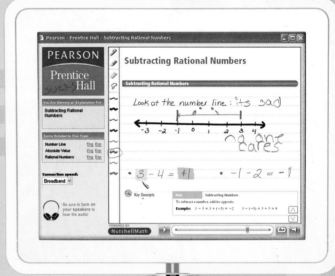

PRENTICE HALL MATHEMATICS

NEW YORK

GEOMETRY

Laurie E. Bass

Randall I. Charles

Basia Hall

Art Johnson

Dan Kennedy

PEARSON

Prentice Hall

Boston, Massachusetts
Upper Saddle River, New Jersey

Authors

Series Authors

Dan Kennedy, Ph.D., is a classroom teacher and the Lupton Distinguished Professor of Mathematics at the Baylor School in Chattanooga, Tennessee. A frequent speaker at professional meetings on the subject of mathematics education reform, Dr. Kennedy has conducted more than 50 workshops and institutes for high school teachers. He is coauthor of textbooks in calculus and precalculus, and from 1990 to 1994 he chaired the College Board's AP Calculus Development Committee. He is a 1992 Tandy Technology Scholar and a 1995 Presidential Award winner.

Randall I. Charles, Ph.D., is Professor Emeritus in the Department of Mathematics and Computer Science at San Jose State University, San Jose, California. He began his career as a high school mathematics teacher, and he was a mathematics supervisor for five years. Dr. Charles has been a member of several NCTM committees and is the former Vice President of the National Council of Supervisors of Mathematics. Much of his writing and research has been in the area of problem solving. He has authored more than 75 mathematics textbooks for kindergarten through college.

Basia Hall currently serves as Manager of Instructional Programs for the Houston Independent School District. With 30 years teaching experience, Ms. Hall has served as a department chair, instructional specialist, instructional supervisor, a school improvement facilitator, and a professional development (TEXTEAMS) trainer. Ms. Hall has developed curriculum for Algebra 1, Geometry, and Algebra 2 and contributed to the development of the Texas Essential Knowledge and Skills. A recipient of the 1992 Presidential Award for Excellence in Mathematics Teaching, Ms. Hall is also a past president of the Texas Association of Supervisors of Mathematics, and she is a state representative for the National Council of Supervisors of Mathematics (NCSM).

Acknowledgments appear on pages 894–896, which constitute an extension of this copyright page.

13-digit ISBN 978-0-202875-2
10-digit ISBN 0-13-202875-1
3 4 5 6 7 8 9 10   12 11 10 09 08

Geometry Authors

Laurie E. Bass is a classroom teacher at the 9–12 division of the Ethical Culture Fieldston School in Riverdale, New York. Ms. Bass has a wide base of teaching experience, ranging from grade 6 through Advanced Placement Calculus. She was the recipient of a 2000 Honorable Mention for the RadioShack National Teacher Awards. She has been a contributing writer of a number of publications, including software-based activities for the Algebra 1 classroom. Among her areas of special interest are cooperative learning for high school students and geometry exploration on the computer. Ms. Bass has been a presenter at a number of local, regional, and national conferences.

Art Johnson, Ed.D., is a professor of mathematics education at Boston University. He is a mathematics educator with 32 years of public school teaching experience, a frequent speaker and workshop leader, and the recipient of a number of awards. Dr. Johnson received the Tandy Prize for Teaching Excellence in 1995, a Presidential Award for Excellence in Mathematics Teaching in 1992, and New Hampshire Teacher of the Year, also in 1992. He was profiled by the Disney Corporation in the American Teacher of the Year Program.

Integrated Algebra and Algebra 2 Authors

Allan E. Bellman is a Lecturer/Supervisor in the School of Education at the University of California, Davis. Before coming to Davis, he was a mathematics teacher for 31 years in Montgomery County, Maryland. He has been an instructor for both the Woodrow Wilson National Fellowship Foundation and the T^3 program. Mr. Bellman has a particular expertise in the use of technology in education and speaks frequently on this topic. He was a 1992 Tandy Technology Scholar.

Sadie Chavis Bragg, Ed.D., is Senior Vice President of Academic Affairs at the Borough of Manhattan Community College of the City University of New York. A former professor of mathematics, she is a past president of the American Mathematical Association of Two-Year Colleges (AMATYC), co-director of the AMATYC project to revise the standards for introductory college mathematics before calculus, and an active member of the Benjamin Banneker Association. Dr. Bragg has coauthored more than 50 mathematics textbooks for kindergarten through college.

William G. Handlin, Sr., is a classroom teacher and Department Chairman of Technology Applications at Spring Woods High School in Houston, Texas. Awarded Life Membership in the Texas Congress of Parents and Teachers for his contributions to the well-being of children, Mr. Handlin is also a frequent workshop and seminar leader in professional meetings throughout the world.

Reviewers

Integrated Algebra Reviewers

Mary Lou Beasley
Southside Fundamental
Middle School
St. Petersburg, Florida

Blanche Smith Brownley
Washington, D.C., Public
Schools
Washington, D.C.

Joseph Caruso
Somerville High School
Somerville, Massachusetts

Belinda Craig
Highland West Junior High
School
Moore, Oklahoma

Jane E. Damaske
Lakeshore Public Schools
Stevensville, Michigan

Stacey A. Ego
Warren Central High School
Indianapolis, Indiana

Earl R. Jones
Formerly, Kansas City
Public Schools
Kansas City, Missouri

Jeanne Lorenson
James H. Blake High School
Silver Spring, Maryland

John T. Mace
Hibbett Middle School
Florence, Alabama

**Ann Marie Palmieri-
Monahan**
Director of Mathematics
Bayonne Board of Education
Bayonne, New Jersey

Marie Schalke
Woodlawn Middle School
Long Grove, Illinois

Julie Welling
LaPorte High School
LaPorte, Indiana

Sharon Zguzenski
Naugatuck High School
Naugatuck, Connecticut

Geometry Reviewers

Marian Avery
Great Valley High School
Malvern, Pennsylvania

Mary Emma Bunch
Farragut High School
Knoxville, Tennessee

Karen A. Cannon
K–12 Mathematics
Coordinator
Rockwood School District
Eureka, Missouri

Johnnie Ebbert
Department Chairman
DeLand High School
DeLand, Florida

Russ Forrer
Math Department Chairman
East Aurora High School
Aurora, Illinois

Andrea Kopco
Midpark High School
Middleburg Heights, Ohio

Gordon E. Maroney III
Camden Fairview High
School
Camden, Arkansas

Charlotte Phillips
Math Coordinator
Wichita USD 259
Wichita, Kansas

Richard P. Strausz
Farmington Public Schools
Farmington, Michigan

Jane Tanner
Jefferson County
International
Baccalaureate School
Birmingham, Alabama

Karen D. Vaughan
Pitt County Schools
Greenville, North Carolina

Robin Washam
Math Specialist
Puget Sound Educational
Service District
Burien, Washington

Algebra 2 Reviewers

Josiane Fouarge
Landry High School
New Orleans, Louisiana

Susan Hvizdos
Math Department Chair
Wheeling Park High School
Wheeling, West Virginia

Kathleen Kohler
Kearny High School
Kearny, New Jersey

Julia Kolb
Leesville Road High School
Raleigh, North Carolina

Deborah R. Kula
Sacred Hearts Academy
Honolulu, Hawaii

Betty Mayberry
Gallatin High School
Gallatin, Tennessee

John L. Pitt
Formerly, Prince William
   County Schools
Manassas, Virginia

Margaret Plouvier
Billings West High School
Billings, Montana

Sandra Sikorski
Berea High School
Berea, Ohio

Tim Visser
Grandview High School
Cherry Creek School District
Aurora, Colorado

Content Consultants

Ann Bell
Mathematics
Prentice Hall Consultant
Franklin, Tennessee

Blanche Brownley
Mathematics
Prentice Hall Consultant
Olney, Maryland

Joe Brumfield
Mathematics
Prentice Hall Consultant
Altadena, California

Linda Buckhalt
Mathematics
Prentice Hall Consultant
Derwood, Maryland

Andrea Gordon
Mathematics
Prentice Hall Consultant
Atlanta, Georgia

Eleanor Lopes
Mathematics
Prentice Hall Consultant
New Castle, Delaware

Sally Marsh
Mathematics
Prentice Hall Consultant
Baltimore, Maryland

Bob Pacyga
Mathematics
Prentice Hall Consultant
Darien, Illinois

Judy Porter
Mathematics
Prentice Hall Consultant
Fuquay-Varena, North Carolina

Rose Primiani
Mathematics
Prentice Hall Consultant
Harbor City, New Jersey

Jayne Radu
Mathematics
Prentice Hall Consultant
Scottsdale, Arizona

Pam Revels
Mathematics
Prentice Hall Consultant
Sarasota, Florida

Barbara Rogers
Mathematics
Prentice Hall Consultant
Raleigh, North Carolina

Michael Seals
Mathematics
Prentice Hall Consultant
Edmond, Oklahoma

Margaret Thomas
Mathematics
Prentice Hall Consultant
Indianapolis, Indiana

New York Mathematics Student Handbook

New York Program Reviewers

John R. Chubb
Mathematics Teacher
Akron High School
Akron, NY

Carole D. Desoe
Math Department Chair
Scarsdale High School
Scarsdale, NY

Lisa Weber
K-12 Mathematics Coordinator
White Plains High School
White Plains, NY

Irene "Sam" Jovell
Senior Mathematics Specialist
Questar III
Castleton, NY
Irene "Sam" Jovell is a former President of NY State Association of Mathematics Supervisors. Sam taught Mathematics at Niskayuna, NY for 33 years. She is also a member of MST Standards writing team.

A. Rose Primiani, Ed.D.
A. Rose Primiani, Ed.D., is Prentice Hall's Program Consultant/Planner for Prentice Hall Mathematics. Formerly a Supervisor of Mathematics for the Yonkers Public Schools and Director of Mathematics for Community School District 10 in New York City.

Table of Contents

Contents in Brief

Chapter 1

Tools of Geometry

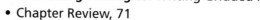

Reasoning and Proof

NEW YORK

Chapter 3

Parallel and Perpendicular Lines

Congruent Triangles

Chapter 5

Relationships Within Triangles

Chapter 6

Quadrilaterals

Chapter 7

Similarity

8

NEW YORK

Right Triangles and Trigonometry

Transformations

Chapter 10

Area

Chapter 11

Surface Area and Volume

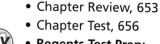
Assessment and Test Prep

Circles

Assessment and Test Prep

Connect Your Learning

through problem solving, activities, and the Web

Applications: Real-World Applications

Activities

Activities: Activity Labs

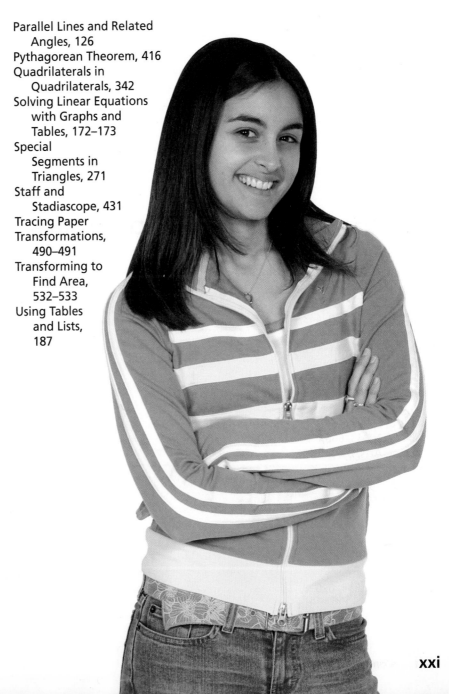

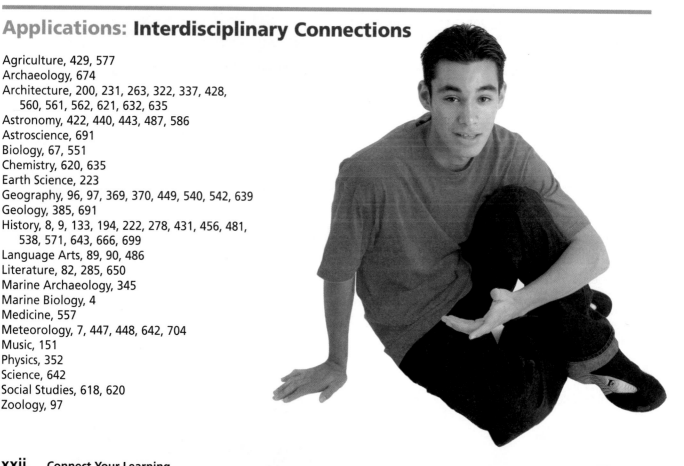

 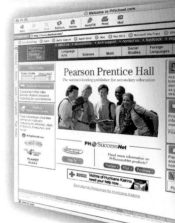

Go Online

Throughout this book you will find links to the Prentice Hall Web site. Use the Web Codes provided with each link to gain direct access to online material. Here's how to *Go Online*:

1. **Go to PHSchool.com**
2. **Enter the Web Code**
3. **Click Go!**

Lesson Web Codes

Lesson Quiz Web Codes: There is an online quiz for every lesson. Access these quizzes with Web Codes aua-0101 through aua-1206 for Lesson 1-1 through Lesson 12-6. See page 9.

Homework Video Tutor Web Codes: For every lesson, there is additional support online to help students complete their homework. Access this homework help online with Web Codes aue-0101 through aue-1206 for Lesson 1-1 through Lesson 12-6. See page 8.

Lesson Quizzes
Web Code format: aua-0105
01 = Chapter 1   05 = Lesson 5

Homework Video Tutor
Web Code format: aue-0605
06 = Chapter 6   05 = Lesson 5

Chapter Web Codes

Chapter	Vocabulary Quizzes	Chapter Tests	Activity Labs
1	auj-0151	aua-0152	
2	auj-0251	aua-0252	aue-0253
3	auj-0351	aua-0352	aue-0353
4	auj-0451	aua-0452	
5	auj-0551	aua-0552	aue-0553
6	auj-0651	aua-0652	
7	auj-0751	aua-0752	
8	auj-0851	aua-0852	aue-0853
9	auj-0951	aua-0952	aue-0953
10	auj-1051	aua-1052	
11	auj-1151	aua-1152	aue-1153
12	auj-1251	aua-1252	
End-of-Course		aua-1254	

Additional Web Codes

Video Tutor Help:
Use Web Code aue-0775 to access engaging online instructional videos to help bring math concepts to life. See page 37.

Data Updates:
Use Web Code aug-9041 to get up-to-date government data for use in examples and exercises. See page 569.

Geometry at Work:
For information about each Geometry at Work feature, use Web Code aub-2031. See page 50.

New York State Learning Standards

To help you master essential mathematical knowledge and deepen your critical thinking skills, the New York State Board of Education established Learning Standards. Here is an overview of the standards for Geometry including Geometric Relationships, Informal and Formal Proofs, Transformational Geometry, and Coordinate Geometry.

Geometry Standard: Geometric Relationships

The standards in this topic cover lines and planes (G.G.1 through G.G.12), and three-dimensional figures (G.G.13 through G.G.16).

When you have mastered these standards you will be able to:

- Use visualization and spatial reasoning to analyze characteristics and properties of geometric shapes.

What It Means To You

Geometry is the study of shapes and figures. The relationships among lines, angles, and planes are important tools for studying more complicated figures.

You will first explore the relationships parallel and perpendicular lines and planes. You will learn to draw conclusions from given conditions. As your understanding develops, you will expand your exploration to include three-dimensional figures like prisms, pyramids, cylinders, and spheres.

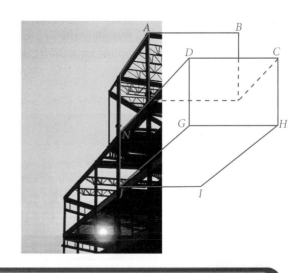

Geometry Standard: Informal and Formal Proofs

There are thirty standards (*G.G.24 through G.G.53*) covering informal and formal proofs.

When you have mastered these standards you will be able to:

- Identify and justify geometric relationships formally and informally.

What It Means To You

Have you ever had a friend say, "Can you prove it?" in response to a statement? You probably tried to think of accepted facts or past events to support your claim. In geometry, you use definitions, postulates (an accepted statement of fact), and theorems (statements that have been proven to be true) to prove that geometric ideas are true. Each step in your proof must be justified and follow a logical order.

As you study geometric figures, you will use your knowledge to develop proofs and theorems, such as those relating similar figures. You will then apply what you have learned to real-world situations.

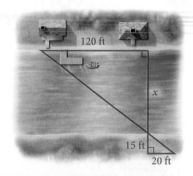

Geometry Standards: Transformational Geometry

The standards in this topic (G.G.54 through G.G.61) cover transformational geometry.

When you have mastered these standards you will be able to:

- Apply transformations and symmetry to analyze problem solving situations
- Justify geometric relationships using transformations.

What It Means To You

If you move on a chess board, you have made a translation of the chess piece. If you read a word in a mirror, its reflection looks backwards. Enlargements and reductions made on photocopiers are dilations. These are all examples of transformations.

You will be studying transformations of figures in the coordinate plane. For example, you will learn to describe the symmetry of a figure and write rules to describe how one figure can be shifted, flipped, or turned to create another figure.

Geometry Standards: Coordinate Geometry

The standards in this topic (G.G.62 through G.G.74) cover coordinate geometry.

When you have mastered these standards you will be able to:

- Apply coordinate geometry to analyze problem solving situations
- Apply the properties of triangles and quadrilaterals in the coordinate plane.

What It Means To You

When you want to know the length of a line segment, you probably think of using a ruler. When the segment is graphed in the coordinate plane, however, there are formulas that can be used to determine length, midpoint, and the slope of a line.

Workout for New York Learning Standards Mastery

Ready to go after Chapter 1

? For help, go to the lesson in green.

1. What is the next term in the sequence?
5, 13, 20, 26, . . .
(Lesson 1-1)

  (1) 44   **(2)** 33
  (3) 32   **(4)** 31

2.

  How many rays are in the next two terms in the sequence?
  (Lesson 1-1)

  (1) 16 and 33 rays
  (2) 17 and 33 rays
  (3) 17 and 34 rays
  (4) 18 and 34 rays

3. Which is the top view of the figure below?
  (Lesson 1-2)

  (1)   **(2)**

  (3)   **(4)**

4. The map below shows Stuart Avenue and Hanover Avenue. Which statement best describes the relationship between the streets?
(Lesson 1-4)

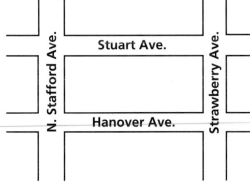

(1) They appear to be parallel.
(2) They appear to be skew.
(3) They appear to intersect.
(4) They appear to form a right angle.

5.

Which is the first step in constructing the perpendicular bisector of $\overline{AB}$?
(Lesson 1-7)

(1) Measure $\overline{AB}$ with a ruler.
(2) Open your compass the length of $\overline{AB}$.
(3) Draw an arc.
(4) Draw a ray.

6. The coordinates of the endpoints of $\overline{AB}$ are
$(-3, 5)$ and $(2, -4)$. What are the coordinates of the midpoint?
(Lesson 1-8)

(1) $(-2.5, -1)$ **(2)** $(-1, 1)$
(3) $(-0.5, 0.5)$ **(4)** $(0.5, -0.5)$

Learning Standards
G.PS.2, G.PS.3, G.CM.11, G.G.18, G.G.66

Workout for New York Learning Standards Mastery

Ready to go after Chapter 2

? For help, go to the lesson in green.

1. Which of the following is the converse of the statement, "If it is a weekend, then I do not have to attend school"?
(Lesson 2-1)

(1) It is a weekend.
(2) I do not have to attend school.
(3) If it is not a weekend, then I do have to attend school.
(4) If I do not have to attend school, then it is a weekend.

2. What is the converse of the statement, "If all sides of a quadrilateral are congruent, then it is a rhombus?"
(Lesson 2-1)

(1) If not all sides of a quadrilateral are congruent, then it is a rhombus.
(2) If not all sides of a quadrilateral are congruent, then it is not a rhombus.
(3) If a quadrilateral is a rhombus, then all of its sides are congruent.
(4) If a quadrilateral is not a rhombus, then not all of its sides are congruent.

3. "If a polygon has exactly five congruent sides, then it is a regular pentagon. If a figure is a regular pentagon, then each interior angle measures 108°."

Which of the following could be a conclusion based on the statements above?
(Lesson 2-3)

(1) If a polygon is a pentagon, then each interior angle measures 108°.
(2) If a polygon has an angle that measures 108°, then it is a pentagon.
(3) If a polygon has exactly five congruent sides, then each interior angle measures 108°.
(4) If a polygon has an angle that does not measure 108°, then it is not a pentagon.

4. Which of these counterexamples shows that the conjecture below is false?

Every square number has exactly three factors.
(Lesson 2-4)

(1) The factors of 2 are 1, 2.
(2) The factors of 4 are 1, 2, 4.
(3) The factors of 8 are 1, 2, 4, 8.
(4) The factors of 16 are 1, 2, 4, 8, 16.

5. Which of these properties justifies the following statement?

If $\angle R \cong \angle S$ and $\angle S \cong \angle T$, then $\angle R \cong \angle T$.
(Lesson 2-4)

(1) Transitive Property
(2) Reflexive Property
(3) Symmetric Property
(4) Substitution Property

6.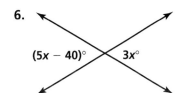

$(5x - 40)°$ $3x°$

What is the value of x?
(Lesson 2-5)

(1) 5
(2) 20
(3) 40
(4) 60

Learning Standards
G.G.26, G.PS.4, G.PS.8

Workout for New York Learning Standards Mastery

Ready to go after Chapter 3

? For help, go to the lesson in green.

1.

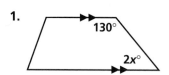

130°

2x°

What is the value of x?
(Lesson 3-1)

(1) 20
(2) 25
(3) 45
(4) 50

2.

A   B

(4x + 15)°

45°

D   C

For what value of x is $\overline{AD} \cong \overline{BC}$?
(Lesson 3-2)

(1) 25
(2) 30
(3) $33\frac{1}{4}$
(4) 135

3.

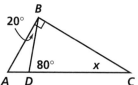

B

20°

80°   x

A   D   C

What is the value of x?
(Lesson 3-4)

(1) 20
(2) $22\frac{1}{2}$
(3) 28
(4) 30

4. A polygon has 11 sides. What is the sum of the measures of the polygon's interior angles? (Lesson 3-5)

(1) 3,240°
(2) 1,980°
(3) 1,978°
(4) 1,620°

5. In hexagon *ABCDEF*, ∠A and ∠B are right angles. If ∠C ≅ ∠D ≅ ∠E ≅ ∠F, what is the measure of ∠F?
(Lesson 3-5)

(1) 45°
(2) 135°
(3) 540°
(4) 720°

6. The outer walls of the Pentagon are formed by two regular pentagons as shown below.

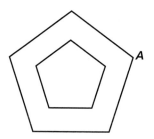

A

What is m∠A?
(Lesson 3-5)

(1) 180°   (2) 144°
(3) 108°   (4)72°

7. Line *k* passes though the points (6, 3) and (−2, −4). What is the slope of a line perpendicular to *k*?
(Lesson 3-7)

(1) $-\frac{8}{7}$ (2) $-\frac{7}{8}$

(3) $\frac{7}{8}$ (4) $\frac{8}{7}$

Learning Standards
G.G.30, G.G.35, G.G.36, G.G.37, G.G.64

Ready to go after Chapter 4

? For help, go to the lesson in green.

1.

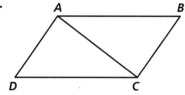

If △ABC ≅ △CDA, which of the following must be true?
(Lesson 4-1)

(1) $\overline{AB} \cong \overline{CA}$
(2) $\overline{BC} \cong \overline{DC}$
(3) ∠CAB ≅ ∠ACD
(4) ∠ABC ≅ ∠CAD

2. The sides of △XYZ are congruent to the sides of △STU. Are the triangles congruent?
(Lesson 4-2)

(1) No; you also need to know that at least one pair of angles are congruent.
(2) Yes; knowing that two sides are congruent is sufficient.
(3) No; we do not know if all three angles are congruent.
(4) Yes; Side-Side-Side Postulate

3.

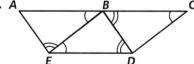

Which two triangles are congruent by the AAS postulate?
(Lesson 4-3)

(1) △ABE and △EBD
(2) △ABE and △BDC
(3) △EBD and △CDB
(4) All three triangles are congruent by AAS.

4.

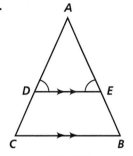

Given: $\overline{DE} \parallel \overline{CB}$,
∠ADE ≅ ∠AED.

Prove: $\overline{AC} \cong \overline{AB}$

Proof: Since $\overline{DE} \parallel \overline{CB}$, ∠ACB ≅ ∠ADE and ∠AED ≅ ∠ABC by the Corresponding Angles Postulate. Since ∠ADE ≅ ∠AED, ∠ACB ≅ ∠ABC by the Transitive Property. Therefore, $\overline{AC}$ is ≅ $\overline{AB}$ by the — **(Lesson 4-5)**

(1) Isosceles Triangle Theorem
(2) Converse of Isosceles Triangle Theorem
(3) Definition of Isosceles Triangle
(4) Definition of Congruent Segments

5. Constructed Response

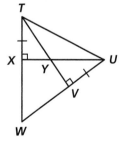

Write a paragraph proof. **(Lesson 4-7)**

Learning Standards
G.G.27, G.G.28, G.G.29, G.G.31

Workout for New York Learning Standards Mastery

Ready to go after Chapter 5

? For help, go to the lesson in green.

1.

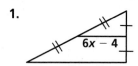

What is the value of x? (**Lesson 5-1**)

(1) 2   (2) 3
(3) 4   (4) 5

2. The two main runways at an airport are 2,744 and 2,014 meters long. They meet at one end.

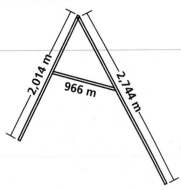

If a taxiway connecting the midpoints of the runways is 966 meters long, how far apart are the other ends? (**Lesson 5-1**)

(1) 1,863 m   (2) 1,932 m
(3) 2,047 m   (4) 2,744 m

3. $\overrightarrow{BD}$ bisects $\angle ABC$.

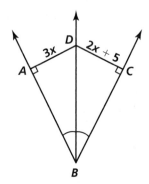

What is the value of x?
(**Lesson 5-3**)

(1) 2   (2) 3
(3) 4   (4) 5

4. What is the symbolic form of the *contrapositive* of "If p, then q"?
(**Lesson 5-4**)

(1) $p \to q$ (2) $q \to p$
(3) $\sim q \to \sim p$ (4) $\sim p \to \sim q$

5.

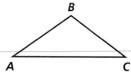

In the diagram, $AB < BC < CA$. From smallest to largest, the angles of $\triangle ABC$ are —
(**Lesson 5-5**)

(1) $\angle C$, $\angle A$, $\angle B$
(2) $\angle B$, $\angle A$, $\angle C$
(3) $\angle A$, $\angle B$, $\angle C$
(4) $\angle B$, $\angle C$, $\angle A$

6. Which list could represent the lengths of the sides of a triangle?
(**Lesson 5-5**)

(1) 1 ft, 2 ft, 4 ft
(2) 4 in., 6 in., 10 in.
(3) 3 m, 5 m, 7m
(4) 7 cm, 10 cm, 25 cm

7. The sides of a triangle are 8 centimeters, 15 centimeters, and x centimeters. Which is a possible value for x?
(**Lesson 5-5**)

(1) 5 cm
(2) 7 cm
(3) 16 cm
(4) 23 cm

Learning Standards
G.G.21, G.G.26, G.G.33, G.G.34, G.G.42

Workout for New York Learning Standards Mastery

Ready to go after Chapter 6

For help, go to the lesson in green.

1. Which type of quadrilateral is *TUVW* if $\overline{TU} \parallel \overline{WV}$, but $\overline{TW}$ is not $\parallel \overline{UV}$?
   (Lesson 6-1)

   (1) kite
   (2) rhombus
   (3) parallelogram
   (4) trapezoid

2.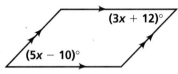

   What is the value of *x*?
   (Lesson 6-2)

   (1) 1
   (2) 8
   (3) 11
   (4) 15

3.

   Which values of *x* and *y* make *DEFG* a parallelogram?
   (Lesson 6-3)

   (1) $x = 3, y = 5$
   (2) $x = 3, y = 3$
   (3) $x = 5, y = 3$
   (4) $x = 7, y = 6$

4.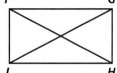

   What is the length of a diagonal of rectangle *FGHI* if $FH = 2x + 3$ and $GI = 5x - 9$?
   (Lesson 6-4)

   (1) 2   (2) 4
   (3) 7   (4) 11

5. The roofline of the building shown below is an isosceles trapezoid.

   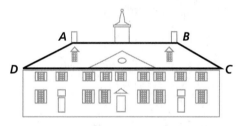

   If m∠B = 150°, what is m∠D?
   (Lesson 6-5)

   (1) 75°   (2) 50°
   (3) 30°   (4) 20°

6.
   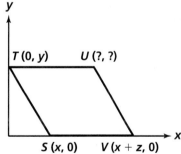

   If *STUV* is a parallelogram, what are the coordinates of point *U*?
   (Lesson 6-6)

   (1) (x, y) (2) (z, y)
   (3) (y, z) (4) $(x + z, y)$

NY

Learning Standards
G.CM.11, G.G.38, G.G.39, G.G.40, G.PS.5

New York Workout

Workout for New York Learning Standards Mastery

Ready to go after Chapter 7

? For help, go to the lesson in green.

1. The ratio of the length of Alan's car to Jan's car is $\frac{7}{8}$. Alan's car is 14 feet long. How long is Jan's car?
(Lesson 7-1)

(1) 14 ft   (2) 15 ft
(3) 16 ft   (4) 18 ft

2.

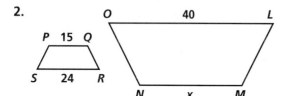

If *PQRS ~ MNOL*, what is the value of *x*?
(Lesson 7-2)

(1) 40   (2) 25
(3) 24   (4) 15

3.

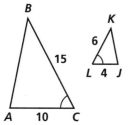

$\triangle ABC \sim \triangle JKL$ by the —?
(Lesson 7-3)

(1) Cross-Product Property
(2) AA Similarity Postulate
(3) SSS Similarity Postulate
(4) SAS Similarity Theorem

4.

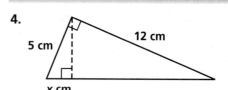

To the nearest centimeter, what is the value of *x*?
(Lesson 7-4)

(1) 1.9   (2) 5
(3) 8.6   (4) 11.1

5.

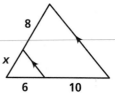

What is the value of *x*?
(Lesson 7-5)

(1) 3   (2) 4.8
(3) 7.5   (4) 13.3

6.

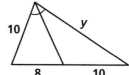

What is the value of *y*?
(Lesson 7-5)

(1) 8   (2) 10
(3) 12.5   (4) 15.5

Learning Standards
G.CN.4, G.G.44, G.G.45, G.G.46, G.G.47, G.RP.2

Workout for New York Learning Standards Mastery

Ready to go after Chapter 8

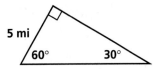 For help, go to the lesson in green.

1. Which of the following is a Pythagorean triple?
(Lesson 8-1)

(1) 3, 5, 6
(2) 8, 14, 17
(3) 12, 35, 37
(4) 20, 23, 29

2.

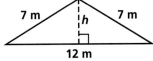

What is the area of the triangle?
(Lesson 8-1)

(1) $\sqrt{13}$ cm²
(2) $6\sqrt{13}$ cm²
(3) $12\sqrt{13}$ cm²
(4) $13\sqrt{12}$ cm²

3. A square tile measures 14 in. by 14 in. Ken cuts the tile along the diagonal. How long is the cut?
(Lesson 8-2)

(1) 7 in.
(2) $7\sqrt{2}$ in.
(3) 14 in.
(4) $14\sqrt{2}$ in.

4.

What is the value of x?
(Lesson 8-2)

(1) 5   **(2)** $\sqrt{5}$
(3) 2   **(4)** $\sqrt{2}$

5.

What is the length of the hypotenuse?
(Lesson 8-2)

(1) 3 mi
(2) 4 mi
(3) $5\sqrt{3}$ mi
(4) 10 mi

6. What is the area of an equilateral triangle with a side that is 6 cm long?
(Lesson 8-2)

(1) $3\sqrt{3}$ cm²
(2) $3\sqrt{5}$ cm²
(3) $9\sqrt{3}$ cm²
(4) 18 cm²

Learning Standards
G.G.48

New York Workout

Workout for New York Learning Standards Mastery

Ready to go after Chapter 9

❓ **For help, go to the lesson in green.**

1.

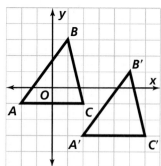

△A'B'C' is —
(Lesson 9-1)

(1) a reflection of △ABC across the x-axis
(2) a reflection of △ABC across the y-axis
(3) a translation of △ABC
(4) a 180° clockwise rotation of △ABC

3.

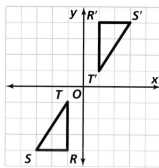

△R'S'T' is —
(Lesson 9-3)

(1) a translation of △RST
(2) a 180° clockwise rotation of △RST
(3) a reflection of △RST across the y-axis
(4) a 90° clockwise rotation of △RST

2.

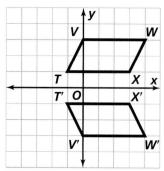

The polygon T'V'W'X' is —
(Lesson 9-2)

(1) a reflection of TVWX across the x-axis
(2) a translation of TVWX
(3) a reflection of TVWX across the y-axis
(4) a 90° clockwise rotation of TVWX

4.

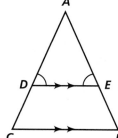

Constructed Response △ADE is a dilation image of △ACB, AD = 6 and DC = 4. Determine the center of dilation and the scale factor.

NY **Learning Standards**
G.G.58, G.G.61

Workout for New York Learning Standards Mastery

Ready to go after Chapter 10

? For help, go to the lesson in green.

1. The coordinates of X are $(-2, 3)$ and the coordinates of Y are $(4, -7)$. What is the approximate distance between X and Y?
(Lesson 1-8)

(1) 11.7   **(2)** 10.2
(2) 7.2   **(4)** 5.1

2.

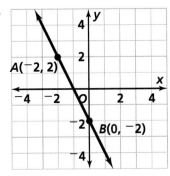

What is the slope of $\overleftrightarrow{AB}$?
(Lesson 3-6)

(1) $\sqrt{13}$ m²
(2) $6\sqrt{13}$ m²
(3) $12\sqrt{13}$ m²
(4) $13\sqrt{12}$ m²

3. What is the symbolic form of the *contrapositive* of "If p, then q"?
(Lesson 5-4)

(1) $p \to q$ **(2)** $q \to p$
(3) $\sim q \to \sim p$ **(4)** $\sim p \to \sim q$

4. The sides of a triangle are 8 centimeters, 15 centimeters, and x centimeters. Which is a possible value for x?
(Lesson 5-5)

(1) 5 cm
(2) 7 cm
(3) 16 cm
(4) 23 cm

5.

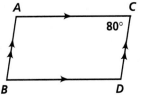

What is the measure of $\angle D$?
(Lesson 6-2)

(1) 80°
(2) 90°
(3) 100°
(4) 110°

6. A digital image that is $3\frac{1}{3}$ inches by $4\frac{2}{3}$ inches on a computer screen is enlarged to make print that is 10 inches by 14 inches. What is the ratio of the width of the image to the width of the print?
(Lesson 7-1)

(1) $\frac{1}{4}$ **2)** $\frac{1}{3}$

(3) $\frac{2}{5}$ **(4)** $\frac{3}{5}$

7.

Constructed Response Is this triangle a right triangle? Explain.
(Lesson 8-1)

Learning Standards
G.CN.4, G.G.26, G.G.33, G.G.38, G.G.48, G.G.67, G.RP.2

New York Workout

Workout for New York Learning Standards Mastery

Ready to go after Chapter 11

? For help, go to the lesson in green.

1.

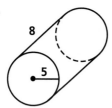

What is the surface area of the cylinder?
(Lesson 11-2)

(1) 130π square units
(2) 100π square units
(3) 80π square units
(4) 50π square units

2.

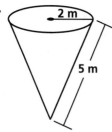

To the nearest tenth, what is the surface area of the cone?
(Lesson 11-3)

(1) 230.3 cm²   (2) 44.0 cm²
(3) 31.4 cm²   (4) 12.6 cm²

3. To the nearest cubic foot, how much water is required to fill the aquarium?
(Lesson 10-5)

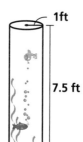

(1) 7.5 cu ft
(2) 15 cu ft
(3) 24 cu ft
(4) 47 cu ft

4.

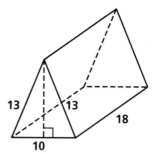

What is the volume of the prism?
(Lesson 11-4)

(1) 2340 cubic units
(2) 1170 cubic units
(3) 1560 cubic units
(4) 1080 cubic units

5. Suppose a square pyramid has a base edge with length 15 inches and height 8 inches. What is the volume?
(Lesson 11-5)

(1) 40 cu in.
(2) 120 cu in.
(3) 600 cu in.
(4) 900 cu in.

6. The volume of a sphere is 2,250 cubic inches. To the nearest square inch, what is the surface area?
(Lesson 11-6)

(1) 830 sq in.
(2) 1,781 sq in.
(3) 2,092 sq in.
(4) 6,750 sq in.

Learning Standards
G.G.12, G.G.13, G.G.14, G.G.15, G.G.16

Workout for New York Learning Standards Mastery

Ready to go after Chapter 12

❓ **For help, go to the lesson in green.**

1.

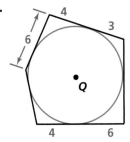

Circle *Q* is inscribed inside a pentagon.
What is the perimeter of the pentagon?
(Lesson 12-1)

(1) 38   **(2)** 30
(3) 19   **(4)** 15

2.

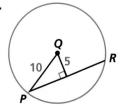

To the nearest tenth, what is the length of $\overline{PR}$?
(Lesson 12-2)

(1) 17.3   **(2)** 8.7
(3) 10.0   **(4)** 5.0

3.

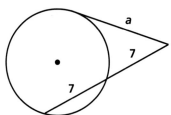

To the nearest whole number, what is the
value of *a*?
(Lesson 11-4)

(1) 17   **(2)** 14
(3) 10   **(4)** 7

4.

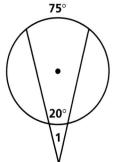

To the nearest tenth, what is m∠1?
(Lesson 12-4)

(1) 20.0°
(2) 27.5°
(3) 45.5°
(4) 55.0°

5. Constructed Response Write the
standard equation of the circle with
center $(-1, -2)$ and radius 5.
(Lesson 12-5)

Learning Standards
G.G.49, G.G.50, G.G.51, G.G.71

New York Workout

Using Your New York

There are many features built into the daily lessons of your New York textbook that will help you learn the important skills and concepts you will need to be successful in this course... and on the Regents Exam! Look through the following pages for some study tips that you will find useful as you complete each lesson.

Mastering the New York Standards

Identifying the NY Standards

Every Chapter opens with a list of the NY Standards you already learned and an overview of the NY Standards that will be covered in the upcoming chapter.

The NY Standards are also correlated to each lesson within the chapter.

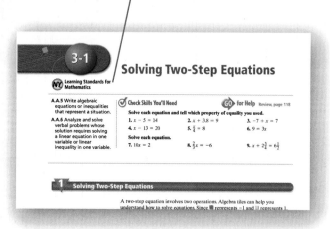

Leveled Exercise Set – including Regents Preparation

Your textbook provides comprehensive exercise sets after every lesson to help you practice what you've learned. Plus, specific New York Regents Exam preparation ensures exam success.

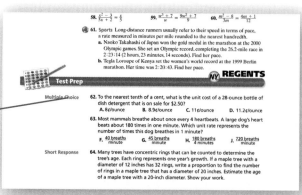

Book for Success

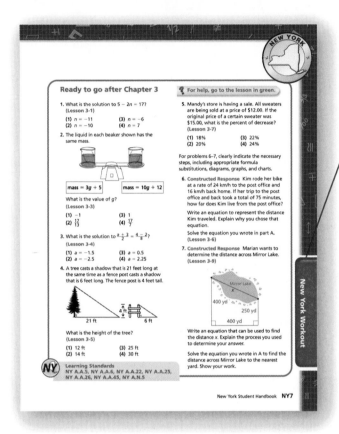

Workout for New York Standards Mastery

In addition to the Regents Exam test prep in the exercise sets, every Chapter has a Workout for New York Standards Mastery to help you succeed on the Regents Exam.

Instant Check System

The Instant Check System is a way for you to monitor your progress on the New York standards quickly and easily in every lesson.

- **Check Skills You'll Need** questions assess prerequisite skills before every lesson.
- **Quick Check** questions after every example ensure you understand before moving to the next example.
- **Check Point Quizzes** throughout the chapter provide a way to assess your cumulative understanding before moving on to more lessons.

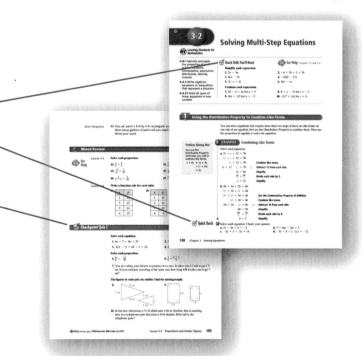

Built-In Help

There are many features built in to every lesson to help support your understanding of the mathematics.

Go for Help

Look for the green labels throughout your book that tell you where to "Go" for help. You'll see this built-in help in the lessons and in the homework exercises.

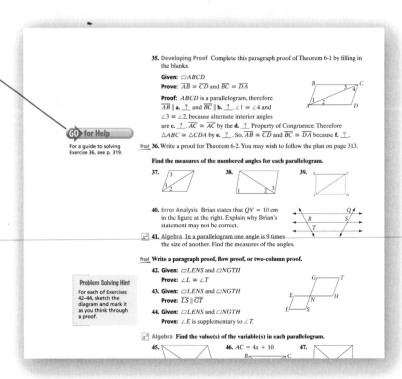

Video Tutor Help

Go online to see engaging videos to help you better understand important math concepts.

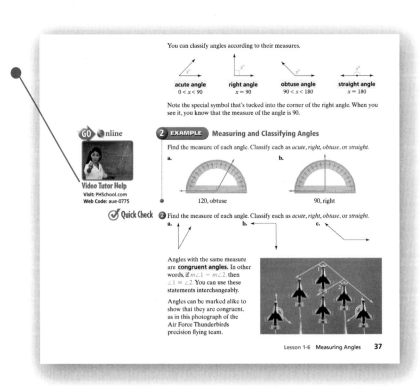

Quick Check

Every lesson includes numerous Examples, each followed by a *Quick Check* question that you can do on your own to see if you understand the skill being introduced. Check your progress with the answers at the back of the book.

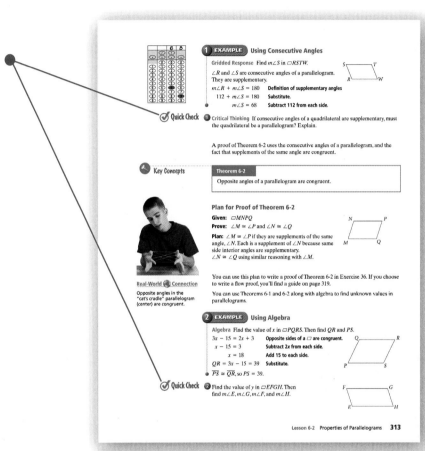

1 EXAMPLE Using Consecutive Angles

Gridded Response Find $m\angle S$ in $\square RSTW$.

$\angle R$ and $\angle S$ are consecutive angles of a parallelogram. They are supplementary.

$m\angle R + m\angle S = 180$ **Definition of supplementary angles**
$112 + m\angle S = 180$ **Substitute.**
$m\angle S = 68$ **Subtract 112 from each side.**

✔ **Quick Check** **1** Critical Thinking If consecutive angles of a quadrilateral are supplementary, must the quadrilateral be a parallelogram? Explain.

A proof of Theorem 6-2 uses the consecutive angles of a parallelogram, and the fact that supplements of the same angle are congruent.

Key Concepts

Theorem 6-2

Opposite angles of a parallelogram are congruent.

Real-World Connection
Opposite angles in the "cat's cradle" parallelogram (center) are congruent.

Plan for Proof of Theorem 6-2

Given: $\square MNPQ$
Prove: $\angle M \cong \angle P$ and $\angle N \cong \angle Q$

Plan: $\angle M \cong \angle P$ if they are supplements of the same angle, $\angle N$. Each is a supplement of $\angle N$ because same side interior angles are supplementary.
$\angle N \cong \angle Q$ using similar reasoning with $\angle M$.

You can use this plan to write a proof of Theorem 6-2 in Exercise 36. If you choose to write a flow proof, you'll find a guide on page 319.

You can use Theorems 6-1 and 6-2 along with algebra to find unknown values in parallelograms.

2 EXAMPLE Using Algebra

Algebra Find the value of x in $\square PQRS$. Then find QR and PS.

$3x - 15 = 2x + 3$ **Opposite sides of a $\square$ are congruent.**
$x - 15 = 3$ **Subtract 2x from each side.**
$x = 18$ **Add 15 to each side.**
$QR = 3x - 15 = 39$ **Substitute.**
$\overline{PS} \cong \overline{QR}$, so $PS = 39$.

✔ **Quick Check** **2** Find the value of y in $\square EFGH$. Then find $m\angle E, m\angle G, m\angle F$, and $m\angle H$.

Lesson 6-2 Properties of Parallelograms   **313**

Understanding Key Concepts

Frequent *Key Concept* boxes summarize important definitions, formulas, theorems, and properties. Use these to review what you've learned.

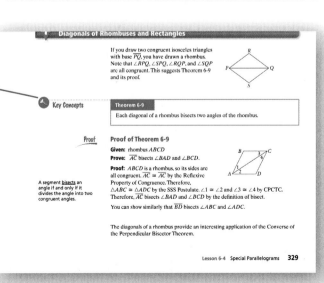

Diagonals of Rhombuses and Rectangles

If you draw two congruent isosceles triangles with base $\overline{PQ}$, you have drawn a rhombus. Note that $\angle RPQ, \angle SPQ, \angle RQP$, and $\angle SQP$ are all congruent. This suggests Theorem 6-9 and its proof.

Key Concepts

Theorem 6-9

Each diagonal of a rhombus bisects two angles of the rhombus.

Proof   **Proof of Theorem 6-9**

Given: rhombus $ABCD$
Prove: $\overline{AC}$ bisects $\angle BAD$ and $\angle BCD$.

Proof: $ABCD$ is a rhombus, so its sides are all congruent. $\overline{AC} \cong \overline{AC}$ by the Reflexive Property of Congruence. Therefore, $\triangle ABC \cong \triangle ADC$ by the SSS Postulate, so $\angle 1 \cong \angle 2$ and $\angle 3 \cong \angle 4$ by CPCTC. Therefore, $\overline{AC}$ bisects $\angle BAD$ and $\angle BCD$ by the definition of bisect. You can show similarly that $\overline{BD}$ bisects $\angle ABC$ and $\angle ADC$.

A segment **bisects** an angle if and only if it divides the angle into two congruent angles.

The diagonals of a rhombus provide an interesting application of the Converse of the Perpendicular Bisector Theorem.

Lesson 6-4 Special Parallelograms   **329**

Online Active Math

Make math come alive with these online activities. Review and practice important math concepts with these engaging online tutorials.

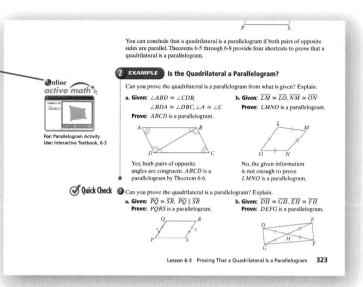

Online active math

For: Parallelogram Activity
Use: Interactive Textbook, 6-3

You can conclude that a quadrilateral is a parallelogram if both pairs of opposite sides are parallel. Theorems 6-5 through 6-8 provide four shortcuts to prove that a quadrilateral is a parallelogram.

2 EXAMPLE Is the Quadrilateral a Parallelogram?

Can you prove the quadrilateral is a parallelogram from what is given? Explain.

a. Given: $\angle ABD \cong \angle CDB$,
$\angle BDA \cong \angle DBC, \angle A \cong \angle C$
Prove: $ABCD$ is a parallelogram.

b. Given: $\overline{LM} \cong \overline{LO}, \overline{NM} \cong \overline{ON}$
Prove: $LMNO$ is a parallelogram.

Yes, both pairs of opposite angles are congruent. $ABCD$ is a parallelogram by Theorem 6-6.

No, the given information is not enough to prove $LMNO$ is a parallelogram.

Quick Check ② Can you prove the quadrilateral is a parallelogram? Explain.

a. Given: $\overline{PQ} \cong \overline{SR}, \overline{PQ} \parallel \overline{SR}$
Prove: $PQRS$ is a parallelogram.

b. Given: $\overline{DH} \cong \overline{GH}, \overline{EH} \cong \overline{FH}$
Prove: $DEFG$ is a parallelogram.

Lesson 6-3 Proving That a Quadrilateral Is a Parallelogram   **323**

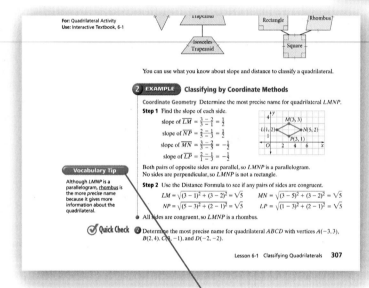

For: Quadrilateral Activity
Use: Interactive Textbook, 6-1

You can use what you know about slope and distance to classify a quadrilateral.

2 EXAMPLE Classifying by Coordinate Methods

Coordinate Geometry Determine the most precise name for quadrilateral $LMNP$.

Step 1 Find the slope of each side.

slope of $\overline{LM} = \frac{3-2}{3-1} = \frac{1}{2}$

slope of $\overline{NP} = \frac{1-2}{3-5} = \frac{1}{2}$

slope of $\overline{MN} = \frac{3-2}{3-5} = -\frac{1}{2}$

slope of $\overline{LP} = \frac{2-1}{1-3} = -\frac{1}{2}$

Both pairs of opposite sides are parallel, so $LMNP$ is a parallelogram. No sides are perpendicular, so $LMNP$ is not a rectangle.

Step 2 Use the Distance Formula to see if any pairs of sides are congruent.

$LM = \sqrt{(3-1)^2 + (3-2)^2} = \sqrt{5}$ $MN = \sqrt{(3-5)^2 + (3-2)^2} = \sqrt{5}$

$NP = \sqrt{(5-3)^2 + (2-1)^2} = \sqrt{5}$ $LP = \sqrt{(1-3)^2 + (2-1)^2} = \sqrt{5}$

All sides are congruent, so $LMNP$ is a rhombus.

Vocabulary Tip

Although $LMNP$ is a parallelogram, rhombus is the more *precise* name because it gives more information about the quadrilateral.

Quick Check ② Determine the most precise name for quadrilateral $ABCD$ with vertices $A(-3,3)$, $B(2,4)$, $C(3,-1)$, and $D(-2,-2)$.

Lesson 6-1 Classifying Quadrilaterals   **307**

Vocabulary Support

Understanding mathematical vocabulary is an important part of studying mathematics. *Vocabulary Tips* and *Vocabulary Builders* throughout the book help focus on the language of math.

Vocabulary Builder

Geometry Vocabulary

To succeed with geometry, you must learn its vocabulary. As a study aid, some students use index cards. Others use special notebooks. With either, you can note

• each geometry term,
• its meaning,
• an example of its use, and
• a drawing, if possible.

A study aid like this can become a valuable tool to use throughout the course.

Parallel Planes – planes that do not intersect

Example – Plane PQRS ∥ Plane WXYZ

EXERCISES

1. Work with a classmate. Plan a study aid that will help both of you learn the geometry vocabulary. Make a sample page. Then share your sample with others. Use the best ideas from the class to revise your method.

2. Begin a vocabulary section of your study aid using Lessons 1-1 through 1-4.

A popular dictionary shows the definitions at the right for *add* and *sum*.

Add (v): To combine two...

Sum (n): The result obtained by...

Practice What You've Learned

There are numerous exercises in each lesson that give you the practice you need to master the concepts in the lesson. Each practice set includes the following sections.

A: Practice by Example

Practice by Example exercises refer you back to the Examples in the lesson, in case you need help with completing these exercises.

B: Apply Your Skills

Apply Your Skills exercises combine skills from earlier lessons to offer you richer skill exercises and multi-step application problems.

C: Challenge

Challenge exercises give you an opportunity to solve problems that extend and stretch your thinking.

Homework Video Tutor

Go online to get help completing your homework! For every lesson, there is a narrated, interactive tutorial to help you review the lesson and understand key concepts.

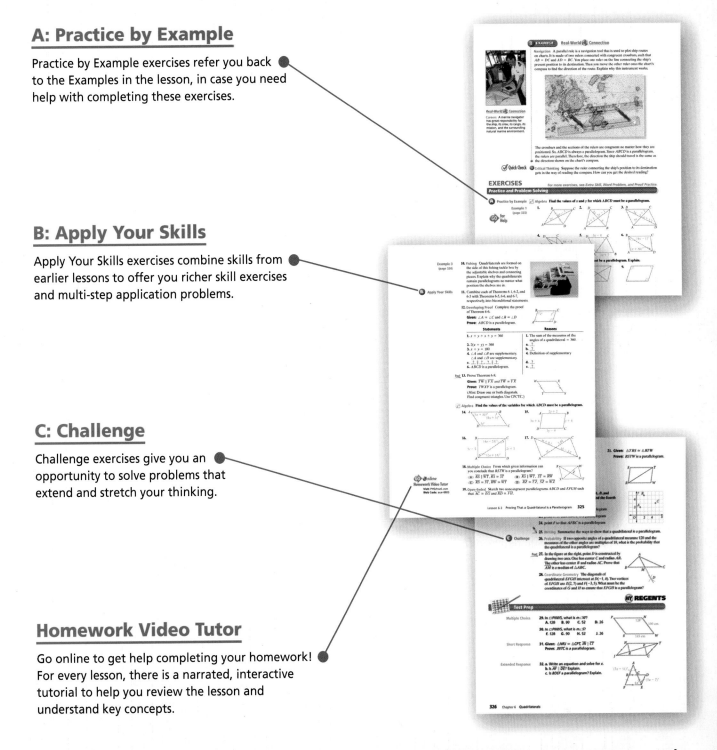

Beginning-Of-Course Diagnostic Test

1. Solve $3 - (2t + 5) = -12$.

2. Solve $5a - 15 + 9a = 3a + 29$.

3. Solve $y - 7 > 3 + 2y$.

4. Solve $15 + 6n < -33$.

5. The figures are made from toothpicks. Use this pattern to determine how many toothpicks are in Figure 10.

Figure 1   Figure 2   Figure 3

6. Find the slope of the line through $(-4, 2)$ and $(5, 8)$.

7. Draw a triangle that has sides of lengths 3.6 cm and 5.2 cm and a 42° angle between these two sides.

8. A fence along a schoolyard is 250 ft long. There is a fence post every 10 ft. How many fence posts are there?

9. Simplify $(-15)^2$.

10. Simplify $\sqrt{256}$.

11. Evaluate the expression $-x(y - 8)^2$ for $x = -2$ and $y = 5$.

12. Simplify $4x - (2 - 3x) + 5$.

Complete.

13. $0.25 \text{ km} = \underline{\ ?\ } \text{ m}$

14. $180 \text{ in.} = \underline{\ ?\ } \text{ yd}$

Solve each system of equations.

15. $\begin{cases} 4x - y = 8 \\ 2x - 4 = 3y \end{cases}$

16. $\begin{cases} 4x - 2y = 9 \\ \qquad y = 2x + 5 \end{cases}$

17. Solve $A = 4\pi r^2$ for r.

18. Simplify the ratio $0.6 : 2.4$.

19. Simplify the ratio $18b^2$ to $45b$.

20. An athletic club has 248 members. Of these, 164 lift weights and 208 perform cardiovascular exercises regularly. All members do at least one of those activities. How many members do both?

21. Solve $|x| - 7 = 6$

Draw graphs of $y = x^2$ and the given function in the same coordinate plane.

22. $y = -2x$

23. $y = x^2 + 3$

24. Solve $x^2 + 6x - 7 = 0$ for x.

25. Maria gave one-half of her jelly beans to Carl. Carl gave one–third of those to Austin. Austin gave one-fourth of those to Carmen. If Carmen received two jelly beans, how many did Maria start with?

26. Write 5.9% as a decimal.

27. Simplify 18% of 300.

28. Max lives 37 mi from his grandmother. What is the range of values this measurement represents?

29. A rectangular prism is 8 in. × 4 in. × 10 in. What is the percent error in calculating the volume of this prism? Round to the nearest percent.

30. Simplify $\sqrt{\frac{3}{5}}$.

31. Simplify $\sqrt{6} \cdot \sqrt{12}$.

A bag contains 4 blue marbles, 6 green marbles, and 2 red marbles.

32. What is the probability of selecting a red marble?

33. A green marble is removed from the bag. What is the probability that the next marble selected will be blue?

What You've Learned

 **NY** **Learning Standards for Mathematics**

In previous courses, you learned

- **A.N.1:** to apply your knowledge of arithmetic to the study of algebra. You also learned about the real number system, including operations on rational and irrational numbers.

- **A.A.5:** to write algebraic expressions, and equations to represent relationships.

- **A.A.7:** to use a variety of techniques to solve equations and inequalities with one or more variables.

 Check Your Readiness **for Help** to the Lesson in green.

Squaring Numbers (Skills Handbook page 753)

Simplify.

1. 3^2 **2.** 4^2 **3.** 11^2

Simplifying Expressions (Skills Handbook page 754)

Simplify each expression. Use 3.14 for π.

4. $2 \cdot 7.5 + 2 \cdot 11$ **5.** $\pi(5)^2$ **6.** $\sqrt{5^2 + 12^2}$

Evaluating Expressions (Skills Handbook page 754)

Evaluate the following expressions for $a = 4$ and $b = -2$.

7. $\dfrac{a + b}{2}$ **8.** $\dfrac{a - 7}{3 - b}$ **9.** $\sqrt{(7 - a)^2 + (2 - b)^2}$

Finding Absolute Value (Skills Handbook page 757)

Simplify each absolute value expression.

10. $|-8|$ **11.** $|2 - 6|$ **12.** $|-5 - (-8)|$

Solving Equations (Skills Handbook page 758)

$\boxed{x^2}$ **Algebra** Solve each equation.

13. $2x + 7 = 13$ **14.** $5x - 12 = 2x + 6$ **15.** $2(x + 3) - 1 = 7x$

Tools of Geometry

◄))**Key Vocabulary**

- acute angle (p. 37)
- angle bisector (p. 46)
- collinear points (p. 17)
- congruent angles (p. 37)
- conjecture (p. 5)
- construction (p. 44)
- coordinate (p. 31)
- counterexample (p. 5)
- line (p. 17)
- midpoint (p. 32)
- net (p. 12)
- obtuse angle (p. 37)
- parallel lines (p. 24)
- perpendicular lines (p. 45)
- plane (p. 17)
- postulate (p. 18)
- ray (p. 23)
- segment (p. 23)

What You'll Learn Next

NY Learning Standards for Mathematics

- **G.PS.2:** In this chapter, you will learn how to make plausible conclusions based on patterns you observe.

- You will learn the foundation blocks for the structure of geometry.

- These foundations will provide you with ways to measure segments and angles.

- **G.PS.3:** You will also learn to use constructions and the coordinate plane to represent geometric figures.

Data Analysis | **Data Analysis Activity Lab** Applying what you learn, on pages 76 and 77 you will connect the speed at which a car wheel turns with the speed of the car.

Patterns and Inductive Reasoning

NY Learning Standards for Mathematics

G.PS.2 Observe and explain patterns to formulate generalizations and conjectures.

 **Check Skills You'll Need**

 for Help Skills Handbook page 753

Here is a list of the counting numbers: 1, 2, 3, 4, 5, . . .
Some are even and some are odd.

1. Make a list of the positive even numbers.

2. Make a list of the positive odd numbers.

3. Copy and extend this list to show the first 10 perfect squares.
$1^2 = 1, 2^2 = 4, 3^2 = 9, 4^2 = 16, \ldots$

4. Which do you think describes the square of any odd number?
It is odd.   It is even.

New Vocabulary • inductive reasoning • conjecture • counterexample

---

1 Using Inductive Reasoning

Inductive reasoning is reasoning that is based on patterns you observe. If you observe a pattern in a sequence, you can use inductive reasoning to tell what the next terms in the sequence will be.

Real-World **Connection**

You can predict growth of the chambered nautilus shell by studying patterns in its cross sections.

1 EXAMPLE **Finding and Using a Pattern**

Find a pattern for each sequence. Use the pattern to show the next two terms in the sequence.

a. 3, 6, 12, 24, . . .

Each term is twice the preceding term. The next two terms are $2 \times 24 = 48$ and $2 \times 48 = 96$.

b.

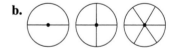

Each circle has one more segment through the center to form equal parts. The next two figures:

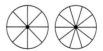

Quick Check **1** Write the next two terms in each sequence.
a. 1, 2, 4, 7, 11, 16, 22, . . .
b. Monday, Tuesday, Wednesday, . . .
c.

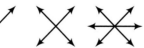

A conclusion you reach using inductive reasoning is called a **conjecture.**

2 EXAMPLE **Using Inductive Reasoning**

Make a conjecture about the sum of the first 30 odd numbers.

Find the first few sums. Notice that each sum is a perfect square.

$1 \qquad\qquad = 1 = 1^2$ ← **The perfect squares form**
$1 + 3 \qquad\quad = 4 = 2^2$ ← **a pattern.**
$1 + 3 + 5 \quad\; = 9 = 3^2$ ←
$1 + 3 + 5 + 7 = 16 = 4^2$ ←

Using inductive reasoning, you can conclude that the sum of the first 30 odd numbers is 30^2, or 900.

✓ **Quick Check** ❷ Make a conjecture about the sum of the first 35 odd numbers. Use your calculator to verify your conjecture.

Not all conjectures turn out to be true. You can prove that a conjecture is false by finding one counterexample. A **counterexample** to a conjecture is an example for which the conjecture is incorrect.

3 EXAMPLE **Finding a Counterexample**

Find a counterexample for each conjecture.
a. The square of any number is greater than the original number.
   The number 1 is a counterexample because $1^2 \not> 1$.
b. You can connect any three points to form a triangle.
   If the three points lie on a line, you cannot form a triangle.

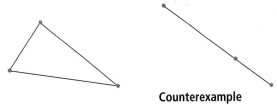

   Counterexample

c. Any number and its absolute value are opposites.
   The conjecture is true for negative numbers, but not positive numbers.
   8 is a counterexample because 8 and $|8|$ are not opposites.

✓ **Quick Check** ❸ Alana makes a conjecture about slicing pizza. She says that if you use only straight cuts, the number of pieces will be twice the number of cuts.

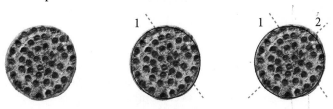

Draw a counterexample that shows you can make 7 pieces using 3 cuts.

4 EXAMPLE Real-World Connection

Business Sales A skateboard shop finds that over a period of five consecutive months, sales of small-wheeled skateboards decreased.

Use inductive reasoning. Make a conjecture about the number of small-wheeled skateboards the shop will sell in June.

The graph shows that sales of small-wheeled skateboards is decreasing by about 3 skateboards each month. By inductive reasoning you might conclude that the shop will sell 42 skateboards in June.

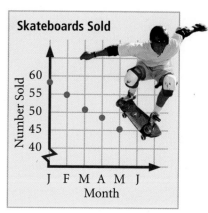

Skateboards Sold

✓ Quick Check **4 a.** Make a conjecture about the number of small-wheeled skateboards the shop will sell in July.

b. **Critical Thinking** How confident would you be in using the graph to make a conjecture about sales in December? Explain.

EXERCISES

For more exercises, see *Extra Skill, Word Problem, and Proof Practice.*

Practice and Problem Solving

 Practice by Example

Example 1
(page 4)

 for Help

Find a pattern for each sequence. Use the pattern to show the next two terms.

1. 5, 10, 20, 40, . . .   **2.** 3, 33, 333, 3333, . . .   **3.** 1, −1, 2, −2, 3, . . .

4. 1, $\frac{1}{2}$, $\frac{1}{4}$, $\frac{1}{8}$, . . .   **5.** 15, 12, 9, 6, . . .   **6.** 81, 27, 9, 3, . . .

7. O, T, T, F, F, S, S, E, . . .   **8.** J, F, M, A, M, . . .   **9.** 1, 2, 6, 24, 120, . . .

10. 2, 4, 8, 16, 32, . . .   **11.** 1, $\frac{1}{4}$, $\frac{1}{9}$, $\frac{1}{16}$, $\frac{1}{25}$, . . .   **12.** 1, $\frac{1}{2}$, $\frac{1}{3}$, $\frac{1}{4}$, . . .

13. George, John, Thomas, James, . . .   **14.** Martha, Abigail, Martha, Dolley, . . .

15. George, Thomas, Abe, Alexander, . . .   **16.** Aquarius, Pisces, Aries, Taurus, . . .

Draw the next figure in each sequence.

17.   **18.**

Example 2
(page 5)

Use the table and inductive reasoning. Make a conjecture about each value.

19. the sum of the first 6 positive even numbers

20. the sum of the first 30 positive even numbers

21. the sum of the first 100 positive even numbers

2	= 2 = 1 · 2
2 + 4	= 6 = 2 · 3
2 + 4 + 6	= 12 = 3 · 4
2 + 4 + 6 + 8	= 20 = 4 · 5
2 + 4 + 6 + 8 + 10	= 30 = 5 · 6

22. Use the pattern in Example 2 to make a conjecture about the sum of the first 100 odd numbers.

Predict the next term in each sequence. Use your calculator to verify your answer.

23. 12345679 × 9 = 111111111
12345679 × 18 = 222222222
12345679 × 27 = 333333333
12345679 × 36 = 444444444
12345679 × 45 = ■

24. 1 × 1 = 1
11 × 11 = 121
111 × 111 = 12321
1111 × 1111 = 1234321
11111 × 11111 = ■

Example 3
(page 5)

Find one counterexample to show that each conjecture is false.

25. The sum of two numbers is greater than either number.

26. The product of two positive numbers is greater than either number.

27. The difference of two integers is less than either integer.

28. The quotient of two proper fractions is a proper fraction.

Example 4
(page 6)

29. Weather The speed with which a cricket chirps is affected by the temperature. If you hear 20 cricket chirps in 14 seconds, what is the temperature?

Chirps per 14 Seconds

5 chirps	45°F
10 chirps	55°F
15 chirps	65°F

75

30. Physical Fitness Dino works out regularly. When he first started exercising, he could do 10 push-ups. After the first month he could do 14 push-ups. After the second month he could do 19, and after the third month he could do 25. Predict the number of push-ups Dino will be able to do after the fifth month of working out. How confident are you of your prediction? Explain.

B **Apply Your Skills**

Find a pattern for each sequence. Use the pattern to show the next two terms.

31. 1, 3, 7, 13, 21, ...

32. 1, 2, 5, 6, 9, ...

33. 0.1, 0.01, 0.001, ...

34. 2, 6, 7, 21, 22, 66, 67, ...

35. 1, 3, 7, 15, 31, ...

36. $0, \frac{1}{2}, \frac{3}{4}, \frac{7}{8}, \frac{15}{16}, \ldots$

37. M, V, E, M, ...

38. AL, AK, AZ, AR, ...

39. H, He, Li, Be, ...

40. Writing Choose two of the sequences in Exercises 31–36 and describe the patterns.

41. Draw two parallel lines on your paper. Locate four points on the paper, each an equal distance from both lines. Describe the figure you get if you continue to locate points, each an equal distance from both lines.

Real-World Connection

Points along the yellow line are equal distances from both sides of the bike trail (Exercise 41).

Draw the next figure in each sequence.

42.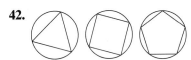

43.

44.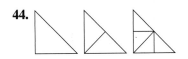

45.

46. Multiple Choice Find the perimeter when 100 triangles are put together in the pattern shown. Assume that all triangle sides are 1 cm long.

Ⓐ 100 cm   Ⓑ 102 cm   Ⓒ 202 cm   Ⓓ 300 cm

47. Math in the Media Read this exerpt from a news article.

> **Top female runners** have been improving about twice as quickly as the fastest men, a new study says. If this pattern continues, women may soon outrun men in competition!
>
> The study is based on world records collected at 10-year intervals, starting in 1905 for men and in the 1920s for women. If the trend continues, the top female and male runners in races ranging from 200 m to 1500 m might attain the same speeds sometime between 2015 and 2055.
>
> Women's marathon records date from 1955 but their rapid fall suggests that the women's record will equal that of men even more quickly.

   a. What conclusion was reached in the study?

   b. How was inductive reasoning used to reach the conclusion?

   c. Explain why the conclusion that women may soon be outrunning men may be incorrect. For which race is the conclusion most suspect? For what reason?

48. Communications The table shows the number of commercial radio stations in the United States for a 50-year period.

   a. Make a line graph of the data.

   b. Use the graph and inductive reasoning to make a conjecture about the number of radio stations in the United States in the year 2010.

   c. How confident are you about your conjecture? Explain.

49. Open-Ended Write two different number-pattern sequences that begin with the same two numbers.

Number of Radio Stations

1950	2,835
1960	4,224
1970	6,519
1980	7,871
1990	9,379
2000	10,577

Source: Federal Communications Commission

50. Error Analysis For each of the past four years, Paulo has grown 2 in. every year. He is now 16 years old and is 5 ft 10 in. tall. He figures that when he is 22 years old he will be 6 ft 10 in. tall. What would you tell Paulo about his conjecture?

for Help

For Exercise 51, you may want to review "Coordinates of a point" in the Glossary.

51. Coordinate Geometry You are given x- and y-coordinates for 14 points.
$A(1,5)$ $B(2,2)$ $C(2,8)$ $D(3,1)$ $E(3,9)$ $F(6,0)$ $G(6,10)$
$H(7,-1)$ $I(7,11)$ $J(9,1)$ $K(9,9)$ $L(10,2)$ $M(10,8)$ $N(11,5)$

   a. Graph each point.

   b. Most of the points fit a pattern. Which points do not?

   c. Describe the figure that fits the pattern.

52. History Leonardo of Pisa (about 1175–1258), also known as Fibonacci (fee buh NAH chee), was born in Italy and educated in North Africa. He was one of the first Europeans known to use modern numerals instead of Roman numerals. The special sequence 1, 1, 2, 3, 5, 8, 13, . . . is known as the Fibonacci sequence. Find the next three terms of this sequence.

Homework Video Tutor

Visit: PHSchool.com
Web Code: aue-0101

53. Time Measurement Leap years have 366 days.

   a. The years 1984, 1988, 1992, 1996, and 2000 are consecutive leap years. Look for a pattern in their dates. Then, make a conjecture about leap years.

   b. Of the years 2010, 2020, 2100, and 2400, which do you think will be leap years?

   c. Research Find out whether your conjecture for part (a) and your answer for part (b) are correct. How are leap years determined?

History When he was in the third grade, German mathematician Karl Gauss (1777–1855) took ten seconds to sum the integers from 1 to 100. Now it's your turn. Find a fast way to sum the integers from 1 to 100; from 1 to n. (*Hint:* Use patterns.)

 55. a. Algebra Write the first six terms of the sequence that starts with 1, and for which the difference between consecutive terms is first 2, and then 3, 4, 5, and 6.

    b. Evaluate $\frac{n^2 + n}{2}$ for $n = 1, 2, 3, 4, 5,$ and 6. Compare the sequence you get with your answer for part (a).

    c. Examine the diagram at the right and explain how it illustrates a value of $\frac{n^2 + n}{2}$.

    d. Draw a similar diagram to represent $\frac{n^2 + n}{2}$ for $n = 5$.

Multiple Choice

56. The sum of the numbers from 1 to 10 is 55. The sum of the numbers from 11 to 20 is 155. The sum of the numbers from 21 to 30 is 255. Based on this pattern, what is the sum of numbers from 91 to 100?
A. 855   **B.** 955   **C.** 1055   **D.** 1155

57. Which of the following conjectures is false?
    F. The product of two even numbers is even.
    G. The sum of two even numbers is even.
    H. The product of two odd numbers is odd.
    J. The sum of two odd numbers is odd.

Short Response

58. a. How many dots would be in each of the next three figures?
    b. Write an expression for the number of dots in the nth figure.

    A   B   C   D

Extended Response

59. a. Describe the pattern. List the next two equations in the pattern.
    b. Guess what the product of 181 and 11 is. Test your conjecture.
    c. State whether the pattern can continue forever. Explain.

$$(101)(11) = 1111$$
$$(111)(11) = 1221$$
$$(121)(11) = 1331$$
$$(131)(11) = 1441$$
$$(141)(11) = 1551$$

Skills Handbook

for Help

60. Measure the sides DE and EF to the nearest millimeter.

61. Measure each angle of DEF to the nearest degree.

62. Draw a triangle that has sides of length 6 cm and 5 cm with a 90° angle between those two sides.

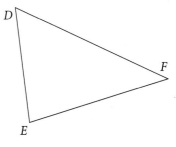

Drawings, Nets, and Other Models

NY Learning Standards for Mathematics

G.PS.3 Use multiple representations to represent and explain problem situations (e.g., spatial, geometric, verbal, numeric, algebraic, and graphical representations).

✓ **Check Skills You'll Need**

GO for Help Lesson 1-1

Draw the next figure in each sequence.

1.

2.

🔊 **New Vocabulary** • isometric drawing • orthographic drawing • foundation drawing • net

1 Drawing Isometric and Orthographic Views

You will study both two-dimensional and three-dimensional figures in geometry. A drawing on a piece of paper is a two-dimensional object. It has length and width. Your textbook is a three-dimensional object. It has length, width, and height. Representing a three-dimensional object on a two-dimensional surface requires special techniques.

You can make an **isometric drawing** on isometric dot paper to show three sides of a figure from a corner view. The simple drawing of a refrigerator at the right is an isometric drawing.

Vocabulary Tip

In Greek, *isos* means "equal" and *metron* means "measure." In an isometric drawing, all 3-D measurements are scaled equally.

1 EXAMPLE Isometric Drawing

Make an isometric drawing of the cube structure at the left.

Isometric drawing:

Step 1

Step 2

Step 3

✓ **Quick Check** ❶ On isometric dot paper, make an isometric drawing of the cube structure.

An **orthographic drawing** is another way to show a three-dimensional figure. It shows a top view, front view, and right-side view.

2 EXAMPLE Orthographic Drawing

Make an orthographic drawing from the isometric drawing at the left.

Isometric drawing:

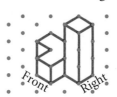

Orthographic drawing:

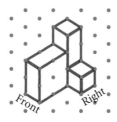

Solid lines show visible edges.

Top

Dashed lines show hidden edges.

Front   Right

✓ **Quick Check** ② Make an orthographic drawing from this isometric drawing.

A **foundation drawing** shows the base of a structure and the height of each part. A foundation drawing of the Sears Tower is shown at the right.

49	89	65
109	109	89
65	89	49

The Sears Tower is made up of nine sections. The numbers tell how many stories tall each section is.

Real-World 🌐 Connection

The foundation drawing shows four heights in the nine sections of the Sears Tower in Chicago, Illinois.

3 EXAMPLE Foundation Drawing

Make a foundation drawing for the isometric drawing at the left.

Isometric drawing:

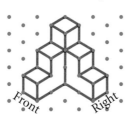

Foundation drawing:

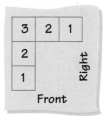

3	2	1
2		
1		

Front / Right

✓ **Quick Check** ③ **a.** How many cubes would you use to make the structure in Example 3?

b. Critical Thinking Which drawing did you use to answer part (a), the foundation drawing or the isometric drawing? Explain.

A **net** is a two-dimensional pattern that you can fold to form a three-dimensional figure. A net shows all of the surfaces of a figure in one view.

4 EXAMPLE Identifying Solids From Nets

Multiple Choice The net at the left shows all the surfaces of a three-dimensional figure. Which figure can you fold from the net?

A

B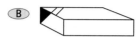

C

D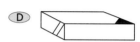

The black triangle is shown on the largest side of the figure instead of the smallest in choices C and D. Those choices cannot be correct. The black triangle will be at the same end as the two diagonal lines when the net is folded into a box. Choice B is correct.

✓ Quick Check 4 The net at the right folds into the cube shown beside it. Draw the cube and show which letters will be on its front and top.

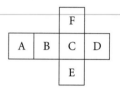

Test-Taking Tip

The number of surfaces on a solid must match the number of regions in its net.

Package designers can use nets to help design containers.

5 EXAMPLE Drawing a Net

Packaging Draw a net for the graham cracker box. Label the net with its dimensions.

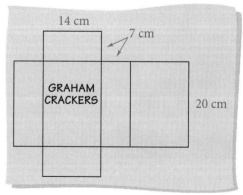

✓ Quick Check 5 Draw a net for the solid shown. Label the net with its dimensions.

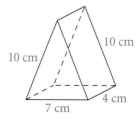

EXERCISES

For more exercises, see *Extra Skill, Word Problem, and Proof Practice.*

Practice and Problem Solving

 Practice by Example

Example 1
(page 10)

On isometric dot paper, make an isometric drawing of each cube structure.

1.   **2.** 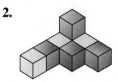 **3.**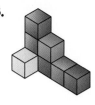

Examples 2, 3
(page 11)

For each figure, make (a) an orthographic drawing, and (b) a foundation drawing.

4. 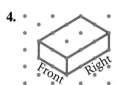 **5.** 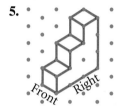 **6.**

How many cubes would you use to make each of the following?

7. the structure in Exercise 4   **8.** the structure in Exercise 5

9. the structure in Exercise 6   **10.** a model of the Sears Tower on page 11

Example 4
(page 12)

Match each three-dimensional figure with its net.

11.   **12.**   **13.**

A. 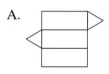   B. 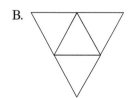   C.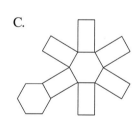

Example 5
(page 12)

Draw a net for each figure. Label the net with its dimensions.

14. 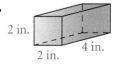 **15.** 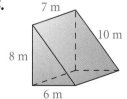 **16.**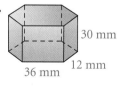

 Apply Your Skills

17. a. Open-Ended Make an isometric drawing of a structure that can be built using 8 cubes.
   b. Make an orthographic drawing of this structure.
   c. Make a foundation drawing for this structure.

For each foundation drawing, make (a) an isometric drawing on dot paper, and (b) an orthographic drawing.

18.

3	3
2	1

Right
Front

19.

1	2	3
	2	1

Right
Front

20.

1		
3	2	
3	2	1

Right
Front

Read the comic strip and complete Exercises 21 and 22.

SHOE by Jeff MacNelly

21. What type of drawing that you've studied in this lesson is a "bird's-eye view"?

22. Writing Photographs of the Washington Monument are typically not taken from a bird's-eye view. Describe a situation in which you would want a photo showing a bird's-eye view.

Visualization **Think about how each net can be folded to form a cube. What is the color of the face that will be opposite the red face?**

23. **24.** **25.** **26.**

27. There are eleven different nets for a cube. Four of them are shown above.
   a. Draw as many of the other seven as you can. (*Hint:* Two nets are the same if you can rotate or flip one to match the other.)
   b. Writing If you were going to make 100 cubes for a mobile, which of the eleven nets would you use? Explain why.

Make an orthographic drawing for each isometric drawing.

28. **29.**  **30.**

31. Draw a net for a cylinder. (*Hint:* The net needs to show three regions: two circles and a rectangle.)

32. There are eight different nets for a pyramid with a square base. Draw as many of them as you can.

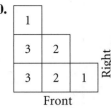

GO Online
Homework Video Tutor
Visit: PHSchool.com
Web Code: aue-0102

33. Visualization Use the orthographic drawing at the right.
  a. Make an isometric drawing of the structure.
  b. Make an isometric drawing of the structure from part (a) after it has been turned on its base 90° counterclockwise.
  c. Make an orthographic drawing of the stucture from part (b).
  d. Turn the structure from part (a) 180°. Repeat parts (b) and (c).

Top

Front   Right

34. The net at the left is folded into a cube. Sketch the cube so that its front face is shaded as shown at the right.

REGENTS

Test Prep

Multiple Choice

35. A three dimensional figure is made with 11 cubes. The top view of the figure shows 5 squares. Which of the following is the greatest possible number of cubes in a stack represented by one of the five squares?
  A. 4   **B.** 5   **C.** 6   **D.** 7

36. Which of the following shows a top, front, and right view of a three-dimensional shape?
  F. an isometric drawing   **G.** an orthographic drawing
  H. a foundation drawing   **J.** a net

Short Response

37. Draw a net for the rectangular box. Label the net with its dimensions.

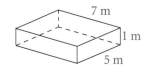
7 m
1 m
5 m

Extended Response

38. Make drawings to show the top view, the front view, and the right-side view of the figure at the right.

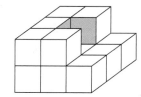

Mixed Review

Lesson 1-1
 for Help

39. Reasoning What is the last digit of 3^{45}? To answer, make a table, look for a pattern, and use inductive reasoning. Explain the pattern.

Find a pattern for each sequence. Use the pattern to show the next two terms.

40. A, C, E, G, . . .   **41.** 2, 6, 12, 20, 30, . . .

42. 4, 16, 64, 256, . . .   **43.** 100, 95, 85, 70, 50, . . .

Skills Handbook $\boxed{x^2}$ **Algebra** **Evaluate each expression for the given values.**

44. $a^2 + b^2$ for $a = 3$ and $b = -5$ **45.** $\frac{1}{2}bh$ for $b = 8$ and $h = 11$

1-3

Points, Lines, and Planes

NY **Learning Standards for Mathematics**

G.CM.4 Explain relationships among different representations of a problem.

G.CM.11 Understand and use appropriate language, representations, and terminology when describing objects, relationships, mathematical solutions, and geometric diagrams.

✔ Check Skills You'll Need

GO for Help Skills Handbook page 760

x^2 **Algebra** Solve each system of equations.

1. $y = x + 5$
  $y = -x + 7$

2. $y = 2x - 4$
  $y = 4x - 10$

3. $y = 2x$
  $y = -x + 15$

4. Copy the diagram of the four points $A, B, C,$ and D. Draw as many different lines as you can to connect pairs of points.

$A\bullet$
$\quad\quad\bullet B$
$C\bullet$
$\quad\bullet D$

🔊 New Vocabulary
- point
- space
- line
- collinear points
- plane
- coplanar
- postulate
- axiom

---

1 Basic Terms of Geometry

Hands-On Activity: How Many Lines Can You Draw?

Many constellations are named for animals and mythological figures. It takes some imagination to join the points representing the stars to get a recognizable figure such as Leo the Lion. How many lines can you draw connecting the 10 points in Leo the Lion?

- Make a table and look for a pattern to help you find out.

1. Mark three points on a circle. Now connect the three points with as many (straight) lines as possible. How many lines can you draw?

2. Mark four points on another circle. How many lines can you draw to connect the four points?

3. Repeat this procedure for five points on a circle and then for six points. How many lines can you draw to connect the points?

4. Use inductive reasoning to tell how many lines you can draw to connect the ten points of the constellation Leo the Lion.

In geometry, some words such as *point, line,* and *plane* are undefined. In order to define these words you need to use words that need further defining. It is important however, to have general descriptions of their meanings.

You can think of a **point** as a location. A point has no size. It is represented by a small dot and is named by a capital letter. A geometric figure is a set of points. **Space** is defined as the set of all points.

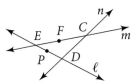

Vocabulary Tip

$\overleftrightarrow{AB}$ ("line *AB*") and $\overleftrightarrow{BA}$ ("line *BA*") name the same line.

You can think of a **line** as a series of points that extends in two opposite directions without end. You can name a line by any two points on the line, such as $\overleftrightarrow{AB}$ (read "line *AB*"). Another way to name a line is with a single lowercase letter, such as line *t* (see above). Points that lie on the same line are **collinear points.**

1 EXAMPLE **Identifying Collinear Points**

a. Are points *E*, *F*, and *C* collinear? If so, name the line on which they lie.

Points *E*, *F*, and *C* are collinear. They lie on line *m*.

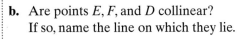

b. Are points *E*, *F*, and *D* collinear? If so, name the line on which they lie.

Points *E*, *F*, and *D* are not collinear.

✓**Quick Check** **1 a.** Are points *F*, *P*, and *C* collinear?
b. Name line *m* in three other ways.
c. Critical Thinking Why do you think arrowheads are used when drawing a line or naming a line such as $\overleftrightarrow{EF}$?

A **plane** is a flat surface that has no thickness. A plane contains many lines and extends without end in the directions of all its lines. You can name a plane by either a single capital letter or by at least three of its noncollinear points. Points and lines in the same plane are **coplanar.**

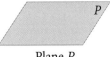

Plane *P*

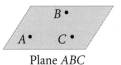

Plane *ABC*

2 EXAMPLE **Naming a Plane**

Each surface of the ice cube represents part of a plane. Name the plane represented by the front of the ice cube.

You can name the plane represented by the front of the ice cube using at least three noncollinear points in the plane. Some names are plane *AEF*, plane *AEB*, and plane *ABFE*.

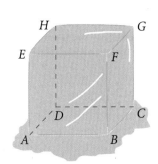

✓**Quick Check** **2** List three different names for the plane represented by the top of the ice cube.

A **postulate** or **axiom** is an accepted statement of fact.

You have used some of the following geometry postulates in algebra. For example, you used Postulate 1-1 when you graphed an equation such as $y = -2x + 8$. You plotted two points and then drew the line through those two points.

 Key Concepts

Postulate 1-1

Through any two points there is exactly one line.

Line t is the only line that passes through points A and B.

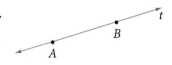

Vocabulary Tip

There is exactly one means "there is one and there is no more than one."

In algebra, one way to solve a system of two equations is to graph the two equations. As the graphs of

$$y = -2x + 8$$

$$y = 3x - 7$$

show, the two lines intersect at a single point, $(3, 2)$. The solution to the system of equations is $(3, 2)$.

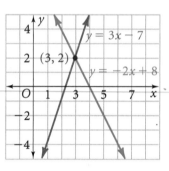

This illustrates Postulate 1-2.

 Key Concepts

Postulate 1-2

If two lines intersect, then they intersect in exactly one point.

$\overleftrightarrow{AE}$ and $\overleftrightarrow{BD}$ intersect at C.

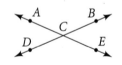

There is a similar postulate about the intersection of planes.

 Key Concepts

Postulate 1-3

If two planes intersect, then they intersect in exactly one line.

Plane RST and plane STW intersect in $\overleftrightarrow{ST}$.

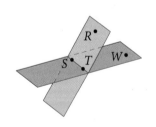

When you know two points in the intersection of two planes, Postulates 1-1 and 1-3 tell you that the line through those points is the line of intersection of the planes.

3 EXAMPLE Finding the Intersection of Two Planes

What is the intersection of
plane *HGFE* and plane *BCGF*?

Plane *HGFE* and plane *BCGF*
intersect in $\overleftrightarrow{GF}$.

Quick Check 3 Name two planes that intersect in $\overleftrightarrow{BF}$.

A three-legged stand will always be stable. As long as
the feet of the stand don't lie in one line, the feet of
the three legs will lie exactly in one plane.

This illustrates Postulate 1-4.

Key Concepts

Postulate 1-4
Through any three noncollinear points there is exactly one plane.

4 EXAMPLE Using Postulate 1-4

a. Shade the plane that contains
   A, B, and *C.*

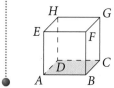

b. Shade the plane that contains
   E, H, and *C.*

Quick Check 4 **a.** Name another point that is in the same
   plane as points *A, B,* and *C.*
   b. Name another point that is coplanar with points *E, H,* and *C.*

EXERCISES

For more exercises, see *Extra Skill, Word Problem, and Proof Practice.*

Practice and Problem Solving

A Practice by Example

Example 1
(page 17)

GO for Help

Are the three points collinear? If so, name the line on which they lie.

1. *A, D, E*   **2.** *B, C, D*

3. *B, C, F*   **4.** *A, E, C*

5. *F, B, D*   **6.** *F, A, E*

7. *G, F, C*   **8.** *A, G, C*

9. Name line *m* in three other ways.

10. Name line *n* in three other ways.

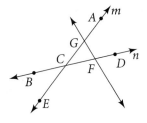

Example 2
(page 17)

Name the plane represented by each surface of the box.

11. the bottom   **12.** the top

13. the front   **14.** the back

15. the left side   **16.** the right side

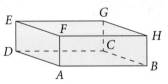

Example 3
(page 19)

**Use the figure at the right for Exercises 17–37.
First, name the intersection of each pair of planes.**

17. planes *QRS* and *RSW*   **18.** planes *UXV* and *WVS*

19. planes *XWV* and *UVR*   **20.** planes *TXW* and *TQU*

Name two planes that intersect in the given line.

21. $\overleftrightarrow{QU}$ **22.** $\overleftrightarrow{TS}$ **23.** $\overrightarrow{XT}$ **24.** $\overrightarrow{VW}$ Exercises 17–37

Example 4
(page 19)

Copy the figure. Shade the plane that contains the given points.

25. *R, V, W*   **26.** *U, V, W*   **27.** *U, X, S*   **28.** *T, U, X*   **29.** *T, V, R*

Name another point in each plane.

30. plane *RVW*   **31.** plane *UVW*   **32.** plane *UXS*   **33.** plane *TUX*   **34.** plane *TVR*

Is the given point coplanar with the other three points?

35. point *Q* with *V, W, S*   **36.** point *U* with *T, V, S*   **37.** point *W* with *X, V, R*

B **Apply Your Skills**

**Postulate 1-4 states that any three noncollinear points lie in
one plane. Find the plane containing the first three points
listed, then decide whether the fourth point is in that plane.
Write *coplanar* or *noncoplanar* to describe the points.**

38. *Z, S, Y, C*   **39.** *S, U, V, Y*

40. *X, Y, Z, U*   **41.** *X, S, V, U*

42. *X, Z, S, V*   **43.** *S, V, C, Y*

44. Photography Photographers and surveyors use a tripod, or three-legged stand,
for their instruments. Use one of the postulates to explain why.

45. Open-Ended Draw a figure with points *B, C, D, E, F,* and *G* that shows $\overleftrightarrow{CD}$,
$\overrightarrow{BG}$, and $\overleftrightarrow{EF}$, with one of the points on all three lines.

If possible, draw a figure to fit each description. Otherwise write *not possible*.

46. four points that are collinear   **47.** two points that are noncollinear

48. three points that are noncollinear   **49.** three points that are noncoplanar

Coordinate Geometry Graph the points and state whether they are collinear.

50. $(3, -3), (2, -3), (-3, 1)$ **51.** $(2, 2), (-2, -2), (3, 2)$

52. $(2, -2), (-2, -2), (3, -2)$ **53.** $(-3, 3), (-3, 2), (-3, -1)$

54. Multiple Choice Which three points are *not* collinear?

    Ⓐ $(0, 0), (0, 2), (0, 4)$ Ⓑ $(0, 0), (3, 0), (5, 0)$

    Ⓒ $(0, 0), (0, 2), (3, 0)$ Ⓓ $(2, -2), (2, 2), (2, 3)$

Real-World **Connection**

Careers The photographer
uses a tripod to help assure
a clear picture.

Use *always*, *sometimes*, or *never* to make a true statement.

55. Intersecting lines are __?__ coplanar.

56. Two planes __?__ intersect in exactly one point.

57. Three points are __?__ coplanar.

58. A plane containing two points of a line __?__ contains the entire line.

59. Four points are __?__ coplanar.

60. Two lines __?__ meet in more than one point.

61. How many planes contain each line and point?

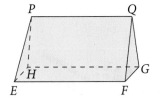

  a. $\overleftrightarrow{EF}$ and point G **b.** $\overrightarrow{PH}$ and point E

  c. $\overleftrightarrow{FG}$ and point P **d.** $\overrightarrow{EP}$ and point G

  e. Make a Conjecture What do you think is true of a line and a point not on the line?

In Exercise 62 and 63, sketch a figure for the given information. Then name the postulate that your figure illustrates.

62. The noncollinear points A, B, and C are all contained in plane N.

63. Planes LNP and MVK intersect in $\overleftrightarrow{NM}$.

64. Optical Illusions The diagram (right) is an optical illusion. Which three points are collinear: A, B, and C or A, B, and D? Are you sure? Use a straightedge to check your answer.

Writing Use postulates to explain each situation.

65. A land surveyor can always find a straight line from the point where she stands to any other point she can see.

66. A carpenter knows that a line can represent the intersection of two flat walls.

67. A furniture maker knows that a three-legged table is always steady, but a four-legged table will sometimes wobble.

Coordinate Geometry Graph the points and state whether they are collinear.

68. $(1, 1), (4, 4), (-3, -3)$ **69.** $(2, 4), (4, 6), (0, 2)$ **70.** $(0, 0), (-5, 1), (6, -2)$

71. $(0, 0), (8, 10), (4, 6)$ **72.** $(0, 0), (0, 3), (0, -10)$ **73.** $(-2, -6), (1, -2), (4, 1)$

C Challenge

74. How many planes contain the same three collinear points? Explain.

75. Navigation Rescue teams use Postulates 1-1 and 1-2 to determine the location of a distress signal. In the diagram, a ship at point A receives a signal from the northeast. A ship at point B receives the same signal from due west. Trace the diagram and find the location of the distress signal. Explain how the two postulates help locate the distress signal.

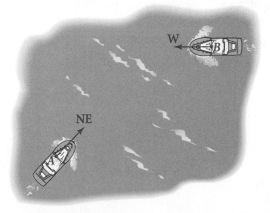

76. a. Open-Ended Suppose two points are in plane P. Explain why it makes sense that the line containing the points would be in the same plane.

   b. Suppose two lines intersect. How many planes do you think contain both lines? You may use the diagram and your answer in part (a) to explain your answer.

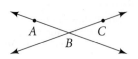

Probability Points are picked at random from A, B, C, and D, which are arranged as shown. Find the probability that the indicated number of points meet the given condition.

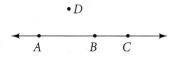

77. 2 points, collinear   **78.** 3 points, collinear   **79.** 3 points, coplanar

Multiple Choice

80. In the figure at the right, which points are collinear with C and H?

   A. B, F
   B. E, F, G
   C. A, D, G, I
   D. A, D, E, H

81. A solid chunk of cheese is to be cut into 4 pieces. What is the least number of slices needed?
   F. 5   **G.** 4   **H.** 3   **J.** 2

82. Ronald is making a table. What is the least number of legs that the table should have so that it will not wobble?
   A. 4   **B.** 3   **C.** 2   **D.** 1

83. At most, how many lines can contain pairs of the points P, Q, and R?
   F. 1   **G.** 2
   H. 3   **J.** 4

Short Response

84. Use the figure at the right.
   a. Name all the planes that form the figure.
   b. Name all the lines that intersect at D.

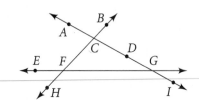

Exercise 84

Mixed Review

Lesson 1-2

Make an orthographic drawing for each figure.

85.
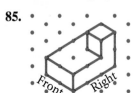

86.

87.

Skills Handbook

Simplify each ratio.

88. 30 to 12   **89.** $\frac{\pi r^2}{2\pi r}$ **90.** $\frac{5}{21} : \frac{5}{7}$

1-4 Segments, Rays, Parallel Lines and Planes

 Learning Standards for Mathematics

G.G.8 Know and apply that if a plane intersects two parallel planes, then the intersection is two parallel lines.

✓ **Check Skills You'll Need**

GO for Help Lesson 1-3

Judging by appearances, will the lines intersect?

1.

2.

3.

Name the plane represented by each surface of the box.

4. the bottom   **5.** the top

6. the front   **7.** the back

8. the left side   **9.** the right side

🔊 **New Vocabulary** • segment • ray • opposite rays • parallel lines
 • skew lines • parallel planes

---

1 Identifying Segments and Rays

Real-World 🌐 Connection

A sunbeam models a ray. The sun is its endpoint.

Many geometric figures, such as squares and angles, are formed by parts of lines called segments or rays. A **segment** is the part of a line consisting of two endpoints and all points between them.

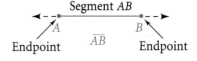

A **ray** is the part of a line consisting of one endpoint and all the points of the line on one side of the endpoint.

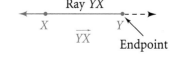

Opposite rays are two collinear rays with the same endpoint. Opposite rays always form a line.

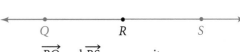

$\overrightarrow{RQ}$ and $\overrightarrow{RS}$ are opposite rays.

1 EXAMPLE **Naming Segments and Rays**

Name the segments and rays in the figure at the right.

• The three segments are $\overline{LP}, \overline{PQ},$ and $\overline{LQ}$.
• The four rays are $\overrightarrow{LP}$ or $\overrightarrow{LQ}, \overrightarrow{PQ}, \overrightarrow{PL},$ and $\overrightarrow{QP}$ or $\overrightarrow{QL}$.

✓ **Quick Check** **①** **Critical Thinking** $\overrightarrow{LP}$ and $\overrightarrow{PL}$ form a line. Are they opposite rays? Explain.

Lines that do not intersect may or may not be coplanar.

Parallel lines are coplanar lines that do not intersect. **Skew lines** are noncoplanar; therefore, they are not parallel and do not intersect.

Vocabulary Tip

You read $\overleftrightarrow{AB} \parallel \overleftrightarrow{EF}$ as "line *AB* is parallel to line *EF*."

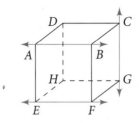

$\overleftrightarrow{AB} \parallel \overleftrightarrow{EF}$
$\overleftrightarrow{AB}$ and $\overleftrightarrow{CG}$ are skew.

Segments or rays are parallel if they lie in parallel lines. They are skew if they lie in skew lines. $\overline{AB}$ and $\overline{CG}$ are skew because $\overleftrightarrow{AB}$ and $\overleftrightarrow{CG}$ are skew.

2 EXAMPLE   Identifying Parallel and Skew Segments

Video Tutor Help

Visit: PhSchool.com
Web Code: aue-0775

a. Name all labeled segments that are parallel to $\overline{DC}$.

$\overline{AB}, \overline{GH},$ and $\overline{JI}$ are parallel to $\overline{DC}$.

b. Name all labeled segments that are skew to $\overline{DC}$.

$\overline{NJ}, \overline{GJ},$ and $\overline{HI}$ are skew to $\overline{DC}$.

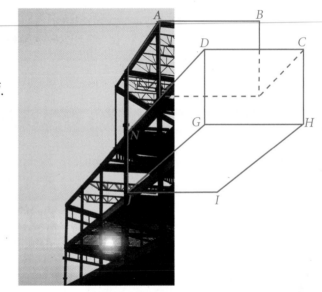

✓ Quick Check   **2** Use the diagram in Example 2.
    a. Name all labeled segments that are parallel to $\overline{GJ}$.
    b. Name all labeled segments that are skew to $\overline{GJ}$.
    c. Name another pair of parallel segments; of skew segments.

Parallel planes are planes that do not intersect. A line and a plane that do not intersect are also parallel.

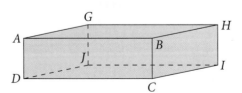

Plane *ABCD* ∥ Plane *GHIJ*.
Plane $ABCD \parallel \overleftrightarrow{GH}$.

3 **EXAMPLE** **Identifying Parallel Planes**

Use the diagram at the right to name
the figures.
a. two pairs of parallel planes

plane *ABHG* ∥ plane *DCIJ*
plane *ADJ* ∥ plane *BCI*

b. a line that is parallel to plane *GHIJ*

$\overleftrightarrow{AB}$ is parallel to *GHIJ*.

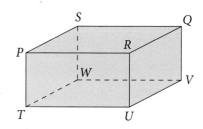

Quick Check **3** Name the figures.
a. three pairs of parallel planes
b. a line that is parallel to plane *QRUV*

EXERCISES

For more exercises, see *Extra Skill, Word Problem, and Proof Practice.*

Practice and Problem Solving

A **Practice by Example**

Example 1
(page 23)

for Help

Use the figure at the right for Exercises 1–3.

1. Name all the labeled segments.

2. Name all the labeled rays.

3. a. Name a pair of opposite rays with *T*
as an endpoint.
b. Name another pair of opposite rays.

Example 2
(page 24)

**Name all segments shown in the diagram that are
parallel to the given segment.**

4. $\overline{AC}$ **5.** $\overline{EF}$ **6.** $\overline{AD}$

**Name all segments shown in the diagram that are
skew to the given segment.**

7. $\overline{AC}$ **8.** $\overline{EF}$ **9.** $\overline{AD}$

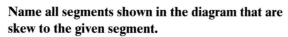

Exercises 4–11

Example 3
(page 25)

Use the diagram above and name a pair of figures to match each description.

10. parallel planes   **11.** a line and a plane that are parallel

Use the figure at the right to name the following.

12. all lines that are parallel to $\overleftrightarrow{AB}$

13. two lines that are skew to $\overleftrightarrow{EJ}$

14. all lines that are parallel to plane *JFAE*

15. the intersection of plane *FAB* and plane *FAE*

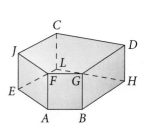

B **Apply Your Skills**

In Exercises 16–23, describe the statement as true or false. If *false*, explain.

16. $\overleftrightarrow{CB} \parallel \overleftrightarrow{HG}$

17. $\overleftrightarrow{ED} \parallel \overleftrightarrow{HG}$

18. plane *AED* ∥ plane *FGH*

19. plane *ABH* ∥ plane *CDF*

20. $\overleftrightarrow{AB}$ and $\overleftrightarrow{HG}$ are skew lines.   21. $\overleftrightarrow{AE}$ and $\overleftrightarrow{BC}$ are skew lines.

22. $\overleftrightarrow{CG}$ and $\overleftrightarrow{AI}$ are skew lines.   23. $\overleftrightarrow{CF}$ and $\overleftrightarrow{AJ}$ are skew lines.

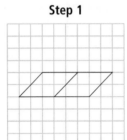

24. The following steps show how to draw planes *A* and *B* intersecting in $\overleftrightarrow{FG}$.

Step 1	**Step 2**	**Step 3**

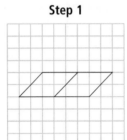
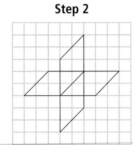
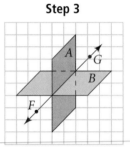

Use similar steps to draw plane *DFE* and plane *DFJ* intersecting in $\overleftrightarrow{DF}$.

Complete Exercises 25–33 with *always*, *sometimes*, or *never* to make a true statement.

25. Two parallel lines are __?__ coplanar.   26. Two skew lines are __?__ coplanar.

27. Two opposite rays __?__ form a line.   28. $\overrightarrow{TQ}$ and $\overleftrightarrow{QT}$ are __?__ the same line.

29. $\overrightarrow{GH}$ and $\overrightarrow{HG}$ are __?__ the same ray.   30. $\overrightarrow{JK}$ and $\overrightarrow{JL}$ are __?__ the same ray.

31. Two planes that do not intersect are __?__ parallel.

32. Two lines that lie in parallel planes are __?__ parallel.

33. Two lines in intersecting planes are __?__ skew.

34. **Multiple Choice** $\overline{FG}$ has endpoints $F(-3, 3)$ and $G(3, 1)$. Which point is also on $\overleftrightarrow{FG}$?

    Ⓐ $(-6, 4)$ Ⓑ $(-1, 2)$ Ⓒ $(0, 2)$ Ⓓ $(6, 0)$

35. **Coordinate Geometry** $\overrightarrow{AB}$ has endpoint $A(2, 3)$ and contains $B(4, 6)$. Give possible coordinates for point *C* so that $\overrightarrow{AB}$ and $\overrightarrow{AC}$ are opposite rays. Graph your answer.

36. **Directional Compass** On a directional compass, the directions north and south can be represented by opposite rays.
    a. Name two other compass directions that can be represented by opposite rays.
    b. What other pairs of opposite directions, if any, can you find?

37. **Open-Ended** Summarize the three ways in which two lines may be related. Give examples from the real world that illustrate the relationships.

GO **Online**

Homework Video Tutor

Visit: PHSchool.com
Web Code: aue-0104

38. Writing The term *skew* is a Middle English word meaning "to escape." Explain how this meaning might be appropriate for skew lines.

39. Critical Thinking Suppose two parallel planes *A* and *B* are each intersected by a third plane *C*.
  a. Make a conjecture about the intersection of planes *A* and *C* and the intersection of planes *B* and *C*.
  b. Find examples in your classroom.

 Challenge

40. a. Draw a line. Draw points *E* and *F* on the line. How many different segments do points *E* and *F* determine? Name the segments.
  b. Draw another line. Draw points *E*, *F*, and *G* on the line. How many segments do points *E*, *F*, and *G* determine? Name them.
  c. Continue to draw lines, labeling one more point each time. Make a table showing the number of points and the number of segments determined. Look for and describe a pattern in the data.
  d. Use your pattern to find how many segments are determined if you label 10 points on a line.
  e. If you label *n* points on a line, how many segments can you name?

Use the figure at the right for Exercises 41 and 42.

41. Do planes *A* and *B* have other lines in common that are parallel to $\overleftrightarrow{CD}$? Explain.

42. Visualization Are there planes that intersect planes *A* and *B* in lines parallel to $\overleftrightarrow{CD}$? Draw a sketch to support your answer.

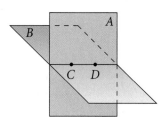

The figure at the right is a pyramid.

43. Name three lines that intersect at one point.

44. What line could be parallel to $\overleftrightarrow{PS}$?

45. Visualization Consider a plane through *V* that is parallel to plane *PQRS*. Can a line in that plane be parallel to $\overleftrightarrow{SR}$? Can it intersect $\overleftrightarrow{SR}$? Can it be skew to $\overleftrightarrow{SR}$? Explain each answer.

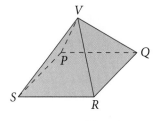

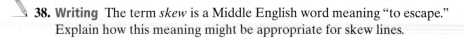

NY REGENTS

Test Prep

Use the figure at the right for Exercises 46–48.

Multiple Choice

46. How many labeled segments are in the figure?
  A. 1 **B.** 4 **C.** 6 **D.** 10

47. Which ray is opposite $\overrightarrow{BC}$?
  F. $\overrightarrow{BE}$ **G.** $\overrightarrow{BD}$ **H.** $\overrightarrow{BA}$ **J.** $\overrightarrow{AB}$

48. What is another name for $\overrightarrow{CA}$?
  A. $\overrightarrow{AC}$ **B.** $\overrightarrow{CB}$ **C.** $\overrightarrow{CE}$ **D.** $\overrightarrow{DC}$

49. Which figure could be the intersection of two planes?
  F. line **G.** ray **H.** point **J.** segment

50. a. Use the diagram to explain how parallel lines and skew lines are alike and how parallel lines and skew lines are different.

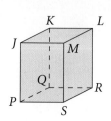

b. Does the diagram suggest other lines that are parallel to $\overleftrightarrow{JM}$, besides $\overleftrightarrow{KL}$, $\overleftrightarrow{QR}$, and $\overleftrightarrow{PS}$? Explain.

Mixed Review

Lesson 1-3

Use the diagram for Exercises 51–58 and name each geometric figure.

51. a line   **52.** a point

53. the intersection of $\overline{DC}$ and $\overleftrightarrow{CG}$

54. two planes that intersect in $\overleftrightarrow{EF}$

55. the plane represented by the top of the box

56. the plane represented by the front of the box

57. the intersection of planes EFG and DFG

58. another point in plane CGH

Draw the following.

59. $\overleftrightarrow{TR}$ **60.** $\overline{PQ}$ **61.** $\overrightarrow{NV}$

Lesson 1-1

Find the next two terms in each sequence.

62. $1, 1.08, 1.16, 1.24, 1.32, \ldots$ **63.** $-1, -2, -4, -7, -11, -16, \ldots$

64. $AB, BC, CD, DE, EF, \ldots$ **65.** $A, D, G, J, M, \ldots$

66. Reasoning Raven conjectured: "If you subtract a number from a given number, the result is always less than the given number." Is her conjecture true? Explain.

Checkpoint Quiz 1   Lessons 1-1 through 1-4

Find the next two terms in each sequence.

1. $19, 21.5, 24, 26.5, \ldots$ **2.** $3.4, 3.45, 3.456, 3.4567, \ldots$

 3. Writing Describe the pattern of each sequence in Exercises 1 and 2.

Use the diagram for Exercises 4–10. In Exercises 4–7, do the points appear to be coplanar? If *yes*, name the plane. If *no*, explain.

4. Points $A, E, F,$ and B **5.** Points $D, C, E,$ and F

6. Points $H, G, F,$ and B **7.** Points $A, E, B,$ and C

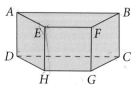

8. Name all the segments parallel to $\overline{HG}$.

9. Name a pair of skew lines.

10. Draw a net for the figure.

Geometry Vocabulary

To succeed with geometry, you must learn its vocabulary. As a study aid, some students use index cards. Others use special notebooks. With either, you can note

- each geometry term,

- its meaning,

- an example of its use, and

- a drawing, if possible.

A study aid like this can become a valuable tool to use throughout the course.

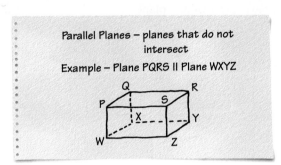

EXERCISES

1. Work with a classmate. Plan a study aid that will help both of you learn the geometry vocabulary. Make a sample page. Then share your sample with others. Use the best ideas from the class to revise your method.

2. Begin a vocabulary section of your study aid using Lessons 1-1 through 1-4.

A popular dictionary shows the definitions at the right for *add* and *sum*.

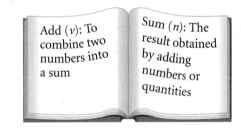

3. The definitions are *circular*. Tell what you think it means for a definition to be circular.

4. Circular defining can involve several statements.

   Show means display. *Display* means illustrate. *Illustrate* means show.

   a. Why are these definitions of *show*, *display*, and *illustrate* circular?
   b. Definitions like these, even though they are circular, can still help you understand vocabulary. Why?

A *mathematical system*, such as geometry, tries to avoid circular definitions. That means some terms must remain undefined.

5. *Point* is a term commonly accepted as undefined in geometry.
   a. Try to write a definition of *point*. Ask classmates whether they think it's a good definition.
   b. What other examples or descriptions—besides those in this book—can help you understand *point*? (You may want to include them in your study aid.)

6. Explain why some vocabulary terms must remain undefined.

Extension

Perpendicular Lines and Planes

FOR USE WITH LESSON 1-4

You can use a parallelogram as a sketch of a plane in space. You can sketch overlapping parallelograms to suggest how two planes intersect in a line. You saw one way to do this on page 26. For another way, sketch one plane and the line of intersection *before* drawing the second plane.

NY G.G.1, G.G.2, G.G.3, G.G.4, G.G.5, G.G.6, G.G.7, G.G.9

Step 1: **Step 2:** **Step 3:**

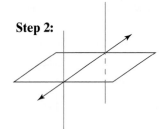
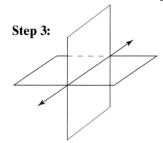

EXERCISES

1. The sketch at the right shows a plane and the line of intersection of a second plane.

   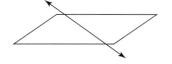

   a. Copy the sketch and complete it to show two intersecting planes that appear to be perpendicular to each other.

   b. Copy the sketch again. This time show two intersecting planes that clearly are not perpendicular to each other.

Draw each of the following.

2. Sketch a plane in space. Then draw two lines that are in the plane and intersect in a point *A*. Draw a third line that is perpendicular to each of the two lines at point *A*.

3. Here are two ways to draw a line intersecting a plane in a point. Try both.
   a. First sketch a plane. Mark the point of intersection. Then draw a line that pierces the plane at that point.
   b. First sketch the line. Mark the point of intersection. Then draw a parallelogram around the point to make it appear that the line pierces the plane at that point.
   c. Now, draw a plane and two lines that intersect the plane at the same point. Make one line appear to be perpendicular to the plane.

4. Draw two lines that appear to be perpendicular to the same plane.

5. Draw two planes that appear to be perpendicular to the same line.

6. Draw line ℓ perpendicular to plane *P* and intersecting plane *P* at point *A*. Then draw a second line that appears to be perpendicular to line ℓ at point A.

7. Draw a line ℓ that appears to be perpendicular to plane *P*. Then draw a second plane *Q* that contains line ℓ. Hint: Before drawing plane *Q*, think of what its intersection with plane *P* would look like.

8. Draw three planes perpendicular to each other. Hint: Think what the intersection of the third plane with the first two planes would look like.

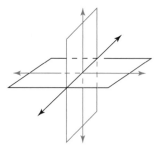

Exercise 8

Measuring Segments

 Learning Standards for Mathematics

G.PS.5 Choose an effective approach to solve a problem from a variety of strategies (numeric, graphic, algebraic).

 **Check Skills You'll Need**

 **for Help** Skills Handbook pages 757 and 758

Simplify each absolute value expression.

1. $|-6|$ **2.** $|3.5|$ **3.** $|7 - 10|$

4. $|-4 - 2|$ **5.** $|-2 - (-4)|$ **6.** $|-3 + 12|$

$\boxed{x^2}$ **Algebra** Solve each equation.

7. $x + 2x - 6 = 6$ **8.** $3x + 9 + 5x = 81$ **9.** $w - 2 = -4 + 7w$

New Vocabulary • coordinate • congruent segments • midpoint

---

1 Finding Segment Lengths

The distance between points C and D on the ruler is 3. You can use the Ruler Postulate to find the distance between points on a number line.

 Key Concepts

> **Postulate 1-5**   **Ruler Postulate**
>
> The points of a line can be put into one-to-one correspondence with the real numbers so that the distance between any two points is the absolute value of the difference of the corresponding numbers.

the length of $\overline{AB}$

$$AB = |a - b|$$

coordinate of A **coordinate** of B

Vocabulary Tip

The <u>congruence</u> symbol ($\cong$) shows that two figures are equal ($=$) in size and similar ($\sim$) in shape.

Two segments with the same length are **congruent ($\cong$) segments.** In other words, if $AB = CD$, then $\overline{AB} \cong \overline{CD}$. You can use these statements interchangeably.

$AB = CD \longrightarrow \overline{AB} \cong \overline{CD}$

As illustrated above, segments can be marked alike to show they are congruent.

1 EXAMPLE Comparing Segment Lengths

Find AB and BC. Are $\overline{AB}$ and $\overline{BC}$ congruent?

A   B   C   D   E

-8 -7 -6 -5 -4 -3 -2 -1 0 1 2 3

$AB = |-8 - (-5)| = |-3| = 3$

$BC = |-5 - (-2)| = |-3| = 3$

● $AB = BC$, so $\overline{AB} \cong \overline{BC}$.

✓ Quick Check ❶ a. Compare CD and DE. Are the segments congruent?

b. **Critical Thinking** To find AB in Example 1, suppose you subtract -8 from -5. Do you get the same result? Why?

Examine the lengths of $\overline{AB}$ and $\overline{BC}$ in Example 1. Notice that $AB + BC = 6$. Notice that $AC = 6$. This suggests the following postulate.

Key Concepts

Postulate 1-6	Segment Addition Postulate

If three points A, B, and C are collinear and B is between A and C, then $AB + BC = AC$.

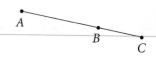

2 EXAMPLE Using the Segment Addition Postulate

Algebra If $DT = 60$, find the value of x. Then find DS and ST.

        $2x - 8$ $3x - 12$
    D   S   T

$DS + ST = DT$	Segment Addition Postulate
$(2x - 8) + (3x - 12) = 60$	Substitute.
$5x - 20 = 60$	Simplify.
$5x = 80$	Add 20 to each side.
$x = 16$	Divide each side by 5.
$DS = 2x - 8 = 2(16) - 8 = 24$	Substitute 16 for x.
● $ST = 3x - 12 = 3(16) - 12 = 36$	

✓ Quick Check ❷ $EG = 100$. Find the value of x. Then find EF and FG.

   $4x - 20$ $2x + 30$
  E   F   G

A **midpoint** of a segment is a point that divides the segment into two congruent segments. A midpoint, or any line, ray, or other segment through a midpoint, is said to *bisect* the segment.

A   B   C

$\overline{AB} \cong \overline{BC}$

3 EXAMPLE Using the Midpoint

Algebra C is the midpoint of $\overline{AB}$. Find AC, CB, and AB.

$AC = CB$	Definition of midpoint
$2x + 1 = 3x - 4$	Substitute.
$2x + 5 = 3x$	Add 4 to each side.
$5 = x$	Subtract 2x from each side.
$AC = 2x + 1 = 2(5) + 1 = 11$	Substitute 5 for x.
$CB = 3x - 4 = 3(5) - 4 = 11$	

● AC and CB are both 11, which is half of 22, the length of $\overline{AB}$.

✓ Quick Check ❸ Z is the midpoint of $\overline{XY}$, and $XY = 27$. Find XZ.

EXERCISES

For more exercises, see *Extra Skill, Word Problem, and Proof Practice.*

Practice and Problem Solving

 Practice by Example

Example 1
(page 32)

GO for Help

Find the length of each segment. Tell whether the segments are congruent.

1. $\overline{AC}$ and $\overline{BD}$ **2.** $\overline{BD}$ and $\overline{CE}$

3. $\overline{AD}$ and $\overline{BE}$ **4.** $\overline{BC}$ and $\overline{CE}$

On a number line, the coordinates of X, Y, Z, and W are –7, –3, 1, and 5, respectively. Find the lengths of the two segments and tell whether they are congruent.

5. $\overline{XY}$ and $\overline{ZW}$ **6.** $\overline{ZX}$ and $\overline{WY}$ **7.** $\overline{YZ}$ and $\overline{XW}$

Example 2
(page 32)

Use the figure at the right for Exercises 8–11.

8. If $RS = 15$ and $ST = 9$, then $RT = \blacksquare$.

9. If $ST = 15$ and $RT = 40$, then $RS = \blacksquare$.

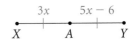

x^2 **10. a. Algebra** If $RS = 3x + 1$, $ST = 2x - 2$, and $RT = 64$, find the value of x.
   b. Find RS and ST.

x^2 **11. a. Algebra** If $RS = 8y + 4$, $ST = 4y + 8$, and $RT = 15y - 9$, find the value of y.
   b. Find RS, ST, and RT.

Example 3
(page 33)

x^2 **12. Algebra** A is the midpoint of $\overline{XY}$.
   a Find XA.
   b. Find AY and XY.

x^2 **Algebra** In Exercises 13–15, use the figure and find PT.

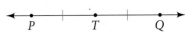

13. $PT = 5x + 3$ and $TQ = 7x - 9$

14. $PT = 4x - 6$ and $TQ = 3x + 4$

15. $PT = 7x - 24$ and $TQ = 6x - 2$

Use the figure at the right for Exercises 16–19.

16. Find the midpoint of $\overline{AB}$.

17. What is the coordinate of the midpoint of $\overline{QB}$?

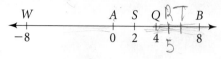

18. What is the coordinate of the midpoint of $\overline{WA}$?

19. What is the coordinate of the midpoint of the segment formed by the two points you found in Exercises 17 and 18?

Suppose the coordinate of *A* is 0, *AR* = 5, and *AT* = 7. What are the possible coordinates of the midpoint of the given segment?

20. $\overline{AR}$ **21.** $\overline{AT}$ **22.** $\overline{RT}$

Visit: PHSchool.com
Web Code: aue-0105

23. Mileage Exit numbers on some highways give mileage from one edge of the state. The highway and its exit numbers resemble a number line. You can find distance between exits in the same way that you find distance on a number line.

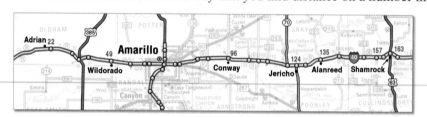

a. Use the map. How many miles is it from Wildorado to Shamrock?

b. Which town is 28 miles from Jericho?

Visualization **Without using your ruler, sketch a segment with the given length. Then use your ruler to see how well you did.**

24. 3 cm   **25.** 3 in.   **26.** 6 in.   **27.** 10 cm   **28.** 65 mm

In Exercises 29–32, describe the statement as *true* or *false*. Explain.

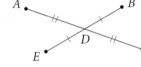

29. $\overline{AB} \cong \overline{CD}$ **30.** $BD < CD$

Exercises 29–33

31. $AC + BD = AD$ **32.** $AC + CD = AD$

33. Suppose $EG = 5$. Find the possible coordinate(s) of point G.

x^2 **Algebra** Use the diagram at the right for Exercises 34 and 35.

34. If $AD = 12$ and $AC = 4y - 36$, find the value of y. Then find AC and DC.

35. If $ED = x + 4$ and $DB = 3x - 8$, find ED, DB, and EB.

36. C is the midpoint of $\overline{AB}$, D is the midpoint of $\overline{AC}$, E is the midpoint of $\overline{AD}$, F is the midpoint of $\overline{ED}$, G is the midpoint of $\overline{EF}$, and H is the midpoint of $\overline{DB}$. If $DC = 16$, find GH.

37. a. Write an algebraic expression that represents GK.

     b. If $GK = 30$, find GH and JK.

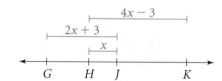

38. Art Project You want to cut pieces of ribbon for an art project. Each piece must be $6\frac{3}{4}$ inches long. For measuring, you have only the old, broken ruler shown at the right.

a. What marks on the ruler would you use to measure the ribbon?

b. **Writing** You also need 5-in. pieces of string. Describe two ways you can use the broken ruler to measure 5 inches.

NY REGENTS

Multiple Choice

39. If $KC = 31$, what is KN?
A. 43   B. 62
C. 74   D. 82

$$\underset{K}{\bullet}\overset{2x + 10}{\rule{2cm}{0.4pt}}\underset{C}{\bullet}\overset{4x + 1}{\rule{2cm}{0.4pt}}\underset{N}{\bullet}$$
Exercises 39–41

40. If $KN = 29$, what is CN?
F. 13   G. 14.5   H. 15.5   J. 16

41. If C is the midpoint of $\overline{KN}$, what is KC?
A. 4.5   B. 9   C. 18   D. 19

42. On a number line, point A has coordinate –6, and B has coordinate 2. Which is the coordinate of point M, the midpoint of segment AB?
F. 4   G. 0   H. –2   J. –3

43. Points X, Y, and Z are collinear with Y between X and Z. Which of the following must be true?
A. $XY = YZ$
B. $XZ - XY = YZ$
C. $XY + XZ = YZ$
D. $XZ = XY - YZ$

Short Response

44. Points L, M, and N are collinear with M between L and N. $LM = 2x + 8$ and MN is one half the length of LM. If $LN = 42$, write and solve an equation to find x.

GO for Help

Lesson 1-4

Complete each statement with *always*, *sometimes*, or *never* to make a true statement.

45. Skew lines are __?__ coplanar.

46. Skew lines __?__ intersect.

47. Opposite rays __?__ form a line.

48. Parallel planes __?__ intersect.

Lesson 1-3

49. Three points are __?__ coplanar.

50. Two points are __?__ collinear.

51. The intersection of two planes is __?__ a line.

52. Intersecting lines are __?__ parallel.

Lesson 1-1

Find the next two terms in each sequence.

53. $5, 10, 15, 20, \ldots$ **54.** $5, 25, 125, 625, \ldots$ **55.** $14, 18, 22, 26, \ldots$

Measuring Angles

 Learning Standards for Mathematics

G.CM.5 Communicate logical arguments clearly, showing why a result makes sense and why the reasoning is valid.

 **Check Skills You'll Need**

GO **for Help** Skills Handbook page 758

Solve each equation.

1. $50 + a = 130$ **2.** $m - 110 = 20$ **3.** $85 - n = 40$

4. $x + 45 = 180$ **5.** $z - 20 = 90$ **6.** $180 - y = 135$

◀)) **New Vocabulary** • angle • acute angle • right angle • obtuse angle
• straight angle • congruent angles • vertical angles
• adjacent angles • complementary angles
• supplementary angles

---

1 Finding Angle Measures

Vocabulary Tip

You may also refer to the angle suggested by the two segments $\overline{BT}$ and $\overline{BQ}$ as $\angle TBQ$.

An **angle** ($\angle$) is formed by two rays with the same endpoint. The rays are the *sides* of the angle. The endpoint is the *vertex* of the angle. The sides of the angle shown here are $\overrightarrow{BT}$ and $\overrightarrow{BQ}$. The vertex is B. You could name this angle $\angle B$, $\angle TBQ$, $\angle QBT$, or $\angle 1$.

1 EXAMPLE **Naming Angles**

Name $\angle 1$ in two other ways.

● $\angle AEC$ and $\angle CEA$ are other names for $\angle 1$.

 **Quick Check** **1** **a.** Name $\angle CED$ two other ways.
b. Critical Thinking Would it be correct to name any of the angles $\angle E$? Explain.

One way to measure an angle is in degrees. To indicate the size or degree measure of an angle, write a lowercase m in front of the angle symbol. The degree measure of angle A is 80. You show this by writing $m\angle A = 80$.

 **Key Concepts**

Postulate 1-7	**Protractor Postulate**

Let $\overrightarrow{OA}$ and $\overrightarrow{OB}$ be opposite rays in a plane. $\overrightarrow{OA}$, $\overrightarrow{OB}$, and all the rays with endpoint O that can be drawn on one side of $\overleftrightarrow{AB}$ can be paired with the real numbers from 0 to 180 so that

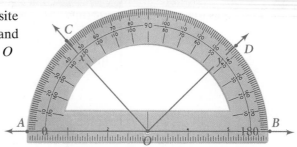

a. $\overrightarrow{OA}$ is paired with 0 and $\overrightarrow{OB}$ is paired with 180.

b. If $\overrightarrow{OC}$ is paired with x and $\overrightarrow{OD}$ is paired with y, then $m\angle COD = |x - y|$.

You can classify angles according to their measures.

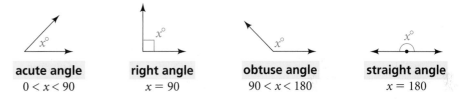

acute angle	**right angle**	**obtuse angle**	**straight angle**
$0 < x < 90$	$x = 90$	$90 < x < 180$	$x = 180$

Note the special symbol that's tucked into the corner of the right angle. When you see it, you know that the measure of the angle is 90.

GO Online

Video Tutor Help

Visit: PHSchool.com
Web Code: aue-0775

2 EXAMPLE **Measuring and Classifying Angles**

Find the measure of each angle. Classify each as *acute*, *right*, *obtuse*, or *straight*.

a.

b.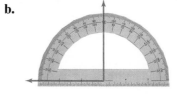

120, obtuse

90, right

✓ **Quick Check** **2** Find the measure of each angle. Classify each as *acute*, *right*, *obtuse*, or *straight*.

a.    **b.**   **c.**

Angles with the same measure are **congruent angles.** In other words, if $m\angle 1 = m\angle 2$, then $\angle 1 \cong \angle 2$. You can use these statements interchangeably.

Angles can be marked alike to show that they are congruent, as in this photograph of the Air Force Thunderbirds precision flying team.

The Angle Addition Postulate is similar to the Segment Addition Postulate.

| Postulate 1-8 | **Angle Addition Postulate** |

If point B is in the interior of $\angle AOC$, then $m\angle AOB + m\angle BOC = m\angle AOC$.

If $\angle AOC$ is a straight angle, then $m\angle AOB + m\angle BOC = 180$.

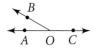

3 EXAMPLE **Using the Angle Addition Postulate**

What is $m\angle TSW$ if $m\angle RST = 50$ and $m\angle RSW = 125$?

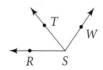

$m\angle RST + m\angle TSW = m\angle RSW$ **Angle Addition Postulate**

$50 + m\angle TSW = 125$ **Substitute.**

$m\angle TSW = 75$ **Subtract 50 from each side.**

✓ Quick Check **3** If $m\angle DEG = 145$, find $m\angle GEF$.

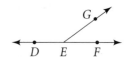

2   Identifying Angle Pairs

Some angle pairs that have special names.

vertical angles

two angles whose sides are opposite rays

adjacent angles

two coplanar angles with a common side, a common vertex, and no common interior points

complementary angles

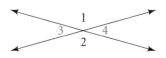

two angles whose measures have sum 90

Each angle is called the *complement* of the other.

supplementary angles

two angles whose measures have sum 180

Each angle is called the *supplement* of the other.

4 EXAMPLE Identifying Angle Pairs

In the diagram identify pairs of numbered angles that are related as follows:

a. complementary
  ∠2 and ∠3

b. supplementary
  ∠4 and ∠5; ∠3 and ∠4

c. vertical
  ∠3 and ∠5

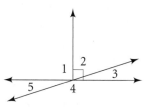

✓ Quick Check ❹ **a.** Name two pairs of adjacent angles in the photo below.
  b. If $m\angle EFD = 27$, find $m\angle AFD$.

When entering the roadway, turn and look for oncoming traffic regardless of what you see in the rear-view mirror.

Whether you draw a diagram or use a given diagram, you can make some conclusions directly from the diagrams. You *can* conclude that angles are

- adjacent angles
- adjacent supplementary angles
- vertical angles

Unless there are marks that give this information, you *cannot* assume

- angles or segments are congruent
- an angle is a right angle
- lines are parallel or perpendicular

Video Tutor Help
Visit: PHSchool.com
Web Code: aue-0775

5 EXAMPLE Making Conclusions From a Diagram

What can you conclude from the information in the diagram?

- ∠1 ≅ ∠2, by the markings.

- ∠2 and ∠3, for example, are adjacent angles.

- ∠4 and ∠5, for example, are adjacent supplementary angles,
  or $m\angle 4 + m\angle 5 = 180$ by the Angle Addition Postulate.

- ∠1 and ∠4, for example, are vertical angles.

✓ Quick Check ❺ Can you make each conclusion from the information in the diagram? Explain.

a. $\overline{TW} \cong \overline{WV}$ **b.** $\overline{PW} \cong \overline{WQ}$

c. $\overline{TV} \perp \overline{PQ}$ **d.** $\overline{TV}$ bisects $\overline{PQ}$.

e. W is the midpoint of $\overline{TV}$.

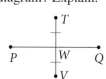

EXERCISES
For more exercises, see *Extra Skill, Word Problem, and Proof Practice.*

Practice and Problem Solving

 Practice by Example

Example 1
(page 36)

 for Help

Name each angle in three ways.

1.

2.
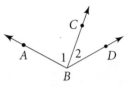

Use the figure at the right. Name the indicated angle in two different ways.

3. ∠1   **4.** ∠2

Example 2
(page 37)

Draw and label a figure to fit each description.

5. an obtuse angle, ∠RST   **6.** an acute acute, ∠BCD

7. a straight angle, ∠EFG   **8.** a right angle, ∠GHI

Use a protractor. Measure and classify each angle.

9.

10. the angle formed by the skis

11.

12.

Example 3
(page 38)

13. Find $m\angle CBD$ if $m\angle ABC = 45$ and $m\angle ABD = 79$.

14. Find $m\angle GFJ$ if $m\angle EFG = 110$.

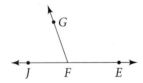

Example 4
(page 39)

Name an angle or angles in the diagram described by each of the following.

15. supplementary to ∠AOD

16. adjacent and congruent to ∠AOE

17. supplementary to ∠EOA

18. complementary to ∠EOD

19. a pair of vertical angles

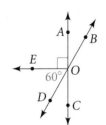

In the diagram above, find the measure of each of the following angles.

20. ∠EOC   **21.** ∠DOC   **22.** ∠BOC   **23.** ∠AOB

Example 5
(page 39)

Can you make each conclusion from the information in the diagram? Explain.

24. $\angle J \cong \angle D$

25. $\angle JAC \cong \angle DAC$

26. $\angle JAE$ and $\angle EAF$ are adjacent and supplementary.

27. $m\angle JCA = m\angle DCA$

28. $m\angle JCA + m\angle ACD = 180$

29. $\overline{AJ} \cong \overline{AD}$ **30.** C is the midpoint of $\overline{JD}$.

31. $\angle EAF$ and $\angle JAD$ are vertical angles.   **32.** $\overrightarrow{AC}$ bisects $\angle JAD$.

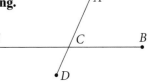

 **Apply Your Skills**

In the diagram, $m\angle ACB = 65$. Find each of the following.

33. $m\angle BCD$ **34.** $m\angle ECD$

Estimation **Estimate the measure of the angle formed by the hands of a clock at each time.**

35. 6:00   **36.** 7:00   **37.** 11:00

38. 4:40   **39.** 5:20   **40.** 10:40

Exercises 33–34

41. Flower Arranging In Japanese flower arranging, you match a stem that is vertical with 0. You match other stems with numbers from 0 to 90, in both directions from the vertical. What numbers would the flowers shown be paired with on a standard protractor?

Real-World Connection

Japanese flower arranging makes precise use of angles to create a mood.

 Algebra **Use the diagram, below right, for Exercises 42–45. Solve for x. Find the angle measures to check your work.**

42. $m\angle AOC = 7x - 2, m\angle AOB = 2x + 8,$
$m\angle BOC = 3x + 14$

43. $m\angle AOB = 4x - 2, m\angle BOC = 5x + 10,$
$m\angle COD = 2x + 14$

44. $m\angle AOB = 28, m\angle BOC = 3x - 2, m\angle AOD = 6x$

45. $m\angle AOB = 4x + 3, m\angle BOC = 7x, m\angle AOD = 16x - 1$

46. Multiple Choice If $m\angle MQV = 90$, which expression can you use to find $m\angle VQP$?
  Ⓐ $m\angle MQP - 90$
  Ⓑ $90 - m\angle MQV$
  Ⓒ $m\angle MQP + 90$
  Ⓓ $90 + m\angle VQP$

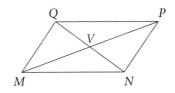

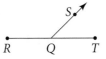

x^2 **47. a. Algebra** Solve for x if $m\angle RQS = 2x + 4$ and $m\angle TQS = 6x + 20$.
  b. What is $m\angle RQS$? $m\angle TQS$?
  c. Show how you can check your answer.

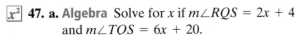

GO Online
Homework Video Tutor
Visit: PHSchool.com
Web Code: aue-0106

C **Challenge**
48. $\overrightarrow{XC}$ bisects $\angle AXB$, $\overrightarrow{XD}$ bisects $\angle AXC$, $\overrightarrow{XE}$ bisects $\angle AXD$, $\overrightarrow{XF}$ bisects $\angle EXD$, $\overrightarrow{XG}$ bisects $\angle EXF$, and $\overrightarrow{XH}$ bisects $\angle DXB$. If $m\angle DXC = 16$, find $m\angle GXH$.

49. **Technology** Leon constructed an angle. Then he constructed a ray from the vertex of the angle to a point in the interior of the angle. He measured all the angles formed. Then he moved the interior ray. What postulate do the two pictures support?

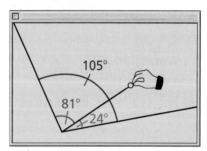 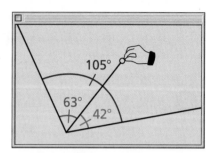

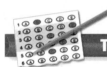 **NY** **REGENTS**

Test Prep

Multiple Choice
50. Two angles are congruent, adjacent, and supplementary. What is the measure of each?
    A. 45   **B.** 90   **C.** 180   **D.** cannot be determined

51. Two angles are congruent and complementary. What is the measure of each?
    F. 45   **G.** 90   **H.** 180   **J.** cannot be determined

52. Two angles are adjacent and supplementary. What is the measure of each?
    A. 45   **B.** 90   **C.** 180   **D.** cannot be determined

53. When 15 is subtracted from the measure of an angle, the result is the measure of a right angle. What is the measure of the original angle?
    F. 75   **G.** 85   **H.** 105   **J.** 115

Short Response
54. You are given that $m\angle ABD + m\angle DBC = m\angle ABC$.
    a. Draw a diagram to show the above.
    b. If $m\angle ABD = 12$ and $\angle ABC$ is obtuse, what are the least and greatest whole number measures possible for $\angle DBC$? Explain.

Mixed Review

 GO for Help

Lesson 1-5   **Use the figure at the right for Exercises 55–56.**

55. If $EG = 75$ and $EF = 28$, what is FG?

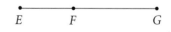

56. If $EG = 49$, $EF = 2x + 3$, and $FG = 4x - 2$, find x. Then find EF and FG.

Lesson 1-2   57. **Writing** Explain the difference between an orthographic drawing and an isometric drawing.

Lesson 1-1   **Find one counterexample to show that each conjecture is false.**

58. The quotient of two integers is not an integer.

59. An even number cannot have 5 as a factor.

Activity Lab

Hands-On

Compass Designs

FOR USE WITH LESSON 1-7

In Lesson 1-7, you use a compass to construct geometric figures. You can construct figures to show geometric relationships, to suggest new relationships, or simply to make appealing geometric designs like the following.

NY G.CN.2: Understand the corresponding procedures for similar problems or mathematical concepts.

Step 1: Open your compass to about 2 inches. Make a circle and mark the point at the center of the circle.

Step 2: Keep the opening of your compass fixed. Place the compass point on the circle. With the pencil end, make a small arc to intersect the circle.

Step 1

Step 3: Place the compass point on the circle at the arc. Mark another arc. Continue around the circle this way to draw four more arcs—six in all.

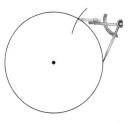

Step 2

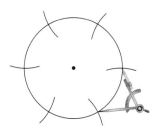

Step 3

Step 4

Step 4: Place your compass point on an arc you marked on the circle. Place the pencil end at the next arc. Draw a large arc that passes through the circle's center and continues to another point on the circle.

Step 5: Draw six large arcs in this manner, each centered at one of the six points marked on the circle. Some people like to call the final design a daisy. You might like to color your daisy.

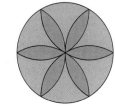

Step 5

EXERCISES

1. In Step 3, did your sixth mark on the circle land precisely on the point where you first placed your compass on the circle?
   a. Survey the class to find how many did.
   b. Because a point has no size, it is very difficult to land on exactly. Try to give a convincing argument that your sixth mark should have landed on your starting point.

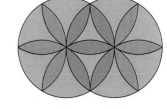

2. Extend your design by using one of the six points on the circle as the center for a new circle. Repeat Steps 2–5 with this circle. Repeat several times to make interlocking circles. Color them as you wish.

Basic Constructions

 Learning Standards for Mathematics

G.G.17 Construct a bisector of a given angle, using a straightedge and compass, and justify the construction.

G.G.18 Construct the perpendicular bisector of a given segment, using a straightedge and compass, and justify the construction.

✓ **Check Skills You'll Need**

GO for Help Lessons 1-5 and 1-6

In Exercises 1–6, sketch each figure.

1. $\overline{CD}$

2. $\overrightarrow{GH}$

3. $\overleftrightarrow{AB}$

4. line m

5. acute $\angle ABC$

6. $\overline{XY} \parallel \overline{ST}$

7. $DE = 20$. Point C is the midpoint of $\overline{DE}$. Find CE.

8. Use a protractor to draw a 60° angle.

9. Use a protractor to draw a 120° angle.

🔊 **New Vocabulary**
- construction
- straightedge
- compass
- perpendicular lines
- perpendicular bisector
- angle bisector

---

1 Constructing Segments and Angles

In a **construction** you use a straightedge and a compass to draw a geometric figure. A **straightedge** is a ruler with no markings on it. A **compass** is a geometric tool used to draw circles and parts of circles called arcs.

Four basic constructions involve constructing congruent segments, congruent angles, and bisectors of segments and angles.

1 EXAMPLE Constructing Congruent Segments

Construct a segment congruent to a given segment.

Given: $\overline{AB}$

Construct: $\overline{CD}$ so that $\overline{CD} \cong \overline{AB}$

Step 1
Draw a ray with endpoint C.

Step 2
Open the compass to the length of $\overline{AB}$.

Step 3
With the same compass setting, put the compass point on point C. Draw an arc that intersects the ray. Label the point of intersection D.

● $\overline{CD} \cong \overline{AB}$

 Quick Check ❶ Use a straightedge to draw $\overline{XY}$. Then construct $\overline{RS}$ so that $RS = 2XY$.

② EXAMPLE **Constructing Congruent Angles**

Construct an angle congruent to a given angle.

Given: ∠A
Construct: ∠S so that ∠S ≅ ∠A

Step 1
Draw a ray with endpoint S.

Step 2
With the compass point on point A, draw an arc that intersects the sides of ∠A. Label the points of intersection B and C.

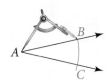

Step 3
With the same compass setting, put the compass point on point S. Draw an arc and label its point of intersection with the ray as R.

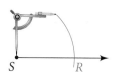

Step 4
Open the compass to the length BC. Keeping the same compass setting, put the compass point on R. Draw an arc to locate point T.

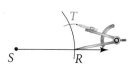

Step 5
Draw $\overrightarrow{ST}$.

● ∠S ≅ ∠A

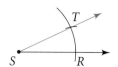

 Quick Check ② Construct ∠F with m∠F = 2m∠B.

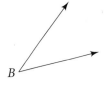

---

 2 **Constructing Bisectors**

Real-World 🌐 Connection

Perpendicular hands signal "Time out."

Perpendicular lines are two lines that intersect to form right angles. The symbol ⊥ means "is perpendicular to." In the diagram at the right, $\overleftrightarrow{AB} \perp \overleftrightarrow{CD}$ and $\overleftrightarrow{CD} \perp \overleftrightarrow{AB}$.

A **perpendicular bisector** of a segment is a line, segment, or ray that is perpendicular to the segment at its midpoint, thereby bisecting the segment into two congruent segments.

As you will learn in Chapter 5, there is just one line that is the perpendicular bisector of a segment in a given plane. Here is a way to construct the perpendicular bisector.

Real-World 🌐 **Connection**

Careers Architects use construction tools to work with their designs.

3 EXAMPLE **Constructing the Perpendicular Bisector**

Construct the perpendicular bisector of a segment.

Given: $\overline{AB}$
Construct: $\overleftrightarrow{XY}$ so that $\overleftrightarrow{XY} \perp \overline{AB}$ at the midpoint M of $\overline{AB}$.

Step 1
Put the compass point on point A and draw a long arc as shown. Be sure the opening is greater than $\frac{1}{2}AB$.

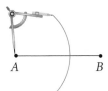

Step 2
With the same compass setting, put the compass point on point B and draw another long arc. Label the points where the two arcs intersect as X and Y.

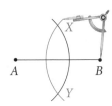

Step 3
Draw $\overleftrightarrow{XY}$. The point of intersection of $\overline{AB}$ and $\overleftrightarrow{XY}$ is M, the midpoint of $\overline{AB}$.

$\overleftrightarrow{XY} \perp \overline{AB}$ at the midpoint of $\overline{AB}$, so $\overleftrightarrow{XY}$ is the perpendicular bisector of $\overline{AB}$.

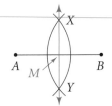

✅ **Quick Check** ❸ Draw $\overline{ST}$. Construct its perpendicular bisector.

An **angle bisector** is a ray that divides an angle into two congruent coplanar angles. Its endpoint is at the angle vertex. Within the ray, a segment with the same endpoint is also an angle bisector. You may say that the ray or segment *bisects* the angle.

4 EXAMPLE **Finding Angle Measures**

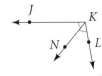

Algebra $\overrightarrow{KN}$ bisects $\angle JKL$ so that $m\angle JKN = 5x - 25$ and $m\angle NKL = 3x + 5$. Solve for x and find $m\angle JKN$.

$m\angle JKN = m\angle NKL$	**Definition of angle bisector**
$5x - 25 = 3x + 5$	**Substitute.**
$5x = 3x + 30$	**Add 25 to each side.**
$2x = 30$	**Subtract 3x from each side.**
$x = 15$	**Divide each side by 2.**
$m\angle JKN = 5x - 25 = 5(15) - 25 = 50$	**Substitute 15 for x.**
$m\angle JKN = 50$	

✅ **Quick Check** ❹ Find $m\angle NKL$ and $m\angle JKL$.

5 EXAMPLE Constructing the Angle Bisector

Construct the bisector of an angle.

Given: $\angle A$

Construct: $\overrightarrow{AX}$, the bisector of $\angle A$

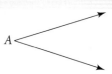

Step 1
Put the compass point on vertex A. Draw an arc that intersects the sides of $\angle A$. Label the points of intersection B and C.

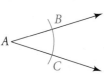

Step 2
Put the compass point on point C and draw an arc. With the same compass setting, draw an arc using point B. Be sure the arcs intersect. Label the point where the two arcs intersect as X.

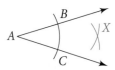

Step 3
Draw $\overrightarrow{AX}$.

$\overrightarrow{AX}$ is the bisector of $\angle CAB$.

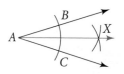

Quick Check 5 **a.** Draw obtuse $\angle XYZ$. Then construct its bisector $\overrightarrow{YP}$.
   b. Explain how you can use your protractor to check your construction.

EXERCISES

For more exercises, see *Extra Skill, Word Problem, and Proof Practice.*

Practice and Problem Solving

A Practice by Example

Example 1
(page 44)

GO for Help

In Exercises 1–8, draw a diagram similar to the given one. Then do the construction. Check your work with a ruler or a protractor.

1. Construct $\overline{XY}$ congruent to $\overline{AB}$.

2. Construct $\overline{VW}$ so that $VW = 2AB$.

3. Construct $\overline{DE}$ so that $DE = TR + PS$.

4. Construct $\overline{QJ}$ so that $QJ = TR - PS$.

Example 2
(page 45)

5. Construct $\angle D$ so that $\angle D \cong \angle C$.

6. Construct $\angle F$ so that $m\angle F = 2m\angle C$.

Example 3
(page 46)

7. Construct the perpendicular bisector of $\overline{AB}$.

8. Construct the perpendicular bisector of $\overline{TR}$.

Example 4
(page 46)

x^2 **9. Algebra** $\overrightarrow{GH}$ bisects $\angle FGI$.
   a. Solve for x and find $m\angle FGH$.
   b. Find $m\angle HGI$.
   c. Find $m\angle FGI$.

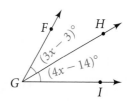

x^2 **Algebra** For Exercises 10–12, $\overrightarrow{BX}$ bisects $\angle ABC$. Solve for x and find $m\angle ABC$.

10. $m\angle ABX = 5x, m\angle XBC = 3x + 10$

11. $m\angle ABC = 4x - 12, m\angle ABX = 24$

12. $m\angle ABX = 4x - 16, m\angle CBX = 2x + 6$

Example 5
(page 47)

13. Draw acute $\angle PQR$. Then construct its bisector.

14. Draw right $\angle TUV$. Then construct its bisector.

B **Apply Your Skills**

15. Use your protractor and draw $\angle W$ with $m\angle W = 120$. Construct $\angle Z \cong \angle W$. Then construct the bisector of $\angle Z$.

Sketch the figure described. Explain how to construct it. Then do the construction.

16. $\overleftrightarrow{XY} \perp \overleftrightarrow{YZ}$

17. $\overrightarrow{ST}$ bisecting right $\angle PSQ$

18. Optics A beam of light and a mirror can be used to study the behavior of light. Light that strikes the mirror is reflected so that the angle of reflection and the angle of incidence are congruent. In the diagram, $\overrightarrow{BC}$ is perpendicular to the mirror, and $\angle ABC$ has a measure of 41°.
  a. Name the angle of reflection and find its measure.
  b. Find $m\angle ABD$.
  c. Find $m\angle ABE$ and $m\angle DBF$.

19. Use a straightedge and protractor.
  a. Draw a mirror and a light beam striking the mirror and reflecting from it.
  b. Construct the bisector of the angle formed by the incoming and reflected light beams. Label the angles of incidence and reflection.

PEANUTS® by Charles M. Schulz

20. Open-Ended Snoopy can draw squares with his compass. You can only draw circles. You can, however, construct a square. Explain how to do this. Use sketches if needed. Then do the construction.

21. Answer these questions about a segment in a plane. Explain each answer.
  a. How many midpoints does the segment have?
  b. How many bisectors does it have? How many lines in the plane are its perpendicular bisectors?
  c. How many lines in space are its perpendicular bisectors?

 **nline**
Homework Video Tutor
Visit: PHSchool.com
Web Code: aue-0107

For Exercises 22–24, copy $\angle 1$ and $\angle 2$.

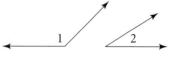

22. Construct $\angle B$ so that $m\angle B = m\angle 1 + m\angle 2$.

23. Construct $\angle C$ so that $m\angle C = m\angle 1 - m\angle 2$.

24. Construct $\angle D$ so that $m\angle D = 2m\angle 2$.

25. Reasoning When $\overrightarrow{BX}$ bisects $\angle ABC$, $\angle ABX \cong \angle CBX$. Lani claims there is always a related equation, $m\angle ABX = \frac{1}{2}m\angle ABC$. Denyse claims the related equation is $2m\angle ABX = m\angle ABC$. Which equation is correct? Explain.

26. Writing Describe how to construct the midpoint of a segment.

27. Construct a 45° angle.

28. a. Draw a large triangle with three acute angles. Construct the bisectors of the three angles. What appears to be true about the three angle bisectors?
   b. Repeat the constructions with a triangle that has one obtuse angle.
   c. Make a Conjecture What appears to be true about the three angle bisectors of any triangle?

Use a ruler to draw segments of 2 cm, 4 cm and 5 cm. Then construct each triangle, if possible. If not possible, explain.

29. with 4-cm, 4-cm, and 5-cm sides   **30.** with 2-cm, 5-cm, and 5-cm sides

31. with 2-cm, 2-cm, and 5-cm sides   **32.** with 2-cm, 2-cm, and 4-cm sides

33. a. Draw a segment, $\overline{XY}$. Construct a triangle with sides congruent to $\overline{XY}$.
   b. Measure the angles of the triangle.
   c. Writing Describe how to construct a 60° angle; a 30° angle.

34. Multiple Choice Which steps best describe how to construct this pattern?

   Ⓐ Use a straightedge to draw the segment and then a compass to draw five half circles.
   Ⓑ Use a straightedge to draw the segment and then a compass to draw six half circles.
   Ⓒ Use a compass to draw five half circles and then a straightedge to join their ends.
   Ⓓ Use a compass to draw six half circles and then a straightedge to join their ends.

Ⓒ Challenge

35. a. Use your compass to draw a circle. Locate three points A, B, and C on the circle.
   b. Construct the perpendicular bisectors of $\overline{AB}$ and $\overline{BC}$.
   c. Critical Thinking Label the intersection of the two perpendicular bisectors as point O. Make a conjecture about point O.

36. Study the figures. Complete the definition of a line perpendicular to a plane:

A line is perpendicular to a plane if it is __?__ to every line in the plane that __?__.

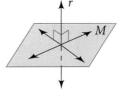
Line $r \perp$ plane M.

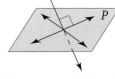
Line t is not $\perp$ plane P.

 **REGENTS**

 **Test Prep**

Multiple Choice

37. What must you do to construct the midpoint of a segment?
   A. Measure half its length.   **B.** Measure twice its length.
   C. Construct an angle bisector.   **D.** Construct a perpendicular bisector.

38. Which of these is the first step in constructing a congruent segment?
  F. Draw a ray.   **G.** Draw a line.
  H. Label two points.   **J.** Measure the segment.

Short Response

39. Explain how to do each construction using a compass and a straightedge.
  a. Draw an acute angle, ∠ABC. Construct an angle congruent to ∠ABC.
  b. Construct an angle whose measure is twice that of ∠ABC.

Extended Response

40. Explain how to do each construction using a compass and a straightedge.
  a. Divide a segment into two congruent segments.
  b. Divide a segment into four congruent segments.
  c. Construct a segment that is 1.25 times as long as a given segment.

Mixed Review

Lesson 1-6

41. Use a protractor to draw a 72° angle.

42. ∠DEF is a straight angle. $m\angle DEG = 80$. Find $m\angle GEF$.

43. $m\angle TUV = 100$ and $m\angle VUW = 80$. Find possible values of $m\angle TUW$.

Lesson 1-5   **Use the number line at the right. Find the length of each segment.**

44. $\overline{AC}$ **45.** $\overline{AD}$

46. $\overline{CD}$ **47.** $\overline{BC}$

```
      A       B       C           D
  ←─●─┼─┼─┼─●─┼─┼─┼─●─┼─┼─┼─●─┼─→
   −7 −6 −5 −4 −3 −2 −1  0  1  2  3  4
```

Lesson 1-4

48. Draw $\overleftrightarrow{RS}$.

Use your drawing from Exercise 48. Answer and explain.

49. Are $\overrightarrow{RS}$ and $\overrightarrow{SR}$ opposite rays?   **50.** Are $\overline{RS}$ and $\overline{SR}$ the same segment?

Geometry at Work

······················· Cabinetmaker

Cabinetmakers not only make cabinets but all types of wooden furniture. The artistry of cabinetmaking can be seen in the beauty and uniqueness of the finest doors, shelves, and tables. The craft is in knowing which types of wood and tools to use, and how to use them.

The carpenter's square is one of the most useful of the cabinetmaker's tools. It can be applied to a variety of measuring tasks. The figure shows how to use a carpenter's square to bisect ∠O.

First, mark equal lengths OA and OC on the sides of the angle. Then position the square so that BA = BC to locate point B. Finally, draw $\overrightarrow{OB}$. $\overrightarrow{OB}$ bisects ∠O.

**For:** Information about cabinetmaking
PHSchool.com **Web Code:** aub-2031

50   Chapter 1   Tools of Geometry

Exploring Constructions

FOR USE WITH LESSON 1-7

Points, lines, and figures are made in geometry software using Draw tools or Construct tools. A figure made by Draw has no constraints. When the figure is manipulated, it moves or changes size freely. A figure made by Construct is dependent upon an existing object. When you manipulate the existing object, the constructed object moves or resizes accordingly.

In this activity you will explore the difference between Draw and Construct. Before you begin, familiarize yourself with the tools of your software.

NY **G.R.1:** Use physical objects, diagrams, charts, tables, graphs, symbols, equations, or objects created using technology as representations of mathematical concepts.

Draw and Construct

- Draw $\overline{AB}$ and Construct the perpendicular bisector $\overleftrightarrow{DC}$.

- Draw $\overline{EF}$ and Construct G, any point on $\overline{EF}$. Draw $\overleftrightarrow{HG}$. Find EG, GF, and $m\angle HGF$. Try to drag G so that $EG = GF$. Try to drag H so that $m\angle HGF = 90$. Were you able to draw the perpendicular bisector of $\overline{EF}$? Explain.

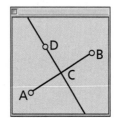
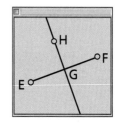

Investigate

- Drag A and B. Observe AC, CB, and $m\angle DCB$. Is $\overleftrightarrow{DC}$ always the perpendicular bisector of $\overline{AB}$ no matter how you manipulate the figure?

- Drag E and F. Observe EG, GF, and $m\angle HGF$. How is the relationship between $\overline{EF}$ and $\overleftrightarrow{HG}$ different from the relationship between $\overline{AB}$ and $\overleftrightarrow{DC}$?

EXERCISES

1. a. Write a description of the general difference between Draw and Construct.
   b. Use your description to explain why the relationship between $\overline{EF}$ and $\overleftrightarrow{HG}$ differs from the relationship between $\overline{AB}$ and $\overleftrightarrow{DC}$.

2. a. Draw $\angle JKL$.
   b. Construct its angle bisector, $\overrightarrow{KM}$.
   c. Manipulate the figure and observe the different angle measures. Is $\overrightarrow{KM}$ always the angle bisector of $\angle JKL$?

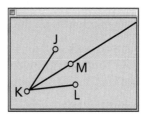

3. a. Draw $\angle NOP$. Draw $\overrightarrow{OQ}$ in the interior of $\angle NOP$. Drag Q until $m\angle NOQ = m\angle QOP$.
   b. Manipulate the figure and observe the different angle measures. Is $\overrightarrow{OQ}$ always the angle bisector of $\angle NOP$?

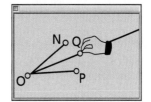

4. For Exercise 2, how can you manipulate the figure on the screen so that it shows a right angle? Justify your answer.

Distance in the Coordinate Plane

Hands-On

FOR USE WITH LESSON 1-8

Much of Manhattan is laid out in a rectangular grid, as shown in this map. In general, the streets are parallel running east and west. The avenues are parallel running north and south.

NY **G.CN.3:** Model situations mathematically, using representations to draw conclusions and formulate new situations.

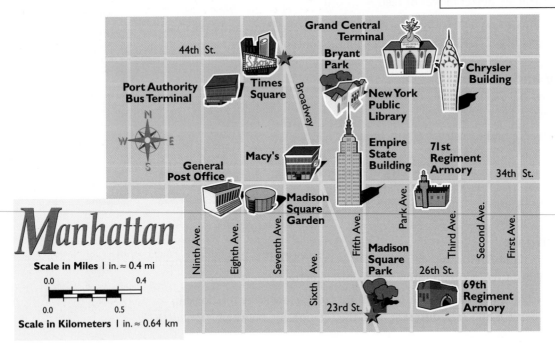

Yvonne's family is at the corner of 44th Street and 7th Avenue. They plan to walk to Madison Square Park at 23rd Street and 5th Avenue. There are several possible routes they can take.

EXERCISES

Use tracing paper to trace the routes on the map. Answer the following questions.

1. Yvonne's father wants to walk east on 44th Street until they reach 5th Avenue. He then plans to walk south on 5th Avenue to Madison Square Park. About how long is his route?

2. Yvonne's mother wants to walk south on 7th Avenue until they reach 23rd Street. She then plans to walk east on 23rd Street to Madison Square Park. About how long is her route?

3. Yvonne notices on the map that Broadway cuts across the grid of streets and leads directly to Madison Square Park. She suggests walking all the way on Broadway. About how long is her route?

4. Whose route is the shortest? Explain.

5. Whose route is the longest? Explain.

6. Give a plausible argument for the following statement: There is no longest route from Times Square to Madison Square Park.

1-8

The Coordinate Plane

 Learning Standards for Mathematics

G.G.66 Find the midpoint of a line segment, given its endpoints.

G.G.67 Find the length of a line segment, given its endpoints.

 **Check Skills You'll Need**

 for Help Skills Handbook pages 753 and 754

Find the square root of each number. Round to the nearest tenth if necessary.

1. 25 **2.** 17 **3.** 123

x^2 **Algebra** Evaluate each expression for $m = -3$ and $n = 7$.

4. $(m - n)^2$ **5.** $(n - m)^2$ **6.** $m^2 + n^2$

x^2 **Algebra** Evaluate each expression for $a = 6$ and $b = -8$.

7. $(a - b)^2$ **8.** $\sqrt{a^2 + b^2}$ **9.** $\dfrac{a + b}{2}$

1 Finding Distance on the Coordinate Plane

You can think of a point as a dot, and a line as a series of points. In coordinate geometry you describe a point by an ordered pair (x, y), called the *coordinates of the point*.

Vocabulary Tip

To help review terms shown here, see "Coordinate plane" in the Glossary.

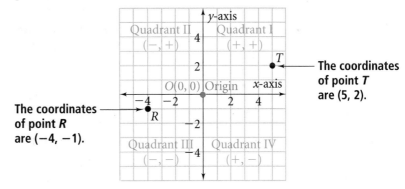

The coordinates of point R are $(-4, -1)$.

The coordinates of point T are $(5, 2)$.

You can use the Ruler Postulate to find the distance between two points if the points are on a horizontal line or a vertical line. To find the distance between two points that are not on a horizontal or vertical line, you can use the Distance Formula.

 Key Concepts

Formula	**The Distance Formula**

The distance d between two points $A(x_1, y_1)$ and $B(x_2, y_2)$ is

$$d = \sqrt{(x_2 - x_1)^2 + (y_2 - y_1)^2}.$$

You will verify this formula in Chapter 8.

For: Distance Activity
Use: Interactive Textbook, 1-8

1 EXAMPLE Finding Distance

Find the distance between $T(5, 2)$ and $R(-4, -1)$ to the nearest tenth.

Let $(5, 2)$ be (x_1, y_1) and $(-4, -1)$ be (x_2, y_2).

$d = \sqrt{(x_2 - x_1)^2 + (y_2 - y_1)^2}$ **Use the Distance Formula.**

$d = \sqrt{(-4 - 5)^2 + (-1 - 2)^2}$ **Substitute.**

$d = \sqrt{(-9)^2 + (-3)^2}$ **Simplify.**

$d = \sqrt{81 + 9} = \sqrt{90}$

90 $\boxed{\sqrt{}}$ **9.4868330** **Use a calculator.**

● To the nearest tenth, $TR = 9.5$.

✓ Quick Check **1 a.** $\overline{AB}$ has endpoints $A(1, -3)$ and $B(-4, 4)$. Find AB to the nearest tenth.
  b. Critical Thinking In Example 1, suppose you let $(-4, -1)$ be (x_1, y_1) and $(5, 2)$ be (x_2, y_2). Do you get the same result? Why?

2 EXAMPLE Real-World 🌐 Connection

Travel Each morning Juanita takes the "Blue Line" subway from Oak Station to Jackson Station. As the map at the left shows, Oak Station is 1 mile west and 2 miles south of City Plaza. Jackson Station is 2 miles east and 4 miles north of City Plaza. Find the distance Juanita travels between Oak Station and Jackson Station.

Let $\text{Oak}(-1, -2)$ be (x_1, y_1) and $\text{Jackson }(2, 4)$ be (x_2, y_2).

$d = \sqrt{(x_2 - x_1)^2 + (y_2 - y_1)^2}$ **Use the Distance Formula.**

$d = \sqrt{(2 - (-1))^2 + (4 - (-2))^2}$ **Substitute.**

$d = \sqrt{3^2 + 6^2}$ **Simplify.**

$d = \sqrt{9 + 36} = \sqrt{45}$

45 $\boxed{\sqrt{}}$ **6.7082039** **Use a calculator.**

● Juanita travels about 6.7 miles between Oak Station and Jackson Station.

✓ Quick Check **2 a.** Find the distance between Elm Station and Symphony Station.
  b. Maple Station is located 6 miles west and 2 miles north of City Plaza. Find the distance between Cedar Station and Maple Station.

2 Finding the Midpoint of a Segment

To find the coordinate of the midpoint of a segment on a number line, find the *average* or *mean* of the coordinates of the endpoints. The coordinate of the midpoint of a segment with endpoints a and b is $\frac{a + b}{2}$.

You can extend this process (see next page) to find the coordinates of the midpoint of a segment in the coordinate plane.

Study the diagram of $\overline{TS}$ with endpoints $T(4, 3)$ and $S(8, 5)$. $\overline{TR}$ is a horizontal segment and $\overline{SR}$ is a vertical segment. The coordinates of R are $(8, 3)$.

The coordinates of M, the midpoint of $\overline{TR}$, are $(6, 3)$. The coordinates of N, the midpoint of $\overline{SR}$, are $(8, 4)$. A vertical line through M and a horizontal line through N meet at P, the midpoint of $\overline{TS}$.

The coordinates of P are $(6, 4)$.

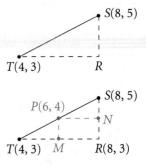

Thus, you find the coordinates of the midpoint of a segment by averaging the x-coordinates and averaging the y-coordinates of the endpoints.

Key Concepts

Formula	The Midpoint Formula

The coordinates of the midpoint M of $\overline{AB}$ with endpoints $A(x_1, y_1)$ and $B(x_2, y_2)$ are the following:

$$M = \left(\frac{x_1 + x_2}{2}, \frac{y_1 + y_2}{2}\right)$$

3 EXAMPLE Finding the Midpoint

Algebra $\overline{QS}$ has endpoints $Q(3, 5)$ and $S(7, -9)$. Find the coordinates of its midpoint M.

Let $(3, 5)$ be (x_1, y_1) and $(7, -9)$ be (x_2, y_2).

x-coordinate of $M = \dfrac{x_1 + x_2}{2} = \dfrac{3 + 7}{2} = \dfrac{10}{2} = 5$

y-coordinate of $M = \dfrac{y_1 + y_2}{2} = \dfrac{5 + (-9)}{2} = \dfrac{-4}{2} = -2$

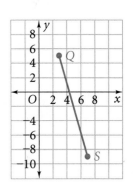

● The coordinates of midpoint M are $(5, -2)$.

✓ Quick Check ③ Find the coordinates of the midpoint of $\overline{XY}$ with endpoints $X(2, -5)$ and $Y(6, 13)$.

4 EXAMPLE Finding an Endpoint

Algebra The midpoint of $\overline{AB}$ is $M(3, 4)$. One endpoint is $A(-3, -2)$. Find the coordinates of the other endpoint B.

Use the Midpoint Formula. Let the coordinates of B be (x_2, y_2).

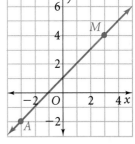

$3 = \dfrac{-3 + x_2}{2}$ ← Midpoint Formula → $4 = \dfrac{-2 + y_2}{2}$

$6 = -3 + x_2$ ← Multiply each side by 2. → $8 = -2 + y_2$

$9 = x_2$ $\qquad\qquad\qquad\qquad\qquad$ $10 = y_2$

● The coordinates of B are $(9, 10)$.

✓ Quick Check ④ The midpoint of $\overline{XY}$ has coordinates $(4, -6)$. X has coordinates $(2, -3)$. Find the coordinates of Y.

EXERCISES

For more exercises, see *Extra Skill, Word Problem, and Proof Practice.*

Practice and Problem Solving

 Practice by Example

Example 1
(page 54)

GO for Help

Find the distance between the points to the nearest tenth.

1. $J(2, -1), K(2, 5)$ **2.** $L(10, 14), M(-8, 14)$ **3.** $N(-1, -11), P(-1, -3)$

4. $A(0, 3), B(0, 12)$ **5.** $C(12, 6), D(-8, 18)$ **6.** $E(6, -2), F(-2, 4)$

7. $Q(12, -12), T(5, 12)$ **8.** $R(0, 5), S(12, 3)$ **9.** $X(-3, -4), Y(5, 5)$

Example 2
(page 54)

Use the map in Example 2 on page 54. Find the distance between the stations.

10. North and South   **11.** Oak and Symphony   **12.** City Plaza and Cedar

Use the map at the right. Find the distances between the stations to the nearest tenth.

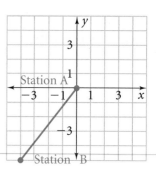

13. Station A and Station B

14. Station B and Station C located at (5, 8)

15. Station B and Station D located at (1, 10)

16. Station E at (2, 12) and Station F at (5, 16)

17. List the stations B, C, D, E, and F in the order of least to greatest distance from Station A.

Example 3
(page 55)

$\boxed{x^2}$ **Algebra Find the coordinates of the midpoint of $\overline{HX}$.**

18. $H(0, 0), X(8, 4)$ **19.** $H(-1, 3), X(7, -1)$

20. $H(13, 8), X(-6, -6)$ **21.** $H(7, 10), X(5, -8)$

22. $H(-6.3, 5.2), X(1.8, -1)$ **23.** $H\left(5\frac{1}{2}, -4\frac{3}{4}\right), X\left(2\frac{1}{4}, -1\frac{1}{4}\right)$

Example 4
(page 55)

$\boxed{x^2}$ **Algebra The coordinates of point T are given. The midpoint of $\overline{ST}$ has coordinates $(5, -8)$. Find the coordinates of point S.**

24. $T(0, 4)$ **25.** $T(5, -15)$ **26.** $T(10, 18)$

27. $T(-2, 8)$ **28.** $T(1, 12)$ **29.** $T(4.5, -2.5)$

An endpoint and a midpoint are given. Find the coordinates of the other endpoint.

30. endpoint $(2, 6)$, midpoint $(5, 12)$ **31.** endpoint $(2, 3)$, midpoint $(3, -4)$

 Apply Your Skills

Find (a) PQ to the nearest tenth and (b) the coordinates of the midpoint of $\overline{PQ}$.

32. $P(3, 2), Q(6, 6)$ **33.** $P(0, -2), Q(3, 3)$ **34.** $P(-4, -2), Q(1, 3)$

35. $P(-5, 2), Q(0, 4)$ **36.** $P(-3, -1), Q(5, -7)$ **37.** $P(-5, -3), Q(-3, -5)$

38. $P(-4, -5), Q(-1, 1)$ **39.** $P(2, 3), Q(4, -2)$ **40.** $P(4, 2), Q(3, 0)$

41. The midpoint of $\overline{TS}$ is the origin. Point T is located in Quadrant II. What quadrant contains point S?

42. Graph the points $A(2, 1), B(6, -1), C(8, 7),$ and $D(4, 9)$. Draw quadrilateral $ABCD$. Use the Midpoint Formula to find the midpoints of $\overline{AC}$ and $\overline{BD}$. What appears to be true?

GO Online
Homework Video Tutor
Visit: PHSchool.com
Web Code: aue-0108

43. Multiple Choice $A(4, -1)$ and $B(-2, 3)$ are points in a coordinate plane. M is the midpoint of $\overline{AB}$. What is the length of $\overline{MB}$ to the nearest tenth of a unit?

    (A) 2.5 units   (B) 3.6 units   (C) 7.2 units   (D) 13.0 units

For each graph, find (a) AB **to the nearest tenth and (b) the coordinates of the midpoint of** $\overline{AB}$.

44. 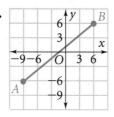   **45.** 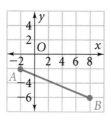   **46.**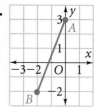

GO for Help

For a guide to solving Exercise 47, see page 60.

 47. Writing An airplane at $T(80, 20)$ needs to fly to both $U(20, 60)$ and $V(110, 85)$. What is the shortest possible distance for the trip? Explain.

 48. Navigation A boat at $X(5, -2)$ needs to travel to $Y(-6, 9)$ or $Z(17, -3)$. Which point is closer? What is the distance to the closer point?

 Communications The cell phone screen at the right shows coordinates of six cities from a grid placed on North America by a long-distance carrier. The carrier finds distance by the Distance Formula. Each grid unit equals $\sqrt{0.1}$ mile. Find the distance between each pair of cities to the nearest mile.

49. Houston and Chicago

50. Denver and New Orleans

51. Boston and San Francisco

52. New Orleans and Houston

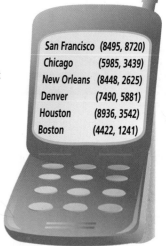

San Francisco (8495, 8720)
Chicago (5985, 3439)
New Orleans (8448, 2625)
Denver (7490, 5881)
Houston (8936, 3542)
Boston (4422, 1241)

Source: Peter H. Dana

Graph $X(-2, 1)$, $Y(2, 3)$, $A(-1, 4)$, $B(0, 2)$, **and** $C(4, 2)$. **For each point described below, give two sets of possible coordinates if they exist. Otherwise, write** *exactly one point* **and give the coordinates, or** *not possible* **and explain.**

53. point D so that $\overleftrightarrow{AD} \parallel \overleftrightarrow{XY}$ **54.** E so that $\overleftrightarrow{EC} \parallel \overleftrightarrow{XY}$

55. point F so that $\overleftrightarrow{FB} \perp \overleftrightarrow{XY}$ **56.** point G so that $\overleftrightarrow{GC} \perp \overleftrightarrow{XY}$

57. point H so that $\overleftrightarrow{HX} \parallel \overleftrightarrow{AY}$, and $\overleftrightarrow{HA} \parallel \overleftrightarrow{XY}$

58. point J so that $\overleftrightarrow{JB} \perp \overleftrightarrow{XY}$, and $\overleftrightarrow{JC} \perp \overleftrightarrow{CY}$

Challenge

59. Open-Ended In a coordinate plane, draw any $\overline{AB}$. Draw another segment that is both congruent and parallel to $\overline{AB}$. Label the new segment $\overline{CD}$ in such a way that $ABCD$ is a quadrilateral.
   a. Find BC and AD. What do you notice?
   b. Write a conjecture that generalizes the result you found in part (a).
   c. Find the midpoint of $\overline{AC}$ and the midpoint of $\overline{BD}$. What do you notice?
   d. Write a conjecture that generalizes the result you found in part (c).
   e. Find the midpoint E of $\overline{AD}$ and the midpoint F of $\overline{BC}$. Find EF and AB. What do you notice?
   f. Write a conjecture that generalizes the result you found in part (e).

Exercise 60

Geometry in 3 Dimensions You can use three coordinates (x, y, z) to locate points in three dimensions. Point P has coordinates $(6, -3.5, 9)$.

60. Give the coordinates of points $A, B, C, D, E, F,$ and G.

61. Draw three axes like those shown. Then graph $R(4, 5, 9)$.

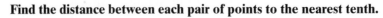

Distance in 3 Dimensions In a three-dimensional coordinate system, the distance between two points (x_1, y_1, z_1) and (x_2, y_2, z_2) can be found using this extension of the Distance Formula.

$$d = \sqrt{(x_2 - x_1)^2 + (y_2 - y_1)^2 + (z_2 - z_1)^2}$$

Find the distance between each pair of points to the nearest tenth.

62. $P(2, 3, 4), B(-2, 4, 9)$ **63.** $Q(0, 12, 15), Y(-8, 20, 12)$

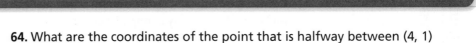

NY REGENTS

Test Prep

Multiple Choice

64. What are the coordinates of the point that is halfway between $(4, 1)$ and $(-22, 8)$?
   A. $(-9, 3.5)$ **B.** $(-9, 4.5)$ **C.** $(-18, 9)$ **D.** $(13, 4.5)$

65. Which point lies the farthest from the origin?
   F. $(0, -7)$ **G.** $(5, 1)$ **H.** $(-4, -3)$ **J.** $(-3, 8)$

66. A segment has endpoints $(14, -8)$ and $(4, 12)$. What are the coordinates of its midpoint?
   A. $(9, 10)$ **B.** $(-5, 10)$ **C.** $(5, -10)$ **D.** $(9, 2)$

67. Susan takes the bus to visit her friend in Valley Stream. She gets on a bus at Gibson and rides to Hewlett. At Hewlett, she transfers to a different bus and rides to Valley Stream. The numbers on the grid represent miles. To the nearest tenth of a mile, how far does Susan travel?
   F. 5.8 miles
   G. 7.3 miles
   H. 9.4 miles
   J. 10.8 miles

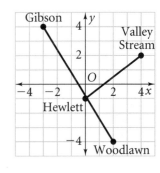

Short Response

68. The midpoint of $\overline{RS}$ is $N(-4, 1)$. One endpoint is $S(0, -7)$.
   a. What are the coordinates of R?
   b. What is the length of $\overline{RS}$ to the nearest tenth of a unit?

69. **a.** Points $P(-4, 6)$, $Q(2, 4)$, and R are collinear. One of the points is the midpoint of the segment formed by the other two points. What are the possible coordinates of R?
   b. $RQ = \sqrt{160}$. Does this information affect your answer to part (a)? Explain.

Lesson 1-7 **Use a straightedge and compass.**

70. Draw $\overline{AB}$. Construct $\overline{PQ}$ so that $PQ = 2AB$.

71. Draw $\overline{LK}$. Construct the perpendicular bisector of $\overline{LK}$.

72. Draw an obtuse $\angle B$. Construct $\angle C$ so that $m\angle C = m\angle B$.

73. Draw an acute $\angle RTS$. Construct the bisector of $\angle RTS$.

Lesson 1-6 **74.** Name $\angle A$ two other ways. **75.** $m\angle PQR = 60$. What is $m\angle RQS$?

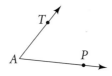

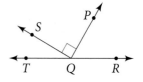

Lesson 1-5 x^2 **76. Algebra** The length of $\overline{AC}$ is 45. If $AB = x + 8$ and $BC = 3x - 3$, find the value of x.

77. Find AB if the coordinate of A is 5 and the coordinate of B is -5.

x^2 **78. Algebra** C is the midpoint of $\overline{EF}$. Find EF.

$$\underset{E}{\overset{2x + 10}{\bullet}} \quad \underset{C}{\overset{5x - 11}{\bullet}} \quad \underset{F}{\bullet}$$

✓ Checkpoint Quiz 2 Lessons 1-5 through 1-8

Use the figure for Exercises 1–3.

1. If $AC = 4x + 5$ and $DC = 3x + 8$, find AC.

2. If $m\angle BCE = 45$ and $m\angle ECD = 65$, find $m\angle BCD$.

3. If $m\angle FCA = 40$, find $m\angle FCD$.

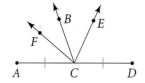

4. $\overleftrightarrow{AX}$ is the perpendicular bisector of $\overline{QS}$ at M.
a. What is $m\angle AMS$?
b. If $QM = 30$, what is QS?

5. $\overrightarrow{PT}$ is the bisector of $\angle APR$. Name two congruent angles.

6. $\overrightarrow{OR}$ is the bisector of right $\angle TOS$. Find $m\angle TOR$.

Use a straightedge to draw three figures like the ones shown at the right. Then do each construction.

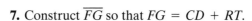

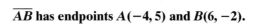

7. Construct $\overline{FG}$ so that $FG = CD + RT$.

8. Construct $\angle HSK$ so that $m\angle HSK = \frac{1}{2}m\angle LSK$.

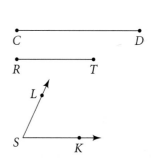

$\overline{AB}$ **has endpoints $A(-4, 5)$ and $B(6, -2)$.**

9. Find AB to the nearest tenth.

10. Find the coordinates of the midpoint of $\overline{AB}$.

GPS Guided Problem Solving

FOR USE WITH PAGE 57, EXERCISE 47

Understanding Word Problems Read the problem below. Then let Rosemarie's thinking guide you through the solution. Check your understanding with the exercises at the bottom of the page.

Writing An airplane at $T(80, 20)$ needs to fly to both $U(20, 60)$ and $V(110, 85)$. What is the shortest possible distance for the trip? Explain.

What Rosemarie Thinks

I'll sketch a graph to understand the problem better.

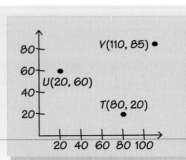

The shortest trip is from T to U to V, or from T to V to U. The total distances for the trips are $TU + UV$ and $TV + VU$.

I have to use the Distance Formula.

I can enter the whole expression for $TU + UV$ on my calculator. I press ENTER and get . . .

Then I press 2nd ENTRY. That shows $TU + UV$ as I had entered it. I'll replace the TU values with the TV values. I press ENTER and get . . .

Now I have to explain. Do I use miles or kilometers? I don't know. I'll use "units."

What Rosemarie Writes

$$TU = \sqrt{(80 - 20)^2 + (20 - 60)^2}$$
$$TV = \sqrt{(80 - 110)^2 + (20 - 85)^2}$$
$$UV = VU = \sqrt{(110 - 20)^2 + (85 - 60)^2}$$

$TU + UV \approx 165.52$

$TV + VU \approx 165.00$

The shortest route for the trip is from T to U to V, or from T to V to U. The T-V-U route is 0.52 unit shorter. The shortest possible distance for the trip is 165 units.

EXERCISES

1. What is the shortest flying distance from $R(65, 72.5)$ to both U and V?

2. An airplane has to fly from T to an airport halfway between U and V. How far must it fly?

60 Guided Problem Solving   Understanding Word Problems

1-9

Perimeter, Circumference, and Area

Learning Standards for Mathematics

G.PS.6 Use a variety of strategies to extend solution methods to other problems.

 Check Skills You'll Need

GO for Help Skills Handbook page 757 and Lesson 1-8

Simplify each absolute value.

1. $|4 - 8|$ **2.** $|10 - (-5)|$ **3.** $|-2 - 6|$

Find the distance between the points to the nearest tenth.

4. $A(2, 3), B(5, 9)$ **5.** $K(-1, -3), L(0, 0)$

6. $W(4, -7), Z(10, -2)$ **7.** $C(-5, 2), D(-7, 6)$

8. $M(-1, -10), P(-12, -3)$ **9.** $Q(-8, -4), R(-3, -10)$

1   Finding Perimeter and Circumference

Hands-On Activity: Finding Perimeter and Area

Draw each figure on centimeter grid paper.

- a rectangle with length 5 cm and width 3 cm
- a rectangle with length 8 cm and height 2 cm
- a rectangle with each side 4 cm

1. To find the perimeter of each rectangle, find the sum of the lengths of the sides. Record the perimeter of each rectangle.

2. To find the area of each rectangle, count the number of square centimeters in its interior. Record the area of each rectangle.

3. Do rectangles with equal perimeters have the same area?

4. Do rectangles with the same area have the same perimeter?

5. Use a piece of string and make a loop. Tie a slip knot. Adjust the loop and fix its total length at 36 cm. Use the loop to approximate different rectangles on your grid paper. Record their lengths, widths, perimeters, and areas. What do you notice?

Vocabulary Tip

You can think of the perimeter of a polygon as the distance around it and the area as the number of square units it encloses.

Online active math

For: Perimeter/Area Activity
Use: Interactive Textbook, 1-9

The perimeter P of a polygon is the sum of the lengths of its sides. The area A of a polygon is the number of square units it encloses. For special figures such as squares, rectangles, and circles, you can use formulas for perimeter (called circumference in circles) and area.

Some formulas for perimeter and area are given in the chart at the top of the next page. You will also find the chart on pages 764 and 765 to be useful at times.

Summary | **Perimeter and Area**

Square with side length s

Rectangle with base b and height h

Circle with radius r and diameter d

Perimeter $P = 4s$

Area $A = s^2$

Perimeter $P = 2b + 2h$

Area $A = bh$

Circumference $C = \pi d$,

or $C = 2\pi r$

Area $= \pi r^2$

The units of measurement for perimeter and circumference include inches, feet, yards, miles, centimeters, meters, and kilometers. When measuring area, use square units such as square inches (in.^{2}), square centimeters (cm^2), square meters (m^2), and square miles (mi^2).

1 EXAMPLE 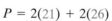 Real-World Connection

Fencing Your pool is 15 ft wide and 20 ft long with a 3-ft wide deck surrounding it. You want to build a fence around the deck. How much fencing will you need?

To find the perimeter of the pool with the deck, first find the width and length of the pool with the deck.

Width of pool and deck $= 15 + 3 + 3 = 21$

Length of pool and deck $= 20 + 3 + 3 = 26$

Perimeter of a rectangle $= 2b + 2h$ **Use the formula for the perimeter of a rectangle.**

$P = 2(21) + 2(26)$ **Substitute.**

$P = 42 + 52$ **Simplify.**

$P = 94$

• You will need 94 ft of fencing.

Vocabulary Tip

For a rectangle, "length" and "width" are sometimes used in place of "base" and "height."

 **Quick Check** ① Suppose you want to frame a picture that is 6 in. by 7 in. with a $\frac{1}{2}$-in. wide frame.
a. Find the perimeter of the picture.
b. Find the perimeter of the outside edge of the frame.

Notice that the formulas for a circle involve π. Since the number π is irrational,

$$\pi = 3.1415926\ldots,$$

you cannot write it as a terminating decimal. For an approximate answer, you can use 3.14 or $\frac{22}{7}$ $\left(3.14 \approx \frac{22}{7}\right)$ for π. You can also use the rounded decimal you get by pressing $\boxed{\pi}$ on your calculator. For an exact answer leave the result in terms of π.

Find the circumference of ⊙A in terms of π. Then find the circumference to the nearest tenth.

$$C = \pi d$$
$$C = 12\pi \qquad \text{This is the exact answer.}$$
$$12 \boxed{\times} \boxed{\pi} \boxed{=} \; 37.699112 \qquad \text{Use a calculator.}$$
$$C \approx 37.7$$

● The circumference of the circle is 12π in., or about 37.7 in.

✓ **Quick Check** **2** **a.** Find the circumference of a circle with a radius of 18 m in terms of π.
b. Find the circumference of a circle with a diameter of 18 m to the nearest tenth.

3 **EXAMPLE** Finding Perimeter in the Coordinate Plane

Algebra Find the perimeter of △ABC.

Find the length of each side. Add the lengths to find the perimeter.

$$AB = |5 - (-1)| = 6$$
$$BC = |6 - (-2)| = 8 \qquad \text{Use the Ruler Postulate.}$$
$$AC = \sqrt{(5 - (-1))^2 + (6 - (-2))^2} \qquad \text{Use the Distance Formula.}$$
$$= \sqrt{6^2 + 8^2} = \sqrt{100} = 10$$
$$AB + BC + AC = 6 + 8 + 10 = 24$$

● The perimeter of △ABC is 24 units.

✓ **Quick Check** **3** Graph quadrilateral KLMN with vertices K(−3, −3), L(1, −3), M(1, 4), and N(−3, 1). Find the perimeter of KLMN.

2 Finding Area

To find area, you should use the same unit for both dimensions.

GO **Online**

Video Tutor Help

Visit: PHSchool.com
Web Code: aue-0775

4 **EXAMPLE** Finding Area of a Rectangle

You are designing a rectangular banner for the front of the museum. The banner will be 4 ft wide and 7 yd high. How much material do you need?

$$7 \text{ yd} = 21 \text{ ft} \qquad \text{Change yards to feet using 1 yd} = 3 \text{ ft.}$$
$$\text{Area} = bh \qquad \text{Use the formula for area of a rectangle.}$$
$$A = 4(21) \qquad \text{Substitute 4 for } b \text{ and 21 for } h.$$
$$A = 84$$

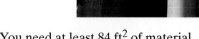

● The area of the banner is 84 square feet (ft²). You need at least 84 ft² of material.

✓ **Quick Check** **4** Find the area of the banner in Example 4 by first changing all units to yards. Compare your answer to the one in Example 4. How do they compare?

5 EXAMPLE Finding Area of a Circle

The diameter of a circle is 10 in. Find the area in terms of π.

radius $= \frac{10}{2}$ or 5 $r = \frac{d}{2}$

Area $= \pi r^2$ **Use the formula for area of a circle.**

$A = \pi(5)^2$ **Substitute 5 for r.**

$A = 25\pi$

● The area of the circle is 25π in.^{2}.

 **Quick Check** ⑤ The diameter of a circle is 5 ft.
a. Find the area in terms of π.
b. Find the area to the nearest tenth.

The following postulates are useful in finding areas of figures with irregular shapes.

 Key Concepts

Postulate 1-9
If two figures are congruent, then their areas are equal.

Postulate 1-10
The area of a region is the sum of the areas of its nonoverlapping parts.

Example 6 applies Postulate 1-10 by summing the areas of the parts of a figure.

6 EXAMPLE Finding Area of an Irregular Shape

Multiple Choice What is the area of the figure at the right?

Ⓐ 12 cm^2 Ⓑ 24 cm^2
Ⓒ 30 cm^2 Ⓓ 36 cm^2

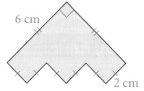

Test-Taking Tip

Marking diagrams on a test can help you understand the problem. If you cannot mark on the test, make a sketch of the diagram on scratch paper.

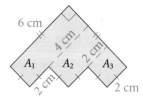

Separate the figure into rectangles.

Area $= bh$ **Use the formula for the area of a rectangle.**

$A_1 = 6 \cdot 2 = 12$ **Find the area of each rectangle.**

$A_2 = 4 \cdot 2 = 8$

$A_3 = 2 \cdot 2 = 4$

Total Area $= 12 + 8 + 4 = 24$ **Add the areas.**

● The area of the figure is 24 cm^2. The correct choice is B.

 **Quick Check** ⑥ Copy the figure in Example 6. Separate it in a different way. Find the area.

EXERCISES

For more exercises, see *Extra Skill, Word Problem, and Proof Practice*.

Practice and Problem Solving

 Practice by Example

Example 1
(page 62)

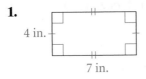 **for Help**

Find the perimeter of each figure.

1.
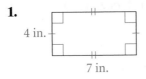
4 in.
7 in.

2.
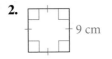
9 cm

Find the perimeter of each rectangle with the given base and height.

3. 21 in., 7 in.   **4.** 16 cm, 23 cm   **5.** 24 m, 36 m

6. Framing A rectangular certificate 8 in. by 10 in. will have a frame $1\frac{1}{2}$ in. wide surrounding it. What is the perimeter of the outside edge of the frame?

7. Fencing A garden that is 5 ft by 6 ft has a walkway 2 ft wide around it. Find the amount of fencing needed to surround the walkway.

Example 2
(page 63)

Find the circumference of each circle in terms of π.

8.

O
15 cm

9.

5 ft
O

10.

O
3.7 in.

11.

$\frac{1}{4}$ m
O

Find the circumference of the circle to the nearest tenth.

12. $r = 9$ in.   **13.** $d = 7.3$ m   **14.** $d = \frac{1}{2}$ yd   **15.** $r = 56$ cm

Example 3
(page 63)

Draw each figure in the coordinate plane. Find the perimeter.

16. $X(0, 2), Y(4, -1), Z(-2, -1)$ **17.** $A(-4, -1), B(4, 5), C(4, -2)$

18. $L(0, 1), M(3, 5), N(5, 5), P(5, 1)$

19. $S(-5, 3), T(7, -2), U(7, -6), V(-5, -6)$

Example 4
(page 63)

Find the area of each rectangle with the given base and height.

20. 4 ft, 4 in.   **21.** 30 in., 4 yd   **22.** 2 ft 3 in., 6 in.

23. 40 cm, 2 m   **24.** 3 m, 190 cm   **25.** 240 cm, 5 m

26. Find the area of a section of road pavement that is 20 ft wide and 100 yd long.

Example 5
(page 64)

Find the area of each circle in terms of π.

27.

20 m

28.

16 ft

29.

$\frac{3}{4}$ in.

30.

0.5 m

31.

6.3 ft

32.

0.1 m

Find the area of each circle to the nearest tenth.

33. $r = 7$ ft   **34.** $d = 8.3$ m   **35.** $d = 24$ cm   **36.** $r = 12$ in.

Example 6
(page 64)

Find the area of the shaded region. All angles are right angles.

37.

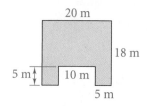

38.

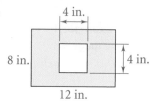

39. a. What is the area of a square whose sides are 12 in. long?
   b. What is the area of a square whose sides are 1 ft long?
   c. Reasoning How many square inches are in a square foot? Explain.

40. a. Count squares to find the area of the
polygon outlined in blue.
   b. Use a formula to find the area of each
square outlined in red.
   c. How does the sum of your results in
part (b) compare to your result in
part (a)? Which postulate does this support?

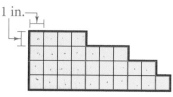

41. Estimation On a postcard from Mexico, Ky sketched the "footprint" of the
pyramid known as El Castillo in the ancient Mayan city Chichen Itza. He said
he estimated the three different lengths on each side to be 22 m, 6 m, and 11 m.
Use those estimates to estimate the area of El Castillo's footprint.

Estimation **Estimate the perimeter and area of each object.**

42. the front cover of this book

43. the front cover of your notebook

44. a classroom bulletin board

45. the top of your desk

46. Writing Choose one exercise from Exercises 42–45 and explain why you chose
your unit of length.

47. The area of an 11-cm wide rectangle is 176 cm^2. What is its length?

48. The perimeter of a rectangle is 40 cm and the base is 12 cm. What is its area?

49. A square and a rectangle have equal area. The rectangle is 64 cm by 81 cm.
What is the perimeter of the square?

50. a. Critical Thinking Can you use the formula for the perimeter of a rectangle
to find the perimeter of any square? Explain.
   b. Can you use the formula for the perimeter of a square to find the perimeter
of any rectangle? Explain.
   c. Use the formula for the perimeter of a square to write a formula for the area
of a square in terms of its perimeter.

51. The surface area of a three-dimensional figure is the
sum of the areas of all of its surfaces. You can find the
surface area by finding the area of a net for the figure.
   a. Draw a net for the solid shown. Label
the dimensions.
   b. What is the area of the net? What is the surface area of the solid?

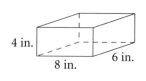

B **Apply Your Skills**

El Castillo
Chichen Itza, Mexico

Real-World **Connection**

Postulate 1-10 can help you
estimate the area of the
"footprint" of El Castillo.

52. Tiling The students in the Art Club are tiling a wall that is 8 ft by 16 ft at the entrance to the community center. They are using tiles that are 6 in. by 6 in. to create a multi-colored design. How many tiles do the students need?

 Algebra Draw each rectangle in the coordinate plane. Find its perimeter and area.

53. $A(-3, 2), B(-2, 2), C(-2, -2), D(-3, -2)$

54. $A(-2, -6), B(-2, -3), C(3, -3), D(3, -6)$

Coordinate Geometry On graph paper, draw polygon *ABCDEFGH* with vertices $A(1, 1)$, $B(10, 1)$, $C(10, 8)$, $D(7, 8)$, $E(7, 5)$, $F(4, 5)$, $G(4, 8)$, and $H(1, 8)$.

55. Find the perimeter of the polygon.

56. Divide the polygon into rectangles. Find the area of the polygon.

57. Biology In the Pacific Northwest, a red fox has a circular home range with a radius of about 718 meters. To the nearest thousand square meters, what is the area of the home range of a red fox?

58. Multiple Choice A rectangle has a base of x units. The area is $(4x^2 - 2x)$ square units. What is the height of the rectangle in terms of x?

Ⓐ $(4 - x)$ units

Ⓑ $(4x^3 - 2x^2)$ units

Ⓒ $(x - 2)$ units

Ⓓ $(4x - 2)$ units

Home Maintenance To determine how much of each item to buy, tell whether you need to know area or perimeter. Explain your choice.

59. wallpaper for a bedroom

60. weatherstripping for a door

61. fence for a garden

62. paint for a basement floor

63. Coordinate Geometry The endpoints of a diameter of a circle are $A(2, 1)$ and $B(5, 5)$. Find the area of the circle in terms of π.

64. Graphing Calculator You want to build a rectangular corral by using the side of a barn for one side and 100 ft of fencing for the other three sides.

a. Make a table on your graphing calculator listing integer values for the base and the corresponding values of the height and area.

b. Make a graph using your table values. Graph the base on the horizontal axis and area on the vertical axis.

c. What are the dimensions of the corral with the greatest area?

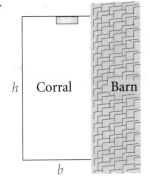

65. How many circles with the given radius are needed for the sum of their areas to equal the area of a circle with the second given radius?

a. 1 in. , 3 in.

b. 2 in. , 6 in.

c. 3 in. , 9 in.

d. Make a Conjecture How many circles with a radius of n in. are needed for the sum of their areas to equal the area of a circle with a radius of $3n$ in.?

 **Algebra** Find the area of each figure.

66. a rectangle with side lengths of $\frac{2a}{5b}$ units and $\frac{3b}{8}$ units

67. a square with perimeter $10n$ units

68. a square with side lengths of $(3m - 4n)$ units

69. Open-Ended The area of a 5 in.-by-5 in. square is the same as the sum of the areas of a 3 in.-by-3 in. square and a 4 in.-by-4 in. square. Find two or more squares whose total area is the same as the area of an 11 in.-by-11 in. square.

70. Track An athletic field is a rectangle, 100 yards by 40 yards, with a semicircle at each of the short sides. A running track 10 yards wide surrounds the field. Find the perimeter of the outside of the running track to the nearest tenth of a yard.

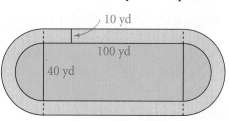

Test Prep

Gridded Response   For Exercises 71 and 72, a rectangular garden has a rectangular walkway around it. The width of the walkway is 8 ft.

71. How many feet greater than the perimeter of the garden is the outside perimeter of the walkway?

72. If the garden is a square with a perimeter of 260 ft, what is the area of the walkway in square feet?

73. You need to tile a 12 ft-by-15 ft floor. The color you want allows you the choices found in the table at the right. How many dollars would it cost to tile the floor with 12 in.-by-12 in. tiles?

Size of Tiles	Cost
12″ × 12″	$3/ft^2
11″ × 11″	$3/ft^2
10″ × 12″	$4/ft^2
6″ × 8″	$4.50/ft^2

74. How many tiles would cover the 12 ft-by-15 ft floor if you choose the 10 in.-by-12 in. tiles?

75. How many dollars would it cost to cover the 12 ft-by-15 ft floor with the tiles that are 6 in. by 8 in.?

Mixed Review

Lesson 1-8

76. The midpoint of $\overline{CD}$ has coordinates $(5, 6)$. Point C has coordinates $(-5, -1)$. Find the coordinates of point D.

Find (a) AB **to the nearest tenth and (b) the coordinates of the midpoint of** $\overline{AB}$.

77. $A(4, 1), B(7, 9)$ **78.** $A(0, 3), B(3, 8)$ **79.** $A(9, 2), B(-3, 9)$

80. $A(0, 1), B(-4, 6)$ **81.** $A(4, 10), B(-2, 3)$ **82.** $A(-1, 1), B(-4, -5)$

Lesson 1-7 $\overleftrightarrow{BG}$ **is the perpendicular bisector of** $\overline{WR}$ **at point** I.

83. What is $m\angle BIR$?   **84.** Name two congruent segments.

85. $\overline{WR}$ has length 124. What is the length of $\overline{IR}$?

Lesson 1-5   **For the given coordinates, find** PQ.

86. $P: 12, Q: -6$ **87.** $P: 3, Q: 9$ **88.** $P: -23, Q: 10$

Comparing Perimeters and Areas

Go Online
PHSchool.com

For: Graphing calculator procedures
Web Code: aue-2104

You can use a graphing calculator or spreadsheet technology to find maximum and minimum values for area and perimeter problems.

You have 32 yards of fencing. You want to make a rectangular pen for the calf you are raising as a 4-H project. What dimensions will give the maximum area? What is the maximum area?

Investigate

Draw some possible rectangular pens and find their areas.

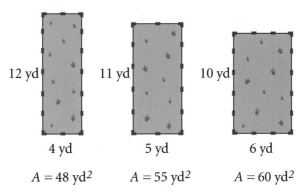

12 yd   11 yd   10 yd

4 yd   5 yd   6 yd

$A = 48$ yd^2 $A = 55$ yd^2 $A = 60$ yd^2

NY **G.R.1:** Use physical objects, diagrams, charts, tables, graphs, symbols, equations, or objects created using technology as representations of mathematical concepts.

Create a graphing calculator table to find area. Let X represent values for the base. The height then is $16 - X$, and the area is $X(16 - X)$. Enter Y1 = $16 - X$ and Y2 = $X(16 - X)$. Set the table so that X starts at 1 and changes by 1.

Scroll down the table. Area is maximum when X (or b) is 8. When $b = 8, h = 8$ and $A = 64$.

You can confirm this result by graphing Y1 = $X(16 - X)$. Trace on the graph to find the maximum area.

A square pen with sides of 8 yd will give maximum area for your calf. The maximum area is 64 yd^2.

X	Y1	Y2
4	12	48
5	11	55
6	10	60
7	9	63
8	8	64
9	7	63
10	6	60

X=4

Y2=X(16-X)

X=8   Y=64

X min = 0   Y min = 0
X max = 18   Y max = 70
X scl = 2   Y scl = 7

EXERCISES

1. **Make a Conjecture** For a fixed perimeter, what rectangular shape will result in a maximum area?

2. Consider that the pen is not restricted to polygon shapes. Determine the area of a circular pen if the circumference is 32 yd. How does this result compare with the maximum square area of 64 yd^2 found in the investigation?

3. You want to make a rectangular garden with an area of 900 ft^2. You want to use a minimum amount of fencing to keep the cost low.
   a. List some possible dimensions for the rectangular garden. Find the perimeter of each rectangle.
   b. Create a graphing calculator table. Use integer values of the base b, and the corresponding values of the height h, to find values for P, the perimeter. What dimensions will give you a garden with the minimum perimeter?

Some tests include gridded-response questions. You find a numerical answer. Then you write the answer at the top of the grid and fill in the corresponding bubbles below. You have to be sure that you use the grid correctly.

1 EXAMPLE

What is the *x*-coordinate of the midpoint of the segment with endpoints $H(5, 1)$ and $K(12, 2)$?

By the Midpoint Formula, you can find that the coordinates of the midpoint are $\left(\frac{17}{2}, \frac{3}{2}\right)$.

The answer to the question is $\frac{17}{2}$.

You can write the answer as 17/2 or 8.5.

The grids at the right show two ways to enter the answer.

Certain tests may have specific directions for entering answers in grids.

If your answer doesn't fit, then you may have misread the directions or the problem itself, or you made a calculation error. Go back and check.

2 EXAMPLE

What is the distance between the points $A(0, 0.14)$ and $B(0.1, 0.2)$? Round your answer to the nearest hundredth.

The answer is 0.12. You grid this as .12.

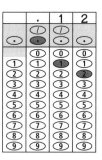

EXERCISES

Write what you would grid for each answer.

1. What is the *y*-coordinate of the midpoint of a segment with endpoints $C(0, -1)$ and $D(-6, 12)$?

2. What is the diameter, in centimeters, of a circle with circumference 0.5π cm?

3. What is the distance between $L(0, 0)$ and $M(0.3, 0.4)$?

4. The endpoints of a diameter of a circle are $R(3, 5)$ and $T(12, 5)$. What is the area of the circle? Round your answer to the nearest tenth.

5. What is the radius of a circle whose area is 10π cm^2? Round your answer to the nearest hundredth of a centimeter.

Chapter Review

Vocabulary Review

acute angle (p. 37)
adjacent angles (p. 38)
angle (p. 36)
angle bisector (p. 46)
axiom (p. 18)
collinear points (p. 17)
compass (p. 44)
complementary angles (p. 38)
congruent angles (p. 37)
congruent segments (p. 31)
conjecture (p. 5)
construction (p. 44)
coordinate (p. 31)
coplanar (p. 17)

counterexample (p. 5)
foundation drawing (p. 11)
inductive reasoning (p. 4)
isometric drawing (p. 10)
line (p. 17)
midpoint (p. 32)
net (p. 12)
obtuse angle (p. 37)
opposite rays (p. 23)
orthographic drawing (p. 11)
parallel lines (p. 24)
parallel planes (p. 24)
perpendicular bisector (p. 45)
perpendicular lines (p. 45)

plane (p. 17)
point (p. 17)
postulate (p. 18)
ray (p. 23)
right angle (p. 37)
segment (p. 23)
skew lines (p. 24)
space (p. 17)
straight angle (p. 37)
straightedge (p. 44)
supplementary angles (p. 38)
vertical angles (p. 38)

Choose the correct term to complete each sentence.

1. Figures that are in the same plane are __?__ .

2. A(n) __?__ is the part of a line consisting of two endpoints and all points between them.

3. Two segments with the same length are __?__ .

4. A(n) __?__ of a segment is a point that divides the segment into two congruent segments.

5. A(n) __?__ is a ray that divides an angle into two congruent angles.

6. A conclusion based upon inductive reasoning is sometimes called a(n) __?__ .

7. A(n) __?__ is an accepted statement of fact.

8. __?__ are coplanar lines that do not intersect.

9. A(n) __?__ is an angle whose measure is between 90 and 180.

Go Online
PHSchool.com
For: Vocabulary quiz
Web Code: auj-0151

Skills and Concepts

1-1 Objectives

▼ To use inductive reasoning to make conjectures

You use **inductive reasoning** when you make conclusions based on patterns you observe. A **conjecture** describes a conclusion reached using inductive reasoning. A **counterexample** to a conjecture is an example for which the conjecture is incorrect.

Find a pattern for each sequence. Describe the pattern and use it to show the next two terms.

10. 40, 35, 30, 25, . . .   **11.** 5, −5, 5, −5, . . .   **12.** 34, 27, 20, 13, 6, . . .

13. 6, 24, 96, 384, . . .   **14.** 2, 4, 8, 16, 32, . . .   **15.** 1, 2, 5, 6, 9, . . .

16. Draw the next figure in the sequence.

1-2 Objectives

▼ To make isometric and orthographic drawings

▼ To draw nets for three-dimensional figures

A **net** is a two-dimensional pattern that you can fold to form a three-dimensional figure. A net shows all surfaces of a figure in one view.

17. The net at the right is for a number cube. What are the three sums of the numbers on opposite surfaces of the cube?

18. On another number cube, the numbers on each pair of opposite surfaces add to 7. Draw a net for this number cube.

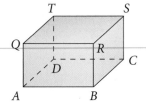

1-3 and 1-4 Objectives

▼ To understand basic terms and postulates of geometry

▼ To identify segments and rays

▼ To recognize parallel lines

Points that lie on the same line are **collinear points.** Points and lines in the same plane are **coplanar. Segments** and **rays** are parts of lines.

Lines that are coplanar and do not intersect are **parallel lines.** Lines in space that are not parallel and do not intersect are **skew.** Planes that do not intersect are **parallel planes.**

Use the figure at the right for Exercises 19–22.

19. Name two intersecting lines.

20. Name three noncollinear points.

21. Name a pair of parallel planes.

22. Name three lines that intersect at D.

Complete with *always, sometimes,* or *never* to make a true statement.

23. A line and a point are __?__ coplanar.   **24.** Two segments are __?__ coplanar.

25. Skew lines are __?__ coplanar.   **26.** Two points are __?__ collinear.

1-5 and 1-6 Objectives

▼ To find the lengths of segments

▼ To find the measures of angles

▼ To identify special angle pairs

Segments with the same length are **congruent segments.** A **midpoint** of a segment divides the segment into two congruent segments.

Two rays with the same endpoint form an **angle.** Angles can be classified as acute, right, obtuse, or straight. Angles with the same measure are **congruent angles.**

27. Find two possible coordinates of Q so that $PQ = 5$.

28. Find the coordinate of the midpoint of $\overline{PH}$.

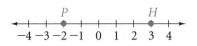

x^2 **Algebra** **Find the value of each variable.**

29.

$$3m + 5 \quad 4m - 10$$

A   B   C

30.

$(3x + 31)°$ $(2x - 6)°$

31. Name a pair of each of the following.
   a. complementary angles
   b. supplementary angles
   c. vertical angles

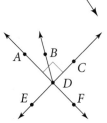

Construction is the process of making geometric figures using a **compass** and a **straightedge.** Four basic constructions involve constructing congruent segments, congruent angles, and bisectors of segments and angles.

Perpendicular lines intersect at right angles. A **perpendicular bisector** of a segment is perpendicular to the segment at its midpoint, and bisects it into two congruent segments. An **angle bisector** is a ray that divides an angle into two congruent angles.

32. Use a protractor to draw a 64° angle. Then construct an angle congruent to your 64° angle.

33. Use a ruler and draw $\overline{PQ}$.
   a. Construct $\overline{AB} \cong \overline{PQ}$.
   b. Construct the perpendicular bisector of $\overline{AB}$.

The x-axis and the y-axis intersect at the origin $(0, 0)$ and determine a coordinate plane. You can find the coordinates of the midpoint M of $\overline{AB}$ with endpoints $A(x_1, y_1)$ and $B(x_2, y_2)$ using the **Midpoint Formula.**

$$M = \left(\frac{x_1 + x_2}{2}, \frac{y_1 + y_2}{2} \right)$$

You can find the distance d between points $A(x_1, y_1)$ and $B(x_2, y_2)$ using the **Distance Formula.**

$$d = \sqrt{(x_2 - x_1)^2 + (y_2 - y_1)^2}$$

Find the distance between the points to the nearest tenth.

34. $A(-1, 5), B(0, 4)$ **35.** $C(-1, -1), D(6, 2)$ **36.** $E(-7, 0), F(5, 8)$

$\overline{GH}$ **has endpoints** $G(-3, 2)$ **and** $H(3, -2)$.

37. Find the coordinates of the midpoint of $\overline{GH}$.

38. Find GH to the nearest tenth.

The perimeter P of a polygon is the sum of the lengths of its sides. The area A of a polygon is the number of square units it encloses.

Formulas:	Square	Rectangle	Circle
	$P = 4s$	$P = 2b + 2h$	$C = \pi d$ or $C = 2\pi r$
	$A = s^2$	$A = bh$	$A = \pi r^2$

Find the perimeter and the area of each figure.

39. 8 cm

40. 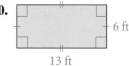 6 ft / 13 ft

41. 3 in. / 5 in.

Find the circumference and the area of each circle to the nearest hundredth.

42. $r = 3$ in.   **43.** $d = 15$ m   **44.** $r = 26$ m

Chapter Test

Go Online
PHSchool.com
For: Chapter Test
Web Code: aua-0152

Describe each pattern and find the next two terms of each sequence.

1. $8, -4, 2, -1, \ldots$ 2. $0, 2, 4, 6, 8, \ldots$

3.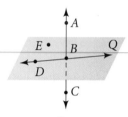

4. **Open-Ended** Write two different sequences whose first three terms are 1, 2, 4. Describe each pattern.

5. Draw a net for a cube.

Use the figure for Exercises 6–9.

6. Name three collinear points.

7. Name four coplanar points.

8. What is the intersection of $\overleftrightarrow{AC}$ and plane Q?

9. How many planes contain each line and each point?
   a. $\overleftrightarrow{BD}$ and point A **b.** $\overleftrightarrow{AB}$ and point C
   c. $\overleftrightarrow{BE}$ and point C **d.** $\overleftrightarrow{BD}$ and point E

10. **Track** The running track is a rectangle with a half circle on each end. If $\overline{FI}$ and $\overline{GH}$ are diameters, find the area inside the track to the nearest tenth.

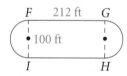

Complete with *always, sometimes,* or *never* to make each statement true.

11. $\overrightarrow{LJ}$ and $\overrightarrow{TJ}$ are ___?___ opposite rays.

12. Four points are ___?___ coplanar.

13. Skew lines are ___?___ coplanar.

14. Two segments that lie in parallel lines are ___?___ parallel.

15. The intersection of two planes is ___?___ a point.

16. **Algebra** $JK = 48$. Find the value of x.

17. **Algebra** $M(x, y)$ is the midpoint of $\overline{CD}$ with endpoints $C(5, 9)$ and $D(17, 29)$.
    a. Find the values of x and y.
    b. Show $MC = MD$.

18. To the nearest tenth, find the perimeter of $\triangle ABC$ with vertices $A(-2, -2)$, $B(0, 5)$, and $C(3, -1)$.

For the given dimensions, find the area of each figure to the nearest hundredth.

19. rectangle   20. square   21. circle
    $b = 4$ m $s = 3.5$ in. $d = 9$ cm
    $h = 2$ cm

Algebra **Use the figure for Exercises 22–24. In Exercises 22 and 23, find the value of each variable.**

22. $m\angle BDK = 3x + 4$, $m\angle JDR = 5x - 10$

23. $m\angle BDJ = 7y + 2$, $m\angle JDR = 2y + 7$

24. Name two complementary angles.

25. **Writing** Why is it useful to have more than one way of naming an angle?

26. Draw an obtuse $\angle ABC$. Use a compass and straightedge to bisect the angle.

Use the figure to complete Exercises 27–30.

27. $\overline{VW}$ is the ___?___ of $\overline{AY}$.

28. If $EY = 3.5$, then $AY = $ ___?___.

29. $\frac{1}{2}$ ___?___ $= AE$

30. ___?___ is the midpoint of ___?___.

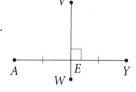

31. **Carpeting** How many square yards of carpet are needed to carpet a room that is 15 ft long and 20 ft wide?

Regents Test Prep

Reading Comprehension Read the passage below, then answer the questions on the basis of what is *stated* or *implied* in the passage.

Instructions for Building the Rainbow Toy Chest

Use $\frac{3}{4}$-in.-thick plywood.
- Cut the top and bottom 18 in. by 42 in.
  Paint the top red and the bottom violet.

Use $\frac{1}{2}$-in.-thick plywood.
- Cut the two sides 18 in. by 60 in.
  Paint: left side brown, right side white.
- Cut the three shelves 15 in. by 41 in.
  Paint: top orange, middle yellow, bottom green.
- Cut the two dividers 24 in. by 15 in.
  Paint: left blue, right indigo.

Use $\frac{3}{4}$-in.-thick particleboard.
- Cut the back 41 in. by 60 in. Paint the back gray.

Assemble the painted pieces using nails and glue.

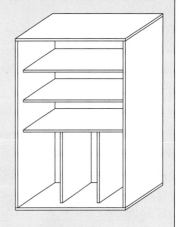

---

1. How many pieces must be cut to make the chest?
   (1) 5   **(2)** 7   **(3)** 9   **(4)** 10

2. What is the shape of each piece?
   (1) square   **(2)** rectangular
   (3) round   **(4)** cannot be determined

3. Which uses thicker wood, a divider or the top?
   (1) divider   **(2)** top
   (3) same thickness   **(4)** cannot be determined

Assume that the toy chest has been assembled.

4. Take the top of the bookshelf as the first horizontal surface and count downward. What is the color of the fourth surface down?
   (1) white   **(2)** yellow   **(3)** green   **(4)** blue

5. Which corners are NOT coplanar?
   (1) the four corners of the right side
   (2) the two top corners of the left side and the two front corners of the bottom
   (3) the two top corners of the right side and the two bottom corners of the left side
   (4) the two front corners of the top and the two back corners of the bottom

6. Where do the violet board and the particleboard meet?
   (1) the front edge of the violet board
   (2) the back edge of the left side
   (3) the right edge of the bottom
   (4) the bottom edge of the back

7. Which edge is skew to the front edge of the orange board?
   (1) the left edge of the orange board
   (2) the top edge of the white board
   (3) the back edge of the yellow board
   (4) the front edge of the middle shelf

8. What are the colors of two boards that are NOT perpendicular?
   (1) blue and gray   **(2)** indigo and violet
   (3) blue and white   **(4)** yellow and brown

9. How tall is the toy chest?

10. What is the area of the top?

11. What is the perimeter of the front?

12. The back edge of each shelf touches the back of the chest. How far "recessed" is the front edge of a shelf from the front edge of a side?

Linear Regression

An automobile contains several data-gathering devices. Among the most visible is the speedometer. How does a speedometer know how fast the car is going? One source of information is the rotation of the wheels.

NY G.PS.1: Use a variety of problem solving strategies to understand new mathematical content.

1 ACTIVITY

1. The tires on the car below have a diameter of 18 in. How far does the car go for each turn of a wheel?

2. Find the number of times that each wheel turns when the car travels 1 mile.

3. Find the revolutions per minute (rpm) of the wheel when the car is going 1 mile per hour (mph).

4. Copy and complete the chart. Find the rpm of the wheel for the speeds listed.

Car Speed (mph)	1	10	20	40	60
Wheel Speed (rpm)	■	■	■	■	■

5. Plot your (rpm, mph) data with rpm on the *x*-axis and mph on the *y*-axis. What do you observe about the points? Describe the relationship between rpm and mph.

6. Find the $\frac{mph}{rpm}$ ratio. This is a *calibration coefficient* that you can use to translate the rpm of the wheel into the mph of the car. Find the speed of the car when the wheels are turning at 1000 rpm.

You know that for a circle with circumference C and diameter d, $C = \pi d$ or $\frac{C}{d} = \pi$.

Here are three ways to approximate π.

- For a circular object, measure C and d. Then find $\frac{C}{d}$.

- For a circular object, have five people measure C and d. Then find $\frac{\text{average } C}{\text{average } d}$.

- Measure C and d for five different objects. Then approximate $\frac{C}{d}$ using the linear regression (LinReg) tool on your graphing calculator.

You can think of $C = \pi d$ as $y = \pi x$, a linear relationship with slope π. When you have a linear relationship $y = ax$ between (x, y) data pairs, LinReg on your calculator will give you a value for a.

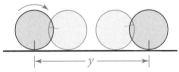 **ACTIVITY**

Gather at least five cylinders with diameters between 2 cm and 20 cm and try the following.

7. Mark a point on a rim of a cylinder. Use the mark as a reference point and record the distance the cylinder rolls in one revolution.

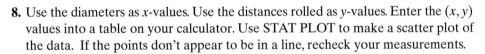

Repeat these steps with your other four cylinders.

8. Use the diameters as x-values. Use the distances rolled as y-values. Enter the (x, y) values into a table on your calculator. Use STAT PLOT to make a scatter plot of the data. If the points don't appear to be in a line, recheck your measurements.

9. Press STAT . In the CALC menu, select LinReg to move LinReg to your home screen. Press ENTER . Record the a value from the screen. The a value should be close to π. Graph Y1 = aX and see how well it fits your scatter plot.

EXERCISES

10. Do a second experiment. Roll each cylinder through five revolutions and record each distance. Again make a table and apply LinReg.
  a. This time, a should be close to 5π. How does the value of $\frac{a}{5}$ compare with the value of a you found in Exercise 9?
  b. Why should this second experiment give you a better approximation of π than the first experiment?

> LinReg
> y=ax+b
> a=15.69259259
> b=-.002962963
>
> ■

11. In Activity 1, suppose you replace the wheels on the car with ones requiring 17-in. tires. Explain how the calibration coefficient would change.

12. Another data-gathering device in an automobile is the odometer. It tells how far the car has traveled. The information it uses also comes from the rotation of the wheels. Explain how this might work.

What You've Learned

NY Learning Standards for Mathematics

In Chapter 1, you learned many of the basic terms and assumptions used in geometry.

- **G.G.66:** to measure angles and segments, do basic constructions, and use the coordinate plane to find the midpoint of a segment and the distance between two points.

- **G.PS.6:** to find the area of and the distance around rectangles and circles.

 Check Your Readiness   **GO** for Help to the Lesson in green.

Evaluating Expressions (Skills Handbook page 754)

x^2 **Algebra** Evaluate each expression for the given value of x.

1. $9x - 13$ for $x = 7$ **2.** $90 - 3x$ for $x = 31$ **3.** $\frac{1}{2}x + 14$ for $x = 23$

Solving Equations (Algebra 1 Review, page 30)

x^2 **Algebra** Solve each equation.

4. $2x - 17 = 4$
5. $3x + 8 = 53$
6. $(10x + 5) + (6x - 1) = 180$
7. $(x + 21) + (2x + 9) = 90$
8. $3x + 4 = 2x - 1$
9. $3(x + 8) = 12$
10. $2(x + 4) = x + 13$
11. $7x + 5 = 5x + 17$
12. $14x = 2(5x + 14)$
13. $2(3x - 4) + 10 = 5(x + 4)$

Segments and Angles (Lessons 1-5 and 1-6)

Use the figure at the right.

14. Name $\angle 1$ in two other ways.

15. Name the vertex of $\angle 2$.

16. If D is the midpoint of $\overline{AB}$, find x.

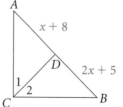

17. If $m\angle ADC$ and $m\angle BDC$ have a sum of 180, name the straight angle.

18. If $\angle 1 \cong \angle 2$, name the bisector of $\angle ACB$.

19. If $m\angle 2 = 45$ and $\angle ACB$ is a right angle, find $m\angle 1$.

20. If $\angle ACB$ is a right angle, $m\angle 1 = 4x$ and $m\angle 2 = 2x + 18$, find $m\angle 1$ and $m\angle 2$.

Reasoning and Proof

◀)) **Key Vocabulary**

What You'll Learn Next

NY **Learning Standards for Mathematics**

- **G.G.25:** In this chapter, you will learn how to write special types of statements known as conditionals, biconditionals, and definitions.

- **G.G.26:** You will use such statements and deductive reasoning to conclude that other statements are true.

- **G.RP.7:** Understanding how deductive reasoning works, you will apply it to form conclusions using algebra.

- **G.RP.7:** You will also use it to study elementary proofs and form your first significant conclusions about geometric relationships.

Activity Lab Applying what you learn, you will do activities involving food on pages 122 and 123.

2-1

Conditional Statements

NY Learning Standards for Mathematics

G.G.25 Know and apply the conditions under which a compound statement is true.

G.G.26 Identify and write the inverse, converse, and contrapositive of a given conditional statement and note the logical equivalences.

✓ **Check Skills You'll Need**   **GO** **for Help** Skills Handbook pages 757 and 758

x^2 **Algebra** Solve each equation.

1. $y + 1 = 0$ **2.** $|x| = 2$ **3.** $17 = 15 + z$

4. $x + 4 = 0$ **5.** $|n| = 0$ **6.** $m - 9 = -10$

◀))) **New Vocabulary**   • conditional   • hypothesis   • conclusion
                          • truth value   • converse

OVERSET

---

1 Conditional Statements

Online
active math

For: Conditionals Activity
Use: Interactive Textbook, 2-1

You have heard *if-then* statements such as this one:
    If you are not completely satisfied, then your money will be refunded.
Another name for an *if-then* statement is a **conditional**. Every conditional has two parts. The part following *if* is the **hypothesis,** and the part following *then* is the **conclusion.**

1 EXAMPLE   Identifying the Hypothesis and the Conclusion

Identify the hypothesis and the conclusion of this conditional statement:
    If Texas won the 2006 Rose Bowl football game, then Texas was college football's 2005 national champion.

Hypothesis: Texas won the 2006 Rose Bowl football game.
● Conclusion: Texas was college football's 2005 national champion.

✓ **Quick Check** ❶ Identify the hypothesis and the conclusion of this conditional statement:
    If T − 38 = 3, then T = 41.

You can write many sentences as conditionals.

2 EXAMPLE   Writing a Conditional

Write each sentence as a conditional.

a. A rectangle has four right angles.
    If a figure is a rectangle, then it has four right angles.

b. A tiger is an animal.
    If something is a tiger, then it is an animal.

✓ **Quick Check** ❷ Write each sentence as a conditional.
    a. An integer that ends with 0 is divisible by 5.
    b. A square has four congruent sides.

A conditional can have a **truth value** of *true* or *false*. To show that a conditional is true, show that every time the hypothesis is true, the conclusion is also true. To show that a conditional is false, you need to find only one counterexample for which the hypothesis is true and the conclusion is false.

3 EXAMPLE Finding a Counterexample

Show that this conditional is false by finding a counterexample:
    If it is February, then there are only 28 days in the month.

To show that this conditional is false, you need to find one counterexample that makes the hypothesis true and the conclusion false.

February in the year 2008 is a counterexample. Because 2008 is a leap year, the month of February has 29 days.

● The conditional is false because February 2008 is a counterexample.

✓ Quick Check ③ Show that this conditional is false by finding a counterexample:
    If the name of a state contains the word *New*, then the state borders an ocean.

You can use a Venn diagram to better understand true conditional statements.

4 EXAMPLE Using a Venn Diagram

Draw a Venn diagram to illustrate this conditional:
    If you live in Chicago, then you live in Illinois.

The set of things that satisfy the hypothesis lies inside the set of things that satisfy the conclusion.

✓ Quick Check ④ Draw a Venn diagram to illustrate this conditional:
    If something is a cocker spaniel, then it is a dog.

Converses

The **converse** of a conditional switches the hypothesis and the conclusion.

5 EXAMPLE Writing the Converse of a Conditional

Write the converse of the following conditional.

<u>Conditional</u>
If two lines intersect to form right angles, **then** they are perpendicular.

<u>Converse</u>
If two lines are perpendicular, **then** they intersect to form right angles.

✓ Quick Check ⑤ Write the converse of the following conditional.
    If two lines are not parallel and do not intersect, then they are skew.

In Example 5, both the original conditional and its converse are true. It is possible for a conditional and its converse to have different truth values.

Test-Taking Tip

Read the question carefully. Multiple-choice items usually ask you to find a true or correct answer. Example 6 asks you to find a converse that is *false*.

6 EXAMPLE Finding the Truth Value of a Converse

Multiple Choice Which conditional statement has a false converse?
- (A) If two planes are parallel, then they have no points in common.
- (B) If a figure is a square, then it has four sides.
- (C) If a point has *x*-coordinate 0, then it lies on the *y*-axis.
- (D) If two angles are congruent, then they have the same measure.

Write the converse of choice B:
  If a figure has four sides, then it is a square.

A rectangle is a counterexample, so the converse is false.
● Choice B is the correct answer.

✓ Quick Check 6 Write the converse of each conditional statement. Determine the truth value of the conditional and its converse. (*Hint:* One of these conditionals is *not* true.)
a. If two lines do not intersect, then they are parallel.
b. If $x = 2$, then $|x| = 2$.

7 EXAMPLE Real-World  Connection

This is an illustration by John Tenniel for *Alice's Adventures in Wonderland.*

Literature In Lewis Carroll's *Alice's Adventures in Wonderland*, the Mad Hatter states: "Why you might just as well say that 'I see what I eat' is the same thing as 'I eat what I see'!" Explain why the Mad Hatter is wrong.

The statement "I see what I eat" can be rewritten as a conditional.
  "If I eat it, then I see it."

The statement "I eat what I see," can be rewritten as a conditional.
  "If I see it, then I eat it."

The two statements are converses of each other. A statement and its converse do not always have the same meaning or the same truth value. The Mad Hatter is wrong to suggest that you can use one just as well as the other.

✓ Quick Check 7 In *Alice's Adventures in Wonderland*, the Dormouse states:
  "... that 'I breathe when I sleep' is the same thing as 'I sleep when I breathe'!"
Use conditionals to explain why this statement is wrong.

You can use symbolic form to represent a conditional and its converse. In symbolic form, the letter *p* stands for the hypothesis and the letter *q* stands for the conclusion.

 Key Concepts

Summary	Conditional Statements and Converses		
Statement	**Example**	**Symbolic Form**	**You Read It**
Conditional	If an angle is a straight angle, then its measure is 180.	$p \rightarrow q$	If *p*, then *q*.
Converse	If the measure of an angle is 180, then it is a straight angle.	$q \rightarrow p$	If *q*, then *p*.

EXERCISES

For more exercises, see *Extra Skill, Word Problem, and Proof Practice.*

Practice and Problem Solving

A Practice by Example

Example 1
(page 80)

GO for Help

1. Identify the hypothesis and the conclusion in the cartoon.

FRANK AND ERNEST   By BOB THAVES

Identify the hypothesis and conclusion of each conditional.

2. If you want to be fit, then you want to get plenty of exercise.

 3. Algebra If $x + 20 = 32$, then $x = 12$.

4. "If you can see the magic in a fairy tale, you can face the future."
— Danielle Steel, novelist

5. "If somebody throws a brick at me, I can catch it and throw it back."
— Harry S Truman

6. "If you can accept defeat and open your pay envelope without feeling guilty,
you're stealing." — George Allen, former NFL coach

7. "If my fans think that I can do everything I say I can do, then they're crazier than I am."—Muhammad Ali

8. ". . . if I could paint that flower in a huge scale, you could not ignore its beauty." — Georgia O'Keeffe, artist

Example 2
(page 80)

Write each sentence as a conditional.

9. Glass objects are fragile.

 10. Algebra $3x - 7 = 14$ implies that $3x = 21$.

11. Whole numbers that have 2 as a factor are even.

12. All obtuse angles have measure greater than 90.

13. Good weather makes a picnic enjoyable.

14. Two skew lines do not lie in the same plane.

Problem Solving Hint

Some conditionals may omit *then*. You can insert it mentally if you wish.

Example 3
(page 81)

Show that each conditional is false by finding a counterexample.

15. If it is not a weekday, then it is Saturday.

16. Odd integers less than 10 are prime.

17. If you live in a country that borders the United States, then you live in Canada.

18. If you play a sport with a ball and a bat, then you play baseball.

Example 4
(page 81)

Draw a Venn diagram to illustrate each statement.

19. If you live in New England, then you live in the United States.

20. If you play the flute, then you are a musician.

21. If an angle has measure 40, then it is acute.

22. Carrots are vegetables.

Example 5
(page 81)

Write the converse of each conditional statement.

23. If you eat your vegetables, then you grow.

24. If a triangle is a right triangle, then it has a 90° angle.

25. If two segments are congruent, then they have the same length.

26. If you do not work, you do not get paid.

Examples 6 and 7
(page 82)

Write the converse of each conditional statement. Determine the truth values of the original conditional and its converse.

27. If you travel from the United States to Kenya, then you have a passport.

28. Coordinate Geometry If a point is in the first quadrant, then its coordinates are positive.

29. Chemistry If a substance is water, then its chemical formula is H_2O.

30. Probability If the probability that an event will occur is 1, then the event is certain to occur.

31. If you are in Indiana, then you are in Indianapolis.

32. If two angles have measure 90, then the angles are congruent.

B **Apply Your Skills**

Write a conditional statement that each Venn diagram illustrates.

33. 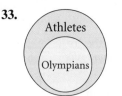   **34.** 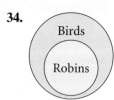   **35.**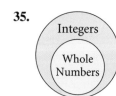

36. Error Analysis Ellen claims that both this conditional and its converse are true.
If x is an integer divisible by 3, then x^2 is an integer divisible by 3.
a. Write the converse of the conditional.
b. Only one of the statements is true. Determine which statement is false and provide a counterexample to support your answer.

37. Multiple Choice Which conditional and its converse are both true?
Ⓐ If $x = 1$, then $2x = 2$.   Ⓑ If $x = 3$, then $x^2 = 6$.
Ⓒ If $x = 2$, then $x^2 = 4$.   Ⓓ If $x^2 = 4$, then $x = 2$.

Jeanette Rankin was one of nine women among 435 members of Congress at the start of World War II.

Write each statement as a conditional.

38. "We're half the people; we should be half the Congress."— Jeanette Rankin, former U.S. Congresswoman, calling for more women in office

39. "A great work is made out of a combination of obedience and liberty." — Nadia Boulanger, orchestra conductor and musical mentor

40. "A problem well stated is a problem half solved." — Charles F. Kettering, inventor

x^2 **Algebra** **Write the converse of each statement. If the converse is true, write _true_; if not true, provide a counterexample.**

41. If $x - 3 = 15$, then $x = 18$.

42. If y is negative, then $-y$ is positive.

43. If $x = -6$, then $|x| = 6$.

44. If $x < 0$, then $x^2 > 0$.

45. If $x = 2$, then $x^2 = 4$.

46. If $x < 0$, then $x^3 < 0$.

47. Advertising Al sees an ad that states, "You want to look good at the beach this summer. Join GoodFit Health Club." Al figures, "I am going to join GoodFit Health Club, so that I will look good at the beach."
   a. Write the statement in the ad as a conditional.
   b. Write Al's statement as a conditional.
   c. Writing Explain why the statement in the ad does not have the same meaning as Al's statement.

Reading Math **Let statements p, q, and r be as follows.**
   p: A figure is square.
   q: A figure has four congruent angles.
   r: A figure has four congruent sides.

Write the words for the symbolic statement shown. Determine the truth value of the statement. If it is false, provide a counterexample.

48. $p \rightarrow q$ **49.** $q \rightarrow p$ **50.** $r \rightarrow q$ **51.** $(q \text{ and } r) \rightarrow p$

Advertising **Advertisements often suggest conditional statements. For example, an ad might imply that if you buy a product, you will be popular.**

52. What conditional is implied in the ad at the right?

53. Open-Ended Find an ad in which a conditional is used or implied.

Write each postulate as a conditional statement.

54. Two intersecting lines meet in exactly one point.

55. Two intersecting planes meet in exactly one line.

56. Two congruent figures have equal areas.

57. Through any two points there is exactly one line.

58. Through any three noncollinear points there is exactly one plane.

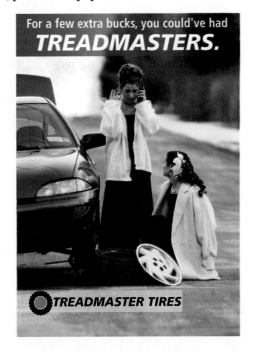

For a few extra bucks, you could've had
TREADMASTERS.

TREADMASTER TIRES

GO Online
Homework Video Tutor
Visit: PHSchool.com
Web Code: aue-0201

Write a statement beginning with *All*, *Some*, or *No* to match each Venn diagram.

59.

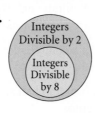

60.

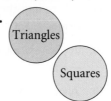

61.

62. Critical Thinking You can write many statements that begin with *All* or *No* as conditionals. Give an example of each. (*Hint:* See Exercises 59–61.)

63. Let *a* represent an integer. Consider the five statements *r*, *s*, *t*, *u*, and *v*:
    r: *a* is even.   *s*: *a* is odd.   *t*: 2*a* is even.   *u*: 2*a* is odd.   *v*: 2*a* + 1 is odd.
How many statements of the form $p \rightarrow q$ can you make from these five statements? Decide whether each of the statements is true or false.

NY REGENTS

Test Prep

Multiple Choice

64. Which is the hypothesis of the following statement?
If $4 < k < 6$, then $-4 > -k > -6$.
  A. $4 < k < 6$ **B.** $4 > k > 6$ **C.** $-4 > -k > -6$ **D.** $-4 < -k < -6$

65. Which is the converse of this statement?
If you can sing, then you can go with Sarah.

  F. You can't sing, then you can't go with Sarah.
  G. If you can't go with Sarah, then you can sing.
  H. If you can't sing, then you can go with Sarah.
  J. If you can go with Sarah, then you can sing.

66. Which statement has a true converse?
  A. If a vehicle is a car, then it has four wheels.
  B. If you go to Asia from the United States, then you cross an ocean.
  C. If you own a dog, then your pet is furry.
  D. If you can stand up, then you can walk.

Short Response

67. Write the converse of the following statement. Determine its truth value.
If Marta is five years old, then she is too young to vote.

Mixed Review

GO for Help

Lesson 1-9

Find the perimeter of each rectangle with the given base and height.

68. 6 in., 12 in.   **69.** 3.5 cm, 7 cm   **70.** $1\frac{3}{4}$ yd, 18 in.   **71.** 11 m, 60 cm

72. Find the area of a circle with diameter 10 in. Leave your answer in terms of π.

Lesson 1-8

Find the distance between the points. Round each answer to the nearest tenth.

73. $A(1, 2), B(4, -2)$ **74.** $M(-5, 1), N(0, 5)$ **75.** $R(0, -6), T(2, 3)$

Lesson 1-1

Find the pattern for each sequence. Use the pattern to show the next two terms.

76. $4, 2, 1, \frac{1}{2}, \ldots$ **77.** $5, 2, -1, -4, \ldots$ **78.** N, M, L, K, ...

2-2

Biconditionals and Definitions

 Learning Standards for Mathematics

G.G.25 Know and apply the conditions under which a compound statement (conjunction, disjunction, conditional, biconditional) is true.

✓ Check Skills You'll Need

GO for Help Lesson 2-1

Identify the hypothesis and the conclusion of each conditional statement.

1. If $x > 10$, then $x > 5$.

2. If you live in Milwaukee, then you live in Wisconsin.

Write each statement as a conditional.

3. Squares have four sides. **4.** All butterflies have wings.

Write the converse of each statement.

5. If the sun shines, then we go on a picnic.

6. If two lines are skew, then they do not intersect.

7. If $x = -3$, then $x^3 = -27$.

◄》 New Vocabulary • biconditional

1 Writing Biconditionals

When a conditional and its converse are true, you can combine them as a true **biconditional.** This is the statement you get by connecting the conditional and its converse with the word *and*. You can write a biconditional more concisely, however, by joining the two parts of each conditional with the phrase *if and only if*.

1 EXAMPLE Writing a Biconditional

Vocabulary Tip

Connect the conditional and its converse with *and*. Then compare with the *if and only if* form.

Consider this true conditional statement. Write its converse. If the converse is also true, combine the statements as a biconditional.

Conditional
If two angles have the same measure, then the angles are congruent.

Converse
If two angles are congruent, then the angles have the same measure.
The converse is also true.

Since both the conditional and its converse are true, you can combine them in a true biconditional by using the phrase *if and only if*.

Biconditional
Two angles have the same measure if and only if the angles are congruent.

✓ Quick Check

1 Consider this true conditional statement. Write its converse. If the converse is also true, combine the statements as a biconditional.

Conditional
If three points are collinear, then they lie on the same line.

Lesson 2-2 Biconditionals and Definitions **87**

You can write a biconditional as two conditionals that are converses of each other.

2 EXAMPLE Separating a Biconditional Into Parts

Algebra Write two statements that form this biconditional about whole numbers:
A number is divisible by 3 if and only if the sum of its digits is divisible by 3.

Here are the two statements. They are converses of each other.

If a number is divisible by 3, then the sum of its digits is divisible by 3.

If the sum of a number's digits is divisible by 3, then the number is divisible by 3.

Quick Check ❷ Write two statements that form this biconditional about integers greater than 1:
A number is prime if and only if it has only two distinct factors, 1 and itself.

Key Concepts

Summary	Biconditional Statements		
A biconditional combines $p \rightarrow q$ and $q \rightarrow p$ as $p \leftrightarrow q$.			
Statement	**Example**	**Symbolic Form**	**You Read It**
Biconditional	An angle is a straight angle if and only if its measure is 180.	$p \leftrightarrow q$	p if and only if q.

2 Recognizing Good Definitions

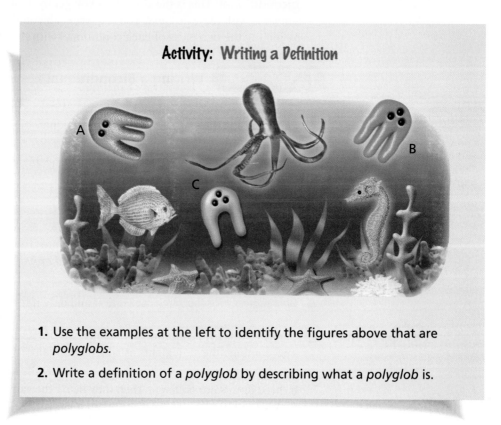

Activity: Writing a Definition

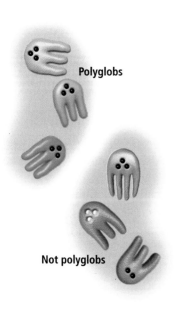

Polyglobs

Not polyglobs

1. Use the examples at the left to identify the figures above that are *polyglobs*.

2. Write a definition of a *polyglob* by describing what a *polyglob* is.

In geometry you start with undefined terms such as point, line, and plane whose meanings you understand intuitively. Then you use those terms to define other terms such as collinear points.

A good definition is a statement that can help you identify or classify an object. A good definition has several important components.

✔ A good definition uses clearly understood terms. The terms should be commonly understood or already defined.

✔ A good definition is precise. Good definitions avoid words such as *large*, *sort of*, and *almost*.

✔ A good definition is reversible. That means that you can write a good definition as a true biconditional.

Real-World **Connection**

The definitions in a dictionary have to be "good definitions."

3 EXAMPLE **Writing a Definition as a Biconditional**

Show that this definition of *perpendicular lines* is reversible. Then write it as a true biconditional.

Definition
Perpendicular lines are two lines that intersect to form right angles.

Conditional
If two lines are perpendicular, then they intersect to form right angles.

Converse
If two lines intersect to form right angles, then they are perpendicular.

The two conditionals—converses of each other—are true, so the definition can be written as a true biconditional.

Biconditional
Two lines are perpendicular if and only if they intersect to form right angles.

☑ **Quick Check** ❸ Show that this definition of *right angle* is reversible. Then write it as a true biconditional.

Definition
A right angle is an angle whose measure is 90.

One way to show that a statement is *not* a good definition is to find a counterexample.

4 EXAMPLE **Real-World** **Connection**

Language Arts Is the given statement a good definition? Explain.
a. An airplane is a vehicle that flies.

The statement is not a good definition because it is not reversible. A helicopter is a counterexample. A helicopter is a vehicle that flies, but a helicopter is not an airplane.

b. A triangle has sharp corners.
The statement is not a good definition because it uses the imprecise word *sharp*, and it is not reversible.

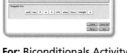

For: Biconditionals Activity
Use: Interactive Textbook, 2-2

☑ **Quick Check** ❹ Is the following statement a good definition? Explain.
A square is a figure with four right angles.

EXERCISES

For more exercises, see *Extra Skill, Word Problem, and Proof Practice.*

Practice and Problem Solving

 Practice by Example

Example 1
(page 87)

 for Help

Each conditional statement below is true. Write its converse. If the converse is also true, combine the statements as a biconditional.

1. If two segments have the same length, then they are congruent.

x^2 **2. Algebra** If $x = 12$, then $2x - 5 = 19$.

3. If a number is divisible by 20, then it is even.

x^2 **4. Algebra** If $x = 3$, then $|x| = 3$.

5. In the United States, if it is July 4th, then it is Independence Day.

x^2 **6. Algebra** If $x = -10$, then $x^2 = 100$.

Example 2
(page 88)

Write the two statements that form each biconditional.

7. A line bisects a segment if and only if the line intersects the segment only at its midpoint.

8. An integer is divisible by 100 if and only if its last two digits are zeros.

9. You live in Washington, D. C., if and only if you live in the capital of the United States.

10. Two lines are parallel if and only if they are coplanar and do not intersect.

11. Two angles are congruent if and only if they have the same measure.

x^2 **12. Algebra** $x^2 = 144$ if and only if $x = 12$ or $x = -12$.

Example 3
(page 89)

Test each statement below to see if it is reversible. If so, write it as a true biconditional. If not, write *not reversible.*

13. A perpendicular bisector of a segment is a line, segment, or ray that is perpendicular to a segment at its midpoint.

14. Parallel planes are planes that do not intersect.

15. A Tarheel is a person who was born in North Carolina.

16. A rectangle is a four-sided figure with at least one right angle.

17. A midpoint of a segment is a point that divides a segment into two congruent segments.

Example 4
(page 89)

Is each statement below a good definition? If not, explain.

18. A cat is an animal with whiskers.

19. A dog is a good pet.

20. A segment is part of a line.

21. Parallel lines do not intersect.

22. A square is a figure with two pairs of parallel sides.

23. An angle bisector is a ray that divides an angle into two congruent angles.

B **Apply Your Skills** **24. Language Arts** Is the following a good definition? Explain.
An obtuse angle is an angle whose measure is greater than 90.

25. Open-Ended Choose a definition from a dictionary or from a glossary. Explain what makes the statement a good definition.

 **26. Writing** Write a definition of *a line parallel to a plane.*

27. Writing Use the figures below to write a good definition of *linear pair*.

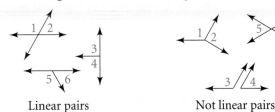

Linear pairs   Not linear pairs

Do angles 1 and 2 form a linear pair? Explain. (*Hint:* See Exercise 27.)

28.   **29.**   **30.**   **31.**

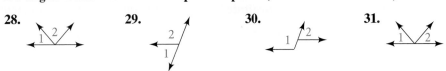

32. Multiple Choice Which conditional and its converse form a true biconditional?

    Ⓐ If $x > 0$, then $|x| > 0$.   Ⓑ If $x^3 = 5$, then $x = 125$.

    Ⓒ If $x = 3$, then $x^2 = 9$.   Ⓓ If $x = 19$, then $2x - 3 = 35$.

🌐 **The American Manual Alphabet** For Exercises 33–37, use the chart below. Decide whether the description of each letter is a good definition. If not, provide a counterexample by giving another letter that could fit the definition.

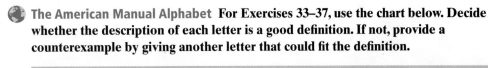

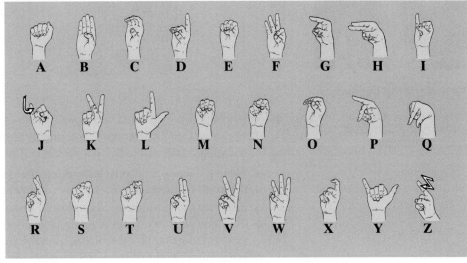

33. The letter D is formed by pointing straight up with the finger beside the thumb and folding the other fingers and the thumb so that they all touch.

34. The letter K is formed by making a V with the two fingers beside the thumb.

35. You have formed the letter Y if and only if the thumb and one finger are extended up and the other fingers are folded into the palm of your hand.

36. You have formed the letter I if and only if the smallest finger is sticking up and the other fingers are folded into the palm of your hand with your thumb folded over them, and your hand is held still.

37. You form the letter B by holding all four fingers tightly together and pointing them straight up while your thumb is folded into the palm of your hand.

Real-World 🌐 Connection

The five letters above form a word to think about.

Vocabulary Tip

The expressions
vice versa and conversely
are synonyms.

Write each statement as a biconditional.

38. Congruent angles are angles with equal measure.

39. When the sum of the digits of an integer is divisible by 9, the integer is divisible by 9 and vice versa.

40. The whole numbers are the nonnegative integers.

Reading Math Let statements *p, q,* and *r,* be as follows.
  p: ∠*A* and ∠*B* are right angles.   *q:* ∠*A* and ∠*B* are supplementary angles.
                    r: ∠*A* and ∠*B* are adjacent angles.
Substitute for *p, q,* and *r,* and write each statement the way you would read it.

41. $p \rightarrow q$ **42.** $q \rightarrow p$ **43.** $p \leftrightarrow q$ **44.** $q \leftrightarrow p$ **45.** $p \leftrightarrow r$ **46.** $r \leftrightarrow q$

C Challenge

47. Reasoning In a band, Amy, Bob, and Carla are the drummer, guitarist, and keyboard player. Use the clues to find the instrument that each one plays.

Carla and the drummer wear different-colored shirts.
The keyboard player is older than Bob.
Amy, the youngest band member, lives next door to the guitarist.

You can solve this type of logic puzzle by eliminating possibilities. Copy the grid below. Put an X in a box once you eliminate it as a possibility.

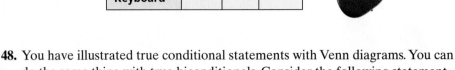

Instrument	Amy	Bob	Carla
Drums			
Guitar			
Keyboard			

Real-World 🌐 **Connection**

Careers Music educators are
well-versed in both traditional
and modern music.

48. You have illustrated true conditional statements with Venn diagrams. You can do the same thing with true biconditionals. Consider the following statement.
  An integer is divisible by 10 if and only if its last digit is 0.

  a. Write the two conditional statements that make up this biconditional.
  b. Illustrate the first conditional from part (a) with a Venn diagram.
  c. Illustrate the second conditional from part (a) with a Venn diagram.
  d. Combine your two Venn diagrams from parts (b) and (c) to form a Venn diagram representing the biconditional statement.
  e. What must be true of the Venn diagram for any true biconditional statement?
  f. Reasoning How does your conclusion in part (e) help to explain why a good definition can be written as a biconditional?

49. Reasoning Alan, Ben, and Cal are seated as shown with their eyes closed. Diane places a hat on each of their heads from a box they know contains 3 red and 2 blue hats. They open their eyes and look forward.

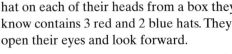

Alan

Ben

Cal

Alan says,
"I cannot deduce what color hat I'm wearing."
Hearing that, Ben says, "I cannot deduce what color I'm wearing, either."
Cal then says, "I know what color I'm wearing!"

How does Cal know the color of his hat?

Multiple Choice

50. Which statement is a good definition?
 A. Skew lines are lines that do not intersect.
 B. Parallel lines are lines that do not intersect.
 C. A square is a rectangle with four congruent sides.
 D. Right angles are angles formed by two intersecting lines.

51. Which statement is NOT true?
 F. If two lines are parallel, then they lie in one plane and do not intersect.
 G. Two lines lie in one plane if and only if the lines are parallel.
 H. If two coplanar lines do not intersect, then the lines are parallel.
 J. Two lines lie in one plane and do not intersect if and only if the two lines are parallel.

52. Which statement is NOT true?
 A. If $x = 1$, then $x^2 = 1$.   **B.** If $x^2 = 1$, then $x = 1$.
 C. If $x = -1$, then $x^2 = 1$.   **D.** $x^2 = 1$ if and only if $x = 1$ or $x = -1$.

Short Response

53. Write the two conditionals that form this biconditional:
 You can go to the movies if and only if you do your homework.

Extended Response

54. Here is a true conditional statement:
 If a person is 18 years old, that person is old enough to vote.
 a. Write the converse.
 b. Determine whether the converse is true or false.
 c. If the converse is false, give a counterexample to show that it is false. If the converse is true, combine the original statement and its converse by writing a biconditional.

Mixed Review

Lesson 2-1

Write each statement as a conditional.

55. Whole numbers that end in zero are even.

56. When $x = -5$, $x^2 = 25$.

57. Sunday is a weekend day.

58. All prime numbers greater than 2 are odd.

Lessons 1-5 and 1-6

59. Draw a segment $\overline{XY}$. Construct a bisector of $\overline{XY}$.

60. Draw an acute angle, $\angle 1$. Construct an angle congruent to $\angle 1$.

61. Draw an obtuse angle, $\angle CAD$. Construct the bisector of $\angle CAD$.

Lesson 1-3

Use the figure at the right to name each of the following.

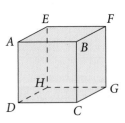

62. two intersecting lines   **63.** two skew lines

64. two parallel lines   **65.** two parallel planes

66. three coplanar points   **67.** two intersecting planes

68. a plane that contains H

69. the intersection of two planes

Deductive Reasoning

 Learning Standards for Mathematics

G.PS.4 Construct various types of reasoning, arguments, justifications and methods of proof for problems.

G.RP.7 Construct a proof using a variety of methods (e.g. deductive analytic transformational).

 Check Skills You'll Need

 **for Help** Lesson 2-1

Write the converse of each statement.

1. If you don't sleep enough, then your grades suffer.

2. If you want to arrive on time, then you must start early.

Write each statement as a conditional.

3. Leap years have 366 days.

4. Students who do not complete their homework will have lower grades.

5. Two lines that are perpendicular meet to form right angles.

6. Every sixteen-year-old is a teenager.

◀)) **New Vocabulary** • deductive reasoning • Law of Detachment • Law of Syllogism

---

1 Using the Law of Detachment

In Chapter 1 you learned that inductive reasoning is based on observing what has happened and then making a conjecture about what will happen. In this lesson, you will study deductive reasoning.

Deductive reasoning (or logical reasoning) is the process of reasoning logically from given statements to a conclusion. If the given statements are true, deductive reasoning produces a true conclusion.

Many people use deductive reasoning in their jobs. A physician diagnosing a patient's illness uses deductive reasoning. A carpenter uses deductive reasoning to determine what materials are needed at a work site.

Real-World  Connection

Careers An auto mechanic uses deductive reasoning as in Example 1.

1 **EXAMPLE** Real-World Connection

Auto Maintenance An auto mechanic knows that if a car has a dead battery, the car will not start. A mechanic begins work on a car and finds the battery is dead. What conclusion can she make?

The mechanic can conclude that the car will not start.

✓ **Quick Check** ❶ **Critical Thinking** Suppose that a mechanic begins work on a car and finds that the car will not start. Can the mechanic conclude that the car has a dead battery? Explain.

In Example 1 the mechanic is using a law of deductive reasoning called the **Law of Detachment.**

Key Concepts

Property	Law of Detachment

If a conditional is true and its hypothesis is true, then its conclusion is true.

In symbolic form:
If $p \rightarrow q$ is a true statement and p is true, then q is true.

Vocabulary Tip

You can read $p \rightarrow q$ as "p implies q."

2 EXAMPLE Using the Law of Detachment

For the given true statements, what can you conclude?

Given: If M is the midpoint of a segment, then it divides the segment into two congruent segments.
M is the midpoint of $\overline{AB}$.

You are given that a conditional and its hypothesis are true. By the Law of Detachment, you can conclude that M divides $\overline{AB}$ into two congruent segments, or $\overline{AM} \cong \overline{MB}$.

 **2** If a baseball player is a pitcher, then that player should not pitch a complete game two days in a row. Vladimir Nuñez is a pitcher. On Monday, he pitches a complete game. What can you conclude?

3 EXAMPLE Real-World Connection

Does the following argument illustrate the Law of Detachment?

Given: If it is snowing, then the temperature is less than or equal to 32°F.
The temperature is 20°F.

You conclude: It must be snowing.

You are given that a conditional and its conclusion are true.
You cannot apply the Law of Detachment and conclude that the hypothesis is true.
You cannot come to any conclusion about whether it is snowing from the information given.

 **3** If possible, use the Law of Detachment to draw a conclusion. If it is not possible to use this law, explain why.

Given: If a road is icy, then driving conditions are hazardous.
Driving conditions are hazardous.

2 Using the Law of Syllogism

Another law of deductive reasoning is the Law of Syllogism. The **Law of Syllogism** allows you to state a conclusion from two true conditional statements when the conclusion of one statement is the hypothesis of the other statement.

 Key Concepts

Property	Law of Syllogism

If $p \rightarrow q$ and $q \rightarrow r$ are true statements, then $p \rightarrow r$ is a true statement.

4 EXAMPLE Using the Law of Syllogism

Vocabulary Tip

2 is a <u>repeated</u> <u>factor</u> of 12 because it appears more than once in the prime factorization of 12.

$12 = 2 \cdot 2 \cdot 3$

Algebra Use the Law of Syllogism to draw a conclusion from the following true statements.

If a number is prime, then it does not have repeated factors.
If a number does not have repeated factors, then it is not a perfect square.

You have two true conditionals where the conclusion of one is the hypothesis of the other. You can use the Law of Syllogism to draw the following conclusion:

● If a number is prime, then it is not a perfect square.

✓ Quick Check **4** If possible, state a conclusion using the Law of Syllogism. If it is not possible to use this law, explain why.
   a. If a number ends in 0, then it is divisible by 10.
     If a number is divisible by 10, then it is divisible by 5.
   b. If a number ends in 6, then it is divisible by 2.
     If a number ends in 4, then it is divisible by 2.

You can use both the Law of Detachment and the Law of Syllogism to draw conclusions.

Nile River

5 EXAMPLE Real-World Connection

Geography Use the Law of Detachment and the Law of Syllogism to draw conclusions from the following true statements.

If a river is more than 4000 mi long, then it is longer than the Amazon.
If a river is longer than the Amazon, then it is the longest river in the world.
The Nile is 4132 mi long.

You can use the first two statements and the Law of Syllogism to conclude:
If a river is more than 4000 mi long, then it is the longest river in the world.

With this, the fact that the Nile is 4132 mi long, and the Law of Detachment, you can also conclude:
● The Nile is the longest river in the world.

Real-World Connection

Over 99% of Egypt's people live close to the Nile River.

✓ Quick Check **5** Use the Law of Detachment and the Law of Syllogism to draw conclusions.
   The Volga River is in Europe.
   If a river is less than 2300 mi long, it is not one of the world's ten longest rivers.
   If a river is in Europe, then it is less than 2300 mi long.

EXERCISES

For more exercises, see *Extra Skill, Word Problem, and Proof Practice*.

Practice and Problem Solving

 Practice by Example

Examples 1 and 2
(pages 94, 95)

Use the Law of Detachment to draw a conclusion.

1. If a student gets an A on a final exam, then the student will pass the course.
   Felicia gets an A on the music theory final exam.

2. If a student wants to go to college, then the student must study hard.
   Rashid wants to go to the University of North Carolina.

3. If two lines are parallel, then they do not intersect.
Line ℓ is parallel to line m.

4. If there is lightning, then it is not safe to be out in the open.
Marla sees lightning from the soccer field.

Example 3
(page 95)

If possible, use the Law of Detachment to draw a conclusion. If not possible, write *not possible.*

5. If a figure is a rectangle, then it has two pairs of parallel sides.
Figure $ABCD$ is a rectangle.

x^2 **6. Algebra** If n is a prime number greater than 2, then n^2 is an odd number.
9^2 is an odd number.

7. If three points are on the same line, then they are collinear.
Points X, Y, and Z are on line m.

8. If an angle is obtuse, then it is not acute.
$\angle XYZ$ is not obtuse.

9. If you are a Golden Gopher, you've attended the University of Minnesota.
(See photo.)

Example 4
(page 96)

Use the Law of Syllogism to draw a conclusion.

10. Zoology If an animal is a red wolf, then its scientific name is *Canis rufus*.
If an animal is named *Canis rufus*, then it is endangered.

11. If two planes intersect, then they intersect in a line.
If two planes are not parallel, then they intersect.

12. If you read a good book, then you enjoy yourself.
If you enjoy yourself, then your time is well spent.

13. If you are studying biology, then you are studying a science.
If you are studying botany, then you are studying biology.

Example 5
(page 96)

Geography **Use the Law of Detachment and the Law of Syllogism to draw conclusions from the following statements.**

14. If a mountain is the highest in Alaska, then it is the highest in the United States.
If an Alaskan mountain is over 20,300 ft high, then it is the highest in Alaska.
Alaska's Mount McKinley is 20,320 ft high.

15. If you live in Lubbock, then you live in Texas. Levon lives in Lubbock.
If you live in Texas, then you live in the 28th state to enter the Union.

B **Apply Your Skills**

For Exercises 16–21, assume that the following statements are true.
    A. If Maria is drinking juice, then it is breakfast time.
    B. If it is lunchtime, then Kira is drinking milk and nothing else.
    C. If it is mealtime, then Curtis is drinking water and nothing else.
    D. If it is breakfast time, then Julio is drinking juice and nothing else.
    E. Maria is drinking juice.

Use only the information given above. For each statement, write *must be true, may be true,* **or** *is not true.* **Explain your reasoning.**

16. Julio is drinking juice.

17. Curtis is drinking water.

18. Kira is drinking milk.

19. Curtis is drinking juice.

20. Maria is drinking water.

21. Julio is drinking milk.

Real-World Connection

Average ocean temperature in Key West is about 80°F, a good snorkeling temperature.

For each of the following, write the first statement as a conditional. If possible, use the Law of Detachment to make a conclusion. If not possible, write *not possible*.

22. All national parks are interesting.
Mammoth Cave is a national park.

23. Weather The temperature is always above 32°F in Key West, Florida.
The temperature is 62°F.

24. Every high school student likes music.
Ling likes music.

25. All squares are rectangles.
ABCD is a square.

26. Writing Give an example of a rule used in your school that could be written as a conditional. Explain how the Law of Detachment is used in applying that rule.

For Exercises 27–31, use the cartoon and deductive reasoning to answer *yes* or *no*. If *no*, explain.

27. Is a person with a red car allowed to park here on Tuesday at 10:00 A.M.?

28. Is a man with a beard allowed to park here on Monday at 10:30 A.M.?

29. Is a woman with a wig allowed to park here on Saturday at 10:00 A.M.?

30. Is a person with a blue car allowed to park here on Tuesday at 9:05 A.M.?

31. Is a person with a convertible with leather seats allowed to park here on Sunday at 6:00 P.M.?

32. Reasoning Assume that the following statements are true.

If Anita goes to the concert, Beth will go.
If Beth goes to the concert, Aisha will go.
If Aisha goes to the concert, Ramon will go.

Only two of the four students went to the concert. Who were they?

Challenge

33. Critical Thinking Consider the following given statements and conclusion.

Given: If an animal is a fish, then it has gills.
A turtle does not have gills.
You conclude: A turtle is not a fish.

This argument does not use the Law of Syllogism or the Law of Detachment, but it does use good deductive reasoning.

a. Draw a Venn diagram to illustrate the given information.
b. Use the Venn diagram to help explain why the argument uses good reasoning.

Multiple Choice

34. What conclusion can you draw from the following two statements?

If a person does not get enough sleep, that person will be tired.
Evan does not get enough sleep.

   A. Evan will get enough sleep.   **B.** Evan will not be tired.
   C. Evan should get enough sleep.   **D.** Evan will be tired.

35. What conclusion can you draw from the following two statements?

If you have a job, then you have an income.
If you have an income, then you must pay taxes.

   F. If you have a job, then you must pay taxes.
   G. If you don't have a job, then you don't pay taxes.
   H. If you pay taxes, then you have a job.
   J. If you have a job, then you don't have to pay taxes.

Short Response

36. Carl reads anything Andrea chooses to read. Bert reads what Carl chooses to read and Carl reads what Bert chooses. Andrea reads whatever Darla chooses to read.
   a. Carl is reading *Hamlet*. Who else, if anyone, must also be reading *Hamlet*?
   b. Exactly three people are reading *King Lear*. Who are they? Explain.

Extended Response

37. Harold, Clara, and Mark each chose a different lunch from three categories: soup, salad, and sandwiches. Each ordered a different drink. Clara will not eat sandwiches. Mark won't eat salad or bread. The person who had the soup also had the iced tea. The person who had the sandwich also had the milk.
   a. Who drank the milk? How do you know?
   b. One person ordered mineral water. What food did the water go with? Explain.

Mixed Review

Lesson 2-2

Is each statement a good definition? If not, find a counterexample.

38. An angle is a figure formed by two rays.

39. A ray is an angle bisector if and only if it divides an angle into two congruent angles.

Lesson 2-1

Show that each conditional is false by finding a counterexample.

40. Geography If the name of a state contains the word *North*, then the state borders Canada.

x^2 **41. Algebra** If you square a fraction, then the result is always greater than the original fraction.

Lesson 1-4

Complete with *always*, *sometimes*, or *never* to make a true statement.

42. Two lines that do not intersect are __?__ parallel.

43. Two lines that intersect are __?__ skew.

44. Two segments that intersect are __?__ coplanar.

1. Identify the hypothesis and the conclusion of this conditional statement:
   If $x > 5$, then $x^2 > 25$.

2. Write this statement as a conditional: Roses are beautiful flowers.

For Exercises 3 and 4, use this conditional statement:
If an integer ends with 0, then the integer is divisible by 2.

3. Write the converse of the statement.

4. Find a counterexample to show that the converse is *not* true.

5. Write the two conditionals that make up this biconditional:
   An angle is an acute angle if and only if its measure is between 0 and 90.

6. Rewrite this definition as a biconditional:
   Points that lie on the same line are collinear.

7. Find a counterexample to show that the following statement is *not* a good definition:
   A computer is a machine with a keyboard and a memory.

Use the Law of Detachment or the Law of Syllogism to draw a conclusion from each pair of statements. If not possible, write *not possible.*

8. If a student is on the basketball team, then that student has passing grades.
   Theresa is on the basketball team.

9. If a student studies geometry, the student studies mathematics.
   If a student studies mathematics, the student's mind is expanded.

10. If you miss the bus, then you will be late for school.
    You are late for school.

A P•int in Time

1500   1600   1700   1800   1900   2000

Hercule Poirot as played by
David Suchet, 1989–1997.

Most people are not detectives, but as a young woman, the English writer Agatha Christie (1890–1976) correctly deduced that many people would like to be. In 1920 she published her first book, a detective novel entitled *The Mysterious Affair at Styles* in which she introduced the eccentric and ultra-logical Belgian detective Hercule Poirot. In this and in many subsequent novels, Poirot solves mysteries not with guns or car chases but with logical reasoning.

 **Go •nline**
PHSchool.com   **For:** Information about Agatha Christie
Web Code: aue-2032

Activity Lab

Mathematical Systems

The geometry you are studying is an example of a *mathematical system*.

NY G.PS.4: Construct various types of reasoning, arguments, justifications and methods of proof for problems.

ACTIVITY

Enter the information at the right into your study aid. (See Vocabulary Builder, page 29.) It tells you about a mathematical system.

You may wish to cross-reference this entry, noting that you learned about:

- undefined terms and definitions on page 29 and in Lesson 2-2,
- postulates in Lessons 1-3 and 1-5, and
- the deductive reasoning framework of geometry in Lesson 2-3.

Mathematical System:
- a vocabulary, and
- a collection of statements in an accepted framework of reasoning.

Example: Euclidean Geometry

Vocabulary	Statements
Undefined terms	Postulates
Defined terms	Theorems

EXERCISES

1. Complete the following:
   a. A mathematical system is made up of a __?__ and a collection of __?__ in an accepted framework of reasoning.
   b. The vocabulary of a mathematical system is made up of __?__ and __?__ .
   c. The statements of a mathematical system are __?__ and __?__ .

A postulate is a statement that you accept as fact without proof. A *theorem* is a statement that you prove true using deductive reasoning. You base the proof on statements that you already know are true.

2. Explain why some statements must be assumed true.

3. If you haven't already done so, begin a "Statements" section of your study aid for the postulates of Lessons 1-3 and 1-5.

Each statement in a mathematical system is either true or false. A statement that is neither true nor false cannot be in the system. One type of such a statement is a *paradox*.

4. The statement, "This sentence is false" is a well-known paradox. Explain.

5. A statement that contradicts common sense is another type of paradox. The Greek logician Zeno (born circa 490 B.C.) is credited with a paradox that says, in effect, a hare can't catch a tortoise in a race if the tortoise gets a head start. Research Zeno's Paradox. Describe it and explain why it is a paradox.

Paper-Folding Constructions

In Chapter 1, you used a compass and straightedge to construct special geometric figures. You can also construct figures by folding paper. Some form of paper that you can see through, such as tracing paper or wax paper, is best.

NY G.G.66: Find the midpoint of a line segment, given its endpoints.

1 ACTIVITY

- Draw a line segment $\overline{AB}$ on your paper.
- Fold and crease the paper so that B is on top of A.
- Unfold. Label the intersection of $\overline{AB}$ and the crease as C.

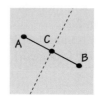

1. Measure lengths AC and CB. What do you notice?

2. Measure the four angles formed by $\overline{AB}$ and the crease. What kind of angles are they?

3. Based on what you found in Exercises 1 and 2, how is the fold line related to $\overline{AB}$?

2 ACTIVITY

- Draw an angle $\angle FGH$ on your paper.
- Fold and crease the paper so that $\overrightarrow{GF}$ lies on top of $\overrightarrow{GH}$.
- Unfold the paper. Label a point J on the fold line and in the interior of $\angle FGH$.

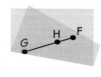
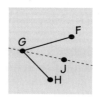

4. Use a protractor. Measure $\angle FGJ$ and $\angle JGH$. What do you notice?

5. Based on what you found in Exercise 4, how is $\overrightarrow{GJ}$ related to $\angle FGH$?

EXERCISES

Fold a line on your paper. Mark a point P on the line and a point Q not on the line. Explain and demonstrate how you can do each of the following.

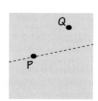

6. Fold a perpendicular to the line at P.

7. Fold a perpendicular to the line through Q.

Reasoning in Algebra

2-4

 Learning Standards for Mathematics

G.PS.8 Determine information required to solve a problem, choose methods for obtaining the information, and define parameters for acceptable solutions.

 **Check Skills You'll Need**

For Exercises 1–5, use the figure at the right.

1. Name ∠1 in two other ways.
2. Name the vertex of ∠2.
3. If ∠1 ≅ ∠2, name the bisector of ∠AOC.
4. If $m\angle AOC = 90$ and $m\angle 1 = 45$, find $m\angle 2$.
5. If $m\angle AOC = 90$, name two perpendicular rays.

GO for Help Lesson 1-6

◀)) **New Vocabulary** • Reflexive Property • Symmetric Property
• Transitive Property

---

1 **Connecting Reasoning in Algebra and Geometry**

In geometry you accept postulates and properties as true. You use deductive reasoning to prove other statements. Some of the properties that you accept as true are the properties of equality from algebra. They are listed below in terms of any numbers $a, b,$ and c.

 Key Concepts

Summary	Properties of Equality
Addition Property	If $a = b$, then $a + c = b + c$.
Subtraction Property	If $a = b$, then $a - c = b - c$.
Multiplication Property	If $a = b$, then $a \cdot c = b \cdot c$.
Division Property	If $a = b$ and $c \neq 0$, then $\frac{a}{c} = \frac{b}{c}$.
Reflexive Property	$a = a$
Symmetric Property	If $a = b$, then $b = a$.
Transitive Property	If $a = b$ and $b = c$, then $a = c$.
Substitution Property	If $a = b$, then b can replace a in any expression.

You also assume that other properties from algebra are true.

 Key Concepts

Property	The Distributive Property
$a(b + c) = ab + ac$	

You use deductive reasoning every time you solve an equation. You can justify each statement that you make with a postulate, a property, or a definition. When you solve problems involving angle measures, you can use the Angle Addition Postulate.

1 EXAMPLE Justifying Steps in Solving an Equation

Algebra Solve for x and justify each step.

Given: $m\angle AOC = 139$

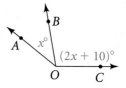

GO for Help

To review the Angle Addition Postulate, go to p. 38.

$m\angle AOB + m\angle BOC = m\angle AOC$	**Angle Addition Postulate**
$x + 2x + 10 = 139$	**Substitution Property**
$3x + 10 = 139$	**Simplify.**
$3x = 129$	**Subtraction Property of Equality**
$x = 43$	**Division Property of Equality**

✓ Quick Check ❶ Fill in each missing reason.

Given: $\overrightarrow{LM}$ bisects $\angle KLN$.

$\overrightarrow{LM}$ bisects $\angle KLN$.	**Given**
$m\angle MLN = m\angle KLM$	**Definition of angle bisector**
$4x = 2x + 40$	?
$2x = 40$	?
$x = 20$	?

You can use the Segment Addition Postulate to justify statements about lengths of segments.

2 EXAMPLE Justifying Steps in Solving an Equation

Algebra Solve for y and justify each step.

Given: $AC = 21$

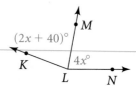

GO for Help

To review the Segment Addition Postulate, go to p. 32.

$AB + BC = AC$	**Segment Addition Postulate**
$2y + (3y - 9) = 21$	**Substitution Property**
$5y - 9 = 21$	**Simplify.**
$5y = 30$	**Addition Property of Equality**
$y = 6$	**Division Property of Equality**

✓ Quick Check ❷ Find AB and BC by substituting $y = 6$ in the expressions in the diagram above. Check that $AB + BC = 21$.

The Reflexive, Symmetric, and Transitive Properties of Equality have corresponding properties of congruence. You can use properties of congruence to justify statements.

Summary	**Properties of Congruence**
Reflexive Property	$\overline{AB} \cong \overline{AB}$ $\angle A \cong \angle A$
Symmetric Property	If $\overline{AB} \cong \overline{CD}$, then $\overline{CD} \cong \overline{AB}$. If $\angle A \cong \angle B$, then $\angle B \cong \angle A$.
Transitive Property	If $\overline{AB} \cong \overline{CD}$ and $\overline{CD} \cong \overline{EF}$, then $\overline{AB} \cong \overline{EF}$. If $\angle A \cong \angle B$ and $\angle B \cong \angle C$, then $\angle A \cong \angle C$.

3 EXAMPLE **Using Properties of Equality and Congruence**

Name the property of equality or congruence that justifies each statement.

a. $\angle K \cong \angle K$
Reflexive Property of Congruence

b. If $2x - 8 = 10$, then $2x = 18$.
Addition Property of Equality

c. If $\overline{RS} \cong \overline{TW}$ and $\overline{TW} \cong \overline{PQ}$, then $\overline{RS} \cong \overline{PQ}$.
Transitive Property of Congruence

d. If $m\angle A = m\angle B$, then $m\angle B = m\angle A$.
Symmetric Property of Equality

✓ Quick Check **3** Name the property of equality or congruence illustrated.
a. $\overline{XY} \cong \overline{XY}$
b. If $m\angle A = 45$ and $45 = m\angle B$, then $m\angle A = m\angle B$.

EXERCISES

For more exercises, see *Extra Skill, Word Problem, and Proof Practice.*

Practice and Problem Solving

A Practice by Example

Examples 1 and 2
(page 104)

 **for Help**

$\boxed{x^2}$ **Algebra** **Fill in the reason that justifies each step.**

1. Solve for x.

$m\angle CDE + m\angle EDF = 180$	**a.** ___?___
$x + (3x + 20) = 180$	**b.** ___?___
$4x + 20 = 180$	**c.** ___?___
$4x = 160$	**d.** ___?___
$x = 40$	**e.** ___?___

2. Solve for n.
Given: $XY = 42$

$XZ + ZY = XY$	**a.** ___?___
$3(n + 4) + 3n = 42$	**b.** ___?___
$3n + 12 + 3n = 42$	**c.** ___?___
$6n + 12 = 42$	**d.** ___?___
$6n = 30$	**e.** ___?___
$n = 5$	**f.** ___?___

x^2 **Algebra** Give a reason for each step.

3. $\frac{1}{2}x - 5 = 10$ Given
     $2\left(\frac{1}{2}x - 5\right) = 20$ **a.** ?
     $x - 10 = 20$ **b.** ?
     $x = 30$ **c.** ?

4. $5(x + 3) = -4$ Given
     $5x + 15 = -4$ **a.** ?
     $5x = -19$ **b.** ?
     $x = -\frac{19}{5}$ **c.** ?

Example 3
(page 105)

Name the property that justifies each statement.

5. $\angle Z \cong \angle Z$

6. $2(3x + 5) = 6x + 10$

7. If $12x = 84$, then $x = 7$.

8. If $\overline{ST} \cong \overline{QR}$, then $\overline{QR} \cong \overline{ST}$.

9. If $m\angle A = 15$, then $3m\angle A = 45$.

10. $XY = XY$

11. If $3x + 14 = 80$, then $3x = 66$.

12. If $KL = MN$, then $MN = KL$.

13. If $2x + y = 5$ and $x = y$,
    then $2x + x = 5$.

14. If $AB - BC = 12$,
    then $AB = 12 + BC$.

15. If $\angle 1 \cong \angle 2$ and $\angle 2 \cong \angle 3$, then $\angle 1 \cong \angle 3$.

B **Apply Your Skills**

Use the given property to complete each statement.

16. Addition Property of Equality
    If $2x - 5 = 10$, then $2x = $? .

17. Subtraction Property of Equality
    If $5x + 6 = 21$, then ? $= 15$.

18. Symmetric Property of Equality
    If $AB = YU$, then ? .

19. Symmetric Property of Congruence
    If $\angle H \cong \angle K$, then ? $\cong \angle H$.

20. Reflexive Property of Congruence
    $\angle PQR \cong$?

21. Distributive Property
    $3(x - 1) = 3x -$?

22. Substitution Property
    If $LM = 7$ and $EF + LM = NP$,
    then ? $= NP$.

23. Transitive Property of Congruence
    If $\angle XYZ \cong \angle AOB$ and
    $\angle AOB \cong \angle WYT$, then ? .

24. **Multiple Choice** Which expression is equivalent to the left side of this
    equation? $-4x + 7y + \frac{1}{3}(12x - 3y) = 180$

    Ⓐ $8x + 4y$ Ⓑ $6y + 8$ Ⓒ $6y$ Ⓓ $8x$

25. **Writing** Jero claims that the statements $\overline{LR} \cong \overline{RL}$ and $\angle CBA \cong \angle ABC$ are
    both true by the Reflexive Property of Congruence. Explain why Jero is correct.

26. Use what you know about transitive properties to complete the following:

    The Transitive Property of Falling Dominoes:

    If domino A causes domino B to fall, and domino B causes domino C to fall,
    then domino A causes domino ? to fall.

Homework Video Tutor
Visit: PHSchool.com
Web Code: aue-0204

27. Algebra Fill in the reason that justifies each step.

Given: C is the midpoint of $\overline{AD}$.

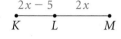

C is the midpoint of $\overline{AD}$.	**a.** ?
$AC = CD$	**b.** ?
$4x = 2x + 12$	**c.** ?
$2x = 12$	**d.** ?
$x = 6$	**e.** ?

GO for Help

For a guide to solving
Exercise 28, see p. 109.

28. Algebra In the figure at the right, $KM = 35$.
  a. Solve for x. Justify each step.
  b. Find the length of $\overline{KL}$.

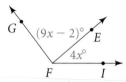

29. Algebra In the figure at the right, $m\angle GFI = 128$.
  a. Solve for x. Justify each step.
  b. Find $m\angle EFI$.

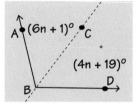

30. Algebra Point C is on the crease when you fold $\overrightarrow{BD}$ onto $\overrightarrow{BA}$. Give the reason that justifies each step. (*Hint:* See page 102, Exercises 4 and 5.)

$\overrightarrow{BC}$ bisects $\angle ABD$.	**a.** ?
$m\angle ABC = m\angle CBD$	**b.** ?
$6n + 1 = 4n + 19$	**c.** ?
$2n = 18$	**d.** ?
$n = 9$	**e.** ?

C Challenge

31. Error Analysis The steps below "show" that $1 = 2$. Find the error.

Given: $a = b$

$a = b$	Given
$ab = b^2$	Multiplication Property of Equality
$ab - a^2 = b^2 - a^2$	Subtraction Property of Equality
$a(b - a) = (b + a)(b - a)$	Distributive Property
$a = b + a$	Division Property of Equality
$a = a + a$	Substitution Property
$a = 2a$	Simplify.
$1 = 2$	Division Property of Equality

Relationships The relationships "is equal to" and "is congruent to" are reflexive, symmetric, and transitive. In a later chapter, you will see that this is also true for the relationship "is similar to." Consider the following relationships among people. State whether each relationship is reflexive, symmetric, transitive, or none of these.

Sample: The relationship "is younger than" is transitive. If Sue is younger than Fred and Fred is younger than Alana, then Sue is younger than Alana. The relationship "is younger than" is not reflexive because Sue is not younger than herself. It is also not symmetric because if Sue is younger than Fred, Fred is not younger than Sue.

Real-World Connection

President Calvin Coolidge, advice columnist Ann Landers, and musician Bill Withers were all born on the Fourth of July. Each one of them "has the same birthday as" either one of the others.

32. has the same birthday as   **33.** is taller than

34. lives in the same state as   **35.** lives in a different state than

36. is the same height as   **37.** is a descendant of

Multiple Choice

38. Which property justifies this statement?

If $4x = 16$, then $16 = 4x$.

A. Multiplication Property of Equality

B. Transitive Property of Equality

C. Reflexive Property of Equality

D. Symmetric Property of Equality

39. The Multiplication Property of Equality justifies which statement below?

F. If $\frac{3}{4}x = 6$, then $\frac{3x}{4} = 6$. **G.** If $\frac{3}{4}x + 5 = 6$, then $\frac{3}{4}x = 1$.

H. If $\frac{3}{4}x = 6$, then $3x = 24$. **J.** If $\frac{3}{4}x - 18 = 6$, then $\frac{3}{4}x = 24$.

40. A transitive property justifies which statement below?

A. If $y - 17 = g$, then $y = g + 17$.

B. If $AM = RS$, then $RS = AM$.

C. If $5(3a - 4) = 120$, then $15a - 20 = 120$.

D. If $\angle J \cong \angle R$ and $\angle R \cong \angle H$, then $\angle J \cong \angle H$.

41. Which equation follows from $\frac{1}{3}m + 1 = 10$ by the Multiplication Property of Equality?

F. $m + 3 = 30$ **G.** $\frac{1}{3}m = 9$ **H.** $\frac{1}{3}m - 9 = 0$ **J.** $m - 27 = 0$

Short Response

42. In the diagram, $x = 2y + 15$ and $x + y = 120$.

a. Use a Property of Equality to explain why $3y + 15 = 120$.

b. Solve for y. Justify each step. Then find the value of x.

Mixed Review

Lesson 2-3

Reasoning **Use logical reasoning to draw a conclusion.**

43. If a student is having difficulty in class, then that student's teacher is concerned. Elena is having difficulty in history class.

44. If a person has a job, then that person is earning money.

If a person is earning money, then that person can save money each week.

Lesson 1-6

Use the diagram at the right and find each measure.

45. $m\angle AOC$ **46.** $m\angle AOD$

47. $m\angle DOB$ **48.** $m\angle BOE$

49. In the diagram, name an obtuse angle and a right angle.

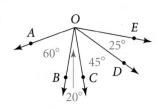

Lesson 1-1

Find the next two terms in each sequence.

50. $19, 21.5, 24, 26.5$ **51.** $3.4, 3.45, 3.456, 3.4567$

52. $-2, 6, -18, 54$ **53.** $8, -4, 2, -1$

Understanding Math Problems Read the problem below. Then let Larissa's thinking guide you through the solution. Check your understanding with the exercises at the bottom of the page.

Algebra In the figure at the right, $KM = 35$.
a. Solve for x. Justify each step.
b. Find the length of $\overline{KL}$.

$$\overset{\displaystyle 2x - 5 \qquad 2x}{\underset{\displaystyle K \qquad L \qquad\quad M}{\bullet\!\!-\!\!-\!\!-\!\!-\!\!\bullet\!\!-\!\!-\!\!-\!\!-\!\!\bullet}}$$

What Larissa Thinks

What information am I given?

Points K, L, and M are collinear. I can use the Segment Addition Postulate to write an equation.

Now I will substitute for KL, LM, and KM. I get an equation that I can solve for x.

I simplify the left side.

I add 5 to each side.

I divide each side by 4.

Part (b) asks me to find KL.
The diagram shows that $KL = 2x - 5$.
I know $x = 10$, so I'll substitute to find KL.

Now, I'll write my answer.

What Larissa Writes

Given: $KM = 35$, $KL = 2x - 5$,
$\qquad\quad LM = 2x$

1. $KL + LM = KM$ 1. Segment Addition Postulate

2. $(2x - 5) + 2x = 35$ 2. Substitute.

3. $4x - 5 = 35$ 3. Simplify.

4. $4x = 40$ 4. Addition Property of Equality

5. $x = 10$ 5. Division Property of Equality

$KL = 2x - 5$
$\quad\ = 2(10) - 5$
$\quad\ = 20 - 5 = 15$

The length of $\overline{KL}$ is 15.

EXERCISES

1. Algebra $\angle 1$ and $\angle 2$ are supplementary; $m\angle 1 = 4y + 15$ and $m\angle 2 = 7y - 11$.

   a. Solve for y. Justify each step.
   b. Find $m\angle 2$.

2. If $\angle 1$ and $\angle 2$ are complementary and have the (algebraic) measures given in Exercise 1, which angle is larger?

Proving Angles Congruent

Learning Standards for Mathematics

G.PS.4 Construct various types of reasoning, arguments, justifications and methods of proof for problems.

✓ **Check Skills You'll Need**

GO for Help Lesson 1-6

x^2 **Algebra** Find the value of each variable.

1.

2.

3.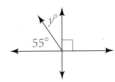

Fill in each blank.

4. Perpendicular lines are two lines that intersect to form __?__.

5. An angle is formed by two rays with the same endpoint. The endpoint is called the __?__ of the angle.

🔊 **New Vocabulary** • theorem • paragraph proof

---

1 Theorems About Angles

Hands-On Activity: Vertical Angles

- Draw two intersecting lines. Number the angles as shown.

- Fold ∠1 onto ∠2.

- Fold ∠3 onto ∠4.

- Make a conjecture about vertical angles.

You can use deductive reasoning to show that a conjecture is true. The set of steps you take is called a proof. The statement that you prove true is a **theorem.** The Investigation above leads to a conjecture that becomes the following theorem.

 Key Concepts

| **Theorem 2-1** | **Vertical Angles Theorem** |

Vertical angles are congruent.

∠1 ≅ ∠2 and ∠3 ≅ ∠4

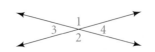

In the proof of a theorem, a "Given" list shows you what you know from the hypothesis of the theorem. You prove the conclusion of the theorem. A diagram records the given information visually.

Here is what the start of many proofs will look like.

what you
know → Given: ~~~~~~

what you → Prove: ~~~~~~
must show

diagram that
← shows what
you know

There are many forms of proofs. A **paragraph proof** is written as sentences in a paragraph. Here is a paragraph proof of Theorem 2-1.

Proof **Given:** ∠1 and ∠2 are   ← what you know →
vertical angles.

Prove: ∠1 ≅ ∠2   ← what you show

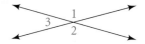

Paragraph Proof: By the Angle Addition Postulate, $m\angle 1 + m\angle 3 = 180$ and $m\angle 2 + m\angle 3 = 180$. By substitution, $m\angle 1 + m\angle 3 = m\angle 2 + m\angle 3$. Subtract $m\angle 3$ from each side. You get $m\angle 1 = m\angle 2$, or $\angle 1 \cong \angle 2$.

You can use the Vertical Angles Theorem to solve for variables and find the measures of angles.

1 EXAMPLE   **Using the Vertical Angles Theorem**

Gridded Response Find the value of x.

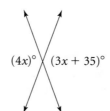

$(4x)°$ ✗ $(3x + 35)°$

$4x = 3x + 35$ **Vertical angles are congruent.**

$x = 35$ **Subtract 3x from each side.**

 **Quick Check** ➊ **a.** Find the measures of the labeled pair of vertical angles in the diagram above.
b. Find the measures of the other pair of vertical angles.
c. Check to see that adjacent angles are supplementary.

The Vertical Angles Theorem is actually a special case of the following theorem. A proof of this theorem is shown on the next page. You can write a proof of another form of this theorem in Exercise 27.

  Key Concepts

Theorem 2-2	Congruent Supplements Theorem

If two angles are supplements of the same angle (or of congruent angles), then the two angles are congruent.

Proof **2** **EXAMPLE** **Proving Theorem 2-2**

Study what you are given, what you are to prove, and the diagram. Write a paragraph proof.

Given: ∠1 and ∠2 are supplementary.
∠3 and ∠2 are supplementary.

Prove: ∠1 ≅ ∠3

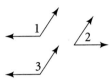

Proof: By the definition of supplementary angles, $m\angle 1 + m\angle 2 = 180$ and $m\angle 3 + m\angle 2 = 180$. By substitution, $m\angle 1 + m\angle 2 = m\angle 3 + m\angle 2$. Subtract $m\angle 2$ from each side. You get $m\angle 1 = m\angle 3$, or $\angle 1 \cong \angle 3$.

 **Quick Check** **2** In the proof above, which Property of Equality allows you to subtract $m\angle 2$ from each side of the equation?

Theorem 2-3 is like the Congruent Supplements Theorem. You can demonstrate its proof in Exercises 7 and 28.

🔑 **Key Concepts**

| Theorem 2-3 | Congruent Complements Theorem |

If two angles are complements of the same angle (or of congruent angles), then the two angles are congruent.

Theorem 2-4

All right angles are congruent.

Theorem 2-5

If two angles are congruent and supplementary, then each is a right angle.

You can complete proofs of Theorems 2-4 and 2-5 in Exercises 14 and 21, respectively.

EXERCISES
For more exercises, see *Extra Skill, Word Problem, and Proof Practice.*

Practice and Problem Solving

A Practice by Example

Example 1
(page 111)

 GO for Help

Find the value of each variable.

1.
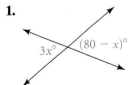
$3x°$ $(80 - x)°$

2.
$3x°$ $y°$ $75°$

3.
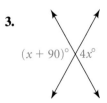
$(x + 90)°$ $4x°$

Find the measures of the labeled angles in each exercise.

4. Exercise 1   **5.** Exercise 2   **6.** Exercise 3

Example 2
(page 112)

7. Developing Proof Complete this proof of one form of Theorem 2-3 by filling in the blanks.

If two angles are complements of the same angle, then the two angles are congruent.

Given: ∠1 and ∠2 are complementary.
     ∠3 and ∠2 are complementary.

Prove: ∠1 ≅ ∠3

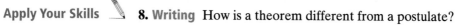

Proof: By the definition of complementary angles,
$m\angle 1 + m\angle 2 =$ **a.** _?_ and $m\angle 3 + m\angle 2 =$ **b.** _?_.
Then $m\angle 1 + m\angle 2 = m\angle 3 + m\angle 2$ by **c.** _?_.
Subtract $m\angle 2$ from each side. You get $m\angle 1 =$ **d.** _?_, or ∠1 ≅ ∠3.

B **Apply Your Skills**

8. Writing How is a theorem different from a postulate?

9. Open-Ended Give an example of vertical angles in your home.

10. Reasoning Explain why this statement is true:
If $m\angle 1 + m\angle 2 = 180$ and $m\angle 3 + m\angle 2 = 180$, then ∠1 ≅ ∠3.

11. Design The two back legs of the director's chair pictured at the left meet in a 72° angle. Find the measure of each angle formed by the two back legs.

 **Algebra** **Find the value of each variable and the measure of each labeled angle.**

12.

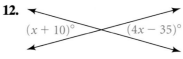
$(x + 10)°$ $(4x - 35)°$

13.

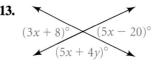
$(3x + 8)°$ $(5x - 20)°$
$(5x + 4y)°$

14. Developing Proof Complete this proof of Theorem 2-4 by filling in the blanks.

All right angles are congruent.

Given: ∠X and ∠Y are right angles.
Prove: ∠X ≅ ∠Y

X   Y

Proof: By the definition of **a.** _?_, $m\angle X = 90$ and $m\angle Y = 90$.
By the Substitution Property, $m\angle X =$ **b.** _?_, or ∠X ≅ ∠Y.

15. Multiple Choice What is the measure of the angle formed by Park St. and Oak St.?

(A) 35°   (B) 45°
(C) 55°   (D) 90°

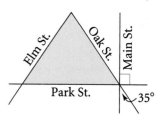

Name two pairs of congruent angles in each figure. Justify your answers.

16.

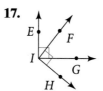

17.

18.

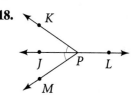

19. Coordinate Geometry ∠DOE contains points $D(2, 3)$, $O(0, 0)$, and $E(5, 1)$. Find the coordinates of a point F so that $\overrightarrow{OF}$ is a side of an angle that is adjacent and supplementary to ∠DOE.

Exercise 11

GO Online
Homework Video Tutor
Visit: PHSchool.com
Web Code: aue-0205

20. Coordinate Geometry $\angle AOX$ contains points $A(1, 3)$, $O(0, 0)$, and $X(4, 0)$.
   a. Find the coordinates of a point B so that $\angle BOA$ and $\angle AOX$ are adjacent complementary angles.
   b. Find the coordinates of a point C so that $\overrightarrow{OC}$ is a side of a different angle that is adjacent and complementary to $\angle AOX$.

21. Developing Proof Complete this proof of Theorem 2-5 by filling in the blanks.

If two angles are congruent and supplementary, then each is a right angle.

Given: $\angle W$ and $\angle V$ are congruent and supplementary.

Prove: $\angle W$ and $\angle V$ are right angles.

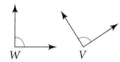

Proof: $\angle W$ and $\angle V$ are congruent, so $m\angle W = m\angle$ **a.** __?__.
$\angle W$ and $\angle V$ are supplementary so $m\angle W + m\angle V =$ **b.** __?__.
Substituting $m\angle W$ for $m\angle V$, you get $m\angle W + m\angle W = 180$, or $2m\angle W = 180$.
By the **c.** __?__ Property of Equality, $m\angle W = 90$.
Since $\angle W \cong \angle V$, $m\angle V = 90$, too. Then both angles are **d.** __?__ angles.

 22. Sports In the photograph, the wheels of the racing wheelchair are tilted so that $\angle 1 \cong \angle 2$. What theorem can you use to justify the statement $\angle 3 \cong \angle 4$?

Exercise 22

$\boxed{x^2}$ **Algebra** **Find the measure of each angle.**

23. $\angle A$ is twice as large as its complement, $\angle B$.

24. $\angle A$ is half as large as its complement, $\angle B$.

25. $\angle A$ is twice as large as its supplement, $\angle B$.

26. $\angle A$ is half as large as twice its supplement, $\angle B$.

Proof **27.** Write a proof for this form of Theorem 2-2.

If two angles are supplements of congruent angles, then the two angles are congruent.

Given: $\angle 1$ and $\angle 2$ are supplementary.
   $\angle 3$ and $\angle 4$ are supplementary.
   $\angle 2 \cong \angle 4$

Prove: $\angle 1 \cong \angle 3$

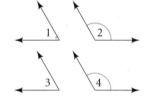

Proof **28.** Write a proof for this form of Theorem 2-3.

If two angles are complements of congruent angles, then the two angles are congruent.

Given: $\angle 1$ and $\angle 2$ are complementary.
   $\angle 3$ and $\angle 4$ are complementary.
   $\angle 2 \cong \angle 4$

Prove: $\angle 1 \cong \angle 3$

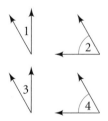

C Challenge

29. Paper Folding After you've done the Activity on page 110, answer these questions.
   a. How is the first fold line you make related to angles 3 and 4?
   b. How is the second fold line you make related to angles 1 and 2?
   c. How are the two fold lines related to each other? Give a convincing argument to support your answer.

x^2 **Algebra** Find the value of each variable and the measure of each labeled angle.

30.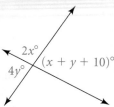
$(y + x)°$
$2x°$
$(y - x)°$

31.
$(x + y + 5)°$
$y°$
$2x°$

32.
$2x°$
$4y°$
$(x + y + 10)°$

Test Prep

Gridded Response

Find the measure of each angle.

33. an angle with measure 8 less than the measure of its complement

34. one angle of a pair of complementary vertical angles

35. an angle with measure three times the measure of its supplement

Use the diagram at the right to find the measure of each of the following angles.

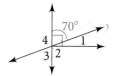
$70°$
4   1
3 2

36. ∠1   **37.** ∠2

38. ∠3   **39.** ∠4

Mixed Review

Lesson 2-4

Use the given property to complete each statement.

40. Subtraction Property of Equality
If $3x + 7 = 19$, then $3x = $ __?__ .

41. Reflexive Property of Congruence
$\overline{AB} \cong$ __?__

42. Substitution Property
If $MN = 3$ and $MN + NP = 15$, then __?__ .

Lesson 2-3

Use deductive reasoning to draw a conclusion. If not possible, write *not possible*.

43. If two lines intersect, then they are coplanar.
Lines m and n are coplanar.

44. If two angles are vertical angles, then they are congruent.
∠1 and ∠2 are vertical angles.

Lesson 2-2

Each conditional statement below is true. Write its converse. If the converse is also true, combine the statements as a biconditional.

45. If $y + 7 = 32$, then $y = 25$.

46. If you live in Australia, then you live south of the equator.

47. If $n > 0$, then $n^2 > 0$.

Writing Short Responses

Short-response questions are usually worth 2 points. To get full credit you must demonstrate a thorough understanding and knowledge of mathematical concepts and techniques.

EXAMPLE

Name a pair of adjacent angles in the diagram at the right. Use the definition of adjacent angles to explain why your angles are adjacent.

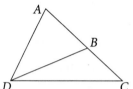

To respond correctly to this problem you have to know that adjacent angles are coplanar angles with a common vertex, a common side, and no common interior points. You have to pick a pair of angles that are adjacent and then use the definition to explain why your angles are adjacent.

2 points	1 points	0 points
∠ABD is adjacent to ∠CBD. They have a common vertex, a common side, and no common interior points.	∠ABD is adjacent to ∠CBD. They don't overlap each other.	∠ABD and ∠ABC

A 2-point response has a correct pair of adjacent angles and an explanation that uses the definition. This 1-point response has a correct pair but an incomplete definition. (An incorrect pair with an explanation that uses the definition is also worth 1 point.) The 0-point response has an incorrect pair and no explanation.

EXERCISES

Score each response to the exercise in the Example above. Explain your reasoning.

1.
∠DBC and ∠ABD are adjacent because they are next to each other and don't overlap.

2.
∠BAD and ∠BCD They aren't next to each other.

3.
∠DBA and ∠DBC Adjacent angles share a vertex and a side and have no interior points in common.

Write a 2-point response to each exercise. Refer to the diagram in the Example.

4. Name a pair of supplementary angles. Use the definition of supplementary angles to explain why your angles are supplementary.

5. ∠ABD and ∠BDC are complementary, $m\angle ABD = 6x$, and $m\angle BDC = 2x + 2$. Use the definition of complementary to explain what x must be.

Chapter Review

Vocabulary Review

biconditional (p. 87)
conclusion (p. 80)
conditional (p. 80)
converse (p. 81)
deductive reasoning (p. 94)

hypothesis (p. 80)
Law of Detachment (p. 94)
Law of Syllogism (p. 95)
paragraph proof (p. 111)
Reflexive Property (p. 105)

Symmetric Property (p. 105)
theorem (p. 110)
Transitive Property (p. 105)
truth value (p. 81)

Choose the correct vocabulary term to complete each sentence.

1. The statement "$\angle A \cong \angle A$" is an example of the __?__ Property of Congruence.

2. In a conditional statement, the part that directly follows *if* is the __?__.

3. "If $\angle A \cong \angle B$ and $\angle B \cong \angle C$, then $\angle A \cong \angle C$" is an example of the __?__ Property of Congruence.

4. When a conditional and its converse are true, they may be written as a single true statement called a __?__.

5. The __?__ of a conditional switches the hypothesis and the conclusion.

6. "If $\angle A \cong \angle B$, then $\angle B \cong \angle A$" is an example of the __?__ Property of Congruence.

7. The part of a conditional statement that follows "then" is the __?__.

8. A conditional has a __?__ of true or false.

9. Reasoning logically from given statements to a conclusion is __?__.

10. A statement that you prove true is a __?__.

Go Online
PHSchool.com
For: Vocabulary quiz
Web Code: auj-0251

Skills and Concepts

2-1 and 2-2 Objectives

▼ To recognize conditional statements
▼ To write converses of conditional statements
▼ To write biconditionals
▼ To recognize good definitions

An *if-then statement* is a **conditional.** The part following *if* is the **hypothesis.** The part following *then* is the **conclusion.** You find the truth value of a conditional by determining whether it is true or false. The symbolic form of a conditional is $p \rightarrow q$.

The **converse** of a conditional switches the hypothesis and the conclusion. The symbolic form of the converse of $p \rightarrow q$ is $q \rightarrow p$.

When a conditional and its converse are true, you can combine them as a true **biconditional.** To write a biconditional, you join the two parts of each conditional with the phrase *if and only if.* The symbolic form of a biconditional is $p \leftrightarrow q$.

For Exercises 11–13, (a) write the converse and (b) determine the truth value of the conditional and its converse. (c) If both statements are true, write a biconditional.

11. If you are a teenager, then you are younger than 20.

12. If an angle is obtuse, then its measure is greater than 90 and less than 180.

13. If a figure is a square, then it has four sides.

14. Write the following sentence as a conditional: All flowers are beautiful.

A good definition is precise. A good definition uses terms that have been previously defined or are commonly accepted.

15. Rico defines a *book* as something you read. Explain why this is not a good definition.

16. Write this definition as a biconditional:
An *oxymoron* is a phrase that contains contradictory terms.

17. Write this biconditional as two statements, a conditional and its converse:
Two angles are complementary if and only if the sum of their measures is 90.

2-3 Objectives

▼ To use the Law of Detachment

▼ To use the Law of Syllogism

Deductive reasoning is the process of reasoning logically from given statements to a conclusion. If the given statements are true, deductive reasoning produces a true conclusion.

The following are two important laws of deductive reasoning:
Law of Detachment: If $p \rightarrow q$ is a true statement and p is true, then q is true.
Law of Syllogism: If $p \rightarrow q$ and $q \rightarrow r$ are true statements, then $p \rightarrow r$ is true.

Use the Law of Detachment to make a conclusion.

18. If you practice table tennis every day, you will become a better player. Lucy practices table tennis every day.

19. Line ℓ and line m are perpendicular. If two lines are perpendicular, they intersect to form right angles.

20. If two angles are supplementary, then the sum of their measures is 180. $\angle 1$ and $\angle 2$ are supplementary.

Use the Law of Syllogism to make a conclusion.

21. If Kate studies, she will get good grades. If Kate gets good grades, she will graduate.

22. If a, then b. If b, then c.

23. If the weather is wet, the Huskies will not play soccer. If the Huskies do not play soccer, Nathan can stop at the ice cream shop.

2-4 Objective

▼ To connect reasoning in algebra and geometry

In algebra, you use deductive reasoning and properties to solve equations. In geometry, each statement in a deductive argument is justified by a property, definition, or postulate. Some of the properties you need are listed below.

Properties of Equality

Addition Property	If $a = b$, then $a + c = b + c$.
Subtraction Property	If $a = b$, then $a - c = b - c$.
Multiplication Property	If $a = b$, then $a \cdot c = b \cdot c$.
Division Property	If $a = b$ and $c \neq 0$, then $\frac{a}{c} \neq \frac{b}{c}$.
Substitution Property	If $a = b$, then b can replace a in any expression.
Distributive Property	$a(b + c) = ab + ac$

Properties of Congruence

Reflexive Property $\overline{AB} \cong \overline{AB}$
$\angle A \cong \angle A$

Symmetric Property If $\overline{AB} \cong \overline{CD}$, then $\overline{CD} \cong \overline{AB}$.
If $\angle A \cong \angle B$, then $\angle B \cong \angle A$.

Transitive Property If $\overline{AB} \cong \overline{CD}$ and $\overline{CD} \cong \overline{EF}$, then $\overline{AB} \cong \overline{EF}$.
If $\angle A \cong \angle B$ and $\angle B \cong \angle C$, then $\angle A \cong \angle C$.

x^2 **24. Algebra** Fill in the reason that justifies each step.

Given: $QS = 42$

$$x + 3 \qquad 2x$$
$$\overline{\underset{Q}{\bullet}\underset{R}{\bullet}\underset{S}{\bullet}}$$

$QR + RS = QS$	**a.** ?
$x + 3 + 2x = 42$	**b.** ?
$3x + 3 = 42$	**c.** ?
$3x = 39$	**d.** ?
$x = 13$	**e.** ?

Use the given property to complete each statement.

25. Addition Property of Equality
If $x = 5$, then $x + 3 = \underline{\ ?\ }$.

26. Division Property of Equality
If $2(AX) = 2(BY)$, then $AX = \underline{\ ?\ }$.

27. Reflexive Property of Equality
$m\angle Y = \underline{\ ?\ }$

28. Symmetric Property of Equality
If $XY = RS$, then $\underline{\ ?\ }$.

29. Transitive Property of Equality
If $x = 5$ and $5 = y$, then $x = \underline{\ ?\ }$.

30. Distributive Property
$2(4x + 5) = 8x + \underline{\ ?\ }$

31. Distributive Property
$3p - 6q = 3(\underline{\ ?\ })$

32. Reflexive Property of Congruence
$\overline{NM} \cong \underline{\ ?\ }$

2-5 Objective

▼ To prove and apply theorems about angles

A statement that you prove true is a **theorem**. A proof written as a paragraph is a **paragraph proof**. You can prove that vertical angles are congruent; that supplements of the same angle are congruent, and that complements of the same angle are congruent.

33. Algebra Find the value of y.

A ∙   ∙ B

$(3y + 20)°$ $(5y - 16)°$
E

C D

34. For the diagram at the left, find each of the following.
  a. $m\angle AEC$
  b. $m\angle BED$
  c. $m\angle AEB$

35. Complete the following paragraph proof.

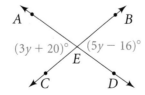

Given: $\angle 1 \cong \angle 4$

Prove: $\angle 2 \cong \angle 3$

Proof: By the Vertical Angles Theorem, $\angle 1 \cong \underline{\ ?\ }$ and $\angle 4 \cong \underline{\ ?\ }$. $\angle 1 \cong \angle 4$ is given, so $\angle 2 \cong \angle 3$ by the $\underline{\ ?\ }$ Property of Congruence.

Chapter Test

For: Chapter Test
Web Code: aua-0252

1. Identify the hypothesis and conclusion:

If $x + 9 = 11$, then $x = 2$.

2. Write this statement as a conditional.

All babies are cute.

3. Find a counterexample to show that this statement is *not* true.

If two angles are complementary, then they are not congruent.

For each statement, (a) write the converse and (b) decide whether the converse is true or false.

4. If a figure is a rectangle, then it has two right angles.

5. If two lines intersect, then they lie in the same plane.

6. If it is snowing in South Carolina, then it is not summer.

Writing Explain why each statement is *not* a good definition.

7. A pencil is a writing instrument.

8. Complementary angles are angles that form a right angle.

9. Vertical angles are angles that are congruent.

For Exercises 10–14, name the property that justifies each statement.

10. If $UV = KL$ and $KL = 6$, then $UV = 6$.

11. If $m\angle 1 + m\angle 2 = m\angle 4 + m\angle 2$, then $m\angle 1 = m\angle 4$.

12. $\angle ABC \cong \angle ABC$

13. If $\frac{1}{2}m\angle D = 45$, then $m\angle D = 90$.

14. If $\angle DEF \cong \angle HJK$, then $\angle HJK \cong \angle DEF$.

15. Find the measure of each angle.

a. $\angle CDM$ **b.** $\angle KDM$

c. $\angle JDK$ **d.** $\angle JDM$

e. $\angle CDB$ **f.** $\angle CDK$

16. The measure of an angle is $2z$. What is the measure of its supplement?

17. The measure of an angle is 52 more than the measure of its complement. What is the measure of the angle?

For each diagram, state two pairs of angles that are congruent. Justify your answers.

18.   **19.**

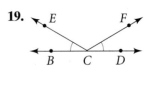

20. Rewrite this biconditional as two conditionals.

A fish is a bluegill if and only if it is a bluish, freshwater sunfish.

21. Complete this statement.

If two angles are complements of congruent angles, the angles are ___?___.

For Exercises 22–26, use the Law of Detachment and the Law of Syllogism to make any possible conclusion. Write *not possible* if you cannot make any conclusion.

22. People who live in glass houses shouldn't throw stones. Lindsay shouldn't throw stones.

23. James wants to be a chemical engineer. If a student wants to be a chemical engineer, that student must graduate from college.

24. $p \rightarrow q$ and $q \rightarrow r$ are true statements.

25. $p \rightarrow q$ and p are true statements.

26. $p \rightarrow q$ and q are true statements.

27. Developing Proof Complete this proof by filling in the blanks.

Given: $\angle FED$ and $\angle DEW$ are complementary.

Prove: $\angle FEW$ is a right angle.

Proof: By the definition of complementary angles, $m\angle FED + m\angle DEW =$ **a.** ___?___.

$m\angle FED + m\angle DEW = m\angle FEW$ by the **b.** ___?___.

$90 = m\angle FEW$ by the **c.** ___?___ Property of Equality.

Then, $\angle FEW$ is a right angle, by the **d.** ___?___.

Regents Test Prep

Multiple Choice

For Exercises 1–15, choose the correct letter.

1. What is the converse of the statement, "If a strawberry is red, then it is ripe"?
   (1) If a strawberry is not red, then it is not ripe.
   (2) If a strawberry is ripe, then it is red.
   (3) A strawberry is ripe if and only if it is red.
   (4) If a strawberry is red, then it is ripe.

2. Which is the intersection of two planes that have a point in common?
   (1) a point   (2) a line
   (3) a plane   (4) a ray

3. Which property justifies this statement?
   If $4AB = 8CD$, then $AB = 2CD$.
   (1) Division Property of Equality
   (2) Reflexive Property of Equality
   (3) Substitution Property of Equality
   (4) Distributive Property

4. Which point lies the farthest from the origin?
   (1) $(0, -7)$ (2) $(-3, 8)$
   (3) $(-4, -3)$ (4) $(5, 1)$

5. What is the length of the segment with endpoints $A(1, 7)$ and $B(-3, -1)$?
   (1) $\sqrt{40}$ (2) 8   (3) $\sqrt{80}$ (4) 40

6. What is the next number in the pattern?
   $1, -4, 9, -16,$
   (1) -35 (2) -25 (3) 25   (4) 35

7. If the measure of an angle is 78 less than the measure of its complement, what is the measure of the angle?
   (1) 6   (2) 12   (3) 51   (4) 84

8. $\angle A$ and $\angle B$ are supplementary and vertical angles. What is $m\angle B$?
   (1) 45   (2) 90   (3) 135   (4) 180

9. What is the midpoint of a segment with endpoints $(0, -4)$ and $(-4, 7)$?
   (1) $(-4, \frac{3}{2})$ (2) $(-2, 3)$
   (3) $(-2, \frac{3}{2})$ (4) $(2, -3)$

10. The measure of an angle is 12 less than twice the measure of its supplement. What is the measure of the angle?
    (1) 28   (2) 34   (3) 64   (4) 116

11. Which statement can be combined with its converse to form a true biconditional?
    (1) If the measure of an angle is 30, then it is an acute angle.
    (2) If a ray is the perpendicular bisector of a segment, then the ray divides the segment into two congruent segments.
    (3) If two lines intersect, then the two lines are not skew.
    (4) If an angle is a straight angle, then its sides are opposite rays.

12. The sum of the measures of a complement and a supplement of an angle is 200. What is the measure of the angle?
    (1) 20   (2) 35   (3) 55   (4) 70

13. Use the Addition Property of Equality to complete this statement:
    If $5x - 12 = 88$, then $5x = \underline{\ ?\ }$.
    (1) 12   (2) 76   (3) 88   (4) 100

14. Which word correctly completes this sentence?
    All $\underline{\ ?\ }$ angles are congruent.
    (1) acute   (2) complementary
    (3) right   (4) supplementary

15. Which point is exactly 5 units from $(-10, 4)$?
    (1) $(-6, -7)$ (2) $(-6, 7)$
    (3) $(6, -7)$ (4) $(6, 7)$

Gridded Response

16. The area of a circle is 10π cm^2. What is the circle's diameter? Round to the nearest hundredth of a centimeter.

17. The measure of an angle is one third the measure of its supplement. What is the measure of the angle?

Short Response

18. $\overline{AB}$ has endpoints $A(3, 6)$ and $B(9, -2)$ and midpoint M. Justify each response.
    a. Find the coordinates of M.
    b. Find AB.

Extended Response

19. Construct a right triangle. Then construct the bisectors of two of its angles.

Activity Lab

The Delicious Side of Division

Applying Reasoning Splitting dessert evenly among brothers and sisters sometimes causes arguments. In some families, one child divides the dessert and another has first choice among the pieces. This encourages the divider to be very, very careful!

Activity 1

Dividing a square cake into an even number of pieces that are alike can be relatively simple. Dividing a square cake into seven same-size pieces is more challenging.

One method is to divide the perimeter of the cake by 7. Then mark the perimeter in seven equal lengths. (Some lengths may go around a corner.) Cut segments from the center of the cake to the marks on the perimeter. This splits the cake into the seven same-size pieces.

Show why this method works.

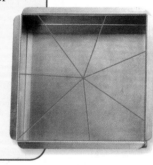

Ground mace

Powdered cocoa

Cooking with Spices

Mace comes from the kernel of an apricot-like fruit that grows mainly in Indonesia. Cocoa comes from the fruit of the cacao tree. Finely ground, both spices add flavor to baked goods.

NY G.CN.6: Recognize and apply mathematics to situations in the outside world.

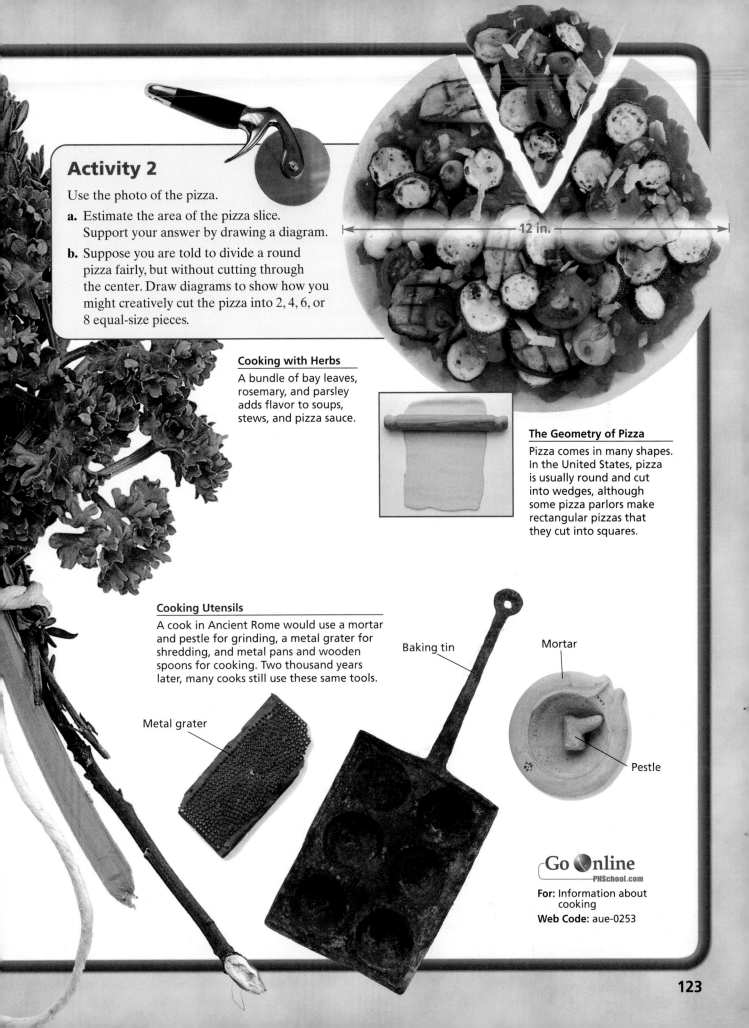

Activity 2

Use the photo of the pizza.

a. Estimate the area of the pizza slice. Support your answer by drawing a diagram.

b. Suppose you are told to divide a round pizza fairly, but without cutting through the center. Draw diagrams to show how you might creatively cut the pizza into 2, 4, 6, or 8 equal-size pieces.

12 in.

Cooking with Herbs

A bundle of bay leaves, rosemary, and parsley adds flavor to soups, stews, and pizza sauce.

The Geometry of Pizza

Pizza comes in many shapes. In the United States, pizza is usually round and cut into wedges, although some pizza parlors make rectangular pizzas that they cut into squares.

Cooking Utensils

A cook in Ancient Rome would use a mortar and pestle for grinding, a metal grater for shredding, and metal pans and wooden spoons for cooking. Two thousand years later, many cooks still use these same tools.

Baking tin

Mortar

Pestle

Metal grater

Go Online
PHSchool.com

For: Information about cooking
Web Code: aue-0253

What You've Learned

NY Learning Standards for Mathematics

- **G.G.8:** In Chapter 1, you learned that two coplanar lines that do not intersect are parallel.

- In Chapters 1 and 2, you learned how to measure angles, recognize congruent angles, and identify angles whose measures have a sum of 180.

- **G.RP.7:** In Chapter 2, you learned how to use deductive reasoning to draw conclusions.

 Check Your Readiness   **for Help** to the Lesson in green.

Evaluating Algebraic Expressions (Skills Handbook page 754)

x^2 **Algebra** Evaluate each expression for the given value of n.

1. $\frac{360}{n}; n = 5$ **2.** $(n - 2)180; n = 9$ **3.** $(n - 2)180; n = 17$

Solving Equations (Algebra 1 Review page 30)

x^2 **Algebra** Solve each equation.

4. $3x + 11 = 7x - 5$ **5.** $(2x + 5) + (3x - 10) = 70$ **6.** $(3x + 2) - (2x - 3) = -19$

Writing and Solving an Equation (Skills Handbook page 758)

Write an equation and solve the problem.

7. The sum of the measures of three angles is 180. One measure is twice the size of each of the other two. Find the measure of each angle.

8. The sum of the measures of three angles is 180. One measure is half the size of each of the other two. Find the measure of each angle.

Drawing Parallel and Perpendicular Lines (Lesson 1-4)

9. Draw a picture of a rectangular box and label its eight corners A through H. Name two lines in your picture that appear to be parallel. Name two lines that appear to be perpendicular.

Drawing and Measuring Angles (Lesson 1-6)

Use a straightedge and draw the given type of angle as best you can. Estimate its measure, and then find its measure with a protractor.

10. acute   **11.** right   **12.** obtuse

Parallel and Perpendicular Lines

◀)) **Key Vocabulary**

- alternate interior angles (p. 127)
- corresponding angles (p. 127)
- equiangular triangle (p. 148)
- equilateral triangle (p. 148)
- exterior angle of a polygon (p. 149)
- flow proof (p. 135)
- isosceles triangle (p. 148)
- polygon (p. 157)
- regular polygon (p. 160)
- remote interior angles (p. 149)
- same-side interior angles (p. 127)
- scalene triangle (p. 148)
- transversal (p. 127)

What You'll Learn Next

(NY) **Learning Standards for Mathematics**

- **G.G.63:** In this chapter, you will use deductive reasoning to make conclusions about parallel and perpendicular lines.

- **G.G.30:** You will use parallel lines to learn about angle measures in triangles and other polygons.

- **G.G.63:** You will also learn ways to think about parallel and perpendicular lines in a coordinate plane.

Activity Lab Applying what you learn, you will do activities involving reflected light and congruence on pages 194 and 195.

125

Parallel Lines and Related Angles

FOR USE WITH LESSON 3-1

Construct

Use geometry software to construct two parallel lines. Check that the lines remain parallel as you manipulate them. Construct a point on each line. Then construct the line through these two points. This line is called a transversal.

Investigate

Measure each of the eight angles formed by the parallel lines and the transversal. Record the measurements. Manipulate the lines. Record the new measurements. What relationships do you notice?

NY G.G.35: Determine if two lines cut by a transversal are parallel, based on the measure of given pairs of angles formed by the transversal and the lines.

EXERCISES

1. When a transversal intersects parallel lines, what are the relationships among the angles formed? Make as many conjectures as possible.

Extend

2. Use your software to construct three or more parallel lines. Construct a line that intersects all three lines.
   a. What relationships exist among the angles formed?
   b. How many different angle measures are there?

3. Construct two parallel lines and a transversal perpendicular to one of the parallel lines. What angle does it make with the second parallel line?

4. Using geometry software, construct two lines and a transversal, making sure that the two lines are *not* parallel. Locate two angles that are on alternate sides of the transversal and in the interior region between the other two lines. Manipulate the lines so that these angles have the same measure.
   a. Make a conjecture as to the relationship between the two lines.
   b. How is this conjecture different from the conjecture(s) you made in Exercise 1?

5. Again, draw two lines and a transversal, making sure that the two lines are *not* parallel. Locate two angles that are on the same side of the transversal and in the interior region between the two lines. Manipulate the lines so that these angles are supplementary.
   a. Make a conjecture as to the relationship between the two lines.
   b. How is this conjecture different from the conjecture(s) you made in Exercise 1?

6. Construct perpendicular lines *a* and *b*. At a point away from the intersection of *a* and *b*, construct line *c* perpendicular to line *a*. Make a conjecture about lines *b* and *c*.

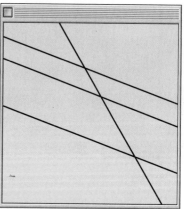

Properties of Parallel Lines

 Learning Standards for Mathematics

G.G.35 Determine if two lines cut by a transversal are parallel, based on the measure of given pairs of angles formed by the transversal and the lines.

✓ **Check Skills You'll Need**

 for Help page 30 or Skills Handbook page 758

x^2 **Algebra** Solve each equation.

1. $x + 2x + 3x = 180$

2. $(w + 23) + (4w + 7) = 180$

3. $90 = 2y - 30$

4. $180 - 5y = 135$

Write an equation and solve the problem.

5. The sum of $m\angle 1$ and twice its complement is 146. Find $m\angle 1$.

6. The measures of two supplementary angles are in the ratio 2 : 3. Find their measures.

🔊 **New Vocabulary**
- transversal
- alternate interior angles
- same-side interior angles
- corresponding angles
- two-column proof
- alternate exterior angles
- same-side exterior angles

---

1 Identifying Angles

A **transversal** is a line that intersects two coplanar lines at two distinct points. The diagram shows the eight angles formed by a transversal t and two lines ℓ and m.

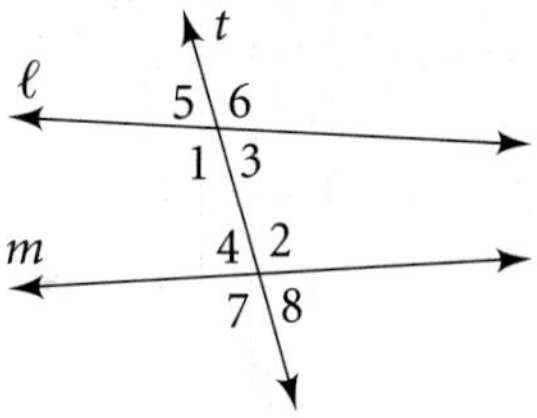

Pairs of the eight angles have special names as suggested by their positions.

∠1 and ∠2 are **alternate interior angles.**

∠1 and ∠4 are **same-side interior angles.**

∠1 and ∠7 are **corresponding angles.**

Video Tutor Help
Visit: PHSchool.com
Web Code: aue-0775

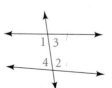 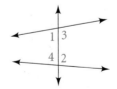 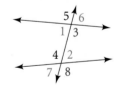

1 EXAMPLE Identifying Angles

Use the diagrams above. Name another pair of alternate interior angles and another pair of same-side interior angles.

● ∠3 and ∠4 are alternate interior angles. ∠2 and ∠3 are same-side interior angles.

✓ **Quick Check** ❶ Name three other pairs of corresponding angles in the diagrams above.

2 EXAMPLE Real-World  Connection

Aviation In the diagram of Lafayette Regional Airport, the black segments are runways and the gray areas are taxiways and terminal buildings. Classify ∠1 and ∠2 as alternate interior angles, same-side interior angles, or corresponding angles.

● ∠1 and ∠2 are corresponding angles.

✓ **Quick Check** ❷ Classify ∠2 and ∠3 as alternate interior angles, same-side interior angles, or corresponding angles.

**Lafayette Regional Airport
Lafayette, Louisiana**

---

2 Properties of Parallel Lines

In the photograph, the vapor trail of the high-flying aircraft suggests a transversal of the parallel trails of the low-flying aircraft.

Vocabulary Tip

<u>Corresponding</u> <u>objects</u> are related in a special way. Here, <u>corresponding angles</u> are angles that are in similar positions on the same side of a transversal.

The same-size angles that appear to be formed by the vapor trails suggest the postulate and theorems below.

🔑 **Key Concepts**

Postulate 3-1	**Corresponding Angles Postulate**

If a transversal intersects two parallel lines, then corresponding angles are congruent.

∠1 ≅ ∠2

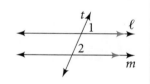

🔑 **Key Concepts**

Theorem 3-1	**Alternate Interior Angles Theorem**

If a transversal intersects two parallel lines, then alternate interior angles are congruent.

∠1 ≅ ∠3

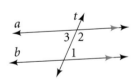

Theorem 3-2	**Same-Side Interior Angles Theorem**

If a transversal intersects two parallel lines, then same-side interior angles are supplementary.

$m\angle 1 + m\angle 2 = 180$

You can display the steps that prove a theorem in a **two-column proof.**

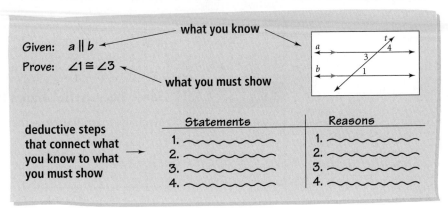

Given: $a \parallel b$ ← what you know

Prove: $\angle 1 \cong \angle 3$ ← what you must show

deductive steps that connect what you know to what you must show →

Statements	Reasons
1. ~~~~~	1. ~~~~~
2. ~~~~~	2. ~~~~~
3. ~~~~~	3. ~~~~~
4. ~~~~~	4. ~~~~~

Proof

Proof of Theorem 3-1

If a transversal intersects two parallel lines, then alternate interior angles are congruent.

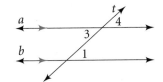

Given: $a \parallel b$

Prove: $\angle 1 \cong \angle 3$

Statements	Reasons
1. $a \parallel b$	1. Given
2. $\angle 1 \cong \angle 4$	2. If lines are $\parallel$, then corresponding angles are congruent.
3. $\angle 4 \cong \angle 3$	3. Vertical angles are congruent.
4. $\angle 1 \cong \angle 3$	4. Transitive Property of Congruence

You will prove Theorem 3-2 in Exercise 29.

Proof **3 EXAMPLE** **Writing a Two-Column Proof**

Study what is given, what you are to prove, and the diagram. Then write a two-column proof.

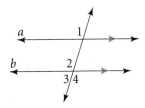

Given: $a \parallel b$

Prove: $\angle 1 \cong \angle 4$

Statements	Reasons
1. $a \parallel b$	1. Given
2. $\angle 1 \cong \angle 2$	2. If lines are $\parallel$, then corresponding angles are congruent.
3. $\angle 2 \cong \angle 4$	3. Vertical angles are congruent.
4. $\angle 1 \cong \angle 4$	4. Transitive Property of Congruence

✓ Quick Check **3** Using the same given information and diagram from Example 3, prove that $\angle 1$ and $\angle 3$ are supplementary.

In Example 3, $\angle 1$ and $\angle 4$ are **alternate exterior angles.** $\angle 1$ and $\angle 3$ are **same-side exterior angles.** Example 3 and its Quick Check prove Theorems 3-3 and 3-4 as stated on the next page.

Key Concepts

| Theorem 3-3 | **Alternate Exterior Angles Theorem** |

If a transversal intersects two parallel lines, then alternate exterior angles are congruent.

$$\angle 1 \cong \angle 3$$

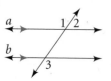

| Theorem 3-4 | **Same-Side Exterior Angles Theorem** |

If a transversal intersects two parallel lines, then same-side exterior angles are supplementary.

$$m\angle 2 + m\angle 3 = 180$$

When you see two parallel lines and a transversal, and you know the measure of one angle, you can find the measures of all the angles. This is illustrated in Example 4.

Test-Taking Tip

You can make marks on any diagrams shown on tests. This can help you keep track of known information.

4 EXAMPLE   **Finding Measures of Angles**

Find $m\angle 1$, and then $m\angle 2$. Which theorem or postulate justifies each answer?

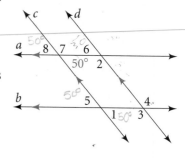

Since $a \parallel b, m\angle 1 = 50$ because corresponding angles are congruent (Corresponding Angles Postulate).

Since $c \parallel d, m\angle 2 = 130$ because same-side interior angles are supplementary (Same-Side Interior Angles Theorem).

✔**Quick Check** ④ Find the measure of each angle. Justify each answer.
   a. $\angle 3$ **b.** $\angle 4$ **c.** $\angle 5$
   d. $\angle 6$ **e.** $\angle 7$ **f.** $\angle 8$

Sometimes you can use algebra to find angle measures.

5 EXAMPLE   **Using Algebra to Find Angle Measures**

Gridded Response Find the value of y in the diagram at the left.

$x = 70$	**Corresponding angles of parallel lines are $\cong$.**
$70 + 50 + y = 180$	**Angle Addition Postulate**
$y = 60$	**Subtraction Property of Equality**

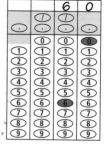

✔**Quick Check** ⑤ Find the values of x and y. Then find the measures of the angles.

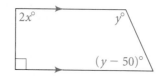

EXERCISES

For more exercises, see *Extra Skill, Word Problem, and Proof Practice.*

Practice and Problem Solving

A Practice by Example

Examples 1, 2
(pages 127, 128)

GO for Help

Name the two lines and the transversal that form each pair of angles. Then classify the pair of angles.

1. ∠2 and ∠3

2. ∠1 and ∠4

3. ∠SPQ and ∠PQR

4. ∠5 and ∠PSR

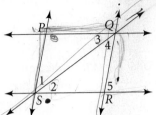

Classify each pair of angles labeled in the same color as *alternate interior angles, same-side interior angles,* or *corresponding angles.*

5. 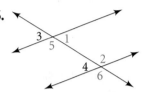 **6.** **7.**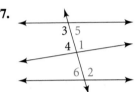

8. The boards securing this barn door suggest two parallel lines and a transversal. Classify ∠1 and ∠2 as alternate interior angles, same-side interior angles, or corresponding angles.

Example 3
(page 129)

9. Developing Proof Supply the missing reasons in this two-column proof.

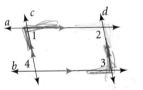

Given: $a \parallel b$, $c \parallel d$

Prove: $\angle 1 \cong \angle 3$

Statements	Reasons
1. $a \parallel b$	**1.** Given
2. ∠3 and ∠2 are supplementary.	**2.** ___
3. $c \parallel d$	**3.** Given
4. ∠1 and ∠2 are supplementary.	**4.** ___
5. $\angle 1 \cong \angle 3$	**5.** ___

Proof **10.** Write a two-column proof for Exercise 9 that does not use ∠2.

Example 4
(page 130)

Find $m\angle 1$, and then $m\angle 2$. Justify each answer.

11. 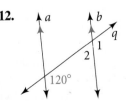 **12.** **13.**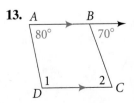

Example 5 x^2 **Algebra** **Find the value of x. Then find the measure of each labeled angle.**
(page 130)

14.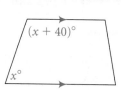

15.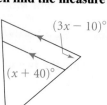

16.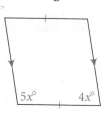

B **Apply Your Skills**

17. In the figure at the right, $f \parallel g$ and $m \parallel n$. Find the measure of each numbered angle.

18. Two pairs of parallel segments form the "pound sign" on your telephone keypad. To find the measures of all the angles in the pound sign, how many angles must you measure? Explain.

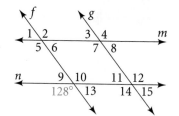

Two lines and a transversal form how many pairs of the following?

19. alternate interior angles

20. corresponding angles

21. same-side exterior angles

22. vertical angles

x^2 **Algebra** **Find the values of the variables.**

Problem Solving Hint

In Exercise 24, turn your book so the other two parallel lines appear horizontal.

23.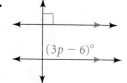

24.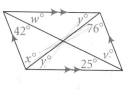

25.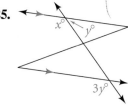

26. Writing Look up the meaning of the prefix *trans*. Explain how the meaning of the prefix relates to the word *transversal*.

27. Open-Ended The letter Z illustrates alternate interior angles. Find at least two other letters that illustrate the pairs of angles presented in this lesson. Draw the letters, mark the angles, and describe them.

Proof **28.** Write a two-column proof.

Given: $a \parallel b$, $\angle 1 \cong \angle 4$

Prove: $\angle 2 \cong \angle 3$

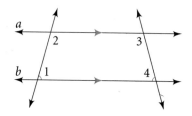

29. Prove Theorem 3-2.

If a transversal intersects two parallel lines, then same-side interior angles are supplementary.

Given: $a \parallel b$

Prove: $\angle 1$ and $\angle 2$ are supplementary.

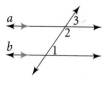

30. Engineering Engineers are laying pipe below ground on opposite sides of the street as shown here. To join the pipe, workers on each side of the street work towards the middle.
a. If one team lays pipe at the angle shown, what should the other team use for $m\angle 1$?
b. Are these two angles alternate interior, same-side interior, or corresponding angles?

GO Online
Homework Video Tutor
Visit: PHSchool.com
Web Code: aue-0301

C Challenge 🌐 **31. History** About 220 B.C., Eratosthenes estimated the circumference of E... He achieved this remark... ... locations in Egypt. ... here and that the ... are parallel. He ... the measures of ... in his simulation.

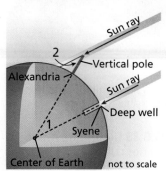

Sun ray
2
Vertical pole
Alexandria
Sun ray
Deep well
1
Syene
Center of Earth
not to scale

Handwritten note: Congruent = distance. Same

... ate interior, ... onding angles.

... hat ∠1 ≅ ∠2?

... e left contains contradictory information.

... ne *B*. Planes *A* and *B* are parallel.

... *es, always,* or *never.* **Justify each answer.**

34. Lines *m* and *n* are __?__ coplanar.

36. Lines *m* and *n* are __?__ skew.

$(60 - 2x)°$

NY REGENTS

A fence on a hill uses vertical posts L and M to hold parallel rails N and P. Use the diagram for Exercises 37–41.

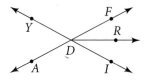

3 M 4
N
L 2
1 5
8 7 6 11 12
P 10 14 13
9
16 15

Multiple Choice **37.** ∠10 and ∠14 are alternate interior angles. Which is the transversal?
A. L **B.** M
C. N **D.** P

38. If $m\angle 1 = 115$, what is $m\angle 16$?
F. 35 **G.** 65 **H.** 85 **J.** 115

39. If $m\angle 10 = x - 24$, what is $m\angle 7$?
A. $156 + x$ **B.** $204 + x$ **C.** $156 - x$ **D.** $204 - x$

40. If $m\angle 1 = 6x$ and $m\angle 12 = 4x$, what is $m\angle 5$?
F. 54 **G.** 60 **H.** 72 **J.** 108

Short Response **41. a.** Describe a plan for showing that ∠1 ≅ ∠5.
b. Explain why ∠1 ≅ ∠5. Justify each step.

Mixed Review

GO for Help

Lesson 2-5 Find the measure of each angle if $m\angle YDF = 121$ and $\overrightarrow{DR}$ bisects ∠FDI.

42. ∠IDA **43.** ∠YDA **44.** ∠RDI

Y F R D A I

Lesson 1-8 **Coordinate Geometry** Find the coordinates of the midpoint of $\overline{AB}$.

45. $A(0, 9), B(1, 5)$ **46.** $A(-3, 8), B(2, -1)$ **47.** $A(10, -1), B(-4, 7)$

Lesson 1-1 Find a pattern for each sequence. Use the pattern to show the next two terms.

48. $4, 8, 12, 16, \ldots$ **49.** $1, -2, 4, -8, \ldots$ **50.** $23, 16, 9, 2, \ldots$

Proving Lines Parallel

NY Learning Standards for Mathematics

G.G.35 Determine if two lines cut by a transversal are parallel, based on the measure of given pairs of angles formed by the transversal and the lines.

 Check Skills You'll Need

 **GO for Help** page 30 and Lesson 2-1

x^2 **Algebra** Solve each equation.

1. $2x + 5 = 27$

2. $8a - 12 = 20$

3. $x - 30 + 4x + 80 = 180$

4. $9x - 7 = 3x + 29$

Write the converse of each conditional statement. Determine the truth value of the converse.

5. If a triangle is a right triangle, then it has a 90° angle.

6. If two angles are vertical angles, then they are congruent.

7. If two angles are same-side interior angles, then they are supplementary.

 **New Vocabulary** • flow proof

---

1 Using a Transversal

On window blinds like those shown here, you move the tilt bar to let in or shut out the light.

When you move the bar, the slats tilt at the same angle. This keeps them parallel and illustrates the converse of the Corresponding Angles Postulate.

 Key Concepts

Postulate 3-2	**Converse of the Corresponding Angles Postulate**

If two lines and a transversal form corresponding angles that are congruent, then the two lines are parallel.

$\ell \parallel m$

In Lesson 3-1, you proved four theorems based on the Corresponding Angles Postulate. You can also prove theorems that are based on its converse. In fact, Theorems 3-5 and 3-6 happen to be converses of the two theorems from Lesson 3-1, Theorems 3-1 and 3-2.

Theorem 3-5	Converse of the Alternate Interior Angles Theorem

If two lines and a transversal form alternate interior angles that are congruent, then the two lines are parallel.

If $\angle 1 \cong \angle 2$, then $\ell \parallel m$.

Theorem 3-6	Converse of the Same-Side Interior Angles Theorem

If two lines and a transversal form same-side interior angles that are supplementary, then the two lines are parallel.

If $\angle 2$ and $\angle 4$ are supplementary, then $\ell \parallel m$.

You have seen two forms of proof—paragraph and two-column. In a third form, called **flow proof,** arrows show the logical connections between the statements. Reasons are written below the statements.

Proof

Proof of Theorem 3-5

If two lines and a transversal form alternate interior angles that are congruent, then the two lines are parallel.

Given: $\angle 1 \cong \angle 2$
Prove: $\ell \parallel m$

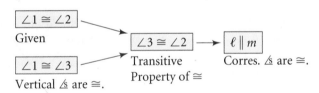

You will write a flow proof of Theorem 3-6 in Exercise 40. Theorems 3-6, 3-5, and Postulate 3-2 now provide you with three ways to prove that two lines are parallel.

1 EXAMPLE Using Theorems 3-5 and 3-6

Which lines, if any, must be parallel if $\angle 1 \cong \angle 2$? Justify your answer with a theorem or postulate.

$\overleftrightarrow{DE} \parallel \overleftrightarrow{KC}$ by Theorem 3-5, the Converse of the Alternate Interior Angles Theorem: If alternate interior angles are congruent, then the lines are parallel.

 **Quick Check** ❶ Which lines, if any, must be parallel if $\angle 3 \cong \angle 4$? Explain.

Theorems 3-3 and 3-4 allow you to make conclusions about alternate exterior angles and same-side exterior angles. Theorems 3-7 and 3-8 on the next page are converses of these two theorems. Like Theorems 3-5 and 3-6, they too can be used to prove that two lines are parallel.

Theorem 3-7	Converse of the Alternate Exterior Angles Theorem

If two lines and a transversal form alternate exterior angles that are congruent, then the two lines are parallel.

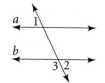

If ∠1 ≅ ∠2, then a ∥ b.

Theorem 3-8	Converse of the Same-Side Exterior Angles Theorem

If two lines and a transversal form same-side exterior angles that are supplementary, then the two lines are parallel.

If ∠1 and ∠3 are supplementary, then a ∥ b.

Here is a two-column proof of Theorem 3-7. You will prove Theorem 3-8 in Exercise 27.

Proof   **Proof of Theorem 3-7**

If two lines and a transversal form alternate exterior angles that are congruent, then the two lines are parallel.

Given: ∠1 ≅ ∠2

Prove: a ∥ b

Statements	Reasons
1. ∠1 ≅ ∠2	**1.** Given
2. ∠1 ≅ ∠4	**2.** Vertical angles are congruent.
3. ∠2 ≅ ∠4	**3.** Transitive Property of Congruence
4. a ∥ b	**4.** If two lines and a transversal form congruent corresponding angles, then the lines are parallel.

You can use algebra along with Postulates 3-1 and 3-2 and Theorems 3-1 through 3-8 to help you solve problems involving parallel lines.

2 EXAMPLE   **Using Algebra**

Algebra Find the value of x for which ℓ ∥ m.

The two angles are corresponding angles.
ℓ ∥ m when 2x + 6 = 40.

2x + 6 = 40

     2x = 34   **Subtract 6 from each side.**

      x = 17   **Divide each side by 2.**

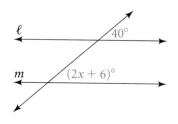

✓ **Quick Check**   **2** Find the value of x for which a ∥ b. Explain how you can check your answer.

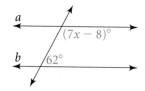

3 EXAMPLE Real-World  Connection

Parking Two workers are painting lines for angled parking spaces. The first worker paints a line so that $m\angle 1 = 60°$. The second worker paints a line so that $m\angle 2 = 60°$. Explain why their lines are parallel.

The angles are congruent alternate exterior angles. By the Converse of the Alternate Exterior Angles Theorem, the lines are parallel.

✓ Quick Check ③ If the second worker uses $\angle 3$, what should $m\angle 3$ be for parallel lines? Explain.

EXERCISES

For more exercises, see *Extra Skill, Word Problem, and Proof Practice*.

Practice and Problem Solving

A **Practice by Example**

Which lines or segments are parallel? Justify your answer.

Example 1
(page 135)

 GO for Help

1.

2.

3.
$$m\angle J + m\angle L = 180$$

4.

Example 2 x^2 **Algebra** **Find the value of x for which $\ell \parallel m$.**
(page 136)

5.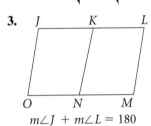

6.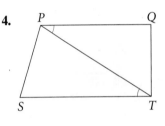

7.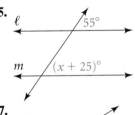

8.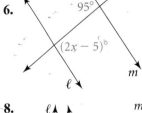

Example 3
(page 137)

9. Drafting An artist uses the drawing tool in the diagram at the right. The artist draws a line, slides the triangle along the flat surface, and draws another line. Explain why the drawn lines must be parallel.

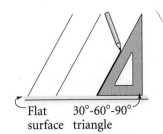

Developing Proof Using the given information, which lines, if any, can you conclude are parallel? Justify each conclusion with a theorem or postulate.

10. ∠2 is supplementary to ∠3.

11. ∠6 is supplementary to ∠7.

12. ∠1 is supplementary to ∠4.

13. $m\angle 7 = 70, m\angle 9 = 110$

14. ∠1 ≅ ∠3

15. ∠9 ≅ ∠12

16. ∠3 ≅ ∠6

17. ∠2 ≅ ∠10

18. ∠1 ≅ ∠8

19. ∠8 ≅ ∠6

20. ∠11 ≅ ∠7

21. ∠5 ≅ ∠10

22. Developing Proof Complete this paragraph proof of Theorem 3-6.

If two lines and a transversal form supplementary same-side interior angles, then the two lines are parallel.

Given: ∠1 and ∠2 are supplementary.

Prove: ℓ ∥ m

Proof: ∠2 is a supplement of **a.** _?_ and ∠3 is a supplement of **b.** _?_ . Since supplements of the same angle are congruent, **c.** _?_ ≅ **d.** _?_ . Since ∠2 and ∠3 are also corresponding angles, ℓ ∥ m by the **e.** _?_ Postulate.

23. Carpentry A T-bevel is a tool used by carpenters to draw congruent angles. By loosening the locking lever, the carpenter can adjust the angle. Explain how the carpenter knows that two lines drawn using the T-bevel are parallel.

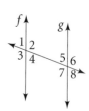

Locking lever

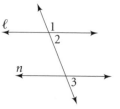

x^2 **Algebra** Find the value of x for which ℓ ∥ m.

24.

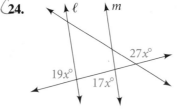

27$x°$

19$x°$ 17$x°$

25.

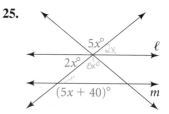

5$x°$ 2$x°$
2$x°$ 8$x°$
$(5x + 40)°$

26. Multiple Choice Lines f and g are parallel. Which statement can you *not* deduce from this information?

(A) ∠1 ≅ ∠8
(B) $m\angle 2 + m\angle 5 = 180$
(C) ∠4 ≅ ∠7
(D) $m\angle 3 + m\angle 8 = 180$

Proof 27. Prove Theorem 3-8.

If two lines and a transversal form same-side exterior angles that are supplementary, then the lines are parallel.

Given: $m\angle 1 + m\angle 3 = 180$

Prove: ℓ ∥ n

Real-World **Connection**

Careers A carpenter must draw angles precisely to ensure good fit.

GO **Online**
Homework Video Tutor
Visit: PHSchool.com
Web Code: aue-0302

x^2 **Algebra** Determine the value of x for which $r \parallel s$. Then find $m\angle 1$ and $m\angle 2$.

28. $m\angle 1 = 80 - x, m\angle 2 = 90 - 2x$

29. $m\angle 1 = 60 - 2x, m\angle 2 = 70 - 4x$

30. $m\angle 1 = 40 - 4x, m\angle 2 = 50 - 8x$

31. $m\angle 1 = 20 - 8x, m\angle 2 = 30 - 16x$

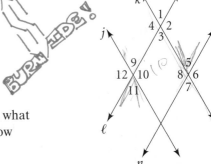

32. Crew If the rowing crew at the left strokes in unison, the oars sweep out angles of equal measure. Explain why the oars on each side of the shell stay parallel.

Open-Ended In each exercise, information is given about the figure below. State another fact about one of the given angles that will guarantee two lines are parallel. Tell which lines will be parallel and why.

33. $\angle 1 \cong \angle 3$

34. $m\angle 8 = 110, m\angle 9 = 70$

35. $\angle 5 \cong \angle 11$

36. $\angle 11$ and $\angle 12$ are supplementary.

37. Reasoning If $\angle 1 \cong \angle 7$ in the diagram, what theorem or postulate can you use to show that $\ell \parallel n$?

Exercise 32

Proof For Exercises 38–39, use the diagram above and write a two-column proof.

38. Given: $\ell \parallel n, \angle 12 \cong \angle 8$
Prove: $j \parallel k$

39. Given: $j \parallel k, m\angle 8 + m\angle 9 = 180$
Prove: $\ell \parallel n$

Proof 40. Rewrite this paragraph proof of Theorem 3-6 as a flow proof.

If two lines and a transversal form supplementary same-side interior angles, then the two lines are parallel.

Given: $\angle 1$ and $\angle 2$ are supplementary.

Prove: $\ell \parallel m$

Proof: It is given that $\angle 1$ and $\angle 2$ are supplementary. $\angle 1$ and $\angle 3$ are also supplementary, so $\angle 2 \cong \angle 3$. Since $\angle 2$ and $\angle 3$ are corresponding angles, $\ell \parallel m$.

C Challenge

Which sides of quadrilateral *PLAN* must be parallel? Explain.

41. $m\angle P = 72, m\angle L = 108, m\angle A = 72, m\angle N = 108$

42. $m\angle P = 59, m\angle L = 37, m\angle A = 143, m\angle N = 121$

43. $m\angle P = 67, m\angle L = 120, m\angle A = 73, m\angle N = 100$

44. $m\angle P = 56, m\angle L = 124, m\angle A = 124, m\angle N = 56$

Proof 45. Prove the following statement is true by following the steps below:

If a transversal intersects two parallel lines, then the bisectors of two corresponding angles are parallel.

a. Draw and label a diagram on paper.

b. State what is given and mark the diagram to keep track of the information.

c. State what you are to prove.

d. Write a plan for proof.

e. Follow your plan and write the proof.

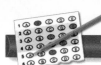
Multiple Choice

46. If *a*, *b*, *c*, and *d* are coplanar lines and *a* ∥ *b*, *b* ⊥ *c*, and *c* ∥ *d*, then which statement must be true?

  A. *d* ⊥ *c*   **B.** *c* ∥ *a*   **C.** *d* ⊥ *a*   **D.** *d* ∥ *b*

Use the diagram for Exercises 47–49.

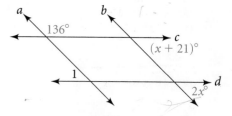

47. For what value of *x* is *c* ∥ *d*?

  F. 21   **G.** 23
  H. 43   **J.** 53

48. If *c* ∥ *d*, what is *m*∠1?

  A. 24   **B.** 44
  C. 136   **D.** 146

Short Response

49. Suppose *a* ∥ *b* in the diagram above.

  a. Write and solve an equation to find the value of *x*.
  b. Write and solve an equation to find whether *c* ∥ *d*. Explain your answer.

Extended Response

50. Two lines, *a* and *b*, are cut by a transversal *t*. ∠1 and ∠2 are any pair of corresponding angles. ∠1 and ∠3 are adjacent angles.
  $m\angle 1 = 2x - 38$, $m\angle 2 = x$, and $m\angle 3 = 6x + 18$.

  a. Draw and label a diagram for the figure described.
  b. Determine whether lines *a* and *b* are parallel. Justify your answer.

Mixed Review

Lesson 3-1

Find *m*∠1, and then *m*∠2. Justify each answer.

51.
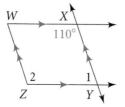

52.
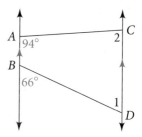

Lesson 2-1

Write the converse of each conditional statement. Determine the truth values of the original conditional and its converse.

53. If you are in Nebraska, you are west of the Mississippi River.

54. If a circle has a diameter of 8 cm, then it has a radius of 4 cm.

55. If a line intersects a pair of parallel lines, then same-side interior angles are supplementary.

56. If you add *ed* to a verb, you form the past tense of a verb.

57. If it is raining, then there are clouds in the sky.

Lesson 1-9

Find the area of each circle. Round to the nearest tenth.

58. *r* = 8 in.   **59.** *d* = 6 cm   **60.** *d* = 9 ft   **61.** *r* = 5 in.

62. *d* = 2.8 m   **63.** *r* = 1.2 m   **64.** *d* = 4.75 ft   **65.** *r* = 0.6 m

3-3

Parallel and Perpendicular Lines

Learning Standards for Mathematics

G.RP.2 Recognize and verify, where appropriate, geometric relationships of perpendicularity, parallelism, congruence, and similarity, using algebraic strategies.

 **Check Skills You'll Need**

 **for Help** Lesson 1-4

Complete each statement with *always*, *sometimes*, or *never*.

1. Two lines in the same plane are ___?___ parallel.

2. Perpendicular lines ___?___ meet at right angles.

3. Two lines in intersecting planes are ___?___ perpendicular.

4. Two lines in parallel planes are ___?___ perpendicular.

1   Relating Parallel and Perpendicular Lines

The two diagrams suggest ways to draw parallel lines. You can draw them
(a) parallel to a given line, or
(b) perpendicular to a given line. Theorems 3-9 and 3-10 guarantee that the lines you draw are indeed parallel.

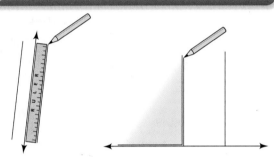

 Key Concepts

Theorem 3-9

If two lines are parallel to the same line, then they are parallel to each other.

$a \parallel b$

Theorem 3-10

In a plane, if two lines are perpendicular to the same line, then they are parallel to each other.

$m \parallel n$

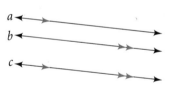

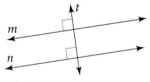

Theorem 3-10 includes the phrase *in a plane*. On the other hand, Theorem 3-9 is true for any three such lines, whether they are coplanar (Exercise 3) or noncoplanar.

Proof **Proof of Theorem 3-10**

Study what is given, what you are to prove, and the diagram. Then write a paragraph proof.

Given: $r \perp t, s \perp t$

Prove: $r \parallel s$

Proof: $\angle 1$ and $\angle 2$ are right angles by the definition of perpendicular, so they are congruent. Since corresponding angles are congruent, $r \parallel s$.

1 EXAMPLE <u>Real-World</u> 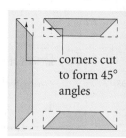 **Connection**

Woodworking To make a frame for a painting, a miter box and a backsaw are used to cut the framing at 45° angles. Explain why cutting the framing at this angle ensures that opposite sides of the frame will be parallel.

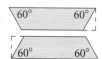

corners cut to form 45° angles

Two adjacent 45° angles form a 90° angle. The opposite sides of the frame are perpendicular to the same side. Thus the opposite sides are parallel because two lines perpendicular to a third line are parallel.

 Quick Check **1** Can you assemble the framing at the right into a frame with opposite sides parallel? Explain.

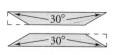

Theorems 3-9 and 3-10 gave conditions by which you can conclude that lines are parallel. Theorem 3-11 provides a way for you to conclude that lines are perpendicular. You will prove Theorem 3-11 in Exercise 11.

Key Concepts

> **Theorem 3-11**
>
> In a plane, if a line is perpendicular to one of two parallel lines, then it is also perpendicular to the other.
>
> $n \perp m$

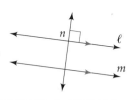

Proof **2 EXAMPLE** **Using Theorem 3-11**

Study what is given, what you are to prove, and the diagram. Then write a paragraph proof.

Given: In a plane, $a \perp b$, $b \perp c$, and $c \perp d$.

Prove: $a \perp d$

Proof: Lines a and c are both perpendicular to line b, so $a \parallel c$ because two lines perpendicular to the same line are parallel. It is given that $c \perp d$. Therefore, $a \perp d$ because a line that is perpendicular to one of two parallel lines is perpendicular to the other (Theorem 3-11).

 Quick Check **2** From what is given in Example 2, can you also conclude $b \parallel d$? Explain.

In the rectangular solid shown here, $\overleftrightarrow{AC}$ and $\overleftrightarrow{BD}$ are parallel. $\overleftrightarrow{EC}$ is perpendicular to $\overleftrightarrow{AC}$, but it is not perpendicular to $\overleftrightarrow{BD}$. This is why Theorem 3-11 states that the lines must be "in a plane."

EXERCISES

For more exercises, see *Extra Skill, Word Problem, and Proof Practice.*

Practice and Problem Solving

 Practice by Example

Example 1
(page 142)

GO for Help

1. A carpenter is building a trellis for vines to grow on. The completed trellis will have two sets of overlapping diagonal slats of wood.
  a. What must be true of $\angle 1$, $\angle 2$, and $\angle 3$ if slats A, B, and C must be parallel?
  b. The carpenter attaches slat D so that it is perpendicular to slat *A*. Is slat D perpendicular to slats B and C? Justify your answer.

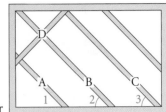

Example 2
(page 142)

Proof **2.** Study what is given, what you are to prove, and the diagram. Then write a proof.

  Given: In a plane, $a \perp b$, $b \perp c$, and $c \parallel d$.

  Prove: $a \parallel d$

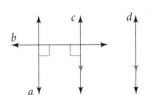

B **Apply Your Skills**

GO for Help

For a guide to solving
Exercise 3, see p. 145.

3. Developing Proof Copy and complete this paragraph proof of Theorem 3-9 for three coplanar lines.

  If two lines are parallel to the same line, then they are parallel to each other.

  Given: $\ell \parallel k$ and $m \parallel k$

  Prove: $\ell \parallel m$

  Proof: $\ell \parallel k$ means that $\angle 2 \cong \angle 1$ by the **a.** _?_ Postulate. $m \parallel k$ means that **b.** _?_ $\cong$ **c.** _?_ for the same reason. By the Transitive Property of Congruence, $\angle 2 \cong \angle 3$. By the **d.** _?_ Postulate, $\ell \parallel m$.

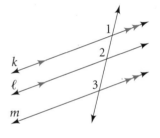

Each of the following statements describes a ladder. What can you conclude about the rungs, one side, or both sides of each ladder? Explain.

4. The rungs are each perpendicular to one side.

5. The rungs are parallel and the top rung is perpendicular to one side.

6. The sides are parallel. The rungs are perpendicular to one side.

7. The rungs are perpendicular to one side. The other side is perpendicular to the top rung.

8. Each side is perpendicular to the top rung.

9. Each rung is parallel to the top rung.

10. The rungs are perpendicular to one side. The sides are not parallel.

Real-World **Connection**

This ladder's rungs are perpendicular to each side. Therefore, the rungs are parallel to each other.

Proof **11.** Prove Theorem 3-11: In a plane, if a line is perpendicular to one of two parallel lines, then it is also perpendicular to the other.

  Given: In a plane, $a \perp b$, and $b \parallel c$.

  Prove: $a \perp c$

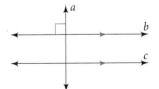

Proof **12.** Prove: If a line is perpendicular to each of two other lines, all in one plane, then the two other lines are parallel.

  Given: $t \perp r$, $t \perp s$

  Prove: $r \parallel s$

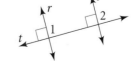

13. Writing Theorem 3-10: In a plane, two lines perpendicular to the same line are parallel. Use the rectangular solid at the right to explain why the words *in a plane* are needed.

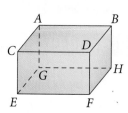

C **Challenge**   *a*, *b*, *c*, and *d* are distinct lines in the same plane. Exercises 14-21 show different combinations of relationships between *a* and *b*, *b* and *c*, and *c* and *d*. For each combination of the three relationships, how are *a* and *d* related?

14. $a \parallel b, b \parallel c, c \parallel d$ **15.** $a \parallel b, b \parallel c, c \perp d$

16. $a \parallel b, b \perp c, c \parallel d$ **17.** $a \perp b, b \parallel c, c \parallel d$

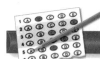

Homework Video Tutor
Visit: PHSchool.com
Web Code: aue-0303

18. $a \parallel b, b \perp c, c \perp d$ **19.** $a \perp b, b \parallel c, c \perp d$

20. $a \perp b, b \perp c, c \parallel d$ **21.** $a \perp b, b \perp c, c \perp d$

Critical Thinking The Reflexive, Symmetric, and Transitive Properties for Congruence (≅) are listed on page 105.

22. Write reflexive, symmetric, and transitive statements for "is parallel to" ($\parallel$). State whether each statement is true or false and justify your answer.

23. Repeat Exercise 22 for "is perpendicular to" ($\perp$).

Test Prep

Multiple Choice   **24.** In a plane, line *e* is parallel to line *f*, line *f* is parallel to line *g*, and line *h* is perpendicular to line *e*. Which of the following MUST be true?
  A. $e \parallel g$ **B.** $h \parallel f$ **C.** $g \parallel h$ **D.** $e \parallel h$

25. In a plane, $\overleftrightarrow{AB}$ is parallel to $\overleftrightarrow{CD}$. $\angle ABC$ is a right angle. What can you conclude?
  I. $\overleftrightarrow{AB} \perp \overleftrightarrow{BC}$ II. $\overleftrightarrow{BC} \perp \overleftrightarrow{CD}$ III. $\overleftrightarrow{BC} \perp \overrightarrow{AD}$
  F. I only   **G.** I and II   **H.** II only   **J.** I, II and III

Mixed Review

Lesson 3-2 $\boxed{x^2}$ **Algebra** Determine the value of *x* for which $a \parallel b$.

26.

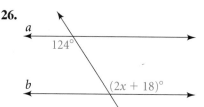

27.

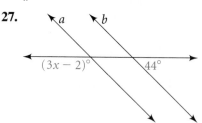

Lesson 2-2   **Each conditional statement below is true. Write its converse. If the converse is also true, combine the statements into a biconditional.**

28. If $x = 7$, then $x^2 = 49$.

29. If two lines in a plane do not meet, then the lines are parallel.

Understanding Proof Problems Read the problem below. Then let Collin's thinking guide you through the solution. Check your understanding with the exercise at the bottom of the page.

Copy and complete the paragraph proof of Theorem 3-9 for three coplanar lines.

If two lines are parallel to the same line, then they are parallel to each other.

Given: $\ell \parallel k$ and $m \parallel k$

Prove: $\ell \parallel m$

Proof: $\ell \parallel k$ means that $\angle 2 \cong \angle 1$ by the **a.** ? Postulate. $m \parallel k$ means that **b.** ? $\cong$ **c.** ? for the same reason. By the Transitive Property of Congruence, $\angle 2 \cong \angle 3$. By the **d.** ? Postulate, $\ell \parallel m$.

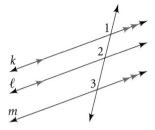

What Collin Thinks

Let's see. $\angle 1$ and $\angle 2$ are corresponding angles. If $\ell \parallel k$, then $\angle 1 \cong \angle 2$ by Postulate 3-1 on page 128.

Now I will use the fact that $m \parallel k$. So I'll ignore line ℓ. That leaves me with $\angle 1$ and $\angle 3$ to consider. These also are corresponding angles.

Now, $\angle 2 \cong \angle 3$ and they are corresponding angles. This fits the postulate in which the hypothesis is "corresponding angles are congruent." By Postulate 3-2 on page 134, I can conclude "the lines are parallel."

What Collin Writes

$\ell \parallel k$ means that $\angle 2 \cong \angle 1$ by the **a. Corresponding Angles Postulate.**

$m \parallel k$ means that **b.** $\underline{\angle 1} \cong$ **c.** $\underline{\angle 3}$ for the same reason.

By the Transitive Property of Congruence, $\angle 2 \cong \angle 3$. By the **d. Converse of the Corresponding Angles Postulate**, $\ell \parallel m$.

EXERCISE

Copy and complete this paragraph proof of Theorem 3-9 for three coplanar lines.

If two lines are parallel to the same line, then they are parallel to each other.

Given: $q \parallel p$ and $p \parallel r$

Prove: $q \parallel r$

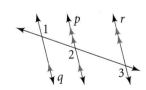

Proof: $q \parallel p$ means that $\angle 1 \cong \angle 2$ by the **a.** ? Theorem. $p \parallel r$ means that **b.** ? $\cong$ **c.** ? by the Corresponding Angles Postulate. By the Transitive Property of Congruence, $\angle 1 \cong \angle 3$. By the **d.** ? Theorem, $q \parallel r$.

Angle Dynamics

Activity 1 suggests one of the most important ideas of Euclidean geometry.

NY G.G.36:
Investigate, justify, and apply theorems about the sum of the measures of the interior and exterior angles of polygons.

1 ACTIVITY

- Draw and cut out a large triangle.

- Number the angles and tear them off.

- Place the three angles adjacent to each other to form one angle, as shown at the right.

1. What kind of angle is formed by the three smaller angles? What is its measure?

2. Make a conjecture about the sum of the measures of the angles of a triangle.

2 ACTIVITY

- Fold a sheet of paper in half three times. Draw a *scalene triangle* (no two sides congruent) on the folded paper. Carefully cut out the triangle. This will give you eight triangles that are all the same size and shape.

- Number the angles of each triangle 1, 2, and 3. Use the same number on the corresponding angles from one triangle to the next.

3. Mark a point *P* in the middle of a blank sheet of paper. How many of your triangles do you think you can fit perfectly about point *P* without gaps or overlaps?

4. What conjecture are you making about the sum of the measures of the angles about a point?

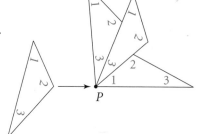

EXERCISES

5. Use the triangles from Activity 2. Based on your conjectures above, what combination of angles 1, 2, and 3 can you place at a point *P* to guarantee a perfect fit of triangles about *P*?

6. There are several ways to arrange the set of six angles 1, 1, 2, 2, 3, 3 to make a perfect fit about *P*. Make three different listings of the numbers 1, 1, 2, 2, 3, 3 around a point. Arrange your triangles about *P* to match your lists. Is any arrangement more appealing than the other two? What happens if your turn over some of the triangles?

7. Follow the steps of Activity 1 using a *quadrilateral* (four-sided figure), such as the one shown at the right. Make a conjecture about the sum of the measures of the angles of a quadrilateral.

8. If you fit the angles of a quadrilateral about a point *P* without gaps or overlaps, what do you think you would discover? Justify your response.

Parallel Lines and the Triangle Angle-Sum Theorem

3-4

 Learning Standards for Mathematics

G.G.30 Investigate, justify, and apply theorems about the sum of the measures of the angles of a triangle.

G.G.32 Investigate, justify, and apply theorems about geometric inequalities, using the exterior angle theorem.

G.G.36 Investigate, justify, and apply theorems about the sum of the measures of the interior and exterior angles of polygons.

✓ Check Skills You'll Need

 for Help Lesson 1-6

Classify each angle as *acute*, *right*, or *obtuse*.

1.    2.    3.

Solve each equation.

4. $30 + 90 + x = 180$ **5.** $55 + x + 105 = 180$

6. $x + 58 = 90$ **7.** $32 + x = 90$

🔊 **New Vocabulary**   • acute triangle   • right triangle   • obtuse triangle
• equiangular triangle   • equilateral triangle
• isosceles triangle   • scalene triangle
• exterior angle of a polygon   • remote interior angles

---

1 | Finding Angle Measures in Triangles

The diagrams at the top of the Activity Lab on page 146 suggest the Triangle-Angle Sum Theorem.

 Key Concepts

| **Theorem 3-12** | 🔑 **Triangle Angle-Sum Theorem** |

The sum of the measures of the angles of a triangle is 180.

$$m\angle A + m\angle B + m\angle C = 180$$

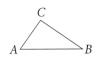

The following proof of Theorem 3-12 relies on the idea that through a point not on a given line you can draw a line parallel to the given line.

Proof **Proof of Theorem 3-12**

Given: $\triangle ABC$

Prove: $m\angle A + m\angle B + m\angle 3 = 180$

Proof: By the Protractor Postulate, you can draw $\overrightarrow{CP}$ so that $m\angle 1 = m\angle A$. Then, $\angle 1$ and $\angle A$ are congruent alternate interior angles, so $\overleftrightarrow{CP} \parallel \overline{AB}$. $\angle 2$ and $\angle B$ are also alternate interior angles, so by the Alternate Interior Angles Theorem, $m\angle 2 = m\angle B$. By substitution, $m\angle A + m\angle B + m\angle 3 = m\angle 1 + m\angle 2 + m\angle 3$, which is equal to 180 by the Angle Addition Postulate.

1 EXAMPLE Applying the Triangle Angle-Sum Theorem

Algebra Find the values of x and y.

To find the value of x, use $\triangle GFJ$.

$39 + 65 + x = 180$	**Triangle Angle-Sum Theorem**
$104 + x = 180$	**Simplify.**
$x = 76$	**Subtract 104 from each side.**

To find the value of y, look at $\angle FJH$. It is a straight angle.

$m\triangle GJF + m\angle GJH = 180$	**Angle Addition Postulate**
$x + y = 180$	**Substitute.**
$76 + y = 180$	**Substitute 76 for x.**
$y = 104$	**Subtract 76 from each side.**

✓ **Quick Check** ❶ Find the value of z in two different ways, each way using the Triangle Angle-Sum Theorem.

In Chapter 1, you classified an angle by its measure. You can also classify a triangle by its angles and sides.

Equiangular
all angles congruent

Acute
all angles acute

Right
one right angle

Obtuse
one obtuse angle

Equilateral
all sides congruent

Isosceles
at least two sides congruent

Scalene
no sides congruent

2 EXAMPLE Classifying a Triangle

Classify the triangle by its sides and its angles.

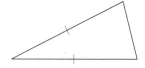

At least two sides are congruent, so the triangle is isosceles. All the angles are acute, so the triangle is acute.

● The triangle is an acute isosceles triangle.

✓ **Quick Check** ❷ Draw and mark a triangle to fit each description. If no triangle can be drawn, write *not possible* and explain why.
  a. acute scalene   **b.** isosceles right   **c.** obtuse equiangular

2 Using Exterior Angles of Triangles

Interior <u>angle of a</u> <u>triangle</u> means the same as "angle of a triangle."

An **exterior angle of a polygon** is an angle formed by a side and an extension of an adjacent side. For each exterior angle of a triangle, the two nonadjacent interior angles are its **remote interior angles.**

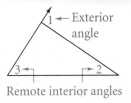
1 ← Exterior angle

Remote interior angles

The diagram at the right suggests a relationship between an exterior angle and its two remote interior angles. Theorem 3-13 states this relationship. You will prove this theorem in Exercise 35.

 Key Concepts

Theorem 3-13	**Triangle Exterior Angle Theorem**

The measure of each exterior angle of a triangle equals the sum of the measures of its two remote interior angles.

$$m\angle 1 = m\angle 2 + m\angle 3$$

Online
active math

For: Triangle Theorems Activity
Use: Interactive Textbook, 3-4

3 EXAMPLE Using the Exterior Angle Theorem

Algebra Find each missing angle measure.

a.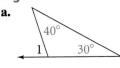
40°
1   30°

$$m\angle 1 = 40 + 30$$
$$m\angle 1 = 70$$

b.
113°
70°
2

$$113 = 70 + m\angle 2$$
$$43 = m\angle 2$$

✓ **Quick Check** **3** Two angles of a triangle measure 45. Find the measure of an exterior angle at each vertex.

4 EXAMPLE Real-World Connection

Test-Taking Tip

To check exterior angle-measure answers, remember that an exterior angle and the adjacent interior angle must be supplementary.

Multiple Choice The lounge chair has different settings that change the angles formed by its parts. Suppose $m\angle 2$ is 32 and $m\angle 3$ is 81. Find $m\angle 1$, the angle formed by the back of the chair and the arm rest.

(A) 67   (B) 81
(C) 113   (D) 180

$$m\angle 1 = m\angle 2 + m\angle 3 \quad \text{**Exterior Angle Theorem**}$$
$$m\angle 1 = 32 + 81 \quad \text{**Substitute.**}$$
$$m\angle 1 = 113 \quad \text{**Simplify.**}$$

The angle formed is a 113° angle.
● The correct choice is C.

✓ **Quick Check** ❹ **a.** Change the setting on the lounge chair so that $m\angle2 = 33$ and $m\angle3 = 97$. Find the new measure of $\angle1$.

b. Explain how you can find $m\angle1$ *without* using the Exterior Angle Theorem.

EXERCISES

For more exercises, see *Extra Skill, Word Problem, and Proof Practice.*

Practice and Problem Solving

A **Practice by Example**

Find $m\angle1$.

Example 1
(page 148)

GO for Help

1.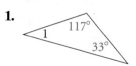

2.

3.

x^2 **Algebra** Find the value of each variable.

4.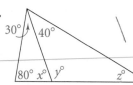

5.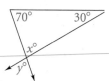

6.

Example 2
(page 148)

Use a protractor and a centimeter ruler to measure the angles and the sides of each triangle. Classify each triangle by its angles and sides.

7.

8.

9.

If possible, draw a triangle to fit each description. Mark the triangle to show known information. If no triangle can be drawn, write *not possible* and explain why.

10. acute equilateral

11. equilateral right

12. obtuse scalene

13. obtuse isosceles

14. isosceles right

15. scalene acute

Example 3
(page 149)

16. a. Which of the numbered angles at the right are exterior angles?

b. Name the remote interior angles for each.

c. How are exterior angles 6 and 8 related?

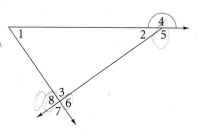

17. a. How many exterior angles at the right are at each vertex of the triangle?

b. How many exterior angles does a triangle have in all?

x^2 **Algebra** Find each missing angle measure.

18.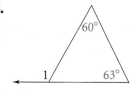

19.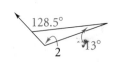

20.

Example 4
(page 149)

21. Music The lid of a grand piano is held open by a prop stick whose length can vary, depending upon the effect desired. The longest prop stick makes angles as shown. What are the values of *x* and *y*?

22. A short prop stick makes the angles shown below. What are the values of *a* and *b*?

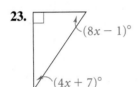

57°

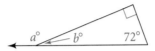

B **Apply Your Skills** x^2 **Algebra** Find the values of the variables and then the measures of the angles. Classify each triangle by its angles. Note that some figures have more than one triangle.

23.
$(8x - 1)°$
$(4x + 7)°$

24.
$(2x + 4)°$
$(2x - 9)°$
$x°$

25.
B
$y°$ $x°$
$54°$ $z°$ $52°$
A D C

26.
E $d°$ $e°$ F
$32°$
$b°$
$c°$ $55°$
H $a°$
G

27. Reasoning What is the measure of each angle of an equiangular triangle? Explain.

28. Writing Is every equilateral triangle isosceles? Is every isosceles triangle equilateral? Explain.

29. Visualization The diagram shows a triangle on a 3-by-3 geoboard. How many triangles with different shapes can be made on this geoboard? Classify each triangle by its sides and angles.

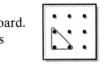

30. Multiple Choice The measure of one angle of a triangle is 115. The other two angles are congruent. What is the measure of each?
Ⓐ 32.5 Ⓑ 65 Ⓒ 57.5 Ⓓ 115

Problem Solving Hint

In Exercise 31, use *x* and 2*x* for the angle measures. In Exercise 32, use 2*x*, 3*x*, and 4*x*.

x^2 **31. Algebra** A right triangle has acute angles whose measures are in the ratio 1 : 2. Find the measures of these angles.

x^2 **32. a. Algebra** The ratio of the angle measures in $\triangle BCR$ is 2 : 3 : 4. Find the angle measures.
b. What type of triangle is $\triangle BCR$?

33. Draw any triangle. Label it △*ABC*. Extend both sides of the triangle to form two exterior angles at vertex *A*. Use the two exterior angles to explain why it does not matter which side of a triangle is extended to form an exterior angle.

Proof **34.** Prove the following theorem.

The acute angles of a right triangle are complementary.

Given: △*ABC* with right angle *C*

Prove: ∠*A* and ∠*B* are complementary.

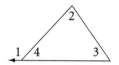

Proof **35.** Prove the Triangle Exterior Angle Theorem.

Given: ∠1 is an exterior angle of the triangle.

Prove: $m\angle 1 = m\angle 2 + m\angle 3$

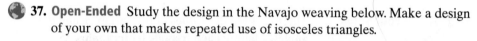

36. Reasoning Two angles of a triangle measure 64 and 48. Find the measure of the largest exterior angle. Explain.

37. Open-Ended Study the design in the Navajo weaving below. Make a design of your own that makes repeated use of isosceles triangles.

38. The measures of the angles of △*RST* are $5\sqrt{x}, 7\sqrt{x}$, and $8\sqrt{x}$.
 a. Find the value of *x*.
 b. Give the measure of each angle.
 c. What type of triangle is △*RST*?

C Challenge

39. Reasoning Sketch a triangle and two exterior angles that have a side of the triangle in common. For what type of triangle, if any, is each statement true? Justify each answer.
 a. The bisectors of the two exterior angles are parallel.
 b. The bisectors of the two exterior angles are perpendicular.
 c. The bisectors of the two exterior angles and the common side of the given triangle form an isosceles triangle.

40. In the figure at the right, $\overline{CD} \perp \overline{AB}$ and $\overline{CD}$ bisects ∠*ACB*. Find $m\angle DBF$.

41. What can you conclude about the bisector of an exterior angle of a triangle if the remote interior angles are congruent? Justify your response.

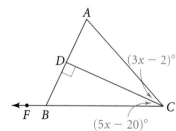

Multiple Choice   Use the diagram at the right for Exercises 42–44.

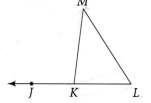

42. $m\angle M = 25$ and $m\angle L = 43$. What is $m\angle JKM$?

    F. 18   **G.** 68   **H.** 117   **J.** 162

43. $m\angle M = 4x$, $m\angle L = 5x$, and $m\angle MKL = 6x$. What is $m\angle JKM$?

    A. 72   **B.** 108   **C.** 120   **D.** 132

44. $m\angle JKM = 15x - 48$, $m\angle L = 5x + 12$, and $m\angle M = 40$. What is $m\angle MKL$?

    F. 9   **G.** 57   **H.** 78   **J.** 97

Mixed Review

Lesson 3-3

Use the diagram at the right for Exercises 45–46.

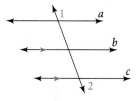

45. If $\angle 1$ and $\angle 2$ are supplementary, what can you conclude about lines a and c? Justify your answer.

46. If $a \parallel c$, what can you conclude about lines a and b? Justify your answer.

Lesson 1-6 $\boxed{x^2}$ **47. Algebra** In the figure at the right, $m\angle AOB = 3x + 20, m\angle BOC = x + 32,$ and $m\angle AOC = 80$. Find the value of x.

Lesson 1-1   **Draw the next figure in each sequence.**

48.

49.

Checkpoint Quiz 1   **Lessons 3-1 through 3-4**

Use the diagram at the right for Exercises 1–9. State the theorem or postulate that justifies each statement.

1. $\angle 1 \cong \angle 3$ **2.** If $\angle 5 \cong \angle 9$, then $d \parallel e$.

3. $m\angle 1 + m\angle 2 = 180$ **4.** If $\angle 4 \cong \angle 7$, then $d \parallel e$.

5. $\angle 1 \cong \angle 4$ **6.** $\angle 7 \cong \angle 9$

7. If $\angle 3 \cong \angle 9$, then $d \parallel e$.   **8.** $\angle 4 \cong \angle 5$

9. If $e \perp b$, then $e \perp c$.

10. Find the measures of the angles of each triangle. Classify each triangle by its angles.

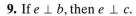
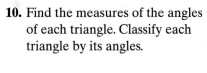

Exploring Spherical Geometry

FOR USE WITH LESSON 3-4

Euclidean geometry is the basis for high school geometry courses. Euclidean geometry is the geometry of flat planes, straight lines, and points. In spherical geometry a "plane" is the curved surface of a sphere and a "line" is a great circle. (A *great circle* is the intersection of a sphere and a plane that contains the center of the sphere.)

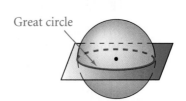

Great circle

1 ACTIVITY

Lines of latitude and longitude are used to identify positions on Earth. Find these lines on a globe. Which of these lines are great circles?

All lines of longitude are great circles. The equator is the only line of latitude that is a great circle. All other lines of latitude are circles smaller than a great circle.

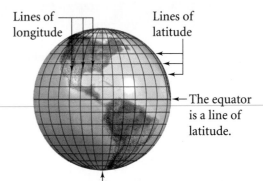

Lines of longitude

Lines of latitude

The equator is a line of latitude.

The lines of longitude all pass through the North and South Poles.

In Euclidean geometry,

Through a point not on a line, there is one and only one line parallel to the given line.

This statement is sometimes called Euclid's Parallel Postulate. Since only great circles are lines in spherical geometry, two lines always intersect. In spherical geometry, the Parallel Postulate is quite different:

Through a point not on a line, there is no line parallel to the given line.

2 ACTIVITY

The diagram at the right shows that any two lines on a sphere intersect at *two* points. Locate the points of intersection of lines of longitude on Earth. What is special about these points?

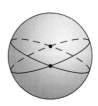

Lines of longitude intersect at the North and South Poles. The poles are on a line that passes through the center of Earth. Thus, the poles lie on Earth's axis and are the endpoints of a diameter of Earth.

One result of Euclid's Parallel Postulate is the Triangle Angle-Sum Theorem of Lesson 3-3. Something quite different happens in spherical geometry as a result of the spherical-geometry Parallel Postulate.

Hold a string taut between any two points on a sphere. The string forms an arc that is part of a great circle. Make three such arcs to form a triangle on the sphere.

Three such triangles are shown below. Find the sum of the measures of the angles of each of the triangles.

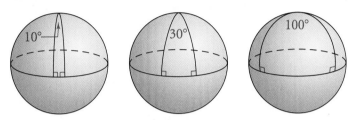

In the first triangle, the sum of the angle measures is 190. In the second triangle, it is 210, and in the third triangle the sum is 280.

EXERCISES

Draw a sketch to illustrate each property of spherical geometry. How does each property compare to what is true in Euclidean geometry?

1. There are pairs of points on a sphere through which more than one line can be drawn.

2. A triangle can have more than one right angle.

3. You can draw two equiangular triangles such that they have different angle measures.

In Exercises 4 and 5, draw a counterexample to show that each of these properties of Euclidean geometry is *not* true in spherical geometry.

4. Two lines that are perpendicular to the same line do not intersect.

5. If two angles of one triangle are congruent to two angles of another triangle, then the third angles are congruent.

6. The figure at the right appears to show parallel lines on a sphere. Explain why this is not so.

7. Explain why a piece of the top circle in the figure is *not* a line segment. (*Hint:* What must be true of line segments in spherical geometry?)

Each of the following statements is true in Euclidean geometry. Does it seem to be true in spherical geometry? Make figures on a globe, ball, or balloon to support your answer.

8. Vertical angles are congruent.

9. Through a point on a line ℓ there exists one and only one line perpendicular to ℓ.

Activity Lab
Exterior Angles of Polygons

FOR USE WITH LESSON 3-5

Construct

Use geometry software. Construct a polygon similar to the one at the right. Extend each side as shown. To measure the exterior angles you will need to mark a point on each ray.

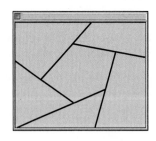

NY G.G.36: Investigate, justify, and apply theorems about the sum of the measures of the interior and exterior angles of polygons.

Investigate

- Measure each exterior angle.
- Calculate the sum of the measures of the exterior angles.
- Manipulate the polygon. Observe the sum of the measures of the exterior angles.

EXERCISES

1. Write a conjecture about the sum of the measures of the exterior angles (one at each vertex) of a convex polygon.

2. Test your conjecture with another polygon.

Extend

3. The figures below show a polygon that is decreasing in size until finally it becomes a point. Describe how you could use this to justify your conjecture in Exercise 1.

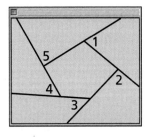 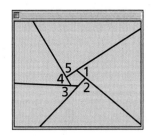 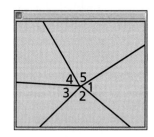

4. The figure at the right shows a square that has been copied several times. Notice that you can use the square to completely cover, or tile, a plane, without gaps or overlaps.

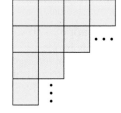

   a. Using geometry software, create several copies of other regular polygons with 3, 5, 6, and 8 sides. Regular polygons have sides of equal length and angles of equal measure.
   b. Which of the polygons you created can tile a plane?
   c. Measure *one* exterior angle of each polygon (including the square).
   d. Write a conjecture about the relationship between the measure of an exterior angle and your ability to tile a plane with a regular polygon.
   e. Test your conjecture with another regular polygon.

5. You can use a square—a rectangle even—to tile a plane. What other four-sided shapes can you use to tile a plane? (*Hint:* See Activity Lab, page 146.)

The Polygon Angle-Sum Theorems

3-5

 Learning Standards for Mathematics

G.G.32 Investigate, justify, and apply theorems about geometric inequalities, using the exterior angle theorem.

G.G.36 Investigate, justify, and apply theorems about the sum of the measures of the interior and exterior angles of polygons.

 **Check Skills You'll Need**   **GO for Help** Lessons 1-6 and 3-4

Find the measure of each angle of quadrilateral *ABCD*.

1. 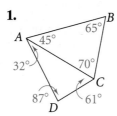   **2.** 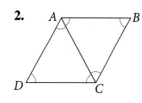   **3.**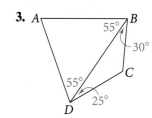

◀)) **New Vocabulary**
- polygon
- convex polygon
- concave polygon
- equilateral polygon
- equiangular polygon
- regular polygon

---

1 **Classifying Polygons**

Real-World Connection

Polygons create striking designs on a soccer ball.

A **polygon** is a closed plane figure with at least three sides that are segments. The sides intersect only at their endpoints, and no adjacent sides are collinear.

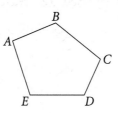   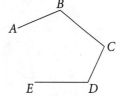   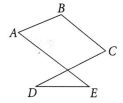

A polygon   Not a polygon;   Not a polygon;
                   not a closed figure   two sides intersect
                                         between endpoints.

To name a polygon, start at any vertex and list the vertices consecutively in a clockwise or counterclockwise direction.

1 **EXAMPLE** **Naming Polygons**

Name the polygon. Then identify its vertices, sides, and angles.

Two names for this polygon are *DHKMGB* and *MKHDBG*.

vertices: *D, H, K, M, G, B*

sides: $\overline{DH}, \overline{HK}, \overline{KM}, \overline{MG}, \overline{GB}, \overline{BD}$

● angles: ∠*D*, ∠*H*, ∠*K*, ∠*M*, ∠*G*, ∠*B*

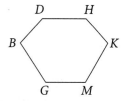

✓ **Quick Check** ❶ Three polygons are pictured at the right. Name each polygon, its sides, and its angles.

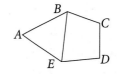

You can classify a polygon by the number of sides it has. The table at the right shows the names of some common polygons.

Polygons are classified as convex or concave.

Sides	Name
3	triangle
4	quadrilateral
5	pentagon
6	hexagon
8	octagon
9	nonagon
10	decagon
12	dodecagon
n	n-gon

A **convex polygon** has no diagonal with points outside the polygon.

A **concave polygon** has at least one diagonal with points outside the polygon.

In this textbook, a polygon is convex unless stated otherwise.

2 EXAMPLE **Real-World Connection**

Tilework The tilework in the photo is a combination of different polygons that form a pleasing pattern. Classify the polygon outlined in red by using the table above. Then classify the polygon as convex or concave.

The polygon outlined in red has 6 sides. Therefore, it is a hexagon.

No diagonal of the hexagon contains points outside the hexagon. The hexagon is convex.

Quick Check ❷ Classify each polygon by its sides. Identify each as convex or concave.

a.

b.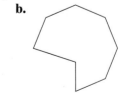

c. the 12-pointed star at the center of the tilework pictured above

Activity: The Sum of Polygon Angle Measures

You can use triangles and the Triangle Angle-Sum Theorem to find the sum of the measures of the angles of a polygon. Record your data in a table like the one begun below.

Polygon	Number of Sides	Number of Triangles Formed	Sum of the Interior Angle Measures
	4	■	■ • 180 = ■

- Sketch polygons with 4, 5, 6, 7, and 8 sides.

- Divide each polygon into triangles by drawing all diagonals that are possible from one vertex.

- Multiply the number of triangles by 180 to find the sum of the measures of the angles of each polygon.

1. Look for patterns in the table. Describe any that you find.

2. **Inductive Reasoning** Write a rule for the sum of the measures of the angles of an n-gon.

Vocabulary Tip

An *n*-gon is a polygon with n sides, where n can be 3, 4, 5, 6, . . .

By dividing a polygon with n sides into $n - 2$ triangles, you can show that the sum of the measures of the angles of any polygon is a multiple of 180.

 Key Concepts

Theorem 3-14	Polygon Angle-Sum Theorem

The sum of the measures of the angles of an n-gon is $(n - 2)180$.

 GO ●**nline**

Video Tutor Help
Visit: PHSchool.com
Web Code: aue-0775

3 **EXAMPLE** **Finding a Polygon Angle Sum**

Find the sum of the measures of the angles of a 15-gon.

For a 15-gon, $n = 15$.

Sum $= (n - 2)180$ **Polygon Angle-Sum Theorem**
     $= (15 - 2)180$ **Substitute.**
     $= 13 \cdot 180$ **Simplify.**
     $= 2340$

● The sum of the measures of the angles of a 15-gon is 2340.

✓ **Quick Check** ❸ **a.** Find the sum of the measures of the angles of a 13-gon.
     b. Critical Thinking The sum of the measures of the angles of a given polygon is 720. How can you use Sum $= (n - 2)180$ to find the number of sides in the polygon?

You will sometimes use algebra with the Polygon Angle-Sum Theorem to find measures of polygon angles.

4 **EXAMPLE**   **Using the Polygon Angle-Sum Theorem**

Algebra Find $m\angle Y$ in pentagon $TVYMR$ at the right.
Use the Polygon Angle-Sum Theorem for $n = 5$.

$$m\angle T + m\angle V + m\angle Y + m\angle M + m\angle R = (5 - 2)180$$
$$90 + 90 + m\angle Y + 90 + 135 = 540 \quad \textbf{Substitute.}$$
$$m\angle Y + 405 = 540 \quad \textbf{Simplify.}$$
$$m\angle Y = 135 \quad \textbf{Subtract 405 from each side.}$$

✓ **Quick Check** ❹ Pentagon $ABCDE$ has 5 congruent angles. Find the measure of each angle.

You can draw exterior angles at any vertex of a polygon. The figures below show that the sum of the measures of the exterior angles, one at each vertex, is 360. This can be proved as a theorem in a way suggested in Exercise 46.

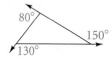

$80 + 150 + 130 = 360$ $115 + 75 + 99 + 71 = 360$ $86 + 59 + 98 + 41 + 76 = 360$

🔑 **Key Concepts**

Theorem 3-15	**Polygon Exterior Angle-Sum Theorem**

The sum of the measures of the exterior angles of a polygon, one at each vertex, is 360.

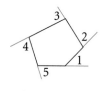

For the pentagon,
$$m\angle 1 + m\angle 2 + m\angle 3 + m\angle 4 + m\angle 5 = 360.$$

An **equilateral polygon** has all sides congruent. An **equiangular polygon** has all angles congruent. A **regular polygon** is both equilateral and equiangular.

5 **EXAMPLE**   **Real-World 🌐 Connection**

⬤nline active math

For: Regular Polygon Activity
Use: Interactive Textbook, 3-5

Packaging The game board at the right has the shape of a regular hexagon. It is packaged in a rectangular box outlined beneath it. The box uses four right triangles made of foam in its four corners. Find $m\angle 1$ in each foam triangle.

Find the measure of an angle of the hexagon first.

- A regular hexagon has 6 sides and 6 congruent angles. The sum of the measures of the interior angles $= (6 - 2)180$, or 720.

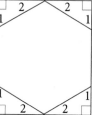

- The measure of one interior angle is $\frac{720}{6}$, or 120.

- The measure of its adjacent exterior angle, $\angle 1$, is $180 - 120$, or 60.

✓ **Quick Check** ❺ a. Find $m\angle 1$ by using the Polygon Exterior Angle-Sum Theorem.
   b. Find $m\angle 2$. Is $\angle 2$ an exterior angle? Explain.

EXERCISES

For more exercises, see *Extra Skill, Word Problem, and Proof Practice.*

Practice and Problem Solving

 Practice by Example

Example 1
(page 157)

GO for Help

Is the figure a polygon? If not, tell why.

1.   **2.** 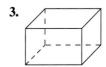   **3.**   **4.**

Name each polygon by its vertices. Then identify its sides and angles.

5. 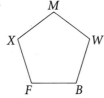   **6.** 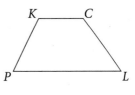   **7.**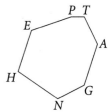

Example 2
(page 158)

Find a polygon in each photograph. Classify the polygon by its number of sides. Tell whether the polygon is convex or concave.

8.    **9.**    **10.**

Example 3
(page 159)

Find the sum of the measures of the angles of each polygon.

11. 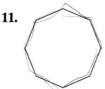   **12.** dodecagon   **13.** decagon

14. 20-gon   **15.** 1002-gon

Example 4
(page 160)

x^2 **Algebra** **Find the missing angle measures.**

16. 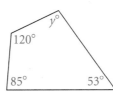   **17.**    **18.**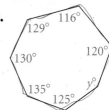

19. 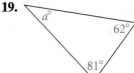   **20.**    **21.**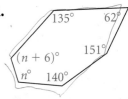

Example 5
(page 160)

Find the measures of an interior angle and an exterior angle of each regular polygon.

22. pentagon   **23.** dodecagon   **24.** 18-gon   **25.** 100-gon

Packaging The nut container at the right has the shape of a regular octagon. It fits in a square box. A cheese wedge fills each corner of the box.

26. Find the measure of each angle of a cheese wedge.

27. Critical Thinking Show how to rearrange the four pieces of cheese to make a regular polygon. What is the measure of each angle of the polygon?

Use a protractor. Sketch each type of regular polygon.

Sample: dodecagon

Use the protractor to equally space 12 points around a circle. (360° ÷ 12 = 30°, so mark a point every 30°.) Connect these points to form a regular dodecagon.

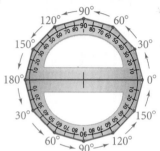

28. triangle **29.** quadrilateral

30. hexagon **31.** octagon

The sum of the measures of the angles of a polygon with *n* sides is given. Find *n*.

32. 180 **33.** 1080 **34.** 1980 **35.** 2880

36. Multiple Choice In the figure at right, $\angle B \cong \angle D$ and $m\angle C = 130$. What is $m\angle B$?

  Ⓐ 25   Ⓑ 50
  Ⓒ 115   Ⓓ 160

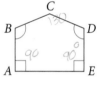

37. Stage Design The diagram at the right shows platforms constructed for a theater-in-the-round stage. Describe the largest platform by the type of regular polygon it suggests. Find the measure of each numbered angle.

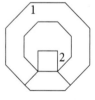

38. Error Analysis Miles said that he measured an angle of a regular polygon to be 130°. Explain why this result is impossible.

39. Critical Thinking A triangle has two congruent angles and an exterior angle with measure 100. Find two possible sets of measures for the angles of the triangle.

The measure of an exterior angle of a regular polygon is given. Find the measure of an interior angle, and find the number of sides.

40. 72 **41.** 36 **42.** 18 **43.** 30 **44.** *x*

45. Probability Find the probability that the measure of an angle of a regular *n*-gon is a positive integer if *n* is an integer and $3 \le n \le 12$.

x^2 **46. Algebra** A polygon has *n* sides. An interior angle of the polygon and an adjacent exterior angle form a straight angle.
  a. What is the sum of the measures of the *n* straight angles?
  b. What is the sum of the measures of the *n* interior angles?
  c. Using your answers above, what is the sum of the measures of the *n* exterior angles?
  d. What theorem do the steps above lead to?

x^2 **Algebra** Find each missing angle measure. Then name the polygon.

47.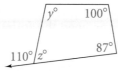

48.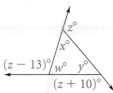

49.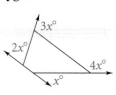

Open-Ended Sketch each figure described in Exercises 50–53.

Problem Solving Hint

In Exercises 51-53, sketch a figure to meet the first condition. Then adjust it to meet the second condition.

50. a quadrilateral that is not equiangular

51. an equiangular quadrilateral that is not regular

52. an equilateral polygon that is not equiangular

53. an equiangular polygon that is not equilateral

54. Critical Thinking Ellen says she has another way to find the sum of the measures of the angles of a polygon. She picks a point inside the polygon, draws a segment to each vertex, counts the number of triangles, multiplies by 180, and then subtracts 360. Does her method work? Explain.

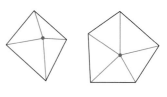

55. Writing Tell what you know about the figure at the right.

56. The measure of an interior angle of a regular polygon is three times the measure of an exterior angle of the same polygon. What is the name of the polygon?

C **Challenge**

57. a. Graphing Calculator Find the measure of an angle of a regular n-gon for $n = 20, 40, 60, 80, \ldots, 200$. Record your results to the nearest tenth as ordered pairs in the form (n, measure of each angle).

   b. Plot the ordered pairs using a window like the one shown at the right.

   c. Data Analysis Based on the graph from part (b), make a statement about the measure of an angle of a regular 1000-gon.

   d. Is there a regular n-gon with an angle of 180°? Explain.

Xmin = 0   Ymin = 160
Xmax = 200 Ymax = 184
Xscl = 20   Yscl = 4

58. a. Explain why the measure of an angle of a regular n-gon is given by the formulas $\frac{180(n-2)}{n}$ and $180 - \frac{360}{n}$.

   b. Use the second formula to explain what happens to the measures in the angles of regular n-gons as n becomes a large number. Explain also what happens to the polygons.

To graph the ordered pairs, use STAT and **STAT PLOT** on your graphing calculator.

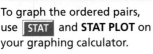

For: Graphing calculator procedures
Web Code: aue-2120

59. Two rays bisect two consecutive angles of a regular decagon and intersect in the decagon's interior. Find the measure of the acute angles formed by the intersecting rays.

Draw, if possible, the concave quadrilateral described. If not possible, explain.

60. with two pairs of congruent adjacent sides

61. with two pairs of congruent opposite sides

62. with three congruent sides

63. with four congruent sides

Gridded Response

For Exercises 64–70, you may need the formula $(n - 2)180$ for the sum of the angle measures in a polygon with n sides.

64. What is the sum of the measures of the angles of a 25-gon?

65. A company is manufacturing a gear that has the shape of a regular polygon. The measure of each angle of the gear is 162. How many sides does the gear have?

66. The car at each vertex of a Ferris wheel holds a maximum of 5 people. The sum of the measures of the angles of the Ferris wheel is 7740. What is the maximum number of people that the Ferris wheel can hold?

67. What is the sum of the measures of the exterior angles, one at each vertex, of an octagon?

68. Exactly three angles of a pentagon are congruent. The other two angles are complementary. What is the measure of one of the three congruent angles?

69. The sum of the measures of the angles of a regular polygon is 4500. How many sides does the polygon have?

70. What is the measure of an exterior angle of a regular polygon with 36 sides?

Mixed Review

Lesson 3-4 x^2 **Find each missing angle measure.**

71.

72.
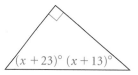

73.
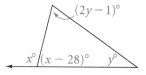

Lesson 2-4 **Name the property that justifies each statement.**

74. $4(2a - 3) = 8a - 12$

75. If $b + c = 7$ and $b = 2$, then $2 + c = 7$.

76. $\overline{RS} \cong \overline{RS}$

77. If $\angle 1 \cong \angle 4$, then $\angle 4 \cong \angle 1$.

78. If $2r = 18$, then $r = 9$.

79. If $AB = BC$ and $BC = 1$, then $AB = 1$.

Lessons 1-4, 1-6 **Identify the following in the diagram.**

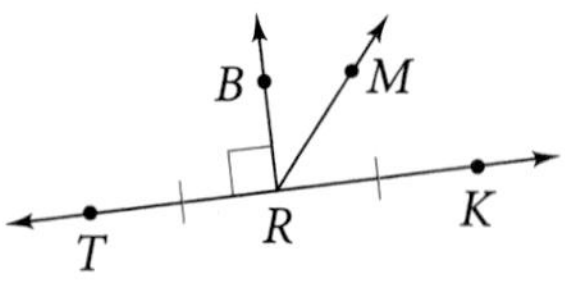

80. a pair of opposite rays

81. two right angles

82. two segments

83. an acute angle

84. an obtuse angle

85. a straight angle

86. a midpoint

Slope

FOR USE WITH LESSON 3-6

The *slope* of a line is the ratio of the vertical change (rise) to the horizontal change (run) between any two points (x_1, y_1) and (x_2, y_2) of the line.

$$\text{slope} = \frac{\text{vertical change}}{\text{horizontal change}} = \frac{\text{rise}}{\text{run}} = \frac{y_2 - y_1}{x_2 - x_1}$$

The slope of a line indicates the steepness of the line and whether it rises or falls from left to right. Line ℓ has slope 1, rising from left to right. Line r has slope -1, falling from left to right. Both form 45° angles with the x-axis.

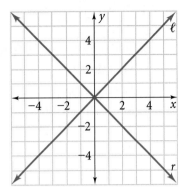

1 EXAMPLE

Find the slope of $\overleftrightarrow{AB}$, which passes through $A(-2, 4)$ and $B(1, -3)$.

Method 1 Use the formula.

$$\text{slope} = \frac{y_2 - y_1}{x_2 - x_1} = \frac{-3 - 4}{1 - (-2)} = \frac{-7}{3}, \text{ or } -\frac{7}{3}$$

Method 2 Use the graph.

$$\text{slope} = \frac{\text{vertical change (rise)}}{\text{horizontal change (run)}} = \frac{-7}{3}, \text{ or } -\frac{7}{3}$$

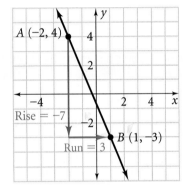

2 EXAMPLE

Compare the rise or fall and the steepness of $\overleftrightarrow{AB}$ in Example 1 with that of line ℓ or r at the top of the page.

The slope of $\overleftrightarrow{AB}$ is negative, so $\overleftrightarrow{AB}$ falls from left to right, the same as line r with slope -1. The slope of $\overleftrightarrow{AB}$ has absolute value $\frac{7}{3}$, which is greater than 1, so $\overleftrightarrow{AB}$ is steeper than line r whose slope has absolute value 1.

EXERCISES

Find the slope of $\overleftrightarrow{AB}$. Compare its rise or fall and its steepness with that of line ℓ or r at the top of the page.

1. $A(4, -6), B(7, 2)$

2. $A(7, -6), B(-5, -8)$

3. $A(-3, 7), B(-1, 4)$

4. $A(-2, -5), B(1, -7)$

5. $A(0, 4), B(4, 0)$

6. $A\left(-3\frac{1}{2}, 3\right), B\left(-7, 2\frac{1}{2}\right)$

7. $A(-1.4, -3.7), B(-2.4, 1.3)$

8. $A(3, -2), B(-6, -2)$

9. $A(5, 9), B(5, -6)$

Open-Ended Graph a line with the given slope m. Compare it with line ℓ or r above.

10. $m = \frac{1}{3}$

11. $m = -1.7$

12. $m = 37$

13. $m = 0$

14. A horizontal line has slope 0. Why?

15. Slope is undefined for a vertical line. Why?

3-6

Lines in the Coordinate Plane

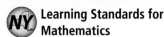 **Learning Standards for Mathematics**

G.PS.8 Determine information required to solve a problem, choose methods for obtaining the information, and define parameters for acceptable solutions.

 Check Skills You'll Need

GO **for Help** page 165

Find the slope of the line that contains each pair of points.

1. $A(-2, 2), B(4, -2)$
2. $P(3, 0), X(0, -5)$
3. $R(-3, -4), S(5, -4)$
4. $K(-3, 3), T(-3, 1)$
5. $C(0, 1), D(3, 3)$
6. $E(-1, 4), F(3, -2)$
7. $G(-8, -9), H(-3, -5)$
8. $L(7, -10), M(1, -4)$

🔊 **New Vocabulary** • slope-intercept form
• standard form of a linear equation • point-slope form

---

1 Graphing Lines

Vocabulary Tip

The *y-intercept* is the y-coordinate of the point where a line crosses the y-axis. The *x-intercept* is the x-coordinate of the point where a line crosses the x-axis.

In algebra, you learned that the graph of a linear equation is a line. The **slope-intercept form** of a linear equation is $y = mx + b$, where m is the slope of the line and b is the y-intercept. Each line at the right has slope 2, but the lines have y-intercepts of 3, -1, and -4.

By Postulate 1-1 (two points determine a line), you need only two points to graph a line. The y-intercept gives you one point. You can use the slope to plot another.

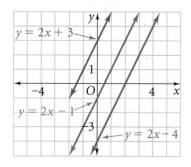

1 **EXAMPLE** **Graphing Lines in Slope-Intercept Form**

Graph the line $y = \frac{3}{4}x + 2$.

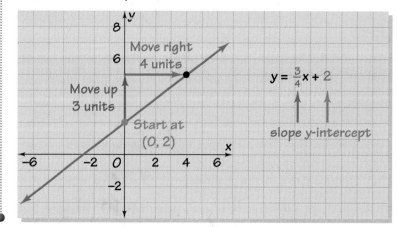

✓ **Quick Check** **1** Graph the line $y = -\frac{1}{2}x - 2$.

Vocabulary Tip

In the <u>standard</u> <u>form</u>, *A*, *B*, and *C* are constants. In Example 2, $6x + 3y = 12$ is in standard form with $A = 6$, $B = 3$, and $C = 12$.

The **standard form of a linear equation** is $Ax + By = C$, where *A*, *B*, and *C* are real numbers and *A* and *B* are not both zero. To graph an equation written in standard form, you can readily find two points for the graph by finding the *x*- and *y*-intercepts.

2 **EXAMPLE** **Graphing Lines Using Intercepts**

Algebra Graph $6x + 3y = 12$.

Step 1 To find the *y*-intercept, substitute 0 for *x*; solve for *y*.

$$6x + 3y = 12$$
$$6(0) + 3y = 12$$
$$3y = 12$$
$$y = 4$$

The *y*-intercept is 4.
A point on the line is $(0, 4)$.

Step 2 To find the *x*-intercept, substitute 0 for *y*; solve for *x*.

$$6x + 3y = 12$$
$$6x + 3(0) = 12$$
$$6x = 12$$
$$x = 2$$

The *x*-intercept is 2.
A point on the line is $(2, 0)$.

Step 3 Plot $(0, 4)$ and $(2, 0)$. Draw the line containing the two points.

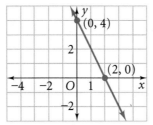

Quick Check **2** Graph $-2x + 4y = -8$.

As an alternative, you can graph an equation in standard form by transforming it into slope-intercept form. Knowing the slope and *y*-intercept beforehand can give you a good mental image of what the graph should look like.

3 **EXAMPLE** **Transforming to Slope-Intercept Form**

Algebra Graph $4x - 2y = 9$.

Step 1 Transform the equation to slope-intercept form.

$$4x - 2y = 9$$
$$-2y = -4x + 9$$
$$\frac{-2y}{-2} = \frac{-4x}{-2} + \frac{9}{-2}$$
$$y = 2x - \frac{9}{2}$$

The *y*-intercept is $-4\frac{1}{2}$ and the slope is 2.

Step 2 Use the *y*-intercept and the slope to plot two points and draw the line containing them.

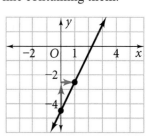

Quick Check **3** Graph $-5x + y = -3$.

A third form for an equation of a line is **point-slope form.** The point-slope form for a nonvertical line through point (x_1, y_1) with slope m is $y - y_1 = m(x - x_1)$.

4 **EXAMPLE** Using Point-Slope Form

Algebra Write an equation of the line through point $P(-1, 4)$ with slope 3.

$y - y_1 = m(x - x_1)$ **Use point-slope form.**

$y - 4 = 3[x - (-1)]$ **Substitute 3 for m and $(-1, 4)$ for (x_1, y_1).**

$y - 4 = 3(x + 1)$ **Simplify.**

✓ Quick Check **4** Write an equation of the line with slope -1 that contains point $P(2, -4)$.

By Postulate 1-1, you need only two points to write an equation of a line.

5 **EXAMPLE** Writing an Equation of a Line Given Two Points

Algebra Write an equation of the line through $A(-2, 3)$ and $B(1, -1)$.

Step 1 Find the slope.

$m = \dfrac{y_2 - y_1}{x_2 - x_1}$

$m = \dfrac{-1 - 3}{1 - (-2)}$ **Substitute $(-2, 3)$ for (x_1, y_1) and $(1, -1)$ for (x_2, y_2).**

$m = -\dfrac{4}{3}$ **Simplify.**

Step 2 Select one of the points. Write an equation in point-slope form.

$y - y_1 = m(x - x_1)$

$y - 3 = -\dfrac{4}{3}[x - (-2)]$ **Substitute $(-2, 3)$ for (x_1, y_1) and $-\dfrac{4}{3}$ for the slope.**

$y - 3 = -\dfrac{4}{3}(x + 2)$ **Simplify.**

✓ Quick Check **5** Write an equation of the line that contains the points $P(5, 0)$ and $Q(7, -3)$.

Recall that the slope of a horizontal line is 0 and the slope of a vertical line is undefined. Thus, horizontal and vertical lines have easily recognized equations.

6 **EXAMPLE** Equations of Horizontal and Vertical Lines

Write equations for the horizontal line and the vertical line that contain $P(3, 2)$.

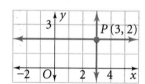

Every point on the horizontal line through $P(3, 2)$ has a y-coordinate of 2. The equation of the line is $y = 2$. It crosses the y-axis at $(0, 2)$.

Every point on the vertical line through $P(3, 2)$ has an x-coordinate of 3. The equation of the line is $x = 3$. It crosses the x-axis at $(3, 0)$.

✓ Quick Check **6** Write equations of the horizontal and vertical lines that contain the point $P(5, -1)$.

EXERCISES

For more exercises, see *Extra Skill, Word Problem, and Proof Practice.*

Practice and Problem Solving

A Practice by Example

Examples 1, 2
(pages 166, 167)

GO for Help

x^2 **Algebra** Graph each line.

1. $y = x + 2$
2. $y = 3x + 4$
3. $y = \frac{1}{2}x - 1$
4. $y = -\frac{5}{3}x + 2$

x^2 **Algebra** Graph each line using intercepts.

5. $2x + 6y = 12$
6. $3x + y = 15$
7. $5x - 2y = 20$
8. $6x - y = 3$
9. $10x + 5y = 40$
10. $1.2x + 2.4y = 2.4$

Example 3
(page 167)

x^2 **Algebra** Write each equation in slope-intercept form and graph the line.

11. $y = 2x + 1$
12. $y - 1 = x$
13. $y + 2x = 4$
14. $8x + 4y = 16$
15. $2x + 6y = 6$
16. $\frac{3}{4}x - \frac{1}{2}y = \frac{1}{8}$

Example 4
(page 168)

x^2 **Algebra** Write an equation in point-slope form of the line that contains the given point and has the given slope.

17. $P(2, 3)$, slope 2
18. $X(4, -1)$, slope 3
19. $R(-3, 5)$, slope -1
20. $A(-2, -6)$, slope -4
21. $V(6, 1)$, slope $\frac{1}{2}$
22. $C(0, 4)$, slope 1

Example 5
(page 168)

Write an equation in point-slope form of the line that contains the given points.

23. $D(0, 5)$, $E(5, 8)$
24. $F(6, 2)$, $G(2, 4)$
25. $H(2, 6)$, $K(-1, 3)$
26. $A(-4, 4)$, $B(2, 10)$
27. $L(-1, 0)$, $M(-3, -1)$
28. $P(8, 10)$, $Q(-4, 2)$

Example 6
(page 168)

Write equations for (a) the horizontal line and (b) the vertical line that contain the given point.

29. $A(4, 7)$
30. $Y(3, -2)$
31. $N(0, -1)$
32. $E(6, 4)$

B Apply Your Skills

Graph each line.

33. $x = 3$
34. $y = -2$
35. $x = 9$
36. $y = 4$
37. $y = 6$

Real-World **Connection**

NASA's Advanced Communications Technology Satellite has a capacity for 250,000 phone calls.

38. Telephone Rates The equation $C = \$.05m + \$4.95$ represents the cost (C) of a long distance telephone call of m minutes.
  a. What is the slope of the line?
  b. What does the slope represent in this situation?
  c. What is the y-intercept (C-intercept)?
  d. What does the y-intercept represent in this situation?

39. Error Analysis A classmate claims that having no slope and having a slope of 0 are the same. Is your classmate correct? Explain.

40. a. What is the slope of the x-axis? Explain.
  b. Write an equation for the x-axis.

41. a. What is the slope of the y-axis? Explain.
  b. Write an equation for the y-axis.

Identify the form of each equation. To graph the line, would you use the given form or change to another form? Explain.

42. $-5x - y = 2$
43. $y = \frac{1}{4}x - \frac{2}{7}$
44. $y + 2 = -(x - 4)$

Critical Thinking Graph three different lines having the given property. Describe how the equations of these lines are alike and how they are different.

45. The lines have slope 2.

46. The lines have y-intercept 2.

 47. Graphing Calculator Graphing calculators use slope-intercept form (rather than standard form or point-slope form) to graph lines. Choose either Exercise 45 or Exercise 46 and write three equations for the lines you graphed. Use the `Y=` window of your graphing calculator to enter your equations. Press `GRAPH`. Do the graphs on the screen confirm the description you wrote previously?

Graph each pair of lines. Then find their point of intersection.

48. $y = -4, x = 6$ **49.** $x = 0, y = 0$ **50.** $x = -1, y = 3$ **51.** $y = 5, x = 4$

52. Building Access By law, the maximum slope of a ramp in new construction is $\frac{1}{12}$. The plan for the new library shows a 3-ft height from the ground to the main entrance. The distance from the sidewalk to the building is 10 ft. Can you design a ramp for the library that complies with the law? Explain.

Real-World Connection

To visualize a slope of $\frac{1}{12}$, think "one foot over, one inch up."

53. Writing Describe the similarities of and the differences between the graphs of the equations $y = 5x - 2$ and $y = -5x - 2$.

54. Open-Ended Write equations for three different lines that contain the point $(5, 6)$.

55. Critical Thinking The x-intercept of a line is 2 and the y-intercept is 4. Use this information to write an equation for the line.

56. The vertices of a triangle are $A(0, 0)$, $B(2, 5)$, and $C(4, 0)$.
    a. Write an equation for the line through A and B.
    b. Write an equation for the line through B and C.
    c. Compare the slopes and y-intercepts of the two lines.

Challenge

Do the three points lie on one line? Justify your answer.

57. $A(5, 6), B(3, 2), C(6, 8)$

58. $D(-2, -2), E(4, -4), F(0, 0)$

59. $G(5, -4), H(2, 3), I(-1, 10)$

60. $J(-2, 9), K(1, -1), L(4, -11)$

A line passes through the given points. Write an equation for the line in point-slope form. Then, rewrite the equation in standard form with integer coefficients.

61. $R(-2, 2), S(0, 8)$ **62.** $T(5, 5), W(7, 6)$ **63.** $X(2, 6), Y(5, 8)$

Multiple Choice

64. Which equation is equivalent to $15x + 3y = 10$?

A. $y = 5x + \frac{10}{3}$ **B.** $y = -5x - \frac{10}{3}$ **C.** $y = 5x - \frac{10}{3}$ **D.** $y = -5x + \frac{10}{3}$

65. Which pair of points $A(-2, 5)$, $B(-1, -2)$, $C(4, -5)$, and $D(7, 0)$, lie on the line with y-intercept closest to the origin?

F. A and B **G.** A and C **H.** B and C **J.** B and D

66. What is the y-intercept of the line whose equation is $y + 9 = 2(x - 3)$?

A. 15   **B.** 9   **C.** -15 **D.** -9

Use the graph at the right for Exercises 67–68.

67. What is the slope of the line?

F. -5 **G.** $-\frac{1}{5}$

H. $\frac{1}{5}$ **J.** 5

68. Which equation is the equation for the line?

A. $5y = x - 6$

B. $y = 5x - 6$

C. $-5x + y = 6$

D. $x + 5y = 6$

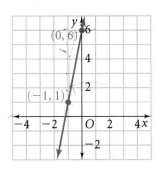

Short Response

69. The slope of line a is $\frac{3}{2}$ and its y-intercept is 12. Line b passes through $(4, 1)$ and $(7, -3)$.

   a. Write an equation for each line.

   b. Graph both lines on the same coordinate plane. From the graph, what is their point of intersection?

Mixed Review

Lesson 3-5

Find the sum of the measures of the angles of each polygon.

70. a nonagon   **71.** a pentagon   **72.** an 11-gon   **73.** a 14-gon

Lesson 2-2

Is each statement a good definition? If not, find a counterexample.

74. A quadrilateral is a polygon with four sides.

75. Skew lines are lines that don't intersect.

76. An acute triangle is a triangle with an acute angle.

Lesson 1-7 x^2 **Algebra** For Exercises 77–80, $\overrightarrow{PQ}$ is the bisector of $\angle MPR$. Solve for a and find the missing angle measure.

77. $m\angle MPQ = 3a$, $m\angle QPR = 2a + 5$, $m\angle MPR = \blacksquare$

78. $m\angle MPQ = 7a$, $m\angle QPR = 4a + 12$, $m\angle MPR = \blacksquare$

79. $m\angle MPQ = 8a - 8$, $m\angle QPR = 5a - 2$, $m\angle QPR = \blacksquare$

80. $m\angle MPQ = 2a + 9$, $m\angle QPR = 4a - 3$, $m\angle MPQ = \blacksquare$

You enter a linear equation in two variables (a linear function) into a graphing calculator in the form

$$Y1 = \text{linear expression}$$

The calculator can then give information about Y1 in both graph and table form.

NY **G.PS.3:** Use multiple representations to represent and explain problem situations (e.g., spatial, geometric, verbal, numeric, algebraic, and graphical representations).

1 ACTIVITY

The windows below use the linear equation $y = 2x - 6$ entered as $Y1 = 2X - 6$.

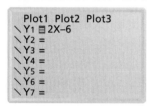
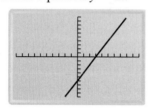

1. How can you find where $2x - 6 = 0$ from the graph? From the table?

2 ACTIVITY

To solve a linear equation from a graph.

let $Y1 = \text{linear expression}$

Then find where $Y1 = 0$ by using the zero feature in CALC.

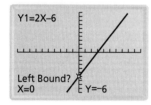
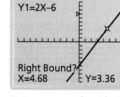
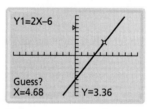

Enter an *x*-value that is to the left of (<) the solution.

Enter an *x*-value that is to the right of (>) the solution.

Enter a guess for the solution.

Try it with these linear equations.

2. $2x - 14 = 0$ **3.** $3x + 1 - (x - 7) = 0$ **4.** $-x + 10 = -14 + x$ **5.** $2.5x + 42.5 = 10$

If you haven't already thought of it, here is a terrific shortcut for solving one linear expression = another linear expression.

Let $Y1 = \text{one linear expression} - (\text{other linear expression})$,

and find where $Y1 = 0$. Be sure to enter the expressions carefully into your calculator. Try it with these.

6. $3x - 17 - 5x = -1 - x + 7x$ **7.** $7 + 4x - (2 + 9x) = x - (6x - 5)$

8. $11 + 3x + 5(7 - x) - 4(3 - (x + 2)) = 3x$

3 ACTIVITY

To help you understand how to solve a linear equation using tables, first think graphically. Look again at the windows in Activity 1.

9. In the graph, to the left of the solution X = 3, are the Y1-values positive or negative? To the right of 3, are the Y1-values positive or negative? Explain.

10. You can say that the Y1-values change __?__ at the solution.

Also, note in the table how Y1 changes sign at the solution X = 3.

Without using your calculator, tell where the Y1-values would change sign in a table for each equation below. State whether the change in Y1 as X increases is from negative to positive or from positive to negative. In each case, tell why. You may assume that the X-values increase by 1 in each step of the table.

11. $y = -x + 4$ **12.** $y = x - 7$ **13.** $y = -2x - 8$ **14.** $y = \frac{1}{2}x + 9$

Now, use a table to solve each equation below. If you do not see the solution, scroll up or down to where you think you'll find Y1 = 0.

15. $0 = 3x - 15$ **16.** $0 = 3x - (x - 4)$ **17.** $0 = 20 + 2x$ **18.** $0 = \frac{1}{3}x - 24$

4 ACTIVITY

Tables become more challenging (and interesting) to use when you see a sign change but you don't see Y1 = 0.

19. For the table of Y1 = $-4x + 11$ at the right, Y1 changes sign between __?__ and __?__.

20. The solution of $-4x + 11 = 0$ is between __?__ and __?__.

X	Y1
0.00	11.00
1.00	7.00
2.00	3.00
3.00	−1.00
4.00	−5.00
5.00	−9.00
6.00	−13.00
X=0	

When you have a solution bounded between two X-values that give a sign change, you can perform a *tabular zoom-in* on the solution. If X is between 2 and 3 (above), you can change TblStart to 2 and ΔTbl to 0.1. Study these three screens that capture the solution $x = 2.75$ of the equation $-4x + 11 = 0$.

```
TABLE SETUP
  TblStart=2
  ΔTble=0.1▮
Indpnt: Auto  Ask
Depnd:  Auto  Ask
```

X	Y1
2.50	1.00
2.60	.60
2.70	.20
2.80	−.20
2.90	−.60
3.00	−1.00
3.10	−1.40
X=3.1	

X	Y1
2.72	.12
2.73	.08
2.74	.04
2.75	0.00
2.76	−.04
2.77	−.08
2.78	−.12
X=2.72	

ΔTble=0.01

Use tables to solve each equation.

21. $5x - 14 = x$ **22.** $2(3x + 11) = 2.5$ **23.** $x = 11.75 - x$ **24.** $3x - 100 = 5(0.9 - x)$

25. Use tables to solve $3x - 11 = 0$. Explain what is different about solving this equation from solving those above.

26. Use tables to solve $x\sqrt{2} - \pi = 0$ to as many decimal places as possible on your graphing calculator. Record your answer. Then solve the equation again using a graph.

3-7

Slopes of Parallel and Perpendicular Lines

 Learning Standards for Mathematics

G.G.63 Determine whether two lines are parallel, perpendicular, or neither, given their equations.

G.G.64 Find the equation of a line, given a point on the line and the equation of a line perpendicular to the given line.

 Check Skills You'll Need

GO for Help page 165 and Lesson 3-6

Find the slope of the line through each pair of points.

1. $F(2, 5), B(-2, 3)$ **2.** $H(0, -5), D(2, 0)$ **3.** $E(1, 1), F(2, -4)$

Find the slope of each line.

4. $y = 2x - 5$ **5.** $x + y = 20$ **6.** $2x - 3y = 6$

7. $x = y$ **8.** $y = 7$ **9.** $y = \frac{2}{3}x + 7$

---

1  **Slope and Parallel Lines**

The relationship between slope and parallel lines is summarized below and proved in Lesson 7-3.

 Key Concepts

Summary	**Slopes of Parallel Lines**

If two nonvertical lines are parallel, their slopes are equal.

If the slopes of two distinct nonvertical lines are equal, the lines are parallel.

Any two vertical lines are parallel.

Real-World 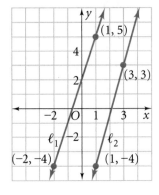 **Connection**

The ramp and rails are parallel because they have the same slope.

You can test whether nonvertical lines are parallel by comparing their slopes.

1 EXAMPLE **Checking for Parallel Lines**

Are lines ℓ_1 and ℓ_2 parallel? Explain.

Find and compare the slopes of the lines.

$$\text{slope of } \ell_1 = \frac{5 - (-4)}{1 - (-2)} = \frac{9}{3} = 3$$

$$\text{slope of } \ell_2 = \frac{3 - (-4)}{3 - 1} = \frac{7}{2}$$

Lines ℓ_1 and ℓ_2 are not parallel because their slopes are not equal.

 Quick Check **1** Line ℓ_3 contains $A(-4, 2)$ and $B(3, 1)$. Line ℓ_4 contains $C(-4, 0)$ and $D(8, -2)$. Are ℓ_3 and ℓ_4 parallel? Explain.

Slope-intercept form allows you to compare slopes easily in order to decide whether lines are parallel.

2 EXAMPLE Determining Whether Lines are Parallel

Multiple Choice Which line is parallel to $4y - 12x = 20$?

- Ⓐ $y = 3x - 1$
- Ⓑ $y = -\frac{1}{3}x - 1$
- Ⓒ $4y = -12x + 20$
- Ⓓ $y = \frac{1}{3}x + 1$

Write $4y - 12x = 20$ in slope-intercept form.

$$4y - 12x = 20$$
$$4y = 12x + 20 \quad \textbf{Add 12x to each side.}$$
$$y = 3x + 5 \quad \textbf{Divide each side by 4.}$$

The line $4y - 12x = 20$ has slope 3. The line $y = 3x - 1$ is the only answer choice with slope 3. The correct choice is A.

✓ **Quick Check** ❷ Are the lines parallel? Explain.

a. $y = -\frac{1}{2}x + 5$ and $2x + 4y = 9$ **b.** $y = -\frac{1}{2}x + 5$ and $2x + 4y = 20$

You can write an equation for a line parallel to a given line.

3 EXAMPLE Writing Equations of Parallel Lines

Write an equation for the line parallel to $y = -4x + 3$ that contains $(1, -2)$.

Step 1 Identify the slope of the given line.

$$y = -4x + 3$$
$$\uparrow$$
$$\text{slope}$$

Step 2 Use point-slope form to write an equation for the new line.

$$y - y_1 = m(x - x_1)$$
$$y - (-2) = -4(x - 1) \quad \textbf{Substitute } -4 \textbf{ for } m \textbf{ and } (1, -2) \textbf{ for } (x_1, y_1).$$
$$y + 2 = -4(x - 1) \quad \textbf{Simplify.}$$

✓ **Quick Check** ❸ Write an equation for the line parallel to $y = -x + 4$ that contains $(-2, 5)$.

2 Slope and Perpendicular Lines

The relationship between perpendicular lines and their slopes is summarized below. These statements will be proved in Lessons 6-6 and 6-7.

Key Concepts

Summary	Slopes of Perpendicular Lines

If two nonvertical lines are perpendicular, the product of their slopes is -1.

If the slopes of two lines have a product of -1, the lines are perpendicular.

Any horizontal line and vertical line are perpendicular.

You can test whether lines are perpendicular by first noting whether either line is vertical or horizontal. If not, check their slopes. If the product of the slopes is -1, the lines are perpendicular.

4 EXAMPLE **Checking for Perpendicular Lines**

Algebra Lines ℓ_1 and ℓ_2 are neither vertical nor horizontal. Are they perpendicular? Explain.

Step 1 Find the slope of each line.

$$m_1 = \text{slope of } \ell_1 = \frac{-2 - 2}{-3 - 0} = \frac{-4}{-3} = \frac{4}{3}$$

$$m_2 = \text{slope of } \ell_2 = \frac{3 - (-3)}{-2 - 6} = \frac{6}{-8} = -\frac{3}{4}$$

Step 2 Find the product of the slopes.

$$m_1 \cdot m_2 = \frac{4}{3} \cdot -\frac{3}{4} = -1$$

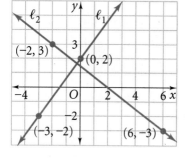

● Lines ℓ_1 and ℓ_2 are perpendicular because the product of their slopes is -1.

✓ **Quick Check** ④ Are ℓ_3 and ℓ_4 perpendicular? Explain.

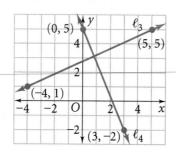

You can write an equation for a line perpendicular to a given line. If the given line is horizontal, write an equation for a vertical line. If the given line is vertical, write an equation for a horizontal line.

5 EXAMPLE **Writing Equations for Perpendicular Lines**

Write an equation for the line through $(-3, 7)$ and perpendicular to $y = -3x - 5$.

Step 1 Identify the slope of the given line.

$$y = -3x - 5$$

slope

Step 2 Find the slope of the line perpendicular to the given line.

Let m be the slope of the perpendicular line.

$-3m = -1$ **The product of the slopes of perpendicular lines is -1.**

$m = \frac{1}{3}$ **Divide each side by -3.**

Step 3 Use point-slope form to write an equation for the new line.

$$y - y_1 = m(x - x_1)$$

$$y - 7 = \frac{1}{3}[x - (-3)]$$ **Substitute $\frac{1}{3}$ for m and $(-3, 7)$ for (x_1, y_1).**

$$y - 7 = \frac{1}{3}(x + 3)$$ **Simplify.**

✓ **Quick Check** ⑤ Write an equation for the line through $(15, -4)$ and perpendicular to $5y - x = 10$.

Vocabulary Tip

Numbers with product -1 are <u>opposite</u> <u>reciprocals</u>. In Example 5, the opposite reciprocal of -3 is $\frac{1}{3}$.

6 EXAMPLE **Real-World** 🌎 **Connection**

The window at the left includes some perpendicular lead strips. The line that contains $\overline{BC}$ has equation $y = -x + 10$. $\overline{AB}$ is perpendicular to $\overline{BC}$. Write an equation for $\overleftrightarrow{AB}$, the line that contains $\overline{AB}$ and point $(-1, 5)$.

The line that contains $\overline{BC}$ has slope -1. Let m be the slope of $\overleftrightarrow{AB}$.

$-1m = -1$ **The product of the slopes is -1.**

$m = 1$

$\overleftrightarrow{AB}$ has slope 1 and can be written in the form $y = 1x + b$, or $y = x + b$.

$y = x + b$ **$\overleftrightarrow{AB}$ has slope 1.**

$5 = -1 + b$ **Substitute 5 for y and -1 for x.**

$6 = b$ **Add 1 to each side.**

● The equation for $\overleftrightarrow{AB}$ is $y = x + 6$.

✓ **Quick Check** **6** If the equation for a line containing a lead strip on a different window is $y = -\frac{2}{3}x + 15$, write an equation for the line perpendicular to it that contains $(2, 8)$.

EXERCISES

For more exercises, see *Extra Skill, Word Problem, and Proof Practice.*

Practice and Problem Solving

A **Practice by Example**

Example 1 (page 174)

GO for Help

In Exercises 1–5, are lines ℓ_1 and ℓ_2 parallel? Explain, using slope.

1.
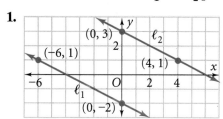

2.
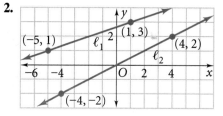

3.
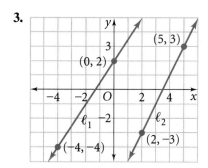

4.
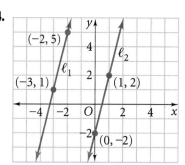

5. Line ℓ_1 contains $A(-3, 6)$ and $B(2, 6)$, and line ℓ_2 contains $C(0, 0)$ and $D(7, 0)$.

Example 2 (page 175)

$\boxed{x^2}$ **Algebra** **Are the lines parallel? Explain.**

6. $y = 2x + 5$
  $y = 2x$

7. $y = \frac{3}{4}x - 10$
  $y = \frac{3}{4}x + 2$

8. $y = -x + 6$
  $x + y = 20$

9. $y - 7x = 6$
  $y + 7x = 8$

10. $3x + 4y = 12$
  $6x + 2y = 6$

11. $2x + 5y = -1$ ·
  $10y = -4x - 20$

Example 3
(page 175)

Write an equation for the line parallel to $\overleftrightarrow{AB}$ that contains point C.

12. $\overleftrightarrow{AB}: y = -2x + 1, C(0, 3)$ **13.** $\overleftrightarrow{AB}: y = \frac{1}{3}x, C(6, 0)$

14. $\overleftrightarrow{AB}: -x + 2y = 4, C(-2, 4)$ **15.** $\overleftrightarrow{AB}: 3x + 2y = 12, C(6, -2)$

Example 4
(page 176)

$\boxed{x^2}$ **Algebra** Are lines ℓ_1 and ℓ_2 perpendicular? Explain using slope.

16.

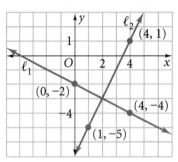

17.

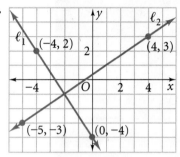

18.

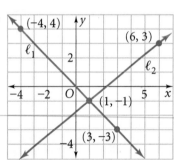

19.

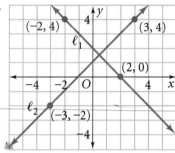

Example 5
(page 176)

Write an equation for the line through P and perpendicular to the given line.

20. $P(6, 6)$; $y = \frac{2}{3}x$ **21.** $P(4, 0)$; $y = \frac{1}{2}x - 5$ **22.** $P(4, 4)$; $y + 2x = -8$

23. P the midpoint of the segment with endpoints $M(1, 9)$ and $N(9, -1)$ in $\overleftrightarrow{MN}$

Example 6
(page 177)

24. Highway Construction Highway planners want to construct a road perpendicular to Route 3 at point O. An equation for the Route 3 line is $y = \frac{2}{3}x$. Find an equation for the line for the new road.

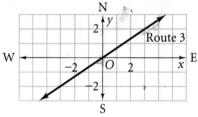

B   **Apply Your Skills** $\boxed{x^2}$ **Algebra**   Are the lines perpendicular? Explain.

25. $y = -x - 7$ **26.** $y = 3$ **27.** $2x - 7y = -42$
    $y - x = 20$ $x = -2$ $4y = -7x - 2$

28. Multiple Choice Which line is perpendicular to $3y + 2x = 12$?

   Ⓐ $6x - 4y = 24$ Ⓑ $y + 3x = -2$

   Ⓒ $2x + 3y = 6$ Ⓓ $y = -2x + 6$

Use slopes to find whether the opposite sides of quadrilateral $ABCD$ are parallel.

29. $A(0, 2), B(3, 4), C(2, 7), D(-1, 5)$ **30.** $A(-3, 1), B(1, -2), C(0, -3), D(-4, 0)$

31. $A(1, 1), B(5, 3), C(7, 1), D(3, 0)$ **32.** $A(1, 0), B(4, 0), C(3, -3), D(-1, -3)$

33. Open-Ended Write equations for two perpendicular lines that have the same y-intercept and do not pass through the origin.

34. Writing Can the y-intercepts of two parallel lines be the same? Explain.

35. Use slope to show that the opposite sides of hexagon *RSTUVW* at the right are parallel.

36. Use slope to determine whether a triangle with vertices $G(3, 2)$, $H(8, 5)$, and $K(0, 10)$ is a right triangle. Explain.

Developing Proof Use slope to explain why each theorem is true for three lines in the coordinate plane.

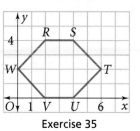
Exercise 35

37. Theorem 3-9: If two lines are parallel to the same line, then they are parallel to each other.

38. Theorem 3-10: In a plane, if two lines are perpendicular to the same line, then they are parallel to each other.

39. Soccer The coordinate system at the right is designed for a soccer field. Each unit represents one yard. Joe is at point $P(35, -20)$. The path of the ball from a corner kick is represented by the equation $y = -\frac{4}{3}x$. To have the best chance for a shot on goal, Joe wants to run toward the ball so that his path meets the path of the ball at a right angle.
  a. Find an equation for the line on which Joe should run.
  b. Critical Thinking Why is point-slope form the best choice for the equation?

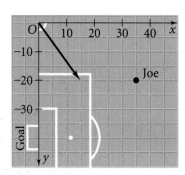

Go Online
PHSchool.com
For: Graphing calculator procedures
Web Code: aue-2122

Determine whether $\overleftrightarrow{AB}$ and $\overleftrightarrow{CD}$ are *parallel*, *perpendicular*, or *neither*.

40. $A\left(-1, \frac{1}{2}\right), B(-1, 2), C(3, 7), D(3, -1)$ **41.** $A(-2, 3), B(-2, 5), C(1, 4), D(2, 4)$

42. $A(2, 4), B(5, 4), C(3, 2), D(0, 8)$ **43.** $A(-3, 2), B(5, 1), C(2, 7), D(1, -1)$

44. Graphing Calculator Use your graphing calculator to find the slope of $\overleftrightarrow{AB}$ in Exercise 43. Enter the *x*-coordinates of *A* and *B* into the L_1 list of your list editor. Enter the *y*-coordinates into the L_2 list. In your **STAT** CALC menu select LinReg $(ax + b)$. **ENTER** to find the slope *a*. Repeat to find the slope of $\overleftrightarrow{CD}$. Are $\overleftrightarrow{AB}$ and $\overleftrightarrow{CD}$ parallel, perpendicular, or neither?

© Challenge

45. Show that the diagonals of the figure at the right are congruent.

46. Show that the diagonals of the figure at the right are perpendicular bisectors of each other.

47. a. Graph the points $P(2, 2)$, $Q(7, 4)$, and $R(3, 5)$.
  b. Find the coordinates of a point *S* that, along with points *P*, *Q*, and *R*, will form the vertices of a quadrilateral whose opposite sides are parallel. Graph the quadrilateral.
  c. Repeat part (b), finding a different point *S* and graphing the new quadrilateral.

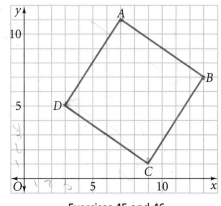
Exercises 45 and 46

48. A triangle has vertices $L(-5, 6)$, $M(-2, -3)$, and $N(4, 5)$. Write an equation for the line perpendicular to $\overline{LM}$ that contains point *N*.

Multiple Choice

49. What is the slope of a line parallel to the line $6x - 4y = 12$?
A. $-\frac{3}{2}$ B. $\frac{3}{2}$ C. $\frac{4}{3}$ D. $-\frac{4}{3}$

50. The slope of a line is 6. What is the slope of a line perpendicular to it?
F. 6 G. -6 H. $\frac{1}{6}$ J. $-\frac{1}{6}$

51. Line f contains the points $(5, -4)$ and $(4, -6)$. What is the slope of a line perpendicular to it?
A. 2 B. $\frac{1}{2}$ C. $-\frac{1}{2}$ D. -2

Short Response

52. Line c contains the points $(2, -2)$ and $(-4, 1)$.
 a. What is the slope of a line perpendicular to line c?
 b. What is the y-intercept of the line perpendicular to line c that contains $(1, 2)$?

Mixed Review

Lesson 3-6 $\boxed{x^2}$ **Algebra** Write an equation for the line containing the given points.

53. $A(0, 3), B(6, 0)$ **54.** $C(-4, 2), D(-1, 7)$ **55.** $E(3, -2), F(-5, -8)$

Lesson 2-4 **Name the property that justifies each statement.**

56. $\angle 4 \cong \angle 4$ **57.** If $m\angle B = 8$, then $2m\angle B = 16$.

58. $-3x + 6 = 3(-x + 2)$ **59.** If $\overline{RS} \cong \overline{MN}$, then $\overline{MN} \cong \overline{RS}$.

Lesson 2-3 **Use the Law of Syllogism to draw a conclusion.**

60. If you are in geometry class, then you are in math class. If you are in math class, then you are at school.

61. If you travel to Switzerland, then you travel to Europe. If you travel to Europe, then you have a passport.

Checkpoint Quiz 2 Lessons 3-5 through 3-7

Use the number of sides to name the polygon. Then find the value of each variable.

1.
$n°$
$(n + 30)°$
$(n - 12)°$ $134°$
$161°$ $(n + 16)°$
$n°$ $126°$

2.
$58°$ $x°$ $w°$
$(2w - 25)°$
$(x + 6)°$
$125°$ $100°$

3.
$50°$
$89°$
$a°$ $m°$ $64°$

$\boxed{x^2}$ **Algebra** Graph each line using intercepts.

4. $4x + y = -8$ **5.** $-2x + 3y = 12$ **6.** $3x + 5y = 30$

Find the slopes of $\overleftrightarrow{RS}$ and $\overleftrightarrow{TV}$. Then determine whether $\overleftrightarrow{RS}$ and $\overleftrightarrow{TV}$ are parallel, perpendicular, or neither. Explain.

7. $R(-2, 6), S(3, 4), T(3, 5), V(0, 0)$ **8.** $R(6, -1), S(7, 0), T(3, -4), V(0, -1)$

9. $R(9, 1), S(5, 6), T(3, 8), V(-2, 4)$ **10.** $R(5, -7), S(-4, -9), T(6, 2), V(-3, 0)$

Constructing Parallel and Perpendicular Lines

Learning Standards for Mathematics

G.G.19 Construct lines parallel (or perpendicular) to a given line through a given point, using a straightedge and compass, and justify the construction.

✓ **Check Skills You'll Need**

GO **for Help** Lesson 1-7

Use a straightedge to draw each figure. Then use a straightedge and compass to construct a figure congruent to it.

1. a segment   **2.** an obtuse angle   **3.** an acute angle

Use a straightedge to draw each figure. Then use a straightedge and compass to bisect it.

4. a segment   **5.** an acute angle   **6.** an obtuse angle

---

1 Constructing Parallel Lines

You can use what you know about parallel lines, transversals, and corresponding angles to construct parallel lines.

1 EXAMPLE **Constructing ℓ ∥ m**

Construct the line parallel to a given line and through a given point that is not on the line.

Given: line ℓ and point N not on ℓ

Construct: line m through N with m ∥ ℓ

Step 1
Label two points H and J on ℓ.
Draw $\overleftrightarrow{HN}$.

Step 2
Construct ∠1 with vertex at N so that ∠1 ≅ ∠NHJ and the two angles are corresponding angles. Label the line you just constructed m.

● m ∥ ℓ

Real-World 🌐 **Connection**

Careers Architects construct parallel and perpendicular lines when they build models of the buildings they design.

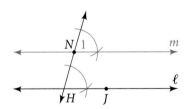

✓ **Quick Check** ❶ **Critical Thinking** Explain why lines ℓ and m must be parallel.

For many constructions, you will find it helpful to first visualize or sketch what the final figure should look like. This will often suggest the construction steps. In Example 2, a sketch is shown at the left of the example.

2 EXAMPLE Constructing a Special Quadrilateral

Construct a quadrilateral with one pair of parallel sides of lengths *a* and *b*.

Given: segments of lengths *a* and *b*

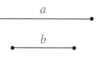

Construct: quadrilateral *ABYZ* with
$AZ = a$, $BY = b$, and $\overline{AZ} \parallel \overline{BY}$

Step 1
Construct $\overline{AZ}$ with length *a*.

Step 2
Draw a point *B* not on $\overleftrightarrow{AZ}$. Then draw $\overrightarrow{AB}$.

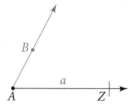

Step 3
Construct a ray parallel to $\overleftrightarrow{AZ}$ through *B*.

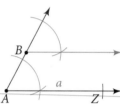

Step 4
Construct *Y* so that $BY = b$. Then draw $\overline{YZ}$.

Quadrilateral *ABYZ* has $AZ = a$, $BY = b$, and $\overline{AZ} \parallel \overline{BY}$.

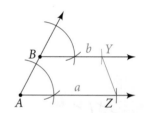

✓ **Quick Check** ❷ Draw two segments. Label their lengths *c* and *d*. Construct a quadrilateral with one pair of parallel sides of lengths *c* and 2*d*.

---

2 Constructing Perpendicular Lines

You can construct perpendicular lines using a compass and a straightedge.

3 EXAMPLE Perpendicular at a Point on a Line

Construct the perpendicular to a given line at a given point on the line.

Given: point *P* on line ℓ

Construct: $\overleftrightarrow{CP}$ with $\overleftrightarrow{CP} \perp \ell$

Step 1
Put the compass point on point *P*. Draw arcs intersecting ℓ in two points. Label the points *A* and *B*.

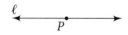

Step 2
Open the compass wider. With the compass tip on *A*, draw an arc above point *P*.

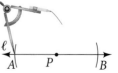

Step 3

Without changing the compass setting, place the compass point on point B. Draw an arc that intersects the arc from Step 2. Label the point of intersection C.

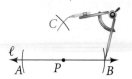

Step 4

Draw $\overleftrightarrow{CP}$.

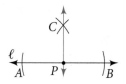

• $\overleftrightarrow{CP} \perp \ell$

✓ **Quick Check** ❸ Use a straightedge to draw $\overleftrightarrow{EF}$. Construct $\overleftrightarrow{FG}$ so that $\overleftrightarrow{FG} \perp \overleftrightarrow{EF}$ at point F.

You will prove in Chapter 5 that the perpendicular segment is the shortest segment from a point to a line. Here is its construction.

❹ EXAMPLE Perpendicular From a Point to a Line

Construct the perpendicular to a given line through a given point not on the line.

Given: line ℓ and point R not on ℓ
Construct: $\overleftrightarrow{RG}$ with $\overleftrightarrow{RG} \perp \ell$

Step 1

Open your compass to a size greater than the distance from R to ℓ. With the compass point on point R, draw an arc that intersects ℓ at two points. Label the points E and F.

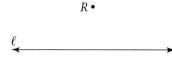

Step 2

Place the compass point on E and make an arc.

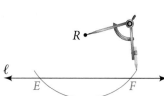

Step 3

Keep the same compass setting. With the compass tip on F, draw an arc that intersects the arc from Step 2. Label the point of intersection G.

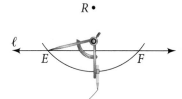

Step 4

Draw $\overleftrightarrow{RG}$.

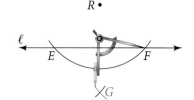

• $\overleftrightarrow{RG} \perp \ell$

Real-World 🌐 Connection

You can draw large circles using a simple, large compass.

✓ **Quick Check** ❹ Draw a line $\overleftrightarrow{CX}$ and a point Z not on $\overleftrightarrow{CX}$. Construct $\overleftrightarrow{ZB}$ so that $\overleftrightarrow{ZB} \perp \overleftrightarrow{CX}$.

EXERCISES

For more exercises, see *Extra Skill, Word Problem, and Proof Practice.*

Practice and Problem Solving

 Practice by Example

Example 1
(page 181)

In Exercises 1–4, draw a figure like the given one. Then construct the line through point J and parallel to $\overleftrightarrow{AB}$.

1.

2.

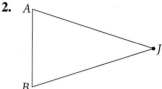

3.

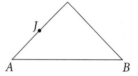

4.

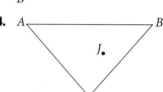

Example 2
(page 182)

For Exercises 5–7, draw two segments. Label their lengths a and b. Construct a quadrilateral with one pair of parallel sides as described.

5. The sides have lengths a and b.

6. The sides have lengths $2a$ and b.

7. The sides have lengths a and $\frac{1}{2}b$.

Example 3
(pages 182, 183)

In Exercises 8–9, draw a figure like the given one. Then construct the line perpendicular to $\overleftrightarrow{AB}$ at point P.

8.

9.

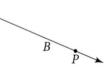

Example 4
(page 183)

In Exercises 10–13, draw a figure like the given one. Then construct the line through point P and perpendicular to $\overleftrightarrow{RS}$.

10.

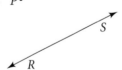

11.

12.

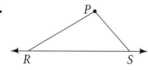

13.

 Apply Your Skills

14. Draw an acute angle. Construct an angle congruent to your angle so that the two angles are alternate interior angles. (*Hint:* Think of the letter Z.)

15. Writing Explain how to use the Converse of the Alternate Interior Angles Theorem to construct a line parallel to a given line through a point not on the line. (*Hint:* See Exercise 14.)

16. Draw obtuse △*ABC* with obtuse ∠*B*.
   a. Construct line ℓ through point *A* so that ℓ ∥ $\overline{BC}$.
   b. Construct line *m* through point *C* so that *m* ∥ $\overline{AB}$.

For Exercises 17–24, use the segments at the left.

17. Draw a line *m*. Construct a segment of length *b* that is perpendicular to line *m*.

18. Construct a rectangle with base *b* and height *c*.

19. Construct a square with sides of length *a*.

20. Construct a rectangle with one side length *a* and a diagonal length *b*.

21. a. Construct a quadrilateral with a pair of parallel sides of length *c*.
   b. Make a Conjecture What appears to be true about the other pair of sides in the quadrilateral you constructed?
   c. Use a protractor, a ruler, or both to check the conjecture you made in part (b).

22. Construct a right triangle with legs of lengths *a* and *b*.

23. Construct a right triangle with legs of lengths *b* and $\frac{1}{2}b$.

24. a. Construct a triangle with sides of lengths *a*, *b*, and *c*.
   b. Construct the midpoint of each side of the triangle.
   c. Form a new triangle by connecting the midpoints.
   d. Make a Conjecture How do the sides of the smaller triangle and the sides of the larger triangle appear to be related?
   e. Use a protractor, a ruler, or both to check the conjecture you made in part (d).

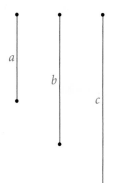

Exercises 17–24

25. Multiple Choice The diagram at the left shows the construction of line $\overleftrightarrow{CP}$ perpendicular to line ℓ through point *P*. Which of the following *must* be true?

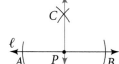

(A) $\overleftrightarrow{CB} \parallel \overleftrightarrow{AB}$ (B) $\overleftrightarrow{AC} \perp \overleftrightarrow{CB}$

(C) $CP = \frac{1}{2}AB$ (D) $\overline{AC} \cong \overline{BC}$

26. Paper Folding You can use paper folding to create a perpendicular to a given line through a given point (Activity Lab, page 102). Fold the paper so that the line folds onto itself and the fold line contains the given point.

Homework Video Tutor
Visit: PHSchool.com
Web Code: aue-0308

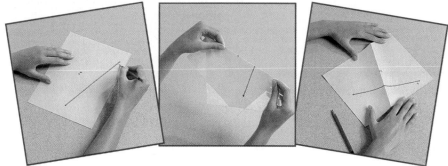

a. Draw a line *m* and a point *W* not on the line. Use paper folding to create the perpendicular to *m* through *W*. Label this fold line *k*.
b. Next, fold the line perpendicular to *k* through *W*. Label this fold line *p*.
c. What is true of *p* and *m*? Justify your answer.

 Challenge

Draw a segment, $\overline{DG}$. Construct a quadrilateral whose diagonals are both congruent to $\overline{DG}$, bisect each other, and meet the additional condition given below. Describe the quadrilateral that you get.

27. The diagonals are not perpendicular.   **28.** The diagonals are perpendicular.

Construct a rectangle whose side lengths a and b meet the given condition.

29. $b = 2a$ **30.** $b = \frac{1}{2}a$ **31.** $b = \frac{1}{3}a$ **32.** $b = \frac{2}{3}a$

Construct a triangle whose side lengths a, b, and c meet the given conditions. If such a triangle is not possible, explain.

33. $a = b = c$ **34.** $a = b = 2c$ **35.** $a = 2b = 2c$ **36.** $a = b + c$

Test Prep

Multiple Choice

37. In the construction shown at the right, the two arcs with centers A and B have the same radius. What must be true of $\overline{PQ}$?
A. $\overline{PQ}$ bisects $\overline{AB}$.   **B.** $\overline{PQ} \parallel \overline{AB}$
C. $\overline{PQ} \cong \overline{AB}$ **D.** $\overline{PQ} \cong \overline{AQ}$

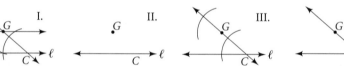

38. Suppose you construct lines ℓ, m, and n so that $\ell \perp m$ and $\ell \parallel n$. Which of the following is true?
F. $m \parallel n$ **G.** $m \parallel \ell$ **H.** $n \perp \ell$ **J.** $n \perp m$

Short Response

39. Use a compass and straightedge to construct the following figure.
a. Draw a line ℓ and a point G not on ℓ. Construct an arc centered at point G to intersect ℓ in two points. Label the points R and T. Draw $\overline{GR}$ and $\overline{GT}$.
b. Classify $\triangle RGT$. Justify your response.

40. These pictures show steps for constructing a line parallel to a given line, but they are not necessarily in order.

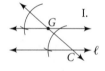  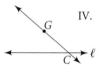

a. List the construction steps in the correct order.
b. For any step that uses a compass, describe the location(s) of the compass point.

Mixed Review

Lesson 3-7

Are the lines parallel? Explain.

41. $y = -4x - 3$ **42.** $y = \frac{1}{2}x + 1$ **43.** $x + 3y = -6$
$\ y = 4x + 3$ $\ y = -2x - 1$ $\ 4x + 12y = -6$

Lesson 1-8

Find the distance between the points to the nearest tenth.

44. $W(8, -2)$ and $Z(2, 6)$

45. $W(-4.5, 1.2)$ and $Z(3.5, -2.8)$

Lesson 1-3

Name the intersection of the planes.

46. plane ABE and plane $EBCD$

47. plane $AFDE$ and plane FCD

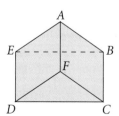

Using Tables and Lists

Tables and lists on your graphing calculator allow you to study relationships both numerically and graphically. The first example reminds you how to build a table.

Go Online
PHSchool.com

For: Graphing calculator procedures
Web Code: aue-2104

1 ACTIVITY

Display a table showing the sums of the measures of the angles of a polygon.

Use the Y= screen and write the polygon angle-sum formula in the form $Y_1 = 180(X - 2)$. Use the **TBLSET** feature so that X starts at 3 and changes by 1.

● Use the **TABLE** feature to see the table of *n*-gon angle sums.

X	Y1
3	180
4	360
5	540
6	720
7	900
8	1080
9	1260

X=3

The second example suggests a powerful way to use lists. First, press Y= CLEAR.

NY **G.R.1:** Use physical objects, diagrams, charts, tables, graphs, symbols, equations, or objects created using technology as representations of mathematical concepts.

2 ACTIVITY

List and plot four ordered pairs for the line $y = 2x - 1$.

On your home screen generate four values for *x* in list **L₁** as follows.

$seq(X, X, -2, 4, 2)$ STO▶ **L₁** ENTER.

Enter the corresponding values for *y* in list **L₂** as follows.

2 **L₁** $- 1$ STO▶ **L₂** ENTER.

Access **STAT PLOT**, press 1, and turn "On" Plot 1. Check that your **Xlist** is **L₁** and your **Ylist** is **L₂**. Then GRAPH in a standard
● viewing window as shown at the right.

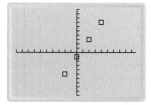

EXERCISES

1. Use lists to plot ordered pairs (x, y) for the relationship $y = 180(x - 2)$. Use the command $seq(X, X, 3, 12)$ to generate L_1 values. Store $180(L_1 - 2)$ into an L_2 list of Y values. Graph the points in an appropriate viewing window.

2. The relationship $y = 180(n - 2)/n$ or $Y_1 = 180(X - 2)/X$ gives the measure of one angle of a regular polygon for each value of *n* or X.
   a. Display a table of regular-polygon angle measures. Use TblStart = 3.
   b. Scroll down your table. What happens to the angle measures as X gets large?
   c. Make L_1, L_2 lists of (n, y) pairs for $n = 3$ to 20.
   d. Plot the (n, y) points. What happens to the angle measures as *n* gets large?

3. The formula $y = 360/n$ gives an exterior-angle measure for a regular *n*-gon. Make L_1, L_2 lists of (n, y) values and plot points. What happens to exterior-angle measures as *n* gets large?

4. Display $180(n - 2)/n$ values and $360/n$ values side by side in a table. What is the sum of the two values in each row? Show that this is the sum for every value of *n*.

Writing Extended Responses

An extended-response question is usually worth a maximum of 4 points and has multiple parts. To get full credit, you need to answer each part and show all your work or justify your reasoning.

EXAMPLE

Algebra Use the triangle at the right.
a. Write an equation that you can use to find the value of x.
b. Show how to solve the equation for x.
c. Find the measure of the smallest interior angle of the triangle.

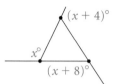

Below are two responses and the amount of credit each received.

4 points	3 points
$x + (x + 4) + (x + 8) = 360$ $3x + 12 = 360$ $3x = 348$ $x = 116$	$180 = 180 - x + 180 - (x + 4) + 180 - (x + 8)$ $180 = 540 - 3x + 12$ $-372 = -3x$ $x = 124$
Exterior angles are 116°, 120°, and 124°. Interior angles are 64°, 60°, 56°. The smallest interior angle has measure 56.	Ext. angles: 124°, 128°, 132° Int. angles: 56°, 52°, 48° The smallest int. angle is 48°.

The 4-point response shows a correct equation and solution. The student examined all the interior angles to find the one with the smallest measure.

There is an error in the 3-point response, but the student completed the problem and answered each part.

If you make an error, you can still get some credit. It is important that you complete each part of the problem or explain how you could do the problem.

EXERCISES

Use the Example above to do each exercise.

1. Where did the student make an error in the 3-point response?

2. When answering an extended-response question, you can describe how you are going to do the problem and then carry out the steps. For the Example exercise, begin with the following observation and write a 4-point response.

   The largest exterior angle has measure $x + 8$, and the smallest interior angle is a supplement of that angle.

3. An even more impressive 4-point response would show a "Check" of the work. What good "Check" of the work can you insert between the last two sentences?

Chapter Review

Vocabulary Review

 acute triangle (p. 148)
alternate exterior angles (p. 129)
alternate interior angles (p. 127)
concave polygon (p. 158)
convex polygon (p. 158)
corresponding angles (p. 127)
equiangular triangle (p. 148)
equiangular polygon (p. 160)
equilateral triangle (p. 148)

equilateral polygon (p. 160)
exterior angle of a polygon (p. 149)
flow proof (p. 135)
isosceles triangle (p. 148)
obtuse triangle (p. 148)
point-slope form (p. 168)
polygon (p. 157)
regular polygon (p. 160)
remote interior angles (p. 149)

right triangle (p. 148)
same-side exterior angles (p. 129)
same-side interior angles (p. 127)
scalene triangle (p. 148)
slope-intercept form (p. 166)
standard form of a
    linear equation (p. 167)
transversal (p. 127)
two-column proof (p. 129)

Choose the correct vocabulary term to complete each sentence.

1. In a triangle, an angle is right, obtuse, or ___?___ .

2. A(n) ___?___ angle has a measure between 90 and 180.

3. When two coplanar lines are cut by a transversal, two angles that are in similar positions on the same side of the transversal are called ___?___ .

4. The measure of a(n) ___?___ angle of a triangle is equal to the sum of the measures of its two remote interior angles.

5. A polygon is ___?___ if no diagonal contains points outside the polygon.

6. A(n) ___?___ polygon has all angles congruent.

7. A(n) ___?___ polygon is both equiangular and equilateral.

8. The linear equation $y - 3 = 4(x + 5)$ is written in ___?___ form.

9. From the ___?___ form of a linear equation, you can easily read the value of the slope and the value of the y-intercept.

10. When two coplanar lines are cut by a transversal, the angles between the two lines and on opposite sides of the transversal are called ___?___ .

Go Online
PHSchool.com
For: Vocabulary quiz
Web Code: auj-0351

Skills and Concepts

3-1 Objectives

▼ To identify angles formed by two lines and a transversal

▼ To prove and use properties of parallel lines

A **transversal** is a line that intersects two coplanar lines at two distinct points.
$\angle 1$ and $\angle 4$ are **corresponding angles.**
$\angle 3$ and $\angle 4$ are **alternate interior angles.**
$\angle 2$ and $\angle 4$ are **same-side interior angles.**

If two parallel lines are cut by a transversal, then

• corresponding angles are congruent.

• alternate interior angles are congruent.

• same-side interior angles are supplementary.

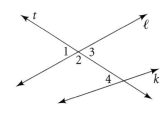

11. Suppose ℓ and k in the diagram above are parallel. If $m\angle 1 = 59$, what are the measures of $\angle 2$, $\angle 3$, and $\angle 4$?

Find $m\angle 1$ and then $m\angle 2$. Justify each answer.

12.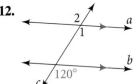

13.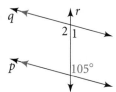

14.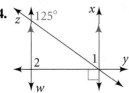

Two lines cut by a transversal are parallel if corresponding angles are congruent, alternate interior angles are congruent, or same-side interior angles are supplementary.

You can construct the line parallel to a given line through a given point not on the line. You can also construct the perpendicular to a given line at a given point on the line or through a given point not on the line.

$\boxed{x^2}$ **Algebra** Find the value of x for which $\ell \parallel m$.

15.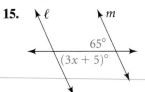

16.

17.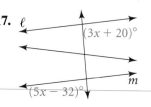

Use the segments at the right for Exercises 18 and 19.

18. Construct a rectangle with side lengths a and b.

19. Construct a quadrilateral with one pair of parallel opposite sides, each side of length $2a$.

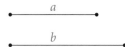

If two lines are perpendicular to the same line, then the two lines are parallel. If a line is perpendicular to one of two parallel lines, then it is perpendicular to both lines. Both of these statements are true only if the lines are coplanar.

Use the diagram at the right for Exercises 20 and 21. State the theorem that justifies each statement.

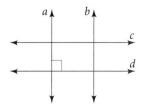

20. If $b \perp c$ and $b \perp d$, then $c \parallel d$.

21. If $c \parallel d$, then $a \perp c$.

The sum of the measures of the angles of a triangle is 180. The measure of each **exterior angle** of a triangle equals the sum of the measures of its two **remote interior angles.**

Find the values of the variables. Then classify each triangle by its sides and angles.

22.

23.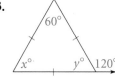

24.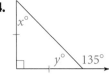

In each of Exercises 25–28, the measures of the three angles of a triangle are given. Find the value of x and then classify the triangle by its angles.

25. $x + 10, x - 20, x + 25$ **26.** $x, 2x, 3x$

27. $20x + 10, 30x - 2, 7x + 1$ **28.** $10x - 3, 14x - 20, x + 3$

29. In a right triangle, what is always true about the angles?

3-5 Objectives

▼ To classify polygons

▼ To find the sums of the measures of the interior and exterior angles of polygons

A **polygon** is a closed plane figure with at least three sides. To name a polygon, start at any vertex and list the vertices consecutively in a clockwise or counterclockwise direction. A polygon is **convex** if no diagonal contains points outside the polygon. Otherwise, it is **concave.**

An **equilateral polygon** has all sides congruent. An **equiangular polygon** has all angles congruent. A **regular polygon** is equilateral and equiangular.

The sum of the measures of the angles of an n-gon is $(n - 2)180$. The sum of the measures of the exterior angles of an n-gon, one at each vertex, is 360.

Find the measure of an interior angle and an exterior angle of each regular polygon.

30. a hexagon   **31.** an octagon   **32.** a decagon   **33.** a 24-gon

34. What is the sum of the measures of the exterior angles for each polygon in Exercises 30–33?

3-6 Objectives

▼ To graph lines given their equations

▼ To write equations of lines

When a linear equation is in **slope-intercept form,** $y = mx + b$, the slope m and the y-intercept b are easily identified. When a linear equation is in **point-slope form,** $(y - y_1) = m(x - x_1)$, point (x_1, y_1) and slope m can easily be identified. The equation $Ax + By = C$, where A and B are not both zero, is in **standard form.** When a linear equation is in standard form, the x- and y-intercepts are readily found.

35. Name the slope and y-intercept of $y = 2x - 1$. Graph the line.

36. Name a point on and the slope of $y - 3 = -2(x + 5)$. Graph the line.

37. Graph $y = -\frac{1}{2}$.   **38.** Graph $3x - 4y = 12$.

39. Write an equation for the vertical line that contains $A(6, -9)$.

3-7 Objectives

▼ To relate slope and parallel lines

▼ To relate slope and perpendicular lines

The slopes of two nonvertical parallel lines are equal. All vertical lines are parallel.

The product of the slopes of two nonvertical perpendicular lines is -1. In a plane, every vertical line is perpendicular to every horizontal line.

Determine whether $\overleftrightarrow{AB}$ and $\overleftrightarrow{CD}$ are *parallel, perpendicular,* or *neither.*

40. $A(-1, -4), B(2, 11), C(1, 1), D(4, 10)$ **41.** $A(2, 8), B(-1, -2), C(3, 7), D(0, -3)$

42. $A(-3, 3), B(0, 2), C(1, 3), D(-2, -6)$ **43.** $A(-1, 3), B(4, 8), C(-6, 0), D(2, 8)$

44. Writing For $B(4, 8)$ and $D(2, 8)$, find the slope of $\overleftrightarrow{BD}$. Explain why the slope of any horizontal line is zero.

Chapter Test

Go Online
PHSchool.com

For: Chapter Test
Web Code: aua-0352

Use a protractor and a centimeter ruler. Classify each triangle by its angles and its sides.

1.

2.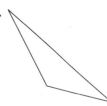

Find $m\angle 1$, **then** $m\angle 2$. **Justify each answer.**

3.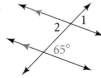

4.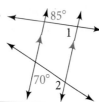

5.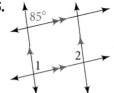

6.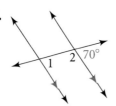

Two lines are parallel and cut by a transversal. Write *yes* or *no* to indicate whether the numbers given could be the measures of a pair of same-side interior angles.

7. 40 and 140

8. 90 and 90

9. 60 and 60

10. 27 and 27

x^2 **Algebra Find the value of** x **for which** $\ell \parallel m$.

11.

12.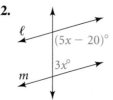

13.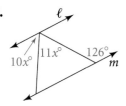

14.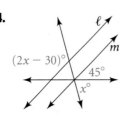

15. Draw a line m and a point T not on the line. Construct the line through T perpendicular to m.

16. Draw an angle, $\angle ABC$. Then construct line m through A so that $m \parallel \overleftrightarrow{BC}$.

17. Open-Ended The letter **F** illustrates a pair of same-side interior angles and a pair of corresponding angles. Find a letter that illustrates alternate interior angles.

18. Open-Ended Describe two corresponding angles formed by lines in your classroom.

19. Supply the reason for each step in the proof.

Given: $\ell \parallel m$ and $\angle 4 \cong \angle 2$
Prove: $n \parallel p$

1. $\ell \parallel m$ **a.** ?

2. $\angle 1 \cong \angle 2$ **b.** ?

3. $\angle 4 \cong \angle 2$ **c.** ?

4. $\angle 1 \cong \angle 4$ **d.** ?

5. $n \parallel p$ **e.** ?

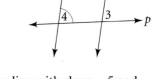

20. Write an equation of the line with slope -5 and containing $A(3, -1)$.

21. Writing Explain how you can determine whether a polygon is concave or convex.

Find the value of each variable.

22.

23.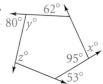

Sketch each pair of lines. Tell whether they are *parallel*, *perpendicular*, or *neither*.

24. $y = 4x + 7$
$y = -\frac{1}{4}x - 3$

25. $y = 3x - 4$
$y = 3x + 1$

26. $y = x + 5$
$y = -5x - 1$

27. $y = -3$
$x = 10$

28. What is the measure of an exterior angle of a regular 12-gon?

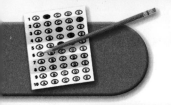

Regents Test Prep

Reading Comprehension Read the passage below. Then answer the questions on the basis of what is *stated* or *implied* in the passage.

Airport Plans Civic leaders in Chicago are discussing a plan to expand O'Hare Airport and build new runways. The airport's present runways, labeled A through G in the diagram, include three pairs that are parallel. Runways B and D intersect to form right angles.

The crisscrossing layout and takeoff/landing patterns prevent the runways from being used to their full capacity. This can cause travel delays. If the planned new runways are built, then four landings and two takeoffs can happen simultaneously. The plan, however, requires an additional 292 acres of land to the south and 141 acres to the north. Taking this land would lead to the demolition of 240 apartment units and more than 300 houses and 70 businesses.

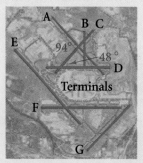

---

1. At present, how many runways are there?
   (1) 3 **(2)** 4 **(3)** 6 **(4)** 7

2. Which list best describes the parallel runways?
   (1) A ∥ C and B ∥ D **(2)** A ∥ E, C ∥ G, and D ∥ F
   (3) A ∥ E and D ∥ F **(4)** A ∥ C, B ∥ D, and E ∥ G

3. Which runways are perpendicular?
   (1) A ⊥ E **(2)** B ⊥ D **(3)** C ⊥ D **(4)** E ⊥ F

4. Why can runways E and G not be used to their full capacity?
   I. crisscrossing layout
   II. crisscrossing takeoff pattern
   III. crisscrossing landing pattern

   (1) I only **(2)** I and II only
   (3) II and III only **(4)** I, II, and III

5. What is the measure of the obtuse angle formed where runways B and C intersect?
   (1) 42 **(2)** 90 **(3)** 132 **(4)** 138

6. What is the measure of the acute angle formed where runways A and D intersect?
   (1) 46 **(2)** 48 **(3)** 86 **(4)** 94

7. If you extend runway C to meet runway E, what is the measure of the acute angle formed?
   (1) 42 **(2)** 48 **(3)** 86 **(4)** 90

8. If you extend runway E to meet runway G, what is the measure of the obtuse angle formed?
   (1) 86 **(2)** 90 **(3)** 94 **(4)** 104

9. At most, how many landings and takeoffs would be possible in one hour on the new runways?
   (1) 4 and 2 **(2)** 40 and 20
   (3) 240 and 120 **(4)** cannot be determined

10. How many acres larger than the current airport will the new airport be?
    (1) 141 **(2)** 161 **(3)** 292 **(4)** 433

11. Each apartment unit, house, or business is one "real estate unit." What is the average number of real estate units per acre that will be demolished in the plan? Explain your answer.

Suppose runways D and B are the *x*- and *y*-axes of a coordinate plane.

12. Which runway has a positive slope?
    (1) A **(2)** B **(3)** C **(4)** D

13. Which runway has a negative slope?
    (1) F **(2)** G **(3)** D **(4)** E

14. Which two runways have slopes whose product is approximately –1?
    (1) A, G **(2)** G, F **(3)** F, D **(4)** D, B

Activity Lab

NY **G.CN.6:**
Recognize and apply mathematics to situations in the outside world.

The Science of Reflection

Applying Parallel Lines When you look at the surface of a still pool of water, your reflection looks back at you. This is because the water acts like a mirror, reflecting a clear, although reversed, image. The type of reflection you see in a mirror depends on the surface of the mirror. Two flat mirrors placed at right angles to each other will seem to magnify the light hitting them by reflecting it directly back to its source.

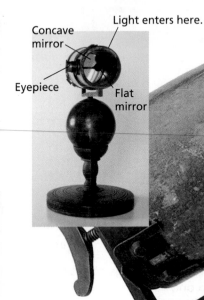

Light enters here.

Concave mirror

Eyepiece

Flat mirror

Eyepiece

Wooden ball mount allows telescope to pivot.

Focusing element

Sir Isaac Newton

Sir Isaac Newton (1642–1727) designed and built the first reflecting telescope. His telescope used mirrors rather than glass lenses to collect and focus light. Most telescopes used by amateur astronomers are reflecting telescopes.

Replica of Newton's telescope

Measuring Distances

Astronauts have placed a cube-corner reflector on the surface of the moon. Each corner provides three perpendicular reflecting planes. By measuring the time it takes a laser beam to bounce back from the reflector, scientists are able to measure the distance from Earth to the moon.

Activity 1

Examine the diagram below. Notice that the mirrors are perpendicular and that each angle of reflection is congruent to the corresponding angle of incidence. Given these two facts, explain why incident and reflected rays must be parallel.

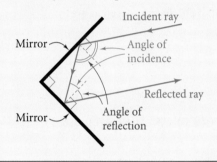

Incident ray
Mirror
Angle of incidence
Reflected ray
Mirror
Angle of reflection

The front view sparkles.

Cut Diamonds
A cut diamond reflects most of the light that falls on its front.

The rear view is dark.

Safety

A bicycle doesn't have the electric taillights that a car does, so it comes equipped with reflectors. Some helmets also have reflectors.

Activity 2

Materials: two flat mirrors

Arrange two flat mirrors so that they form a right angle. Look at your face in one of the mirrors and wink. Then look into the seam where the two mirrors meet, find your face, and wink. How does the reflection in the perpendicular mirrors differ from the reflection in the single mirror?

Hall of Mirrors

The Hall of Mirrors at France's Palace of Versailles is brightly lit even on a cloudy day, because each of its windows is placed opposite a mirror of the same size and shape.

PHSchool.com

For: Information about mirrors and reflections
Web Code: aue-0353

What You've Learned

NY **Learning Standards for Mathematics**

In Chapter 1, you learned the meanings of congruent segments and congruent angles.

● **G.RP.7:** In Chapter 2, you used deductive reasoning to prove angles congruent.

● **G.RP.2:** In Chapter 3, you developed relationships involving congruent angles, parallel lines, perpendicular lines, and polygons.

Check Your Readiness   **GO** **for Help** to the Lesson in green.

The Distance Formula (Lesson 1-8)

Find the lengths of the sides of △ABC.

1. $A(3, 1), B(-1, 1), C(-1, -2)$ **2.** $A(-3, 2), B(-3, -6), C(8, 6)$ **3.** $A(-1, -2), B(6, 1), C(2, 5)$

Proving Angles Congruent (Lesson 2-5)

Draw a conclusion based on the information given.

4. $\angle A$ is supplementary to $\angle B$; $\angle C$ is supplementary to $\angle B$

5. $\angle A$ is supplementary to $\angle B$; $\angle A \cong \angle B$

6. $\angle 1$ is complementary to $\angle 2$

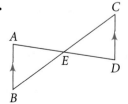

7. $\overrightarrow{FA} \perp \overrightarrow{FC}; \overrightarrow{FB} \perp \overrightarrow{FD}$

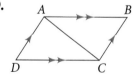

Parallel Lines and the Triangle Angle-Sum Theorem (Lesson 3-3)

What can you conclude from each diagram?

8.

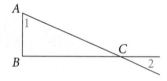

9.

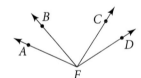

10.

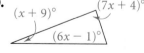

$(x + 9)°$ $(7x + 4)°$ $(6x - 1)°$

Congruent Triangles

◀)) Key Vocabulary

- base of an isosceles triangle (p. 228)
- base angles of an isosceles triangle (p. 228)
- congruent polygons (p. 198)
- corollary (p. 229)
- CPCTC (corresponding parts of congruent triangles are congruent) (p. 221)
- hypotenuse (p. 235)
- legs of a right triangle (p. 235)
- legs of an isosceles triangle (p. 228)
- vertex angle of an isosceles triangle (p. 228)

What You'll Learn Next

(NY) Learning Standards for Mathematics

- **G.R.8:** In this chapter, you will learn the meaning of congruent polygons.

- **G.G.38:** You will learn how to prove two triangles congruent by five different methods.

- **G.G.31:** By learning how to prove triangles congruent, you will discover properties of an isosceles triangle.

- **G.G.29:** You will also learn how to draw other conclusions, once two triangles have been proved congruent.

Data Analysis **Activity Lab** You will do activities involving games and congruent triangles on pages 254 and 255.

Congruent Figures

 Learning Standards for Mathematics

G.G.27 Write a proof arguing from a given hypothesis to a given conclusion.

 **Check Skills You'll Need**

 **for Help** Algebra 1 Review, page 30

x^2 **Algebra** Solve each equation.

1. $x + 6 = 25$ **2.** $x + 7 + 13 = 33$

3. $5x = 540$ **4.** $x + 10 = 2x$

5. For the triangle at the right, use the Triangle Angle-Sum Theorem to find the value of y.

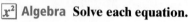 **New Vocabulary** • congruent polygons

---

1   Congruent Figures

Congruent figures have the same size and shape. When two figures are congruent, you can move one so that it fits exactly on the other one. Three ways to make such a move—a slide, a flip, and a turn—are shown below. You will learn much more about slides, flips, and turns in Chapter 9.

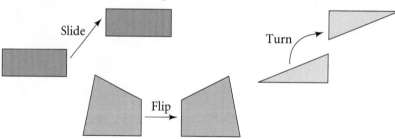

Congruent polygons have congruent corresponding parts—their matching sides and angles. Matching vertices are corresponding vertices. When you name congruent polygons, always list corresponding vertices in the same order.

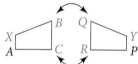

C corresponds to R.
$\angle B$ corresponds to $\angle Q$.
$\overline{AX}$ corresponds to $\overline{PY}$.
$ACBX \cong PRQY$

---

1 EXAMPLE   Naming Congruent Parts

$\triangle TJD \cong \triangle RCF$. List the congruent corresponding parts.

Sides: $\overline{TJ} \cong \overline{RC}$ $\overline{JD} \cong \overline{CF}$ $\overline{DT} \cong \overline{FR}$

Angles: $\angle T \cong \angle R$ $\angle J \cong \angle C$ $\angle D \cong \angle F$

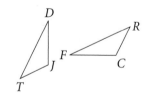

 Quick Check ❶ $\triangle WYS \cong \triangle MKV$. List the congruent corresponding parts. Use three letters to name each angle.

2 EXAMPLE **Real-World** 🌐 **Connection**

Spacecraft The fins of the Space Shuttle suggest congruent pentagons. Find $m\angle B$.

In the congruent pentagons, B corresponds to E, so you know that $\angle B \cong \angle E$. You can find $m\angle B$ by first finding $m\angle E$.

Use the Polygon Angle-Sum Theorem. It tells you that the sum of the measures of the angles of pentagon *SPACE* is $(5 - 2)180$, or 540.

$$m\angle S + m\angle P + m\angle A + m\angle C + m\angle E = 540 \quad \text{Polygon Angle-Sum Theorem}$$
$$88 + 90 + 90 + 132 + m\angle E = 540 \quad \text{Substitute.}$$
$$400 + m\angle E = 540 \quad \text{Simplify.}$$
$$m\angle E = 140 \quad \text{Subtract 400 from each side.}$$

● $m\angle B = m\angle E$, so $m\angle B = 140$.

 **Quick Check** ❷ It is given that $\triangle WYS \cong \triangle MKV$. If $m\angle Y = 35$, what is $m\angle K$? Explain.

Two triangles are congruent when they have three pairs of congruent corresponding sides and three pairs of congruent corresponding angles.

3 EXAMPLE **Finding Congruent Triangles**

Decide whether the triangles are congruent. Justify your answer.

$\overline{AC} \cong \overline{EC}$	**Given**
$\overline{AB} \cong \overline{ED}$	**AB = 3 = ED**
$\overline{BC} \cong \overline{DC}$	**BC = 4 = DC**
$\angle A \cong \angle E$	**Given**
$\angle B \cong \angle D$	**All right angles are congruent.**
$\angle BCA \cong \angle DCE$	**Vertical angles are congruent.**

● $\triangle ABC \cong \triangle EDC$ by the definition of congruent triangles.

 **Quick Check** ❸ Can you conclude $\triangle JKL \cong \triangle MNL$? Justify your answer.

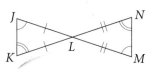

The next theorem follows from the Triangle Angle-Sum Theorem. In Exercise 45, you will show why this theorem is true.

 Key Concepts

Theorem 4-1

If two angles of one triangle are congruent to two angles of another triangle, then the third angles are congruent.

$\angle C \cong \angle F$

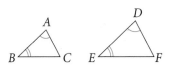

Example 4 shows a typical setup—Given, Prove, and diagram—that requires a proof from you. The form of proof you use is generally a matter of preference.

Proof **4 EXAMPLE** **Proving Triangles Congruent**

Given: $\overline{PQ} \cong \overline{PS}, \overline{QR} \cong \overline{SR}, \angle Q \cong \angle S, \angle QPR \cong \angle SPR$
Prove: $\triangle PQR \cong \triangle PSR$

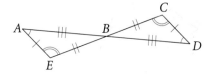

Statements	Reasons
1. $\overline{PQ} \cong \overline{PS}, \overline{QR} \cong \overline{SR}$	**1.** Given
2. $\overline{PR} \cong \overline{PR}$	**2.** Reflexive Property of $\cong$
3. $\angle Q \cong \angle S, \angle QPR \cong \angle SPR$	**3.** Given
4. $\angle QRP \cong \angle SRP$	**4.** Theorem 4-1
● **5.** $\triangle PQR \cong \triangle PSR$	**5.** Definition of $\cong$ triangles

✓ **Quick Check** **4** **Given:** $\angle A \cong \angle D, \angle E \cong \angle C, \overline{AE} \cong \overline{DC},$
$\overline{EB} \cong \overline{CB}, \overline{BA} \cong \overline{BD}$
Prove: $\triangle AEB \cong \triangle DCB$

EXERCISES

For more exercises, see *Extra Skill, Word Problem, and Proof Practice.*

Practice and Problem Solving

A Practice by Example

Example 1
(page 198)

GO for Help

1. Building Builders use the King Post truss, below left, for the top of a simple structure. In this truss, $\triangle ABC \cong \triangle ABD$. List the congruent corresponding parts.

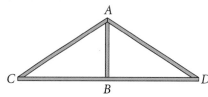
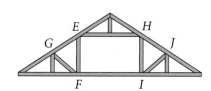

2. The Attic Frame truss, above right, provides open space in the center for storage. In this truss, $\triangle EFG \cong \triangle HIJ$. List the congruent corresponding parts.

$\triangle LMC \cong \triangle BJK$. **Complete the congruence statements.**

3. $\overline{LC} \cong \underline{\ ?\ }$ **4.** $\overline{KJ} \cong \underline{\ ?\ }$

5. $\overline{JB} \cong \underline{\ ?\ }$ **6.** $\angle L \cong \underline{\ ?\ }$

7. $\angle K \cong \underline{\ ?\ }$ **8.** $\angle M \cong \underline{\ ?\ }$

9. $\triangle CML \cong \underline{\ ?\ }$ **10.** $\triangle KBJ \cong \underline{\ ?\ }$

11. $\triangle MLC \cong \underline{\ ?\ }$ **12.** $\triangle JKB \cong \underline{\ ?\ }$

Real-World Connection

Exposed beams show the congruent triangles used in Tudor architecture.

13. The last piece of the jigsaw puzzle must be put into place. Name the corners that correspond to corners $A, B, C,$ and D.

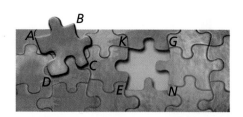

$POLY \cong SIDE$. **List each of the following.**

14. four pairs of congruent sides   **15.** four pairs of congruent angles

Example 2
(page 199)

In the two lifeguard chairs, $ABCD \cong FGHI$. Find the measure of the angle or the length of the side.

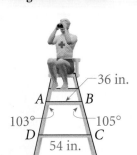

16. $\overline{AD}$ **17.** $\overline{HI}$

18. $\angle FGH$ **19.** $\angle ADC$

20. $\overline{FG}$ **21.** $\overline{BC}$

22. $\angle DCB$ **23.** $\angle IFG$

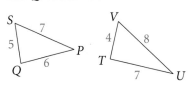

Example 3
(page 199)

In Exercises 24–27, can you conclude the figures are congruent? Justify each answer.

24. $\triangle TRK$ and $\triangle TUK$

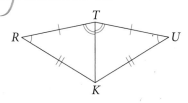

25. $\triangle SPQ$ and $\triangle TUV$

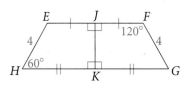

26. $\triangle XYZ$ and $\triangle XYP$

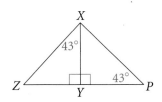

27. $HEJK$ and $GFJK$

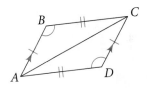

Example 4 *Proof* **28. Given:** $\overline{AB} \parallel \overline{DC}, \angle B \cong \angle D,$
(page 200) $\overline{AB} \cong \overline{DC}, \overline{BC} \cong \overline{AD}$

Prove: $\triangle ABC \cong \triangle CDA$

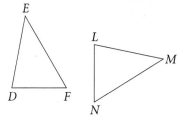

B **Apply Your Skills**

29. Multiple Choice If $\triangle DEF \cong \triangle LMN$, which of the following must be a correct congruence statement?

Ⓐ $\overline{DE} \cong \overline{LN}$ Ⓑ $\angle N \cong \angle F$
Ⓒ $\overline{FE} \cong \overline{NL}$ Ⓓ $\angle M \cong \angle F$

x^2 **Algebra** Find the values of the variables.

30.

$\triangle ABC \cong \triangle KLM$

31.

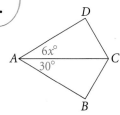

$\triangle ACD \cong \triangle ACB$

GO for Help

To review the Triangle Angle-Sum Theorem, go to Lesson 3-4.

x^2 **Algebra** $\triangle ABC \cong \triangle DEF$. **Find the measures of the given angles or the lengths of the given sides.**

32. $m\angle A = x + 10, m\angle D = 2x$

33. $m\angle B = 3y, m\angle E = 21$

34. $BC = 3z + 2, EF = z + 6$

35. $AC = 7a + 5, DF = 5a + 9$

36. Parquet Floor Explain why it is important that $PACH \cong OLDE$.

37. Sports Cards The 225 cards in Tracy's sports card collection are rectangles of three different sizes. Describe how Tracy could quickly sort the cards.

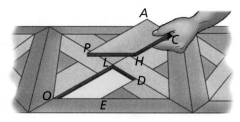

Exercise 36

Write a congruence statement for each pair of triangles.

38.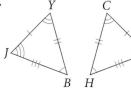

39.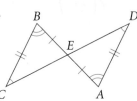

E is the midpoint of $\overline{CD}$.

40.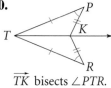

$\overrightarrow{TK}$ bisects $\angle PTR$.

41. Complete in two different ways:

$\triangle JLM \cong \underline{\ ?\ }$

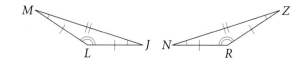

42. Writing Die-cast toys are a popular collector's item. Explain why the two die-cast toys that Pearl is studying at the left have congruent shapes.

43. Open-Ended Write a congruence statement for two triangles. List the congruent sides and angles.

Proof **44. Given:** $\overline{PR} \parallel \overline{TQ}, \overline{PR} \cong \overline{TQ}$
$\overline{PS} \cong \overline{QS}, \overline{PQ}$ bisects $\overline{RT}$.
Prove: $\triangle PRS \cong \triangle QTS$

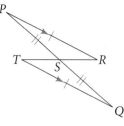

45. Prove Theorem 4-1.
Given: $\angle A \cong \angle D, \angle B \cong \angle E$
Prove: $\angle C \cong \angle F$

Exercise 42

Coordinate Geometry **Vertices of** $\triangle GHJ$ **are** $G(-2, -1), H(-2, 3),$ **and** $J(1, 3)$.

46. $\triangle KLM \cong \triangle GHJ$. Find $KL, LM,$ and KM.

47. If L and M have coordinates $L(3, -3)$ and $M(6, -3)$, how many pairs of coordinates are possible for K? Find one such pair.

48. a. How many quadrilaterals (convex and concave) with different shapes or sizes can you make on a three-by-three geoboard? One is shown at the right.
b. How many quadrilaterals of each type are there?

Gridded Response

Use the diagrams at the right for Exercises 49–51.
$ABCDE \cong PFKYM$.

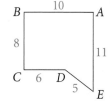
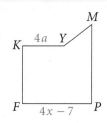

49. What is the value of a?

50. What is the value of x?

51. What is the perimeter of $PFKYM$?

52. $\triangle HLN \cong \triangle GST$, $m\angle H = 66$, and $m\angle S = 42$. What is $m\angle T$?

Mixed Review

Lesson 3-8

Constructions For Exercises 53 and 54, construct the geometric figure.

53. a square

54. a rectangle whose length is twice its width

Lesson 3-4

55. Find $m\angle A$ in the figure at the right.

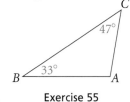

Exercise 55

Lesson 2-4

Use the given property to complete each statement.

56. Symmetric Property of Equality

If $PQ = RS$, then ? .

57. Reflexive Property of Congruence

$\angle 1 \cong$?

58. Addition Property of Equality

If $m\angle A - 4 = 8$, then $m\angle A =$? .

59. Transitive Property of Congruence

If $\overline{AB} \cong \overline{DE}$ and $\overline{DE} \cong \overline{GH}$, then ? .

Geometry at Work

Die Casting

Two centuries ago, people manufactured articles by hand. Each article produced was slightly different from every other. In 1800, inventor Eli Whitney recognized that he could speed up manufacturing by using congruent parts. Whitney made a die, or mold, for each part of a musket he was producing for the U.S. Army. This allowed workers to rapidly cast the parts and assemble them into standard-sized muskets. It ushered in the era of mass production.

Today, die makers are highly skilled industrial workers who shape dies out of metal, plastic, rubber, and other materials. Machines create and assemble the congruent die-cast parts into standard-sized objects, like the die-cast toy cars at the left. Other workers supply a final inspection and skilled hand finishing.

For: Information about die casting
Web Code: aub-2031

Building Congruent Triangles

Are there shortcuts for finding congruent triangles? Explore this with your classmates by building and comparing triangles.

NY G.PS.7: Work in collaboration with others to propose, critique, evaluate, and value alternative approaches to problem solving.

1 ACTIVITY

- Cut three strips of paper with lengths 7 in., 8 in., and 9 in.
- Use the strips to form a triangle on top of a piece of tracing paper. Mark each vertex of the triangle on your tracing paper.
- Remove the strips. Use a straightedge to carefully draw your triangle.

 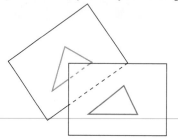

1. Is your triangle congruent to your classmates' triangles? Place your paper on top of a classmate's paper. Try to make your triangle fit exactly on the other triangle.

2. Make a conjecture: What seems to be true about two triangles in which three sides of one are congruent to three sides of the other?

3. As a class, choose three different lengths and repeat the steps above.
 a. Are the triangles congruent?
 b. Does this support your conjecture from Exercise 2?

2 ACTIVITY

- Use a straightedge to carefully draw a triangle on tracing paper. Label it △ABC.
- Use a protractor. Carefully measure ∠A and ∠B. Use a ruler to measure the side between them, $\overline{AB}$.
- Write the measurements on an index card and swap cards with a classmate. Draw a triangle using only your classmate's measurements.

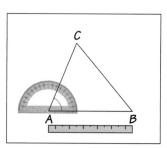

4. Compare your new triangle to your classmate's original △ABC. Are the triangles congruent? Is your classmate's new triangle congruent to your original △ABC?

5. Make a conjecture: What seems to be true when two triangles have congruent sides between two pairs of congruent angles?

6. Make a conjecture: At most, how many triangle measurements must you know to guarantee that all triangles built with those measurements will be congruent?

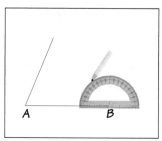

Triangle Congruence by SSS and SAS

 Learning Standards for Mathematics

G.G.28 Determine the congruence of two triangles by using one of the five congruence techniques (SSS, SAS, ASA, AAS, HL), given sufficient information about the sides and/or angles of two congruent triangles.

Check Skills You'll Need

 for Help Lesson 1-6

What can you conclude from each diagram?

1.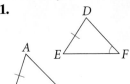

2.

3.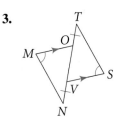

---

1 Using the SSS and SAS Postulates

In Lesson 4-1 you learned that if two triangles have three pairs of congruent corresponding angles and three pairs of congruent corresponding sides, then the triangles are congruent.

If you know this,

$\angle A \cong \angle X$ $\overline{AB} \cong \overline{XY}$

$\angle B \cong \angle Y$ $\overline{AC} \cong \overline{XZ}$

$\angle C \cong \angle Z$ $\overline{BC} \cong \overline{YZ}$

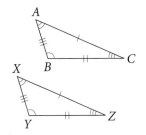

then you know this.

$\triangle ABC \cong \triangle XYZ$

However, you do not need to know that all six corresponding parts are congruent in order to conclude that two triangles are congruent. It is enough to know only that corresponding sides are congruent.

 Key Concepts

Postulate 4-1	**Side-Side-Side (SSS) Postulate**

If the three sides of one triangle are congruent to the three sides of another triangle, then the two triangles are congruent.

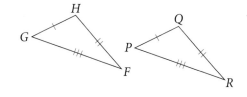

$\triangle GHF \cong \triangle PQR$

As described on page 101, postulates are the self-evident truths in a mathematical system. Postulate 4-1 is perhaps the most self-evident truth about triangles. It agrees with the notion that triangles are rigid figures, a property demonstrated in bicycle frames and steel bridges.

1 EXAMPLE **Using SSS**

Bridge Design The bridge girders are the same size, as marked. Write a flow proof.

Given: $\overline{AB} \cong \overline{CB}, \overline{AD} \cong \overline{CD}$

Prove: $\triangle ABD \cong \triangle CBD$

Plan: To prove the triangles congruent by the SSS Postulate, the three pairs of sides need to be congruent.

Proof:

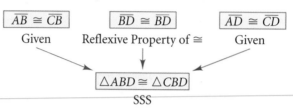

$\overline{AB} \cong \overline{CB}$	$\overline{BD} \cong \overline{BD}$	$\overline{AD} \cong \overline{CD}$
Given	Reflexive Property of $\cong$	Given

$\triangle ABD \cong \triangle CBD$
SSS

✓ Quick Check **1** **Given:** $\overline{HF} \cong \overline{HJ}, \overline{FG} \cong \overline{JK}$,
H is the midpoint of $\overline{GK}$.

Prove: $\triangle FGH \cong \triangle JKH$

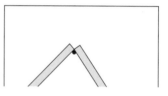

🔑 **Key Concepts**

Postulate 4-2	**Side-Angle-Side (SAS) Postulate**

If two sides and the included angle of one triangle are congruent to two sides and the included angle of another triangle, then the two triangles are congruent.

$\triangle BCA \cong \triangle FDE$

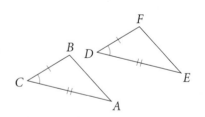

The word *included* is used frequently when referring to the angles and the sides of a triangle.

$\overline{BX}$ is included between $\angle B$ and $\angle X$.

$\angle N$ is included between $\overline{NB}$ and $\overline{NX}$.

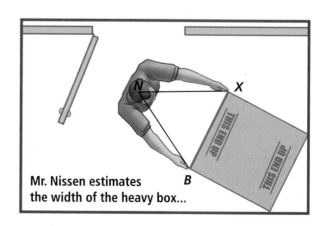

Mr. Nissen estimates the width of the heavy box...

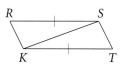

...to help decide whether the box will fit through the doorway.

Mr. Nissen kept his arms at a fixed angle as he moved from the box to the doorway. The triangle he used beside the box is congruent to the triangle he used beside the doorway. He knows the two triangles are congruent because two sides and the included angle of one are congruent to two sides and the included angle of the other.

2 EXAMPLE Using SAS

$\overline{RS} \cong \overline{TK}$. What other information do you need to prove $\triangle RSK \cong \triangle TKS$ by SAS?

You are given $\overline{RS} \cong \overline{TK}$. Also, $\overline{KS} \cong \overline{KS}$ by the Reflexive Property of Congruence. Therefore, if you know $\angle RSK \cong \angle TKS$, you can prove $\triangle RSK \cong \triangle TKS$ by SAS.

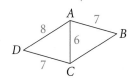

Quick Check ❷ What other information do you need to prove $\triangle ABC \cong \triangle CDA$ by SAS?

In Lesson 1-7, you learned to construct congruent segments using a compass open to a fixed angle. SAS tells you that the triangles outlined here are congruent, so $\overline{AB} \cong \overline{CD}$.

3 EXAMPLE Are the Triangles Congruent?

From the information given, can you prove $\triangle RED \cong \triangle CAT$? Explain.

Given: $\overline{RE} \cong \overline{CA}, \overline{RD} \cong \overline{CT}, \angle R \cong \angle T$

No, there is not enough information to prove $\triangle RED \cong \triangle CAT$. $\angle T$ is not included between $\overline{CA}$ and $\overline{CT}$. $\triangle RED$ may or may not be congruent to $\triangle CAT$.

Quick Check ❸ From the information given, can you prove $\triangle AEB \cong \triangle DBC$? Explain.

Given: $\overline{EB} \cong \overline{CB}, \overline{AE} \cong \overline{DB}$

Practice and Problem Solving

A **Practice by Example**

Example 1
(page 206)

GO for Help

1. Developing Proof Copy and complete the flow proof.

Given: $\overline{JK} \cong \overline{LM}, \overline{JM} \cong \overline{LK}$

Prove: $\triangle JKM \cong \triangle LMK$

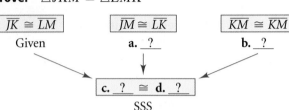

$\overline{JK} \cong \overline{LM}$	$\overline{JM} \cong \overline{LK}$	$\overline{KM} \cong \overline{KM}$
Given	**a.** ?	**b.** ?

c. ? $\cong$ **d.** ?

SSS

Proof 2. Given: $\overline{IE} \cong \overline{GH}, \overline{EF} \cong \overline{HF}$,
F is the midpoint of $\overline{GI}$.

Prove: $\triangle EFI \cong \triangle HFG$

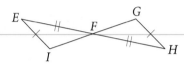

3. Given: $\overline{WZ} \cong \overline{ZS} \cong \overline{SD} \cong \overline{DW}$

Prove: $\triangle WZD \cong \triangle SDZ$

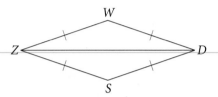

Example 2
(page 207)

Is the information you are given enough for you to prove that the two triangles are congruent? Explain.

4.

The vertical beam $\overline{OB}$ is perpendicular to the porch roof. $P, O,$ and R are equally spaced.

5.

The diagonal legs have equal lengths and are joined at their midpoints.

6.

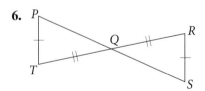

7.

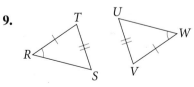

Example 3
(page 207)

What other information, if any, do you need to prove the two triangles congruent by SSS or SAS?

8.

9.

Copy the triangle. Start at any vertex and label the triangle as △WVU.

10. What sides include ∠V?

11. What angle is included between $\overline{WV}$ and $\overline{WU}$?

12. What angles include $\overline{UV}$?

13. What side is included between ∠W and ∠U?

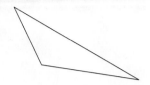

Name the indicated part(s) of △XYZ without drawing △XYZ.

14. the angle included between $\overline{XY}$ and $\overline{XZ}$ **15.** the sides that include ∠Z

From the information given in the diagram, can you prove that the two triangles are congruent? Explain.

16.

17.

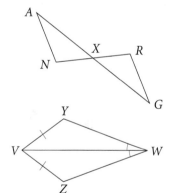

Proof 18. Given: X is the midpoint of $\overline{AG}$ and of $\overline{NR}$.
   Prove: △ANX ≅ △GRX

19. Multiple Choice What additional information do you need to prove the two triangles congruent by SAS?

Ⓐ $\overline{YW} \cong \overline{ZW}$
Ⓑ ∠Y ≅ ∠Z
Ⓒ ∠WVY ≅ ∠ZWV
Ⓓ no additional information needed

Is there enough information to prove the two triangles congruent? If so, write the congruence statement and name the postulate you would use. If not, write *not possible* and tell what other information you would need.

20.

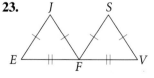

21.

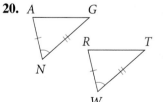

22.

23.

24.

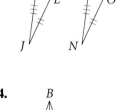

25.

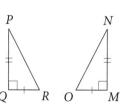

GO Online
Homework Video Tutor
Visit: PHSchool.com
Web Code: aue-0402

Real-World Connection

Four isosceles triangles cap the Smith Tower in Seattle.

Proof 26. Given: $\overline{GK}$ bisects $\angle JGM$, $\overline{GJ} \cong \overline{GM}$.

Prove: $\triangle GJK \cong \triangle GMK$

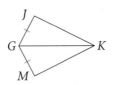

27. Given: $\overline{AE}$ and $\overline{BD}$ bisect each other.

Prove: $\triangle ACB \cong \triangle ECD$

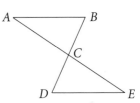

From the information given, can you prove the two triangles congruent? Explain.

28. $\triangle ABC$ and $\triangle DEF$ with $\angle A \cong \angle D, \angle B \cong \angle E, \angle C \cong \angle F$

29. $\triangle GHI$ and $\triangle JKL$ with $\overline{GH} \cong \overline{JK}, \overline{HI} \cong \overline{KL}, \angle I \cong \angle L$

30. $\triangle MNP$ and $\triangle QRS$ with $\overline{MN} \cong \overline{QR}, \angle N \cong \angle R, \overline{NP} \cong \overline{RS}$

Constructions Use a straightedge to draw $\triangle JKL$. Construct $\triangle MNP \cong \triangle JKL$ using the given postulate.

31. SSS   **32.** SAS

33. a. Open-Ended List three real-life uses of congruent triangles.
   b. Writing For each, tell whether you think congruence is necessary and why.

What can you prove about $\triangle ISP$ and $\triangle OSP$ given the information in the diagram and the information below?

34. $\overline{SP}$ is the bisector of $\angle ISO$.

35. $\overline{SP}$ is a bisector of $\overline{IO}$.

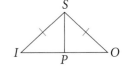

In $ABCD$, $\overline{AD} \parallel \overline{BC}$ and $\overline{AD} \cong \overline{BC}$.
Can you prove the two triangles congruent? Explain.

36. $\triangle ADB$ and $\triangle CBD$ **37.** $\triangle ABC$ and $\triangle CDA$

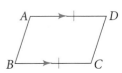

Proof 38. Given: $\overline{FG} \parallel \overline{KL}, \overline{FG} \cong \overline{KL}$

Prove: $\triangle FGK \cong \triangle KLF$

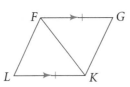

39. Given: $\overline{AB} \perp \overline{CM}, \overline{AB} \perp \overline{DB}, M$ is the midpoint of $\overline{AB}, \overline{CM} \cong \overline{DB}$.

Prove: $\triangle AMC \cong \triangle MBD$

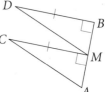

Challenge Proof 40. Given: $\overline{HK} \cong \overline{LG}, \overline{HF} \cong \overline{LJ}, \overline{FG} \cong \overline{JK}$

Prove: $\triangle FGH \cong \triangle JKL$

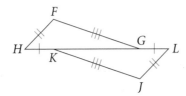

41. Given: $\angle N \cong \angle L, \overline{MN} \cong \overline{OL}, \overline{NO} \cong \overline{LM}$

Prove: $\overline{MN} \parallel \overline{LO}$

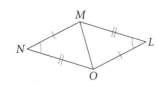

42. Error Analysis A friend conjectures that there should be an AAA Congruence Postulate since there is a SSS Congruence Postulate. Give a counterexample to disprove your friend's conjecture.

43. Critical Thinking Four sides of polygon *ABCD* are congruent, respectively, to the four sides of polygon *EFGH*.
  a. Must the two quadrilaterals be congruent? Explain.
  b. Is a quadrilateral a rigid figure? Explain.
  c. What could you "add" to a quadrilateral to make it a rigid figure? Explain.

Multiple Choice

Use the figures at the right for Exercises 44–46.

44. Suppose $\overline{TM} \cong \overline{GL}$ and $\angle M \cong \angle G$.
What additional information is needed
to prove $\triangle MTD \cong \triangle GLS$ by SAS?

  A. $\angle T \cong \angle L$ **B.** $\angle T \cong \angle S$ **C.** $\overline{TD} \cong \overline{SL}$ **D.** $\overline{MD} \cong \overline{SG}$

45. Suppose $\overline{TD} \cong \overline{SG}$ and $\overline{MD} \cong \overline{SL}$. What additional information is needed to prove the two triangles congruent by SAS?

  F. $\angle T \cong \angle S$ **G.** $\angle D \cong \angle S$ **H.** $\angle S \cong \angle L$ **J.** $\angle D \cong \angle G$

46. Suppose *TD* = 10 cm, *DM* = 9 cm, *TM* = 11 cm, *SL* = 11 cm, and
SG = 9 cm. What else do you need to know in order to prove that the two
triangles are congruent by SSS?

  A. *LG* = 9 cm   **B.** *TD* = *SL*   **C.** *GL* = 10 cm   **D.** *TM* = *SG*

Short Response

47. In the diagram, $\overline{WB} \cong \overline{BZ}$ and $\angle W \cong \angle ABZ$.
  a. State another conclusion you can make.
    Name the property that justifies your conclusion.
  b. Based on the information given in the diagram, can you
    prove the two triangles congruent? Justify your answer.

Lesson 4-1

ABCD ≅ *EFGH*. **Name the angle or side that corresponds to the given part.**

  48. $\angle A$ **49.** $\overline{EF}$ **50.** $\overline{BC}$ **51.** $\angle G$

Lesson 2-2

52. The following two statements are about lines with defined slopes. Combine them into a single biconditional.

If the product of the slopes of two lines is −1, then the lines are perpendicular.
If two lines are perpendicular, then the product of their slopes is −1.

53. Write the two conditional statements that form this biconditional:
$x = 2$ if and only if $2x = 4$.

Lesson 2-1

Write the converse of the statement. Decide whether the statement and its converse are true or false.

  54. If $x = 3$ then $2x = 6$.   **55.** If $x = 3$ then $x^2 = 9$.

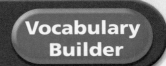

High-use academic words are words that you will see often in textbooks and on tests. These words are not math vocabulary terms, but knowing them will help you to succeed in mathematics.

Words to Learn: Direction Words

Some words tell you what to do in a problem. You need to understand what these words are asking so that you give the correct answer.

Word	Meaning
Explain	State facts and details to make an idea easy to understand.
Justify	Think of this as "Explain why." Give reasons to support a statement, conclusion, or action.

EXERCISES

Use the meanings of the words listed above to help you do the following.

1. *Explain* how to prepare your favorite snack.

2. In the student council meeting, you make a motion to recommend that the school establish a community service requirement.
   a. One reason, you claim, is that this requirement will help students become more responsible adults. What might you say to *justify* this claim.
   b. Determine another way in which you could *justify* your motion.
   c. Write an argument that would *justify* opposition to your motion.

Use the triangles at the right for Exercises 3–5.

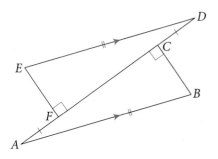

3. Are $\overline{AB}$ and $\overline{ED}$ parallel in the diagram at the right? Explain.

4. Are triangles *ABC* and *DEF* congruent? Justify your answer.

5. Can you conclude that $\overline{BC} \cong \overline{EF}$? Justify your answer.

6. **Word Knowledge** Think about the word *conjecture*.
   a. Choose the letter for how well you know the word.
      A. I know its meaning.
      B. I've seen it, but I don't know its meaning.
      C. I don't know it.
   b. **Research** Look up *conjecture* in a dictionary or online. Write its definition.
   c. Write a sentence involving mathematics that uses the word *conjecture*.

Triangle Congruence by ASA and AAS

4-3

 Learning Standards for Mathematics

G.G.27 Write a proof arguing from a given hypothesis to a given conclusion.

Check Skills You'll Need

GO for Help Lesson 4-2

In △*JHK*, which side is included between the given pair of angles?

1. ∠*J* and ∠*H*   **2.** ∠*H* and ∠*K*

In △*NLM*, which angle is included between the given pair of sides?

3. $\overline{LN}$ and $\overline{LM}$ **4.** $\overline{NM}$ and $\overline{LN}$

Give a reason to justify each statement.

5. $\overline{PR} \cong \overline{PR}$

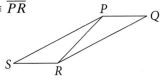

6. ∠*A* ≅ ∠*D*

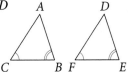

---

1 **Using the ASA Postulate and the AAS Theorem**

In Lesson 4-2 you learned that two triangles are congruent if two pairs of sides are congruent and the included angles are congruent (SAS). The Activity Lab on page 204 suggests that two triangles are also congruent if two pairs of angles are congruent and the included sides are congruent (ASA).

 Key Concepts

Postulate 4-3	**Angle-Side-Angle (ASA) Postulate**

If two angles and the included side of one triangle are congruent to two angles and the included side of another triangle, then the two triangles are congruent.

△*HGB* ≅ △*NKP*

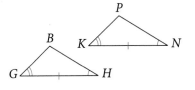

---

1 **EXAMPLE** **Using ASA**

Multiple Choice Which triangle is congruent to △*CAT* by the ASA Postulate?

A △*DOG*   **B** △*INF*

C △*GDO*   **D** △*FNI*

∠*C* ≅ ∠*G*, $\overline{CA} \cong \overline{GD}$, and ∠*A* ≅ ∠*D*.

△*CAT* ≅ △*GDO* by ASA. Choice C is correct.

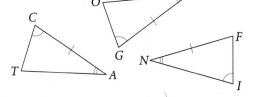

Test-Taking Tip

In a triangle congruence statement, remember to list corresponding vertices in the same order.

 **Quick Check** **1** Can you conclude that △*INF* is congruent to either of the other two triangles? Explain.

Here is how you can use the ASA Postulate in a proof.

2 EXAMPLE **Real-World** 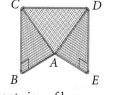 **Connection**

Lacrosse Study what you are given and what you are to prove about the lacrosse goal. Then write a proof that uses ASA.

Given: $\angle CAB \cong \angle DAE$, $\overline{AB} \cong \overline{AE}$, $\angle ABC$ and $\angle AED$ are right angles.

Prove: $\triangle ABC \cong \triangle AED$

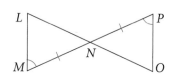

Front view of lacrosse goal

Proof: $\angle ABC \cong \angle AED$ because all right angles are congruent. You are given that $\overline{AB} \cong \overline{AE}$ and $\angle CAB \cong \angle DAE$. Thus, $\triangle ABC \cong \triangle AED$ by ASA.

✓ Quick Check ❷ Write a proof.

Given: $\overline{NM} \cong \overline{NP}$, $\angle M \cong \angle P$

Prove: $\triangle NML \cong \triangle NPO$

You can use the ASA Postulate to prove the Angle-Angle-Side Congruence Theorem. A flow proof is shown below.

Key Concepts

Theorem 4-2	**Angle-Angle-Side (AAS) Theorem**

If two angles and a nonincluded side of one triangle are congruent to two angles and the corresponding nonincluded side of another triangle, then the triangles are congruent.

$$\triangle CDM \cong \triangle XGT$$

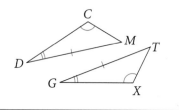

Proof **Proof of the Angle-Angle-Side Theorem**

Given: $\angle A \cong \angle X$, $\angle B \cong \angle Y$, $\overline{BC} \cong \overline{YZ}$

Prove: $\triangle ABC \cong \triangle XYZ$

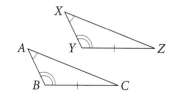

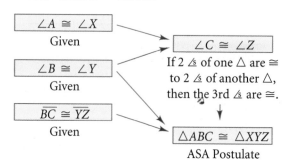

A proof starts in your head with a plan. It may help to jot down your plan. You will find this especially helpful as proofs become more complex. As you become skilled at proof, you will find that a good plan looks a lot like a paragraph proof.

③ EXAMPLE Planning a Proof

Study what you are given and what you are to prove.
Then plan a proof that uses AAS.

Given: $\angle S \cong \angle Q$,
$\overline{RP}$ bisects $\angle SRQ$.

Prove: $\triangle SRP \cong \triangle QRP$

Plan: $\triangle SRP \cong \triangle QRP$ by AAS
if $\overline{SP} \cong \overline{QP}$ or $\overline{RP} \cong \overline{RP}$.

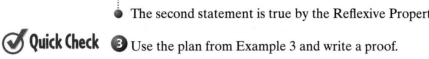

● The second statement is true by the Reflexive Property of Congruence.

✓ Quick Check ③ Use the plan from Example 3 and write a proof.

Proof **④ EXAMPLE** Writing a Proof

Video Tutor Help

Visit: PhSchool.com
Web Code: aue-0775

Study what you are given and what you are to prove.
Then write a proof that uses AAS.

Given: $\overline{XQ} \parallel \overline{TR}$, $\overline{XR}$ bisects $\overline{QT}$.

Prove: $\triangle XMQ \cong \triangle RMT$

Plan: $\overline{XQ} \parallel \overline{TR}$ gives two pairs of congruent
alternate interior angles. $\overline{XR}$ bisects $\overline{QT}$
means $\overline{QM} \cong \overline{TM}$. Use AAS.

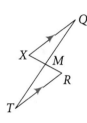

Statements	Reasons
1. $\overline{XQ} \parallel \overline{TR}$	1. Given
2. $\angle Q \cong \angle T, \angle X \cong \angle R$	2. _?_
3. $\overline{XR}$ bisects $\overline{QT}$.	3. Given
4. $\overline{QM} \cong \overline{TM}$	4. Definition of segment bisector
5. $\triangle XMQ \cong \triangle RMT$	5. AAS

✓ Quick Check ④ **a.** Supply the reason that justifies Step 2.
b. Critical Thinking Explain how you could prove $\triangle XMQ \cong \triangle RMT$ by ASA.

EXERCISES

For more exercises, see *Extra Skill, Word Problem, and Proof Practice.*

Practice and Problem Solving

Ⓐ Practice by Example

Example 1
(page 213)

Name two triangles that are congruent by the ASA Postulate.

1. 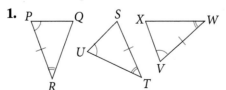 **2.**

Answer each question without drawing the triangle.

3. Which side is included between $\angle R$ and $\angle S$ in $\triangle RST$?

4. Which angles include $\overline{NO}$ in $\triangle NOM$?

Example 2
(page 214)

5. Developing Proof Complete the proof by filling in the blanks.

Given: ∠LKM ≅ ∠JKM,
    ∠LMK ≅ ∠JMK

Prove: △LKM ≅ △JKM

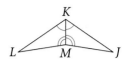

Proof: ∠LKM ≅ ∠JKM and ∠LMK ≅ ∠JMK are
given. $\overline{KM} ≅ \overline{KM}$ by the **a.** _?_ Property of Congruence.
△LKM ≅ △JKM by the **b.** _?_ Postulate.

Proof **6. Given:** ∠BAC ≅ ∠DAC
    $\overline{AC} ⊥ \overline{BD}$

Prove: △ABC ≅ △ADC

7. Given: $\overline{QR} ≅ \overline{TS}, \overline{QR} \parallel \overline{ST}$

Prove: △QRT ≅ △TSQ

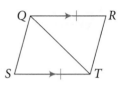

Example 3
(page 215)

8. Developing Proof Complete the proof plan by filling in the blanks.

Given: ∠UWT and ∠UWV are right angles,
    ∠T ≅ ∠V.

Prove: △UWT ≅ △UWV

Plan: △UWT ≅ △UWV by AAS if ∠T ≅ ∠V,
∠UWT ≅ **a.** _?_ , and $\overline{UW}$ ≅ **b.** _?_ .
∠UWT ≅ ∠UWV because all **c.** _?_ angles are congruent.
$\overline{UW} ≅ \overline{UW}$ by the **d.** _?_ Property of Congruence.

Proof **9.** Use your plan from Exercise 8 and write a proof.

Example 4
(page 215)

10. Developing Proof Complete the proof by filling in the blanks.

Given: ∠N ≅ ∠S, line ℓ bisects $\overline{TR}$ at Q.

Prove: △NQT ≅ △SQR

Statements	Reasons
1. ∠N ≅ ∠S	**1.** Given
2. ∠NQT ≅ ∠SQR	**2.** _?_
3. ℓ bisects $\overline{TR}$ at Q.	**3.** _?_
c. _?_	**4.** Definition of bisect
5. △NQT ≅ △SQR	**5.** _?_

Proof **11. Given:** ∠V ≅ ∠Y,
    $\overline{WZ}$ bisects ∠VWY.

Prove: △VWZ ≅ △YWZ

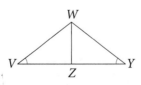

12. Given: $\overline{PQ} ⊥ \overline{QS}, \overline{RS} ⊥ \overline{QS},$
    T is the midpoint of $\overline{PR}$.

Prove: △PQT ≅ △RST

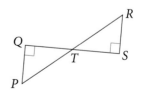

Write a congruence statement for each pair of triangles. Name the postulate or theorem that justifies your statement.

13.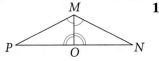

14.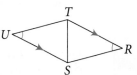

15.

16. Multiple Choice For the triangles shown, $\angle D \cong \angle T$, $\angle E \cong \angle U$, and $\overline{EO} \cong \overline{UX}$. Which of the following statements is true?

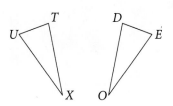

Ⓐ $\triangle TUX \cong \triangle DOE$ by ASA
Ⓑ $\triangle UTX \cong \triangle DEO$ by AAS
Ⓒ $\triangle TXU \cong \triangle ODE$ by ASA
Ⓓ $\triangle TUX \cong \triangle DEO$ by AAS

17. Writing Anita says that you can rewrite any proof that uses the AAS Theorem as a proof that uses the ASA Postulate. Do you agree with Anita? Explain.

18. $\angle E \cong \angle I$ and $\overline{FE} \cong \overline{GI}$. What else must you know to prove $\triangle FDE \cong \triangle GHI$ by AAS? By ASA?

19. Reasoning Can you prove the triangles at the right congruent using ASA or AAS? Justify your answer.

Proof **20. Given:** $\angle N \cong \angle P, \overline{MO} \cong \overline{QO}$
  Prove: $\triangle MON \cong \triangle QOP$

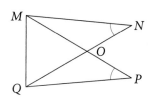

21. Given: $\angle F \cong \angle H, \overline{FG} \parallel \overline{JH}$
  Prove: $\triangle FGJ \cong \triangle HJG$

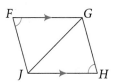

Proof **22. Given:** $\overline{AE} \parallel \overline{BD}, \overline{AE} \cong \overline{BD},$
    $\angle E \cong \angle D$
  Prove: $\triangle AEB \cong \triangle BDC$

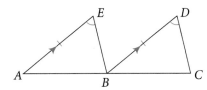

23. Given: $\overline{DH}$ bisects $\angle BDF,$
    $\angle 1 \cong \angle 2.$
  Prove: $\triangle BDH \cong \triangle FDH$

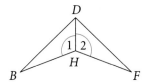

24. Constructions Using a straightedge, draw a triangle. Label it $\triangle JKL$. Construct $\triangle MNP \cong \triangle JKL$ so you know that the triangles are congruent by ASA.

Proof **25. Given:** $\overline{AB} \parallel \overline{DC}, \overline{AD} \parallel \overline{BC}$
  Prove: $\triangle ABC \cong \triangle CDA$

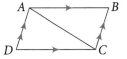

26. Reasoning If possible, draw two noncongruent triangles that have two pairs of congruent angles and one pair of congruent sides. If this is not possible, explain why.

27. a. Open-Ended Draw a triangle. Draw a second triangle that shares a common side with the first one and is congruent to it.

 b. Think about how you drew your second triangle. What postulate or theorem did you use to make the second triangle congruent to the first one?

Real-World 🌐 Connection

The two triangles above are congruent if just one additional condition is met.

Use the figure at the right. Name as many pairs of congruent triangles as you can for the information given.

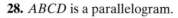

 28. *ABCD* is a parallelogram.

 29. *ABCD* is a rectangle.

30. Reasoning △*JKL* ≅ △*MNP*. What additional information about $\overline{KQ}$ and $\overline{NR}$ will allow you to conclude that △*JKQ* ≅ △*MNR*? Explain.

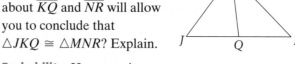

31. Probability Here are six congruence statements about the triangles at the right.

∠*A* ≅ ∠*X*   ∠*B* ≅ ∠*Y*   ∠*C* ≅ ∠*Z*

$\overline{AB} \cong \overline{XY}$ $\overline{AC} \cong \overline{XZ}$ $\overline{BC} \cong \overline{YZ}$

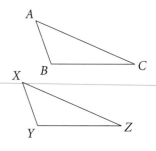

There are 20 ways to choose a group of three statements from these six. What is the probability that three statements chosen at random from the six will guarantee that the triangles are congruent?

32. △*RST* at the right has *RS* = 5, *RT* = 9, and *m*∠*T* = 30. Show that there is no SSA congruence rule by constructing △*UVW* with *UV* = 5, *UW* = 9, and *m*∠*W* = 30, but with △*UVW* ≇ △*RST*.

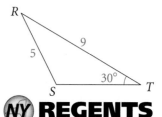

NY REGENTS

Test Prep

Multiple Choice

33. Which of the following is NOT a method used to prove triangles congruent?
 A. AAS   **B.** ASA   **C.** SAS   **D.** SSA

34. Suppose $\overline{RT} \cong \overline{ND}$ and ∠*R* ≅ ∠*N*. What additional information is needed to prove △*RTJ* ≅ △*NDF* by ASA?
 F. ∠*T* ≅ ∠*D*   **G.** ∠*R* ≅ ∠*N*   **H.** ∠*J* ≅ ∠*D*   **J.** ∠*T* ≅ ∠*F*

Short Response

35. $\overline{PQ}$ bisects ∠*RPS* and ∠*RQS*. Justify each answer.
 a. Which pairs of angles, if any, are congruent?
 b. By what theorem or postulate can you prove that △*PRQ* ≅ △*PSQ*?

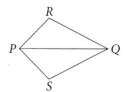

Extended Response

36. $\overline{LJ} \parallel \overline{KG}$ and *M* is the midpoint of $\overline{LG}$.
 a. Why is $\overline{LM} \cong \overline{GM}$?
 b. Can the two triangles be proved congruent by ASA? Explain.
 c. Can the two triangles be proved congruent by AAS? Explain.

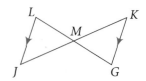

GO for Help

Lesson 4-2

In Exercises 37 and 38, decide whether you can use the SSS Postulate or the SAS Postulate to prove the triangles congruent. If so, write the congruence statement and name the postulate. If not, write *not possible*.

37.

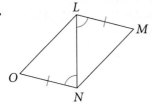

38.

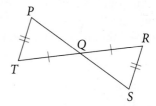

39. For any △ABC, which sides are *not* included between ∠A and ∠B?

Lesson 3-2

40. State the theorem or postulate that justifies the statement: If ∠1 ≅ ∠3, then a ∥ b.

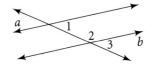

Lesson 1-9

Photography You want to arrange class-trip photos without overlap to make a 2 ft-by-3 ft poster. You collect 3 in.-by-5 in. and 4 in.-by-6 in. photos. What is the greatest number of each type of photo that you can fit on your poster?

41. 3 in.-by-5 in.

42. 4 in.-by-6 in.

43. What percent more paper is used for a large photo than a regular photo?

Checkpoint Quiz 1   **Lessons 4-1 through 4-3**

1. △RST ≅ △JKL. List the three pairs of congruent corresponding sides and the three pairs of congruent corresponding angles.

State the postulate or theorem you can use to prove the triangles congruent. If the triangles cannot be proven congruent, write *not possible*.

2.

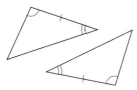

3.

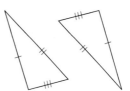

4.

5.

6.

7.

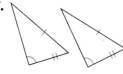

Use the information given in the diagram. Tell why each statement is true.

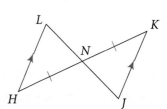

8. ∠H ≅ ∠K

9. ∠HNL ≅ ∠KNJ

10. △HNL ≅ △KNJ

Exploring AAA and SSA

Technology

FOR USE WITH LESSON 4-3

So far, four statements allow you to conclude that two triangles are congruent. You can refer to them as SSS, SAS, ASA, and AAS. It is good mathematics to wonder about the other two possibilities, AAA and SSA.

NY G.R.1: Use physical objects, diagrams, charts, tables, graphs, symbols, equations, or objects created using technology as representations of mathematical concepts.

Construct

Use geometry software to construct $\overrightarrow{AB}$ and $\overrightarrow{AC}$. Construct $\overline{BC}$ to create $\triangle ABC$. Construct a line parallel to $\overline{BC}$ that intersects $\overrightarrow{AB}$ and $\overrightarrow{AC}$ at points D and E to form $\triangle ADE$.

Investigate

Are the three angles of $\triangle ABC$ congruent to the three angles of $\triangle ADE$? Manipulate the figure to change the positions of $\overline{DE}$ and $\overline{BC}$. Do the corresponding angles of the triangles remain congruent? Are the two triangles congruent? Can the two triangles be congruent?

In Exercise 1, you will be asked to make a conjecture about this investigation.

Construct

Construct $\overrightarrow{AB}$. Draw a circle with center C that intersects $\overrightarrow{AB}$ in two points. Construct $\overline{AC}$. Construct a point E on the circle and construct $\overline{CE}$.

Investigate

Move point E around the circle until E is on $\overrightarrow{AB}$ and forms $\triangle ACE$. Then move E on the circle to the other point on $\overrightarrow{AB}$ to form another $\triangle ACE$.

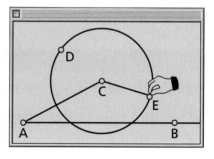

Compare the measures of $\overline{AC}, \overline{CE},$ and $\angle A$ in one triangle with the measures of $\overline{AC}, \overline{CE},$ and $\angle A$ in the other triangle. Are two sides and a nonincluded angle of one triangle congruent to two sides and a nonincluded angle of the other triangle? Are the two triangles congruent? Do you get the same results if you change the size of $\angle A$ and the size of the circle?

EXERCISES

1. **Make a Conjecture** Based on your first investigation above, is there an AAA congruence theorem? Explain.

   For Exercises 2–4, use what you learned in your second investigation above.

2. **Make a Conjecture** Do you think there is an SSA congruence theorem? Why?

3. Manipulate the figure so that $\angle A$ is obtuse. Decide whether the circle can intersect $\overrightarrow{AB}$ twice to form two triangles. Could there be an SSA congruence theorem if the congruent angles are obtuse? Explain.

4. Suppose you are given $\overline{CE}, \overline{AC},$ and $\angle A$. What must be true about $CE, AC,$ and $m\angle A$ so that you can construct exactly one $\triangle ACE$? (*Hint:* Consider cases.)

220 Activity Lab Exploring AAA and SSA

4-4

Using Congruent Triangles: CPCTC

G.G.29 Identify corresponding parts of congruent triangles.

✓ Check Skills You'll Need

GO for Help Lesson 4-1

In the diagram, $\triangle JRC \cong \triangle HVG$.

1. List the congruent corresponding angles.
2. List the congruent corresponding sides.

You are given that $\triangle TIC \cong \triangle LOK$.

3. List the congruent corresponding angles.
4. List the congruent corresponding sides.

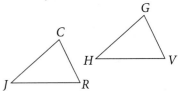

🔊 **New Vocabulary** • CPCTC

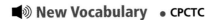

1 Proving Parts of Triangles Congruent

With SSS, SAS, ASA, and AAS, you know how to use three parts of triangles to show that the triangles are congruent. Once you have triangles congruent, you can make conclusions about their other parts because, by definition, corresponding parts of congruent triangles are congruent. You can abbreviate this as **CPCTC.**

Proof → **1 EXAMPLE** Real-World 🌐 Connection

Real-World 🌐 Connection

Shapes formed by the ribs, stretchers, and shaft are congruent whether an umbrella is open or closed.

Umbrella Frames In an umbrella frame, the stretchers are congruent and they open to angles of equal measure.

Given: $\overline{SL} \cong \overline{SR}$,
$\angle 1 \cong \angle 2$

Prove that the angles formed by the shaft and the ribs are congruent.

Prove: $\angle 3 \cong \angle 4$

Proof: It is given that $\overline{SL} \cong \overline{SR}$ and $\angle 1 \cong \angle 2$. $\overline{SC} \cong \overline{SC}$ by the Reflexive Property of Congruence. $\triangle LSC \cong \triangle RSC$ by SAS, so $\angle 3 \cong \angle 4$ by CPCTC.

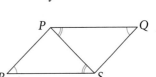

✓ **Quick Check** 1 **Given:** $\angle Q \cong \angle R$,
$\angle QPS \cong \angle RSP$

Prove: $\overline{SQ} \cong \overline{PR}$

You can use congruent triangles and CPCTC to measure distances, such as the distance across a river, indirectly.

History According to legend, one of Napoleon's officers used congruent triangles to estimate the width of a river. On the riverbank, the officer stood up straight and lowered the visor of his cap until the farthest thing he could see was the edge of the opposite bank. He then turned and noted the spot on his side of the river that was in line with his eye and the tip of his visor.

Given: ∠DEG and ∠DEF are right angles; ∠EDG ≅ ∠EDF.

The officer then paced off the distance to this spot and declared that distance to be the width of the river! Use congruent triangles to prove that he was correct.

Prove: $\overline{EF} \cong \overline{EG}$

Statements	Reasons
1. ∠EDG ≅ ∠EDF	**1.** Given
2. $\overline{DE} \cong \overline{DE}$	**2.** Reflexive Property of Congruence
3. ∠DEG and ∠DEF are right angles.	**3.** Given
4. ∠DEG ≅ ∠DEF	**4.** All right angles are congruent.
5. △DEF ≅ △DEG	**5.** ASA Postulate
6. $\overline{EF} \cong \overline{EG}$	**6.** CPCTC

 **Quick Check** **2** About how wide was the river if the officer stepped off 20 paces and each pace was about $2\frac{1}{2}$ ft long?

EXERCISES

For more exercises, see *Extra Skill, Word Problem, and Proof Practice.*

Practice and Problem Solving

 Practice by Example

Example 1
(page 221)

 for Help

1. Developing Proof State why the two triangles are congruent. Give the congruence statement. Then list all the other parts of the triangles that are congruent by CPCTC.

Proof **2. Given:** ∠ABD ≅ ∠CBD, ∠BDA ≅ ∠BDC
Prove: $\overline{AB} \cong \overline{CB}$

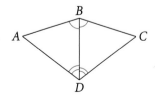

3. Given: $\overline{OM} \cong \overline{ER}$, $\overline{ME} \cong \overline{RO}$
Prove: ∠M ≅ ∠R

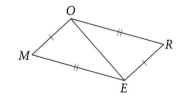

Example 2
(page 222)

4. Developing Proof Two cars of the same model have hood braces that are identical, connect to the body of the car in the same place, and fit into the same slot in the hood.

Given: $\overline{CA} \cong \overline{VE}, \overline{AR} \cong \overline{EH}, \overline{RC} \cong \overline{HV}$

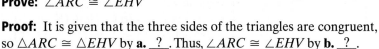

Complete the proof that the hood braces hold the hoods open at the same angle.

Prove: $\angle ARC \cong \angle EHV$

Proof: It is given that the three sides of the triangles are congruent, so $\triangle ARC \cong \triangle EHV$ by **a.** ? . Thus, $\angle ARC \cong \angle EHV$ by **b.** ? .

B **Apply Your Skills** 🌐 **5. Earth Science** Some distances are best measured indirectly.

Sinkhole Swallows House

The large sinkhole in this photo occurred suddenly in 1981 in Winter Park, Florida, following a severe drought. Increased water consumption lowers the water table. Sinkholes form when caverns in the underlying limestone dry up and collapse.

A geometry class indirectly measured the distance across a sinkhole. The distances they measured are shown in the diagram. Explain how to use their measurements to find the distance across the sinkhole.

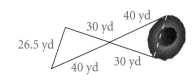

Proof **6. Given:** $\angle SPT \cong \angle OPT$, $\overline{SP} \cong \overline{OP}$

Prove: $\angle S \cong \angle O$

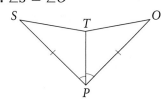

7. Given: $\overline{YT} \cong \overline{YP}, \angle C \cong \angle R$, $\angle T \cong \angle P$

Prove: $\overline{CT} \cong \overline{RP}$

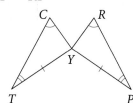

GO **Online**
Homework Video Tutor
Visit: PHSchool.com
Web Code: aue-0404

Copy and mark the figure to show the given information. Explain how you would use SSS, SAS, ASA, or AAS with CPCTC to prove $\angle P \cong \angle Q$.

8. Given: $\overline{PK} \cong \overline{QK}$, $\overline{KL}$ bisects $\angle PKQ$.

9. Given: $\overline{KL}$ is the perpendicular bisector of $\overline{PQ}$.

10. Given: $\overline{KL} \perp \overline{PQ}, \overline{KL}$ bisects $\angle PKQ$.

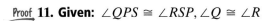

Proof **11. Given:** $\angle QPS \cong \angle RSP, \angle Q \cong \angle R$

Prove: $\overline{PQ} \cong \overline{SR}$

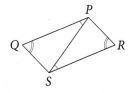

12. Writing Karen cut this pattern for the stained glass shown here so that $AB = CB$ and $AD = CD$. Must $\angle A$ be congruent to $\angle C$? Explain.

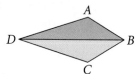

13. Constructions The construction of a line perpendicular to line ℓ through point P on ℓ is shown here.
 a. Which lengths or distances are equal by construction?
 b. Explain why you can conclude that $\overleftrightarrow{CP}$ is perpendicular to ℓ. (*Hint:* Do the construction. Then draw $\overline{CA}$ and $\overline{CB}$.)

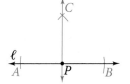

GO for Help

For a guide to solving Exercise 14, see page 226.

14. Error Analysis The proof is incorrect. Find the error and tell how you would correct the proof.

Given: $\angle A \cong \angle C$, $\overline{BD}$ bisects $\angle ABC$.
Prove: $\overline{AB} \cong \overline{CB}$

Statements	Reasons
1. $\angle A \cong \angle C$	1. Given
2. $\overline{BD}$ bisects $\angle ABC$.	2. Given
3. $\angle 1 \cong \angle 2$	3. Definition of bisect
4. $\overline{AD} \cong \overline{CD}$	4. Definition of bisect
5. $\triangle ABD \cong \triangle CBD$	5. AAS Theorem
6. $\overline{AB} \cong \overline{CB}$	6. CPCTC

Proof 15. Given: $\overline{BA} \cong \overline{BC}$, $\overline{BD}$ bisects $\angle ABC$.

Prove: $\overline{BD} \perp \overline{AC}$, $\overline{BD}$ bisects $\overline{AC}$.

16. Given: $\ell \perp \overline{AB}$, ℓ bisects $\overline{AB}$ at C, P is on ℓ.

Prove: $PA = PB$

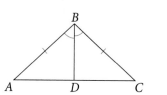

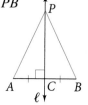

17. Constructions In the construction of the bisector of $\angle A$ below, $\overline{AB} \cong \overline{AC}$ because they are radii of the same circle. $\overline{BX} \cong \overline{CX}$ because both arcs had the same compass setting. Tell why you can conclude that $\overrightarrow{AX}$ bisects $\angle BAC$.

Problem Solving Hint

In the third diagram, what two triangles must be congruent, and why?

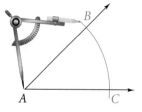

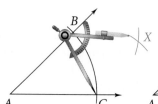

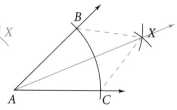

Proof 18. Given: $\overline{BE} \perp \overline{AC}$, $\overline{DF} \perp \overline{AC}$, $\overline{BE} \cong \overline{DF}$, $\overline{AF} \cong \overline{EC}$

Prove: $\overline{AB} \cong \overline{DC}$

19. Given: $\overline{JK} \parallel \overline{QP}$, $\overline{JK} \cong \overline{QP}$

Prove: $\overline{KQ}$ bisects $\overline{JP}$.

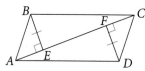

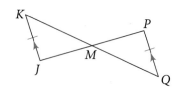

C Challenge *Proof* **20. Given:** $\overline{PR} \parallel \overline{MG}, \overline{MP} \parallel \overline{GR}$

Prove: Each diagonal of *PRGM* divides
PRGM into two congruent triangles.

Proof **21. Given:** $\overline{PR} \parallel \overline{MG}, \overline{MP} \parallel \overline{GR}$

Prove: $\overline{PR} \cong \overline{MG}, \overline{MP} \cong \overline{GR}$

(*Hint:* See Exercise 20.)

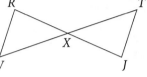

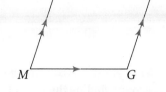

Multiple Choice

22. In the diagram, $\triangle RXW \cong \triangle JXT$. Which
statement is NOT necessarily true?

A. $\angle J \cong \angle R$ **B.** $\angle W \cong \angle T$

C. $\overline{WX} \cong \overline{JX}$ **D.** $\overline{RW} \cong \overline{JT}$

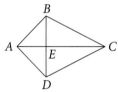

23. Which is true by CPCTC?

F. $\overline{AC}$ bisects $\overline{BD}$ **G.** $\angle BAC \cong \angle DCA$

H. $\angle ABE \cong \angle EDC$ **J.** $\overline{BC} \cong \overline{DC}$

24. Which is *not* true by CPCTC?

A. $\overline{BE} \cong \overline{DE}$ **B.** $\angle BAC \cong \angle DAC$

C. $\angle BCA \cong \angle DCE$ **D.** $\overline{AB} \cong \overline{AD}$

$\triangle ABC \cong \triangle ADC$

Exercises 23–24

Short Response

25. In the diagram, $\overline{KB}$ bisects $\angle VKT$ and $\overline{KV} \cong \overline{KT}$.

a. What do you need to show in order to
conclude $\angle KBV \cong \angle KBT$? State whether it is
possible to show this and justify your answer.

b. Show that $\overline{VB} \cong \overline{TB}$.

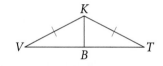

 GO for Help

Lesson 4-3

What postulate or theorem can you use to prove the triangles congruent?

26.  **27.**

Lesson 2-5

28. The measure of an angle is 10 more than the measure of its supplement. Find
the measures of both angles.

Lesson 2-3

**If possible, use the Law of Detachment to draw a conclusion. If it is not possible to
draw a conclusion, write *not possible.***

29. If two nonvertical lines are parallel, then their slopes are equal. Line *m* is
nonvertical and parallel to line *n*.

30. If a convex polygon is a quadrilateral, then the sum of its angle measures is 360.
Convex polygon *ABCDE* has five sides.

31. If a quadrilateral is a square, then it has four congruent sides. Quadrilateral
ABCD has four congruent sides.

Analyzing Errors Read the problem below. Then let Lucas's thinking guide you through finding the error. Check your understanding with the exercise at the bottom of the page.

Error Analysis This proof is incorrect. Find the error and tell how you would correct the proof.

Given: $\angle A \cong \angle C$, $\overline{BD}$ bisects $\angle ABC$.
Prove: $\overline{AB} \cong \overline{CB}$

Statements	Reasons
1. $\angle A \cong \angle C$	1. Given
2. $\overline{BD}$ bisects $\angle ABC$.	2. Given
3. $\angle 1 \cong \angle 2$	3. Definition of bisect
4. $\overline{AD} \cong \overline{CD}$	4. Definition of bisect
5. $\triangle ABD \cong \triangle CBD$	5. AAS Theorem
6. $\overline{AB} \cong \overline{CB}$	6. CPCTC

What Lucas Thinks

An error could be anywhere. I'll check some easier parts first.

The first two lines match the given information. The last line matches what is to be proved.

In the last line, $\overline{AB}$ and $\overline{CB}$ are corresponding parts from $\triangle ABD$ and $\triangle CBD$. If the triangles are congruent, then CPCTC is correct.

$\triangle ABD \cong \triangle CBD$ by AAS. I see AAS in lines 1, 3, and 4, and they are in the right order. But I'm still not sure that lines 3 and 4 are correct.

I see the error!

Now I have to correct the proof. I still have two pairs of congruent angles. I know $\overline{BD} \cong \overline{BD}$. I can use this for line 4. AAS still works. I'm done!

What Lucas Writes

The error is in line 4. You cannot say $\overline{AD} \cong \overline{CD}$ by the definition of bisect. $\overline{BD}$ is given to be an angle bisector, not a segment bisector.

Replace line 4 with:

4. $\overline{BD} \cong \overline{BD}$ 4. $\cong$ is reflexive.

EXERCISES

1. In the proof above, let the Given and Statement 2 be "$\overline{BD}$ bisects $\overline{AC}$." Find the error in this proof and tell how you might correct the proof.

2. A proof states "$\triangle RST \cong \triangle VTS$ by SSA." Find the error. Show a counterexample.

In Chapter 3, you learned that isosceles triangles have two congruent sides. Folding one of the sides onto the other will suggest another important property of isosceles triangles.

NY G.CM.9: Formulate mathematical questions that elicit, extend, or challenge strategies, solutions, and/or conjectures of others.

1 ACTIVITY

• Construct an isosceles triangle on tracing paper.

• Name your triangle △ABC, with A and B opposite the congruent sides.

• Fold the paper so the two congruent sides fit precisely one on top of the other. Crease the paper. Label the intersection of the fold line and $\overline{AB}$ as point D.

1. What do you notice about ∠A and ∠B? Compare your results with others. Make a conjecture about the angles opposite the congruent sides in an isosceles triangle.

2. Study the fold line $\overline{CD}$ and the base $\overline{AB}$.
   a. As best you can tell, what type of angle is ∠CDA? ∠CDB?
   b. How do $\overline{AD}$ and $\overline{BD}$ seem to be related?
   c. Use your answers to parts (a) and (b) to complete the conjecture:
      The fold line $\overline{CD}$ is the __?__ of the base $\overline{AB}$ of isosceles △ABC.

2 ACTIVITY

In Activity 1, you made a conjecture about angles opposite the congruent sides of a triangle. You can also fold paper to study whether the converse is true.

• On tracing paper, draw an acute angle with F as its vertex and one side $\overline{FG}$. Construct ∠G as shown, so that ∠G ≅ ∠F.

• Fold the paper so the two congruent angles, ∠F and ∠G, fit precisely one on top of the other.

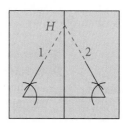

3. Explain why sides 1 and 2 meet at point H on the fold line. Make a conjecture about sides $\overline{FH}$ and $\overline{GH}$ opposite congruent angles in a triangle.

4. Write your conjectures from Steps 1 and 3 as a biconditional.

Isosceles and Equilateral Triangles

Learning Standards for Mathematics

G.G.31 Investigate, justify, and apply the isosceles triangle theorem and its converse.

Check Skills You'll Need

1. Name the angle opposite $\overline{AB}$.
2. Name the angle opposite $\overline{BC}$.
3. Name the side opposite $\angle A$.
4. Name the side opposite $\angle C$.
$\boxed{x^2}$ 5. **Algebra** Find the value of x.

GO for Help   Lesson 3-4

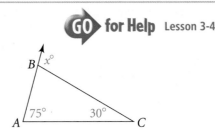

New Vocabulary

- legs of an isosceles triangle
- base of an isosceles triangle
- vertex angle of an isosceles triangle
- base angles of an isosceles triangle   • corollary

---

1   The Isosceles Triangle Theorems

Vocabulary Tip

Isosceles is derived from the Greek *isos* for equal and *skelos* for leg.

Isosceles triangles are common in the real world. You can find them in structures such as bridges and buildings. The congruent sides of an isosceles triangle are its **legs.** The third side is the **base.** The two congruent sides form the **vertex angle.** The other two angles are the **base angles.**

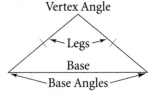

An isosceles triangle has a certain type of *symmetry* about a line through its vertex angle. The theorems below suggest this symmetry, which you will study in greater detail in Lesson 9-4.

Key Concepts

Theorem 4-3   **Isosceles Triangle Theorem**
If two sides of a triangle are congruent, then the angles opposite those sides are congruent. $\angle A \cong \angle B$

Theorem 4-4   **Converse of Isosceles Triangle Theorem**
If two angles of a triangle are congruent, then the sides opposite the angles are congruent. $\overline{AC} \cong \overline{BC}$

Theorem 4-5
The bisector of the vertex angle of an isosceles triangle is the perpendicular bisector of the base. $\overline{CD} \perp \overline{AB}$ and $\overline{CD}$ bisects $\overline{AB}$.

In the following proof of the Isosceles Triangle Theorem, you use a special segment, the bisector of the vertex angle. You will prove Theorems 4-4 and 4-5 in the Exercises.

Proof of the Isosceles Triangle Theorem

Proof →

Begin with isosceles $\triangle XYZ$ with $\overline{XY} \cong \overline{XZ}$. Draw $\overline{XB}$, the bisector of the vertex angle $\angle YXZ$.

Given: $\overline{XY} \cong \overline{XZ}$, $\overline{XB}$ bisects $\angle YXZ$.

Prove: $\angle Y \cong \angle Z$

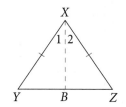

Proof: You are given that $\overline{XY} \cong \overline{XZ}$. By the definition of angle bisector, $\angle 1 \cong \angle 2$. By the Reflexive Property of Congruence, $\overline{XB} \cong \overline{XB}$. Therefore, by the SAS Postulate, $\triangle XYB \cong \triangle XZB$, and $\angle Y \cong \angle Z$ by CPCTC.

Real-World Connection

This A-shaped roof has congruent legs and congruent base angles.

1 EXAMPLE **Using the Isosceles Triangle Theorems**

Explain why each statement is true.

a. $\angle WVS \cong \angle S$

$\overline{WV} \cong \overline{WS}$ so $\angle WVS \cong \angle S$ by the Isosceles Triangle Theorem.

b. $\overline{TR} \cong \overline{TS}$

$\angle R \cong \angle WVS$ and $\angle WVS \cong \angle S$, so $\angle R \cong \angle S$ by the Transitive Property of $\cong$. $\overline{TR} \cong \overline{TS}$ by the Converse of the Isosceles Triangle Theorem.

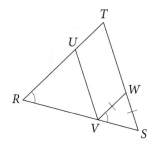

✓ **Quick Check** ❶ Can you deduce that $\triangle RUV$ is isosceles? Explain.

2 EXAMPLE **Using Algebra**

Multiple Choice Find the value of y.

- Ⓐ 17
- Ⓑ 27
- Ⓒ 54
- Ⓓ 90

Test-Taking Tip

Remember that the acute angles of a right triangle are complementary.

By Theorem 4-5, you know that $\overline{MO} \perp \overline{LN}$, so $\angle MON = 90$. $\triangle MLN$ is isosceles, so $\angle L \cong \angle N$ and $m\angle N = 63$.

$m\angle N + 90 + y = 180$ **Triangle Angle-Sum Theorem**
$63 + 90 + y = 180$ **Substitute for $m\angle N$.**
$y = 27$ **Subtract 153 from each side.**

The correct answer is B.

✓ **Quick Check** ❷ Suppose $m\angle L = 43$. Find the value of y.

A **corollary** is a statement that follows immediately from a theorem. Corollaries to the Isosceles Triangle Theorem and its converse appear on the next page.

Key Concepts

Corollary	**Corollary to Theorem 4-3**

If a triangle is equilateral, then the triangle is equiangular.

$$\angle X \cong \angle Y \cong \angle Z$$

Corollary	**Corollary to Theorem 4-4**

If a triangle is equiangular, then the triangle is equilateral.

$$\overline{XY} \cong \overline{YZ} \cong \overline{ZX}$$

Online active math

For: Isosceles Triangle Activity
Use: Interactive Textbook, 4-5

3 EXAMPLE **Real-World Connection**

Landscaping A landscaper uses rectangles and equilateral triangles for the path around the hexagonal garden. Find the value of x.

In a rectangle, an angle measure is 90; in an equilateral triangle, it is 60.

$$x + 90 + 60 + 90 = 360$$
$$x = 120$$

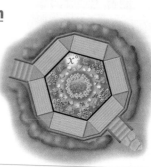

Quick Check ③ What is the measure of the angle at each outside corner of the path?

EXERCISES

For more exercises, see *Extra Skill, Word Problem, and Proof Practice.*

Practice and Problem Solving

A Practice by Example

Example 1
(page 229)

for Help

Complete each statement. Explain why it is true.

1. $\overline{VT} \cong$ ___?___

2. $\overline{UT} \cong$ ___?___ $\cong \overline{YX}$

3. $\overline{VU} \cong$ ___?___

4. $\angle VYU \cong$ ___?___

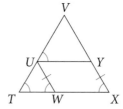

Example 2
(page 229)

x^2 **Algebra** **Find the values of x and y.**

5.

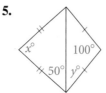

6.

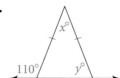

7.

Example 3
(page 230)

8. A square and a regular hexagon are placed so that they have a common side. Find $m\angle SHA$ and $m\angle HAS$.

9. Five fences meet at a point to form angles with measures $x, 2x, 3x, 4x,$ and $5x$ around the point. Find the measure of each angle.

Find each value.

10. If $m\angle L = 58$, then $m\angle LKJ = $ ■.

11. If $JL = 5$, then $ML = $ ■.

12. If $m\angle JKM = 48$, then $m\angle J = $ ■.

13. If $m\angle J = 55$, then $m\angle JKM = $ ■.

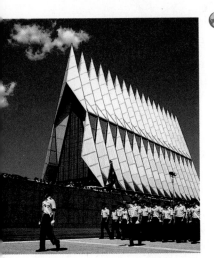

Exercise 14

14. **Architecture** Seventeen spires, pictured at the left, cover the Cadet Chapel at the Air Force Academy in Colorado Springs, Colorado. Each spire is an isosceles triangle with a 40° vertex angle. Find the measure of each base angle.

15. **Developing Proof** Here is another way to prove the Isosceles Triangle Theorem. Supply the missing parts.

Begin with isosceles $\triangle HKJ$ with $\overline{KH} \cong \overline{KJ}$. Draw **a.** ? , a bisector of the base $\overline{HJ}$.

Given: $\overline{KH} \cong \overline{KJ}$, **b.** ? bisects $\overline{HJ}$.

Prove: $\angle H \cong \angle J$

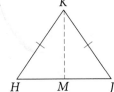

Statements	Reasons
1. $\overline{KM}$ bisects $\overline{HJ}$.	**c.** ?
2. $\overline{HM} \cong \overline{JM}$	**d.** ?
3. $\overline{KH} \cong \overline{KJ}$	3. Given
4. $\overline{KM} \cong \overline{KM}$	**e.** ?
5. $\triangle KHM \cong \triangle KJM$	**f.** ?
6. $\angle H \cong \angle J$	**g.** ?

Proof 16. Supply the missing parts in this statement of the Converse of the Isosceles Triangle Theorem. Then write a proof.

Begin with $\triangle PRQ$ with $\angle P \cong \angle Q$. Draw **a.** ? , the bisector of $\angle PRQ$.

Given: $\angle P \cong \angle Q$, **b.** ? bisects $\angle PRQ$.

Prove: $\overline{PR} \cong \overline{QR}$

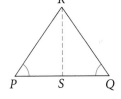

17. **Graphic Arts** A former logo for the National Council of Teachers of Mathematics is shown at the right. Trace the logo onto paper.
a. Highlight an obtuse isosceles triangle in the design. Then find its angle measures.
b. How many different sizes of angles can you find in the logo? What are their measures?

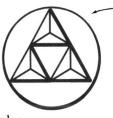

The triangles in the logo have these congruent sides and angles.

18. **Writing** Explain how each corollary on page 230 follows from its theorem. First, write one explanation and then write the second similar to the first.

19. **Multiple Choice** The perimeter of the triangle at the right is 20. Find x.

Ⓐ 3   Ⓑ 5   Ⓒ 6   Ⓓ 7

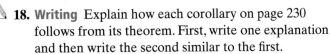

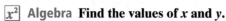 **Algebra** Find the values of *x* and *y*.

20.

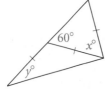

21.

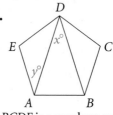

ABCDE is a regular pentagon.

22.

23. Write the Isosceles Triangle Theorem and its converse as a biconditional.

24. Critical Thinking An exterior angle of an isosceles triangle has measure 100. Find two possible sets of measures for the angles of the triangle.

25. a. Communications In the diagram at the right, what type of triangles are formed by the cables of the same height and the ground?

    b. What are the two different base lengths of the triangles?

    c. How is the tower related to each of the triangles?

26. Critical Thinking Curtis defines the base of an isosceles triangle as its "bottom side." Is his definition a good one? Explain.

27. Reasoning What are the measures of the base angles of an isosceles right triangle? Explain.

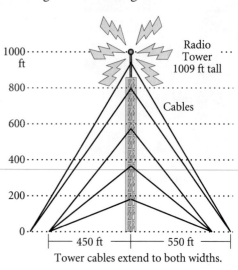

 28. Given: $\overline{AE} \cong \overline{DE}$, $\overline{AB} \cong \overline{DC}$
    Prove: $\triangle ABE \cong \triangle DCE$

 29. Prove Theorem 4-5. Use the diagram next to it on page 228.

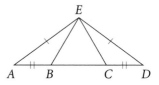

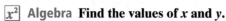 **Algebra** Find the values of *m* and *n*.

30.

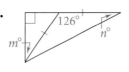

31.

32.

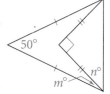

C **Challenge**

Coordinate Geometry For each pair of points, there are six points that could be the third vertex of an isosceles right triangle. Find the coordinates of each point.

33. $(4, 0)$ and $(0, 4)$ **34.** $(0, 0)$ and $(5, 5)$ **35.** $(2, 3)$ and $(5, 6)$

 36. Algebra A triangle has angle measures $x + 15$, $3x - 35$, and $4x$.
    a. Find the value of *x*.   **b.** Find the measure of each angle.
    c. What type of triangle is it? Why?

37. State the converse of Theorem 4-5. If the converse is true, write a paragraph proof. If the converse is false, give a counterexample.

Real-World  **Connection**

The circle is a basic shape for many tile designs.

38. Crafts The design in Step 3 is used in Hmong crafts and in Islamic and Mexican tiles. To create it, the artist starts by drawing a circle and four equally spaced diameters.

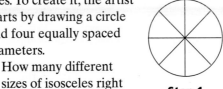

Step 1　　　**Step 2**　　　**Step 3**

　　a. How many different sizes of isosceles right triangles can you find in Step 2? Trace an example of each onto your paper.
　　b. How many times does a triangle of each size in part (a) appear in the Step 2 diagram?

Reasoning **What measures are possible for the base angles of each type of triangle? Explain.**

39. an isosceles obtuse triangle　　　　**40.** an isosceles acute triangle

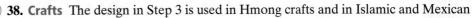
Multiple Choice

41. In isosceles △*ABC*, the vertex angle is ∠*A*. What can be proved?
　A. *AB* = *CB*　　　**B.** ∠*A* ≅ ∠*B*
　C. *m*∠*B* = *m*∠*C*　**D.** $\overline{BC}$ ≅ $\overline{AC}$

42. In the diagram at the right, *m*∠1 = 40. What is *m*∠2?
　F. 40　　　**G.** 50　　　**H.** 80　　　**J.** 100

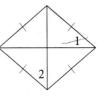

43. In an isosceles triangle, the measure of the vertex angle is 4*x*. The measure of each base angle is 2*x* + 10. What is the measure of the vertex angle?
　A. 10　　　**B.** 20　　　**C.** 50　　　**D.** 80

Short Response

44. In the figure at the right, *m*∠*APB* = 60.
　a. What is *m*∠*PAB*? Explain.
　b. ∠*PAB* and ∠*QAB* are complementary. What is *m*∠*AQB*? Show your work.

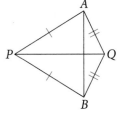

Mixed Review

Lesson 4-4

 for Help

45. *m*∠*R* = 59, *m*∠*T* = 93 = *m*∠*H*, *m*∠*V* = 28, and *RT* = *GH*. What, if anything, can you conclude about *RC* and *GV*? Explain.

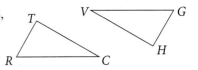

Lessons 4-2, 4-3

Which congruence statement, SSS, SAS, ASA, or AAS, would you use to conclude that the two triangles are congruent?

46. 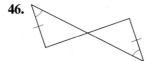　　　　　　　**47.**

Lesson 3-5

48. How many sides are in a regular polygon whose exterior angles measure 15°?

Systems of Linear Equations

You can solve a system of equations in two variables by finding the intersection of their graphs or by using substitution to form a one-variable equation.

1 EXAMPLE

x^2 **Algebra** Solve the system: $y = 3x + 5$
$$y = x + 1$$

$y = x + 1$ **Start with one equation.**

$3x + 5 = x + 1$ **Substitute $3x + 5$ for y.**

$2x = -4$ **Solve for x.**

$x = -2$

Substitute -2 for x in either equation and solve for y.

$y = x + 1$
$= (-2) + 1 = -1$

Since $x = -2$ and $y = -1$, the solution is $(-2, -1)$. This is the point of intersection of the two lines.

The graph of a linear system with *infinitely many solutions* is one line, and the graph of a linear system with *no solution* is two parallel lines.

2 EXAMPLE

x^2 **Algebra** Solve the system: $x + y = 3$
$$4x + 4y = 8$$

$x + y = 3$

$x = 3 - y$ **Solve the first equation for x.**

$4(3 - y) + 4y = 8$ **Substitute $3 - y$ for x in the second equation.**

$12 - 4y + 4y = 8$ **Solve for y.**

$12 = 8$ **False!**

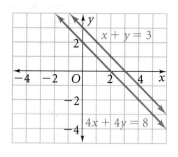

Since $12 = 8$ is a false statement, the system has no solution.

EXERCISES

Solve each system of equations.

1. $y = x - 4$
$y = 3x + 2$

2. $2x - y = 8$
$x + 2y = 9$

3. $3x + y = 4$
$-6x - 2y = 12$

4. $2x - 3 = y + 3$
$2x + y = -3$

5. $y = x + 1$
$x = y - 1$

6. $x - y = 4$
$3x - 3y = 6$

7. $y = -x + 2$
$2y = 4 - 2x$

8. $y = 2x + 1$
$y = 3x - 7$

9. $x - y = 2$
$x + y = 1$

10. What is the solution of the system, $x = 4$, $y = 3$? Interpret the solution using the coordinate plane.

Congruence in Right Triangles

Learning Standards for Mathematics

G.G.28 Determine the congruence of two triangles by using one of the five congruence techniques (SSS, SAS, ASA, AAS, HL), given sufficient information about the sides and/or angles of two congruent triangles.

✓ Check Skills You'll Need

GO for Help Lessons 4-2 and 4-3

Tell whether the abbreviation identifies a congruence statement.

1. SSS   **2.** SAS   **3.** SSA

4. ASA   **5.** AAS   **6.** AAA

Can you conclude that the two triangles are congruent? Explain.

7.    **8.**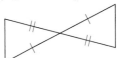

🔊 **New Vocabulary** • hypotenuse • legs of a right triangle

---

1   The Hypotenuse-Leg Theorem

In a right triangle, the side opposite the right angle is the longest side and is called the **hypotenuse.** The other two sides are called **legs.**

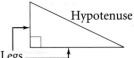

Right triangles provide a special case for which there is an SSA congruence rule. (See Lesson 4-3, Exercise 32.) It occurs when hypotenuses are congruent and one pair of legs are congruent.

 Key Concepts

Theorem 4-6	**Hypotenuse-Leg (HL) Theorem**

If the hypotenuse and a leg of one right triangle are congruent to the hypotenuse and a leg of another right triangle, then the triangles are congruent.

Proof → **Proof of the HL Theorem**

Given: $\triangle PQR$ and $\triangle XYZ$ are right triangles, with right angles Q and Y respectively. $\overline{PR} \cong \overline{XZ}$, and $\overline{PQ} \cong \overline{XY}$.

Prove: $\triangle PQR \cong \triangle XYZ$

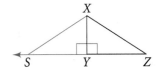

Proof: On $\triangle XYZ$ at the right, draw $\overrightarrow{ZY}$. Mark point S as shown so that $YS = QR$. Then, $\triangle PQR \cong \triangle XYS$ by SAS. By CPCTC, $\overline{PR} \cong \overline{XS}$. It is given that $\overline{PR} \cong \overline{XZ}$, so $\overline{XS} \cong \overline{XZ}$ by the Transitive Property of Congruence.

By the Isosceles Triangle Theorem, $\angle S \cong \angle Z$, so $\triangle XYS \cong \triangle XYZ$ by AAS. Therefore, $\triangle PQR \cong \triangle XYZ$ by the Transitive Property of Congruence.

To use the HL Theorem, you must show that three conditions are met.

- There are two right triangles.
- The triangles have congruent hypotenuses.
- There is one pair of congruent legs.

For: Right Triangles Activity
Use: Interactive Textbook, 4-6

1 EXAMPLE Real-World Connection

Tent Design On the tent, $\angle CPA$ and $\angle MPA$ are right angles and $\overline{CA} \cong \overline{MA}$. Can you use one pattern to cut fabric for both flaps of the tent? Explain.

Check whether the two right triangles meet the three conditions for the HL Theorem.

- You are given that $\angle CPA$ and $\angle MPA$ are right angles. $\triangle CPA$ and $\triangle MPA$ are right triangles.
- The hypotenuses of the triangles are $\overline{CA}$ and $\overline{MA}$. You are given that $\overline{CA} \cong \overline{MA}$.
- $\overline{PA}$ is a leg of both $\triangle CPA$ and $\triangle MPA$. $\overline{PA} \cong \overline{PA}$ by the Reflexive Property of Congruence.

$\triangle CPA \cong \triangle MPA$ by the HL Theorem. The triangles are the same shape and size. You can use one pattern for both flaps.

Quick Check ❶ Which two triangles are congruent by the HL Theorem? Write a correct congruence statement.

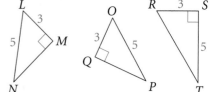

Proof 2 EXAMPLE Using the HL Theorem

Given: $\overline{CD} \cong \overline{EA}$, $\overline{AD}$ is the perpendicular bisector of $\overline{CE}$.

Prove: $\triangle CBD \cong \triangle EBA$

Proof:

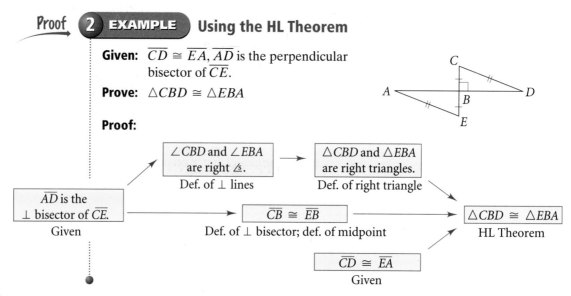

Quick Check ❷ Prove that the two triangles you named in Quick Check 1 are congruent.

Proof → **3** **EXAMPLE** Using the HL Theorem

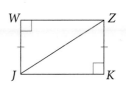

Given: $\overline{WJ} \cong \overline{KZ}$, $\angle W$ and $\angle K$ are right angles.
Prove: $\triangle JWZ \cong \triangle ZKJ$

Statements	Reasons
1. $\angle W$ and $\angle K$ are right angles.	1. Given
2. $\triangle JWZ$ and $\triangle ZKJ$ are right triangles.	2. Definition of right triangle
3. $\overline{JZ} \cong \overline{JZ}$	3. Reflexive Property of Congruence
4. $\overline{WJ} \cong \overline{KZ}$	4. Given
5. $\triangle JWZ \cong \triangle ZKJ$	5. HL Theorem

✓ **Quick Check** **3** **Given:** $\angle PRS$ and $\angle RPQ$ are right angles, $\overline{SP} \cong \overline{QR}$.
Prove: $\triangle PRS \cong \triangle RPQ$

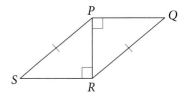

EXERCISES

For more exercises, see *Extra Skill, Word Problem, and Proof Practice.*

Practice and Problem Solving

A Practice by Example

Example 1
(page 236)

GO for Help

Write a short paragraph to explain why the two triangles are congruent.

1.

2.

What additional information do you need to prove the triangles congruent by HL?

3. $\triangle BLT$ and $\triangle RKQ$

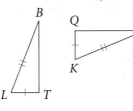

4. $\triangle XRV$ and $\triangle TRV$

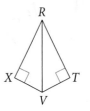

Example 2
(page 236)

5. Developing Proof Complete the flow proof.
Given: $\overline{PS} \cong \overline{PT}$, $\angle PRS \cong \angle PRT$
Prove: $\triangle PRS \cong \triangle PRT$

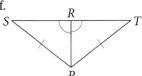

$\angle PRS$ and $\angle PRT$ are $\cong$ and supplementary ⦟.
Given (by diagram)
→ $\angle PRS$ and $\angle PRT$ are right ⦟.
a. ?
→ $\triangle PRS$ and $\triangle PRT$ are right △.
b. ?

$\overline{PS} \cong \overline{PT}$
c. ?

$\overline{PR} \cong \overline{PR}$
d. ?

→ $\triangle PRS \cong \triangle PRT$
e. ?

Proof 6. Given: $\overline{AD} \cong \overline{CB}$, $\angle D$ and $\angle B$ are right angles.

Prove: $\triangle ADC \cong \triangle CBA$

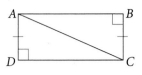

Example 3
(page 237)

7. Developing Proof Complete the two-column proof.

Given: $\overline{JL} \perp \overline{LM}$, $\overline{LJ} \perp \overline{JK}$, $\overline{MJ} \cong \overline{KL}$

Prove: $\triangle JLM \cong \triangle LJK$

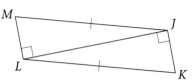

Statements	Reasons
1. $\overline{JL} \perp \overline{LM}$ and $\overline{LJ} \perp \overline{JK}$	**a.** ?
2. $\angle JLM$ and $\angle LJK$ are right angles.	**b.** ?
c. ?	3. Definition of a right triangle
4. $\overline{MJ} \cong \overline{KL}$	**d.** ?
e. ?	5. Reflexive Property of Congruence
6. $\triangle JLM \cong \triangle LJK$	**f.** ?

Proof 8. Given: $\overline{HV} \perp \overline{GT}$, $\overline{GH} \cong \overline{TV}$, I is the midpoint of $\overline{HV}$.

Prove: $\triangle IGH \cong \triangle ITV$

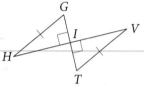

B **Apply Your Skills**

 9. Antiques To repair an antique clock, a 12-toothed wheel has to be made by cutting right triangles out of a regular polygon that has twelve 4-cm sides. The hypotenuse of each triangle is a side of the regular polygon, and the shorter leg is 1 cm long. Explain why the 12 triangles must be congruent.

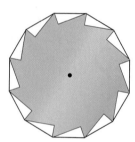

x^2 **Algebra In Exercises 10 and 11, for what values of**
x and y are the triangles congruent by HL?

10.

11.

Real-World **Connection**

Interest in antiques and shifts in fashion have stabilized the need for dial-clock repair skills.

12. Critical Thinking While working for a landscape architect, you are told to lay out a flower bed in the shape of a right triangle with sides of 3 yd and 7 yd. Explain what else you need to know in order to make the flower bed.

Proof 13. Given: $\overline{RS} \cong \overline{TU}$, $\overline{RS} \perp \overline{ST}$, $\overline{TU} \perp \overline{UV}$, T is the midpoint of $\overline{RV}$.

Prove: $\triangle RST \cong \triangle TUV$

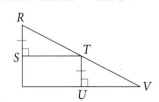

14. Given: $\overline{JM} \cong \overline{WP}$, $\overline{JP} \parallel \overline{MW}$, $\overline{JP} \perp \overline{PM}$

Prove: $\triangle JMP \cong \triangle WPM$

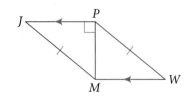

GO **Online**
Homework Video Tutor

Visit: PHSchool.com
Web Code: aue-0406

Proof 15. Study Exercise 5. There is a different set of steps that will prove $\triangle PRS \cong \triangle PRT$. Decide what they are. Then write a proof using these steps.

Constructions Copy the triangle and construct a triangle congruent to it using the method stated.

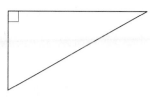

16. by SAS   **17.** by HL

18. by ASA   **19.** by SSS

Proof **20. Given:** $\overline{EB} \cong \overline{DB}$, $\angle A$ and $\angle C$ are right angles, and B is the midpoint of $\overline{AC}$.
Prove: $\triangle BEA \cong \triangle BDC$

21. Given: $\overline{LO}$ bisects $\angle MLN$, $\overline{OM} \perp \overline{LM}$, and $\overline{ON} \perp \overline{LN}$.
Prove: $\triangle LMO \cong \triangle LNO$

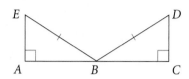

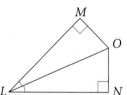

22. Open-Ended You are the DJ for the school dance. To set up, you have placed one speaker in the corner of the platform. What measurement(s) could you make with a tape measure to make sure that a matching speaker is in the other corner at exactly the same angle? Explain why your method works.

23. a. Coordinate Geometry Use grid paper. Graph the points $E(-1, -1)$, $F(-2, -6)$, $G(-4, -4)$, and $D(-6, -2)$. Connect the points with segments.
   b. Find the slope for each of $\overline{DG}$, $\overline{GF}$, and $\overline{GE}$.
   c. Use your answer to part (b) to describe $\angle EGD$ and $\angle EGF$.
   d. Use the Distance Formula to find DE and FE.
   e. Write a paragraph to prove that $\triangle EGD \cong \triangle EGF$.

Exercise 22

24. Critical Thinking "A HA!" exclaims Francis. "There is an HA Theorem . . . , something like the HL Theorem!" Explain what Francis is saying and why he is correct or incorrect.

C Challenge

Geometry in 3 Dimensions Use the figure at the right for Exercises 25 and 26.

Proof **25. Given:** $\overline{BE} \perp \overline{EA}$, $\overline{BE} \perp \overline{EC}$, $\triangle ABC$ is equilateral.
Prove: $\triangle AEB \cong \triangle CEB$

26. Given: $\triangle AEB \cong \triangle CEB$, $\overline{BE} \perp \overline{EA}$, and $\overline{BE} \perp \overline{EC}$. Can you prove that $\triangle ABC$ is equilateral? Explain.

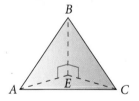

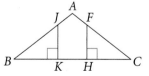

Test Prep

Multiple Choice In Exercises 27 and 28, which additional congruence statement could you use to prove that $\triangle BJK \cong \triangle CFH$ by HL?

27. Given: $\overline{BJ} \cong \overline{CF}$
   A. $\overline{JK} \cong \overline{FH}$ **B.** $\angle B \cong \angle C$
   C. $\overline{AJ} \cong \overline{AF}$ **D.** $\angle BJK \cong \angle CFH$

28. Given: $\overline{BK} \cong \overline{CH}$
   F. $\overline{JK} \cong \overline{FH}$ **G.** $\angle B \cong \angle C$ **H.** $\overline{JB} \cong \overline{FC}$ **J.** $\angle BJK \cong \angle CFH$

29. Which congruence statement can be used to prove that the two triangles are congruent?

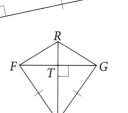

   A. SAS   **B.** SSS

   C. ASA   **D.** HL

Short Response

30. a. Use the diagram at the right to name all the pairs of triangles you could prove congruent by using the HL Theorem.

   b. Suppose you need to prove $\triangle RFW \cong \triangle RGW$. What specifically do you need to prove before you can use the HL Theorem?

Mixed Review

Lesson 4-5

For Exercises 31 and 32, what type of triangle must $\triangle XYZ$ be?

31. $\triangle XYZ \cong \triangle ZYX$ **32.** $\triangle XYZ \cong \triangle ZXY$

Lesson 3-7

33. Connect $A(3, 3)$, $B(5, 5)$, $C(9, 1)$, and $D(9, -3)$ in order. Are any sides of the figure parallel? Are any sides perpendicular? Explain.

Lesson 3-1

State the postulate or theorem that justifies each statement.

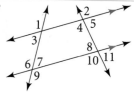

34. $\angle 5 \cong \angle 8$ **35.** $m\angle 4 + m\angle 8 = 180$

36. $\angle 6 \cong \angle 9$ **37.** $\angle 4 \cong \angle 10$

38. $\angle 1 \cong \angle 6$ **39.** $\angle 6$ and $\angle 3$ are supplementary.

✓ Checkpoint Quiz 2   Lessons 4-4 Through 4-6

1. In the diagram at the right, $\triangle PQR \cong \triangle SRQ$ by SAS. What other pairs of sides and angles can you conclude are congruent by CPCTC?

2. Complete the plan for a proof.

   Given: Isosceles $\triangle JKL$ with $\overline{JK} \cong \overline{JL}$; $\overline{KM}$ and $\overline{LM}$ are bisectors of the base angles.

   Prove: $\triangle KML$ is isosceles.

   Plan: Since $\triangle JKL$ is isosceles, $\angle JKL \cong \angle JLK$ by the **a.** __?__ Theorem. Since $\overline{KM}$ and $\overline{LM}$ are angle bisectors, $\angle MKL$ **b.** __?__ $\angle MLK$. Therefore, $\triangle KML$ is isosceles by the **c.** __?__ Theorem.

Exercise 3

3. Six triangles are pictured in the diagram at the left. Which of the triangles are isosceles? Explain.

4. Why are these triangles congruent?   **5.** Explain why $\overline{GW} \cong \overline{ST}$.

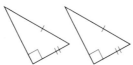

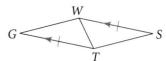

4-7

Using Corresponding Parts of Congruent Triangles

NY Learning Standards for Mathematics

G.G.29 Identify corresponding parts of congruent triangles.

✓ Check Skills You'll Need

1. How many triangles will the next two figures in this pattern have?

GO for Help Lessons 1-1 and 4-3

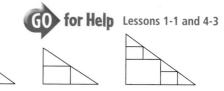

2. Can you conclude that the triangles are congruent? Explain.
   a. △AZK and △DRS   **b.** △SDR and △JTN   **c.** △ZKA and △NJT

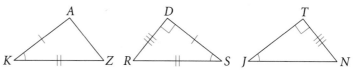

---

1 Using Overlapping Triangles in Proofs

Some triangle relationships are difficult to see because the triangles overlap. Overlapping triangles may have a common side or angle. You can simplify your work with overlapping triangles by separating and redrawing the triangles.

Vocabulary Tip

Overlapping triangles share part or all of one or more sides.

1 EXAMPLE Identifying Common Parts

Separate and redraw △DFG and △EHG. Identify the common angle.

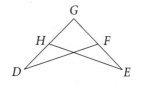

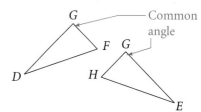

✓ Quick Check

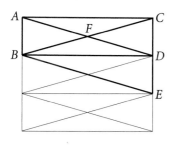

1. **Engineering** The diagram at the left shows triangles from the scaffolding that workers used when they repaired and cleaned the Statue of Liberty.
   a. Name the common side in △ADC and △BCD.
   b. Name another pair of triangles that share a common side. Name the common side.

In overlapping triangles, a common side or angle is congruent to itself by the Reflexive Property of Congruence.

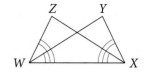

Proof ➋ EXAMPLE   Using Common Parts

Given: ∠ZXW ≅ ∠YWX, ∠ZWX ≅ ∠YXW

Write a plan and then a proof to show
that the two "outside" segments are congruent.

Prove: $\overline{ZW} \cong \overline{YX}$

Plan: First, separate the
overlapping triangles.
$\overline{ZW} \cong \overline{YX}$ by CPCTC if
△ZXW ≅ △YWX. Show
this congruence by ASA.

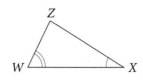
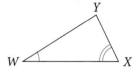

Proof:

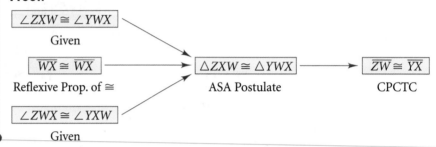

∠ZXW ≅ ∠YWX		
Given		
$\overline{WX} \cong \overline{WX}$	△ZXW ≅ △YWX	$\overline{ZW} \cong \overline{YX}$
Reflexive Prop. of ≅	ASA Postulate	CPCTC
∠ZWX ≅ ∠YXW		
Given		

✓ **Quick Check** ➋ Write a plan and then a proof.

Given: △ACD ≅ △BDC
Prove: $\overline{CE} \cong \overline{DE}$

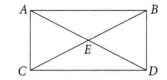

❷ Using Two Pairs of Congruent Triangles

Sometimes you can prove one pair of triangles congruent and then use their
congruent corresponding parts to prove another pair congruent.

Proof ➌ EXAMPLE   Using Two Pairs of Triangles

Given: In the quilt, E is the midpoint of $\overline{AC}$ and $\overline{DB}$.
Prove: △GED ≅ △JEB

Write a plan and then a proof.

Plan: △GED ≅ △JEB by ASA if ∠D ≅ ∠B. These angles are congruent by
CPCTC if △AED ≅ △CEB. These triangles are congruent by SAS.

Proof: E is the midpoint of $\overline{AC}$ and $\overline{DB}$, so $\overline{AE} \cong \overline{CE}$ and $\overline{DE} \cong \overline{BE}$.
∠AED ≅ ∠CEB because vertical angles are congruent. Therefore,
△AED ≅ △CEB by SAS. ∠D ≅ ∠B by CPCTC, and ∠GED ≅ ∠JEB
because they are vertical angles. Therefore, △GED ≅ △JEB by ASA.

✓ **Quick Check** ➌ Write a plan and then a proof.

Given: $\overline{PS} \cong \overline{RS}$, ∠PSQ ≅ ∠RSQ
Prove: △QPT ≅ △QRT

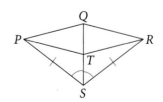

When triangles overlap, you can keep track of information by drawing other diagrams that separate the overlapping triangles.

Proof → **4 EXAMPLE** **Separating Overlapping Triangles**

Given: $\overline{CA} \cong \overline{CE}$, $\overline{BA} \cong \overline{DE}$

Write a plan and then a proof to show that two small segments inside the triangle are congruent.

Prove: $\overline{BX} \cong \overline{DX}$

Plan: $\overline{BX} \cong \overline{DX}$ by CPCTC if $\triangle BXA \cong \triangle DXE$. This congruence holds by AAS if $\angle ABX \cong \angle EDX$. These are congruent by CPCTC in $\triangle BAE$ and $\triangle DEA$, which are congruent by SAS.

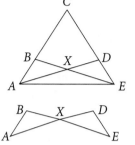

Proof:

Statements	Reasons
1. $\overline{BA} \cong \overline{DE}$	1. Given
2. $\overline{CA} \cong \overline{CE}$	2. Given
3. $\angle CAE \cong \angle CEA$	3. Isosceles Triangle Theorem
4. $\overline{AE} \cong \overline{AE}$	4. Reflexive Property of Congruence
5. $\triangle BAE \cong \triangle DEA$	5. SAS
6. $\angle ABE \cong \angle EDA$	6. CPCTC
7. $\angle BXA \cong \angle DXE$	7. Vertical angles are congruent.
8. $\triangle BXA \cong \triangle DXE$	8. AAS
9. $\overline{BX} \cong \overline{DX}$	9. CPCTC

Real-World Connection

The Japanese paper-folding art of origami involves many overlapping triangles.

✓ **Quick Check** **4** Plan a proof. Separate the overlapping triangles in your plan. Then follow your plan and write a proof.

Given: $\angle CAD \cong \angle EAD$, $\angle C \cong \angle E$
Prove: $\overline{BD} \cong \overline{FD}$

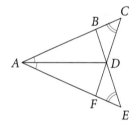

EXERCISES

For more exercises, see *Extra Skill, Word Problem, and Proof Practice.*

Practice and Problem Solving

A Practice by Example

Example 1
(page 241)

GO for Help

In each diagram, the red and blue triangles are congruent. Identify their common side or angle.

1.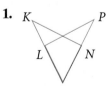

2.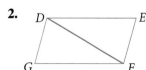

3.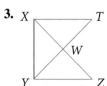

Separate and redraw the indicated triangles. Identify any common angles or sides.

4. △PQS and △QPR

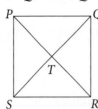

5. △ACB and △PRB

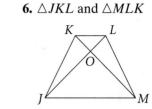

6. △JKL and △MLK

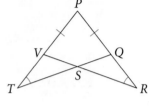

Example 2
(page 242)

7. Developing Proof Complete the flow proof.

Given: ∠T ≅ ∠R, $\overline{PQ}$ ≅ $\overline{PV}$

Prove: ∠PQT ≅ ∠PVR

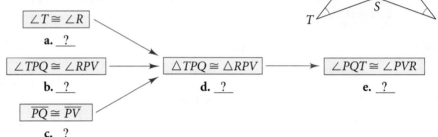

<u>Proof</u> **Write a plan and then a proof.**

8. Given: $\overline{RS}$ ≅ $\overline{UT}$, $\overline{RT}$ ≅ $\overline{US}$
Prove: △RST ≅ △UTS

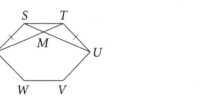

9. Given: $\overline{QD}$ ≅ $\overline{UA}$, ∠QDA ≅ ∠UAD
Prove: △QDA ≅ △UAD

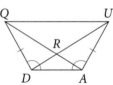

Examples 3, 4
(pages 242 and 243)

10. Given: ∠1 ≅ ∠2, ∠3 ≅ ∠4
Prove: △QET ≅ △QEU

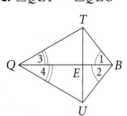

11. Given: $\overline{AD}$ ≅ $\overline{ED}$,
D is the midpoint of $\overline{BF}$.

Prove: △ADC ≅ △EDG

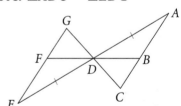

B **Apply Your Skills**

Open-Ended **Draw the diagram described.**

12. Draw a vertical segment on your paper. On the right side of the segment draw two triangles that share the given segment as a common side.

13. Draw two triangles that have a common angle.

14. Draw two regular pentagons, each with its five diagonals.
  a. In one, shade two triangles that share a common angle.
  b. In the other, shade two triangles that share a common side.

15. Draw two regular hexagons and their diagonals. For these diagrams, do parts (a) and (b) of the preceding exercise.

GO **online**
Homework Video Tutor
Visit: PHSchool.com
Web Code: aue-0407

16. Multiple Choice In the diagram, $\angle V \cong \angle S$, $\overline{VU} \cong \overline{ST}$, and $\overline{PS} \cong \overline{QV}$. Which two triangles can you prove congruent by SAS?

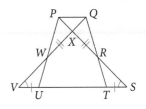

Ⓐ $\triangle WVU \cong \triangle RST$

Ⓑ $\triangle PSU \cong \triangle QVT$

Ⓒ $\triangle PWX \cong \triangle QRX$

Ⓓ none of these

Proof **17. Given:** $\overline{AC} \cong \overline{EC}, \overline{CB} \cong \overline{CD}$

Prove: $\angle A \cong \angle E$

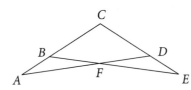

18. Given: $\overline{QT} \perp \overline{PR}, \overline{QT}$ bisects $\overline{PR}$, $\overline{QT}$ bisects $\angle VQS$.

Prove: $\overline{VQ} \cong \overline{SQ}$

Clothes Design **The figure at the right is part of a clothing design pattern. In the figure, $\overline{AB} \parallel \overline{DE} \parallel \overline{FG}, \overline{AB} \perp \overline{BC}$, and $\overline{GC} \perp \overline{AC}$. $\triangle DEC$ is isosceles with base $\overline{DC}$, and $m\angle A = 56$.**

19. Find the measures of all the numbered angles in the figure.

20. $\overline{AB} \cong \overline{FC}$. Name two congruent triangles and tell how you can prove them congruent.

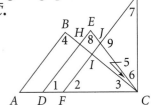

Proof **21. Given:** $\overline{TE} \cong \overline{RI}, \overline{TI} \cong \overline{RE}$, $\angle TDI$ and $\angle ROE$ are right $\angle$s.

Prove: $\overline{TD} \cong \overline{RO}$

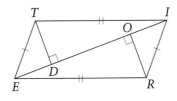

22. Given: $\overline{AB} \perp \overline{BC}, \overline{DC} \perp \overline{BC}$, $\overline{AC} \cong \overline{DB}$

Prove: $\overline{AE} \cong \overline{DE}$

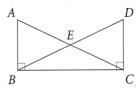

🅲 Challenge

23. Reasoning Draw a quadrilateral $ABCD$ with $\overline{AB} \parallel \overline{DC}$ and $\overline{AD} \parallel \overline{BC}$, and its diagonals $\overline{AC}$ and $\overline{DB}$ intersecting at E. Label your diagram to indicate the parallel sides.

✏️ **a.** List all the pairs of congruent segments that you can find in your diagram.

✏️ **b. Writing** Explain how you know that the segments you listed are congruent.

Proof **Name a pair of overlapping congruent triangles in each diagram. State whether the triangles are congruent by SSS, SAS, ASA, AAS, or HL. Plan and write a proof.**

24. Given:

$\overline{AC} \cong \overline{BC}$, $\angle A \cong \angle B$

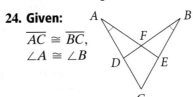

25. Given:

$\overline{WY} \perp \overline{YX}$, $\overline{ZX} \perp \overline{YX}$, $\overline{WX} \cong \overline{ZY}$

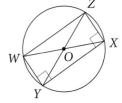

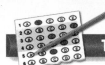

Multiple Choice

Use the diagram at the right for Exercises 26–28.

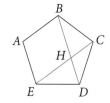

26. If $m\angle KJM = 25$, what is $m\angle LKJ$?

    A. 25   **B.** 30   **C.** 65   **D.** 85

27. If $m\angle KJM = 30$ and $x = 7.4$, what is the perimeter of $\triangle LKJ$?

    F. 44.4   **G.** 22.2   **H.** 14.8   **J.** 7.4

28. If $m\angle LJK = 47$, what is $m\angle LJM$?

    A. 23.5   **B.** 25   **C.** 43   **D.** 47

Short Response

29. The pentagon at the right is equilateral and equiangular.

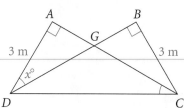

    a. What two triangles must be congruent to prove $\overline{HB} \cong \overline{HE}$?

    b. Write a proof to show $\overline{HB} \cong \overline{HE}$.

Extended Response

30. a. In the figure at the right, why is $\triangle ACD \cong \triangle BDC$?

    b. Copy the figure. Mark each angle that has measure x.

    c. What is the value of x? Explain how you found your answer.

    d. What is $m\angle AGB$?

    e. What is CD? Explain your answer.

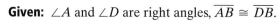

Mixed Review

GO for Help

Lesson 4-6

31. Complete the plan for a proof.

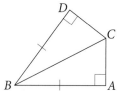

    Given: $\angle A$ and $\angle D$ are right angles, $\overline{AB} \cong \overline{DB}$.

    Prove: $\triangle ABC \cong \triangle DBC$

    Plan: $\triangle ABC$ and $\triangle DBC$ are **a.** ? triangles with legs that are given to be **b.** ? . The hypotenuse is congruent to itself by the **c.** ? Property of Congruence. $\triangle ABC \cong \triangle DBC$ by the **d.** ? Theorem.

Lesson 3-8

Constructions Draw a line p and a point M not on p. Construct the described line.

32. line n through M so that $n \perp p$ **33.** line r through M so that $r \parallel p$

Lesson 3-6

Write an equation in point-slope form of the line that contains the given point and has the given slope.

34. $P(2, -6)$; slope $\frac{1}{2}$ **35.** $Q(0, 5)$; slope 1

36. $R(-3, 6)$; slope -2 **37.** $S(0, 0)$; slope $-\frac{1}{3}$

Write an equation in point-slope form of the line that contains the given points.

38. $A(1, 4), B(0, 2)$ **39.** $E(3, -5), F(6, 0)$ **40.** $X(-4, -3), Y(2, -8)$

Writing Flow Proofs

Proofs can get long and complex. While a two-column proof may appear more organized, a flow proof can show the logic flow better and thus be easier to follow.

To write a flow proof, sketch the logic "paths" of a proof on scratch paper. Then organize your work into a neat, easy-to-follow flow diagram, as in this Example.

NY G.R.2: Recognize, compare, and use an array of representational forms.

EXAMPLE

Given: $\overline{KJ} \cong \overline{LM}, \angle KJN \cong \angle LMN, \overline{JN} \cong \overline{MN}$

Prove: $\overline{JL} \cong \overline{MK}$

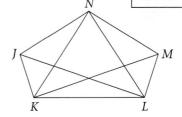

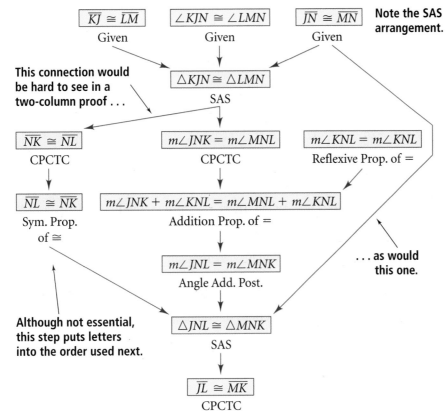

Note the SAS arrangement.

This connection would be hard to see in a two-column proof . . .

. . . as would this one.

Although not essential, this step puts letters into the order used next.

EXERCISES

Write a flow proof. Make the flow of logic as easy to follow as you can.

1. Given: $\overline{AB} \cong \overline{DC}$, E is the midpoint of $\overline{AD}$, and $\overline{CE} \cong \overline{BE}$.
Prove: $\overline{AC} \cong \overline{DB}$

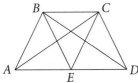

2. Given: $\overline{RQ} \cong \overline{RS}$, $\overline{RP} \cong \overline{RT}$, $\overline{QP} \cong \overline{ST}$
Prove: $\overline{QT} \cong \overline{SP}$

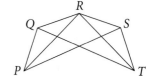

A test item may ask you to use a data set to determine a relationship between quantities. Often the relationship is linear and may be expressed as a **linear function,** a function that can be represented by a linear equation in two variables. These are the equations you studied in Lesson 3-6.

EXAMPLE

Each row in the table shows the measure b of a base angle and the measure v of the vertex angle of an isosceles triangle. Which function equation describes the relationship between b and v?

(A) $v = 16b$ (B) $v = 170 - b$ (C) $v = 180 - b$

(D) $v = -2b + 180$ (E) $v = 2b - 20$

b	v
10	160
20	140
30	120
40	100
50	80

Method 1

The choices suggest a linear relationship. For each change in b of 10, the change in v is -20. The slope must be -2. The correct choice is D.

Method 2

For any isosceles triangle, $b + b + v = 180$ by the Triangle Angle-Sum Theorem. Simplifying, $2b + v = 180$, or $v = -2b + 180$. The correct choice is D.

EXERCISES

1. The Example above gives you two methods for solving. There is a third method that works for every test item of this type. What is it? (*Hint:* What is true about the (b, v) values in the table and an equation that describes the b-v relationship?) For the Example, which method do you prefer? Explain.

2. Each row in the table shows the measure a of a base angle of an isosceles triangle. It also shows the measure b of an angle formed by the bisectors of a base angle and the vertex angle. Which equation describes the relationship between a and b? Explain how you found your answer.

(A) $2b - a = 180$ (B) $2b + a = 180$ (C) $b - 2a = 75$

(D) $b + 2a = 115$ (E) $b + a = 105$

a	b
10	95
20	100
30	105
40	110
50	115

3. What might the values of x and y in each row of the table represent?

(A) The measure x of an angle and the measure y of its complement.

(B) The measure x of an angle and the measure y of its supplement.

(C) The measure x of a base angle of an isosceles triangle and the measure y of the vertex angle.

(D) The measure x of a base angle of an isosceles triangle and the measure y of an exterior angle at the vertex.

(E) The measure x of the vertex angle of an isosceles triangle and the sum y of the measures of the two base angles.

x	y
60	120
45	90
30	60
15	30

Chapter Review

Vocabulary Review

base of an isosceles triangle (p. 228)	CPCTC (corresponding parts of congruent triangles are congruent) (p. 221)	legs of an isosceles triangle (p. 228)
base angles of an isosceles triangle (p. 228)		vertex angle of an isosceles triangle (p. 228)
congruent polygons (p. 198)	hypotenuse (p. 235)	
corollary (p. 229)	legs of a right triangle (p. 235)	

Choose the correct term to complete each sentence.

1. The two congruent sides of an isosceles triangle are the __?__.

2. The two congruent sides of an isosceles triangle form the __?__.

3. If you know that two triangles are congruent, then the corresponding sides and angles of the triangles are congruent because __?__.

4. The side opposite the right angle of a right triangle is the __?__.

5. The angles of an isosceles triangle that are not the vertex angle are called the __?__.

6. A __?__ to a theorem is a statement that follows immediately from the theorem.

Go Online
PHSchool.com

For: Vocabulary quiz
Web Code: auj-0451

7. The __?__ are the two sides of a right triangle that are not the hypotenuse.

8. __?__ have congruent corresponding parts.

9. The side of an isosceles triangle that is not a leg is called the __?__.

Skills and Concepts

4-1 Objectives

▼ To recognize congruent figures and their corresponding parts

Congruent polygons have congruent corresponding parts. When you name congruent polygons, always list corresponding vertices in the same order.

Two triangles are congruent when they have three pairs of congruent corresponding sides and three pairs of congruent corresponding angles.

RSTUV ≅ KLMNO. Complete the congruence statements.

10. $\overline{TS} \cong$ __?__

11. $\angle N \cong$ __?__

12. $\overline{LM} \cong$ __?__

13. $VUTSR \cong$ __?__

WXYZ ≅ PQRS. Find the measure of the angle or the length of the side.

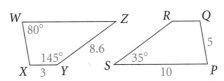

14. $\angle P$ **15.** $\overline{QR}$ **16.** $\overline{WX}$ **17.** $\angle Z$ **18.** $\angle X$

4-2 and 4-3 Objectives

▼ To prove two triangles congruent using the SSS and SAS Postulates

▼ To prove two triangles congruent using the ASA Postulate and the AAS Theorem

If three sides of one triangle are congruent to three sides of another triangle, then the two triangles are congruent by the **Side-Side-Side (SSS) Postulate.**

If two sides and the included angle of one triangle are congruent to two sides and the included angle of another triangle, then the two triangles are congruent by the **Side-Angle-Side (SAS) Postulate.**

If two angles and the included side of one triangle are congruent to two angles and the included side of another triangle, then the two triangles are congruent by the **Angle-Side-Angle (ASA) Postulate.**

If two angles and a nonincluded side of one triangle are congruent to two angles and the corresponding nonincluded side of another triangle, then the two triangles are congruent by the **Angle-Angle-Side (AAS) Theorem.**

Which postulate or theorem, if any, could you use to prove the two triangles congruent? If the triangles *cannot* be proven congruent, write *not possible.*

19.

20.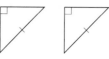

21.

22.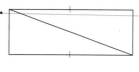

23.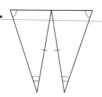

24.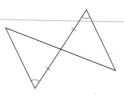

Write a congruence statement for each pair of triangles. Name the postulate or theorem that justifies your statement. If the triangles *cannot* be proven congruent, write *not possible.*

25.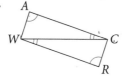

26.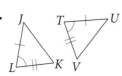

27.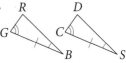

4-4 Objectives

▼ To use triangle congruence and CPCTC to prove that parts of two triangles are congruent

Once you know that triangles are congruent, you can make conclusions about corresponding segments and angles because, by definition, **corresponding parts of congruent triangles are congruent (CPCTC).** You can use congruent triangles in the proofs of many theorems.

Explain how you can use SSS, SAS, ASA, or AAS with CPCTC to prove the statement true.

28. $\overline{TV} \cong \overline{YW}$

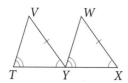

29. $\overline{BE} \cong \overline{DE}$

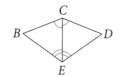

30. $\overline{KN} \cong \overline{ML}$

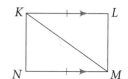

4-5 and 4-6 Objectives

▼ To use and apply properties of isosceles triangles

▼ To prove triangles congruent using the HL Theorem

If two sides of a triangle are congruent, then the angles opposite those sides are also congruent by the **Isosceles Triangle Theorem.** If two angles of a triangle are congruent, then the sides opposite the angles are congruent by the **Converse of the Isosceles Triangle Theorem.**

The bisector of the vertex angle of an isosceles triangle is the perpendicular bisector of the base.

If the hypotenuse and a leg of one right triangle are congruent to the hypotenuse and a leg of another right triangle, then the triangles are congruent by the **Hypotenuse-Leg (HL) Theorem.**

x^2 **Algebra** Find the values of x and y.

31.

32.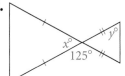

33.

Write a proof for each of the following.

34. Given: $\overline{PS} \perp \overline{SQ}, \overline{RQ} \perp \overline{QS}, \overline{PQ} \cong \overline{RS}$
   Prove: $\triangle PSQ \cong \triangle RQS$

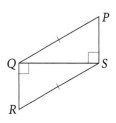

35. Given: $\overline{LN} \perp \overline{KM}, \overline{KL} \cong \overline{ML}$
   Prove: $\triangle KLN \cong \triangle MLN$

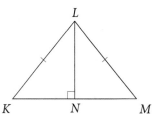

4-7 Objectives

▼ To identify congruent overlapping triangles

▼ To prove two triangles congruent by first proving two other triangles congruent

You can prove overlapping triangles congruent. You can also use the common or shared sides and angles of triangles in congruence proofs.

Name a pair of overlapping congruent triangles in each diagram. State whether the triangles are congruent by SSS, SAS, ASA, AAS, or HL.

36.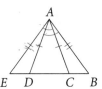

37.

38.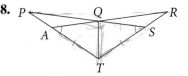

39.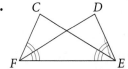

Write a congruence statement for each pair of triangles.

1.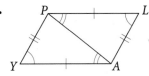

2.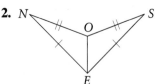

Which postulate, if any, could you use to prove the two triangles congruent? If not enough information is given, write *not possible.*

3.

4.

5.

6.

7.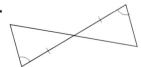

8.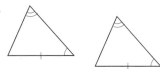

9. Writing Explain why you cannot use AAA to prove two triangles congruent.

10. Open-Ended Draw a picture to represent $\triangle CEO \cong \triangle HDF$. Name all of the pairs of corresponding congruent parts.

11. If two game boards have the same area, are the game boards congruent? Explain your answer.

x^2 **12. Algebra** Find the value of the variable.

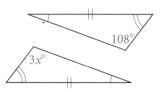

Write a proof for each of the following.

13. Given: $\overline{AT} \cong \overline{GS}$,
$\overline{AT} \parallel \overline{GS}$
Prove: $\triangle GAT \cong \triangle TSG$

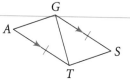

14. Given: $\overline{LN}$ bisects $\angle OLM$ and $\angle ONM$.
Prove: $\triangle OLN \cong \triangle MLN$

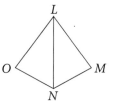

Name a pair of overlapping congruent triangles in each diagram. State whether the triangles are congruent by SSS, SAS, ASA, AAS, or HL.

15. Given: $\overline{CE} \cong \overline{DF}$,
$\overline{CF} \cong \overline{DE}$

16. Given: $\overline{RT} \cong \overline{QT}$,
$\overline{AT} \cong \overline{ST}$

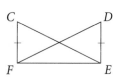

17. Open-Ended Draw two parallel lines and draw two parallel transversals through your parallel lines. Then draw a third transversal to create two congruent triangles. Label your triangles and write the congruence statement.

Regents Test Prep

Multiple Choice

For Exercises 1–8, choose the correct letter.

1. What is $m\angle CDF$?

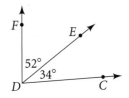

(1) 18   **(2)** 86   **(3)** 94   **(4)** 274

2. Which angles could an obtuse triangle have?

   I. a right angle
   II. two acute angles
   III. an obtuse angle
   IV. two vertical angles

(1) I and II   **(2)** II and III
(3) III and IV   **(4)** I and IV

3. What is the area in square units of a rectangle with vertices $(-2, 5)$, $(3, 5)$, $(3, -1)$, and $(-2, -1)$?

(1) 56   **(2)** 30   **(3)** 25   **(4)** 24

4. Quadrilateral $ABCD \cong QRST$. Which segment is congruent to $\overline{TS}$?

(1) $\overline{AB}$ **(2)** $\overline{BC}$ **(3)** $\overline{CB}$ **(4)** $\overline{DC}$

5. By which postulate or theorem are the triangles congruent?

(1) SAS   **(2)** SSS   **(3)** ASA   **(4)** AAS

6. Which condition(s) will allow you to prove that $\ell \parallel m$?

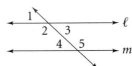

   I. $\angle 1 \cong \angle 4$
   II. $\angle 2 \cong \angle 5$
   III. $m\angle 2 + m\angle 4 = 180$
   IV. $\angle 3 \cong \angle 4$

(1) III only   **(2)** I and III only
(3) II and IV only   **(4)** I, II, III, and IV

7. Which equation represents a line perpendicular to the line shown?

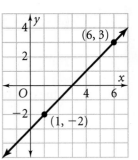

(1) $y = x - 3$
(2) $3x + 2y = 10$
(3) $y = -5x$
(4) $x + y = -7$

8. Which postulate or theorem justifies the statement $\triangle JLV \cong \triangle PMK$?

(1) AAS
(2) SAS
(3) ASA
(4) SSS

Gridded Response

9. An isosceles triangle has two angles measuring 54.5 and 71. What is the measure of the third angle?

10. What is the number of feet in the circumference of a circle with a diameter of 10 ft? Use 3.14 for π.

11. What is the measure of the complement of a $56°$ angle?

12. What is the measure of the supplement of a $35°$ angle?

Short Response

Explain your work.

13. Draw an angle. Then construct another angle congruent to the first.

14. Construct the perpendicular bisector of a segment $\overline{MN}$.

Extended Response

15. Find CD and the coordinates of the midpoint of $\overline{CD}$ if the endpoints are $C(5, 7)$ and $D(10, -5)$. Explain your work.

Data Analysis

NY G.PS.5: Choose an effective approach to solve a problem from a variety of strategies (numeric, graphic, algebraic).

Probability

You and your friends are planning a neighborhood carnival. The carnival games will stress fun, not winning. All children will win points at every game. At the end of the day, the children can use their points to attend a picnic and talent show at the high school.

 **1** ACTIVITY

1. Draw three segments that are 4, 5, and 6 cm long. Then use a compass and straightedge to construct a triangle with sides those lengths (Figure 1).

Figure 1

Figure 2

2. Construct two more (red) triangles with sides of 4, 5, and 6 cm. Cut out the three triangles and arrange them as shown in Figure 2.

3. Explain why the three triangles you constructed are congruent. Also, explain why the triangle (blue) bordered by the three triangles is congruent to each of the three triangles.

2 ACTIVITY

For the carnival, you use a large version of the design from Activity 1 to make the surface (Figure 3) for a beanbag-toss.

4. The white triangle is 25% of the surface. If you were to randomly touch your pencil point to the surface, the probability of touching the white triangle is 25%. What is the probability of touching red? Of touching blue?

Figure 3

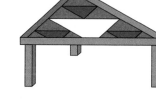

5. Design a simulation of the random tossing of beanbags onto the game surface. Use the randInt function on your calculator to generate random integers from 1 to 16. Let each integer 1–16 represent a "hit" on one of the colors.

   a. Why does it make sense to use the integers 1–16?

   b. Copy and complete the table. In the second column, use the values found in Exercise 4. In the third column, represent each color with a range of integers corresponding to the probabilities.

Color	Probability	Use Integers
White	$\frac{4}{16}$	1, 2, 3, 4
Red		
Blue		

6. Carry out your simulation from Exercise 5. Generate 50 random numbers and count the number of hits for each color. Divide each count by 50 to find the percent of hits on each color.

7. Compare the simulation results with the probabilities you found in Exercise 4. What might cause the differences you found between your theoretical and experimental results?

8. If you run the simulation with 100 random numbers instead of 50, would you expect the experimental results to be closer to or further from the theoretical results? Explain.

3 ACTIVITY

To set up a dart game, you measure ∠E to be 20° and draw △DEF. Then you construct triangles congruent to △DEF to complete a pinwheel design.

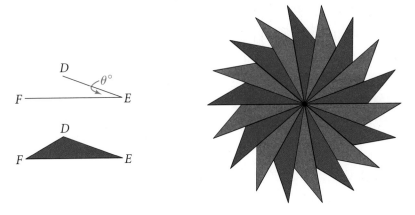

9. What other integer values could you have used for θ (the Greek letter *theta*) to make a design of the type shown? Explain.

10. Of the values you listed in Exercise 9, choose one from 24 to 60 to be θ. Draw △DEF that has one angle with measure θ. Construct a design like the one above by constructing triangles congruent to △DEF. (*Hint:* Find an efficient way to use the construction suggested on page 210, Exercise 32.) Color your design.

EXERCISES

11. In the beanbag-toss game of Activity 2, players should win more points for landing beanbags on colors with lesser probabilities. Based on the probabilities from Activity 2, what would be a fair number of points to award for landing on each color? Explain why your numbers are fair.

12. For the dartboard design in Activity 3, the probability of a dart landing randomly on either color is 50%. Explain how you could re-color the design to make probabilities that are unequal. How would you award points for your new color scheme?

13. In the design at the right θ = 72.
  a. How would you color this with two colors?
  b. How would you award points?

What You've Learned

NY Learning Standards for Mathematics

- **G.RP.2:** In Chapter 1, you learned how to identify segments, lines, and angles. You also learned the meaning of some important terms such as bisector, congruence, midpoint, perpendicular, and parallel.

- **G.PS.4:** In Chapters 2 and 3, you made conjectures about angles, parallel lines, and perpendicular lines, and learned how to use deductive reasoning to prove the conjectures true.

- **G.G.28:** In Chapter 4, you learned how to prove triangles congruent.

 Check Your Readiness **for Help** to the Lesson in green.

Inequalities (Previous Course)

x^2 **Algebra** Solve each inequality.

1. $3x + 10 \leq 22$ **2.** $4x - 1 > 2x + 14$ **3.** $30 - 5x \geq x + 24$

Basic Constructions (Lesson 1-7)

Use a compass and straightedge for the following.

4. Construct the perpendicular bisector of a segment.

5. Construct the angle bisector of an angle.

Distance Formula (Lesson 1-8)

Find the distance between each pair of points.

6. $(1, 4), (4, 8)$ **7.** $(-6, 2), (-1, 14)$ **8.** $(-3, -2), (5, -6)$

Midpoint Formula (Lesson 1-8)

Find the midpoint of the segments whose endpoints are given.

9. $(4, 11), (6, 3)$ **10.** $(-8, -3), (2, -4)$ **11.** $(-7, 15), (-2, -10)$

Slope (Algebra 1 Review, page 165)

Find the slope of the line containing each pair of points.

12. $(8, 3), (7, 12)$ **13.** $(3, -2), (0, 6)$ **14.** $(-5, 4), (-2, 4)$

Relationships Within Triangles

What You'll Learn Next

NY Learning Standards for Mathematics

- **G.G.21:** In this chapter, you will learn about geometric relationships within triangles.

- **G.G.21:** You will learn about three lines that pass through one point and find the four sets of such lines that exist for every triangle.

- **G.G.26:** You will learn about two other types of statements that are related to a conditional, as well as another type of reasoning – indirect reasoning.

- **G.G.24:** You will apply indirect reasoning to deduce information about inequalities in triangles.

 Activity Lab Applying what you learn, you will do activities involving balance on pages 302 and 303.

LESSONS

- **5-1** Midsegments of Triangles
- **5-2** Bisectors in Triangles
- **5-3** Concurrent Lines, Medians, and Altitudes
- **5-4** Inverses, Contrapositives, and Indirect Reasoning
- **5-5** Inequalities in Triangles

🔊 Key Vocabulary

- altitude of a triangle (p. 275)
- centroid (p. 274)
- circumcenter of a triangle (p. 273)
- circumscribed about (p. 273)
- concurrent (p. 273)
- contrapositive (p. 280)
- coordinate proof (p. 260)
- distance from a point to a line (p. 266)
- equivalent statements (p. 281)
- incenter of a triangle (p. 273)
- indirect proof (p. 281)
- indirect reasoning (p. 281)
- inscribed in (p. 273)
- inverse (p. 280)
- median of a triangle (p. 274)
- midsegment (p. 259)
- negation (p. 280)
- orthocenter of a triangle (p. 275)
- point of concurrency (p. 273)

Activity Lab

Investigating Midsegments

FOR USE WITH LESSON 5-1

Construct

Use geometry software to draw and label △*ABC*.
Construct the midpoints *D* and *E* of $\overline{AB}$ and $\overline{AC}$.
Connect the midpoints with a *midsegment*.

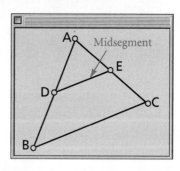

Investigate

- Measure the lengths of $\overline{DE}$ and $\overline{BC}$. Calculate $\frac{DE}{BC}$.
- Measure the slopes of $\overline{DE}$ and $\overline{BC}$.
- Manipulate the triangle and observe the lengths and slopes of $\overline{DE}$ and $\overline{BC}$.

NY G.G.42: Investigate, justify, and apply theorems about geometric relationships, based on the properties of the line segment joining the midpoints of two sides of the triangle.

EXERCISES

1. Make conjectures about the lengths and slopes of midsegments.

2. Construct the midpoint *F* of $\overline{BC}$. Then construct the other two midsegments of △*ABC*. Test whether these midsegments support your conjectures in Exercise 1.

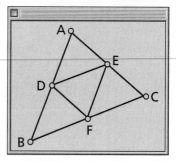

3. △*ABC* and the three midsegments form four small triangles.
   a. Measure the sides of the four small triangles and list those that you find are congruent.
   b. Use a postulate from Chapter 4 to make a conjecture about the four small triangles.

For the remaining exercises, assume your conjectures in Exercises 1 and 3 are true.

4. What can you say about the areas of the four small triangles in the window above?

5. How does △*ABC* compare to each small triangle
   a. in area?
   b. in perimeter?

6. Construct the three midsegments of △*DEF*. Label this triangle △*GHI*. How does △*ABC* compare to △*GHI*
   a. in area?
   b. in perimeter?
   c. Suppose you construct the midsegment triangle inside △*GHI*. Predict how △*ABC* would compare to this third midsegment triangle in area and perimeter.

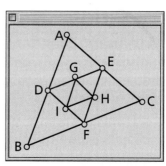

Exercise 6

Extend

7. • Draw quadrilateral *RSTU*.
   • Construct the midpoints of each side of the quadrilateral.
   • Join consecutive midpoints to form quadrilateral *YXWV*.
   • Manipulate the figure and observe the shape of quadrilateral *YXWV*.

   Make a conjecture about the sides of quadrilateral *YXWV*.

5-1

Midsegments of Triangles

Learning Standards for Mathematics

G.G.21 Investigate and apply the concurrence of medians, altitudes, angle bisectors, and perpendicular bisectors of triangles.

G.G.42 Investigate, justify, and apply theorems about geometric relationships, based on the properties of the line segment joining the midpoints of two sides of the triangle.

Lesson Preview

 Check Skills You'll Need

 for Help Lesson 1-8 and page 165

Find the coordinates of the midpoint of each segment.

1. $\overline{AB}$ with $A(-2, 3)$ and $B(4, 1)$

2. $\overline{CD}$ with $C(0, 5)$ and $D(3, 6)$

3. $\overline{EF}$ with $E(-4, 6)$ and $F(3, 10)$

4. $\overline{GH}$ with $G(7, 10)$ and $H(-5, -8)$

Find the slope of the line containing each pair of points.

5. $A(-2, 3)$ and $B(3, 1)$ **6.** $C(0, 5)$ and $D(3, 6)$

7. $E(-4, 6)$ and $F(3, 10)$ **8.** $G(7, 10)$ and $H(-5, -8)$

 New Vocabulary • midsegment • coordinate proof

---

1 Using Properties of Midsegments

Hands-On Activity: Midsegments of Triangles

Draw, label, and cut out a large scalene triangle. Do the same with other right, acute, and obtuse triangles. Label the vertices *A*, *B*, and *C*.

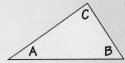

• For each triangle fold *A* onto *C* to find the midpoint of $\overline{AC}$. Do the same for $\overline{BC}$. Label the midpoints *L* and *N*, then draw $\overline{LN}$.

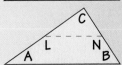

• Fold each triangle on $\overline{LN}$.

• Fold *A* to *C*. Fold *B* to *C*.

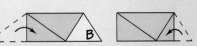

1. How does *LN* compare to *AB*? Explain.

2. Make a conjecture about how the segment joining the midpoints of two sides of a triangle is related to the third side of the triangle.

In $\triangle ABC$ above, $\overline{LN}$ is a triangle midsegment. A **midsegment** of a triangle is a segment connecting the midpoints of two sides.

Theorem 5-1   **Triangle Midsegment Theorem**

If a segment joins the midpoints of two sides of a triangle, then the segment is parallel to the third side, and is half its length.

One way to prove the Triangle Midsegment Theorem is to use coordinate geometry and algebra. This style of proof is called a **coordinate proof.** You begin the proof by placing a triangle in a convenient spot on the coordinate plane. You then choose variables for the coordinates of the vertices.

Proof   **Proof of Theorem 5-1**

Given: R is the midpoint of $\overline{OP}$.
S is the midpoint of $\overline{QP}$.

Prove: $\overline{RS} \parallel \overline{OQ}$ and $RS = \frac{1}{2}OQ$

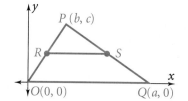

- Use the Midpoint Formula to find the coordinates of R and S.

$$R: \left(\frac{0+b}{2}, \frac{0+c}{2}\right) = \left(\frac{b}{2}, \frac{c}{2}\right)$$

$$S: \left(\frac{a+b}{2}, \frac{0+c}{2}\right) = \left(\frac{a+b}{2}, \frac{c}{2}\right)$$

Vocabulary Tip

The Midpoint Formula
$$M = \left(\frac{x_1 + x_2}{2}, \frac{y_1 + y_2}{2}\right)$$

The Distance Formula
$$d = \sqrt{(x_2 - x_1)^2 + (y_2 - y_1)^2}$$

- To prove that $\overline{RS}$ and $\overline{OQ}$ are parallel, show that their slopes are equal. Because the y-coordinates of R and S are the same, the slope of $\overline{RS}$ is zero. The same is true for $\overline{OQ}$. Therefore, $\overline{RS} \parallel \overline{OQ}$.

- Use the Distance Formula to find RS and OQ.

$$RS = \sqrt{\left(\frac{a+b}{2} - \frac{b}{2}\right)^2 + \left(\frac{c}{2} - \frac{c}{2}\right)^2}$$

$$= \sqrt{\left(\frac{a}{2} + \frac{b}{2} - \frac{b}{2}\right)^2 + 0^2} = \sqrt{\left(\frac{a}{2}\right)^2} = \frac{a}{2} = \frac{1}{2}a$$

$$OQ = \sqrt{(a - 0)^2 + (0 - 0)^2}$$

$$= \sqrt{a^2 + 0^2} = a$$

Therefore, $RS = \frac{1}{2}OQ$.

1 EXAMPLE   **Finding Lengths**

In $\triangle EFG$, H, J, and K are midpoints. Find HJ, JK, and FG.

$HJ = \frac{1}{2}EG$ or $\frac{1}{2}(100)$; $HJ = 50$

$JK = \frac{1}{2}EF$ or $\frac{1}{2}(60)$; $JK = 30$

HK or $40 = \frac{1}{2}FG$; $FG = 80$

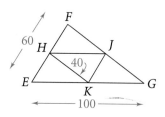

 **Quick Check**   **1** $AB = 10$ and $CD = 18$. Find EB, BC, and AC.

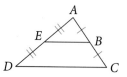

In △DEF, A, B, and C are midpoints. Name pairs of parallel segments.

The midsegments are $\overline{AB}$, $\overline{BC}$, and $\overline{CA}$.

By the Triangle Midsegment Theorem,
• $\overline{AB} \parallel \overline{DF}$, $\overline{BC} \parallel \overline{ED}$, and $\overline{AC} \parallel \overline{EF}$.

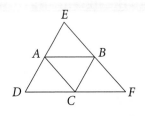

✓ Quick Check **2** **Critical Thinking** Find $m\angle VUZ$. Justify your answer.

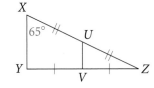

You can use the Triangle Midsegment Theorem to find lengths of segments that might be difficult to measure directly.

3 **EXAMPLE** **Real-World**  **Connection**

Indirect Measurement Dean plans to swim the length of the lake, as shown in the photo. How far would Dean swim?

Here is what Dean does to find the distance he would swim across the lake.

Step 1: He measures his stride and adjusts it so that it averages about 3 ft.

Step 2: Then he begins at the left edge of the lake (first diagram). He paces 35 strides along the edge of the lake and sets a stake.

Step 3: He paces 35 more strides in the same direction and sets another stake.

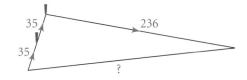

Step 4: He paces to where his swim will end at the other side of the lake, counting 236 strides.

Step 5: Then (second diagram) he paces 118 strides, or half the distance, back towards the second stake.

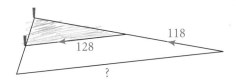

Step 6: He paces to the first stake, counting 128 strides.

Step 7: He converts strides to feet.

$$128 \text{ strides} \times \frac{3 \text{ ft}}{1 \text{ stride}} = 384 \text{ ft}$$

Step 8: He uses Theorem 5-1. The distance across the lake is twice the length of the midsegment.

$$2(384 \text{ ft}) = 768 \text{ ft}$$

• Dean would swim approximately 768 ft.

✓ Quick Check **3** **a.** $\overline{CD}$ is a new bridge being built over a lake as shown. Find the length of the bridge.
b. How long is the bridge in miles?

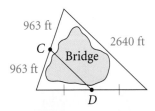

963 ft
2640 ft
C
Bridge
963 ft
D

Practice and Problem Solving

A Practice by Example

Example 1
(page 260)

Mental Math Find the value of *x*.

1.

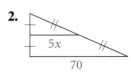
x
18

2.

5*x*
70

3.

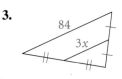
84
3*x*

4.

x − 1
45

5.

x − 1
5

6.

5*x* − 2
4

Points *E, D,* and *H* are midpoints of △*TUV*.
UV = 80, *TV* = 100, and *HD* = 80.

7. Find *HE*.   **8.** Find *ED*.

9. Find *TU*.   **10.** Find *TE*.

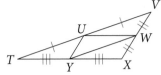

Example 2
(page 261)

Identify pairs of parallel segments in each diagram.

11.

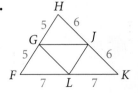

12.

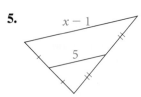

13. a. In the figure at the right, identify
pairs of parallel segments.
  b. If *m*∠*QST* = 40, find *m*∠*QPR*.

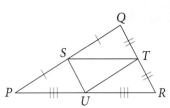

Name the segment that is parallel to the given segment.

14. $\overline{AB}$ **15.** $\overline{BC}$

16. $\overline{EF}$ **17.** $\overline{CA}$

18. $\overline{GE}$ **19.** $\overline{FG}$

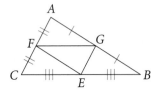

Example 3
(page 261)

20. Indirect Measurement Kate wants to paddle
her canoe across the lake. To determine how
far she must paddle, she paced out a triangle,
counting the number of strides, as shown.
  a. If Kate's strides average 3.5 ft, what is the
  length of the longest side of the triangle?
  b. What distance must Kate paddle across
  the lake?

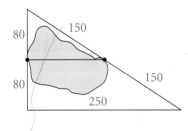

Apply Your Skills

Problem Solving Hint

The highlighted segment is halfway up the face of the Rock and Roll Hall of Fame.

21. a. Architecture The triangular face of the Rock and Roll Hall of Fame in Cleveland, Ohio, is isosceles. The length of the base is 229 ft 6 in. What is the length of the highlighted segment?

✏️ **b. Writing** Explain your reasoning.

X is the midpoint of $\overline{UV}$. Y is the midpoint of $\overline{UW}$.

22. If $m\angle UXY = 60$, find $m\angle V$.

23. If $m\angle W = 45$ find $m\angle UYX$.

24. If $XY = 50$, find VW.

25. If $VW = 110$, find XY.

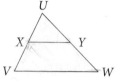

26. Coordinate Geometry The coordinates of the vertices of a triangle are $E(1, 2)$, $F(5, 6)$, and $G(3, -2)$.
  a. Find the coordinates of H, the midpoint of $\overline{EG}$, and J, the midpoint of $\overline{FG}$.
  b. Verify that $\overline{HJ} \parallel \overline{EF}$.   **c.** Verify that $HJ = \frac{1}{2}EF$.

$\overline{IJ}$ is a midsegment of $\triangle FGH$. $IJ = 7$, $FH = 10$, and $GH = 13$. Find the perimeter of each triangle.

27. $\triangle IJH$ **28.** $\triangle FGH$

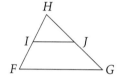

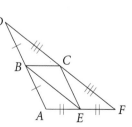

29. Multiple Choice Marita is designing a kite to look like the one on the left. Its diagonals are to measure 64 cm and 90 cm. She will use ribbon to connect the midpoints of its sides. How much ribbon will Marita need?
  (A) 77 cm   (B) 122 cm   (C) 154 cm   (D) 308 cm

Exercise 29

x^2 **Algebra** Find the value of each variable.

30.

31.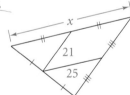

32.

33.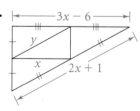

GO Online
Homework Video Tutor
Visit: PHSchool.com
Web Code: aue-0501

Use the figure at the right for Exercises 34–36.

34. If $DF = 24$, $BC = 6$, and $DB = 8$, find the perimeter of $\triangle ADF$.

x^2 **35. Algebra** If $BE = 2x + 6$ and $DF = 5x + 9$, find the value of x, then find DF.

x^2 **36. Algebra** If $EC = 3x - 1$ and $AD = 5x + 7$, find the value of x, then find EC.

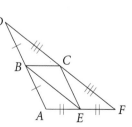

C Challenge

37. Open-Ended Explain how you could use the Triangle Midsegment Theorem as the basis for this construction. Draw $\overline{CD}$. Draw point A not on $\overline{CD}$. Construct $\overline{AB}$ so that $\overline{AB} \parallel \overline{CD}$ and $AB = \frac{1}{2}CD$.

38. Coordinate Geometry In $\triangle GHJ$, $K(2, 3)$ is the midpoint of $\overline{GH}$, $L(4, 1)$ is the midpoint of $\overline{HJ}$, and $M(6, 2)$ is the midpoint of $\overline{GJ}$. Find the coordinates of G, H, and J.

Proof **39.** Complete the prove statement and then write a proof.

Given: S, T, and U are midpoints.

Prove: $\triangle YST \cong \triangle TUZ \cong \triangle SVU \cong \underline{\ ?\ }$.

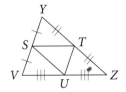

Test Prep

Gridded Response

Q and P are midpoints of two sides of $\triangle RST$.

40. What is RS?

41. What is TQ?

42. What is TS?

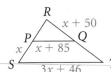

43. What is $m\angle ABC$?

44. What is $m\angle D$?

45. What is $m\angle A$?

46. What is $m\angle CBE$?

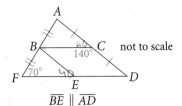

$\overline{BE} \parallel \overline{AD}$

Mixed Review

Lesson 4-7

Name a pair of overlapping congruent triangles in each diagram. State whether the triangles are congruent by SSS, SAS, ASA, AAS, or HL.

47.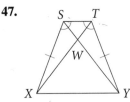

48.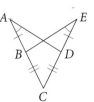

49.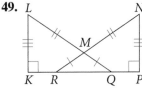

Lesson 3-6 $\boxed{x^2}$ **Algebra** Graph each line.

50. $y = x + 2$

51. $y = 3x - 2$

52. $y = -x - 5$

Lesson 3-2 $\boxed{x^2}$ **Algebra** Determine the value of x for which $\ell \parallel m$.

53.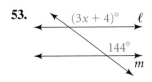

54.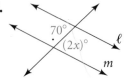

55.

Bisectors in Triangles

 Learning Standards for Mathematics

G.G.21 Investigate and apply the concurrence of medians, altitudes, angle bisectors, and perpendicular bisectors of triangles.

✓ **Check Skills You'll Need**

 **for Help** Lesson 1-7

Use a compass and a straightedge for the following.

1. Draw a triangle, $\triangle XYZ$. Construct $\triangle STV$ so that $\triangle STV \cong \triangle XYZ$.

2. Draw acute $\angle P$. Construct $\angle Q$ so that $\angle Q \cong \angle P$.

3. Draw $\overline{AB}$. Construct a line $\overleftrightarrow{CD}$ so that $\overleftrightarrow{CD} \perp \overline{AB}$ and $\overleftrightarrow{CD}$ bisects $\overline{AB}$.

4. Draw acute $\angle E$. Construct the bisector of $\angle E$.

$\overrightarrow{TM}$ **bisects $\angle STU$ so that $m\angle STM = 5x + 4$ and $m\angle MTU = 6x - 2$.**

x^2 **5. Algebra** Find the value of x.   6. Find $m\angle STU$.

🔊 **New Vocabulary**   • distance from a point to a line

---

1   Perpendicular Bisectors and Angle Bisectors

Triangles play a key role in relationships involving perpendicular bisectors and angle bisectors.

In the diagram below on the left, $\overleftrightarrow{CD}$ is the perpendicular bisector of $\overline{AB}$. $\overleftrightarrow{CD}$ is perpendicular to $\overline{AB}$ at its midpoint. In the diagram on the right, $\overline{CA}$ and $\overline{CB}$ are drawn to complete the triangles, $\triangle CAD$ and $\triangle CBD$.

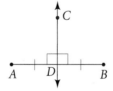 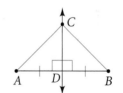

You should recognize from your work in Chapter 4 that $\triangle CAD \cong \triangle CBD$. Thus, you can conclude that $\overline{CA} \cong \overline{CB}$, that $CA = CB$, or simply that C is equidistant from points A and B.

This suggests a proof of Theorem 5-2 below. Its converse is also true and is stated as Theorem 5-3. You will prove these theorems in the exercises.

 Key Concepts

Theorem 5-2	**Perpendicular Bisector Theorem**

If a point is on the perpendicular bisector of a segment, then it is equidistant from the endpoints of the segment.

Theorem 5-3	**Converse of the Perpendicular Bisector Theorem**

If a point is equidistant from the endpoints of a segment, then it is on the perpendicular bisector of the segment.

1 EXAMPLE **Real-World** **Connection**

National Landmarks Find the set of points on the map of Washington, D.C. that are equidistant from the Jefferson Memorial and the White House.

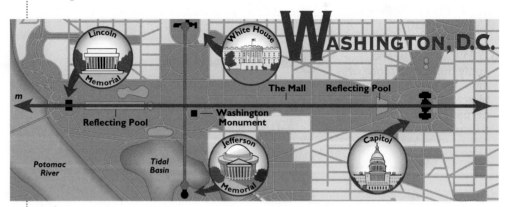

The red segment connects the Jefferson Memorial and the White House. All points on the perpendicular bisector *m* of this segment are equidistant from the Jefferson Memorial and the White House.

✓ Quick Check **1** Use the information given in the diagram. $\overleftrightarrow{CD}$ is the perpendicular bisector of $\overline{AB}$. Find *CA* and *DB*. Explain your reasoning.

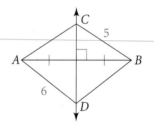

The **distance from a point to a line** is the length of the perpendicular segment from the point to the line. In the diagram, $\overrightarrow{AD}$ is the bisector of $\angle CAB$. If you measure the lengths of the perpendicular segments from *D* to the two sides of the angle, you will find that the lengths are equal so *D* is equidistant from the sides.

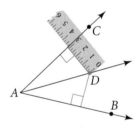

🔑 Key Concepts

Theorem 5-4	**Angle Bisector Theorem**

If a point is on the bisector of an angle, then the point is equidistant from the sides of the angle.

Theorem 5-5	**Converse of the Angle Bisector Theorem**

If a point in the interior of an angle is equidistant from the sides of the angle, then the point is on the angle bisector.

You will use congruent triangles to prove these theorems in the exercises.

You can combine Theorems 5-4 and 5-5 into a biconditional: A point in the interior of an angle is equidistant from the sides of the angle if and only if it is on the angle bisector.

2 EXAMPLE Using the Angle Bisector Theorem

Multiple Choice What is the length of $\overline{FD}$?

Ⓐ 8 Ⓑ 16 Ⓒ 30 Ⓓ 40

From the diagram you can see that F is on the bisector of $\angle ACE$. Therefore, $FB = FD$.

$FB = FD$	
$5x = 2x + 24$	**Substitute.**
$3x = 24$	**Subtract 2x.**
$x = 8$	**Divide by 3.**
$FB = 5x = 5(8) = 40$	**Substitute.**
$FD = 40$	**Substitute.**

● The correct answer is D.

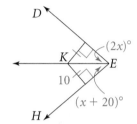

Test-Taking Tip

Answer the question asked. Choice A is the value of x, but the question asks for FD.

✓ **Quick Check**

❷ **a.** According to the diagram, how far is K from $\overrightarrow{EH}$? From $\overrightarrow{ED}$?

b. What can you conclude about $\overrightarrow{EK}$?

c. Find the value of x.

d. Find $m\angle DEH$.

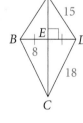

EXERCISES

For more exercises, see *Extra Skill, Word Problem, and Proof Practice.*

Practice and Problem Solving

 Practice by Example

Example 1
(page 266)

Use the figure at the right for Exercises 1–4.

1. From the information given in the figure, how is $\overline{AC}$ related to $\overline{BD}$?

2. Find AB.   **3.** Find BC.   **4.** Find ED.

5. On a piece of paper, mark a point H for home and a point S for school. Describe the set of points equidistant from H and S

Example 2
(page 267)

x^2 **6. Algebra** Find $x, JK,$ and JM.

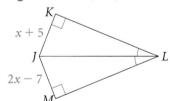

7. Algebra Find $y, ST,$ and TU.

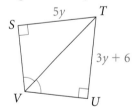

Use the figure at the right for Exercises 8–11.

8. According to the diagram, how far is L from $\overrightarrow{HK}$? From $\overrightarrow{HF}$?

9. How is $\overrightarrow{HL}$ related to $\angle KHF$? Explain.

10. Find the value of y.

11. Find $m\angle KHL$ and $m\angle FHL$.

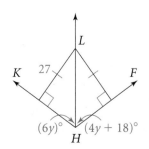

B **Apply Your Skills** x^2 **Algebra** Use the figure, below right, for Exercises 12–16.

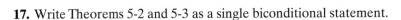

12. Find the value of x.

13. Find TW.

14. Find WZ.

15. What kind of triangle is $\triangle TWZ$? Explain.

16. If R is on the perpendicular bisector of $\overline{TZ}$,
then R is __?__ from T and Z, or __?__ = __?__.

17. Write Theorems 5-2 and 5-3 as a single biconditional statement.

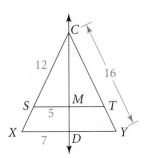

$\overleftrightarrow{CD}$ **is the perpendicular bisector of both $\overline{XY}$ and $\overline{ST}$,
and $CY = 16$. Find each length.**

18. CT **19.** TY

20. SX **21.** CX

22. MT **23.** ST

24. DY **25.** XY

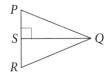

GO Online
Homework Video Tutor
Visit: PHSchool.com
Web Code: aue-0502

26. What kind of triangles are $\triangle SCT$ and $\triangle XCY$?
Explain.

27. **Error Analysis** To prove that $\triangle PQR$ is isosceles, a
student began by stating that since Q is on the segment
perpendicular to $\overline{PR}$, Q is equidistant from the endpoints
of $\overline{PR}$. What additional information does the student
need in order to make that statement?

 Writing Determine whether point A must be on the bisector of $\angle TXR$. Explain.

28.   **29.**   **30.**

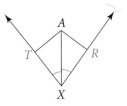

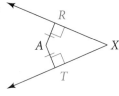

31. **Baseball** What is the common
name for the part of a baseball
field that is equidistant from
the foul lines and 60 ft 6 in.
from home plate?

32. a. Constructions Draw a large
triangle, $\triangle CDE$. Construct the
angle bisectors of each angle.
   b. Make a Conjecture What
appears to be true about the
angle bisectors?
   c. Test your conjecture with
another triangle.

Real-World **Connection**

On a baseball field, second
base is equidistant from the
foul lines and 127 feet from
home plate.

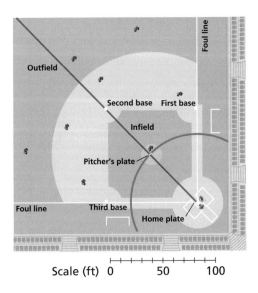

Scale (ft) 0   50   100

Real-World 🌐 **Connection**

The picture hangs straight when the hook is on the perpendicular bisector of the picture's top edge.

33. a. Constructions Draw a large acute scalene triangle, △*PQR*. Construct the perpendicular bisectors of each side.
  b. Make a Conjecture What appears to be true about the perpendicular bisectors?
  c. Test your conjecture with another triangle.

Coordinate Geometry **Find two points on the perpendicular bisector of $\overline{AB}$. Verify your results by showing each point is equidistant from *A* and *B*.**

34. $A(0, 0), B(0, 4)$ **35.** $A(0, 2), B(6, 2)$ **36.** $A(3, 3), B(3, -3)$

37. $A(3, 0), B(0, 3)$ **38.** $A(3, 0), B(1, 4)$ **39.** $A(3, 0), B(2, 5)$

Proof **40.** Prove the Perpendicular Bisector Theorem.

  Given: $\overleftrightarrow{CD} \perp \overline{AB}, \overleftrightarrow{CD}$ bisects $\overline{AB}$.
  Prove: $DA = DB$

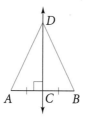

41. Prove the Converse of the Perpendicular Bisector Theorem.

  Given: $AP = AQ$ with $\overline{AB} \perp \overline{PQ}$ at *B*.
  Prove: $\overline{AB}$ is the perpendicular bisector of *PQ*.

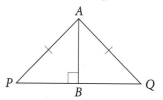

42. Coordinate Geometry You are given points $A(6, 8), O(0, 0)$, and $B(10, 0)$.
  a. Write equations of lines ℓ and m such that $\ell \perp \overleftrightarrow{OA}$ at *A* and $m \perp \overleftrightarrow{OB}$ at *B*.
  b. Find the intersection *C* of lines ℓ and m.
  c. Show that $CA = CB$.
  d. Explain why *C* is on the bisector of $\angle AOB$.

Proof **43.** Prove the Angle Bisector Theorem.

  Given: $\overline{PB} \perp \overrightarrow{AB}, \overline{PC} \perp \overrightarrow{AC},$
  $\overrightarrow{AP}$ bisects $\angle BAC$.
  Prove: $PB = PC$

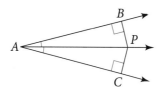

44. Prove the Converse of the Angle Bisector Theorem.

  Given: $\overline{SP} \perp \overrightarrow{QP}, \overline{SR} \perp \overrightarrow{QR},$
  $SP = SR$
  Prove: $\overrightarrow{QS}$ bisects $\angle PQR$.

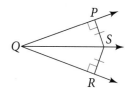

45. Multiple Choice Which equation represents the perpendicular bisector of the segment with endpoints $A(0, 0)$ and $B(6, 0)$?
  A $y = 3$ **B** $x = 6$ **C** $y = x + 6$ **D** $x = 3$

Coordinate Geometry **Write an equation of the perpendicular bisector of $\overline{AB}$.**

46. $A(0, 0), B(0, 4)$ **47.** $A(1, -1), B(3, 1)$ **48.** $A(-2, 0), B(2, 8)$

C **Challenge**

49. *A*, *B*, and *C* are three noncollinear points. Describe and sketch a line in plane *ABC* such that points *A*, *B*, and *C* are equidistant from the line. Justify your response.

50. Reasoning M is the intersection of the perpendicular bisectors of two sides of $\triangle ABC$.
  a. Explain why M is equidistant from A, B, and C.
  b. Line ℓ is perpendicular to plane ABC at M. Explain why a point E on ℓ is equidistant from A, B, and C. (*Hint:* See page 49, Exercise 36. Explain why $\triangle EAM \cong \triangle EBM \cong \triangle ECM$.)

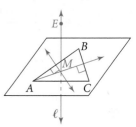

Multiple Choice

Use the figure at the right for Exercises 51–53.

51. What is *TK*?
  A. 4 **B.** 5 **C.** 15 **D.** 25

52. If $m\angle CTR = 27$, what is $m\angle K$?
  F. 27 **G.** 54 **H.** 63 **J.** 76

53. Suppose $RK = 8$. What is the perimeter of $\triangle TPK$?
  A. 25 **B.** 33 **C.** 50 **D.** 66

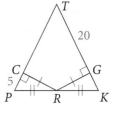

Short Response

54. In the figure at the right, explain why $\overline{MV}$ is the angle bisector of $\angle KVR$.

Extended Response

55. In the figure at the right, explain why $\overline{MV}$ is the perpendicular bisector of $\overline{KR}$.

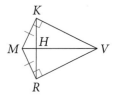

Mixed Review

Lesson 5-1 $\boxed{x^2}$ **Algebra** Find the value of x.

56.

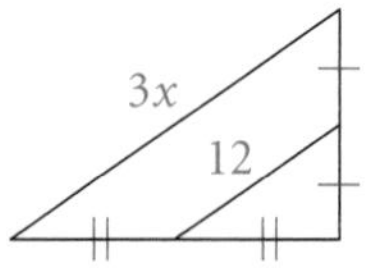

57.
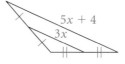

58.

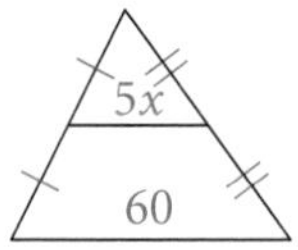

Lesson 2-4

Name the property that justifies each statement.

59. $AB = AB$

60. If $2x = 30$, then $x = 15$.

61. If $x = 30 - x$, then $2x = 30$.

62. $3(4x - 1) = 12x - 3$

63. If $m\angle 3 = m\angle 4$ and $m\angle 4 = m\angle 5$, then $m\angle 3 = m\angle 5$.

64. If $\angle 3 \cong \angle 4$ and $\angle 4 \cong \angle 5$, then $\angle 3 \cong \angle 5$.

Lesson 1-8

Coordinate Geometry Find C the midpoint of $\overline{AB}$. Then show that $AC = CB = \frac{1}{2}AB$.

65. $A(0, 5)$, $B(6, 8)$ **66.** $A(-2, 8)$, $B(2, -1)$ **67.** $A(5, 3)$, $B(6, 7)$

Activity Lab

Special Segments in Triangles

FOR USE WITH LESSON 5-3

Construct

Use geometry software.

- Construct a triangle and the three perpendicular bisectors of its sides.

- Construct a triangle and its three angle bisectors.

- An *altitude* of a triangle is the perpendicular segment from a vertex to the line containing the opposite side. Construct a triangle. Through a vertex of the triangle construct a line that is perpendicular to the line containing the side opposite that vertex. Next construct the altitudes from the other two vertices.

- A *median* of a triangle is the segment joining the midpoint of a side and the opposite vertex. Construct a triangle. Construct the midpoint of one side. Draw the median. Then construct the other two medians.

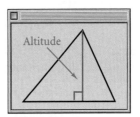

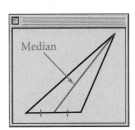

NY G.G.21:
Investigate and apply the concurrence of medians, altitudes, angle bisectors, and perpendicular bisectors of triangles.

EXERCISES

1. In the constructions above, what property do the perpendicular bisectors, angle bisectors, lines containing altitudes, and medians seem to have?

2. Manipulate the triangles. Does the property still hold as you manipulate the triangles?

3. List your conjectures about the perpendicular bisectors, angle bisectors, lines containing altitudes, and medians of a triangle.

Extend

4. Copy the table. Think about acute triangles, right triangles, and obtuse triangles. Use *inside*, *on*, or *outside* to describe the location of the intersection of the segments or lines for each type of triangle.

	Perpendicular Bisectors	Angle Bisectors	Lines Containing the Altitudes	Medians
Acute Triangle				
Right Triangle				
Obtuse Triangle				

5. What observations, if any, can you make about these special segments for isosceles triangles? Equilateral triangles?

6. Your Exercise 3 conjecture should identify some special points. One of these points is equidistant from the three vertices of its triangle, no matter what shape the triangle has.
   a. Use your software. Find which special segments locate this extra-special point.
   b. Draw the circle suggested by the properties of this point.

5-3

Concurrent Lines, Medians, and Altitudes

 Learning Standards for Mathematics

G.G.21 Investigate and apply the concurrence of medians, altitudes, angle bisectors, and perpendicular bisectors of triangles.

G.G.43 Investigate, justify, and apply theorems about the centroid of a triangle, dividing each median into segments whose lengths are in the ratio 2:1.

✓ Check Skills You'll Need

GO **for Help** Lesson 1-7

For Exercises 1–2, draw a large triangle. Construct each figure.

1. an angle bisector

2. a perpendicular bisector of a side

3. Draw $\overline{GH}$. Construct $\overleftrightarrow{CD} \perp \overline{GH}$ at the midpoint of $\overline{GH}$.

4. Draw $\overleftrightarrow{AB}$ with a point E not on $\overleftrightarrow{AB}$. Construct $\overleftrightarrow{EF} \perp \overleftrightarrow{AB}$.

🔊 New Vocabulary

- concurrent • point of concurrency
- circumcenter of a triangle • circumscribed about
- incenter of a triangle • inscribed in
- median of a triangle • centroid • altitude of a triangle
- orthocenter of a triangle

---

1 Properties of Bisectors

Hands-On Activity: Paper Folding Bisectors

- Draw and cut out five different triangles: two acute, two right, and one obtuse.

- Step 1: Use paper folding to create the angle bisectors of each angle of an acute triangle. What do you notice about the angle bisectors?

- Step 2: Repeat Step 1 with a right triangle and an obtuse triangle. Does your discovery from Step 1 still hold true?

Folding an Angle Bisector

1. Make a conjecture about the bisectors of the angles of a triangle.

Folding a Perpendicular Bisector

- Step 3: Use paper folding to create the perpendicular bisector of each side of an acute triangle. What do you notice about the perpendicular bisectors?

- Step 4: Repeat Step 3 with a right triangle. What do you notice?

2. Make a conjecture about the perpendicular bisectors of the sides of a triangle.

---

272 Chapter 5 Relationships Within Triangles

When three or more lines intersect in one point, they are **concurrent.** The point at which they intersect is the **point of concurrency.** For any triangle, four different sets of lines are concurrent. Theorems 5-6 and 5-7 tell you about two of them.

 Key Concepts

> **Theorem 5-6**
>
> The perpendicular bisectors of the sides of a triangle are concurrent at a point equidistant from the vertices.
>
> **Theorem 5-7**
>
> The bisectors of the angles of a triangle are concurrent at a point equidistant from the sides.

You will prove these theorems in the exercises.

This figure shows $\triangle QRS$ with the perpendicular bisectors of its sides concurrent at C. The point of concurrency of the perpendicular bisectors of a triangle is called the **circumcenter of the triangle.**

Points Q, R, and S are equidistant from C, the circumcenter. The circle is **circumscribed about** the triangle.

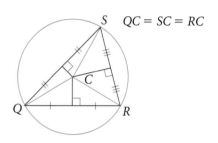

$QC = SC = RC$

1 EXAMPLE **Finding the Circumcenter**

Coordinate Geometry Find the center of the circle that you can circumscribe about $\triangle OPS$.

Two perpendicular bisectors of sides of $\triangle OPS$ are $x = 2$ and $y = 3$. These lines intersect at $(2, 3)$. This point is the center of the circle.

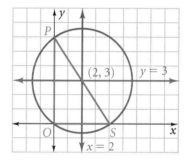

✓ **Quick Check** **1 a.** Find the center of the circle that you can circumscribe about the triangle with vertices $(0, 0)$, $(-8, 0)$, and $(0, 6)$.

b. Critical Thinking In Example 1, explain why it is not necessary to find the third perpendicular bisector.

This figure shows $\triangle UTV$ with the bisectors of its angles concurrent at I. The point of concurrency of the angle bisectors of a triangle is called the **incenter of the triangle.**

Points X, Y, and Z are equidistant from I, the incenter. The circle is **inscribed in** the triangle.

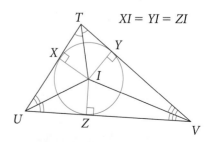

$XI = YI = ZI$

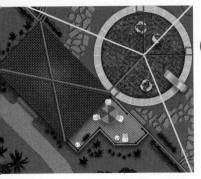

2 EXAMPLE Real-World 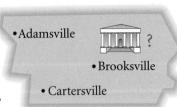 Connection

Pools The Jacksons want to install the largest possible circular pool in their triangular backyard. Where would the largest possible pool be located?

Locate the center of the pool at the point of concurrency of the angle bisectors. This point is equidistant from the sides of the yard. If you choose any other point as the center of the pool, it will be closer to at least one of the sides of the yard, and the pool will be smaller.

 Quick Check **2 a.** The towns of Adamsville, Brooksville, and Cartersville want to build a library that is equidistant from the three towns. Trace the diagram and show where they should build the library.

b. What theorem did you use to find the location?

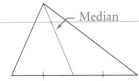

2 Medians and Altitudes

A **median of a triangle** is a segment whose endpoints are a vertex and the midpoint of the opposite side.

— Median

 Key Concepts

Theorem 5-8

The medians of a triangle are concurrent at a point that is two thirds the distance from each vertex to the midpoint of the opposite side.

$$DC = \frac{2}{3}DJ \qquad EC = \frac{2}{3}EG \qquad FC = \frac{2}{3}FH$$

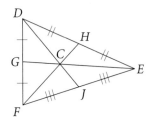

Test-Taking Tip

If you don't remember the meaning of a term, like centroid, the diagram may give a clue.

In a triangle, the point of concurrency of the medians is the **centroid.** The point is also called the center of gravity of a triangle because it is the point where a triangular shape will balance. (See DK Activity Lab, page 303.) You will prove Theorem 5-8 in Chapter 6.

3 EXAMPLE Finding Lengths of Medians

Gridded Response In $\triangle ABC$ at the left, D is the centroid and $DE = 6$. Find BE.

Since D is a centroid, $BD = \frac{2}{3}BE$ and $DE = \frac{1}{3}BE$.

$\frac{1}{3}BE = DE$

$\frac{1}{3}BE = 6$ **Substitute 6 for *DE*.**

$BE = 18$

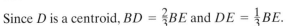

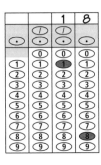

 Quick Check **3** Find *BD*. Check that $BD + DE = BE$.

An **altitude of a triangle** is the perpendicular segment from a vertex to the line containing the opposite side. Unlike angle bisectors and medians, an altitude of a triangle can be a side of a triangle or it may lie outside the triangle.

Acute Triangle:
Altitude is inside.

Right Triangle:
Altitude is a side.

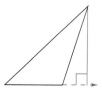

Obtuse Triangle:
Altitude is outside.

4 EXAMPLE **Identifying Medians and Altitudes**

Is $\overline{ST}$ a median, an altitude, or neither? Explain.

$\overline{ST}$ is a segment extending from vertex S to the side opposite S. Also, $\overline{ST} \perp \overline{VU}$.

● $\overline{ST}$ is an altitude of $\triangle VSU$.

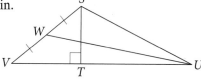

✓ Quick Check ● **4** Is $\overline{UW}$ a median, an altitude, or neither? Explain.

The lines containing the altitudes of a triangle are concurrent at the **orthocenter of the triangle.** A proof of this theorem appears in Chapter 6.

 Key Concepts

Theorem 5-9
The lines that contain the altitudes of a triangle are concurrent.

EXERCISES

For more exercises, see *Extra Skill, Word Problem, and Proof Practice.*

Practice and Problem Solving

 A Practice by Example

Example 1
(page 273)

 GO for Help

Coordinate Geometry Find the center of the circle that you can circumscribe about each triangle.

1.

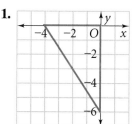

2.

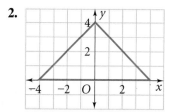

Coordinate Geometry Find the center of the circle that you can circumscribe about $\triangle ABC$.

3. $A(0,0)$
   $B(3,0)$
   $C(3,2)$

4. $A(0,0)$
   $B(4,0)$
   $C(4,-3)$

5. $A(-4,5)$
   $B(-2,5)$
   $C(-2,-2)$

6. $A(-1,-2)$
   $B(-5,-2)$
   $C(-1,-7)$

7. $A(1,4)$
   $B(1,2)$
   $C(6,2)$

Example 2
(page 274)

Name the point of concurrency of the angle bisectors.

8.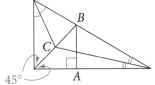

9.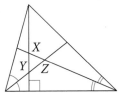

10. City Planning Copy the diagram of Altgeld Park. Show where park officials should place a drinking fountain so that it is equidistant from the tennis court, the playground, and the volleyball court.

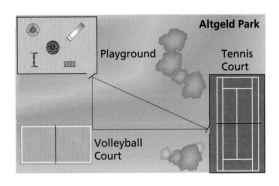

Altgeld Park

Playground

Tennis Court

Volleyball Court

Example 3
(page 274)

In △TUV, Y is the centroid.

11. If $YW = 9$, find TY and TW.

12. If $YU = 9$, find ZY and ZU.

13. If $VX = 9$, find VY and YX.

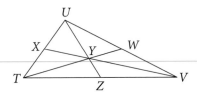

Example 4
(page 275)

Is $\overline{AB}$ a median, an altitude, or neither? Explain.

14.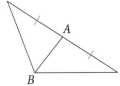

15.

16.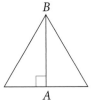

B **Apply Your Skills**

Constructions Draw the triangle. Then construct the inscribed circle and the circumscribed circle.

17. right triangle, △DEF

18. obtuse triangle, △STU

In Exercises 19–22, name each figure in △BDF.

19. an angle bisector

20. a median

21. a perpendicular bisector

22. an altitude

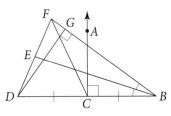

23. Critical Thinking A centroid separates a median into two segments. What is the ratio of the lengths of those segments?

24. Writing Ivars found a yellowed parchment inside an antique book. It read:
From the spot I buried Olaf's treasure, equal sets of paces did I measure; each of three directions in a line, there to plant a seedling Norway pine. I could not return for failing health; now the hounds of Haiti guard my wealth.—Karl
After searching Caribbean islands for five years, Ivars found one with three tall Norway pines. How might Ivars find where Karl buried Olaf's treasure?

The figures below show how to construct medians and altitudes by paper folding.

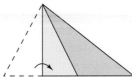 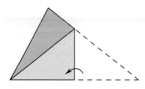 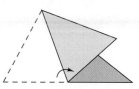

To find an altitude, fold the triangle so that a side overlaps itself and the fold contains the opposite vertex.

To find a median, fold one vertex to another vertex. This locates the midpoint of a side.

Then fold so that the fold contains the midpoint and the opposite vertex.

Problem Solving Hint

Paper-folding an altitude is the same as paper-folding the perpendicular to a line through a point not on the line.

25. Cut out a large triangle. Paper-fold very carefully to construct the three medians of the triangle and demonstrate Theorem 5-8.

26. Cut out a large acute triangle. Paper-fold very carefully to construct the three altitudes of the triangle and demonstrate Theorem 5-9.

27. Multiple Choice C is the centroid of $\triangle DEF$. If $GF = 6x^2 + 9y$, which expression represents CF?

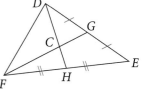

Ⓐ $2x^2 + 9y$ Ⓑ $2x^2 + 3y$

Ⓒ $6x^2 + 9y$ Ⓓ $4x^2 + 6y$

28. Is $\overline{AB}$ a perpendicular bisector, an angle bisector, a median, an altitude, or none of these? Explain.

a.   **b.** 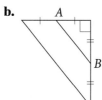   **c.**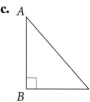

29. Developing Proof Complete this proof of Theorem 5-6 by filling in the blanks.

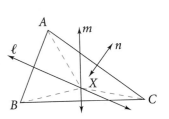

Given: Lines $\ell, m,$ and n are perpendicular bisectors of the sides of $\triangle ABC$. X is the intersection of lines ℓ and m.

Prove: Line n contains point X, and $XA = XB = XC$.

Proof: Since ℓ is the perpendicular bisector of **a.** _?_, $XA = XB$. Since m is the perpendicular bisector of **b.** _?_, $XB = $ **c.** _?_. Thus $XA = XB = XC$. Since $XA = XC$, X is on line n by the Converse of the **d.** _?_ Theorem.

Proof **30.** Prove Theorem 5-7.

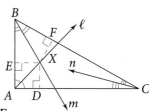

Given: Rays $\ell, m,$ and n are bisectors of the angles of $\triangle ABC$. X is the intersection of rays ℓ and m and $\overline{XD} \perp \overline{AC}, \overline{XE} \perp \overline{AB}, \overline{XF} \perp \overline{BC}$.

Prove: Ray n contains point X, and $XD = XE = XF$.

31. What kind of triangle has its circumcenter on one of its sides? Explain.

Homework Video Tutor

Visit: PHSchool.com
Web Code: aue-0503

32. Coordinate Geometry Complete the following steps to locate the centroid.

a. Find the coordinates of midpoints L, M, and N.

b. Find equations of $\overleftrightarrow{AM}$, $\overleftrightarrow{BN}$, and $\overleftrightarrow{CL}$.

c. Find the coordinates of P, the intersection of $\overleftrightarrow{AM}$ and $\overleftrightarrow{BN}$. This is the centroid.

d. Show that point P is on $\overleftrightarrow{CL}$.

e. Use the Distance Formula to show that point P is $\frac{2}{3}$ of the distance from each vertex to the midpoint of the opposite side.

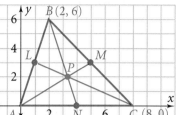

C Challenge

For Exercises 33 and 34, points of concurrency have been drawn for two triangles. Match the points with the lines and segments listed in I–IV.

33.
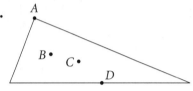

34.
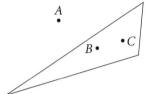

I. perpendicular bisectors of sides II. angle bisectors
III. medians IV. lines containing altitudes

35. In an isosceles triangle, show that the circumcenter, incenter, centroid, and orthocenter can be four different points but all four must be collinear.

36. History In 1765 Leonhard Euler proved that for any triangle, three of the four points of concurrency are collinear. The line that contains these three points is known as Euler's Line. Use Exercises 33 and 34 to determine which point of concurrency does not necessarily lie on Euler's Line.

NY REGENTS

Test Prep

Multiple Choice

Use the figure at the right for Exercises 37–39.

37. What is RD if $RL = 54$ cm?
A. 81 cm B. 108 cm
C. 162 cm D. 216 cm

38. What is WL if $WJ = 210$ mm?
F. 70 mm G. 105 mm
H. 140 mm J. 157.5 mm

39. What is x if $WL = 15x$ and $LJ = 5x + 3$?
A. 0.3 B. 0.4 C. 0.6 D. 1.2

Short Response

40. Name all types of triangles for which the centroid, circumcenter, incenter, and orthocenter are all inside the triangle. Classify the triangles according to the sides as well as the angles.

Extended Response

41. The point of concurrency of the three altitudes of a triangle lies outside the triangle. Where are its circumcenter, incenter, and centroid located in relation to the triangle? Draw and label a diagram to support each of your answers.

Lesson 5-2   Determine whether point *B* must be on the bisector of ∠*T*. **Explain.**

42.

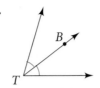

43.

44.

Lesson 3-4   Classify each △*JKL* by its angles.

45. $m\angle J = 37, m\angle K = 53, m\angle L = 90$ **46.** $m\angle J = 47, m\angle K = 98, m\angle L = 35$

Lesson 1-4   In the figure at the right, *ABCD* is a square. Identify each of the following.

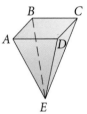

47. a line skew to $\overleftrightarrow{ED}$ **48.** a line skew to $\overleftrightarrow{EB}$

49. two intersecting planes   **50.** two parallel segments

51. the intersection of plane *ABC* and plane *BCE*

Checkpoint Quiz 1   **Lessons 5-1 through 5-3**

x^2 **Algebra** Find the value of *x*.

1.

2.

3. a. $\overline{AB}$ is a midsegment of △*XYZ*.
      $AB = 52$. Find *YZ*.
   b. $AX = 26$ and $BZ = 36$. Find
      the perimeter of △*XYZ*.

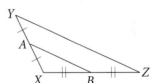

Use the diagram. What can you conclude about each of the following? **Explain.**

4. ∠*CDB*

5. △*ABD* and △*CBD*

6. $\overline{AD}$ and $\overline{DC}$

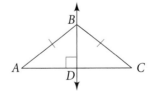

Use the figure at the right.

7. What can you conclude about $\overrightarrow{XY}$? Explain.

8. Find *XZ*. Justify your response.

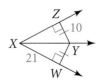

 Writing For a given triangle, describe how you can construct the following.

9. a median   **10.** an altitude

Inverses, Contrapositives, and Indirect Reasoning

 Learning Standards for Mathematics

G.G.24 Determine the negation of a statement and establish its truth value.

G.G.26 Identify and write the inverse, converse, and contrapositive of a given conditional statement and note the logical equivalences.

✓ **Check Skills You'll Need**

 **GO** **for Help** Lessons 2-1 and 2-2

Write the converse of each statement.

1. If it snows tomorrow, then we will go skiing.

2. If two lines are parallel, then they do not intersect.

3. If $x = -1$, then $x^2 = 1$.

Write two conditional statements that make up each biconditional.

4. A point is on the bisector of an angle if and only if it is equidistant from the sides of the angle.

5. A point is on the perpendicular bisector of a segment if and only if it is equidistant from the endpoints of the segment.

6. You will pass a geometry course if and only if you are successful with your homework.

◀ᴼ))) **New Vocabulary** • negation • inverse • contrapositive
• equivalent statements • indirect reasoning
• indirect proof

---

1 Writing the Negation, Inverse, and Contrapositive

The statement, "Knoxville is the capital of Tennessee," is false. The **negation** of a statement has the opposite truth value. The negation, "Knoxville is not the capital of Tennessee," is true.

1 EXAMPLE **Writing the Negation of a Statement**

Write the negation of each statement.

a. Statement: $\angle ABC$ is obtuse.

Negation: $\angle ABC$ is not obtuse.

b. Statement: Lines m and n are not perpendicular.

Negation: Lines m and n are perpendicular.

✓ **Quick Check** ❶ Write the negation of each statement.
a. $m\angle XYZ > 70$.
b. Today is not Tuesday.

Vocabulary Tip

The prefix <u>contra</u> is Latin for "against."

The **inverse** of a conditional statement negates both the hypothesis and the conclusion. The **contrapositive** of a conditional switches the hypothesis and the conclusion and negates both.

"If you don't stand for something, you'll fall for anything."

—Maya Angelou, poet and author

2 EXAMPLE Writing the Inverse and Contrapositive

Write the inverse and the contrapositive of the conditional statement.

Conditional: If a figure is a square, then it is a rectangle.

Negate both.

Inverse: If a figure is not a square, then it is not a rectangle.

Conditional: If a figure is a square, then it is a rectangle.

Switch and negate both.

● Contrapositive: If a figure is not a rectangle, then it is not a square.

 **2** Write (a) the inverse and (b) the contrapositive of Maya Angelou's statement under the photo at the left.

You know that a conditional statement and its converse can have different truth values. A conditional statement and its inverse can also have different truth values. The contrapositive of a conditional statement, however, always has the same truth value as the conditional. A conditional statement and its contrapositive are equivalent. **Equivalent statements** have the same truth value.

🔑 **Key Concepts**

Summary	Negation, Inverse, and Contrapositive Statements		
Statement	**Example**	**Symbolic Form**	**You Read It**
Conditional	If an angle is a straight angle, then its measure is 180.	$p \rightarrow q$	If p, then q.
Negation (of p)	An angle is not a straight angle.	$\sim p$	Not p.
Inverse	If an angle is not a straight angle, then its measure is not 180.	$\sim p \rightarrow \sim q$	If not p, then not q.
Contrapositive	If an angle's measure is not 180, then it is not a straight angle.	$\sim q \rightarrow \sim p$	If not q, then not p.

2 Using Indirect Reasoning

Suppose your brother tells you, "Susan called a few minutes ago." You think through these three steps.

Step 1 You have two friends named Susan.

Step 2 You know that one of them is at band practice.

Step 3 You conclude that the other Susan must have been the caller.

This type of reasoning is called indirect reasoning. In **indirect reasoning,** all possibilities are considered and then all but one are proved false. The remaining possibility must be true.

A proof involving indirect reasoning is an **indirect proof.** In an indirect proof, a statement and its negation often are the only possibilities.

 **Key Concepts**

Summary	Writing an Indirect Proof

Step 1 State as an assumption the opposite (negation) of what you want to prove.

Step 2 Show that this assumption leads to a contradiction.

Step 3 Conclude that the assumption must be false and that what you want to prove must be true.

In the first step of an indirect proof you assume as true the opposite of what you want to prove.

3 EXAMPLE The First Step of an Indirect Proof

Write the first step of an indirect proof.

a. Prove: Quadrilateral $QRWX$ does not have four acute angles.

Assume that quadrilateral $QRWX$ has four acute angles.

b. Prove: An integer n is divisible by 5.

Assume that the integer n is not divisible by 5.

 **3** You want to prove each statement true. Write the first step of an indirect proof.
a. The shoes cost no more than $20.   **b.** $m\angle A > m\angle B$

To do an indirect proof, you have to be able to identify a contradiction.

4 EXAMPLE Identifying Contradictions

Identify the two statements that contradict each other.

   I. $\triangle ABC$ is acute.   II. $\triangle ABC$ is scalene.   III. $\triangle ABC$ is equiangular.

A triangle can be acute and scalene. I and II do not contradict each other.

An equiangular triangle is an acute triangle. I and III do not contradict each other.

An equiangular triangle must be equilateral, so it cannot be scalene. II and III contradict each other.

Jack, 18, is a good driver. Jack concludes he will get a good insurance rate. His insurance bill is a contradiction.

 **4** Identify the two statements that contradict each other.
   I. $\overline{FG} \parallel \overline{KL}$ II. $\overline{FG} \perp \overline{KL}$ III. $\overline{FG} \cong \overline{KL}$

Proof **5 EXAMPLE** Indirect Proof

Read the conditional statement. Think about what is given and what you are to prove. Then give the steps of an indirect proof.

If Jaeleen spends more than $50 to buy two items at a bicycle shop, then at least one of the items costs more than $25.

Given: The cost of two items is more than $50.

Prove: At least one of the items costs more than $25.

Step 1 Assume as true the opposite of what you want to prove. That is, assume that neither item costs more than $25.

Step 2 This means that each item costs $25 or less. This, in turn, means that the two items together cost $50 or less. This contradicts the given information that the amount spent is more than $50.

Step 3 Conclude that the assumption is false. One item must cost more than $25.

 **Quick Check** **5** **Critical Thinking** You plan to write an indirect proof showing that ∠X is an obtuse angle. In the first step you assume that ∠X is an acute angle. What have you overlooked?

EXERCISES

For more exercises, see *Extra Skill, Word Problem, and Proof Practice*.

Practice and Problem Solving

A **Practice by Example**

Example 1
(page 280)

 GO for Help

Write the negation of each statement.

1. Two angles are congruent.

2. You are not sixteen years old.

3. The angle is not obtuse.

4. The soccer game is on Friday.

5. The figure is a triangle.

6. $m\angle A < 90$

Example 2
(page 281)

Write (a) the inverse and (b) the contrapositive of each conditional statement.

7. If you eat all of your vegetables, then you will grow.

8. If a figure is a square, then all of its angles are right angles.

9. If a figure is a rectangle, then it has four sides.

Example 3
(page 282)

Write the first step of an indirect proof.

10. It is raining outside.

11. ∠J is not a right angle.

12. △PEN is isosceles.

13. At least one angle is obtuse.

14. $\overline{XY} \cong \overline{AB}$

15. $m\angle 2 > 90$

Example 4
(page 282)

Identify the two statements that contradict each other.

16. I. △PQR is equilateral.

    II. △PQR is a right triangle.

    III. △PQR is isosceles.

17. I. In right △ABC, $m\angle A = 60$.

    II. In right △ABC, $\angle A \cong \angle C$.

    III. In right △ABC, $m\angle B = 90$.

18. I. ℓ and m are skew.

    II. ℓ and m do not intersect.

    III. ℓ ∥ m

19. I. Each of the two items that Val bought costs more than $10.

    II. Val spent $34 for the two items.

    III. Neither of the two items that Val bought costs more than $15.

Example 5
(page 282)

20. Developing Proof Fill in the blanks to prove the following statement.
If the Debate and Chess Clubs together have fewer than 20 members and the Chess Club has 10 members, then the Debate Club has fewer than 10 members.

Given: The total membership of the Debate Club and the Chess Club is fewer than 20. The Chess Club has 10 members.

Prove: The Debate Club has fewer than 10 members.

Proof: Assume that the Debate Club has 10 or more members. This means that together the two clubs have **a.** ? members. This contradicts the given information that **b.** ? . The assumption is false. Therefore it is true that **c.** ? .

21. Developing Proof Fill in the blanks to prove the following statement.
In a given triangle, $\triangle LMN$, there is at most one right angle.

Given: $\triangle LMN$

Prove: $\triangle LMN$ has at most one right angle.

Proof: Assume that $\triangle LMN$ has more than one **a.** ? . That is, assume that both $\angle M$ and $\angle N$ are **b.** ? . If $\angle M$ and $\angle N$ are both right angles, then $m\angle M = m\angle N =$ **c.** ? . By the Triangle Angle-Sum Theorem, $m\angle L + m\angle M + m\angle N =$ **d.** ? . Use substitution to find $m\angle L +$ **e.** ? $+$ **f.** ? $= 180$. When you solve for $m\angle L$, you find that $m\angle L =$ **g.** ? . This means that there is no $\triangle LMN$, which contradicts the given statement. So the assumption that $\triangle LMN$ has **h.** ? must be false. Therefore, $\triangle LMN$ has **i.** ? .

GO for Help

For a guide to solving Exercise 21, see page 287.

B **Apply Your Skills**

GO Online
Homework Video Tutor
Visit: PHSchool.com
Web Code: aue-0504

Write an indirect proof.

Proof **22. Given:** $\triangle ABC$ with $BC > AC$
Prove: $\angle A \not\equiv \angle B$

Proof **23. Given:** $\triangle XYZ$ is isosceles.
Prove: Neither base angle is a right angle.

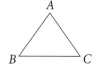

Write (a) the inverse and (b) the contrapositive of each statement. Give the truth value of each.

24. If you live in El Paso, then you live in Texas.

25. If four points are collinear, then they are coplanar.

Open-Ended Write a true conditional statement for each given condition. If such a statement is not possible, tell why.

26. The inverse is false.   **27.** The inverse is true.

28. The contrapositive is false.   **29.** The contrapositive is true.

Writing For Exercises 30–33, write a convincing argument that uses indirect reasoning.

30. Fresh skid marks appear behind a green car at the scene of an accident. Show that the driver of the green car applied the brakes.

31. Ice is forming on the sidewalk in front of Toni's house. Show that the temperature of the sidewalk surface must be 32°F or lower.

32. An obtuse triangle cannot contain a right angle.

33. In a plane, a line has no more than one perpendicular at any of its points.

Real-World Connection

Water freezes at 32°F. Sidewalk "salt" lowers the freezing point of water.

Write the conditional statement illustrated by each Venn diagram. Then write its contrapositive.

34.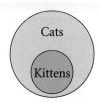

35.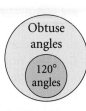

36.

37. Error Analysis Angie saw an ad that stated "If you don't drink Muscle Rex, then you won't build muscles." Angie bought Muscle Rex and drank it, and nothing happened. She sent an e-mail to the company asking for her money back. The company would not refund her money. They claimed that her reasoning was faulty. Using one or more of the terms *converse*, *inverse*, or *contrapositive*, explain why Angie's reasoning was faulty.

38. Earl lives near a noisy construction site at which work ends promptly at 5:00 each workday. Earl thinks, "Today is Tuesday. If it were before 5:00, I would hear construction noise, but I don't hear any. So it must be later than 5:00."
a. What does Earl prove?
b. What assumption does he make?
c. What fact would contradict the assumption?

39. Literature In Arthur Conan Doyle's story "The Sign of the Four," Sherlock Holmes talks to his friend Watson about how a culprit enters a room that has only four entrances: a door, a window, a chimney, and a hole in the roof.
  "You will not apply my precept," he said, shaking his head. "How often have I said to you that when you have eliminated the impossible, whatever remains, however improbable, must be the truth? We know that he did not come through the door, the window, or the chimney. We also know that he could not have been concealed in the room, as there is no concealment possible. Whence, then, did he come?"
How did the culprit enter the room? Explain.

40. Open-Ended Describe a real-life situation in which you used an indirect argument to convince someone of your point of view. Outline your argument.

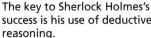

The key to Sherlock Holmes's success is his use of deductive reasoning.

C Challenge **Proof 41. Given:** $\triangle ABC$ is scalene, $m\angle ABX = 36$, and $m\angle CBX = 36$.
Prove: $\overline{XB}$ is not perpendicular to $\overline{AC}$.

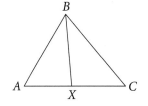

Test Prep

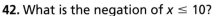

Multiple Choice

42. What is the negation of $x \le 10$?
  A. $x \le -10$ **B.** $-x \le 10$ **C.** $-x > 10$ **D.** $x > 10$

43. What is the negation of $y > 8$?
  F. $y \le -8$ **G.** $y \le 8$ **H.** $-y > 8$ **J.** $y > 8$

44. What is the inverse of $p \rightarrow q$?
  A. $q \rightarrow p$ **B.** $\sim q \rightarrow \sim p$ **C.** $p \rightarrow q$ **D.** $\sim p \rightarrow \sim q$

45. What is the contrapositive of the following statement?
If two parallel lines are cut by a transversal, then the corresponding angles are congruent.

  F. If two lines are cut by a transversal and the corresponding angles are congruent, then the two lines are parallel.

  G. If two nonparallel lines are cut by a transversal, then the corresponding angles are not congruent.

  H. If two lines are cut by a transversal and the corresponding angles are not congruent, then the two lines are not parallel.

  J. If two parallel lines are cut by a transversal, then the corresponding angles are not congruent.

Short Response

46. Use indirect reasoning to give a convincing argument that an obtuse triangle has at most one obtuse angle.

Mixed Review

Lesson 5-3

Is $\overline{XY}$ a perpendicular bisector, an angle bisector, an altitude, a median, or none of these? Explain.

47.   **48.**   **49.**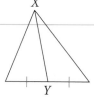

Lesson 3-1

Classify each pair of angles as *alternate interior angles*, *same-side interior angles*, or *corresponding angles*.

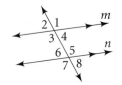

50. $\angle 1$ and $\angle 5$ **51.** $\angle 4$ and $\angle 5$

52. $\angle 3$ and $\angle 5$ **53.** $\angle 3$ and $\angle 7$

Lesson 2-4

Use the given property to complete each statement.

54. Addition Property of Equality
If $5x - 10 = 25$, then $5x = \underline{\ \ ?\ \ }$.

55. Symmetric Property of Equality
If $m\angle ABC = 45$, then $\underline{\ \ ?\ \ }$.

Geometry at Work

················· **Industrial Designer**

Industrial designers work on two-dimensional surfaces to develop products that have three-dimensional appeal to consumers. They use computer-aided design (CAD) software to create two-dimensional screen images and manipulate them for three-dimensional effects. Fashion designers use CAD to study their creations on electronic human forms from various angles and distances.

 **For:** Information about industrial design
  PHSchool.com **Web Code:** aub-2031

Understanding Proof Problems Read the problem below. Then let Resa's thinking guide you through the solution. Check your understanding with the exercises at the bottom of the page.

Fill in the blanks to prove the following statement.

In a given triangle, $\triangle LMN$, there is at most one right angle.

Given: $\triangle LMN$

Prove: $\triangle LMN$ has at most one right angle.

Proof: Assume that $\triangle LMN$ has more than one **a. _?_**. That is, assume that both $\angle M$ and $\angle N$ are **b. _?_**. If $\angle M$ and $\angle N$ are both right angles, then $m\angle M = m\angle N = $ **c. _?_**. By the Triangle Angle-Sum Theorem, $m\angle L + m\angle M + m\angle N = $ **d. _?_**. Use substitution to find $m\angle L + $ **e. _?_** $ + $ **f. _?_** $ = 180$. When you solve for $m\angle L$, you find that $m\angle L = $ **g. _?_**. This means that there is no $\triangle LMN$, which contradicts the information you are given. So the assumption that $\triangle LMN$ has **h. _?_** must be false. Therefore, $\triangle LMN$ has **i. _?_**.

What Resa Thinks

To prove $\triangle LMN$ has at most one right angle, I will assume that it does not have *at most* one. This means that it has more than one.

Now I'll copy the next line in the proof. Huh? Where did $\angle M$ and $\angle N$ come from? Oh, I see. "more than one" means "at least two," so I have to choose at least two angles and assume they are right angles.
The measure of a right angle is 90.

The Triangle Angle-Sum Theorem says that the sum of the measures of the angles in a triangle is 180. "Use substitution." I can substitute 90 for $m\angle M$ and for $m\angle N$. Solving for $m\angle L$ is easy.

But wait! How can $m\angle L = 0$?
Oh! I remember. In an indirect proof, I'm looking for a contradiction. Well, I've got one! The rest is easy.

What Resa Writes

Assume that $\triangle LMN$ has more than one **a. right angle.**

That is, assume that both $\angle M$ and $\angle N$ are **b. right angles.**

If $\angle M$ and $\angle N$ are both right angles, then $m\angle M = m\angle N = $ **c. 90.**

$m\angle L + m\angle M + m\angle N = $ **d. 180.**
Substitute to find
$m\angle L + $ **e. 90** $ + $ **f. 90** $ = 180$.
$m\angle L = $ **g. 0.**

So the assumption that $\triangle LMN$ has **h. more than one right angle** must be false. Therefore $\triangle LMN$ has **i. at most one right angle.**

EXERCISES

1. Given: $\triangle ABC$ is a right triangle.

　Prove: $\triangle ABC$ is not an obtuse triangle.

2. Given: $\triangle ABC$ is an obtuse triangle.

　Prove: $\triangle ABC$ is not a right triangle.

Solving Inequalities

The solutions of an inequality are all the numbers that make the inequality true. The following chart reviews the Properties of Inequality.

Property	Properties of Inequality
	For all real numbers $a, b, c,$ and d:
Addition Property	If $a > b$ and $c \geq d$, then $a + c > b + d$.
Multiplication Property	If $a > b$ and $c > 0$, then $ac > bc$.
	If $a > b$ and $c < 0$, then $ac < bc$.
Transitive Property	If $a > b$ and $b > c$, then $a > c$.
Comparison Property	If $a = b + c$ and $c > 0$, then $a > b$.

You use the Addition and Multiplication Properties of Inequality to solve inequalities.

EXAMPLE

Algebra Solve $-6x + 7 > 25$.

$-6x + 7 - 7 > 25 - 7$ Add -7 to each side (or subtract 7 from each side).

$\dfrac{-6x}{-6} < \dfrac{18}{-6}$ Multiply each side by $-\frac{1}{6}$ (or divide each side by -6). Remember to reverse the inequality symbol.

$x < -3$ Simplify.

EXERCISES

$\boxed{x^2}$ **Algebra** Solve each inequality.

1. $7x - 13 \leq -20$ **2.** $3x + 8 > 16$ **3.** $-2x - 5 < 16$

4. $8y + 2 \geq 14$ **5.** $5a + 1 \leq 91$ **6.** $-x - 2 > 17$

7. $-4z - 10 < -12$ **8.** $9x - 8 \geq 82$ **9.** $6n + 3 \leq -18$

10. $c + 13 > 34$ **11.** $3x - 5x + 2 < 12$ **12.** $x - 19 < -78$

13. $-n - 27 \leq 92$ **14.** $-9t + 47 < 101$ **15.** $8x - 4 + x > -76$

16. $2(y - 5) > -24$ **17.** $8b + 3 \geq 67$ **18.** $-3(4x - 1) \geq 15$

19. $r - 9 \leq -67$ **20.** $\frac{1}{2}(4x - 7) \geq 19$ **21.** $5x - 3x + 2x < -20$

22. $9x - 10x + 4 < 12$ **23.** $-3x - 7x \leq 97$ **24.** $8y - 33 > -1$

25. $4a + 17 \geq 13$ **26.** $-4(5z + 2) > 20$ **27.** $x + 78 \geq -284$

28. $6c \geq -12 - 24$ **29.** $27 - 12 < 3x$ **30.** $8y - 4y + 11 \leq -33$

31. $5x - 2x + 13 > -8$ **32.** $4(5a + 3) \leq -8$ **33.** $8c + 2c + 7 < -10 - 3$

Inequalities in Triangles

 Learning Standards for Mathematics

G.G.33 Investigate, justify, and apply the triangle inequality theorem.

G.G.34 Determine either the longest side of a triangle given the three angle measures or the largest angle given the lengths of three sides of a triangle.

✓ **Check Skills You'll Need**

GO for Help Lessons 1-8 and 5-4

Graph the triangles with the given vertices. List the sides in order from shortest to longest.

1. $A(5, 0), B(0, 8), C(0, 0)$

2. $P(2, 4), Q(-5, 1), R(0, 0)$

3. $G(3, 0), H(4, 3), J(8, 0)$

4. $X(-4, 3), Y(-1, 1), Z(-1, 4)$

Recall the steps for indirect proof.

5. You want to prove $m\angle A > m\angle B$.

Write the first step of an indirect proof.

6. In an indirect proof, you deduce that $AB \geq AC$ is false. What conclusion can you make?

---

1 Inequalities Involving Angles of Triangles

When you empty a container of juice into two glasses, it is difficult to be sure that the glasses get equal amounts. You can be sure, however, that each glass holds less than the original amount in the container. This is a simple application of the Comparison Property of Inequality.

 Key Concepts

Property	Comparison Property of Inequality
	If $a = b + c$ and $c > 0$, then $a > b$.

 Proof

Proof of the Comparison Property

Given: $a = b + c, c > 0$

Prove: $a > b$

Statements	Reasons
1. $c > 0$	**1.** Given
2. $b + c > b + 0$	**2.** Addition Property of Inequality
3. $b + c > b$	**3.** Simplify.
4. $a = b + c$	**4.** Given
5. $a > b$	**5.** Substitute a for $b + c$ in Statement 3.

The Comparison Property of Inequality allows you to prove the following corollary to the Exterior Angle Theorem for triangles (Theorem 3-13).

Key Concepts

Corollary	Corollary to the Triangle Exterior Angle Theorem

The measure of an exterior angle of a triangle is greater than the measure of each of its remote interior angles.

$$m\angle 1 > m\angle 2 \text{ and } m\angle 1 > m\angle 3$$

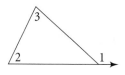

Proof

Proof of the Corollary

Given: ∠1 is an exterior angle of the triangle.

Prove: $m\angle 1 > m\angle 2$ and $m\angle 1 > m\angle 3$.

Proof: By the Exterior Angle Theorem, $m\angle 1 = m\angle 2 + m\angle 3$. Since $m\angle 2 > 0$ and $m\angle 3 > 0$, you can apply the Comparison Property of Inequality and conclude that $m\angle 1 > m\angle 2$ and $m\angle 1 > m\angle 3$.

1 **EXAMPLE** **Applying the Corollary**

In the diagram, $m\angle 2 = m\angle 1$ by the Isosceles Triangle Theorem. Explain why $m\angle 2 > m\angle 3$.

By the corollary to the Exterior Angle Theorem, $m\angle 1 > m\angle 3$. So, $m\angle 2 > m\angle 3$ by substitution.

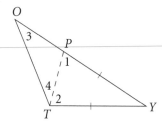

 Quick Check **1** Explain why $m\angle OTY > m\angle 3$.

You will prove the following inequality theorem in the exercises.

Key Concepts

Theorem 5-10

If two sides of a triangle are not congruent, then the larger angle lies opposite the longer side.

If $XZ > XY$, then $m\angle Y > m\angle Z$.

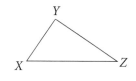

Real-World  **Connection**

Careers Landscape architects blend structures with decorative plantings.

2 **EXAMPLE** **Real-World** **Connection**

Deck Design A landscape architect is designing a triangular deck. She wants to place benches in the two larger corners. Which corners have the larger angles?

Corners B and C have the larger angles. They are opposite the two longer sides of 27 ft and 21 ft.

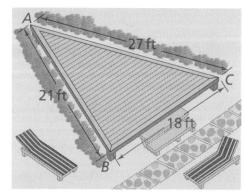

Quick Check **2** List the angles of $\triangle ABC$ in order from smallest to largest.

Theorem 5-10 on the preceding page states that the larger angle is opposite the longer side. The converse is also true.

Key Concepts

Theorem 5-11

If two angles of a triangle are not congruent, then the longer side lies opposite the larger angle.

If $m\angle A > m\angle B$, then $BC > AC$.

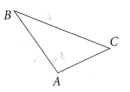

Proof

Indirect Proof of Theorem 5-11

Given: $m\angle A > m\angle B$

Prove: $BC > AC$

Step 1 Assume $BC \not> AC$. That is, assume $BC < AC$ or $BC = AC$.

Step 2 If $BC < AC$, then $m\angle A < m\angle B$ (Theorem 5-10). This contradicts the given fact that $m\angle A > m\angle B$. Therefore, $BC < AC$ must be false.

If $BC = AC$, then $m\angle A = m\angle B$ (Isosceles Triangle Theorem). This also contradicts $m\angle A > m\angle B$. Therefore, $BC = AC$ must be false.

Step 3 The assumption $BC \not> AC$ is false, so $BC > AC$.

3 EXAMPLE Using Theorem 5-11

Test-Taking Tip

Don't be distracted! Choice B lists the sides in order, but from longest to shortest, not shortest to longest.

Multiple Choice Which choice shows the sides of $\triangle TUV$ in order from shortest to longest?

Ⓐ $\overline{TV}, \overline{UV}, \overline{UT}$ Ⓑ $\overline{UT}, \overline{UV}, \overline{TV}$

Ⓒ $\overline{UV}, \overline{UT}, \overline{TV}$ Ⓓ $\overline{TV}, \overline{UT}, \overline{UV}$

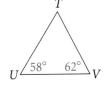

By the Triangle Angle-Sum Theorem, $m\angle T = 60$.

$58 < 60 < 62$, so $m\angle U < m\angle T < m\angle V$. By Theorem 5-11, $TV < UV < UT$. The correct choice is A.

✓ Quick Check ③ List the sides of the $\triangle XYZ$ in order from shortest to longest. Explain your listing.

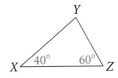

The lengths of three segments must be related in a certain way to form a triangle.

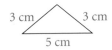

3 cm, 3 cm, 5 cm

2 cm, 2 cm, 6 cm

Notice that only one of the sets of three segments above can form a triangle. The sum of the smallest two lengths must be greater than the greatest length. This is Theorem 5-12 (see next page). You will prove it in the exercises.

 **Key Concepts**

| Theorem 5-12 | **Triangle Inequality Theorem** |

The sum of the lengths of any two sides of a triangle is greater than the length of the third side.

$$XY + YZ > XZ$$
$$YZ + ZX > YX$$
$$ZX + XY > ZY$$

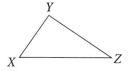

For: Triangle Inequality Activity
Use: Interactive Textbook, 5-5

4 EXAMPLE **Using the Triangle Inequality Theorem**

Can a triangle have sides with the given lengths? Explain.

a. 3 ft, 7 ft, 8 ft

$3 + 7 > 8$
$8 + 7 > 3$
$3 + 8 > 7$ Yes

The sum of any two lengths is greater than the third length.

b. 3 cm, 6 cm, 10 cm

$3 + 6 \ngtr 10$ No

The sum of 3 and 6 is not greater than 10, contradicting Theorem 5-12.

 **Quick Check** **4** Can a triangle have sides with the given lengths? Explain.
a. 2 m, 7 m, and 9 m   **b.** 4 yd, 6 yd, and 9 yd

5 EXAMPLE **Finding Possible Side Lengths**

Algebra A triangle has sides of lengths 8 cm and 10 cm. Describe the lengths possible for the third side.

Let x represent the length of the third side. By the Triangle Inequality Theorem,

| $x + 8 > 10$ | $x + 10 > 8$ | $8 + 10 > x$ |
| $x > 2$ | $x > -2$ | $x < 18$ |

The third side must be longer than 2 cm and shorter than 18 cm.

 **Quick Check** **5** A triangle has sides of lengths 3 in. and 12 in. Describe the lengths possible for the third side.

EXERCISES

For more exercises, see *Extra Skill, Word Problem, and Proof Practice.*

Practice and Problem Solving

A Practice by Example

Example 1
(page 290)

Explain why $m\angle 1 > m\angle 2$.

1.

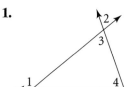

2.

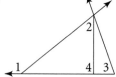

3.

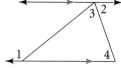

Example 2 (page 290)	**List the angles of each triangle in order from smallest to largest.**

4.

5.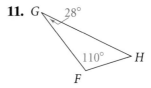

6.

(figure with H, 4, G, 6, I, right angle)

7. $\triangle ABC$, where $AB = 8$, $BC = 5$, and $CA = 7$

8. $\triangle DEF$, where $DE = 15$, $EF = 18$, and $DF = 5$

9. $\triangle XYZ$, where $XY = 12$, $YZ = 24$, and $ZX = 30$

Example 3 (page 291)	**List the sides of each triangle in order from shortest to longest.**

10.

(triangle O, 45°, 75°, M, N)

11.

(triangle G, 28°, 110°, F, H)

12.

(triangle T, U, 30°, V, right angle)

13. $\triangle ABC$, with $m\angle A = 90$, $m\angle B = 40$, and $m\angle C = 50$

14. $\triangle DEF$, with $m\angle D = 20$, $m\angle E = 120$, and $m\angle F = 40$

15. $\triangle XYZ$, with $m\angle X = 51$, $m\angle Y = 59$, and $m\angle Z = 70$

Example 4 (page 292)	**Can a triangle have sides with the given lengths? Explain.**

16. 2 in., 3 in., 6 in. **17.** 11 cm, 12 cm, 15 cm **18.** 8 m, 10 m, 19 m

19. 1 cm, 15 cm, 15 cm **20.** 2 yd, 9 yd, 10 yd **21.** 4 m, 5 m, 9 m

Example 5 (page 292)	$\boxed{x^2}$ **Algebra** **The lengths of two sides of a triangle are given. Describe the lengths possible for the third side.**

22. 8 ft, 12 ft **23.** 5 in., 16 in. **24.** 6 cm, 6 cm

25. 18 m, 23 m **26.** 4 yd, 7 yd **27.** 20 km, 35 km

B Apply Your Skills

28. Error Analysis The Shau family is crossing Kansas on Highway 70. A sign reads "Wichita 90 miles, Topeka 110 miles." Avi says, "I didn't know that it was only 20 miles from Wichita to Topeka." Explain to Avi why the distance between the two cities doesn't have to be 20 miles.

 29. Writing Explain why the distance between the two peaks in the photograph is greater than the difference of the distances from the hiker to each of the peaks.

30. The Hinge Theorem The hypothesis of the Hinge Theorem is stated below. The conclusion is missing.

Suppose two sides of one triangle are congruent to two sides of another triangle. If the included angle of the first triangle is larger than the included angle of the second triangle, then ? .

a. Draw a diagram to illustrate the hypothesis.

b. The conclusion of the Hinge Theorem concerns the sides opposite the two angles mentioned in the hypothesis. Write the conclusion.

c. Draw a diagram to illustrate the converse.

d. Converse of the Hinge Theorem Write the conclusion to this theorem.

Suppose two sides of one triangle are congruent to two sides of another triangle. If the third side of the first triangle is greater than the third side of the second triangle, then ? .

Exercise 29

31. Shortcuts Explain how the student in the photograph is applying the Triangle Inequality Theorem.

x^2 **32. Algebra** Find the longest side of $\triangle ABC$, if $m\angle A = 70, m\angle B = 2x - 10$, and $m\angle C = 3x + 20$.

33. Developing Proof Fill in the blanks for a proof of Theorem 5-10: If two sides of a triangle are not congruent, then the larger angle lies opposite the longer side.

Given: $\triangle TOY$, with $YO > YT$.

Prove: a. _?_ > **b.** _?_

Mark P on $\overline{YO}$ so that $\overline{YP} \cong \overline{YT}$. Draw $\overline{TP}$.

Statements	Reasons
1. $\overline{YP} \cong \overline{YT}$	1. Ruler Post.
2. $m\angle 1 = m\angle 2$	**c.** _?_
3. $m\angle OTY = m\angle 4 + m\angle 2$	**d.** _?_
4. $m\angle OTY > m\angle 2$	**e.** _?_
5. $m\angle OTY > m\angle 1$	**f.** _?_
6. $m\angle 1 > m\angle 3$	**g.** _?_
7. $m\angle OTY > m\angle 3$	**h.** _?_

Real-World Connection

Some recall the Triangle Inequality Theorem as "The shortest path between two points is the straight path."

Proof **34.** Prove this corollary to Theorem 5-11:
The perpendicular segment from a point to a line is the shortest segment from the point to the line.

Given: $\overline{PT} \perp \overline{TA}$

Prove: $PA > PT$

Critical Thinking Determine which segment is shortest in each diagram.

35.

36.

37.

Homework Video Tutor

Visit: PHSchool.com
Web Code: aue-0505

C Challenge

38. Probability A student has two straws, one 6 cm long and the other 9 cm long. She picks a third straw at random from a group of four straws whose lengths are 3 cm, 5 cm, 11 cm, and 15 cm. What is the probability that the straw she picks will allow her to form a triangle?

For Exercises 39 and 40, x and y are whole numbers, $1 < x < 5$, and $2 < y < 9$.

39. The sides of a triangle are 5 cm, x cm, and y cm. List possible (x, y) pairs.

40. Probability What is the probability that you can draw an isosceles triangle that has sides 5 cm, x cm, and y cm, with x and y chosen at random?

Proof **41.** Prove Theorem 5-12: The sum of the lengths of any two sides of a triangle is greater than the length of the third side.

Given: $\triangle ABC$

Prove: $AC + CB > AB$

(*Hint:* On $\overrightarrow{BC}$ mark a point D not on $\overline{BC}$, so that $DC = AC$. Draw $\overline{DA}$ and use Theorem 5-11 with $\triangle ABD$.)

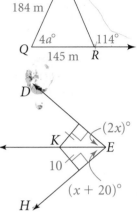

Multiple Choice

42. For △PQR, which is the best estimate for PR?
  A. 137 m   **B.** 145 m
  C. 163 m   **D.** 187 m

43. Two sides of a triangle measure 13 and 15.
  Which length is NOT possible for the third side?
  F. 2   **G.** 8
  H. 14   **J.** 20

44. Which statement is true for the figure at the right?
  A. JN > JB
  B. JN > BN
  C. The shortest side is $\overline{JB}$.
  D. The longest side is $\overline{BN}$.

45. For △ABC, what must be true about an exterior angle at A?
  F. It is larger than ∠A.   **G.** It is smaller than ∠A.
  H. It is larger than ∠B.   **J.** It is smaller than ∠C.

46. Which lengths can be lengths for the sides of a triangle?
  A. 1, 2, 5   **B.** 3, 2, 5   **C.** 5, 2, 5   **D.** 7, 2, 5

47. For △JKL, LJ < JK < KL. What must be true about angles J, K, and L?
  F. m∠L < m∠J < m∠K   **G.** m∠L > m∠J > m∠K
  H. m∠J < m∠L < m∠K   **J.** m∠J > m∠L > m∠K

Short Response

48. In △PQR, PQ > PR > QR. One angle measures 170°. List all possible whole number values for m∠P.

49. In △ABC, m∠A > m∠C > m∠B.
  a. Of $\overline{AB}$ and $\overline{AC}$, one measures 5 inches and the other measures 9 inches. Which measures 9 inches? Explain.
  b. Based on your conclusion for part (a), find all possible whole-number measures for the third side. Explain.

Mixed Review

Lesson 5-4

Write the negation of each statement.

50. $m∠A ≤ m∠B$

51. $m∠X > m∠B$

52. The angle is a right angle.

53. The triangle is not obtuse.

Lesson 2-5

Use the diagram. Find the measure of each angle.

54. ∠ADH   **55.** ∠GDH

56. ∠CDH   **57.** ∠ADG

Lesson 1-9

Find to the nearest tenth of a square unit the area of each circle with the given radius r or diameter d.

58. r = 1.6 ft   **59.** d = 35 mm   **60.** r = 0.5 m   **61.** d = 20 mi

Indirect Proofs in Space

Definition: Line ℓ and plane P that intersect in point A are **perpendicular** if and only if ℓ is perpendicular to every line in P that contains A.

NY G.G.1, G.G.2, G.G.3, G.G.4, G.G.5, G.G.6, G.G.7, G.G.9

You can use this definition to give convincing arguments—some indirect—about perpendicular lines and planes. You will also need the plane geometry fact that through a point on a line ℓ there is exactly one line perpendicular to ℓ.

EXAMPLE

For a point A in plane P, explain why there cannot be two lines perpendicular to P through A.

Assume there are two lines a and b both perpendicular to P through A. Lines a and b determine a plane Q (page 22, Exercise 76) that intersects P in line c. So in plane Q, $a \perp c$ and $b \perp c$, which is a contradiction.

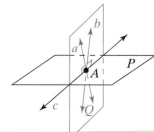

Thus, there can be at most one line perpendicular to P through A.

EXERCISES

Draw a diagram (see page 30) to illustrate each fact about perpendicular lines and planes. Then give a convincing argument that the statement is true.

1. For a point A *not* in plane P, explain why there cannot be two lines perpendicular to P through A.

2. Suppose line ℓ is perpendicular to plane P at point O. Explain why any line perpendicular to line ℓ at point O must be in plane P.

3. Given line ℓ and point A, tell why there is only one plane perpendicular to ℓ through A.
   a. Case 1: Assume that point A is on line ℓ.
   b. Case 2: Assume that point A is not on line ℓ.

4. Explain why two planes that are perpendicular to the same line must be parallel.

5. Definition: Planes P and R that intersect in line ℓ are **perpendicular** if and only if there is a plane S perpendicular to ℓ and intersecting P and R to form a right angle. The drawing at the right shows the right angle formed by such a plane S (not shown).
   a. Copy and complete the drawing to show plane S.
   b. Then give a convincing argument: If a line is perpendicular to a plane, then every plane containing the line is perpendicular to the given plane.

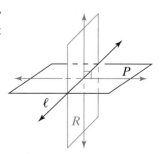

6. Two lines r and s are both perpendicular to plane P. Tell why r and s must be coplanar.

7. Suppose line ℓ is perpendicular to each of lines m and n in plane P at their point of intersection O. Tell why line ℓ must be perpendicular to each of the other lines in P that contain O. (*Hint:* Try direct, rather than indirect, reasoning.)

8. Explain why two planes P and Q are perpendicular to each other if and only if one plane contains a line t that is perpendicular to the other plane. Note that this is a biconditional.

Chapter 5

Chapter Review

Vocabulary Review

altitude of a triangle (p. 275)
centroid (p. 274)
circumcenter of a triangle (p. 273)
circumscribed about (p. 273)
concurrent (p. 273)
contrapositive (p. 280)
coordinate proof (p. 260)
distance from a point to a line (p. 266)
equivalent statements (p. 281)
incenter of a triangle (p. 273)
indirect proof (p. 281)
indirect reasoning (p. 281)
inscribed in (p. 273)
inverse (p. 280)
median of a triangle (p. 274)
midsegment (p. 259)
negation (p. 280)
orthocenter of a triangle (p. 275)
point of concurrency (p. 273)

Choose the correct vocabulary term to complete each sentence.

1. A *(centroid, median of a triangle)* is a segment whose endpoints are a vertex and the midpoint of the side opposite the vertex.

2. The length of the perpendicular segment from a point to a line is the *(midsegment, distance from the point to the line).*

3. If T is a point on the perpendicular bisector of $\overline{FG}$, then $TF = TG$ because of the *(Perpendicular Bisector Theorem, Angle Bisector Theorem).*

4. The *(altitude, median)* of a triangle is a perpendicular segment from a vertex to the line containing the side opposite the vertex.

5. The notation $\sim q \rightarrow \sim p$ is the *(inverse, contrapositive)* of $p \rightarrow q$.

6. To write a(n) *(indirect proof, negation)*, you start by assuming that the opposite of what you want to prove is true.

7. In $\triangle ABC$, $AB + BC > AC$ because of the *(Comparison Property of Inequality, Triangle Inequality Theorem).*

8. The *(circumcenter, incenter)* of a triangle is the point of concurrency of the angle bisectors of the triangle.

9. The *(Angle Bisector Theorem, Triangle Inequality Theorem)* says that if a point is on the bisector of an angle, then it is equidistant from the sides of the angle.

10. A point where three lines intersect is a *(point of concurrency, incenter).*

Go Online
PHSchool.com
For: Vocabulary quiz
Web Code: auj-0551

Skills and Concepts

5-1 and 5-2 Objectives

- To use properties of midsegments to solve problems
- To use properties of perpendicular bisectors and angle bisectors

A **midsegment** of a triangle is a segment that connects the midpoints of two sides. A midsegment is parallel to the third side, and is half its length.

In a **coordinate proof,** a figure is drawn on a coordinate plane and formulas are used to prove properties of the figure.

The **distance from a point to a line** is the length of the perpendicular segment from the point to the line. The Perpendicular Bisector Theorem together with its converse states that a point is on the perpendicular bisector of a segment if and only if it is equidistant from the endpoints of the segment. The Angle Bisector Theorem together with its converse states that a point is on the bisector of an angle if and only if it is equidistant from the sides of the angle.

x^2 **Algebra** Find the value of x.

11.

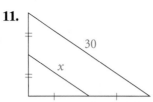

12.

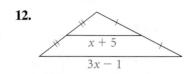

Use the figure to find each segment length or angle measure.

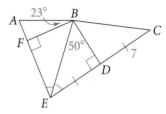

13. $m\angle BEF$

14. FE

15. EC

16. $m\angle CEA$

5-3 Objectives

▼ To identify properties of perpendicular bisectors and angle bisectors

▼ To identify properties of medians and altitudes of a triangle

When three or more lines intersect in one point, they are **concurrent**.

The **median of a triangle** is a segment whose endpoints are a vertex and the midpoint of the opposite side. The **altitude of a triangle** is a perpendicular segment from a vertex to the line containing the opposite side.

For any given triangle, special segments and lines are concurrent:

• the perpendicular bisectors of the sides at the circumcenter, the center of the circle that can be **circumscribed about** the triangle

• the bisectors of the angles at the incenter, the center of the circle that can be **inscribed in** the triangle

• the medians at the **centroid**

• the lines containing the altitudes at the **orthocenter of the triangle.**

Graph $\triangle ABC$ with vertices $A(2, 3)$, $B(-4, -3)$, and $C(2, -3)$. Find the coordinates of each point of concurrency.

17. circumcenter   **18.** centroid   **19.** orthocenter

Determine whether $\overline{AB}$ is a perpendicular bisector, an angle bisector, a median, an altitude, or none of these. Explain.

20.

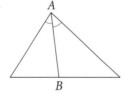

21.

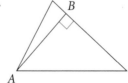

22.

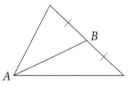

5-4 Objectives

▼ To write the negation of a statement and the inverse and contrapositive of a conditional statement

▼ To use indirect reasoning

The **negation** of a statement has the opposite truth value. The **inverse** of a conditional statement is the negation of both the hypothesis and the conclusion. The **contrapositive** of a conditional statement switches the hypothesis and the conclusion and negates both. Statements that always have the same truth value are **equivalent statements.**

To use **indirect reasoning,** consider all possibilities and then prove all but one false. The remaining possibility must be true.

The three steps of an **indirect proof** are:

Step 1 State as an assumption the opposite (negation) of what you want to prove.

Step 2 Show that this assumption leads to a contradiction.

Step 3 Conclude that the assumption must be false and that what you want to prove must be true.

Write the inverse and the contrapositive of each statement.

23. If it is snowing, then it is cold outside.

24. If an angle is obtuse, then its measure is greater than 90 and less than 180.

25. If a figure is a square, then its sides are congruent.

26. If you are in Australia, then you are south of the equator.

Write a convincing argument that uses indirect reasoning.

27. The product of two numbers is even. Show that at least one of the two numbers must be even.

28. Show that a right angle cannot be formed by the intersection of nonperpendicular lines.

29. Show that a triangle can have at most one obtuse angle.

30. Show that an equilateral triangle cannot have an obtuse angle.

5-5 Objectives

▼ To use inequalities involving angles of triangles

▼ To use inequalities involving sides of triangles

If two sides of a triangle are not congruent, then the larger angle lies opposite the longer side. The converse is also true. If two angles are not congruent, then the longer side lies opposite the larger angle.

The measure of an exterior angle of a triangle is greater than the measure of each of its remote interior angles. The sum of the lengths of any two sides of a triangle is greater than the length of the third side.

List the angles and sides in order from smallest to largest.

31.

32.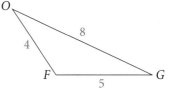

Is it possible for a triangle to have sides with the given lengths? Explain.

33. 5 in., 8 in., 15 in.   **34.** 10 cm, 12 cm, 20 cm

35. 20 m, 22 m, 24 m   **36.** 3 ft, 6 ft, 8 ft

37. 1 yd, 1 yd, 3 yd   **38.** 5 km, 6 km, 7 km

Two side lengths of a triangle are given. Write an inequality to show the range of values, x, for the length of the third side.

39. 4 in., 7 in.   **40.** 8 m, 15 m

41. 2 cm, 8 cm   **42.** 12 ft, 13 ft

Chapter Test

Go Online
PHSchool.com

For: Chapter Test
Web Code: aua-0552

Write (a) the inverse and (b) the contrapositive of each statement.

1. If a polygon has eight sides, then it is an octagon.

2. If it is a leap year, then it is an even-numbered year.

3. If it is snowing, then it is not summer.

4. What can you conclude from the diagram? Justify your answer.

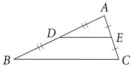

Identify the two statements that contradict each other.

5. I. $\triangle PQR$ is a right triangle.
    II. $\triangle PQR$ is an obtuse triangle.
    III. $\triangle PQR$ is scalene.

6. I. $\angle DAS \cong \angle CAT$
    II. $\angle DAS$ and $\angle CAT$ are vertical.
    III. $\angle DAS$ and $\angle CAT$ are adjacent.

List the angles of $\triangle ABC$ from smallest to largest.

7. $AB = 9, BC = 4, AC = 12$

8. $AB = 10, BC = 11, AC = 9$

9. $AB = 3, BC = 9, AC = 7$

10. **Open-Ended** Write three lengths that cannot be the lengths of the three sides of a triangle. Explain your answer.

List the sides of each triangle in order from shortest to longest.

11.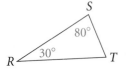

12.

x^2 **Algebra** Find the value of x in each figure.

13.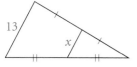

14.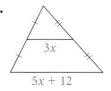

15. **Writing** Use indirect reasoning to explain why the following statement is true: If an isosceles triangle is obtuse, then the obtuse angle is the vertex angle.

x^2 **Algebra** Find the values of x and KM in each figure.

16.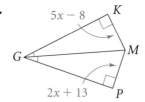

17.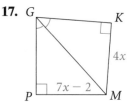

Coordinate Geometry Find the center of the circle that circumscribes $\triangle ABC$.

18. $A(0, 0), B(0, -3), C(-5, -3)$

19. $A(0, 5), B(-4, 5), C(-4, -3)$

20. $A(3, -1), B(-2, -1), C(3, -8)$

21. $\triangle ABC$ has vertices $A(2, 5), B(2, -3), C(10, -3)$. What point of concurrency is at $(6, 1)$?

22. **Given:** $\overleftrightarrow{PQ}$ is the perpendicular bisector of $\overline{AB}$. $\overleftrightarrow{QT}$ is the perpendicular bisector of $\overline{AC}$.

    Prove: $QC = QB$

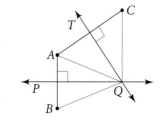

23. What can you conclude about point Y? Explain.

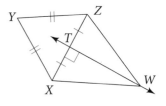

24. In the figure, $WK = KR$. What can you conclude about point A? Explain.

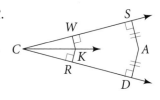

Regents Test Prep

Reading Comprehension **Read the passage below. Then answer the questions on the basis of what is *stated* or *implied* in the passage.**

The Kitchen Triangle Architects, builders, and anyone who has ever prepared a full meal in a kitchen know that in most kitchen activities you go back and forth between the sink, the stove, and the refrigerator. Those three locations form what is called the "work triangle" or the "kitchen triangle."

A large apartment complex made this table of kitchen triangle dimensions for four of its units.

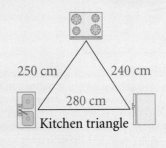

Kitchen Triangle Distances (cm)

Apt.	Stove to Sink	to Fridge	to Stove
A	250	280	240
B	145	320	165
C	115	220	330
D	310	152	270

To have an efficient kitchen, it should be easy to traverse each side of the kitchen triangle.

1. Why are the sink, stove, and refrigerator the vertices of the kitchen triangle?
   (1) Architects and builders say they are.
   (2) You walk between these three appliances in most kitchen activities.
   (3) Most apartments have these appliances.
   (4) A sink, stove, and refrigerator form an efficient kitchen.

2. In apartment A, which appliance is at the vertex of the largest angle?
   (1) the sink   (2) the stove
   (3) the refrigerator (4) cannot be determined

3. In apartment D, which appliance is at the vertex of the smallest angle?
   (1) the sink   (2) the stove
   (3) the refrigerator (4) cannot be determined

4. The description of one kitchen triangle is incorrect. Which one? Justify your answer.

5. In which kitchen is it least efficient for two persons to work? Justify your answer.

In kitchen triangle A, a microwave is halfway between the stove and sink, and an electric can opener is halfway between the fridge and sink.

6. What is the distance between the microwave and the can opener? Justify your answer.
   (1) 120 cm   (2) 125 cm
   (3) 240 cm   (4) 265 cm

7. To go from the microwave to the can opener, you must go in the direction you would travel from the stove to the refrigerator. Why?

8. Which kitchen triangle has a midsegment triangle with smallest perimeter?
   (1) A   (2) D
   (3) C   (4) cannot be determined

9. For which kitchen triangle do you need the fewest steps to go from one vertex to another and then to the third?
   (1) A   (2) C
   (3) D   (4) cannot be determined

10. In Apartment D, how far is it from stove to sink by way of the refrigerator?
    (1) 270 cm (2) 310 cm (3) 422 cm (4) 462 cm

Activity Lab

Point of Balance

Applying Theorems About Triangles For an object to spin smoothly, its weight must be evenly distributed around its central axis. A toy top, an ice skater, and a carousel all spin around a central axis. If the spin goes slightly out of alignment, the spinning object begins to wobble and may eventually tip over.

Gyroscopes

A gyroscope is a wheel mounted in a set of rings so that the axis of the wheel can turn in any direction. When the wheel spins rapidly, a small gyroscope becomes so stable that it can balance on the tip of a pencil.

 G.G.21:
Investigate and apply the concurrence of medians, altitudes, angle bisectors, and perpendicular bisectors of triangles.

Amusing Rides

Carousels were traditionally used to train young princes in tournament riding. The more skilled the rider, the more likely he was to spear one or more of the small gold rings that hung along the outer edge. It seemed like so much fun that eventually the rest of the population wanted to ride the carousel.

The Spin

Figure skaters at the Olympic Games compete in a short program (33.3% of the total score) and a long program (66.7%). A short program can last a maximum of 2 min 40 s and must include three spins.

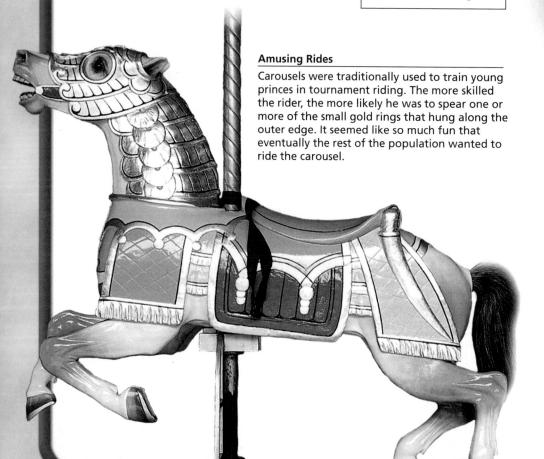

The Bicycle as a Gyroscope

The spinning wheels of a bicycle act as gyroscopes and help to keep the bike upright. Like all gyroscopes, the wheels tend to remain spinning in the same plane, giving the bicycle stability.

Earth as a Gyroscope

The axis of Earth is remarkably stable, but our planet does wobble slightly. It takes 25,800 years for each circular wobble. This wobble is largely caused by the gravitational pull of the sun and moon on Earth's equatorial bulge.

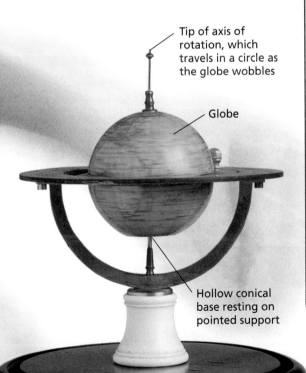

Tip of axis of rotation, which travels in a circle as the globe wobbles

Globe

Hollow conical base resting on pointed support

Activity

Materials: cardboard, straightedge, scissors, compass, graph paper, pencil

Draw a large triangle on the cardboard. Construct its medians. Locate the centroid, the point at which the medians meet. Cut out the triangle.

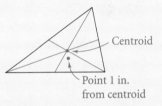

Centroid

Point 1 in. from centroid

a. Balance your triangle by placing the centroid on the point of your pencil and pushing down slightly to dent the cardboard. Give your triangle a gentle spin and watch it as it moves. Describe the movement.

b. Locate a point 1 in. from the centroid. Support your triangle on your pencil at this point. Give your triangle a gentle spin and watch it as it moves. Describe any differences from its movement in part (a).

c. Suppose the coordinates of the vertices of a triangle are $(0, 0)$, $(16, 0)$, and $(20, 18)$. Find the coordinates of the centroid. Test your work by making a graph of the triangle. Glue the graph to a piece of cardboard and then cut out the triangle. Try to balance the triangle at the coordinates you calculated.

d. Draw a large convex quadrilateral on the cardboard. Devise a way to locate its centroid. Cut out the quadrilateral and test its balance with spins as described in parts (a) and (b). Describe what you find.

Go Online
PHSchool.com

For: Information about objects that spin
Web Code: aue-0553

What You've Learned

NY Learning Standards for Mathematics

● **G.G.35:** In Chapter 3, you learned that parallel lines produce pairs of angles that are congruent or supplementary.

● **G.G.28:** In Chapter 4, you learned how to prove triangles congruent. You also learned how to use CPCTC to draw additional conclusions.

● **G.G.21:** In Chapter 5, you learned that properties of special segments of a triangle can provide additional information about a triangle.

INSTANT CHECK SYSTEM™ Check Your Readiness

GO for Help to the Lesson in green.

Properties of Parallel Lines (Lesson 3-1)

x^2 **Algebra** Use properties of parallel lines to find the value of x.

1.
$(x + 9)°$
$(2x - 21)°$

2.
$(3x - 14)°$
$(2x - 16)°$

3.
$5x°$
$(176 - 3x)°$

Proving Lines Parallel (Lesson 3-2)

x^2 **Algebra** Determine whether $\overleftrightarrow{AB}$ (or $\overline{AB}$) is parallel to $\overleftrightarrow{CD}$ (or $\overline{CD}$).

4.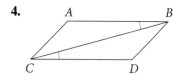
A B
C D

5.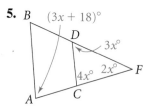
B $(3x + 18)°$
D
$3x°$
$4x°$ $2x°$ F
A C

6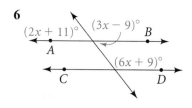
$(2x + 11)°$ $(3x - 9)°$ B
A
$(6x + 9)°$
C D

Using Slope to Determine Parallel and Perpendicular Lines (Lesson 3-7)

x^2 **Algebra** Determine whether each pair of lines is parallel, perpendicular, or neither.

7. $y = -2x; y = -2x + 4$ **8.** $y = -\frac{3}{5}x + 1; y = \frac{5}{3}x - 3$ **9.** $2x - 3y = 1; 3x - 2y = 8$

Proving Triangles Congruent (Lessons 4-2 and 4-3)

Determine the postulate or theorem that makes each pair of triangles congruent.

10.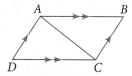
A B
D C

11.

12.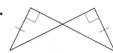

Quadrilaterals

🔊 **Key Vocabulary**

What You'll Learn Next

(NY) Learning Standards for Mathematics

- **G.G.38:** In this chapter, you will learn properties of parallelograms and other special quadrilaterals.

- **G.G.41:** You will learn properties of quadrilaterals that allow you to classify quadrilaterals.

- **G.G.69:** You will use these properties to help you place figures in the coordinate plane.

Data Analysis **Activity Lab** Applying what you learn, you will model two investment strategies on pages 362 and 363.

Classifying Quadrilaterals

G.CM.11 Understand and use appropriate language, representations, and terminology when describing objects, relationships, mathematical solutions, and geometric diagrams.

Find the distance between the points to the nearest tenth.

1. $M(2, -5), N(-7, 1)$ **2.** $P(-1, -3), Q(-6, -9)$ **3.** $C(-4, 6), D(5, -3)$

Find the slope of the line through each pair of points.

4. $X(0, 6), Y(4, 9)$ **5.** $R(3, 8), S(6, 0)$ **6.** $A(4, 3), B(2, 1)$

 **New Vocabulary** • **parallelogram** • **rhombus** • **rectangle** • **square** • **kite** • **trapezoid** • **isosceles trapezoid**

1   Classifying Special Quadrilaterals

Seven important types of quadrilaterals are defined below.

 Key Concepts

Definitions	**Special Quadrilaterals**

A **parallelogram** is a quadrilateral with both pairs of opposite sides parallel.

A **rhombus** is a parallelogram with four congruent sides.

A **rectangle** is a parallelogram with four right angles.

A **square** is a parallelogram with four congruent sides and four right angles.

A **kite** is a quadrilateral with two pairs of adjacent sides congruent and no opposite sides congruent.

A **trapezoid** is a quadrilateral with exactly one pair of parallel sides. The **isosceles trapezoid** at the right is a trapezoid whose nonparallel opposite sides are congruent.

Real-World Connection

A "kite" is not the only special quadrilateral used to make a kite!

1 EXAMPLE Classifying a Quadrilateral

Judging by appearance, classify *DEFG*
in as many ways as possible.

DEFG is a quadrilateral because it has four sides.

It is a parallelogram because both pairs of opposite sides are parallel.

● It is a rectangle because it has four right angles.

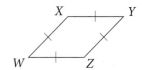

 **Quick Check** **1** **a.** Judging by appearance, classify *WXYZ* at the right
in as many ways as possible.

b. Critical Thinking Which name gives the most
information about *WXYZ*? Explain.

For: Quadrilateral Activity
Use: Interactive Textbook, 6-1

The diagram below shows the relationships among special quadrilaterals.

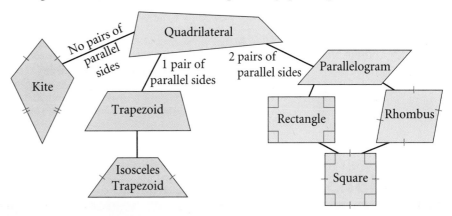

You can use what you know about slope and distance to classify a quadrilateral.

2 EXAMPLE Classifying by Coordinate Methods

Coordinate Geometry Determine the most precise name for quadrilateral *LMNP*.

Step 1 Find the slope of each side.

$$\text{slope of } \overline{LM} = \frac{3-2}{3-1} = \frac{1}{2}$$

$$\text{slope of } \overline{NP} = \frac{2-1}{5-3} = \frac{1}{2}$$

$$\text{slope of } \overline{MN} = \frac{3-2}{3-5} = -\frac{1}{2}$$

$$\text{slope of } \overline{LP} = \frac{2-1}{1-3} = -\frac{1}{2}$$

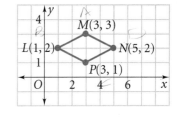

Both pairs of opposite sides are parallel, so *LMNP* is a parallelogram.
No sides are perpendicular, so *LMNP* is not a rectangle.

Step 2 Use the Distance Formula to see if any pairs of sides are congruent.

$$LM = \sqrt{(3-1)^2 + (3-2)^2} = \sqrt{5} \qquad MN = \sqrt{(3-5)^2 + (3-2)^2} = \sqrt{5}$$

$$NP = \sqrt{(5-3)^2 + (2-1)^2} = \sqrt{5} \qquad LP = \sqrt{(1-3)^2 + (2-1)^2} = \sqrt{5}$$

● All sides are congruent, so *LMNP* is a rhombus.

Vocabulary Tip

Although *LMNP* is a
parallelogram, rhombus is
the more *precise* name
because it gives more
information about the
quadrilateral.

 **Quick Check** **2** Determine the most precise name for quadrilateral *ABCD* with vertices *A*(−3, 3),
B(2, 4), *C*(3, −1), and *D*(−2, −2).

Lesson 6-1 Classifying Quadrilaterals **307**

You can use the definitions of special quadrilaterals and algebra to find lengths of sides.

3 EXAMPLE **Using the Properties of Special Quadrilaterals**

Algebra Find the values of the variables for the kite.

$KB = JB$	Definition of kite
$3x - 5 = 2x + 4$	Substitute.
$x - 5 = 4$	Subtract $2x$ from each side.
$x = 9$	Add 5 to each side.
$KT = x + 6 = 15$	Substitute 9 for x.
$KT = JT$	Definition of kite
$15 = 2y + 5$	Substitute.
$10 = 2y$	Subtract 5 from each side.
$5 = y$	Divide each side by 2.

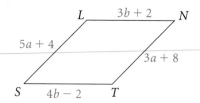

✓ Quick Check ❸ Find the values of the variables for the rhombus. Then find the lengths of the sides.

EXERCISES

For more exercises, see *Extra Skill, Word Problem, and Proof Practice.*

Practice and Problem Solving

 Practice by Example

Example 1
(page 307)

 for Help

These quadrilaterals are made from a toy building set. Judging by appearance, classify each quadrilateral in as many ways as possible.

1.   **2.**    **3.**

4.   **5.**    **6.**

Determine the most precise name for each quadrilateral.

7.   **8.** 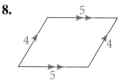   **9.**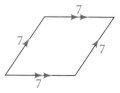

10. 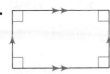   **11.**    **12.**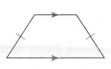

Example 2
(page 307)

Coordinate Geometry Graph and label each quadrilateral with the given vertices. Then determine the most precise name for each quadrilateral.

13. $A(3, 5), B(7, 6), C(6, 2), D(2, 1)$ **14.** $W(-1, 1), X(0, 2), Y(1, 1), Z(0, -2)$

15. $J(2, 1), K(5, 4), L(7, 2), M(2, -3)$ **16.** $R(-2, -3), S(4, 0), T(3, 2), V(-3, -1)$

17. $N(-6, -4), P(-3, 1),$ **18.** $E(-3, 1), F(-7, -3),$
     $Q(0, 2), R(-3, 5)$ $G(6, -3), H(2, 1)$

Example 3
(page 308)

x^2 **Algebra** Find the values of the variables. Then find the lengths of the sides.

19. kite   **20.** kite   **21.** kite

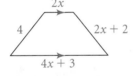   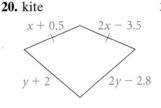   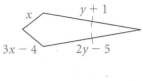

22. isosceles trapezoid   **23.** rhombus   **24.** square

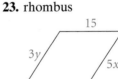   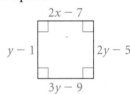

B **Apply Your Skills** x^2 **Algebra** In each figure, find the measures of the angles and the lengths of the sides.

25. isosceles trapezoid $DEFG$ **26.** rhombus $HKJI$

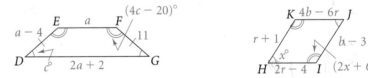

27. Art American artist Charles Demuth created *My Egypt*, the oil painting pictured at the left. It is in an art style called Cubism, in which subjects are made of cubes and other geometric forms. Identify the types of special quadrilaterals you see in the painting.

28. Multiple Choice $K(-3, 0), I(0, 2),$ and $T(3, 0)$ are the vertices of a kite. Which point could be the fourth vertex?
    Ⓐ $E(0, 2)$ Ⓑ $E(0, 0)$ Ⓒ $E(0, -2)$ Ⓓ $E(0, -10)$

Draw each figure on graph paper. If not possible, explain.

29. a parallelogram that is neither   **30.** an isosceles trapezoid with vertical
     a rectangle nor a rhombus   and horizontal congruent sides

31. a trapezoid with only one right angle   **32.** a trapezoid with two right angles

33. a rhombus that is not a square   **34.** a kite with two right angles

Exercise 27

35. Copy the Venn diagram. Add the labels *Rectangles*, *Rhombuses*, and *Trapezoids* to the diagram in the appropriate places.

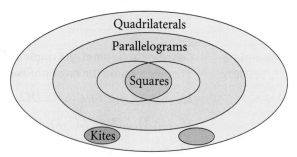

State whether each statement is *true* or *false*. Justify your response. You may find the diagram from Exercise 35 helpful.

36. All squares are rectangles.

37. A trapezoid is a parallelogram.

38. A rhombus can be a kite.

39. Some parallelograms are squares.

40. Every quadrilateral is a parallelogram.

41. All rhombuses are squares.

42. Paper Folding Fold a nonsquare, rectangular piece of paper in half horizontally and then vertically, as shown at the right. Draw and then cut along the line connecting the two opposite corners containing a fold. What quadrilateral do you find when you unfold the paper? Why doesn't it matter what size rectangle you start with?

43. Identify a parallelogram, rhombus, rectangle, square, kite, and trapezoid in your classroom. State whether your trapezoid is isosceles.

44. Writing Describe the difference between a rhombus and a kite.

Name each type of special quadrilateral that can meet the given condition. Make sketches to support your answers.

45. exactly one pair of congruent sides

46. two pairs of parallel sides

47. four right angles

48. adjacent sides that are congruent

49. Error Analysis Lauren argues, "A parallelogram has two pairs of parallel sides, so it certainly has one pair of parallel sides. Therefore a parallelogram must also be a trapezoid." What is the error in Lauren's argument?

Name the type(s) of special quadrilateral(s) it appears that you can form by joining the triangles in each pair. Make sketches to support your answers.

Sample two congruent scalene triangles

parallelogram

50. two congruent scalene right triangles

51. two congruent equilateral triangles

52. two congruent isosceles right triangles

53. two congruent isosceles acute triangles

Real-World Connection

To make this butterfly, an origami square was folded first on its diagonals.

Problem Solving Hint

In Exercises 50–53, if you flip one of the two triangles, you may find different quadrilaterals.

Identify a parallelogram, rhombus, rectangle, square, kite and trapezoid at each site. State whether your trapezoid is isosceles.

54. home   **55.** somewhere other than school and home

 **Challenge**   **Reasoning** A scrap of paper covers part of each quadrilateral. Name all the special quadrilaterals that each could be. Explain each choice.

56.   **57.**    **58.**   **59.**

NY REGENTS

Multiple Choice

60. Which statement is NEVER true?
   A. Square *ABCD* is a rhombus.
   B. Parallelogram *PQRS* is a square.
   C. Trapezoid *GHJK* is a parallelogram.
   D. Square *WXYZ* is a parallelogram.

61. Which statement is true for some, but not all, rectangles?
   F. Opposite sides are parallel.
   G. It is a parallelogram.
   H. Adjacent sides are perpendicular.
   J. All sides are congruent.

62. A parallelogram has four congruent sides. Which name best describes the figure?
   A. trapezoid   **B.** parallelogram   **C.** rhombus   **D.** square

63. Which name best describes a parallelogram with four congruent angles?
   F. kite   **G.** rhombus   **H.** rectangle   **J.** square

Short Response

64. *A*(−3, 1), *B*(−1, −2), and *C*(2, 1) are three vertices of quadrilateral *ABCD*. Could *ABCD* be a rectangle? Explain.

Mixed Review

Lesson 5-5   **Can a triangle have sides with the given lengths? Explain.**

65. 8 mm, 6 mm, 3 mm   **66.** 5 ft, 20 ft, 7 ft   **67.** 3 m, 5 m, 8 m

Lesson 4-1   **Quadrilaterals *RSTV* and *NMQP* are congruent. Find the length of the side or the measure of the angle.**

68. $\overline{MN}$ **69.** $\overline{VT}$

70. $\overline{ST}$ **71.** $\angle S$

72. $\angle V$ **73.** $\angle R$

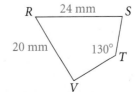

Lesson 3-7   **74.** Write an equation for the line parallel to $y = -3x - 5$ that contains point (0, 4).

6-2

Properties of Parallelograms

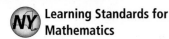

Learning Standards for Mathematics

G.G.38 Investigate, justify, and apply theorems about parallelograms involving their angles, sides, and diagonals.

✓ **Check Skills You'll Need**

GO for Help Lessons 4-1 and 4-3

Use the figure at the right.

1. Name the postulate or theorem that justifies the congruence $\triangle EFG \cong \triangle GHE$.

2. Complete each statement.
   a. $\angle FEG \cong$ _?_ **b.** $\angle EFG \cong$ _?_
   c. $\angle FGE \cong$ _?_ **d.** $\overline{EF} \cong$ _?_
   e. $\overline{FG} \cong$ _?_ **f.** $\overline{GE} \cong$ _?_

3. What other relationship exists between $\overline{FG}$ and $\overline{EH}$?

◀)) **New Vocabulary** • consecutive angles

Properties: Sides and Angles

You can use what you know about parallel lines and transversals to prove some theorems about parallelograms.

 Key Concepts

Theorem 6-1

Opposite sides of a parallelogram are congruent.

Proof **Proof of Theorem 6-1**

Given: ▱ABCD
Prove: $\overline{AB} \cong \overline{CD}, \overline{BC} \cong \overline{DA}$

Vocabulary Tip

Read "▱" as "parallelogram." The plural is "▱."

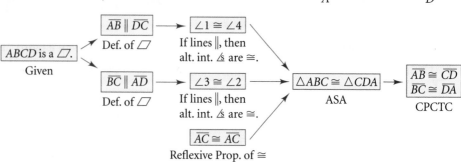

Angles of a polygon that share a side are **consecutive angles.** A parallelogram has opposite sides parallel. Its consecutive angles are same-side interior angles so they are supplementary. In ▱ABCD, consecutive angles B and C are supplementary, as are consecutive angles C and D.

1 EXAMPLE Using Consecutive Angles

Gridded Response Find $m\angle S$ in $\square RSTW$.

$\angle R$ and $\angle S$ are consecutive angles of a parallelogram.
They are supplementary.

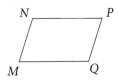

$m\angle R + m\angle S = 180$	**Definition of supplementary angles**
$112 + m\angle S = 180$	**Substitute.**
$m\angle S = 68$	**Subtract 112 from each side.**

 Quick Check ❶ **Critical Thinking** If consecutive angles of a quadrilateral are supplementary, must the quadrilateral be a parallelogram? Explain.

A proof of Theorem 6-2 uses the consecutive angles of a parallelogram, and the fact that supplements of the same angle are congruent.

🔑 **Key Concepts**

Theorem 6-2
Opposite angles of a parallelogram are congruent.

Plan for Proof of Theorem 6-2

Given: $\square MNPQ$

Prove: $\angle M \cong \angle P$ and $\angle N \cong \angle Q$

Plan: $\angle M \cong \angle P$ if they are supplements of the same angle, $\angle N$. Each is a supplement of $\angle N$ because same side interior angles are supplementary.
$\angle N \cong \angle Q$ using similar reasoning with $\angle M$.

You can use this plan to write a proof of Theorem 6-2 in Exercise 36. If you choose to write a flow proof, you'll find a guide on page 319.

You can use Theorems 6-1 and 6-2 along with algebra to find unknown values in parallelograms.

Real-World 🌐 Connection

Opposite angles in the "cat's cradle" parallelogram (center) are congruent.

2 EXAMPLE Using Algebra

Algebra Find the value of x in $\square PQRS$. Then find QR and PS.

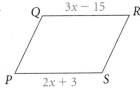

$3x - 15 = 2x + 3$	**Opposite sides of a $\square$ are congruent.**
$x - 15 = 3$	**Subtract 2x from each side.**
$x = 18$	**Add 15 to each side.**
$QR = 3x - 15 = 39$	**Substitute.**

$\overline{PS} \cong \overline{QR}$, so $PS = 39$.

 **Quick Check** ❷ Find the value of y in $\square EFGH$. Then find $m\angle E, m\angle G, m\angle F,$ and $m\angle H$.

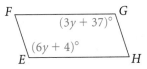

The diagonals of parallelograms have a special property.

 Key Concepts

Theorem 6-3

The diagonals of a parallelogram bisect each other.

 Proof

Proof of Theorem 6-3

Given: $\square ABCD$

Prove: $\overline{AC}$ and $\overline{BD}$ bisect each other at E.

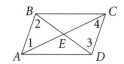

Statements	Reasons
1. $ABCD$ is a parallelogram.	1. Given
2. $\overline{AB} \parallel \overline{DC}$	2. Definition of parallelogram
3. $\angle 1 \cong \angle 4; \angle 2 \cong \angle 3$	3. Parallel lines form $\cong$ alt. int. $\angle$s.
4. $\overline{AB} \cong \overline{DC}$	4. Opposite sides of a $\square$ are $\cong$.
5. $\triangle ABE \cong \triangle CDE$	5. ASA
6. $\overline{AE} \cong \overline{CE}; \overline{BE} \cong \overline{DE}$	6. CPCTC
7. $\overline{AC}$ and $\overline{BD}$ bisect each other at E.	7. Definition of bisector

You can use Theorem 6-3 to find unknown lengths in parallelograms.

Real-World Connection

The railing braces are diagonals of parallelograms so they bisect each other.

3 **EXAMPLE** **Using Algebra**

Solve a system of linear equations to find the values of x and y in $\square ABCD$. Then find $AE, EC, BE,$ and ED.

① $3y - 7 = 2x$	The diagonals of a
② $y = x + 1$	parallelogram bisect each other.
$3(x + 1) - 7 = 2x$	Substitute $x + 1$ for y in equation ①.
$3x + 3 - 7 = 2x$	Distribute.
$3x - 4 = 2x$	Simplify.
$3x = 2x + 4$	Add 4 to each side.
$x = 4$	Subtract $2x$ from each side.
$3y - 7 = 2(4) = 8$	Substitute 4 for x in equations ① and ②.
$y = 4 + 1 = 5$	

● $AE = EC = 8$ and $BE = ED = 5$.

✓ **Quick Check** **3** Find the values of a and b.

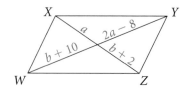

In Exercise 52, you will use Theorem 6-1, opposite sides of a parallelogram are congruent, to prove the following theorem.

Theorem 6-4

If three (or more) parallel lines cut off congruent segments on one transversal, then they cut off congruent segments on every transversal.

$$\overline{BD} \cong \overline{DF}$$

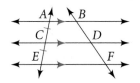

4 EXAMPLE **Real-World**  **Connection**

Measurement Show how to separate a blank card into three strips that are the same height by using lined paper, a straightedge, and Theorem 6-4.

The lines of the paper are parallel and equally spaced. Place a corner of the top edge of the card on the first line of the paper. Place the corner of the bottom edge on the fourth line. Mark the points where the second and third lines intersect the card. The marks will be equally spaced because the edge of the card is a transversal for the equally spaced parallel lines of the paper. Repeat for the other side of the card. Connect the marks using a straightedge.

✓ Quick Check **4** In the figure at the right, $\overleftrightarrow{DH} \parallel \overleftrightarrow{CG} \parallel \overleftrightarrow{BF} \parallel \overleftrightarrow{AE}$, $AB = BC = CD = 2$, and $EF = 2.5$. Find EH.

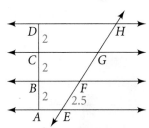

EXERCISES

For more exercises, see *Extra Skill, Word Problem, and Proof Practice.*

Practice and Problem Solving

A **Practice by Example** $\boxed{x^2}$ **Algebra** **Find the value of x in each parallelogram.**

Example 1
(page 313)

GO **for Help**

1.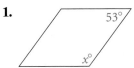

2.

3.

4.

5.

6.

Example 2 $\boxed{x^2}$ **Algebra** **Find the value of x and the length of each side.**
(page 313)

7.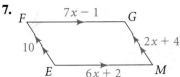

8.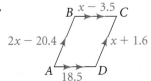

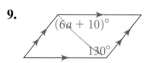 **Algebra** Find the value of a.

9.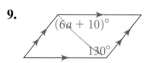

10.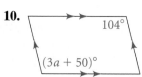

11.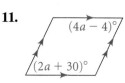

Find the value of a and the measure of each angle in each parallelogram.

12.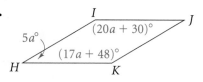

13.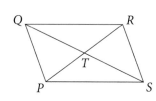

Example 3
(page 314)

x^2 **Algebra** Find the values of x and y in $\square PQRS$.

14. $PT = 2x, TR = y + 4, QT = x + 2, TS = y$

15. $PT = x + 2, TR = y, QT = 2x, TS = y + 3$

16. $PT = y, TR = x + 3, QT = 2y, TS = 3x - 1$

17. $PT = 2x, TR = y + 3, QT = 3x, TS = 2y$

18. $PT = 8x, TR = 6y, QT = 2x + 2, TS = 2y$

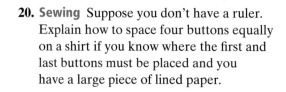

Example 4
(page 315)

19. Find ED and FD in the figure at the right.

20. Sewing Suppose you don't have a ruler.
Explain how to space four buttons equally
on a shirt if you know where the first and
last buttons must be placed and you
have a large piece of lined paper.

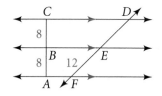

In the figure, the horizontal lines are parallel and $PQ = QR = RS$.
Find each length.

21. ZU **22.** XZ

23. XU **24.** TZ

25. TU **26.** XV

27. YX **28.** YV

29. WX **30.** WV

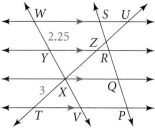

B **Apply Your Skills** x^2 **Algebra** Use the given information to find the lengths of all four sides of $\square ABCD$.

31. The perimeter is 48 in. AB is 5 in. less than BC.

32. The perimeter is 92 cm. AD is 7 cm more than twice AB.

33. Multiple Choice What is the value of x in
the parallelogram?

    Ⓐ 15   Ⓑ 45

    Ⓒ 60   Ⓓ 135

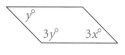

 34. Writing Explain how to find the measures of the remaining three angles of a
parallelogram if you already know the measure of one of the angles.

35. Developing Proof Complete this paragraph proof of Theorem 6-1 by filling in the blanks.

Given: ▱ABCD

Prove: $\overline{AB} \cong \overline{CD}$ and $\overline{BC} \cong \overline{DA}$

Proof: ABCD is a parallelogram, therefore $\overline{AB} \parallel$ **a.** ___?___ and $\overline{BC} \parallel$ **b.** ___?___. ∠1 ≅ ∠4 and ∠3 ≅ ∠2, because alternate interior angles are **c.** ___?___. $\overline{AC} \cong \overline{AC}$ by the **d.** ___?___ Property of Congruence. Therefore △ABC ≅ △CDA by **e.** ___?___ . So, $\overline{AB} \cong \overline{CD}$ and $\overline{BC} \cong \overline{DA}$ because **f.** ___?___ .

For a guide to solving Exercise 36, see p. 319.

Proof 36. Write a proof for Theorem 6-2. You may wish to follow the plan on page 313.

Find the measures of the numbered angles for each parallelogram.

37.

38.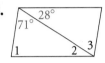

39.

40. Error Analysis Brian states that $QV = 10$ cm in the figure at the right. Explain why Brian's statement may not be correct.

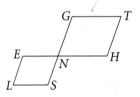

x^2 41. Algebra In a parallelogram one angle is 9 times the size of another. Find the measures of the angles.

Proof Write a paragraph proof, flow proof, or two-column proof.

42. Given: ▱LENS and ▱NGTH
   Prove: ∠L ≅ ∠T

43. Given: ▱LENS and ▱NGTH
   Prove: $\overline{LS} \parallel \overline{GT}$

44. Given: ▱LENS and ▱NGTH
   Prove: ∠E is supplementary to ∠T.

Problem Solving Hint

For each of Exercises 42–44, sketch the diagram and mark it as you think through a proof.

x^2 Algebra Find the value(s) of the variable(s) in each parallelogram.

45.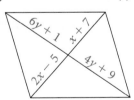

46. $AC = 4x + 10$

47.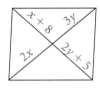

Proof Write a paragraph proof, flow proof, or two-column proof.

48. Given: ▱RSTW and ▱XYTZ
   Prove: ∠R ≅ ∠X

49. Given: ▱RSTW and ▱XYTZ
   Prove: $\overline{XY} \parallel \overline{RS}$

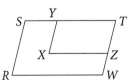

50. Given: ▱ABCD and $\overline{AC}$ bisects ∠DAB.
   Prove: $\overline{AC}$ bisects ∠DCB.

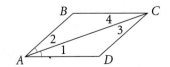

GO Online
Homework Video Tutor
Visit: PHSchool.com
Web Code: aue-0602

51. a. Open-Ended Sketch two parallelograms whose corresponding sides are congruent but whose corresponding angles are not congruent.

b. Critical Thinking Is there an SSSS congruence theorem for parallelograms? Explain.

Proof **52.** Prove Theorem 6-4.

Given: $\overleftrightarrow{AB} \parallel \overleftrightarrow{CD} \parallel \overleftrightarrow{EF}$ and $\overline{AC} \cong \overline{CE}$

Prove: $\overline{BD} \cong \overline{DF}$

(*Hint:* Draw lines through B and D parallel to $\overleftrightarrow{AE}$ and intersecting $\overleftrightarrow{CD}$ at G and $\overleftrightarrow{EF}$ at H.)

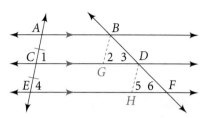

Proof **53. a.** Prove that if two sides and the included angle of one parallelogram are congruent to corresponding parts of another parallelogram, then the parallelograms are congruent. (*Hint:* Prove that all the corresponding parts of the parallelograms are congruent.)

b. Is there a theorem similar to SAS for trapezoids? Explain.

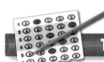

Test Prep

Gridded Response   Use the parallelogram at the right for Exercises 54–57. Find the indicated segment length or angle measure.

54. *JM*   **55.** *ML*   **56.** $m\angle L$ **57.** $m\angle J$

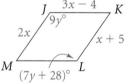

58. The measures of three angles in a parallelogram are 20, 160, and 20. Find the measure of the fourth angle.

59. The measures of two angles in a parallelogram are 32 and 32. Find the measure of one of the other two angles.

60. Two consecutive angles in a parallelogram have measures $x + 5$ and $4x - 10$. Find the measure of the smaller angle.

Mixed Review

GO for Help

Lesson 6-1   Determine the most precise name for each figure.

61.

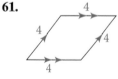

62.

Lesson 4-6   **63.** What additional information do you need to prove $\triangle ADC \cong \triangle ABC$ by the HL Theorem?

Lesson 3-1   In the figure at the right, $\overleftrightarrow{PQ} \parallel \overleftrightarrow{RS}$. Find each measure.

64. $m\angle 1$ **65.** $m\angle 2$ **66.** $m\angle 3$ **67.** $m\angle 4$

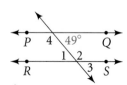

Understanding Proof Problems Let Ravi's thinking guide you through a flow proof for Theorem 6-2. He follows the plan on page 313, also shown below. Check your understanding with the exercise at the bottom of the page.

Theorem 6-2 Opposite angles of a parallelogram are congruent.

Given: ▱MNPQ

Prove: ∠M ≅ ∠P and ∠N ≅ ∠Q

Plan: ∠M ≅ ∠P if they are supplements of the same angle, ∠N. Each is a supplement of ∠N because same-side interior angles are supplementary. ∠N ≅ ∠Q using similar reasoning with ∠M.

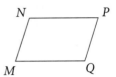

What Ravi Thinks

The second sentence of the plan is

"Each is a supplement of ∠N because same-side interior angles are supplementary."

This tells me where to begin. Opposite sides of a parallelogram are parallel while the other sides act as transversals. The opposite sides and transversals form same-side interior angles.

Same-side interior angles are supplementary and supplements of the same angle are congruent.

This gives me half the proof.

The last sentence of the plan is

"∠N ≅ ∠Q using similar reasoning with ∠M."

This tells me I can repeat the first part of the proof, but with different angles.

I have found a clever way to add this second part to my flow proof. It's in red.

What Ravi Writes

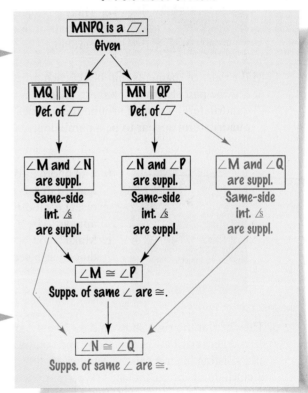

EXERCISE

For Exercise 42 on page 317, write a proof that follows the plan shown below.

Given: ▱LENS and ▱NGTH

Prove: ∠L ≅ ∠T

Plan: ∠L ≅ ∠T if their opposite angles, ∠ENS and ∠HNG, are congruent. ∠ENS ≅ ∠HNG because they are vertical angles.

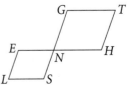

Geo-Models

By definition, a quadrilateral is a parallelogram if both pairs of opposite sides are parallel. You can use a geoboard or geopaper to explore other conditions that seem to force a quadrilateral to be a parallelogram.

NY G.CN.1: Understand and make connections among multiple representations of the same mathematical idea.

1. On geopaper or a geoboard, make pairs of segments that are congruent and parallel.

   a. The figure at the right shows one pair. Imagine the endpoints connected to form a quadrilateral. Does the quadrilateral appear to be a parallelogram?

   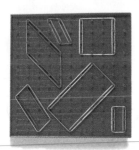
   b. The photo at the left shows other pairs of congruent parallel segments. A large elastic band wraps around each pair to form a quadrilateral. Does each quadrilateral appear to be a parallelogram?
   c. Make a conjecture about a quadrilateral that has one pair of opposite sides congruent and parallel.

2. Make pairs of segments that bisect each other.

   a. For the pair shown at the right, imagine a quadrilateral wrapped around them. Does the quadrilateral appear to be a parallelogram?

   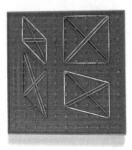
   b. The photo at the left shows four more quadrilaterals wrapped around diagonals that bisect each other. Does each quadrilateral appear to be a parallelogram?
   c. Make a conjecture about a quadrilateral whose diagonals bisect each other.

3. The chart at the right shows a 2-by-2 geoboard and a 3-by-3 geoboard at the top. Copy and complete the chart, listing in each cell the number of different quadrilaterals of the indicated type that you can find on each size geoboard. Two quadrilaterals are different if they are not congruent. One cell has been completed for you.

4. How many different squares can you find on a 4-by-4 geoboard?

5. The vertices of a cube form a 2-by-2-by-2 *lattice* in space. How many different squares can you find whose vertices are points of this 2-by-2-by-2 lattice?

6. How many different squares can you find in a 3-by-3-by-3 lattice?

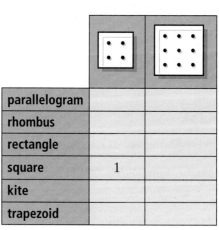

parallelogram		
rhombus		
rectangle		
square	1	
kite		
trapezoid		

6-3

Proving That a Quadrilateral Is a Parallelogram

Learning Standards for Mathematics

G.G.38 Investigate, justify, and apply theorems about parallelograms involving their angles, sides, and diagonals.

G.G.41 Justify that some quadrilaterals are parallelograms, rhombuses, rectangles, squares, or trapezoids.

✔ **Check Skills You'll Need**

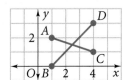

GO for Help Lessons 1-8 and 3-7

Use the figure at the right.

1. Find the coordinates of the midpoints of $\overline{AC}$ and $\overline{BD}$. What is the relationship between $\overline{AC}$ and $\overline{BD}$?

2. Find the slopes of $\overline{BC}$ and $\overline{AD}$. How do they compare?

3. Are $\overline{AB}$ and $\overline{DC}$ parallel? Explain.

4. What type of figure is $ABCD$?

---

1 ▸ Is the Quadrilateral a Parallelogram?

Theorems 6-5 and 6-6 are converses of Theorems 6-1 and 6-2, respectively, from the previous lesson. They provide two ways to conclude that a quadrilateral is a parallelogram.

 Key Concepts

> **Theorem 6-5**
>
> If both pairs of opposite sides of a quadrilateral are congruent, then the quadrilateral is a parallelogram.

Real-World Connection

The frame remains a parallelogram as it is raised and lowered, and the backboard stays vertical.

Proof of Theorem 6-5

Given: $\overline{WX} \cong \overline{ZY}$ and $\overline{XY} \cong \overline{WZ}$

Prove: $WXYZ$ is a parallelogram.

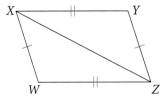

Proof: Draw diagonal $\overline{XZ}$. Since opposite sides of $WXYZ$ are congruent, $\triangle WXZ \cong \triangle YZX$ by SSS. Using CPCTC, $\angle WXZ \cong \angle YZX$, so $\overline{WX} \parallel \overline{ZY}$. Also, $\angle WZX \cong \angle YXZ$, so $\overline{WZ} \parallel \overline{XY}$. $WXYZ$ is a parallelogram by definition.

You will complete a proof of Theorem 6-6 in Exercise 12.

 Key Concepts

> **Theorem 6-6**
>
> If both pairs of opposite angles of a quadrilateral are congruent, then the quadrilateral is a parallelogram.

Theorem 6-7 is the converse of Theorem 6-3 of the previous lesson.

 **Key Concepts**

> **Theorem 6-7**
>
> If the diagonals of a quadrilateral bisect each other, then the quadrilateral is a parallelogram.

Proof

Proof of Theorem 6-7

Given: $\overline{AC}$ and $\overline{BD}$ bisect each other at E.

Prove: $ABCD$ is a parallelogram.

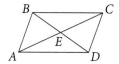

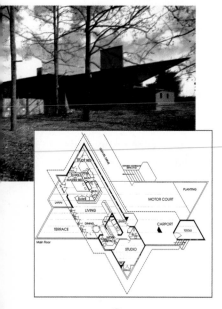

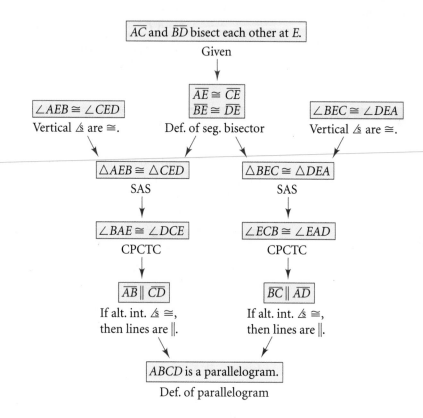

Real-World  **Connection**

Frank Lloyd Wright, a famous architect, used parallelograms in designs of many houses, such as the Kraus House in Kirkwood, Missouri.

Theorem 6-8 suggests that if you keep two objects of the same length parallel, such as cross-country skis, then the quadrilateral determined by their endpoints must be a parallelogram. You will prove Theorem 6-8 in Exercise 13.

 **Key Concepts**

> **Theorem 6-8**
>
> If one pair of opposite sides of a quadrilateral is both congruent and parallel, then the quadrilateral is a parallelogram.

You can use algebra and Theorems 6-7 and 6-8 to find values for which quadrilaterals are parallelograms.

1 EXAMPLE Finding Values for Parallelograms

Test-Taking Tip

Read test questions carefully. Here the value of y is choice C, but you are asked for the value of x.

Multiple Choice For what value of x must $MLPN$ be a parallelogram?

Ⓐ 1 Ⓑ 3 Ⓒ 9 Ⓓ 27

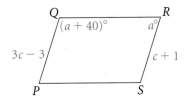

Diagonals of a parallelogram bisect each other, so $2y - 7 = y + 2$ and $3x = y$.

$2y - 7 = y + 2$ **Write the equation with one variable.**

$y - 7 = 2$ **Collect the variables on one side.**

$y = 9$ **Solve.**

Substitute 9 for y in the second equation. Find x.

$3x = y$

$3x = 9$ **Substitute 9 for y.**

$x = 3$ **Solve.**

For $x = 3$, $MLPN$ is a parallelogram. Answer B is correct.

Quick Check ❶ Find the values of a and c for which $PQRS$ must be a parallelogram.

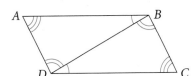

You can conclude that a quadrilateral is a parallelogram if both pairs of opposite sides are parallel. Theorems 6-5 through 6-8 provide four shortcuts to prove that a quadrilateral is a parallelogram.

2 EXAMPLE Is the Quadrilateral a Parallelogram?

For: Parallelogram Activity
Use: Interactive Textbook, 6-3

Can you prove the quadrilateral is a parallelogram from what is given? Explain.

a. Given: $\angle ABD \cong \angle CDB$, $\angle BDA \cong \angle DBC$, $\angle A \cong \angle C$
Prove: $ABCD$ is a parallelogram.

b. Given: $\overline{LM} \cong \overline{LO}, \overline{NM} \cong \overline{ON}$
Prove: $LMNO$ is a parallelogram.

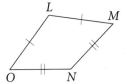

Yes, both pairs of opposite angles are congruent. $ABCD$ is a parallelogram by Theorem 6-6.

No, the given information is not enough to prove $LMNO$ is a parallelogram.

Quick Check ❷ Can you prove the quadrilateral is a parallelogram? Explain.

a. Given: $\overline{PQ} \cong \overline{SR}$, $\overline{PQ} \parallel \overline{SR}$
Prove: $PQRS$ is a parallelogram.

b. Given: $\overline{DH} \cong \overline{GH}, \overline{EH} \cong \overline{FH}$
Prove: $DEFG$ is a parallelogram.

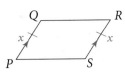
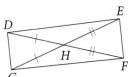

3 EXAMPLE **Real-World Connection**

Navigation A parallel rule is a navigation tool that is used to plot ship routes on charts. It is made of two rulers connected with congruent crossbars, such that $AB = DC$ and $AD = BC$. You place one ruler on the line connecting the ship's present position to its destination. Then you move the other ruler onto the chart's compass to find the direction of the route. Explain why this instrument works.

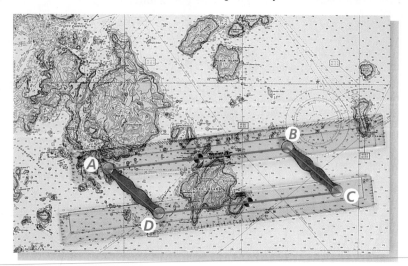

The crossbars and the sections of the rulers are congruent no matter how they are positioned. So, $ABCD$ is always a parallelogram. Since $ABCD$ is a parallelogram, the rulers are parallel. Therefore, the direction the ship should travel is the same as the direction shown on the chart's compass.

✓ Quick Check ③ **Critical Thinking** Suppose the ruler connecting the ship's position to its destination gets in the way of reading the compass. How can you get the desired reading?

EXERCISES

For more exercises, see *Extra Skill, Word Problem, and Proof Practice.*

Practice and Problem Solving

Ⓐ **Practice by Example**  **Algebra** Find the values of x and y for which $ABCD$ must be a parallelogram.

Example 1
(page 323)

1.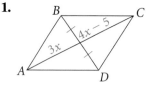

2.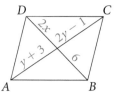

3.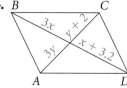

4.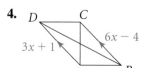

5.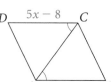

6.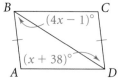

Example 2
(page 323)

Determine whether the quadrilateral must be a parallelogram. Explain.

7.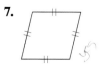

8.

9.

Example 3
(page 324)

10. Fishing Quadrilaterals are formed on the side of this fishing tackle box by the adjustable shelves and connecting pieces. Explain why the quadrilaterals remain parallelograms no matter what position the shelves are in.

 **Apply Your Skills**

11. Combine each of Theorems 6-1, 6-2, and 6-3 with Theorems 6-5, 6-6, and 6-7, respectively, into biconditional statements.

12. Developing Proof Complete the proof of Theorem 6-6.

Given: $\angle A \cong \angle C$ and $\angle B \cong \angle D$
Prove: $ABCD$ is a parallelogram.

Statements	Reasons
1. $x + y + x + y = 360$	**1.** The sum of the measures of the angles of a quadrilateral = 360.
2. $2(x + y) = 360$	**a.** ?
3. $x + y = 180$	**b.** ?
4. $\angle A$ and $\angle B$ are supplementary. $\angle A$ and $\angle D$ are supplementary.	**4.** Definition of supplementary
c. ? $\parallel$? , ? $\parallel$?	**d.** ?
6. $ABCD$ is a parallelogram.	**e.** ?

Proof **13.** Prove Theorem 6-8.

Given: $\overline{TW} \parallel \overline{YX}$ and $\overline{TW} \cong \overline{YX}$
Prove: $TWXY$ is a parallelogram.
(*Hint:* Draw one or both diagonals. Find congruent triangles. Use CPCTC.)

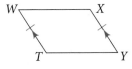

x^2 **Algebra** **Find the values of the variables for which *ABCD* must be a parallelogram.**

14.

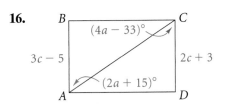

15.

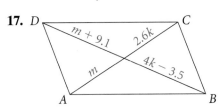

16.

17.

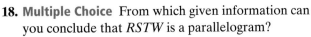

18. Multiple Choice From which given information can you conclude that $RSTW$ is a parallelogram?

- **A** $\overline{RS} \parallel \overline{WT}, \overline{RS} \cong \overline{ST}$
- **B** $\overline{RS} \parallel \overline{WT}, \overline{ST} \cong \overline{RW}$
- **C** $\overline{RS} \cong \overline{ST}, \overline{RW} \cong \overline{WT}$
- **D** $\overline{RZ} \cong \overline{TZ}, \overline{SZ} \cong \overline{WZ}$

19. Open-Ended Sketch two noncongruent parallelograms $ABCD$ and $EFGH$ such that $\overline{AC} \cong \overline{EG}$ and $\overline{BD} \cong \overline{FH}$.

GO O**nline**
Homework Video Tutor
Visit: PHSchool.com
Web Code: aue-0603

Proof **20. Given:** ∠JKN ≅ ∠LMN
∠LKN ≅ ∠JMN
Prove: JKLM is a parallelogram.

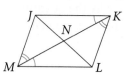

21. Given: △TRS ≅ △RTW
Prove: RSTW is a parallelogram.

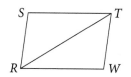

Coordinate Geometry Given points A, B, and C in the coordinate plane as shown, find the fourth point described below.

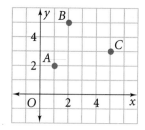

22. point D so that ABCD is a parallelogram

23. point E so that ABEC is a parallelogram

24. point F so that AFBC is a parallelogram

 25. Writing Summarize the ways to show that a quadrilateral is a parallelogram.

C Challenge **26. Probability** If two opposite angles of a quadrilateral measure 120 and the measures of the other angles are multiples of 10, what is the probability that the quadrilateral is a parallelogram?

Proof **27.** In the figure at the right, point D is constructed by drawing two arcs. One has center C and radius AB. The other has center B and radius AC. Prove that $\overline{AM}$ is a median of △ABC.

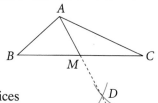

28. Coordinate Geometry The diagonals of quadrilateral EFGH intersect at D(−1, 4). Two vertices of EFGH are E(2, 7) and F(−3, 5). What must be the coordinates of G and H to ensure that EFGH is a parallelogram?

NY REGENTS

Test Prep

Multiple Choice

29. In ▱PNWS, what is m∠W?
   A. 128   **B.** 90   **C.** 52   **D.** 26

30. In ▱PNWS, what is m∠S?
   F. 128   **G.** 90   **H.** 52   **J.** 26

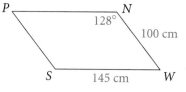

Short Response

31. Given: △NRJ ≅ △CPT, $\overline{JN} \parallel \overline{CT}$
Prove: JNTC is a parallelogram.

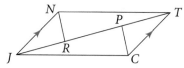

Extended Response

32. a. Write an equation and solve for x.
   b. Is $\overline{AF} \parallel \overline{DE}$? Explain.
   c. Is BDEF a parallelogram? Explain.

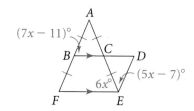

Lesson 6-2 x^2 **Algebra** Find the value of each variable in each parallelogram.

33.

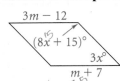

34.

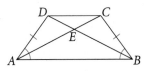

35.

Lesson 4-7

36. Explain how you can use overlapping congruent triangles to prove $\overline{AC} \cong \overline{BD}$.

Lesson 2-2

Write the two conditional statements that make up each biconditional.

37. The diagonals of a quadrilateral bisect each other if and only if the quadrilateral is a parallelogram.

38. Two lines are parallel if and only if the two lines and a transversal form corresponding angles that are congruent.

39. Two nonvertical lines are perpendicular if and only if the product of their slopes is -1.

✓ Checkpoint Quiz 1　　　　　　　　　Lessons 6-1 through 6-3

Find the measures of the numbered angles for each parallelogram.

1.

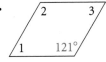

2.

3.

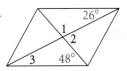

Classify each quadrilateral in as many ways as possible.

4.

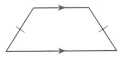

5.

6.

x^2 **Algebra** Find the values of the variables for which $ABCD$ is a parallelogram.

7.

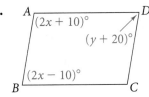

8.

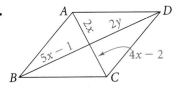

9. In the figure at the right, $\overleftrightarrow{AB} \parallel \overleftrightarrow{CD} \parallel \overleftrightarrow{EF}$. Find AE.

10. What is the most precise name for a quadrilateral with vertices $(3, 5)$, $(-1, 4)$, $(3, -5)$, and $(7, 4)$?

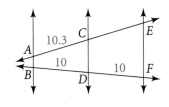

Diagonals of Parallelograms

Construct

Use geometry software to construct a parallelogram.

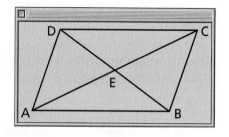

NY **G.PS.3:** Use multiple representations to represent and explain problem situations.

- Construct segments $\overline{AB}$ and $\overline{BC}$.
- Construct a line through C parallel to $\overline{AB}$ and a line through A parallel to $\overline{BC}$. Label the point where the two lines intersect as D.
- Hide the lines and construct $\overline{AD}$ and $\overline{CD}$.
- Construct the diagonals of $\square ABCD$ and label their point of intersection E.

Set up the following measurements. Use them where indicated.

M1: $m\angle DAB$, or some other angle of the parallelogram, so that you can tell when the parallelogram is a rectangle

M2: AB and AD, or some other adjacent sides of the parallelogram, so you can tell when it is a rhombus

M3: $m\angle AED$, or some other angle at the intersection of the diagonals, so you can tell when the diagonals are perpendicular

M4: AC and DB, the diagonal lengths, so you can tell when they are equal

M5: the angles at each vertex of the parallelogram, so you can tell when the diagonals bisect the angles of the parallelogram

Investigate

- Manipulate the parallelogram to get a rectangle (M1). Make note of what appear to be any special properties of the diagonals of a rectangle. Manipulate the rectangle to check whether the properties hold (M3–5).
- Manipulate the parallelogram to get a rhombus (M2). Make note of what appear to be any special properties of the diagonals of a rhombus. Manipulate the rhombus to check whether the properties hold (M3–5).

EXERCISES

Make as many conjectures as you can about each of the following.

1. the diagonals of rectangles   **2.** the diagonals of rhombuses   **3.** the diagonals of squares

Extend

4. Manipulate the diagonals so they are perpendicular (M3). Make a conjecture about the type of parallelogram that is determined by perpendicular diagonals.

5. Manipulate the diagonals so they are congruent (M4). Make a conjecture about the type of parallelogram that is determined by congruent diagonals.

6. Manipulate the diagonals so they bisect the angles of the parallelogram (M5). Make a conjecture about the type of parallelogram that is determined by diagonals that bisect the angles.

7. Construct a trapezoid and then its two diagonals. Manipulate the trapezoid until the diagonals are the same length. Make a conjecture about the type of trapezoid that is determined by congruent diagonals.

Special Parallelograms

 Learning Standards for Mathematics

G.G.39 Investigate, justify, and apply theorems about special parallelograms (rectangles, rhombuses, squares) involving their angles, sides, and diagonals.

G.G.41 Justify that some quadrilaterals are parallelograms, rhombuses, rectangles, squares, or trapezoids.

✓ **Check Skills You'll Need**

GO for Help Lesson 6-2

PACE is a parallelogram and $m\angle PAC = 109$. Complete each of the following.

1. $EC = \blacksquare$ **2.** $EP = \blacksquare$

3. $m\angle CEP = \blacksquare$ **4.** $PR = \blacksquare$

5. $RE = \blacksquare$ **6.** $CP = \blacksquare$

7. $m\angle EPA = \blacksquare$ **8.** $m\angle ECA = \blacksquare$

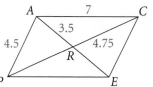

9. Draw a rhombus that is not a square. Draw a rectangle that is not a square. Explain why each is not a square.

---

1 Diagonals of Rhombuses and Rectangles

If you draw two congruent isosceles triangles with base $\overline{PQ}$, you have drawn a rhombus. Note that $\angle RPQ$, $\angle SPQ$, $\angle RQP$, and $\angle SQP$ are all congruent. This suggests Theorem 6-9 and its proof.

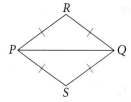

🔑 **Key Concepts**

Theorem 6-9

Each diagonal of a rhombus bisects two angles of the rhombus.

Proof **Proof of Theorem 6-9**

Given: rhombus *ABCD*

Prove: $\overline{AC}$ bisects $\angle BAD$ and $\angle BCD$.

Vocabulary Tip

A segment <u>bisects</u> an angle if and only if it divides the angle into two congruent angles.

Proof: *ABCD* is a rhombus, so its sides are all congruent. $\overline{AC} \cong \overline{AC}$ by the Reflexive Property of Congruence. Therefore, $\triangle ABC \cong \triangle ADC$ by the SSS Postulate. $\angle 1 \cong \angle 2$ and $\angle 3 \cong \angle 4$ by CPCTC. Therefore, $\overline{AC}$ bisects $\angle BAD$ and $\angle BCD$ by the definition of bisect.

You can show similarly that $\overline{BD}$ bisects $\angle ABC$ and $\angle ADC$.

The diagonals of a rhombus provide an interesting application of the Converse of the Perpendicular Bisector Theorem.

In the rhombus at the right, points B and D are equidistant from A and C. By the Converse of the Perpendicular Bisector Theorem, they are on the perpendicular bisector of $\overline{AC}$. This proves the next theorem. In Exercise 22, you will prove it a second way.

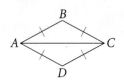

Key Concepts

Theorem 6-10

The diagonals of a rhombus are perpendicular.

$$\overline{AC} \perp \overline{BD}$$

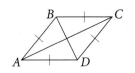

pantograph

You can use Theorems 6-9 and 6-10 to find angle measures in rhombuses.

1 EXAMPLE **Finding Angle Measures**

$MNPQ$ is a rhombus and $m\angle N = 120$.
Find the measures of the numbered angles.

$m\angle 1 = m\angle 3$	**Isosceles Triangle Theorem**
$m\angle 1 + m\angle 3 + 120 = 180$	**Triangle Angle-Sum Theorem**
$2(m\angle 1) + 120 = 180$	**Substitute.**
$2(m\angle 1) = 60$	**Subtract 120 from each side.**
$m\angle 1 = 30$	**Divide each side by 2.**

Therefore, $m\angle 1 = m\angle 3 = 30$. By Theorem 6-9, $m\angle 1 = m\angle 2$ and $m\angle 3 = m\angle 4$.
Therefore, $m\angle 1 = m\angle 2 = m\angle 3 = m\angle 4 = 30$.

Real-World  **Connection**

The diagonals of the rhombus formed by the pantograph stay perpendicular when the pantograph lifts or lowers.

✓ Quick Check ❶ Find the measures of the numbered angles in the rhombus.

The diagonals of a rectangle, another parallelogram, also have a special property.

Key Concepts

Theorem 6-11

The diagonals of a rectangle are congruent.

Proof **Proof of Theorem 6-11**

Given: Rectangle $ABCD$

Prove: $\overline{AC} \cong \overline{BD}$

Proof: $ABCD$ is a rectangle, so it is also a parallelogram.
$\overline{AB} \cong \overline{DC}$ because opposite sides of a parallelogram are congruent.
$\overline{BC} \cong \overline{BC}$ by the Reflexive Property of Congruence.
$\angle ABC$ and $\angle DCB$ are right angles by the definition of rectangle.
$\angle ABC \cong \angle DCB$ because all right angles are congruent.
$\triangle ABC \cong \triangle DCB$ by SAS. $\overline{AC} \cong \overline{BD}$ by CPCTC.

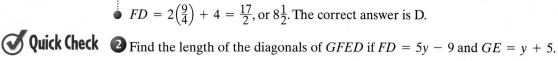

② EXAMPLE Finding Diagonal Length

Multiple Choice Find the length of $\overline{FD}$ in rectangle $GFED$ if $FD = 2y + 4$ and $GE = 6y - 5$.

Ⓐ $2\frac{1}{2}$ Ⓑ $3\frac{1}{4}$ Ⓒ $4\frac{1}{2}$ Ⓓ $8\frac{1}{2}$

$2y + 4 = 6y - 5$ **Diagonals of a rectangle are congruent.**

$9 = 4y$ **Subtract 2y from each side and add 5 to each side.**

$\frac{9}{4} = y$ **Divide each side by 4.**

● $FD = 2\left(\frac{9}{4}\right) + 4 = \frac{17}{2}$, or $8\frac{1}{2}$. The correct answer is D.

 Quick Check ❷ Find the length of the diagonals of *GFED* if $FD = 5y - 9$ and $GE = y + 5$.

---

❷ Is the Parallelogram a Rhombus or a Rectangle?

The following theorems are the converses of Theorems 6-9, 6-10, and 6-11, respectively. You will prove these theorems in the Exercises.

🔑 **Key Concepts**

> **Theorem 6-12**
>
> If one diagonal of a parallelogram bisects two angles of the parallelogram, then the parallelogram is a rhombus.
>
> **Theorem 6-13**
>
> If the diagonals of a parallelogram are perpendicular, then the parallelogram is a rhombus.
>
> **Theorem 6-14**
>
> If the diagonals of a parallelogram are congruent, then the parallelogram is a rectangle.

You can use Theorems 6-12, 6-13, and 6-14 to classify parallelograms.

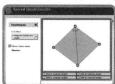

For: Quadrilateral Activity
Use: Interactive Textbook, 6-4

③ EXAMPLE Identifying Special Parallelograms

Can you conclude that the parallelogram is a rhombus or a rectangle? Explain.

a.

Yes. A diagonal bisects two angles. By Theorem 6-12, this parallelogram is a rhombus.

b.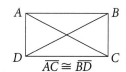

Yes. The diagonals are congruent. By Theorem 6-14, this parallelogram is a rectangle.

 **Quick Check** ❸ A parallelogram has angles of 30°, 150°, 30°, and 150°. Can you conclude that it is a rhombus or a rectangle? Explain.

You can use properties of diagonals to construct special parallelograms.

4 EXAMPLE **Real-World** **Connection**

Community Service Builders use properties of diagonals to "square up" rectangular shapes like building frames and playing-field boundaries.

Suppose you are on the volunteer building team at the right. You are helping to lay out a rectangular patio. Explain how to use properties of diagonals to locate the four corners.

To locate the corners, you can use two theorems:

- Theorem 6-7: If the diagonals of a quadrilateral bisect each other, then the quadrilateral is a parallelogram.

- Theorem 6-14: If the diagonals of a parallelogram are congruent, then the parallelogram is a rectangle.

First, cut two pieces of rope that will be the diagonals of the foundation rectangle. Cut them the same length because of Theorem 6-14. Join them at their midpoints because of Theorem 6-7. Then pull the ropes straight and taut. The ends of the ropes will be the corners of a rectangle.

Real-World  **Connection**

A well-planned volunteer effort can frame a small house in a day.

✓ **Quick Check** **4** Kate thinks that they can adapt this method slightly to stake off a square play area. Is she right? Explain.

EXERCISES

For more exercises, see *Extra Skill, Word Problem, and Proof Practice.*

Practice and Problem Solving

 Practice by Example

Example 1
(page 330)

 for Help

Find the measures of the numbered angles in each rhombus.

1.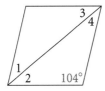

2.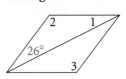

3.

4.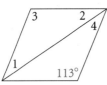

5.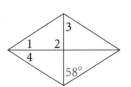

6.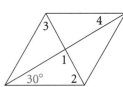

7.

8.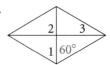

9.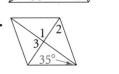

Example 2
(page 331)

x^2 **Algebra** *LMNP* **is a rectangle. Find the value of x and the length of each diagonal.**

10. $LN = x$ and $MP = 2x - 4$

11. $LN = 5x - 8$ and $MP = 2x + 1$

12. $LN = 3x + 1$ and $MP = 8x - 4$

13. $LN = 9x - 14$ and $MP = 7x + 4$

14. $LN = 7x - 2$ and $MP = 4x + 3$

15. $LN = 3x + 5$ and $MP = 9x - 10$

Example 3
(page 331)

Is the parallelogram a rhombus or a rectangle? Justify your answer.

16. 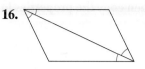 **17.** 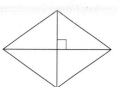 **18.**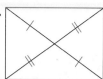

Example 4
(page 332)

19. Hardware You can use a simple device called a turnbuckle to "square up" structures that are parallelograms. For the gate pictured at the right, you tighten or loosen the turnbuckle on the diagonal cable so that the cable stays congruent to the other diagonal. Explain why a frame that normally is rectangular will, when it sags, keep the shape of a parallelogram.

Turnbuckle

20. Carpentry A carpenter is building a bookcase. How can she use a tape measure to check that the bookshelf is rectangular? Justify your answer and name any theorems used.

B Apply Your Skills

21. Reasoning Suppose the diagonals of a parallelogram are both perpendicular and congruent. What type of special quadrilateral is it? Explain your reasoning.

22. Developing Proof Complete the flow proof of Theorem 6-10.

Given: *ABCD* is a rhombus.
Prove: $\overline{AC} \perp \overline{BD}$

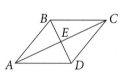

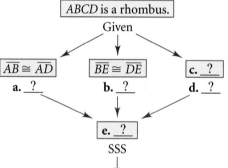

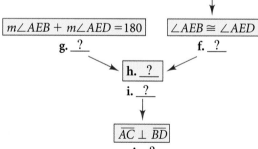

Proof **23.** Prove Theorem 6-13.

Given: $\square ABCD; \overline{AC} \perp \overline{BD}$ at *E*.
Prove: *ABCD* is a rhombus.

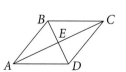

GO **O**nline
Homework Video Tutor
Visit: PHSchool.com
Web Code: aue-0604

24. Multiple Choice A diagonal of a parallelogram bisects one angle of the parallelogram. What kind of quadrilateral must the figure be?
Ⓐ rhombus　Ⓑ rectangle　Ⓒ square　Ⓓ cannot tell

Using Symbols Create your own distinctive symbols for a parallelogram, rhombus, rectangle, and square. Then copy the properties in Exercises 25–34. After each property, use your symbols to list the quadrilaterals having that property.

25. All sides are ≅.

26. Opposite sides are ≅.

27. Opposite sides are ∥.

28. Opposite ⩘ are ≅.

29. All ⩘ are right ⩘.

30. Consecutive ⩘ are supplementary.

31. Diagonals bisect each other.

32. Diagonals are ≅.

33. Diagonals are ⊥.

34. Each diagonal bisects opposite ⩘.

Which, if any, of the properties in Exercises 25–34 can the following type of quadrilateral have? Draw diagrams to illustrate.

35. a trapezoid

36. a kite

37. a quadrilateral that is not a special quadrilateral

 38. Writing Summarize the properties of squares that follow from a square being **(a)** a parallelogram, **(b)** a rhombus, and **(c)** a rectangle.

Proof 39. Prove Theorem 6-12.

Given: $ABCD$ is a parallelogram; $\overline{AC}$ bisects ∠BAD and ∠BCD.

Prove: $ABCD$ is a rhombus.

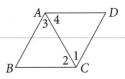

Proof 40. Prove Theorem 6-14.

Given: $\square ABCD$; $\overline{AC} \cong \overline{BD}$

Prove: $ABCD$ is a rectangle.

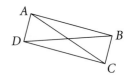

GO for Help

To review what makes a good definition, see Lesson 2-2.

Reasoning Decide whether each of these is a good definition. Justify your answer.

41. A rectangle is a quadrilateral with four right angles.

42. A rhombus is a quadrilateral with four congruent sides.

43. A square is a quadrilateral with four right angles and four congruent sides.

x^2 **Algebra** Find the value(s) of the variable(s) for each parallelogram.

44. $RZ = 2x + 5$, $SW = 5x - 20$

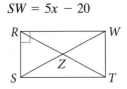

45. $m∠1 = 3y - 6$

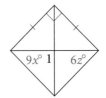

46. $BD = 4x - y + 1$

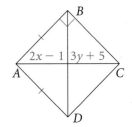

Open-Ended Given two segments with lengths a and b ($a \neq b$), what special quadrilaterals can you sketch that meet these conditions? Show each sketch.

47. Both diagonals have length a.

48. The two diagonals have lengths a and b.

49. One diagonal has length a, one side of the quadrilateral has length b.

Algebra *ABCD* is a rectangle. Find the length of each diagonal.

50. $AC = 2(x - 3)$ and $BD = x + 5$ **51.** $AC = 2(5a + 1)$ and $BD = 2(a + 1)$

52. $AC = \frac{3y}{5}$ and $BD = 3y - 4$ **53.** $AC = \frac{3c}{9}$ and $BD = 4 - c$

Constructions Explain how to construct each figure, given its diagonals.

54. parallelogram   **55.** rectangle   **56.** rhombus

57. square   **58.** kite   **59.** trapezoid

 **Challenge**

Determine whether the quadrilateral can be a parallelogram. If not, write *impossible*. Explain.

60. The diagonals are congruent, but the quadrilateral has no right angles.

61. Each diagonal is 3 cm long and two opposite sides are 2 cm long.

62. Two opposite angles are right angles, but the quadrilateral is not a rectangle.

Proof **63.** In Theorem 6-12, replace "two angles" with "one angle." Write a paragraph that proves this new statement true or show a counterexample to prove it false.

 Test Prep

Multiple Choice

64. The diagonals of a quadrilateral are perpendicular bisectors of each other. What name best describes the quadrilateral?
    A. rectangle   **B.** parallelogram   **C.** quadrilateral   **D.** rhombus

65. The diagonals of a quadrilateral bisect both pairs of opposite angles. What name best describes the quadrilateral?
    F. parallelogram   **G.** quadrilateral   **H.** rectangle   **J.** rhombus

Short Response

66. Given: *QRST* is a rhombus, $\overline{QS}$ intersects $\overline{RT}$ at *P*, *QR* = 9 cm, and *QP* = 4.5 cm. Find $m\angle RST$. Explain your work.

Mixed Review

Lesson 6-3

Can you conclude that the quadrilateral is a parallelogram? Explain.

67. 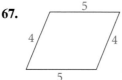   **68.**   **69.**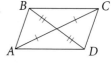

Lesson 5-1

In △*PQR*, points *S*, *T*, and *U* are midpoints. Complete each statement.

70. $TQ = \underline{\ ?\ }$ **71.** $PQ = \underline{\ ?\ }$ **72.** $TU = \underline{\ ?\ }$

73. $\overline{SU} \parallel \underline{\ ?\ }$ **74.** $\overline{TU} \parallel \underline{\ ?\ }$ **75.** $\overline{PQ} \parallel \underline{\ ?\ }$

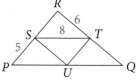

Lesson 3-4 **76. Algebra** Find the value of *c*.

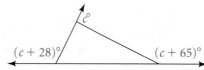

6-5

Trapezoids and Kites

Learning Standards for Mathematics

G.G.40 Investigate, justify, and apply theorems about trapezoids (including isosceles trapezoids) involving their angles, sides, medians, and diagonals.

G.G.41 Justify that some quadrilaterals are parallelograms, rhombuses, rectangles, squares, or trapezoids.

✓ **Check Skills You'll Need**

GO▶ for Help Lesson 6-1

x^2 **Algebra** **Find the values of the variables. Then find the lengths of the sides.**

1.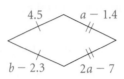

2.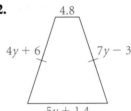

3.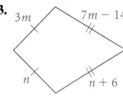

◀)) **New Vocabulary** • base angles of a trapezoid

---

1 Properties of Trapezoids and Kites

The parallel sides of a trapezoid are its bases. The nonparallel sides are its legs. Two angles that share a base of a trapezoid are **base angles** of the trapezoid.

The following theorem is about each pair of base angles. You will be asked to prove it in Exercise 38.

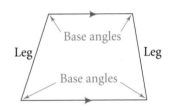

🔑 **Key Concepts**

Theorem 6-15
The base angles of an isosceles trapezoid are congruent.

The bases of a trapezoid are parallel. Therefore the two angles that share a leg are supplementary. This fact and Theorem 6-15 allow you to solve problems involving the angles of a trapezoid.

 EXAMPLE **Finding Angle Measures in Trapezoids**

$ABCD$ is an isosceles trapezoid and $m\angle B = 102$. Find $m\angle A, m\angle C,$ and $m\angle D$.

$m\angle A + m\angle B = 180$ **Two angles that share a leg are supplementary.**

$m\angle A + 102 = 180$ **Substitute.**

$m\angle A = 78$ **Subtract 102 from each side.**

● By Theorem 6-15, $m\angle C = m\angle B = 102$ and $m\angle D = m\angle A = 78$.

Real-World 🌐 Connection

In the isosceles trapezoids at the top of this electric tea kettle, each pair of base angles are congruent.

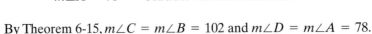

❶ In the isosceles trapezoid, $m\angle S = 70$. Find $m\angle P, m\angle Q$, and $m\angle R$.

---

② **EXAMPLE**   **Real-World** 🌐 **Connection**

Architecture The second ring of the ceiling shown at the left is made from congruent isosceles trapezoids that create the illusion of circles. What are the measures of the base angles of these trapezoids?

Each trapezoid is part of an isosceles triangle whose base angles are the acute base angles of the trapezoid. The isosceles triangle has a vertex angle that is half as large as one of the 20 angles at the center of the ceiling.

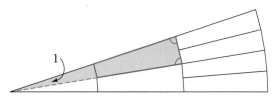

The measure of each angle at the center of the ceiling is $\frac{360}{20}$ or 18.

The measure of $\angle 1$ is $\frac{18}{2}$, or 9.

The measure of each acute base angle is $\frac{180 - 9}{2}$, or 85.5.

● The measure of each obtuse base angle is $180 - 85.5$, or 94.5.

✓ **Quick Check** ❷ A glass ceiling like the one above has 18 angles meeting at the center instead of 20. What are the measures of the base angles of the trapezoids in its second ring?

You are looking up at Harbour Centre Tower in Vancouver, Canada.

Like the diagonals of parallelograms, the diagonals of an isosceles trapezoid have a special property.

🔑 **Key Concepts**

Theorem 6-16

The diagonals of an isosceles trapezoid are congruent.

Proof

Proof of Theorem 6-16

Given: Isosceles trapezoid $ABCD$ with $\overline{AB} \cong \overline{DC}$
Prove: $\overline{AC} \cong \overline{DB}$

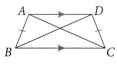

It is given that $\overline{AB} \cong \overline{DC}$. Because the base angles of an isosceles trapezoid are congruent, $\angle ABC \cong \angle DCB$. By the Reflexive Property of Congruence, $\overline{BC} \cong \overline{BC}$. Then, by the SAS Postulate, $\triangle ABC \cong \triangle DCB$. Therefore, $\overline{AC} \cong \overline{DB}$ by CPCTC.

Another special quadrilateral that is not a parallelogram is a kite. The diagonals of a kite, like the diagonals of a rhombus, are perpendicular. A proof of this for a kite (next page) is quite like its proof for a rhombus (at the top of page 330).

 **Key Concepts**

Theorem 6-17
The diagonals of a kite are perpendicular.

Proof **Proof of Theorem 6-17**

Given: Kite $RSTW$ with $\overline{TS} \cong \overline{TW}$ and $\overline{RS} \cong \overline{RW}$
Prove: $\overline{TR} \perp \overline{SW}$

Both T and R are equidistant from S and W. By the Converse of the Perpendicular Bisector Theorem, T and R lie on the perpendicular bisector of $\overline{SW}$. Since there is exactly one line through any two points (Postulate 1-1), $\overline{TR}$ must be the perpendicular bisector of $\overline{SW}$. Therefore, $\overline{TR} \perp \overline{SW}$.

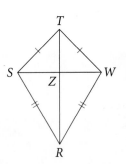

You can use Theorem 6-17 to find angle measures in kites.

3 EXAMPLE **Finding Angle Measures in Kites**

Find $m\angle 1, m\angle 2$, and $m\angle 3$ in the kite.

$m\angle 1 = 90$	**Diagonals of a kite are perpendicular.**
$90 + m\angle 2 + 32 = 180$	**Triangle Angle-Sum Theorem**
$122 + m\angle 2 = 180$	**Simplify.**
$m\angle 2 = 58$	**Subtract 122 from each side.**

$\triangle ABD \cong \triangle CBD$ by SSS.

● By CPCTC, $m\angle 3 = m\angle DBC = 32$.

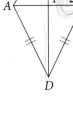

✓ Quick Check ❸ Find $m\angle 1, m\angle 2$, and $m\angle 3$ in the kite.

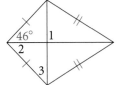

EXERCISES

For more exercises, see *Extra Skill, Word Problem, and Proof Practice.*

Practice and Problem Solving

A Practice by Example

Example 1
(page 336)

 for Help

Each trapezoid is isosceles. Find the measure of each angle.

1.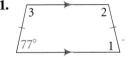

2.

3.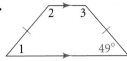

4.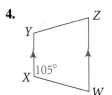

5.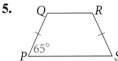

6.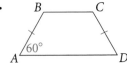

Example 2
(page 337)

7. Design Each patio umbrella is made of eight panels that are congruent isosceles triangles with parallel stripes. A sample panel is shown at the right. The vertex angle of the panel measures 42.
   a. Classify the quadrilaterals shown as blue stripes on the panel.
   b. Find the measures of the quadrilaterals' interior angles.

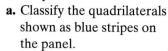

Example 3
(page 338)

Find the measures of the numbered angles in each kite.

8.

9.

10.

11.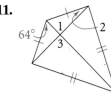

12.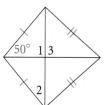

13.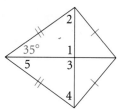

14.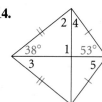

15.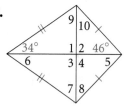

16.

Ⓑ Apply Your Skills

17. Open-Ended Sketch two kites that are not congruent, but with the diagonals of one congruent to the diagonals of the other.

18. The perimeter of a kite is 66 cm. The length of one of its sides is 3 cm less than twice the length of another. Find the length of each side of the kite.

19. Critical Thinking If $KLMN$ is an isosceles trapezoid, is it possible for $\overline{KM}$ to bisect $\angle LMN$ and $\angle LKN$? Explain.

x^2 Algebra **Find the value of the variable in each isosceles trapezoid.**

20.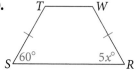

21.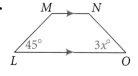

22.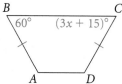

GO Online
Homework Video Tutor
Visit: PHSchool.com
Web Code: aue-0605

23.
$TV = 2x - 1$
$US = x + 2$

24.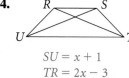
$SU = x + 1$
$TR = 2x - 3$

25.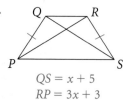
$QS = x + 5$
$RP = 3x + 3$

$\boxed{x^2}$ **Algebra** Find the value(s) of the variable(s) in each kite.

26.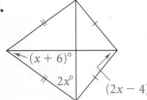

27.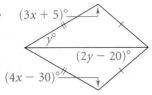

28.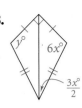

112°

Exercises 29–30

🌐 **Bridge Design** A quadrilateral is formed by the beams of the bridge at the left.

29. Classify the quadrilateral. Explain your reasoning.

30. Find the measures of the other interior angles of the quadrilateral.

Critical Thinking Can two angles of a kite be as follows? Explain.

31. opposite and acute

32. consecutive and obtuse

33. opposite and supplementary

34. consecutive and supplementary

35. opposite and complementary

36. consecutive and complementary

✏️ **37. Writing** A kite is sometimes defined as a quadrilateral with two pairs of consecutive sides congruent. Compare this to the definition you learned in Lesson 6-1. Are parallelograms, trapezoids, rhombuses, rectangles, or squares special kinds of kites according to the changed definition? Explain.

38. Developing Proof The plan suggests a proof of Theorem 6-15. Write a proof that follows the plan.

Given: Isosceles trapezoid $ABCD$ with $\overline{AB} \cong \overline{DC}$.

Prove: $\angle B \cong \angle C$ and $\angle BAD \cong \angle D$

Plan: Begin by drawing $\overline{AE} \parallel \overline{DC}$ to form parallelogram $AECD$ so that $\overline{AE} \cong \overline{DC} \cong \overline{AB}$. $\angle B \cong \angle C$ because $\angle B \cong \angle 1$ and $\angle 1 \cong \angle C$. Also, $\angle BAD \cong \angle D$ because they are supplements of the congruent angles, $\angle B$ and $\angle C$.

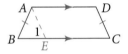

Proof Write a proof. Use the given figure with additional lines as needed.

39. Given: Isosceles trapezoid $TRAP$ with $\overline{TR} \cong \overline{PA}$

Prove: $\angle RTA \cong \angle APR$

ⓒ **Challenge**

40. Given: Isosceles trapezoid $TRAP$ with $\overline{TR} \cong \overline{PA}$; $\overline{BI}$ is the perpendicular bisector of $\overline{RA}$ intersecting $\overline{RA}$ at B and $\overline{TP}$ at I.

Prove: $\overline{BI}$ is the perpendicular bisector of $\overline{TP}$.

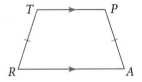

For a trapezoid, consider the segment joining the midpoints of the two given segments. How are its length and the lengths of the two parallel sides of the trapezoid related? Justify your answer.

41. the two nonparallel sides

42. the diagonals

43. $\overleftrightarrow{BN}$ is the perpendicular bisector of $\overline{AC}$ at N. Describe the set of points, D, for which $ABCD$ is a kite.

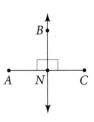

Proof 44. Prove that the angles formed by the noncongruent sides of a kite are congruent. (*Hint:* Draw a diagonal of the kite.)

Multiple Choice

45. Which statement is true for every trapezoid?
- **A.** Exactly two sides are congruent.
- **B.** Exactly two sides are parallel.
- **C.** Opposite angles are supplementary.
- **D.** The diagonals bisect each other.

46. Which statement is true for every kite?
- **F.** Opposite sides are congruent.
- **G.** At least two sides are parallel.
- **H.** Opposite angles are supplementary.
- **J.** The diagonals are perpendicular.

47. Two consecutive angles of a trapezoid are right angles. Three of the following statements about the trapezoid could be true. Which statement CANNOT be true?
- **A.** The two right angles are base angles.
- **B.** The diagonals are not congruent.
- **C.** Two of the sides are congruent.
- **D.** No two sides are congruent.

48. Quadrilateral *EFGH* is a kite. What is the value of *x*?
- **F.** 15
- **G.** 70
- **H.** 85
- **J.** 160

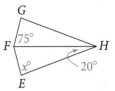

Short Response

49. In the trapezoid at the right, $BE = 2x - 8$, $DE = x - 4$, and $AC = x + 2$.
- **a.** Write and solve an equation for *x*.
- **b.** Find the length of each diagonal.

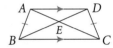

50. Diagonal $\overline{RB}$ of kite *RHBW* forms an equilateral triangle with two of the sides. $m\angle BWR = 40$. Draw and label a diagram showing the diagonal and the measures of all the angles. Which angles of the kite are largest?

Mixed Review

Lesson 6-4

Find the indicated angle measures for the rhombus.

51. $m\angle 1$ **52.** $m\angle 2$ **53.** $m\angle 3$

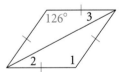

Lesson 5-2 x^2 **Algebra** Find the values indicated.

54. a. *a*
    b. *FG*
    c. *GH*

55. a. *x*
    b. *CD*
    c. *BC*

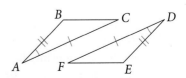

Lesson 4-2

56. State the postulate that justifies the statement $\triangle ABC \cong \triangle DEF$.

Quadrilaterals in Quadrilaterals

Construct

- Use geometry software to construct a quadrilateral *ABCD*.

- Construct the midpoint of each side of *ABCD*.

- Construct segments joining the midpoints, in order, to form quadrilateral *EFGH*.

Investigate

- Measure the lengths of the sides of *EFGH* and their slopes.

- Measure the angles of *EFGH*.

What kind of quadrilateral does *EFGH* appear to be?

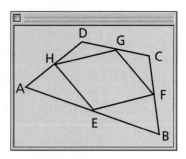

NY G.R.1: Use physical objects, diagrams, charts, tables, graphs, symbols, equations, or objects created using technology as representations of mathematical concepts.

EXERCISES

1. Manipulate quadrilateral *ABCD*.
  a. Make a conjecture about the quadrilateral whose vertices are the midpoints of the sides of a quadrilateral.
  b. Does your conjecture hold when *ABCD* is concave?
  c. Can you manipulate *ABCD* so that your conjecture doesn't hold?

Extend

2. Draw the diagonals of *ABCD*.
  a. Describe *EFGH* when the diagonals are perpendicular.
  b. Describe *EFGH* when the diagonals are congruent.
  c. Describe *EFGH* when the diagonals are both perpendicular and congruent.

3. Construct the midpoints of *EFGH* and use them to construct quadrilateral *IJKL*. Construct the midpoints of *IJKL* and use them to construct quadrilateral *MNOP*. For *MNOP* and *EFGH*, compare the ratios of the lengths of the sides, perimeters, and areas. How are the sides of *MNOP* and *EFGH* related?

4. Writing During the investigation, you made a conjecture as to the type of quadrilateral *EFGH* appears to be. Prove your conjecture. Include in your proof the Midsegment Theorem, "If a segment joins the midpoints of two sides of a triangle, then the segment is parallel to the third side and half its length."

5. Describe the quadrilateral formed by joining the midpoints, in order, of the sides of each of the following. Justify each response.
  a. parallelogram   **b.** rectangle   **c.** rhombus   **d.** square
  e. trapezoid   **f.** isosceles trapezoid   **g.** kite

6-6

Placing Figures in the Coordinate Plane

Learning Standards for Mathematics

G.PS.5 Choose an effective approach to solve a problem from a variety of strategies (numeric, graphic, algebraic).

✓ **Check Skills You'll Need**

 **for Help** Lesson 6-1

Draw a quadrilateral with the given vertices. Then determine the most precise name for each quadrilateral.

1. $H(-5, 0), E(-3, 2), A(3, 2), T(5, 0)$

2. $S(0, 0), A(4, 0), N(3, 2), D(-1, 2)$

3. $R(0, 0), A(5, 5), I(8, 4), N(7, 1)$

4. $W(-3, 0), I(0, 3), N(3, 0), D(0, -3)$

---

1 Naming Coordinates

When working with a figure in the coordinate plane, it generally is good practice to place a vertex at the origin and one side on an axis. You can also use multiples of 2 to avoid fractions when finding midpoints.

1 EXAMPLE  **Real-World**  **Connection**

T-Shirt Design An art class creates T-shirt designs by drawing quadrilaterals and connecting their midpoints to form other quadrilaterals. Tiana claims that everyone's inner quadrilateral will be a parallelogram. Is she correct? Explain.

Draw quadrilateral $OACE$ with one vertex at the origin and one side on the x-axis. Since you are finding midpoints, use coordinates that are multiples of 2. Find the coordinates of the midpoints $T, W, V,$ and U.

$$T = \text{midpoint of } \overline{OA} = \left(\frac{2a + 0}{2}, \frac{2b + 0}{2}\right) = (a, b)$$

$$W = \text{midpoint of } \overline{AC} = \left(\frac{2a + 2c}{2}, \frac{2b + 2d}{2}\right) = (a + c, b + d)$$

$$V = \text{midpoint of } \overline{CE} = \left(\frac{2c + 2e}{2}, \frac{2d + 0}{2}\right) = (c + e, d)$$

$$U = \text{midpoint of } \overline{OE} = \left(\frac{0 + 2e}{2}, \frac{0 + 0}{2}\right) = (e, 0)$$

Find the slopes of the sides of $TWVU$.

$$\text{slope of } \overline{TW} = \frac{b - (b + d)}{a - (a + c)} = \frac{d}{c} \qquad \text{slope of } \overline{VU} = \frac{d - 0}{(c + e) - e} = \frac{d}{c}$$

The slopes are equal, so $\overline{TW} \parallel \overline{VU}$. Similarly, $\overline{WV} \parallel \overline{TU}$.

Since both pairs of opposite sides of $TWVU$ are parallel, $TWVU$ is a parallelogram, and Tiana is correct.

✓ **Quick Check** ❶ Find the slopes of $\overline{WV}$ and $\overline{TU}$. Verify that $\overline{WV}$ and $\overline{TU}$ are parallel.

Centering figures at the origin can also be helpful when placing figures in the coordinate plane.

2 EXAMPLE **Naming Coordinates**

Algebra In the diagram, rectangle *KLMN* is centered at the origin with sides parallel to the axes. Find the missing coordinates.

L has coordinates (a, b), so the coordinates of the other vertices are $K(-a, b)$, $M(a, -b)$, and
● $N(-a, -b)$.

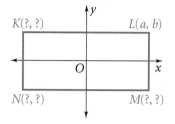

✓ **Quick Check** ❷ Use the properties of parallelogram *OPQR* to find the missing coordinates. Do not use any new variables.

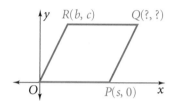

EXERCISES

For more exercises, see *Extra Skill, Word Problem, and Proof Practice.*

Practice and Problem Solving

 A **Practice by Example**

Example 1
(page 343)

GO **for Help**

1. Claim: The midpoint of the hypotenuse of a right triangle is equidistant from the vertices of the triangle. Follow Steps 1 and 2 to place a right triangle in the coordinate plane. Complete Steps 3–5 to verify the claim.

Given: Right $\triangle ABC$ with M the midpoint of hypotenuse $\overline{AB}$

Prove: $MA = MB = MC$

Step 1: Draw right $\triangle ABC$ on a coordinate plane. Locate the right angle, $\angle C$, at the origin and leg $\overline{CA}$ on the positive x-axis.

Step 2: You seek a midpoint, so label coordinates using multiples of 2. The coordinates of point A are **a.** ? . The coordinates of point B are **b.** ? .

Step 3: By the Midpoint Formula, the coordinates of midpoint M are **c.** ? .

Step 4: By the Distance Formula, $MA = $ **d.** ? , $MB = $ **e.** ? , and $MC = $ **f.** ? .

Step 5: Conclusion: **g.** ?

Example 2
(page 344)

x^2 **Algebra** **Give coordinates for points *W* and *Z* without using any new variables.**

2. rectangle

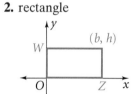

3. square

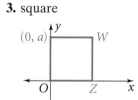

4. square

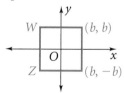

5. parallelogram

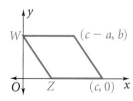

6. rhombus

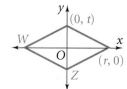

7. isosceles trapezoid

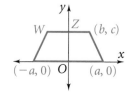

8. Writing Choose values for r and t in Exercise 6. Find the slope and length of each side. State why the figure satisfies the definition of a rhombus.

9. What property of a rhombus makes it convenient to place its diagonals on the x- and y-axes? (See Exercise 6.)

Here are coordinates for eight points in the coordinate plane ($q > p > 0$).
$$A(0, 0), \ B(p, 0), \ C(q, 0), \ D(p + q, 0), \ E(0, q), \ F(p, q), \ G(q, q), \ H(p + q, q)$$
Which four points, if any, are the vertices for each type of figure?

10. parallelogram   **11.** rhombus   **12.** rectangle

13. square   **14.** trapezoid   **15.** isosceles trapezoid

Refer to the diagrams in Exercises 2–7. Use the coordinates given below in place of the ones shown. Then give the coordinates for points W and Z without using any new variables. The new diagram for Exercise 16 is shown here.

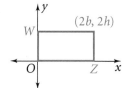

16. Ex. 2, $(2b, 2h)$ **17.** Ex. 3, $(0, 2a)$

18. Ex. 4, $(2b, 2b), (2b, -2b)$ **19.** Ex. 5, $(2c, 0), (2c - 2a, b)$

20. Ex. 6, $(2r, 0), (0, 2t)$ **21.** Ex. 7, $(-2a, 0), (2a, 0), (2b, 2c)$

22. Multiple Choice The coordinates of three vertices of a rectangle are $(-2a, 0)$, $(2a, 0)$, and $(2a, 2b)$. What are the coordinates of the midpoint of the diagonal joining one of these points with the fourth vertex?
Ⓐ $(0, b)$ Ⓑ $(0, 2b)$ Ⓒ $(2a, b)$ Ⓓ $(-2a, 2b)$

Give the coordinates for point P without using any new variables.

23. isosceles trapezoid   **24.** trapezoid with a right $\angle$ **25.** kite

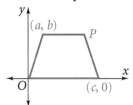   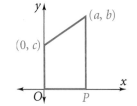   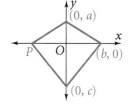

26. a. Draw a square whose diagonals of length $2b$ lie on the x- and y-axes.
   b. Give the coordinates of the vertices of the square.
   c. Compute the length of a side of the square.
   d. Find the slopes of two adjacent sides of the square.
   e. Do the slopes show that the sides are perpendicular? Explain.

27. Make two drawings of an isosceles triangle with base length $2b$ and height $2c$.
   a. In one drawing, place the base on the x-axis with a vertex at the origin.
   b. In the second, place the base on the x-axis with its midpoint at the origin.
   c. Find the lengths of the legs of the triangle as placed in part (a).
   d. Find the lengths of the legs of the triangle as placed in part (b).
   e. How do the results of parts (c) and (d) compare?

 28. Marine Archaeology Marine archaeologists sometimes use a coordinate system on the ocean floor. They record the coordinates of points where artifacts are found. Assume that each diver searches a square area and can go no farther than b units from the starting point. Draw a model for the region one diver can search. Assign coordinates to the vertices without using any new variables.

Real-World Connection

Careers Underwater archaeologists also spend time on land studying historical data to locate submerged sites or ships.

29. Coordinate Proof Follow the steps below to prove:

If two nonvertical lines are perpendicular, the product of their slopes is -1.

Step 1: Two nonvertical lines, ℓ_1 and ℓ_2, intersect. Which coordinate point might be the easiest to work with as the point of intersection?

Step 2: To work with the slope of a line, you need two points on the line. Choose one point $A(a, b)$ on ℓ_1. What are the coordinates of C?

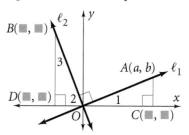

Step 3: Notice that $\angle 1$ and $\angle 3$ are both complements of $\angle 2$. Why?

Step 4: This means that the two triangles pictured have congruent angles. Thus, if any pair of sides are congruent, the two triangles are congruent. Congruent triangles are desirable, so what would be a good choice for the coordinates of point D?

Step 5: If you made a choice for D so that $\triangle ACO \cong \triangle ODB$ what must be the coordinates of point B?

Step 6: Now, complete the proof that the product of slopes is -1.

Multiple Choice

30. The vertices of a rhombus are located at $(a, 0)$, $(0, b)$, $(-a, 0)$, and $(0, -b)$, where $a, b > 0$. What is the midpoint of the side that is in Quadrant II?

A. $\left(\frac{a}{2}, \frac{b}{2}\right)$ **B.** $\left(-\frac{a}{2}, \frac{b}{2}\right)$ **C.** $\left(-\frac{a}{2}, -\frac{b}{2}\right)$ **D.** $\left(\frac{a}{2}, -\frac{b}{2}\right)$

31. The vertices of a kite are located at $(0, a)$, $(b, 0)$, $(0, -c)$, and $(-b, 0)$, where $a, b, c, d > 0$. What is the slope of the side in Quadrant IV?

F. $\frac{c}{b}$ **G.** $\frac{b}{c}$ **H.** $-\frac{b}{c}$ **J.** $-\frac{c}{b}$

32. The vertices of a square are located at $(a, 0)$, (a, a), $(0, a)$, and $(0, 0)$. What is the length of a diagonal?

A. a **B.** $2a$ **C.** $a\sqrt{2}$ **D.** $2\sqrt{a}$

Short Response

33. The vertices of a rectangle are $(2b, 0)$, $(2b, 2a)$, $(0, 2a)$, and $(0, 0)$. What are the coordinates of the midpoint of each diagonal? What can you conclude from your answers?

Mixed Review

 for Help

Lesson 6-5 $\boxed{x^2}$ **34. Algebra** Find the measure of each angle and the value of x in the isosceles trapezoid.

Lesson 5-3 **Find the center of the circle that circumscribes $\triangle ABC$.**

35. $A(1, 1)$, $B(5, 3)$, $C(5, 1)$ **36.** $A(-5, 0)$, $B(-1, -8)$, $C(-1, 0)$

Lesson 4-3 **37. Given:** $\angle ACD \cong \angle ACB$ and $\angle D \cong \angle B$

Prove: $\triangle ADC \cong \triangle ABC$

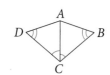

x^2 **Algebra** Find the value(s) of the variable(s).

1.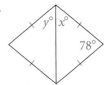

2.

3.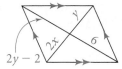

4.

5.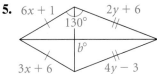

In Exercises 6–8, decide whether the statement is *true* or *false*. If true, explain why. If false, show a counterexample.

6. A quadrilateral with congruent diagonals is an isosceles trapezoid or rectangle.

7. A quadrilateral with congruent and perpendicular diagonals is a square.

8. Each diagonal of a kite bisects two angles of the kite.

Give the coordinates for point *A* without using any new variables.

9.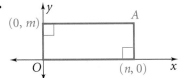

10.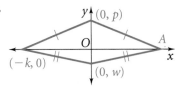

A Point in Time

400   200 B.C.   0   A.D. 200   2000

Many walls in ancient Egypt were decorated with reliefs. The relief in the photo was created in the year 255 B.C. First, the artist sketched the scene on papyrus overlaid with a grid. Next, the wall was marked with a grid the size of the intended sculpture. To draw each line, a tightly stretched string that had been dipped in red ochre was plucked, like a guitar string.

Using the grid squares as guides, the artist transferred the drawing to the wall. Then, a sculptor cut the background away, leaving the scene slightly raised. Finally, an artist painted the scene.

Go Online
PHSchool.com   **For:** Information about Egyptian reliefs
Web Code: aue-2032

6-7 Proofs Using Coordinate Geometry

 Learning Standards for Mathematics

G.G.66 Find the midpoint of a line segment, given its endpoints.

G.G.67 Find the length of a line segment, given its endpoints.

 Check Skills You'll Need

 **for Help** Lesson 6-6

1. Graph the rhombus with vertices $A(2, 2)$, $B(7, 2)$, $C(4, -2)$, and $D(-1, -2)$. Then, connect the midpoints of consecutive sides to form a quadrilateral. What do you notice about the quadrilateral?

x^2 **Algebra** Give the coordinates of B without using any new variables.

2. rectangle

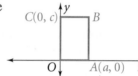

3. isosceles triangle

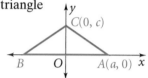

🔊 **New Vocabulary** • midsegment of a trapezoid

---

1 Building Proofs in the Coordinate Plane

In Lesson 5-1, you learned about midsegments of triangles. A trapezoid also has a midsegment. The **midsegment of a trapezoid** is the segment that joins the midpoints of the nonparallel opposite sides. It has two unique properties.

 Key Concepts

Theorem 6-18	Trapezoid Midsegment Theorem

(1) The midsegment of a trapezoid is parallel to the bases.
(2) The length of the midsegment of a trapezoid is half the sum of the lengths of the bases.

$$\overline{MN} \parallel \overline{TP}, \overline{MN} \parallel \overline{RA}, \text{ and } MN = \tfrac{1}{2}(TP + RA).$$

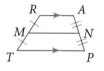

Formulas for slope, midpoint, and distance are used in a proof of Theorem 6-18.

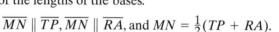 **EXAMPLE** Planning a Coordinate Geometry Proof

Developing Proof Plan a coordinate proof of Theorem 6-18.

Given: $\overline{MN}$ is the midsegment of trapezoid $TRAP$.
Prove: $\overline{MN} \parallel \overline{TP}, \overline{MN} \parallel \overline{RA},$ and $MN = \tfrac{1}{2}(TP + RA)$.

Plan: Place the trapezoid in the coordinate plane with a vertex at the origin and a base along the x-axis. Since midpoints will be involved, use multiples of 2 to name coordinates. To show lines are parallel, check for equal slopes. To compare lengths, use the Distance Formula.

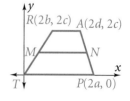

✓ Quick Check ❶ Complete the coordinate proof of Theorem 6-18.
   a. Find the coordinates of midpoints *M* and *N*. How do the multiples of 2 help?
   b. Find and compare the slopes of $\overline{MN}, \overline{TP}$, and $\overline{RA}$.
   c. Find and compare the lengths *MN, TP*, and *RA*.
   d. In parts (b) and (c), how does placing a base along the *x*-axis help?

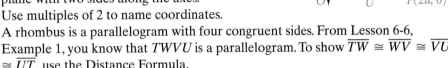

Algebra The rectangular flag at the left is constructed by connecting the midpoints of its sides. Use coordinate geometry to prove that the quadrilateral formed by connecting the midpoints of the sides of a rectangle is a rhombus.

Given: *MNPO* is a rectangle.
   T, W, V, U are midpoints of its sides.

Prove: *TWVU* is a rhombus.

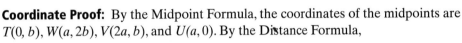

Plan: Place the rectangle in the coordinate plane with two sides along the axes. Use multiples of 2 to name coordinates. A rhombus is a parallelogram with four congruent sides. From Lesson 6-6, Example 1, you know that *TWVU* is a parallelogram. To show $\overline{TW} \cong \overline{WV} \cong \overline{VU} \cong \overline{UT}$, use the Distance Formula.

Coordinate Proof: By the Midpoint Formula, the coordinates of the midpoints are $T(0, b), W(a, 2b), V(2a, b)$, and $U(a, 0)$. By the Distance Formula,

$$TW = \sqrt{(a - 0)^2 + (2b - b)^2} = \sqrt{a^2 + b^2}$$

$$WV = \sqrt{(2a - a)^2 + (b - 2b)^2} = \sqrt{a^2 + b^2}$$

$$VU = \sqrt{(a - 2a)^2 + (0 - b)^2} = \sqrt{a^2 + b^2}$$

$$UT = \sqrt{(0 - a)^2 + (b - 0)^2} = \sqrt{a^2 + b^2}$$

$\overline{TW} \cong \overline{WV} \cong \overline{VU} \cong \overline{UT}$, so parallelogram *TWVU* is a rhombus.

Real-World 🌐 Connection

A flag's length, called the *fly*, usually is greater than its width, called the *hoist*.

✓ Quick Check ❷ **Critical Thinking** Explain why the proof using $M(0, 2b), N(2a, 2b), P(2a, 0)$, and $O(0, 0)$ is easier than a proof using $M(0, b), N(a, b), P(a, 0)$, and $O(0, 0)$.

EXERCISES

For more exercises, see *Extra Skill, Word Problem, and Proof Practice.*

Practice and Problem Solving

Ⓐ Practice by Example
   Example 1
   (page 348)
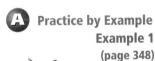

1. *W* and *Z* are the midpoints of $\overline{OR}$ and $\overline{ST}$, respectively. In parts (a)–(c), find the coordinates of *W* and *Z*.

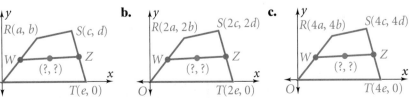

   d. You are to plan a coordinate proof involving the midpoint of $\overline{WZ}$. Which of the figures (a)–(c) would you prefer to use? Explain.

Developing Proof **Complete the plan for each coordinate proof.**

2. The diagonals of a parallelogram bisect each other (Theorem 6-3).

Given: Parallelogram $ABCD$

Prove: $\overline{AC}$ bisects $\overline{BD}$, and $\overline{BD}$ bisects $\overline{AC}$.

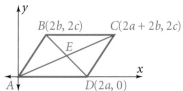

Plan: Place the parallelogram in the coordinate plane with a vertex at the **a.** _?_ and a side along the **b.** _?_ . Since midpoints will be involved, use multiples of **c.** _?_ to name coordinates. To show segments bisect each other, show the midpoints have the same **d.** _?_ .

3. The diagonals of an isosceles trapezoid are congruent (Theorem 6-16).

Given: Trapezoid $EFGH$ with $\overline{FE} \cong \overline{GH}$

Prove: $\overline{EG} \cong \overline{HF}$

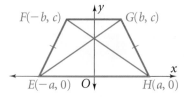

Plan: The trapezoid is isosceles, so place one base on the x-axis so that the **a.** _?_ bisects its bases. To show the diagonals are congruent, use the **b.** _?_ Formula.

4. The median to the hypotenuse of a right triangle is half the hypotenuse.

Given: $\triangle MNO$ is a right triangle with right $\angle MON$.
P is the midpoint of $\overline{MN}$.

Prove: $OP = \frac{1}{2}MN$

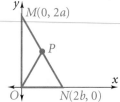

Plan: Place the right triangle in the coordinate plane with the vertex of the **a.** _?_ at the origin and the **b.** _?_ along each axis. Since midpoints will be involved, use **c.** _?_ to name coordinates for points **d.** _?_ and **e.** _?_ . Use the **f.** _?_ Formula to find the coordinates of P. To compare lengths, use the **g.** _?_ Formula.

5. The segments joining the midpoints of consecutive sides of an isosceles trapezoid form a rhombus.

Given: Trapezoid $TRAP$ with $\overline{TR} \cong \overline{PA}$; D, E, F, and G are midpoints of the indicated sides.

Prove: $DEFG$ is a rhombus.

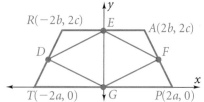

Plan: The trapezoid is **a.** _?_ , so place one base on the **b.** _?_ so that the **c.** _?_ bisects its bases. Use multiples of 2 to name coordinates since **d.** _?_ will be involved. A rhombus is a parallelogram with four **e.** _?_ . To show opposite sides are parallel, show that their **f.** _?_ are the same. To show sides are congruent, use **g.** _?_ .

Example 2
(page 349)

Developing Proof **Follow the plans above to complete the coordinate proofs.**

6. (Exercise 3) The diagonals of an isosceles trapezoid are congruent.

Proof: By the Distance Formula, $EG =$ **a.** _?_ and $HF =$ **b.** _?_ . Therefore, $\overline{EG} \cong \overline{HF}$ by the definition of congruence.

7. (Exercise 4) The median from the vertex of the right angle of a right triangle is half as long as the hypotenuse.

Proof: By the Distance Formula, $OP =$ **a.** _?_ and $MN =$ **b.** _?_ . Therefore, $OP = \frac{1}{2}MN$.

8. (Exercise 5) The segments joining the midpoints of consecutive sides of an isosceles trapezoid form a rhombus.

Proof: The midpoints have coordinates **a.** $D(\underline{\ ?\ }, \underline{\ ?\ })$, $E(\underline{\ ?\ }, \underline{\ ?\ })$, $F(\underline{\ ?\ }, \underline{\ ?\ })$, and $G(\underline{\ ?\ }, \underline{\ ?\ })$. By the Distance Formula, $DE = $ **b.** $\underline{\ ?\ }$, $EF = $ **c.** $\underline{\ ?\ }$, $FG = $ **d.** $\underline{\ ?\ }$, and $GD = $ **e.** $\underline{\ ?\ }$. The slope of $DE = $ **f.** $\underline{\ ?\ }$ and the slope of $FG = $ **g.** $\underline{\ ?\ }$. The slope of $EF = $ **h.** $\underline{\ ?\ }$ and that of $GD = $ **i.** $\underline{\ ?\ }$. Thus, $DEFG$ is a parallelogram with congruent **j.** $\underline{\ ?\ }$, so **k.** $\underline{\ ?\ }$ is a rhombus by the definition of rhombus.

9. **Developing Proof** Use coordinate geometry to prove that the diagonals of a rectangle bisect each other.

   Proof: The midpoint of $\overline{AC}$ is **a.** $\underline{\ ?\ }$. The midpoint of $\overline{BD}$ is **b.** $\underline{\ ?\ }$. The midpoints are **c.** $\underline{\ ?\ }$, so the diagonals bisect each other.

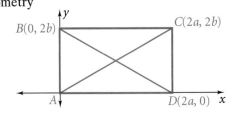

B Apply Your Skills

10. **Open-Ended** Give an example of a statement that you think is easier to prove with a coordinate geometry proof than with a paragraph, flow, or two-column proof. Explain your choice.

Problem Solving Hint
Lines with undefined slope, like $\overline{KN}$ and $\overline{LM}$, are vertical lines. All vertical lines are parallel.

Proof 11. Prove: The midpoints of the sides of a kite determine a rectangle.

   Given: Kite $DEFG$ with $DE = EF$ and $DG = GF$; K, L, M, and N are midpoints of the sides.

   Prove: $KLMN$ is a rectangle.

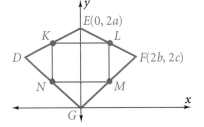

State whether each type of conclusion shown here could be reached using coordinate methods. Give a reason for each answer.

12. $\overline{AB} \cong \overline{CD}$

13. $\overline{AB} \parallel \overline{CD}$

14. $\overline{AB} \perp \overline{CD}$

15. $\overline{AB}$ bisects $\overline{CD}$.

16. $\overline{AB}$ bisects $\angle CAD$.

17. $\angle A \cong \angle B$

18. $\angle A$ is a right angle.

19. $AB + BC = AC$

20. $\triangle ABC$ is isosceles.

21. $\angle A$ and $\angle B$ are supplementary.

22. $\overline{AB}, \overline{CD}$, and $\overline{EF}$ are concurrent.

23. Quadrilateral $ABCD$ is a rhombus.

24. **Multiple Choice** $DEFG$ is a rhombus. What is the slope of its diagonal $\overline{DF}$?

    Ⓐ 0   Ⓑ $\left(\dfrac{a}{b}\right)$

    Ⓒ 1   Ⓓ undefined

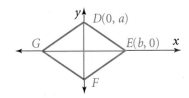

GO Online
Homework Video Tutor
Visit: PHSchool.com
Web Code: aue-0607

A and B have coordinates -2 and 10 on a number line. Find the coordinates of the points that separate $\overline{AB}$ into the given number of congruent segments.

25. 4   26. 6   27. 10   28. 50   29. n

The endpoints of $\overline{AB}$ are $A(-3, 5)$ and $B(9, 15)$. Find the coordinates of the points that separate $\overline{AB}$ into the given number of congruent segments.

30. 4 **31.** 6 **32.** 10 **33.** 50 **34.** n

Proof **35.** You learned in Lesson 5-3 (Theorem 5-8) that the centroid of a triangle, the point where the medians meet, is $\frac{2}{3}$ of the distance from each vertex to the midpoint of the opposite side. Complete the following steps to prove this theorem.

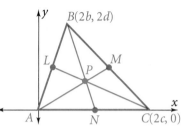

a. Find the coordinates of points L, M, and N, the midpoints of the sides of $\triangle ABC$.

b. Find equations of $\overleftrightarrow{AM}, \overleftrightarrow{BN}$, and $\overleftrightarrow{CL}$.

c. Find the coordinates of point P, the intersection of $\overrightarrow{AM}$ and $\overrightarrow{BN}$.

d. Show that point P is on $\overleftrightarrow{CL}$.

e. Use the Distance Formula to show that point P is $\frac{2}{3}$ of the distance from each vertex to the midpoint of the opposite side.

Problem Solving Hint

To show three lines are concurrent, you can show

1) one point is on all three lines

2) the intersection point of two lines is on the third line, or

3) the intersection of one pair of lines is the same as the intersection of another pair.

Proof **36.** Complete the following steps to prove Theorem 5-9. You are given $\triangle ABC$ with altitudes p, q, and r. Show that p, q, and r intersect in a point (called the orthocenter of the triangle).

a. The slope of $\overline{BC}$ is $\frac{c}{-b}$. What is the slope of line p?

b. Show that the equation of line p is $y = \frac{b}{c}(x - a)$.

c. What is the equation of line q?

d. Show that lines p and q intersect at $\left(0, \frac{-ab}{c}\right)$.

e. The slope of $\overline{AC}$ is $\frac{c}{-a}$. What is the slope of line r?

f. Show that the equation of line r is $y = \frac{a}{c}(x - b)$.

g. Show that lines r and q intersect at $\left(0, \frac{-ab}{c}\right)$.

h. Give the coordinates of the orthocenter of $\triangle ABC$.

C Challenge

The endpoints of $\overline{AB}$ are as given. Find the coordinates of the points that separate $\overline{AB}$ into n congruent segments.

37. A has coordinate a and B has coordinate b on a number line.

38. A has coordinates (a, c) and B has coordinates (b, d) in the coordinate plane.

39. Use the diagram at the right.

a. Explain using area why $\frac{1}{2}ad = \frac{1}{2}bc$ and hence $ad = bc$. (*Hint*: Area of triangle $= \frac{1}{2} \cdot$ base $\cdot$ height)

b. Use slope and part (a) to show: If $\frac{a}{b} = \frac{c}{d}$, then $ad = bc$.

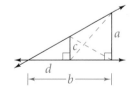

40. Physics For a mobile to be in balance, you suspend each part of the mobile at its center of mass. The center of mass, or centroid, is the point around which the weight of an object appears to be evenly distributed. You learned a method for finding the centroid of a triangle in Lesson 5-3. Now use it to help you find the centroid of a quadrilateral. (*Hint*: Where is the centroid of a segment?)

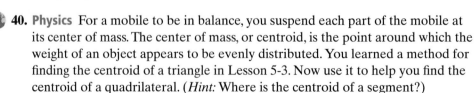

Real-World Connection

Carefully balanced metal shapes hang from wires in the mobiles of Alexander Calder (1898–1976).

Proof **41.** Write a coordinate proof of the theorem:

If the slopes of two lines have product -1, the lines are perpendicular.

a. First, argue that neither line can be horizontal or vertical.
b. Then, tell why the lines must intersect. (*Hint:* Use indirect reasoning.)
c. Knowing that they do intersect, place the lines in the coordinate plane, choose a point on ℓ_1, find a related point on ℓ_2, and complete the proof.

NY REGENTS

Test Prep

Multiple Choice

42. Two points on a line are $(-7, 10)$ and $(9, 2)$. Two points on a line parallel to that line are $(1, -3)$ and $(x, 4)$. What is the value of x?
A. -13 **B.** 13 **C.** 15 **D.** -15

43. Two points on a line are $(-4, 0)$ and $(8, 8)$. Two points on a line perpendicular to that line are $(8, -1)$ and $(6, y)$. What is the value of y?
F. 3 **G.** 2 **H.** $-\frac{7}{3}$ **J.** -4

Short Response

44. The endpoints of a segment are $(7, -3)$ and (a, b). The midpoint is $(3, 4)$.
a. What are the coordinates of the other endpoint? Show your work.
b. What is the length of the segment? Show your work.

Extended Response

45. Given: $\triangle ABC$; P, Q, and R are the midpoints of $\overline{AB}$, $\overline{AC}$, and $\overline{BC}$, respectively.
a. Place $\triangle ABC$ in the coordinate plane by writing coordinates for A, B, and C.
b. What are the coordinates of P, Q, and R?
c. Use coordinate geometry to prove $\triangle APQ \cong \triangle RQP$ by SSS.

Mixed Review

Lesson 6-6

46. Rectangle $LMNP$ at the right is centered at the origin. Give coordinates for point P without using any new variables.

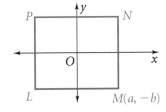

Lesson 5-4

Write (a) the inverse and (b) the contrapositive of each statement.

47. If the sum of the angles of a polygon is not $360°$, then the polygon is not a quadrilateral.

48. If $x = 51$, then $2x = 102$.

49. If $a = 5$, then $a^2 = 25$.

50. If $b < -4$, then b is negative.

51. If $c > 0$, then c is positive.

Lesson 4-4

Explain how you can use SSS, SAS, ASA, or AAS with CPCTC to prove each statement true.

52. $\overline{AB} \cong \overline{CB}$

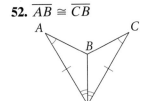

53. $\angle 1 \cong \angle 2$

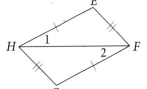

54. $\angle K \cong \angle M$

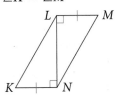

Quadratics

The graph of the quadratic function $y = ax^2$ ($a > 1$) is a vertical stretch of $y = x^2$ by the factor a. Each y-value for $y = ax^2$ is a times the corresponding y-value for $y = x^2$.

For $0 < a < 1$, the function $y = ax^2$ is a vertical shrink of $y = x^2$ by the factor a.

> **NY** **G.G.70:** Solve systems of equations involving one linear equation and one quadratic equation graphically.

1 EXAMPLE

Graph $y = x^2$, $y = 3x^2$, and $y = \frac{1}{3}x^2$. Compare the graphs.

First, draw $y = x^2$. For $y = 3x^2$, draw each point to be 3 times as high as the corresponding point in $y = x^2$.

To draw $y = \frac{1}{3}x^2$, draw each point one-third as high as the corresponding point in $y = x^2$.

The graph of $y = 3x^2$ is a stretch of the graph of $y = x^2$ by the factor 3. The graph of $y = \frac{1}{3}x^2$ is a shrink of the graph of $y = x^2$ by the factor $\frac{1}{3}$.

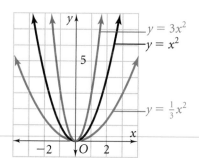

The graph of $y = -ax^2$ is a *reflection* (mirror image) of the graph of $y = ax^2$ across the x-axis. Each y-value for $y = -ax^2$ is the opposite of the corresponding y-value for $y = ax^2$.

2 EXAMPLE

Graph $y = 3x^2$ and $y = -3x^2$. Compare the graphs.

Draw $y = 3x^2$ as in Example 1. For $y = -3x^2$, draw each point so its y-value is the opposite of the corresponding y-value for $y = 3x^2$.

The graph of $y = -3x^2$ is the reflection across the x-axis of the graph of $y = 3x^2$.

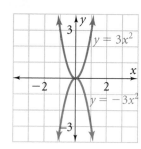

EXERCISES

Draw graphs of $y = x^2$ and the given function in the same coordinate plane. Compare the graphs.

1. $y = 2x^2$ **2.** $y = \frac{1}{2}x^2$ **3.** $y = 0.75x^2$ **4.** $y = \frac{3}{2}x^2$

For the given value of a, draw graphs of $y = ax^2$ and $y = -ax^2$ in the same coordinate plane. Compare the graphs.

5. $a = 4$ **6.** $a = \frac{1}{4}$ **7.** $a = \frac{2}{3}$ **8.** $a = \frac{5}{2}$

The graph of $y = x^2 + k$ is a vertical shift, or *translation*, of $y = x^2$ by $|k|$ units.
If $k > 0$, the shift is upward. If $k < 0$, the shift is downward.

3 EXAMPLE

Graph $y = x^2$, $y = x^2 + 2$, and $y = x^2 - 3$. Compare the graphs.

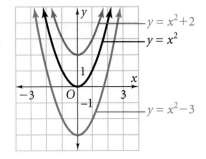

Draw $y = x^2$ as in Example 1. For $y = x^2 + 2$, draw each point to be 2 units higher than the corresponding point in $y = x^2$.

For $y = x^2 - 3$ draw each point to be 3 units lower than the corresponding point in $y = x^2$.

The graph of $y = x^2 + 2$ is 2 units above the graph of $y = x^2$.
The graph of $y = x^2 - 3$ is 3 units below the graph of $y = x^2$.

EXERCISES

Draw graphs of $y = x^2$ and the given function in the same coordinate plane. Compare the graphs.

9. $y = x^2 + 1$ **10.** $y = x^2 - 1$ **11.** $y = x^2 + \frac{5}{2}$ **12.** $y = x^2 - 3.5$

It is possible for a quadratic equation in two variables to be part of a system of equations. You can solve such systems graphically.

4 EXAMPLE

Solve each system of equations.

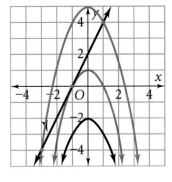

A: $y = 2x + 2$ B: $y = 2x + 2$ C: $y = 2x + 2$
   $y = -x^2 - 2$ $y = -x^2 + 1$ $y = -x^2 + 5$

For System A, the line intersects the parabola in no points. The system has no solutions.

For System B, the line intersects the parabola in one point. The system has one solution, $(-1, 0)$.

For System C, the line intersects the parabola in two points. The system has two solutions, $(-3, -4)$ and $(1, 4)$.

EXERCISES

Solve each system graphically.

13. $y = x + 2$ **14.** $y = -12$ **15.** $y = -4x + 4$ **16.** $y = -x$
   $y = x^2$ $y = -3x^2$ $y = x^2 - 1$ $y = -2x^2 + 3$

State whether the system has 0, 1, or 2 solutions.

17. $y = 3x + \frac{9}{4}$ **18.** $y = \frac{3}{4}x - \frac{1}{2}$ **19.** $y = -2x + 4$ **20.** $y = 12x - 17$
   $y = -x^2$ $y = \frac{1}{4}x^2$ $y = -x^2 + 4$ $y = 3x^2 - 4$

Drawing a Diagram

Sometimes, if a problem does not have a diagram, you should draw one. A diagram helps you to see the given information so that you can use the information to make inferences.

1 EXAMPLE

Side $\overline{AB}$ of square $ABCD$ has endpoints $A(0,0)$ and $B(5,2)$. What are the possible locations of point C?

You need to draw a diagram to do this problem. Point C can be in two different locations.

First draw the segment $\overline{AB}$, and then begin the two possible squares that have $\overline{AB}$ as a side. To locate C from B you can either go up 5 units and left 2 units, or you can go down 5 units and right 2 units. The coordinates of point C can be either $(3,7)$ or $(7,-3)$.

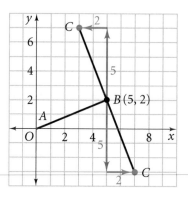

2 EXAMPLE

The midpoints of the sides of a rhombus are joined to form a quadrilateral. What special quadrilateral is formed? Explain.

Draw a rhombus and join the midpoints of the sides. The midpoint quadrilateral is a parallelogram since both pairs of sides are parallel to a diagonal of the rhombus. The diagonals of the rhombus are perpendicular, so adjacent sides of the parallelogram are perpendicular to each other.

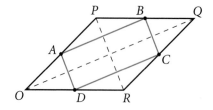

The quadrilateral formed is a rectangle.

EXERCISES

Draw a diagram to help you answer each question.

1. Angie says that if the diagonals of a quadrilateral are perpendicular, then the quadrilateral is a kite or a rhombus. Provide a counterexample to Angie's claim by sketching a quadrilateral with perpendicular diagonals that is neither a rhombus nor a kite.

2. Three vertices of parallelogram $ABCD$ are $A(0,0)$, $B(5,2)$, and $C(8,5)$. What are the possible locations of point D?

3. Three vertices of a parallelogram are $(0,0)$, $(5,2)$ and $(8,5)$. What are the possible locations of the fourth vertex?

4. The diagonal $\overline{AC}$ of square $ABCD$ has endpoints $A(0,0)$ and $C(6,4)$. What are the coordinates of the other two vertices of the square?

5. The midpoints of the sides of an isosceles trapezoid are joined to form a quadrilateral. What special quadrilateral is formed? Explain.

Chapter Review

Vocabulary Review

base angles of a trapezoid (p. 336)
consecutive angles (p. 312)
isosceles trapezoid (p. 306)
kite (p. 306)

midsegment of a trapezoid (p. 348)
parallelogram (p. 306)
rectangle (p. 306)

rhombus (p. 306)
square (p. 306)
trapezoid (p. 306)

To complete each definition, find the appropriate word in the second column.

1. A(n) ? is a parallelogram with four right angles.

2. A(n) ? is a quadrilateral with two pairs of adjacent sides congruent and no opposite sides congruent.

3. Angles of a polygon that share a common side are ? .

4. A(n) ? is a quadrilateral with exactly one pair of parallel sides.

5. A(n) ? is a parallelogram with four congruent sides.

6. The ? of a trapezoid is the segment that joins the midpoints of the nonparallel opposite sides.

7. A(n) ? is a quadrilateral with both pairs of opposite sides parallel.

8. A(n) ? is a parallelogram with four congruent sides and four right angles.

9. A(n) ? is a trapezoid whose nonparallel opposite sides are congruent.

10. Two angles that share a base of a trapezoid are its ? .

A. parallelogram

B. trapezoid

C. square

D. base angles

E. isosceles trapezoid

F. rectangle

G. consecutive angles

H. kite

I. rhombus

J. midsegment

Go Online
PHSchool.com
For: Vocabulary quiz
Web Code: auj-0651

Skills and Concepts

6-1 Objective

▼ To define and classify special types of quadrilaterals

Special quadrilaterals are defined by their characteristics.

A **parallelogram** is a quadrilateral with both pairs of opposite sides parallel.
A **rhombus** is a parallelogram with four congruent sides.
A **rectangle** is a parallelogram with four right angles.
A **square** is a parallelogram with four congruent sides and four right angles.
A **kite** is a quadrilateral with two pairs of adjacent sides congruent and no opposite sides congruent.
A **trapezoid** is a quadrilateral with exactly one pair of parallel sides.
An **isosceles trapezoid** is a trapezoid whose nonparallel opposite sides are congruent.

Draw and label each quadrilateral with the given vertices. Then determine the most precise name for each quadrilateral.

11. $N(-1, 2), M(3, 4), K(5, 0), L(1, -2)$ 12. $P(-4, 2), Q(-1, 3), R(7, 0), S(4, -1)$

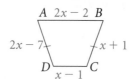

$\boxed{x^2}$ **Algebra** Find the values of the variables and the lengths of the sides.

13. isosceles trapezoid $ABCD$

14. kite $KLMN$

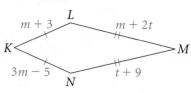

6-2 and 6-3 Objectives

▼ To use relationships among sides and among angles of parallelograms

▼ To use relationships involving diagonals of parallelograms or transversals

▼ To determine whether a quadrilateral is a parallelogram

Opposite sides and opposite angles of a parallelogram are congruent. The diagonals of a parallelogram bisect each other.

If three (or more) parallel lines cut off congruent segments on one transversal, then they cut off congruent segments on every transversal.

A quadrilateral is a parallelogram if any one of the following is true.
Both pairs of opposite sides are congruent.
Both pairs of opposite angles are congruent.
The diagonals bisect each other.
One pair of opposite sides is both congruent and parallel.

Find the measures of the numbered angles for each parallelogram.

15.

16.

17.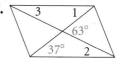

Determine whether the quadrilateral must be a parallelogram.

18.

19.

20.

21.

$\boxed{x^2}$ **Algebra** Find the values of the variables for which $ABCD$ must be a parallelogram.

22.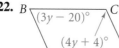

23.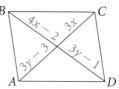

6-4 and 6-5 Objectives

▼ To use properties of diagonals of rhombuses and rectangles

▼ To determine whether a parallelogram is a rhombus or a rectangle

▼ To verify and use properties of trapezoids and kites

Each diagonal of a rhombus bisects two angles of the rhombus. The diagonals of a rhombus are perpendicular.

The diagonals of a rectangle are congruent.

If one diagonal of a parallelogram bisects two angles of the parallelogram, then the parallelogram is a rhombus. If the diagonals of a parallelogram are perpendicular, then the parallelogram is a rhombus. If the diagonals of a parallelogram are congruent, then the parallelogram is a rectangle.

The parallel sides of a trapezoid are its bases and the nonparallel sides are its legs. Two angles that share a base of a trapezoid are **base angles** of the trapezoid.

The base angles of an isosceles trapezoid are congruent. The diagonals of an isosceles trapezoid are congruent.

The diagonals of a kite are perpendicular.

Find the measures of the numbered angles in each quadrilateral.

24.

25.

26.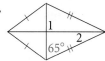

Find AC for each quadrilateral.

27.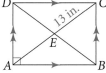

28.

29.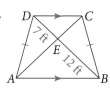

6-6 and 6-7 Objectives

▼ To name coordinates of special figures by using their properties

▼ To prove theorems using figures in the coordinate plane

In coordinate proofs, it generally is good practice to center the figure on the origin or place a vertex at the origin and one side of the figure on an axis.

The segment that joins the midpoints of the nonparallel sides of a trapezoid is the **midsegment** of the trapezoid. It is parallel to the bases and half as long as the sum of the lengths of the bases.

The formulas for slope, midpoint, and distance are used in coordinate proofs.

Give the coordinates of point P without using any new variable.

30. rectangle

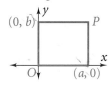

31. square

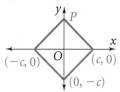

32. parallelogram

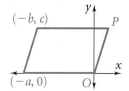

Complete each coordinate proof.

33. The diagonals of a square are perpendicular.

Given: Square $FGHI$ with vertices $I(0, 0)$, $H(a, 0)$, $G(a, a)$, and $F(0, a)$

Prove: $\overline{FH} \perp \overline{GI}$

The slope of $\overline{FH}$ is **a.** ? . The slope of $\overline{GI}$ is **b.** ? . $\overline{FH} \perp \overline{GI}$ because **c.** ? .

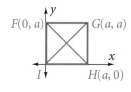

34. The diagonals of a rectangle are congruent.

Given: Rectangle $ABCD$

Prove: $\overline{AC} \cong \overline{BD}$

Vertex A is at the origin with coordinates $(0, 0)$. Name B as $(a, 0)$, C as (**a.** ? , b), and D as **b.** ? . $AC = $ **c.** ? and $BD = $ **d.** ? . So, $AC = $ **e.** ? , and $\overline{AC} \cong \overline{BD}$.

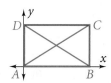

Chapter Test

Go Online
PHSchool.com
For: Chapter Test
Web Code: aua-0652

Graph each quadrilateral *ABCD*. Then determine the most precise name for it.

1. $A(1, 2), B(11, 2), C(7, 5), D(4, 5)$

2. $A(3, -2), B(5, 4), C(3, 6), D(1, 4)$

3. $A(1, -4), B(1, 1), C(-2, 2), D(-2, -3)$

4. **Open-Ended** Write the coordinates of four points that determine each figure with the given conditions. One vertex is at the origin and one side is 3 units long.
   a. square   **b.** parallelogram
   c. rectangle   **d.** trapezoid

Find *AN* for each parallelogram.

5.
9 in.

6.
6.5 cm

7. Sketch two noncongruent parallelograms *ABCD* and *EFGH* such that $\overline{AC} \cong \overline{BD} \cong \overline{EG} \cong \overline{FH}$.

 **Algebra** **Find the values of the variables for each parallelogram.**

8.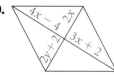

9.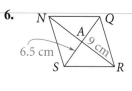

10.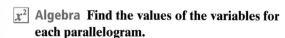

11.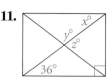

Does the information allow you to prove that *ABCD* is a parallelogram? Explain.

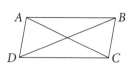

12. $\overline{AC}$ bisects $\overline{BD}$.

13. $\overline{AB} \cong \overline{DC}; \overline{AB} \parallel \overline{DC}$ 14. $\overline{AB} \cong \overline{DC}; \overline{BC} \cong \overline{AD}$

15. $\angle DAB \cong \angle BCD$ and $\angle ABC \cong \angle CDA$

16. **Writing** Explain why a square cannot be a kite.

 17. **Algebra** Determine the values of the variables for which *ABCD* is a parallelogram.

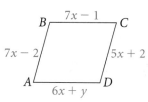

Find the measures of ∠1 and ∠2.

18.

19.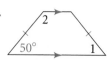

Give the coordinates for points *S* and *T* without using any new variables. Then find the midpoint and the slope of $\overline{ST}$.

20. rectangle

21. parallelogram
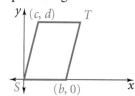

22. You want a garden border halfway between the front of your house and the city sidewalk, which are parallel to each other. You've measured those two edges. Find the length of the garden border.

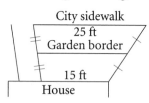

23. Complete this coordinate proof that the diagonals of a square are congruent.

Given: Square *ABCD* with vertices
$A(0, 0), B(a, 0), C(a, a)$, and $D(0, a)$
Prove: $\overline{AC} \cong \overline{BD}$

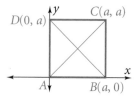

$AC =$ **a.** __?__ , and $BD =$ **b.** __?__ . So, **c.** __?__ $= BD$ and $\overline{AC} \cong \overline{BD}$.

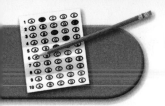

Regents Test Prep

Multiple Choice

For Exercises 1–11, choose the correct letter.

1. What is a name for the quadrilateral below?

    I. square
    II. rectangle
    III. rhombus
    IV. parallelogram

(1) I only   **(2)** IV only
(3) II and IV   **(4)** I, II, and IV

2. An isosceles triangle has two angles measuring 48 and 84. What is the measure of the third angle?

(1) 84   **(2)** 51   **(3)** 49   **(4)** 48

3. $\overline{DE}$ is a midsegment of $\triangle ABC$. What is DE?

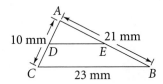

(1) 11.5 mm   **(2)** 11 mm
(3) 10.5 mm   **(4)** 10 mm

4. Which can you use to prove that two lines are parallel?

(1) supplementary corresponding angles
(2) congruent alternate interior angles
(3) congruent vertical angles
(4) congruent same-side interior angles

5. How can you prove that the two triangles are congruent?

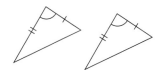

(1) ASA   **(2)** SSS   **(3)** SAS   **(4)** CPCTC

6. What is the measure of $\angle MOQ$?

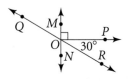

(1) 15   **(2)** 30   **(3)** 45   **(4)** 60

7. What is the midpoint of the segment whose endpoints are $M(6, -11)$ and $N(-18, 7)$?

(1) $(-6, -2)$ **(2)** $(6, 2)$
(3) $(-12, 9)$ **(4)** $(12, -9)$

8. Which statement is true for the figure?

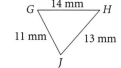

(1) $m\angle J < m\angle G < m\angle H$
(2) $m\angle H < m\angle G < m\angle J$
(3) $m\angle J < m\angle H < m\angle G$
(4) $m\angle H < m\angle J < m\angle G$

9. Which equation represents the line that is parallel to the graph of $2y - 5x = 10$?

(1) $y = -2.5x + 5$ **(2)** $(y - 1) = 2.5(x + 6)$
(3) $y = -0.4x$ **(4)** $(y - 4) = 0.4(x + 5)$

10. Which quadrilateral does NOT always have perpendicular diagonals?

(1) square   **(2)** rhombus
(3) kite   **(4)** isosceles trapezoid

11. In which point do the bisectors of the angles of a triangle meet?

(1) centroid   **(2)** circumcenter
(3) ncenter   **(4)** orthocenter

Gridded Response

12. In the parallelogram, DB is 15. What is DE?

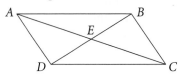

13. Algebra What is the value of x?

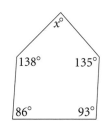

Short Response

14. In rectangle $ABCD$, $AC = 5(x - 2)$ and $BD = 3(x + 2)$. What is the value of x?

Extended Response

15. A parallelogram has vertices $L(-2, 5)$, $M(3, 3)$, $N(1, 0)$. What are possible coordinates for its fourth vertex? Explain.

Interpreting Data

Art is a harmony parallel with nature.

—Paul Cezanne

Although *parallel* has well-defined meaning as part of your geometry vocabulary, it has a more general everyday use. It refers to things—even people—that move, develop, or exist in certain, similar ways.

• Two books written in different languages telling the same story have parallel text.

• Two computers working on different data with the same program are parallel-processing.

• History teaches that events of today can parallel events of the past.

NY G.CM.5: Communicate logical arguments clearly, showing why a result makes sense and why the reasoning is valid.

1 ACTIVITY

Suppose you and your twin sibling begin your working careers at the same time. Also, you both decide to save some money for at least the first 15 years of work.

1. Your twin saves $1000 per year for the first 10 years and $4000 per year for the 5 years after that. Graph your twin's accumulated savings for each year of the 15-year period.

2. You decide to save $4000 per year for the first 5 years and then $1000 per year for the 10 years after that. Graph your accumulated savings for each year. Use a different color on the same coordinate plane as your twin's graph.

3. Do the two of you end up with same or different total savings? Why?

4. Use what you know about coordinate geometry. Explain why the resulting graph suggests a parallelogram.

In Activity 1 you compared accumulated savings, but you weren't asked to consider the interest that the savings could earn.

The table below shows the average annual rates of return for money invested in three different funds. Money invested 10 years ago in Growth Fund A earned a yearly average of 4.8% interest over 10 years.

Fund	Rate of Return		
	5-year	10-year	15-year
Growth Fund A	3.9%	4.8%	6.7%
Small Cap Fund B	11.3%	−5.2%	1.4%
Balanced Fund C	4.6%	2.1%	3.1%

5. The *risk* of an investment is the amount that the investment return may vary, up or down, from what you expect. Which fund in the table appears to have the greatest risk? Explain your answer.

6. You and your twin decide to invest in the same fund. Choose an interest rate based on the rates of return in the table. Use this annual interest rate to make a new graph that shows the value of your savings and the value of your twin's savings over 15 years.

7. Explain how interest changed your graph so that it is no longer a parallelogram.

8. Explain how this new graph might influence how you choose to save some of your income once you begin your working career.

EXERCISES

9. "Poetry seems to exist in a parallel universe outside daily life in America," said Rita Dove, former Poet Laureate of the United States. What might she mean?

Rita Dove

10. The formula for calculating the value V of an investment of P dollars at an annual interest rate of r percent (in decimal form) for n years is
$$V = P(1 + r)^n.$$

   a. What is the value of $1000 invested at 5% for 10 years? For 20 years?

   b. How much should be invested at 5% for the value to be $1000 in 10 years? In 20 years?

   c. In how many years will $1000 invested at 5% grow to at least $2000? (*Hint:* Use tables on a graphing calculator.)

11. Many people diversify their investments. They put their money into several funds to try and maximize interest while minimizing risk. In Activity 2, suppose that you divide your money equally between the three funds. Based on the 15-year rates of return, what average annual rate of return would you expect?

What You've Learned

NY Learning Standards for Mathematics

- **G.G.28:** In Chapter 4, you learned several different methods for providing triangles congruent. You learned how to use CPCTC to find additional information about congruent triangles.

- **G.G.42:** In Chapter 5, you learned that a midsegment of a triangle is parallel to a side of the triangle and half its length.

- **G.G.39:** In Chapter 6, you learned the properties of special parallelograms and strategies for working with figures in the coordinate plane.

 Check Your Readiness

GO for Help to the Lesson in green.

Simplifying Ratios (Skills Handbook page 756)

Simplify each ratio.

1. 12 : 18

2. $\frac{55}{11}$

3. $\frac{20a^2}{15a^5}$

4. $\frac{3x^2 - 12x}{x^3 - x^2}$

Polygon Angle-Sum Theorems (Lesson 3-4)

Determine the measure of an angle of each regular polygon.

5. pentagon

6. octagon

7. decagon

8. 27-gon

Congruent Figures (Lesson 4-1)

$\triangle PAC \cong \triangle DHL$. Complete the congruence statements.

9. $\overline{PC} \cong$ __?__

10. $\angle H \cong$ __?__

11. $\angle PCA \cong$ __?__

12. $\triangle HDL \cong$ __?__

Triangle Congruence (Lessons 4-2 and 4-3)

State why each pair of triangles are congruent.

13.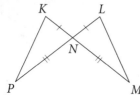

14.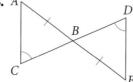

15.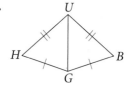

Midsegments of Triangles (Lesson 5-1)

Use the figure at the right for Exercises 16–17.

16. If $BC = 12$, then $BF =$ __?__ and $DE =$ __?__ .

17. If $EF = 4.7$, then $AD =$ __?__ and $AC =$ __?__ .

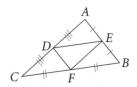

Similarity

◀)) **Key Vocabulary**

What You'll Learn Next

NY) Learning Standards for Mathematics

- G.RP.2: In this chapter, you will learn that similar polygons are polygons that have the same shape but not necessarily the same size.

- G.G.45: You will learn how to prove triangles similar.

- G.G.45: Through proving triangles similar, you will find additional relationships within triangles.

You will also learn how the perimeters and areas of similar figures are related.

Activity Lab Applying what you learn on pages 412 and 413, you will explore the precision of indirect measurements.

Ratios and Proportions

G.CN.4 Understand how concepts, procedures, and mathematical results in one area of mathematics can be used to solve problems in other areas of mathematics.

 **Check Skills You'll Need**

GO for Help Skills Handbook p. 756 and Lesson 5-1

Simplify each ratio.

1. $\frac{2}{4}$ 2. $\frac{8}{12}$ 3. $\frac{6}{8}$ 4. $\frac{10}{10}$

5. 20 : 30   6. 8 to 2   7. 2 to 8   8. 12 : 9

9. Draw a triangle. Then draw its three midsegments to form a smaller triangle. How do the lengths of the sides of the smaller triangle compare to the lengths of the sides of the larger triangle?

🔊 **New Vocabulary**   • ratio   • proportion   • extended proportion
• **Cross-Product Property**   • **scale drawing**   • **scale**

---

1 Using Ratios and Proportions

Vocabulary Tip

You can read *a* : *b* as the ratio *a* to *b*.

A **ratio** is a comparison of two quantities. You can write the ratio of *a* to *b* or *a* : *b* as the quotient $\frac{a}{b}$ when $b \neq 0$. Unless otherwise stated, the terms and expressions appearing in ratios in this book are assumed to be nonzero.

1 EXAMPLE   **Real-World Connection**

Photography A photo that is 8 in. wide and $5\frac{1}{3}$ in. high is enlarged to a poster that is 2 ft wide and $1\frac{1}{3}$ ft high. What is the ratio of the width of the photo to the width of the poster?

$$\frac{\text{width of photo}}{\text{width of poster}} = \frac{8 \text{ in.}}{2 \text{ ft}} = \frac{8 \text{ in.}}{24 \text{ in.}} = \frac{8}{24} = \frac{1}{3}$$

● The ratio of the width of the photo to the width of the poster is 1 : 3 or $\frac{1}{3}$.

✓ **Quick Check**   ① What is the ratio of the height of the photo to the height of the poster?

You can read both $\frac{a}{b} = \frac{c}{d}$
and $a : b = c : d$
as _a is to b as c is to d_.

A **proportion** is a statement that two ratios are equal. You can write a proportion in these forms:

$$\frac{a}{b} = \frac{c}{d} \text{ and } a : b = c : d$$

When three or more ratios are equal, you can write an **extended proportion.** For example, you could write the following:

$$\frac{6}{24} = \frac{4}{16} = \frac{1}{4}$$

Two equations are equivalent when either can be deduced from the other using the Properties of Equality. Several equations are equivalent to a proportion. Some of them are important enough to be called Properties of Proportions.

Key Concepts

Property	Properties of Proportions
$\frac{a}{b} = \frac{c}{d}$ is equivalent to	(1) $ad = bc$ (2) $\frac{b}{a} = \frac{d}{c}$
	(3) $\frac{a}{c} = \frac{b}{d}$ (4) $\frac{a+b}{b} = \frac{c+d}{d}$

Multiplying both sides of $\frac{a}{b} = \frac{c}{d}$ by bd results in the first property, called the **Cross-Product Property.** You may state this property as "The product of the extremes is equal to the product of the means."

$$\begin{array}{c} \text{means} \\ \downarrow \quad \downarrow \\ a : b = c : d \\ \uparrow \qquad \qquad \uparrow \\ \text{extremes} \\ \frac{a}{b} = \frac{c}{d} \\ ad = bc \end{array}$$

2 EXAMPLE **Properties of Proportions**

Algebra If $\frac{x}{y} = \frac{5}{6}$, complete each statement.

a. $6x = \blacksquare$

$6x = 5y$

b. $\frac{y}{x} = \frac{\blacksquare}{\blacksquare}$

$\frac{y}{x} = \frac{6}{5}$

c. $\frac{x}{5} = \frac{\blacksquare}{\blacksquare}$

$\frac{x}{5} = \frac{y}{6}$

d. $\frac{x+y}{y} = \frac{\blacksquare}{\blacksquare}$

$\frac{x+y}{y} = \frac{11}{6}$

Quick Check **2 Critical Thinking** Write two proportions that are equivalent to $\frac{m}{4} = \frac{n}{11}$.

You solve a proportion by finding the value of the variable.

3 EXAMPLE **Solving for a Variable**

Algebra Solve each proportion.

a. $\frac{x}{5} = \frac{12}{7}$

$7x = 5(12)$ ← **Cross-Product Property** → $4(y + 3) = 8y$

$7x = 60$

$x = \frac{60}{7}$

b. $\frac{y+3}{8} = \frac{y}{4}$

$4(y + 3) = 8y$

$4y + 12 = 8y$

$12 = 4y$

$y = 3$

Quick Check **3** Solve each proportion.

a. $\frac{5}{z} = \frac{20}{3}$

b. $\frac{18}{n+6} = \frac{6}{n}$

In a **scale drawing,** the **scale** compares each length in the drawing to the actual length. The lengths used in a scale can be in different units. A scale might be written as 1 in. to 100 mi, 1 in. = 12 ft, or 1 mm : 1 m. You can use proportions to find the actual dimensions represented in a scale drawing.

4 EXAMPLE Real-World Connection

Multiple Choice Use a ruler to measure the length ℓ of the bedroom in the scale drawing. What is the length of the actual bedroom?

Ⓐ 10 ft Ⓑ 14 ft Ⓒ $18\frac{2}{7}$ ft Ⓓ $24\frac{2}{5}$ ft

An inch ruler shows that $\ell = \frac{7}{8}$ in.

$$\frac{1}{16} = \frac{\frac{7}{8}}{\ell} \qquad \frac{\text{drawing length (in.)}}{\text{actual length (ft)}}$$

$$\ell = 16\left(\frac{7}{8}\right) \qquad \text{Cross-Product Property}$$

$$\ell = 14$$

The actual bedroom is 14 ft long. The correct answer is B.

Scale: 1 in. = 16 ft

Test-Taking Tip

If you use a ruler and your answer does not match any answer choice, first check that you measured correctly.

✓ **Quick Check** ④ Find the width of the actual bedroom.

EXERCISES

For more exercises, see *Extra Skill, Word Problem, and Proof Practice.*

Practice and Problem Solving

Ⓐ **Practice by Example**

Example 1
(page 366)

for Help

1. The base of the pyramid at the right is a square whose sides measure 0.675 m. The intent was for the sides to measure 675 m. What is the ratio of the length of a base side in the small pyramid to the length of a base side in the intended pyramid?

2. Models The Leaning Tower of Pisa in Italy is about 185 ft tall. A model of the Leaning Tower is 6 in. tall. What is the ratio of the height of the model to the height of the real tower?

"We had a little problem with the decimal point."

Example 2
(page 367)

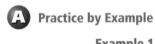 **Algebra** If $\frac{a}{b} = \frac{3}{4}$, complete each statement.

3. $4a = \blacksquare$

4. $\frac{b}{a} = \frac{\blacksquare}{\blacksquare}$

5. $\frac{a}{3} = \frac{\blacksquare}{\blacksquare}$

6. $\frac{4}{3} = \frac{\blacksquare}{\blacksquare}$

7. $\frac{4}{b} = \frac{\blacksquare}{\blacksquare}$

8. $3b = \blacksquare$

9. $\frac{a + b}{b} = \frac{\blacksquare}{\blacksquare}$

10. $\frac{a}{a + b} = \frac{\blacksquare}{\blacksquare}$

11. $\frac{a + 3}{3} = \frac{\blacksquare}{\blacksquare}$

Example 3
(page 367)

x^2 **Algebra** Solve each proportion.

12. $\frac{x}{2} = \frac{8}{4}$ **13.** $\frac{9}{5} = \frac{3}{x}$ **14.** $\frac{1}{3} = \frac{x}{12}$

15. $\frac{5}{x} = \frac{8}{11}$ **16.** $\frac{4}{x} = \frac{5}{9}$ **17.** $\frac{5}{6} = \frac{6}{x}$

18. $\frac{x+3}{3} = \frac{10+4}{4}$ **19.** $\frac{x+7}{7} = \frac{15}{5}$ **20.** $\frac{3}{5} = \frac{6}{x+3}$

Example 4
(page 368)

21. Geography The scale for this map of Louisiana is 1 in. = 40 mi. On the map, the distance from Lake Charles to Baton Rouge is about $3\frac{1}{8}$ in. About how far apart are the two cities?

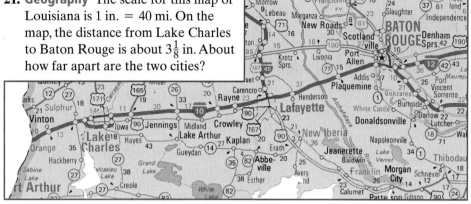

Map Reading Measure the map distance. Then find the actual distance.

22. Morgan City to Rayne   **23.** Vinton to New Roads   **24.** Kaplan to Plaquemine

25. Design You want to make a scale drawing of your bedroom to help you arrange your furniture. You decide on a scale of 3 in. = 2 ft. Your bedroom is a 12 ft-by-15 ft rectangle. What should be its dimensions in your scale drawing?

B **Apply Your Skills**

For each rectangle, find the ratio of the longer side to the shorter side.

26.

30 m

65 m

27.

50 in.

40 in.

28.

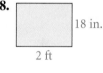

18 in.

2 ft

29. Multiple Choice The diameter of a dinner plate is 1 ft. In a dollhouse set, the diameter of a dinner plate is $1\frac{1}{4}$ in. What is the ratio of the diameter of the dollhouse plate to the diameter of the full-size plate?

   Ⓐ $\frac{5}{48}$ Ⓑ $\frac{4}{5}$ Ⓒ $\frac{5}{4}$ Ⓓ $\frac{48}{5}$

Complete each statement.

30. If $\frac{x}{7} = \frac{y}{3}$, then $\frac{x}{y} = \blacksquare$.

31. If $4m = 9n$, then $\frac{m}{n} = \blacksquare$.

32. If $\frac{30}{t} = \frac{18}{r}$, then $\frac{t}{r} = \blacksquare$.

33. If $\frac{a+5}{5} = \frac{b+2}{2}$, then $\frac{a}{5} = \blacksquare$.

34. Writing Use a map in your classroom or a map from a textbook. Explain how to use a ruler and the scale of the map to approximate an actual distance. Give an example. (If you do not have access to another map, use the map above.)

GO Online

Homework Video Tutor

Visit: PHSchool.com
Web Code: aue-0701

x^2 **Algebra** Solve each proportion.

35. $\frac{y}{10} = \frac{15}{25}$ **36.** $\frac{9}{24} = \frac{12}{n}$ **37.** $\frac{11}{14} = \frac{b}{21}$ **38.** $\frac{5}{x-3} = \frac{10}{x}$

39. $\frac{8}{n+4} = \frac{4}{n}$ **40.** $\frac{2b-1}{5} = \frac{b}{12}$ **41.** $\frac{x-3}{3} = \frac{2}{x+2}$ **42.** $\frac{3-4x}{1+5x} = \frac{1}{2+3x}$

43. Models The sandwich shop at the left is 40 ft tall. The shop is an enlargement of an actual milk bottle. The scale used in construction is 5 ft = 2 cm. Find the height of the actual milk bottle.

44. Geography Students at the University of Minnesota in Minneapolis built a model globe 42 ft in diameter using a scale of 1 : 1,000,000. About how tall is Mount Everest on the model? (Mount Everest is about 29,000 ft tall.)

Complete each extended proportion.

45. $\frac{8}{12} = \frac{6}{\blacksquare} = \frac{12}{\blacksquare}$

46. $\frac{\blacksquare}{15} = \frac{15}{25} = \frac{\blacksquare}{20}$

47. $\frac{14}{\blacksquare} = \frac{\blacksquare}{12} = \frac{35}{20}$

Games Choose a scale and make a scale drawing of the playing region.

48. A pool table is 5 ft by 10 ft.

49. A bowling lane is 3.5 ft by 60 ft.

50. A basketball court is 92 ft by 50 ft.

51. A football field is 160 ft by 120 yd.

52. Error Analysis One rectangle has length 3 in. and width 4 ft. Another rectangle has length 3 ft and width 4 yd. Elaine claims that the two rectangles are similar because their corresponding angles are congruent and their corresponding sides are in proportion. Explain why Elaine's reasoning is incorrect.

If $\frac{a}{b} = \frac{c}{d}$**, complete each statement.**

53. $\frac{a + b}{c + d} = \blacksquare$

54. $\frac{a + c}{b + d} = \blacksquare$

55. $\frac{a + 2b}{b} = \blacksquare$

C Challenge $\boxed{x^2}$ **Algebra Justify the indicated property of proportions.**

56. Property (2): If $\frac{a}{b} = \frac{c}{d}$, then $\frac{b}{a} = \frac{d}{c}$.

57. Property (3): If $\frac{a}{b} = \frac{c}{d}$, then $\frac{a}{c} = \frac{b}{d}$.

58. Property (4): If $\frac{a}{b} = \frac{c}{d}$, then $\frac{a + b}{b} = \frac{c + d}{d}$.

Solve each extended proportion for x and y with x > 0 and y > 0.

59. $\frac{x}{6} = \frac{x + 10}{18} = \frac{4x}{y}$

60. $\frac{x}{5} = \frac{9}{y} = \frac{y}{25}$

61. $\frac{1}{x} = \frac{4}{x + 9} = \frac{7}{y}$

Real-World  **Connection**

This sandwich shop is on Museum Wharf in Boston, Massachusetts.

Test Prep

Multiple Choice Solve each proportion.

62. $\frac{21}{x} = \frac{7}{3}$ A. 3   B. 7   C. 9   D. 14

63. $\frac{4}{x - 1} = \frac{1}{x}$ F. −3   G. $-\frac{1}{3}$ H. $\frac{1}{3}$ J. 3

64. $\frac{x}{x + 6} = \frac{2}{3}$ A. 4   B. 6   C. 8   D. 12

65. $\frac{3}{8} = \frac{x + 3}{9}$ F. $3\frac{3}{8}$ G. 3   H. $\frac{3}{8}$ J. $\frac{1}{3}$

Short Response **66.** A map of Long Island has the scale 2.75 cm = 16 km. On the map, Target Rock is 23.2 cm from Lake Montauk.
    a. Write a proportion that you can solve to determine the actual distance from Target Rock to Lake Montauk.
    b. Find the actual distance. Round your answer to the nearest kilometer.

GO for Help

Lesson 6-7

Give the coordinates for point *P* without using any new variables.

67. square

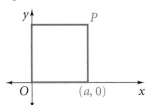

68. parallelogram

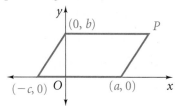

Lesson 6-1

Graph each quadrilateral *ABCD*. Classify *ABCD* in as many ways as possible.

69. $A(-1, -2), B(3, -2), C(1, 4), D(-3, 4)$

70. $A(2, -1), B(6, 2), C(8, 2), D(10, -1)$

71. $A(-7, 1), B(-5, 3), C(0, -2), D(-2, -4)$

72. $A(1, 1), B(-4, 4), C(1, 7), D(6, 4)$

Lesson 5-4

In each exercise, identify two statements that contradict each other.

73. I. $\triangle PQR$ is isosceles.
    II. $\triangle PQR$ is an obtuse triangle.
    III. $\triangle PQR$ is scalene.

74. I. $\angle 1 \cong \angle 2$
    II. $\angle 1$ and $\angle 2$ are complementary.
    III. $m\angle 1 + m\angle 2 = 180$

Write (a) the inverse and (b) the contrapositive of each statement.

75. If an angle is acute, then it has measure between 0 and 90.

76. If two lines are parallel, then they are coplanar.

77. If two angles are complementary, then both angles are acute.

A P●int in Time

1500 1600 1700 1800 1900 2000

In 1675, Danish astronomer Ole Römer used proportions to estimate the speed of light. He carefully measured the movements of Jupiter's moons. With Earth at point *B*, a moon emerged from behind Jupiter 16.6 min later than when Earth was at point *A*. He reasoned that it must have taken 16.6 min for the light to travel from point *A* to point *B*. Using proportions, Römer estimated the speed of light to be 150,000 mi/s. This estimate is about 81% of today's accepted value of 186,282 mi/s.

Go Online
PHSchool.com
For: Information about the speed of light
Web Code: aue-2032

Solving Quadratic Equations

There are several ways to solve a quadratic equation in standard form:

$$ax^2 + bx + c = 0, a \neq 0.$$

Page 355 reviewed graphical and algebraic methods. Another method—one that works for all quadratic equations—is the Quadratic Formula:

$$x = \frac{-b \pm \sqrt{b^2 - 4ac}}{2a}$$

1 EXAMPLE

Solve for x: $7x^2 + 6x - 1 = 0$. **The equation is in standard form.**

$a = 7, b = 6, c = -1$

$$x = \frac{-6 \pm \sqrt{6^2 - 4(7)(-1)}}{2(7)}$$ **Substitute in the Quadratic Formula.**

$$x = \frac{-6 \pm \sqrt{36 + 28}}{14}$$ **Simplify.**

$$x = \frac{-6 \pm \sqrt{64}}{14}$$

$$x = \frac{-6 + 8}{14} \text{ or } x = \frac{-6 - 8}{14}$$

$$x = \tfrac{1}{7} \text{ or } x = -1$$

Sometimes you may need a calculator to approximate the solutions.

2 EXAMPLE

Solve for x: $-3x^2 - 5x + 1 = 0$.

$a = -3, b = -5, c = 1$

$$x = \frac{-(-5) \pm \sqrt{(-5)^2 - 4(-3)(1)}}{2(-3)}$$ **Substitute in the Quadratic Formula.**

$$x = \frac{5 \pm \sqrt{25 + 12}}{-6}$$ **Simplify.**

$$x = \frac{5 + \sqrt{37}}{-6} \text{ or } x = \frac{5 - \sqrt{37}}{-6}$$

$$x \approx -1.85 \text{ or } x \approx 0.18$$ **Use a calculator and round.**

EXERCISES

Solve each quadratic equation by the method of your choice. Round answers that are not integers to the nearest hundredth.

1. $x^2 + 5x - 14 = 0$ **2.** $4x^2 - 13x + 3 = 0$ **3.** $2x^2 + 7x + 3 = 0$

4. $5x^2 + 2x - 2 = 0$ **5.** $6x^2 + 10x = 5$ **6.** $1 = 2x^2 - 6x$

7. $x^2 - 6x = 27$ **8.** $2x^2 - 10x + 11 = 0$ **9.** $8x^2 - 2x - 3 = 0$

Similar Polygons

7-2

 Learning Standards for Mathematics

G.RP.2 Recognize and verify, where appropriate, geometric relationships of perpendicularity, parallelism, congruence, and similarity, using algebraic strategies.

✓ **Check Skills You'll Need**

 **for Help** Lessons 4-1 and 7-1

1. $\triangle ABC \cong \triangle HIJ$. Name three pairs of congruent sides.

x^2 **Algebra** Solve each proportion.

2. $\frac{3}{4} = \frac{x}{8}$ **3.** $\frac{2}{x} = \frac{8}{24}$ **4.** $\frac{x}{9} = \frac{1}{3}$ **5.** $\frac{10}{25} = \frac{2}{x}$

◀)) **New Vocabulary** • similar • similarity ratio • golden rectangle • golden ratio

---

1 Similar Polygons

Two figures that have the same shape but not necessarily the same size are similar (∼). Two polygons are **similar** if (1) corresponding angles are congruent and (2) corresponding sides are proportional. The ratio of the lengths of corresponding sides is the **similarity ratio.**

1 **EXAMPLE** **Understanding Similarity**

$ABCD \sim EFGH$. Complete each statement.

a. $m\angle E = \blacksquare$
  $m\angle E = m\angle A = 53$ **Corresponding angles are ≅.**

b. $\frac{AB}{EF} = \frac{AD}{\blacksquare}$
  $\frac{AB}{EF} = \frac{AD}{EH}$ **Corresponding sides are proportional.**

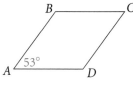

✓ **Quick Check** ❶ Complete: $m\angle B = \blacksquare$ and $\frac{GH}{CD} = \frac{FG}{\blacksquare}$

2 **EXAMPLE** **Determining Similarity**

Determine whether the triangles are similar. If they are, write a similarity statement and give the similarity ratio.

Three pairs of angles are congruent.
Also, corresponding sides are proportional.

$\frac{AC}{FD} = \frac{18}{24} = \frac{3}{4}$ $\frac{AB}{FE} = \frac{15}{20} = \frac{3}{4}$ $\frac{BC}{ED} = \frac{12}{16} = \frac{3}{4}$

$\triangle ABC \sim \triangle FED$ with a similarity ratio of $\frac{3}{4}$ or 3 : 4.

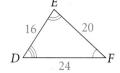

✓ **Quick Check** ❷ Sketch $\triangle XYZ$ and $\triangle MNP$ with $\angle X \cong \angle M$, $\angle Y \cong \angle N$, and $\angle Z \cong \angle P$. Also, $XY = 12$, $YZ = 14$, $ZX = 16$, $MN = 18$, $NP = 21$, and $PM = 24$. Can you conclude that the two triangles are similar? Explain.

You can use proportions to find unknown lengths in similar polygons.

③ EXAMPLE Using Similar Figures

Algebra $LMNO \sim QRST$
Find the value of x.

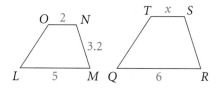

Write a proportion.

$\dfrac{LM}{QR} = \dfrac{ON}{TS}$ **Corresponding sides of $\sim$ polygons are proportional.**

$\dfrac{5}{6} = \dfrac{2}{x}$ **Substitute.**

$5x = 12$ **Cross-Product Property**

$x = 2.4$ **Solve for x.**

✓ Quick Check ③ Use the figures above. Find SR to the nearest tenth.

2 Applying Similar Polygons

You can use similar polygons to find measures when using enlarged or reduced images.

④ EXAMPLE Real-World 🌐 Connection

Technology If you have a vision problem, a magnification system can help you read. You choose a level of magnification. Then you place an image under the viewer. A similar, magnified image appears on the video screen.

The video screen pictured is 16 in. wide by 12 in. tall. What is the largest complete video image possible for a block of text that is 6 in. wide by 4 in. tall?

The 6-in. width can be magnified at most to a rectangle with a longer side of 16 in. Let the shorter side of the video image have length x.

$\dfrac{6}{16} = \dfrac{4}{x}$ **Corresponding sides of $\sim$ polygons are proportional.**

$6x = 64$ **Cross-Product Property**

$x = 10\frac{2}{3}$ **Solve for x.**

The 4-in. height of the block of text enlarges to $10\frac{2}{3}$ in. This is less than the 12-in. height of the video screen, so the entire block of text fits on the screen.

The largest complete video image possible for the block of text is 16 in. by $10\frac{2}{3}$ in.

✓ Quick Check ④ On the video screen in Example 4, what is the largest complete image possible for a photograph that is 3 in. wide by 5 in. tall?

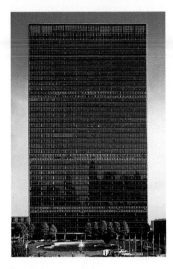

A **golden rectangle** is a rectangle that can be divided into a square and a rectangle that is similar to the original rectangle. A pattern of repeated golden rectangles is shown at the right. Each golden rectangle that is formed is copied and divided again. Each golden rectangle is similar to the original rectangle.

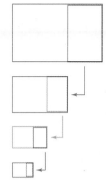

In any golden rectangle, the length and width are in the **golden ratio** which is about 1.618 : 1. You will derive this ratio in Exercise 51.

The golden rectangle is considered pleasing to the human eye. It has appeared in architecture and art since ancient times. It has intrigued artists including Leonardo da Vinci (1452–1519). Da Vinci illustrated *The Divine Proportion*, a book about the golden rectangle.

You can apply the golden ratio to real-life design problems.

5 EXAMPLE **Real-World** 🌐 **Connection**

Art An artist plans to paint a picture. He wants the canvas to be a golden rectangle with its longer horizontal sides 30 cm wide. How high should the canvas be?

Let h be the height of the canvas.

$\frac{30}{h} = \frac{1.618}{1}$	**Write a proportion.**
$1.618h = 30$	**Cross-Product Property**
$h = \frac{30}{1.618}$	**Solve for h.**
$h = 18.541409$	**Use a calculator.**

● The canvas should be about 18.5 cm high.

✓ **Quick Check** ⑤ A golden rectangle has shorter sides of length 20 cm. Find the length of the longer sides.

EXERCISES

For more exercises, see *Extra Skill, Word Problem, and Proof Practice.*

Practice and Problem Solving

🅐 **Practice by Example**

Example 1
(page 373)

GO for Help

JDRT ~ JHYX. **Complete the congruence and proportion statements.**

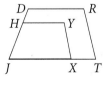

1. $\angle D \cong \underline{\ ?\ }$

2. $\angle Y \cong \underline{\ ?\ }$

3. $\angle T \cong \underline{\ ?\ }$

4. $\frac{JD}{JH} = \frac{DR}{\blacksquare}$

5. $\frac{RT}{YX} = \frac{\blacksquare}{JX}$

6. $\frac{\blacksquare}{DR} = \frac{YX}{RT}$

Example 2
(page 373)

Are the polygons similar? If they are, write a similarity statement and give the similarity ratio. If they are not, explain.

7.

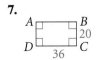

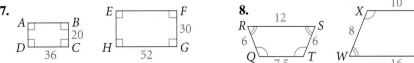

8.

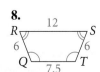

Are the polygons similar? If they are, write a similarity statement and give the similarity ratio. If they are not, explain.

9.
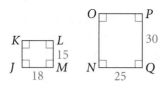

10.
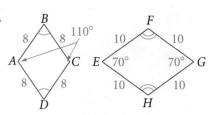

11.
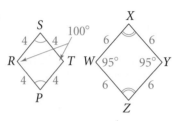

12.
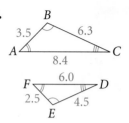

Example 3 x^2 **Algebra** **The polygons are similar. Find the value of each variable.**
(page 374)

13.
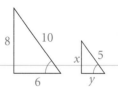

14.
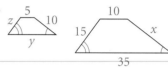

15.
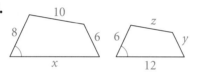

16.
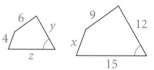

Example 4 **17. Drawing** You want to draw an enlargement of a design that is painted on a
(page 374)   3 in.-by-5 in. card. You plan to draw on an $8\frac{1}{2}$ in.-by-11 in. piece of paper. What
             are the dimensions of the largest complete enlargement you can draw?

18. A map has dimensions 9 in. by 15 in. You want to reduce the map so that it will
fit on a 4 in.-by-6 in. index card. What are the dimensions of the largest possible
complete map that you can fit on the index card?

Example 5 **19. Electrical Equipment** A switch plate for a standard wall switch has the shape
(page 375)   of a golden rectangle. The longer side of the switch plate is about 114 mm.
             How long is the shorter side? Round your answer to the nearest millimeter.

20. Design You want the banner you are creating from one piece of cloth to be a
golden rectangle. The cloth will be cut from a bolt that is 54 in. wide. What are
the dimensions of the largest banner that you can make?

B **Apply Your Skills**

$\triangle DFG \sim \triangle HKM$. **Use the diagram to find the following.**

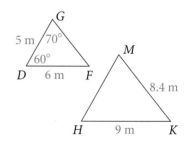

21. the similarity ratio of $\triangle DFG$ to $\triangle HKM$

22. the similarity ratio of $\triangle HKM$ to $\triangle DFG$

23. $m\angle F$ **24.** $m\angle K$ **25.** $m\angle M$

26. $\frac{DF}{HK}$ **27.** HM **28.** GF

29. Writing Are two congruent figures similar? Explain.

30. a. Reading Math What two symbols combine to form the congruence symbol?
   b. Explain why the congruence symbol makes sense.

31. Multiple Choice An art class is painting a rectangular mural for a community
festival. The students planned the mural with a diagram that is 80 in. long and
16 in. high. The mural is 4 ft high. What is its length?

   Ⓐ 5 ft 4 in.   Ⓑ 10 ft   Ⓒ 20 ft   Ⓓ 26 ft 8 in.

x^2 **Algebra** **Find the values of the variables.**

32.

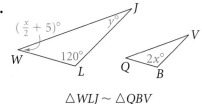

$\triangle WLJ \sim \triangle QBV$

33.

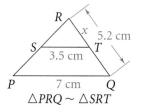

$\triangle PRQ \sim \triangle SRT$

The polygons in each exercise are similar. Find the similarity ratio of the first to the second.

34.

35.

36. Exercise 13   **37.** Exercise 14   **38.** Exercise 15   **39.** Exercise 16

🌐 **Logo Design** A company logo is a rhombus with 4-cm sides and angles of 60° and
120°. Find the angle measures and side lengths if the logo is changed as follows.

40. reduced by 50%   **41.** reduced to 50%   **42.** reduced by 20%

43. reduced to 20%   **44.** reduced by 75%   **45.** reduced to 75%

Find the other side length of the golden rectangle to the nearest tenth of an inch.

46. The shorter side is 10 in.   **47.** The longer side is 10 in.

🌐 **48. Money** From 1861 to 1928, U. S. paper currency measured 7.42 in. by 3.13 in.
The dimensions of a current bill are
shown here. Are the
old and new bills
similar rectangles?
Explain.

Go Online
Homework Video Tutor

Visit: PHSchool.com
Web Code: aue-0702

49. Critical Thinking Are all circles similar? Explain.

50. Open-Ended Draw two polygons with sides in the ratio 2 : 1 that are not similar.

 Challenge **51. a. Algebra** You know that the golden ratio is about 1.618 : 1. You can use the definition of a golden rectangle to derive this ratio. Let *ABCD* at the left be a golden rectangle with width 1. By the definition of golden rectangle, *ABCD* ~ *BCFE*. Fill in the reasons in the following argument.

1. $\frac{AB}{BC} = \frac{BC}{CF}$	1. ?
2. $\frac{x}{1} = \frac{1}{x-1}$	2. ?
3. $x^2 - x = 1$	3. ?
4. $x^2 - x - 1 = 0$	4. ?

b. Using the quadratic formula to solve the equation in part (a), you get $x = \frac{1 \pm \sqrt{5}}{2}$. Explain why $x = \frac{1 - \sqrt{5}}{2}$ does not make sense in this situation.

c. The golden ratio is the ratio $x : 1$, or $\frac{1 + \sqrt{5}}{2} : 1$. Use a calculator to find the value of $x = \frac{1 + \sqrt{5}}{2}$ to the nearest ten-thousandth.

52. a. In the Fibonacci Sequence (see Lesson 1-1), each term after the first two is the sum of the two preceding terms. The first seven terms of the Fibonacci Sequence are 1, 1, 2, 3, 5, 8, and 13. Find the next seven terms.

b. Start with the second term. Here is the ratio of each term to the prior term.

$$\frac{1}{1} = 1 \qquad \frac{2}{1} = 2 \qquad \frac{3}{2} = 1.5 \qquad \frac{5}{3} = 1.6667$$

Find the next nine ratios. Round to the nearest ten thousandth.

c. Compare the ratios that you found in part (b) to the golden ratio.

Multiple Choice Use the figures at the right for Exercises 53–54.

53. What is $m\angle Q$?

   A. 20   **B.** 30

   C. 50   **D.** 100

54. What is the similarity ratio of △*ABC* to △*PQR*?

   F. 5 : 8   **G.** 7 : 10

   H. 10 : 7   **J.** 8 : 5

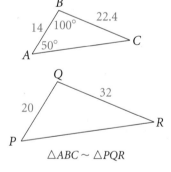

△*ABC* ~ △*PQR*

Short Response **55. a.** Find $m\angle C$.

   b. Write a proportion and solve it to find *CP*.

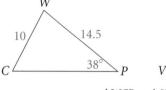 

△*WCP* ~ △*HNV*

56. Quadrilateral *ABCD* ~ quadrilateral *JKLM* with a similarity ratio of 2 : 3.

   a. If $BC = 8$ cm, find *KL*.

   b. If $m\angle BCD = 38$, find $m\angle KLM$.

Lesson 7-1   If $\frac{x}{7} = \frac{y}{9}$, complete each of the following using properties of proportions.

 for Help

57. $9x = \blacksquare$ **58.** $\frac{x}{y} = \frac{\blacksquare}{\blacksquare}$ **59.** $\frac{x+7}{7} = \frac{\blacksquare}{\blacksquare}$

Lesson 6-3   **Can you conclude that the quadrilateral is a parallelogram? Explain.**

60.   **61.**   **62.**

Lesson 4-5   **Use the marked △CEA for Exercises 63–66.**

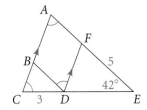

63. Name the isosceles triangles in the figure.

64. $\overline{CD} \cong \underline{\ ?\ } \cong \underline{\ ?\ }$

Find the value of each of the following.

65. AE **66.** $m\angle A$

 Checkpoint Quiz 1   **Lessons 7-1 through 7-2**

1. Models A table is 4 ft high. A small model of the table is 6 in. high. What is the ratio of the height of the model table to the height of the real table?

2. If $\frac{a}{b} = \frac{9}{10}$, complete this statement: $\frac{a}{9} = \frac{\blacksquare}{\blacksquare}$

3. Are the two polygons similar? If so, give the similarity ratio of the first polygon to the second. If not, explain.

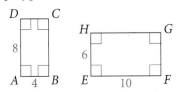

Solve each proportion.

4. $\frac{y}{6} = \frac{18}{54}$ **5.** $\frac{5}{7} = \frac{x-2}{4}$

6. The scale of a scale drawing is 2 in. = 5 ft. A room is 5 in. long on the scale drawing. Find the actual length of the room.

△ABC ~ △DBF at the right. Complete each statement.

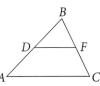

7. $m\angle A = m\angle \underline{\ ?\ }$ **8.** $\frac{AB}{DB} = \frac{BC}{\blacksquare}$

9. A postcard is 6 in. by 4 in. A printing shop will enlarge it so that the longer side is any length up to 3 ft. Find the dimensions of the biggest enlargement.

x^2 **10. Algebra** The polygons at the right are similar. Find the value of x.

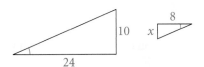

Fractals

Fractals are objects that have three important properties:

- You can form them by repeating steps—a process called *iteration*.

- They require infinitely many iterations. In practice, you can continue until the objects become too small to draw. Even then the steps could continue in your mind.

- At each stage, a portion of the object is a reduced copy of the entire object at the previous stage. This property is called *self-similarity*.

> **NY G.R.8:** Use mathematics to show and understand mathematical phenomena (e.g., use investigation, discovery, conjecture, reasoning, arguments, justification and proofs to validate that the two base angles of an isosceles triangle are congruent).

1 EXAMPLE

The segment below of length 1 unit is Stage 0 of a fractal tree. Draw Stage 1 and Stage 2 of the tree. For each stage, draw two branches from the top third of each segment.

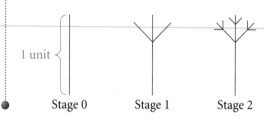

1 unit

Stage 0   Stage 1   Stage 2

Amazingly, some fractals are used to describe natural formations such as mountain ranges and clouds. In 1904, Swedish mathematician Helge von Koch created the Koch Curve, a fractal that is used to model coastlines.

2 EXAMPLE

The segment at the right of length 1 unit is Stage 0 of a Koch Curve. Draw Stages 1–4 of the curve. For each stage, replace the middle third of each segment with two segments, both equal in length to the middle third.

- For Stage 1, replace the middle third with two segments, both $\frac{1}{3}$ unit long.

- For Stage 2, replace the middle third of each segment with two segments, both $\frac{1}{9}$ unit long.

- Continue with a third and fourth iteration.

Stages 0–4 are shown at the right.

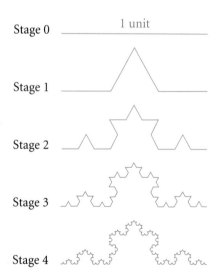

Stage 0   1 unit

Stage 1

Stage 2

Stage 3

Stage 4

You can construct a Koch Curve on each side of an equilateral triangle and get a Koch Snowflake.

The equilateral triangle at the right is Stage 0 of a Koch Snowflake. Draw Stage 1.

- Draw an equilateral triangle on the middle third of each side.

- Erase the middle third of each side to get Stage 1 of the snowflake. (Continued in Exercises 5–9.)

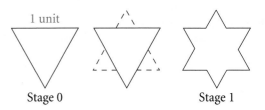

Stage 0   Stage 1

EXERCISES

1. Draw Stage 3 of the fractal tree in Example 1.

Use the Koch Curve in Example 2 for Exercises 2–4.

2. Complete the table to find the length of the Koch Curve at each stage.

3. Examine the results of Exercise 2 and look for a pattern. Use this pattern to predict the length of the Koch Curve at Stage 3; at Stage 4.

Stage	0	1	2
Length	1	■	■

4. Suppose you are able to complete a Koch Curve to Stage n.
  a. Write an expression for the length of the curve.
  b. What happens to the length of the curve as n gets large?

5. Draw Stage 2 of the Koch Snowflake in Example 3.

Stage 3 of the Koch Snowflake is shown at the right. Use it and the earlier stages to answer Exercises 6–8.

6. At each stage, is the snowflake equilateral?

7. a. Complete the table to find the perimeter at each stage.

Stage	Number of Sides	Length of a Side	Perimeter
0	3	1	3
1	■	$\frac{1}{3}$	■
2	48	■	■
3	■	■	■

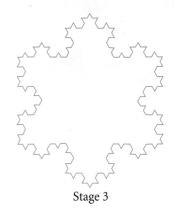

Stage 3

  b. Predict the perimeter at Stage 4.
  c. Will there be a stage at which the perimeter is greater than 100 units? Explain.

8. Exercises 4 and 7 suggest that there is no bound on the perimeter of the Koch snowflake. Is this true about the area of the Koch snowflake? Explain.

9. To draw the Sierpinski Triangle fractal, start with an equilateral triangle. For each stage, connect the midpoints of all of the triangles "pointed upwards." Stages 0–2 are shown at the right. Draw Stage 3.

Stage 0   Stage 1   Stage 2

Proving Triangles Similar

Learning Standards for Mathematics

G.G.44 Establish similarity of triangles, using the following theorems: AA, SAS, and SSS.

G.G.45 Investigate, justify, and apply theorems about similar triangles.

 **Check Skills You'll Need**

 for Help Lessons 4-2 and 4-3

Name the postulate or theorem you can use to prove the triangles congruent.

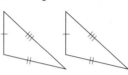

New Vocabulary • indirect measurement

---

1 The AA Postulate and the SAS and SSS Theorems

Real-World Connection

The gables on the historic House of the Seven Gables in Salem, Massachusetts, suggest similar triangles.

Activity: Triangles with Two Pairs of Congruent Angles

- Draw two triangles of different sizes, each with a 50° angle and a 60° angle.

- Measure the sides of each triangle to the nearest millimeter.

- Find the ratio of the lengths of each pair of corresponding sides.

1. What conclusion can you make about the two triangles?

2. Complete this conjecture:

   If two angles of one triangle are congruent to two angles of another triangle, then the triangles are ___?___.

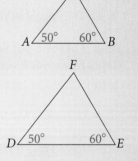

In this lesson, you will show triangles are similar without using the definition of similar triangles. The two triangles shown above suggest the following postulate.

 Key Concepts

Postulate 7-1	**Angle-Angle Similarity (AA ∼) Postulate**

If two angles of one triangle are congruent to two angles of another triangle, then the triangles are similar.

$\triangle TRS \sim \triangle PLM$

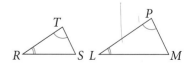

1 **EXAMPLE** **Using the AA ~ Postulate**

Explain why the triangles are similar.
Write a similarity statement.

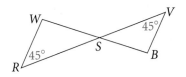

$\angle RSW \cong \angle VSB$ because vertical angles are
congruent. $\angle R \cong \angle V$ because their measures
are equal. $\triangle RSW \sim \triangle VSB$ by the Angle-Angle
Similarity Postulate.

✓ **Quick Check** ❶ **Critical Thinking** In Example 1, you have enough information to write a similarity
statement. Do you have enough information to find the similarity ratio? Explain.

The next two theorems follow from the AA Similarity Postulate.

🔑 **Key Concepts**

Theorem 7-1	**Side-Angle-Side Similarity (SAS ~) Theorem**

If an angle of one triangle is congruent to an angle of a second triangle,
and the sides including the two angles are proportional, then the triangles
are similar.

Proof **Proof of Theorem 7-1**

Given: $\angle A \cong \angle Q$, $\dfrac{AB}{QR} = \dfrac{AC}{QS}$

Prove: $\triangle ABC \sim \triangle QRS$

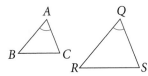

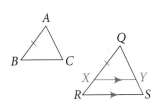

Choose X on $\overline{QR}$ so that $QX = AB$. (See figures at left.) Draw $\overline{XY} \parallel \overline{RS}$.
Then $\angle QXY \cong \angle R$. By the AA ~ Postulate, $\triangle QXY \sim \triangle QRS$. Thus $\dfrac{QX}{QR} = \dfrac{QY}{QS}$.
Combining this proportion, the given proportion, and the fact that $AB = QX$
shows that $\dfrac{AC}{QS} = \dfrac{QY}{QS}$. Hence, $AC = QY$.

Then $\triangle ABC \cong \triangle QXY$ by the SAS Congruence Postulate. $\angle B \cong \angle QXY$ by
CPCTC, and $\angle B \cong \angle R$ by the Transitive Property. Thus, $\triangle ABC \sim \triangle QRS$ by the
AA ~ Postulate.

🔑 **Key Concepts**

Theorem 7-2	**Side-Side-Side Similarity (SSS ~) Theorem**

If the corresponding sides of two triangles are proportional, then the triangles
are similar.

Proof **Proof of Theorem 7-2**

Given: $\dfrac{AB}{QR} = \dfrac{BC}{RS} = \dfrac{AC}{QS}$

Prove: $\triangle ABC \sim \triangle QRS$

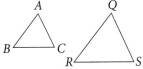

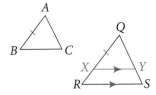

Draw $\overline{XY}$ as in the proof of Theorem 7-1. By the AA ~ Postulate,
$\triangle QXY \sim \triangle QRS$. Thus $\dfrac{QX}{QR} = \dfrac{XY}{RS} = \dfrac{QY}{QS}$. Combining this proportion, the given
proportion, and the fact that $AB = QX$ shows that $BC = XY$ and $AC = QY$.

Then $\triangle ABC \cong \triangle QXY$ by the SSS Congruence Postulate and, as in the proof
above, $\triangle ABC \sim \triangle QRS$ by the AA ~ Postulate.

Lesson 7-3 Proving Triangles Similar **383**

2 EXAMPLE Using Similarity Theorems

Explain why the triangles must be similar.
Write a similarity statement.

$\angle QRP \cong \angle XYZ$ because they are right angles.

$\frac{QR}{XY} = \frac{3}{4}$ and $\frac{PR}{ZY} = \frac{6}{8} = \frac{3}{4}$.

Therefore, $\triangle QRP \sim \triangle XYZ$
by the SAS $\sim$ Theorem.

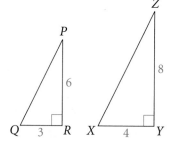

✓ Quick Check ② Explain why the triangles must be similar.
Write a similarity statement.

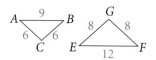

---

2 Applying AA, SAS, and SSS Similarity

You can apply the AA Similarity Postulate and the SAS and SSS Similarity Theorems to find the lengths of sides in similar triangles.

3 EXAMPLE Finding Lengths in Similar Triangles

Multiple Choice If the triangles are similar, find DE.

Ⓐ 10   Ⓑ 15   Ⓒ 32
Ⓓ The triangles are not similar.

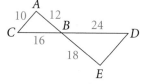

$\angle ABC \cong \angle EBD$ because vertical angles are congruent.

$\frac{AB}{EB} = \frac{12}{18} = \frac{2}{3}$ and $\frac{CB}{DB} = \frac{16}{24} = \frac{2}{3}$

Therefore, $\triangle ABC \sim \triangle EBD$ by the SAS $\sim$ Theorem.

$\frac{CA}{DE} = \frac{2}{3}$ **Corresponding sides of $\sim$ triangles are proportional.**

$\frac{10}{DE} = \frac{2}{3}$ **Substitute.**

$2DE = 30$ **Cross-Product Property**

$DE = 15$ **The correct answer is B.**

Test-Taking Tip

In similarity and congruence problems, try to name the figures to show the correct correspondence between vertices.

✓ Quick Check ③ Find the value of x in the figure at the right.

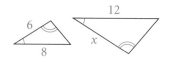

You can use similar triangles and measurements to find distances that are difficult to measure directly. This is called **indirect measurement.**

One method of indirect measurement uses the fact that light reflects off a mirror at the same angle at which it hits the mirror. A second method uses the similar triangles that are formed by certain figures and their shadows.

Both methods are illustrated in Example 4.

Geology Ramon places a mirror on the ground 40.5 ft from the base of a geyser. He walks backwards until he can see the top of the geyser in the middle of the mirror. At that point, Ramon's eyes are 6 ft above the ground and he is 7 ft from the image in the mirror. Use similar triangles to find the height of the geyser.

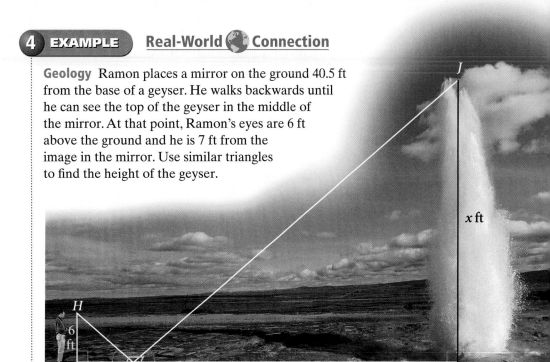

∠HVT ≅ ∠JVS follows from the fact that the angle of incidence equals the angle of reflection. See p. 48, Exercise 18.

△HTV ∼ △JSV	**AA ∼ Postulate**
$\frac{HT}{JS} = \frac{TV}{SV}$	**Corresponding sides of ∼ triangles are proportional.**
$\frac{6}{x} = \frac{7}{40.5}$	**Substitute.**
$243 = 7x$	**Cross-Product Property**
$34.7 \approx x$	**Solve for x.**

● The geyser is about 35 ft high.

 **Quick Check** ④ In sunlight, a cactus casts a 9-ft shadow. At the same time a person 6 ft tall casts a 4-ft shadow. Use similar triangles to find the height of the cactus.

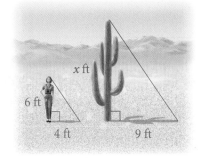

EXERCISES

For more exercises, see Extra Skill, Word Problem, and Proof Practice.

Practice and Problem Solving

 Practice by Example

Examples 1 and 2
(pages 383 and 384)

Can you conclude the triangles are similar? If so, write a similarity statement and name the postulate or theorem you used. If not, explain.

1.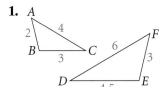

2.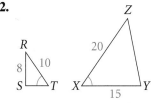

3. If possible, find the similarity ratio for each pair of similar triangles in Exercises 1 and 2. If not possible, explain.

Are the triangles similar? If so, write a similarity statement and name the postulate or theorem you used. If not, explain.

4.

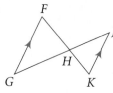

5.

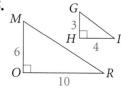

6.

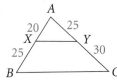

7.

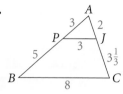

8.

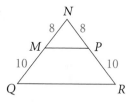

9.

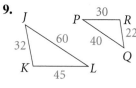

Example 3 x^2 **Algebra** **Explain why the triangles are similar. Then find the value of x.**
(page 384)

10.

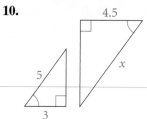

11.

12.

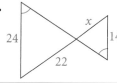

13.

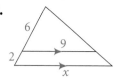

14.

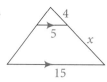

15.

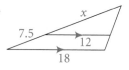

Example 4 **Indirect Measurement** **Explain why the triangles are similar. Then find the**
(page 385) **distance represented by x.**

16.

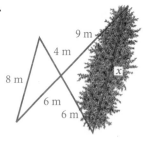

17.

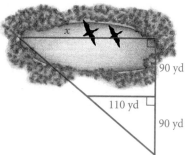

18.

19.

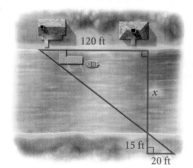

B **Apply Your Skills**   **Tall Buildings** Use the news article for Exercises 20 and 21.

20. Writing Explain how Hannelore Kraus could use indirect measurement to estimate the length of the shadow of the building at a particular time of day.

21. Indirect Measurement Suppose Ms. Kraus is 1.75 m tall. When her shadow is 1 m long, about how long would the shadow of the proposed building be?

22. a. Classify *RSTW*.
  b. Must any of the triangles shown be similar? Explain.

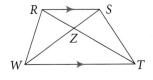

23. a. Critical Thinking Are two isosceles triangles always similar? Explain.
  b. Are two isosceles right triangles always similar? Explain.

Can you conclude that the triangles are similar? If so, write a similarity statement and name the postulate or theorem you used. If not, explain.

24.
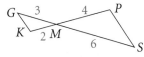

25.
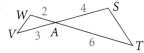

26.
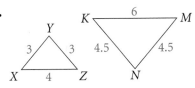

27.
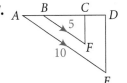

28. Indirect Measurement In sunlight, a vertical yardstick casts a 1-ft shadow at the same time that a nearby tree casts a 15-ft shadow. How tall is the tree?

29. Open-Ended Name something with a height that would be difficult to measure directly. Describe how you could measure it indirectly.

Find the similarity ratio of the larger to the smaller triangle in each exercise.

30. Ex. 10   **31.** Ex. 11   **32.** Ex. 12   **33.** Ex. 13   **34.** Ex. 14

35. Ex. 15   **36.** Ex. 16   **37.** Ex. 17   **38.** Ex. 18   **39.** Ex. 19

40. Constructions Draw any △*ABC*. Use a straightedge and a compass to construct △*RST* so that △*ABC* ~ △*RST* with similarity ratio 1 : 3.

Proof **41. Given:** $RT \cdot TQ = MT \cdot TS$
  Prove: △*RTM* ~ △*STQ*

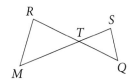

C **Challenge** *Proof* **42.** Write a proof of the following:
Any two nonvertical parallel lines have equal slopes.

  Given: nonvertical lines ℓ_1 and ℓ_2,
    $\ell_1 \parallel \ell_2$, $\overline{EF}$ and $\overline{BC} \perp$ to the *x*-axis
  Prove: $\dfrac{BC}{AC} = \dfrac{EF}{DF}$

Hint: Use the *x*-axis as a transversal to show that $\angle BAC \cong \angle EDF$.

Proof 43. Use the diagram in Exercise 42 on the preceding page.
Prove: Any two nonvertical lines with equal slopes are parallel.

44. There are three similar triangles in the
figure at the right. Find them and
prove that they are similar.

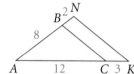

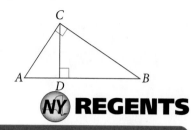

Multiple Choice

45. Complete the statement △ABC ~ __?__ , and identify the reason why
the triangles are similar.
  A. △AKN; SSS ~
  B. △AKN; SAS ~
  C. △ANK; SAS ~
  D. △ANK; AA ~

46. Complete the statement △ABC ~ __?__ , and identify the reason why
the triangles are similar.
  F. △LGC; SSS ~
  G. △GLC; SSS ~
  H. △LGC; AA ~
  J. △GLC; AA ~

Short Response

47. Suppose △VLQ ~ △PSX.
  a. Explain how you would find $m\angle X$ if $m\angle V = 48$ and $m\angle L = 80$.
  b. Find $m\angle X$.

Extended Response

48. Hank is 6 ft tall. Hank measured the shadow of a tree and found it to be
30 ft long. He then measured his own shadow. It was 10 ft long.
  a. Draw and label a diagram that you could use to find the height of the
tree. Write a similarity statement and justify your answer.
  b. Write a proportion and solve it to find the height of the tree.

Mixed Review

Lesson 7-2

TRAP ~ EZYD. **Complete each statement.**

49. $\angle T \cong$ __?__

50. $\angle D \cong$ __?__

51. $\angle A \cong$ __?__

52. $\dfrac{AP}{YD} = \dfrac{RA}{\blacksquare}$

53. $\dfrac{TR}{RA} = \dfrac{\blacksquare}{ZY}$

54. $\dfrac{DE}{PT} = \dfrac{\blacksquare}{AR}$

Lesson 6-6

Give possible coordinates of points *W* and *Z* without using any new variables.

55.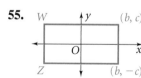

Rectangle

56.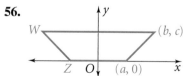

Isosceles trapezoid

Lesson 5-5

57. A triangle has sides with lengths 9 m and 15 m. Write an inequality that shows
the range of possible lengths for the third side.

388 Chapter 7 Similarity

Abbreviations and Symbols

Many cellular phones and other modern communication devices can send and receive short text messages. Because of their small screen sizes and the need to send information quickly, people have found ways to communicate using new abbreviations and symbols.

1 EXAMPLE

Decode the short text message shown at the right.

GTG
TLK2UL8TR

In words:
● Got to go. Talk to you later.

Abbreviations and symbols have always been part of the written vocabulary of mathematics. As with electronic text messaging, the purposes of "mathematical text messaging" include speed and economy in the recording of information.

2 EXAMPLE

Write this statement using words. Then write it again using symbols from the diagram.

In ~ rt. ⧌, corres. ⊥ seg. are prop. to corres. sides.

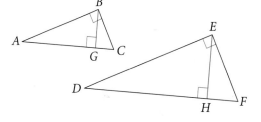

In words:
In similar right triangles, corresponding perpendicular segments are proportional to corresponding sides.

In symbols from the diagram:
● $\triangle ABC \sim \triangle DEF, \overline{BG} \perp \overline{AC}, \overline{EH} \perp \overline{DF} \rightarrow \dfrac{BG}{EH} = \dfrac{AC}{DF}$

EXERCISES

Write each short text message using words.

1. SUP   **2.** BCNU   **3.** THNQ   **4.** ILBL8   **5.** RURDY

Write each math statement using words.

6. Alt. int. ⧌ formed by 2 ‖ lines and a transv. are ≅.

7. A ▱ is a rhom. ↔ 1 diag. bis. opp. ⧌.

Write each math statement using symbols from the diagram.

8. Parallel lines intercept proportional segments on all transversals.

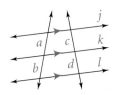

9. The altitude to the hypotenuse of a right triangle divides the triangle into two triangles that are similar to the original triangle and to each other.

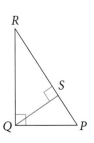

Simplifying Radicals

You can multiply and divide numbers that are under radical signs.

1 EXAMPLE

Simplify the expressions $\sqrt{2} \cdot \sqrt{8}$ and $\sqrt{294} \div \sqrt{3}$.

$$\sqrt{2} \cdot \sqrt{8} = \sqrt{2 \cdot 8} \qquad \leftarrow \text{Rewrite using one radical sign.} \rightarrow \qquad \sqrt{294} \div \sqrt{3} = \sqrt{\frac{294}{3}}$$

$$= \sqrt{16} \qquad \leftarrow \text{Simplify the expression under the radical.} \rightarrow \qquad = \sqrt{98}$$

$$= 4 \qquad \leftarrow \text{Factor out perfect squares and simplify.} \rightarrow \qquad = \sqrt{49 \cdot 2}$$

$$= 7\sqrt{2}$$

A radical expression is in simplest form when all the following are true.

- The number under the radical sign has no perfect square factors other than 1.
- The number under the radical sign does not contain a fraction.
- A denominator does not contain a radical expression.

2 EXAMPLE

Write $\sqrt{\frac{4}{3}}$ in simplest form.

$$\sqrt{\frac{4}{3}} = \frac{\sqrt{4}}{\sqrt{3}} \qquad \text{Rewrite the single radical as a quotient.}$$

$$= \frac{2}{\sqrt{3}} \qquad \text{Simplify the numerator.}$$

$$= \frac{2}{\sqrt{3}} \cdot \frac{\sqrt{3}}{\sqrt{3}} \qquad \text{Multiply by a form of 1 to rationalize the denominator.}$$

$$= \frac{2\sqrt{3}}{3} \qquad \text{Simplify.}$$

Check using a calculator: $\sqrt{\frac{4}{3}} \approx 1.1547005$ and $\frac{2\sqrt{3}}{3} \approx 1.1547005$.

EXERCISES

Simplify each expression.

1. $\sqrt{5} \cdot \sqrt{10}$ **2.** $\sqrt{243}$ **3.** $\sqrt{128} \div \sqrt{2}$ **4.** $\sqrt{\frac{125}{4}}$

5. $\sqrt{6} \cdot \sqrt{8}$ **6.** $\frac{\sqrt{36}}{\sqrt{3}}$ **7.** $\frac{\sqrt{144}}{\sqrt{2}}$ **8.** $\sqrt{3} \cdot \sqrt{12}$

9. $\sqrt{72} \div \sqrt{2}$ **10.** $\sqrt{169}$ **11.** $28 \div \sqrt{8}$ **12.** $\sqrt{300} \div \sqrt{5}$

13. $\sqrt{12} \cdot \sqrt{2}$ **14.** $\frac{\sqrt{24}}{\sqrt{3}}$ **15.** $\sqrt{\frac{75}{3}}$ **16.** $\sqrt{18} \cdot \sqrt{2}$

17. $\sqrt{68}$ **18.** $\sqrt{3} \cdot \sqrt{15}$ **19.** $\frac{\sqrt{20}}{\sqrt{5}}$ **20.** $45 \div \sqrt{3}$

21. $\sqrt{\frac{25}{20}}$ **22.** $\sqrt{\frac{8}{28}}$ **23.** $\frac{\sqrt{6} \cdot \sqrt{3}}{\sqrt{9}}$ **24.** $\frac{\sqrt{3} \cdot \sqrt{15}}{\sqrt{2}}$

25. Find and simplify the difference of the golden ratio (see page 378, Exercise 51) and its multiplicative inverse.

Similarity in Right Triangles

NY Learning Standards for Mathematics

G.G.45 Investigate, justify, and apply theorems about similar triangles.

G.G.47 Investigate, justify, and apply theorems about mean proportionality.

x^2 **Algebra** Solve each proportion.

1. $\frac{x}{8} = \frac{18}{24}$ 2. $\frac{2}{3} = \frac{x}{7}$ 3. $\frac{15}{4} = \frac{18}{x}$ 4. $\frac{51}{x} = \frac{17}{13}$

5. $\frac{4}{10} = \frac{x}{5}$ 6. $\frac{3}{m} = \frac{9}{8}$ 7. $\frac{w}{2} = \frac{20}{9}$ 8. $\frac{9}{6} = \frac{27}{a}$

9. Draw a right triangle. Label the triangle $\triangle ABC$ with right angle $\angle C$. Draw the altitude to the hypotenuse. Label the altitude $\overline{CD}$. Name the two smaller right triangles that are formed.

◀)) **New Vocabulary** • geometric mean

---

1 Using Similarity in Right Triangles

Hands-On Activity: Similarity in Right Triangles

- Draw one diagonal on a rectangular sheet of paper. Cut the paper on the diagonal to make two congruent right triangles.

- In one of the triangles, use paper folding to locate the altitude to the hypotenuse. Cut the triangle along the altitude to make two smaller right triangles.

- Label the angles of the three triangles as shown.

- Compare the angles of the three triangles by placing the angles on top of one another.

1. Which angles have the same measure as $\angle 1$?

2. Which angles have the same measure as $\angle 2$?

3. Which angles have the same measure as $\angle 3$?

4. Based on your results, what is true about the three triangles?

5. Use the diagram at the right to complete the similarity statement.
$\triangle RST \sim \triangle\ \underline{?}\ \sim \triangle\ \underline{?}$

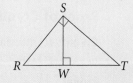

In a right triangle, the altitude to the hypotenuse yields three similar triangles.

Key Concepts

> **Theorem 7-3**
>
> The altitude to the hypotenuse of a right triangle divides the triangle into two triangles that are similar to the original triangle and to each other.

Proof

Proof of Theorem 7-3

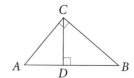

Given: Right triangle, $\triangle ABC$, with $\overline{CD}$ the altitude to the hypotenuse

Prove: $\triangle ABC \sim \triangle ACD \sim \triangle CBD$

Proof: Both smaller triangles are right triangles. Each also shares an angle with $\triangle ABC$. Thus each smaller triangle is similar to $\triangle ABC$ by the AA ~ Postulate. Since both smaller triangles are similar to $\triangle ABC$, their corresponding angles are congruent. Thus they are similar to each other.

Vocabulary Tip

In the proportion
$\frac{a}{b} = \frac{c}{d}$,
b and c are the <u>means</u>.

Proportions in which the means are equal occur frequently in geometry. For any two positive numbers a and b, the **geometric mean** of a and b is the positive number x such that $\frac{a}{x} = \frac{x}{b}$. Note that $x = \sqrt{ab}$.

1 EXAMPLE **Finding the Geometric Mean**

Algebra Find the geometric mean of 4 and 18.

$\frac{4}{x} = \frac{x}{18}$ **Write a proportion.**

$x^2 = 72$ **Cross-Product Property**

$x = \sqrt{72}$ **Take the square root.**

$x = 6\sqrt{2}$ **Write in simplest radical form.**

● The geometric mean of 4 and 18 is $6\sqrt{2}$.

 **Quick Check** ❶ Find the geometric mean of 15 and 20.

Two important corollaries of Theorem 7-3 involve a geometric mean.

Key Concepts

> **Corollary**   **Corollary 1 to Theorem 7-3**
>
> The length of the altitude to the hypotenuse of a right triangle is the geometric mean of the lengths of the segments of the hypotenuse.

Proof

Proof of Corollary 1

Given: Right triangle, $\triangle ABC$, with $\overline{CD}$ the altitude to the hypotenuse

Prove: $\frac{AD}{CD} = \frac{CD}{DB}$

Proof: By Theorem 7-3, $\triangle ACD \sim \triangle CBD$. Since corresponding sides of similar triangles are proportional, $\frac{AD}{CD} = \frac{CD}{DB}$.

Corollary	Corollary 2 to Theorem 7-3

The altitude to the hypotenuse of a right triangle separates the hypotenuse so that the length of each leg of the triangle is the geometric mean of the length of the adjacent hypotenuse segment and the length of the hypotenuse.

Proof

Proof of Corollary 2

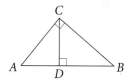

Given: Right triangle, $\triangle ABC$, with $\overline{CD}$ the altitude to the hypotenuse

Prove: $\dfrac{AD}{AC} = \dfrac{AC}{AB}$, $\dfrac{DB}{CB} = \dfrac{CB}{AB}$

Proof: By Theorem 7-3, $\triangle ACD \sim \triangle ABC$. Their corresponding sides are proportional, so $\dfrac{AD}{AC} = \dfrac{AC}{AB}$. Similarly, $\triangle CBD \sim \triangle ABC$ and $\dfrac{DB}{CB} = \dfrac{CB}{AB}$.

Online active math

For: Similarity Activity
Use: Interactive Textbook, 7-4

2 EXAMPLE Applying Corollaries 1 and 2

Algebra Solve for x and y.

Use Corollary 2 to solve for x:

$\dfrac{4}{x} = \dfrac{x}{4+5}$ ← **Write a proportion.** →

$x^2 = 36$ ← **Cross-Product Property** →

$x = 6$ ← **Take the square root.** →

Use Corollary 1 to solve for y:

$\dfrac{4}{y} = \dfrac{y}{5}$

$y^2 = 20$

$y = 2\sqrt{5}$

✓ Quick Check ❷ Solve for x and y.

3 EXAMPLE Real-World 🌐 Connection

Recreation The 300-m path to the information center and the 400-m path to the canoe rental dock meet at a right angle at the parking lot. Marla walks straight from the parking lot to the lake as shown, where a sign tells her that she is 320 m from the dock. How far is Marla from the information center?

$\dfrac{x}{300} = \dfrac{300}{x+320}$ **Corollary 2**

$x^2 + 320x = 90{,}000$ **Cross-Product Property**

$x^2 + 320x - 90{,}000 = 0$ **Standard Form Quadratic Equation**

$x = 180$ **Solve. Use a calculator (see p. 372).**

Real-World 🌐 Connection

Paddling a canoe burns about 175 calories per hour.

● Marla is 180 m from the information center.

✓ Quick Check ❸ How far did Marla walk from the parking lot to the lake?

EXERCISES

For more exercises, see *Extra Skill, Word Problem, and Proof Practice.*

Practice and Problem Solving

 Practice by Example

 Algebra Find the geometric mean of each pair of numbers.

Example 1
(page 392)

1. 4 and 9   **2.** 4 and 10   **3.** 4 and 12   **4.** 3 and 48

5. 7 and 56   **6.** 5 and 125   **7.** 9 and 24   **8.** 7 and 9

Example 2
(page 393)

x^2 **Algebra** Refer to the figure to complete each proportion.

9. $\dfrac{r}{h} = \dfrac{h}{\blacksquare}$ **10.** $\dfrac{c}{a} = \dfrac{a}{\blacksquare}$ **11.** $\dfrac{\blacksquare}{b} = \dfrac{b}{s}$

12. $\dfrac{r}{\blacksquare} = \dfrac{\blacksquare}{c}$ **13.** $\dfrac{r}{h} = \dfrac{\blacksquare}{s}$ **14.** $\dfrac{s}{b} = \dfrac{\blacksquare}{c}$

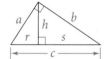

x^2 **Algebra** Solve for x.

15. 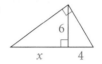   **16.** 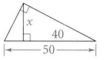   **17.**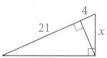

18.   **19.**   **20.**

Example 3
(page 393)

21. a. Civil Engineering Study the plan at the right. A service station will be built on the highway, and a road will connect it with Cray. How far from Blare should the service station be located so that the proposed road will be perpendicular to the highway?
  b. How long will the new road be?

 Apply Your Skills

22. Complete:
$\triangle JKL \sim \triangle\underline{\ ?\ } \sim \triangle\underline{\ ?\ }$

23. a. The altitude to the hypotenuse of a right triangle divides the hypotenuse into segments 2 cm and 8 cm long. Find the length h of the altitude.
  b. Drawing Use the value you found for h in part (a), along with the lengths 2 cm and 8 cm, to draw the right triangle accurately.
  c. Writing Explain how you drew the triangle in part (b).

24. a. Open-Ended Draw a right triangle so that the altitude from the right angle to the hypotenuse bisects the hypotenuse.
  b. How does the length of the altitude compare with the lengths of the segments of the hypotenuse? Explain.

Vocabulary Tip

<u>Respectively</u> in Exercise 25 means you match the lists in the order named:
$A(4, 2)$, $D(4, 6)$, $B(4, 15)$.

25. Coordinate Geometry $\overline{CD}$ is the altitude to the hypotenuse of right $\triangle ABC$. The coordinates of A, D, and B are $(4, 2)$, $(4, 6)$, and $(4, 15)$, respectively. Find all possible coordinates of point C.

x^2 **Algebra** Find the geometric mean of each pair of numbers.

26. 3 and 16   **27.** 4 and 49   **28.** $\sqrt{8}$ and $\sqrt{2}$ **29.** $\sqrt{28}$ and $\sqrt{7}$

30. $\frac{1}{2}$ and 2   **31.** 5 and 1.25   **32.** 1 and 1000   **33.** 11 and 1331

x^2 Algebra Find the values of the variables.

34.

35.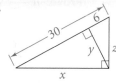

36.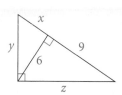

x^2 37. Algebra The altitude to the hypotenuse of a right triangle divides the hypotenuse into segments with lengths in the ratio 1 : 2. The length of the altitude is 8. How long is the hypotenuse?

38. Multiple Choice To estimate the height of a totem pole, Jorge uses a small square of plastic. He holds the square up to his eyes and walks backward from the pole. He stops when the bottom of the pole lines up with the bottom edge of the square and the top of the pole lines up with the top edge of the square. Jorge's eye level is about 2 m from the ground. He is about 3 m from the pole. Which is the best estimate for the height of the totem pole?

    Ⓐ 4.5 m   Ⓑ 5 m   Ⓒ 6.5 m   Ⓓ 9 m

Exercise 38

For a right triangle, denote lengths as follows: ℓ_1 and ℓ_2 the legs, h the hypotenuse, a the altitude, and h_1 and h_2 the hypotenuse segments determined by the altitude. For the two given measures, find the other four. Use simplest radical form.

39. $h = 2, h_1 = 1$ **40.** $h_1 = 4, h_2 = 9$ **41.** $a = 6, h_1 = 6$ **42.** $\ell_1 = 2, h_2 = 3$

43. $h = 13, \ell_2 = 12$ **44.** $\ell_1 = 4, h_1 = 3$ **45.** $a = 8, h_1 = 16$ **46.** $h_1 = 3, \ell_2 = 6\sqrt{3}$

Proof 47. Given: isosceles right $\triangle ABC$
Prove: $AB = x\sqrt{2}$

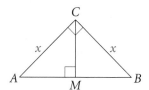

48. Given: equilateral $\triangle ABC$
Prove: $h = 10\sqrt{3}$

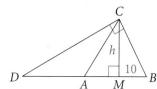

x^2 Algebra Find the value of x.

49.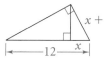

50.

51.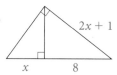

C Challenge

52. a. Lauren thinks she has found a new corollary: The product of the lengths of the two legs of a right triangle is equal to the product of the lengths of the hypotenuse and the altitude to the hypotenuse. Draw a figure for this corollary. Write the *Given* information and what you are to *Prove*.
b. Critical Thinking Is Lauren's corollary true? Explain.

53. In the diagram $c = q + r$. Also, Corollary 2 to Theorem 7-3 gives you two more equations involving $a, b, c, q,$ and r. The result is a system of three equations in five variables.
a. Reduce the system to one equation in three variables by eliminating q and r.
b. State in words what the one resulting equation tells you.

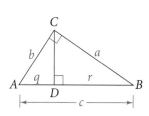

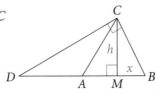

Proof 54. **Given:** equilateral $\triangle ABC$
Prove: $h = x\sqrt{3}$

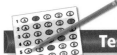
Multiple Choice

55. What is the geometric mean of 12 and 18?
  A. 1.5   **B.** $\sqrt{6}$ **C.** 15   **D.** $6\sqrt{6}$

56. What is the geometric mean of 2 and 36?
  F. 17   **G.** $6\sqrt{2}$ **H.** 38   **J.** $2\sqrt{6}$

57. Solve for m.
  A. 7   **B.** 15
  C. 20   **D.** 25

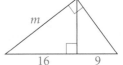

58. The altitude to the hypotenuse of a right triangle divides the hypotenuse into segments of lengths 5 and 15. What is the length of the altitude?
  F. 3   **G.** 10   **H.** $5\sqrt{3}$ **J.** $5\sqrt{5}$

Short Response

59. a. Explain how you could solve for x.
  b. What is the value of x?

Mixed Review

for Help

Lesson 7-3

If the triangles are similar, **(a)** write a similarity statement and **(b)** name the postulate or theorem you used. If the triangles are not similar, write *not similar*.

60.

61.

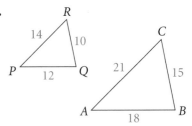

62.

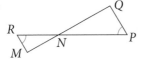

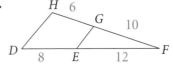

Lesson 6-2 $\boxed{x^2}$ **Algebra** Find the values of x and y in $\square RSTV$.

63. $RP = 2x, PT = y + 2, VP = y, PS = x + 3$

64. $RP = 4x, PT = 3y - 3, VP = 2x + 3, PS = y + 6$

65. $RV = 2x + 3, VT = 5x, TS = y + 5, SR = 4y - 1$

Lesson 5-1

Find the value of x.

66.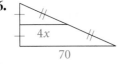

67.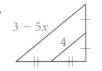

Exploring Proportions in Triangles

FOR USE WITH LESSON 7-5

Construct

- Use geometry software. Draw $\triangle ABC$ and construct point D on $\overline{AB}$.
- Construct a line through D parallel to $\overline{AC}$.
- Construct the intersection E of the parallel line with $\overline{BC}$.

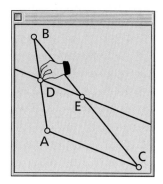

Investigate

- Measure $\overline{BD}, \overline{DA}, \overline{BE}$, and $\overline{EC}$.
- Calculate the ratios $\frac{BD}{DA}$ and $\frac{BE}{EC}$.
- Manipulate $\triangle ABC$ and observe the ratios $\frac{BD}{DA}$ and $\frac{BE}{EC}$.
(Save your observations for Exercise 1.)

> **NY** **G.G.46:**
> Investigate, justify, and apply theorems about proportional relationships among the segments of the sides of the triangle, given one or more lines parallel to one side of a triangle and intersecting the other two sides of the triangle.

Construct

- Use geometry software. Draw $\triangle ABC$. Construct the bisector of $\angle A$. Construct point D, the intersection of the bisector and $\overline{CB}$.

Investigate

- Measure $\overline{AC}, \overline{AB}, \overline{CD}$, and $\overline{DB}$.
- Calculate the ratios $\frac{AC}{AB}$ and $\frac{CD}{DB}$.
- Manipulate $\triangle ABC$ and observe the ratios $\frac{AC}{AB}$ and $\frac{CD}{DB}$.
(Save your observations for Exercise 2.)

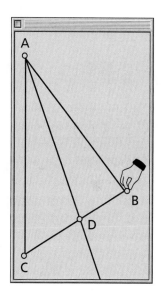

EXERCISES

1. Suppose a line parallel to one side of a triangle intersects the other two sides. Make a conjecture about the four segments formed.

2. The bisector of an angle of a triangle divides the opposite side into two segments. Make a conjecture about the two segments and the other two sides of the triangle.

Extend

- Construct $\overleftrightarrow{AB} \parallel \overleftrightarrow{CD}$. Then construct lines $\overleftrightarrow{AC}$ and $\overleftrightarrow{BD}$.
- Construct point E on $\overleftrightarrow{AC}$.
- Construct $\overleftrightarrow{EF} \parallel \overleftrightarrow{AB}$ with F the intersection of $\overleftrightarrow{EF}$ and $\overleftrightarrow{BD}$.
- Measure $\overline{AC}, \overline{CE}, \overline{BD}$, and $\overline{DF}$.
- Calculate the ratios $\frac{AC}{CE}$ and $\frac{BD}{DF}$.
- Manipulate the locations of A and B. Observe $\frac{AC}{CE}$ and $\frac{BD}{DF}$.

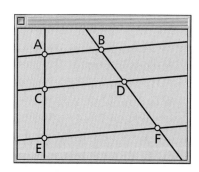

3. Suppose three parallel lines intersect two transversals. Make a conjecture about the segments of the transversals.

4. Suppose that four or more parallel lines intersect two transversals. Make a conjecture about the segments of the transversals.

7-5

Proportions in Triangles

NY Learning Standards for Mathematics

G.G.45 Investigate, justify, and apply theorems about similar triangles.

G.G.46 Investigate, justify, and apply theorems about proportional relationships among the segments of the sides of the triangle, given one or more lines parallel to one side of a triangle and intersecting the other two sides of the triangle.

✓ **Check Skills You'll Need**

GO for Help Lesson 7-2

The two triangles in each diagram are similar. Find the value of *x* in each.

1.

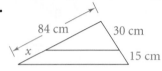

84 cm   30 cm   *x*   15 cm

2.

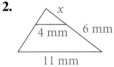

x   4 mm   6 mm   11 mm

3.

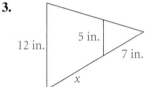

12 in.   5 in.   7 in.   *x*

4.

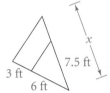

x   7.5 ft   3 ft   6 ft

---

1 Using the Side-Splitter Theorem

You can use similar triangles to prove the following theorem.

 Key Concepts

Theorem 7-4	**Side-Splitter Theorem**

If a line is parallel to one side of a triangle and intersects the other two sides, then it divides those sides proportionally.

Proof

Proof of Theorem 7-4

Given: $\triangle QXY$ with $\overleftrightarrow{RS} \parallel \overleftrightarrow{XY}$

Prove: $\dfrac{XR}{RQ} = \dfrac{YS}{SQ}$

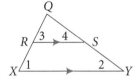

Statements	**Reasons**
1. $\overleftrightarrow{RS} \parallel \overleftrightarrow{XY}$	1. Given
2. $\angle 1 \cong \angle 3, \angle 2 \cong \angle 4$	2. If lines are ∥, then corr. ⧊ are ≅.
3. $\triangle QXY \sim \triangle QRS$	3. AA ~ Postulate
4. $\dfrac{XQ}{RQ} = \dfrac{YQ}{SQ}$	4. Corr. sides of ~ ⧊ are proportional.
5. $XQ = XR + RQ, YQ = YS + SQ$	5. Segment Addition Postulate
6. $\dfrac{XR + RQ}{RQ} = \dfrac{YS + SQ}{SQ}$	6. Substitute.
7. $\dfrac{XR}{RQ} = \dfrac{YS}{SQ}$	7. A Property of Proportions

1 EXAMPLE Using the Side-Splitter Theorem

Gridded Response Find the value of x.

$\dfrac{TS}{SR} = \dfrac{TU}{UV}$ **Side-Splitter Theorem**

$\dfrac{x}{16} = \dfrac{5}{10}$ **Substitute.**

$x = \dfrac{5}{10} \cdot 16$ **Solve for x.**

$x = 8$

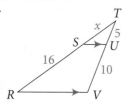

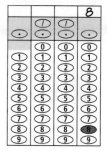

✓ Quick Check ❶ Use the Side-Splitter Theorem to find the value of x.

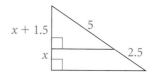

The following corollary to the Side Splitter Theorem says that parallel lines divide all transversals proportionally. You will prove this corollary in Exercise 35.

Key Concepts

Corollary	**Corollary to Theorem 7-4**

If three parallel lines intersect two transversals, then the segments intercepted on the transversals are proportional.

$\dfrac{a}{b} = \dfrac{c}{d}$

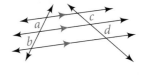

2 EXAMPLE Real-World 🌐 Connection

Sail Making Sail makers sometimes use a computer to create a pattern for a sail. After they cut out the panels of the sail, they sew them together to form the sail.

The edges of the panels in the sail at the right are parallel. Find the lengths x and y.

$\dfrac{2}{x} = \dfrac{1.7}{1.7}$ **Side-Splitter Theorem**

$x = 2$

$\dfrac{3}{2} = \dfrac{y}{1.7}$ **Corollary to the Side-Splitter Theorem**

$\dfrac{3}{2}(1.7) = y$ **Solve for y.**

$2.55 = y$

Length x is 2 ft and length y is 2.55 ft.

Real-World 🌐 Connection

You windsurf with a large sail in light winds and a small sail in strong winds.

✓ Quick Check ❷ Solve for x and y.

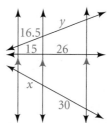

Using the Triangle-Angle-Bisector Theorem

You can use the Side-Splitter Theorem to prove the following relationship.

Key Concepts

Theorem 7-5   **Triangle-Angle-Bisector Theorem**

If a ray bisects an angle of a triangle, then it divides the opposite side into two segments that are proportional to the other two sides of the triangle.

Proof

Proof of Theorem 7-5

Given: $\triangle ABC$, $\overrightarrow{AD}$ bisects $\angle CAB$.

Prove: $\frac{CD}{DB} = \frac{CA}{BA}$

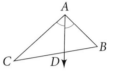

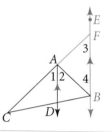

Draw $\overleftrightarrow{BE} \parallel \overleftrightarrow{DA}$. Extend $\overline{CA}$ to meet $\overleftrightarrow{BE}$ at point F.

Proof: By the Side-Splitter Theorem, $\frac{CD}{DB} = \frac{CA}{AF}$.
By the Corresponding Angles Postulate, $\angle 3 \cong \angle 1$.
Since $\overrightarrow{AD}$ bisects $\angle CAB$, $\angle 1 \cong \angle 2$. By the Alternate Interior Angles Theorem, $\angle 2 \cong \angle 4$. Using the Transitive Property of Congruence, you know that $\angle 3 \cong \angle 4$.
By the Converse of the Isosceles Triangle Theorem, $BA = AF$. Substituting BA for AF, $\frac{CD}{DB} = \frac{CA}{BA}$.

Problem Solving Hint

Drawing $\overleftrightarrow{BE} \parallel \overline{DA}$ sets up $\triangle BCF$ for the Side-Splitter Theorem, as well as congruent $\angle$1, 2, 3, and 4.

3 EXAMPLE **Using the Triangle-Angle-Bisector Theorem**

Algebra Find the value of x.

$\frac{PS}{SR} = \frac{PQ}{RQ}$ **Triangle-Angle-Bisector Theorem**

$\frac{x}{6} = \frac{8}{5}$ **Substitute.**

$5x = 48$ **Cross-Product Property**

$x = 9.6$ **Solve for x.**

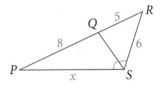

✓ **Quick Check** ❸ Find the value of y.

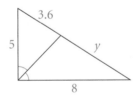

EXERCISES

For more exercises, see *Extra Skill, Word Problem, and Proof Practice.*

Practice and Problem Solving

 Practice by Example x^2 **Algebra Solve for x.**

Example 1
(page 399)

1.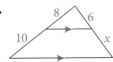

2.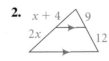

3.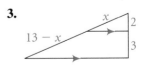

400 Chapter 7 Similarity

Example 2
(page 399)

Use the figure at the right to complete each proportion.

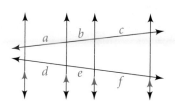

4. $\dfrac{a}{b} = \dfrac{\blacksquare}{e}$

5. $\dfrac{b}{\blacksquare} = \dfrac{e}{f}$

6. $\dfrac{f}{e} = \dfrac{c}{\blacksquare}$

7. $\dfrac{a}{b + c} = \dfrac{\blacksquare}{e + f}$

$\boxed{x^2}$ **Algebra** Solve for x.

8.
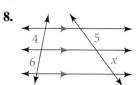

9.
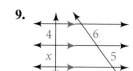

10.
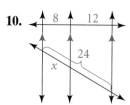

Example 3
(page 400)

$\boxed{x^2}$ **Algebra** Solve for x.

11.

12.

13.

14.

15.
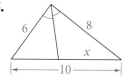

16.
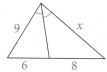

Use the figure at the right to complete each proportion.

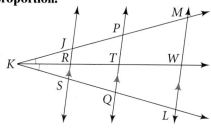

17. $\dfrac{RS}{\blacksquare} = \dfrac{JR}{KJ}$

18. $\dfrac{KJ}{JP} = \dfrac{KS}{\blacksquare}$

19. $\dfrac{QL}{PM} = \dfrac{SQ}{\blacksquare}$

20. $\dfrac{PT}{\blacksquare} = \dfrac{TQ}{KQ}$

21. $\dfrac{KL}{LW} = \dfrac{\blacksquare}{MW}$

22. $\dfrac{\blacksquare}{KP} = \dfrac{LQ}{KQ}$

23. $\dfrac{\blacksquare}{SQ} = \dfrac{JK}{KS}$

24. $\dfrac{KL}{KM} = \dfrac{\blacksquare}{MW}$

B **Apply Your Skills**

Real-World Connection

Careers Some urban planners specialize in environmental issues or historic preservation.

Urban Design In Washington, D.C., 17th, 18th, 19th, and 20th Streets are parallel streets that intersect Pennsylvania Avenue and I Street.

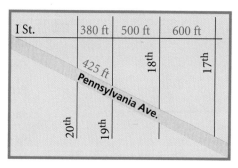

25. How long (to the nearest foot) is Pennsylvania Avenue between 19th Street and 18th Street?

26. How long (to the nearest foot) is Pennsylvania Avenue between 18th Street and 17th Street?

27. The sides of a triangle are 5 cm, 12 cm, and 13 cm long. Find the lengths, to the nearest tenth, of the segments into which the bisector of each angle divides the opposite side.

28. **Open-Ended** In a triangle, the bisector of an angle divides the opposite side into two segments with lengths 6 cm and 9 cm. How long could the other two sides of the triangle be? (*Caution:* Make sure the three sides satisfy the Triangle Inequality Theorem.)

For a guide to solving
Exercise 29, see p. 405.

29. Surveying The perimeter of the triangular
lot at the right is 50 m. The surveyor's tape
bisects an angle. Find the lengths x and y.

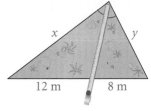

30. Critical Thinking Sharell draws $\triangle ABC$. She finds
that the bisector of $\angle C$ bisects the opposite side.
  a. Sketch $\triangle ABC$ and the bisector.
  b. Writing What type of triangle is $\triangle ABC$? Explain your reasoning.

x^2 **Algebra** Solve for x.

31.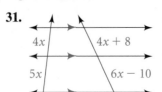

32.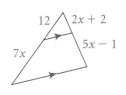

33.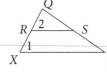

34. Developing Proof Copy and complete this two-column proof of the Converse
of the Side-Splitter Theorem: If a line divides two sides of a triangle
proportionally, then it is parallel to the third side.

Given: $\dfrac{XR}{RQ} = \dfrac{YS}{SQ}$

Prove: $\overline{RS} \parallel \overline{XY}$

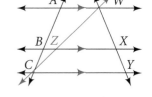

Statements	Reasons
1. $\dfrac{XR}{RQ} = \dfrac{YS}{SQ}$	**a.** ?
2. $\dfrac{XR + RQ}{RQ} = \dfrac{YS + SQ}{SQ}$	**b.** ?
3. $\dfrac{XQ}{RQ} = \dfrac{YQ}{SQ}$	**c.** ?
4. $\angle Q \cong \angle Q$	**d.** ?
5. $\triangle XQY \sim \triangle RQS$	**e.** ?
6. $\angle 1 \cong \angle 2$	**f.** ?
7. $\overline{RS} \parallel \overline{XY}$	**g.** ?

Proof 35. Follow the steps below. Write a proof of the Corollary
to the Side-Splitter Theorem found on page 399.

Given: $\overleftrightarrow{AW} \parallel \overleftrightarrow{BX} \parallel \overleftrightarrow{CY}$

Prove: $\dfrac{AB}{BC} = \dfrac{WX}{XY}$

Begin by drawing $\overleftrightarrow{WC}$, intersecting $\overline{BX}$ at point Z.
  a. Apply the Side-Splitter Theorem to $\triangle ACW$: $\dfrac{\blacksquare}{\blacksquare} = \dfrac{WZ}{ZC}$.
  b. Apply the Side-Splitter Theorem to $\triangle CWY$: $\dfrac{WZ}{ZC} = \dfrac{\blacksquare}{\blacksquare}$.
  c. Substitute to prove the corollary.

**Determine whether the red segments are parallel. Explain each answer. You can
use the theorem proved in Exercise 34.**

36.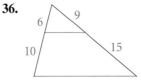

37.

38.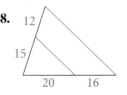

GO Online
Homework Video Tutor
Visit: PHSchool.com
Web Code: aue-0705

39. Oil Spills Describe how you could use the figure at the right to find the length of the oil spill indirectly. What measurements and calculations would you use?

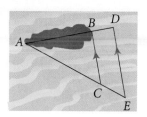

40. An angle bisector of a triangle divides the opposite side of the triangle into segments 5 cm and 3 cm long. A second side of the triangle is 7.5 cm long. Find all possible lengths for the third side of the triangle.

Geometry in 3 Dimensions In the figure at the right, $\overleftrightarrow{FG} \parallel \overleftrightarrow{AB}, \overleftrightarrow{GH} \parallel \overleftrightarrow{BC}, AF = 2, FE = 4,$ and $BG = 3$.

41. Find GE.

42. If $EH = 5$, find HC.

43. If $FG = 3$, find the perimeter of $\triangle ABE$.

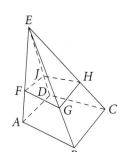

Proof **44.** One side of a triangle is k times as long as a second side. The bisector of their angle cuts the third side into two segments. Prove that one of those segments is k times as long as the other.

C Challenge Proof **45.** Use the definition in part (a) to prove the statements in parts (b) and (c).
  a. Write a definition for a midsegment of a parallelogram.
  b. A parallelogram midsegment is parallel to two sides of the parallelogram.
  c. A parallelogram midsegment bisects the diagonals of a parallelogram.

46. State the converse of the Triangle-Angle-Bisector Theorem. Give a convincing argument that the converse is true or a counterexample to prove that it is false.

47. In $\triangle ABC$, the bisectors of $\angle A, \angle B,$ and $\angle C$ cut the opposite sides into lengths a_1 and a_2, b_1 and $b_2,$ and c_1 and c_2, respectively, labeled in this order counterclockwise around $\triangle ABC$. Find the perimeter of $\triangle ABC$ for each of the following.

  a. $a_1 = \frac{5}{3}, a_2 = \frac{10}{3}, b_1 = \frac{15}{4}$
  b. $a_1 = \frac{10}{3}, b_1 = \frac{10}{9}, c_1 = \frac{8}{7}$

NY REGENTS

Test Prep

Multiple Choice **48.** Use the figure at the right. What is x?
  A. 5   **B.** 10
  C. 15   **D.** 20

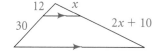

49. The legs of a right triangle have lengths 7 and 24. The bisector of the right angle divides the hypotenuse into two segments. What is the length of the shorter segment of the hypotenuse to the nearest tenth?
  F. 5.6   **G.** 8.0   **H.** 19.4   **J.** 25

Short Response **50.** What is n? Show your work.

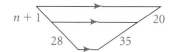

Extended Response **51.** The bisectors of an angle of a triangle divide the opposite side of the triangle into segments 4 cm and 5 cm long. A second side of the triangle is 6 cm long.
  a. Draw two diagrams you can use to find the two possible different lengths for the third side.
  b. Use each diagram in part (a) to write a proportion. Solve for each possible length of the third side of the triangle. Show your work.

Mixed Review

Lesson 7-4

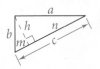

Refer to the figure to complete each proportion.

52. $\frac{n}{h} = \frac{h}{\blacksquare}$

53. $\frac{\blacksquare}{b} = \frac{b}{c}$

54. $\frac{n}{a} = \frac{a}{\blacksquare}$

55. $\frac{m}{h} = \frac{\blacksquare}{n}$

Lesson 6-4 $\boxed{x^2}$ **Algebra** *RSTV* is a rectangle. Find the lengths of the diagonals $\overline{RT}$ and $\overline{SV}$.

56. $RT = 5x + 8, SV = x + 32$

57. $RT = 42 - x, SV = 9x - 8$

58. $RT = 8x - 4, SV = 6x + 9$

59. $RT = 3x + 5, SV = 5x + 4$

Lesson 5-3 Find the center of the circle that you can circumscribe about each $\triangle ABC$.

60. $A(0,0)$
$B(6,0)$
$C(0,-6)$

61. $A(2,5)$
$B(-2,5)$
$C(-2,-1)$

62. $A(-2,0)$
$B(5,5)$
$C(-2,5)$

For $\triangle ABC$, name the point of intersection associated with each set of segments.

63. the medians

64. the angle bisectors

65. the altitudes

Checkpoint Quiz 2   **Lessons 7-3 through 7-5**

Determine whether the triangles are similar. If so, write the similarity statement. Also, write the postulate or theorem that proves they are similar.

1.   **2.**

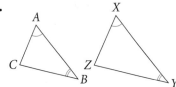

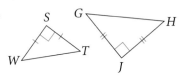

$\boxed{x^2}$ **Algebra** The polygons are similar. Find the value of each variable.

3.   **4.**

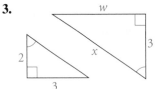

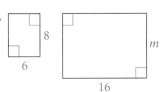

$\boxed{x^2}$ **Algebra** Find the value of each variable.

5.   **6.**   **7.**

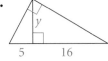

8.   **9.**   **10.**

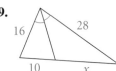

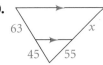

Understanding Word Problems Read the problem below. Then let Kyla's thinking guide you through the solution. Check your understanding with the exercises at the bottom of the page.

Surveying The perimeter of the triangular lot at the right is 50 m. The surveyor's tape bisects an angle. Find the lengths x and y.

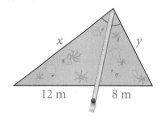

What Kyla Thinks

The perimeter of the triangular lot is 50. The perimeter is the distance around the lot. I can write and simplify an equation.

The surveyor's tape bisects the top angle. By the Triangle-Angle-Bisector Theorem, the tape divides the bottom side into two lengths that are proportional to the other two sides.

I can write and simplify a proportion.

I have a system of two equations (✔ and ✔) in two unknowns. I can substitute $\frac{2}{3}x$ for y in the first equation and solve for x.

Now I can solve for y.

What Kyla Writes

$$x + y + 8 + 12 = 50$$
$$x + y + 20 = 50$$
$$x + y = 30 \quad ✔$$

$$\frac{12}{8} = \frac{x}{y}$$
$$12y = 8x$$
$$y = \frac{2}{3}x \quad ✔$$

$$x + \frac{2}{3}x = 30$$
$$\frac{5}{3}x = 30$$
$$x = 18$$

Since $x + y = 30$, y must be 12.

EXERCISES

Refer to the exercise above. If the surveyor's tape forms segments of the following lengths, what are x and y?

1. 10 m and 10 m   **2.** 8 m and 12 m   **3.** 15 m and 5 m   **4.** 2 m and 18 m

5. Again, refer to the problem Kyla solved. If the cuts measure a and b, and $a + b = 20$, show that $x = \frac{3}{2}a$ and $y = \frac{3}{2}b$.

Your choices in a multiple-choice question include a correct answer. A strategy is to test each choice in the original problem. You may find mental math to be particularly useful for this.

EXAMPLE

Algebra What is the value of x in the diagram at the right?

- (A) 0
- (B) 3
- (C) 6
- (D) 10
- (E) 12

The triangles are similar. The proportion $\frac{3}{5} = \frac{x}{x+4}$ is apparent in the diagram. You can test answer choices using mental math.

Let $x = 0$.	Let $x = 3$.	Let $x = 6$.
$\frac{x}{x+4} = \frac{0}{0+4} = 0 \neq \frac{3}{5}$	$\frac{x}{x+4} = \frac{3}{3+4} = \frac{3}{7} \neq \frac{3}{5}$	$\frac{x}{x+4} = \frac{6}{6+4} = \frac{6}{10} = \frac{3}{5}$
A is not the answer.	B is not the answer.	Yes! C is likely the answer.

To help make sure, let $x = 10$.

$\frac{x}{x+4} = \frac{10}{10+4} = \frac{10}{14} = \frac{5}{7} \neq \frac{3}{5}$

D is not the answer.

Finally, let $x = 12$.

$\frac{x}{x+4} = \frac{12}{12+4} = \frac{12}{16} = \frac{3}{4} \neq \frac{3}{5}$

E is not the answer.

● You can mark C as your answer.

EXERCISES

1. What number is a solution to $\frac{x}{12} = \frac{x-3}{8}$?

- (A) 2
- (B) 3
- (C) 6
- (D) 8
- (E) 9

2. What number is a solution to $\frac{8}{13} = \frac{20}{a+12}$?

- (A) 20.5
- (B) 31
- (C) 33
- (D) 34
- (E) 44.5

x^2 **3. Algebra** What is the value of x?

- (A) 4
- (B) 6
- (C) 8
- (D) 10
- (E) 12

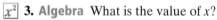

x^2 **4. Algebra** What is the value of y?

- (A) 2
- (B) 4
- (C) 6
- (D) 8
- (E) 10

5. $AB = 2$, $CD = 4$, $FE = 10$, and $AE = 20$. What is AC?

- (A) 2
- (B) 4
- (C) 6
- (D) 8
- (E) 10

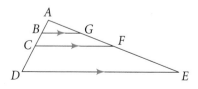

Chapter Review

Vocabulary Review

Cross-Product Property (p. 367)　　**golden rectangle** (p. 375)　　　**scale** (p. 368)
extended proportion (p. 367)　　**indirect measurement** (p. 384)　**scale drawing** (p. 368)
geometric mean (p. 392)　　　　**proportion** (p. 367)　　　　　**similar** (p. 373)
golden ratio (p. 375)　　　　　**ratio** (p. 366)　　　　　　　**similarity ratio** (p. 373)

Go Online
PHSchool.com

For: Vocabulary quiz
Web Code: auj-0751

Choose the correct term to complete each sentence.

1. Two polygons are __?__ if corresponding angles are congruent and corresponding sides are proportional.

2. The __?__ states that the product of the extremes is equal to the product of the means.

3. A __?__ is a rectangle that can be divided into a square and a rectangle that is similar to the original rectangle.

4. The ratio of the lengths of corresponding sides of two similar figures is the __?__.

5. A __?__ is a statement that two ratios are equal.

6. Finding distances using similar triangles is called __?__.

7. The length and width of a golden rectangle are in the __?__.

Skills and Concepts

7-1 Objectives

▼ To write ratios and solve proportions

A ratio is a comparison of two quantities by division. You can write the ratio of a to b or $a : b$ as the quotient $\frac{a}{b}$ when $b \neq 0$.

A **proportion** is a statement that two ratios are equal. According to the **Properties of Proportions,** $\frac{a}{b} = \frac{c}{d}$ is equivalent to

$\quad$ (1) $ad = bc$ $\quad$ (2) $\frac{b}{a} = \frac{d}{c}$ $\quad$ (3) $\frac{a}{c} = \frac{b}{d}$ $\quad$ (4) $\frac{a+b}{b} = \frac{c+d}{d}$

Property 1, above, illustrates the **Cross-Product Property,** which states that the product of the extremes is equal to the product of the means.

When three or more ratios are equal, you can write an **extended proportion.**

In a **scale drawing,** the **scale** compares each length in the drawing to the actual length being represented.

🌐 **Dollhouses** **Dollhouse furnishings come in different sizes depending on the size of the dollhouse. For each exercise, write a ratio of the size of the dollhouse item to the size of the larger item.**

8. dollhouse sofa: $1\frac{1}{2}$ in. long;
$\quad$ real sofa: 6 ft long

9. dollhouse piano: $1\frac{3}{4}$ in. high
$\quad$ real piano: 3 ft 6 in. high

If $\frac{p}{q} = \frac{2}{5}$, tell whether each equation must be true. Explain.

10. $2q = 5p$ $\qquad$ **11.** $\frac{5}{2} = \frac{q}{p}$ $\qquad$ **12.** $5q = 2p$ $\qquad$ **13.** $\frac{p}{2} = \frac{q}{5}$

Similar polygons have congruent corresponding angles and proportional corresponding sides. The ratio of the lengths of corresponding sides is the **similarity ratio.**

A **golden rectangle** is a rectangle that can be divided into a square and a rectangle that is similar to the original rectangle. In any golden rectangle, the length and the width are in the **golden ratio,** which is $\frac{1 + \sqrt{5}}{2} : 1$, or about 1.618 : 1.

If two angles of one triangle are congruent to two angles of another triangle, then the triangles are similar by the **Angle-Angle Similarity Postulate** (AA ~).

If an angle of one triangle is congruent to an angle of a second triangle, and the sides including the two angles are proportional, then the triangles are similar by the **Side-Angle-Side Similarity Theorem** (SAS ~).

If the corresponding sides of two triangles are proportional, then the triangles are similar by the **Side-Side-Side Similarity Theorem** (SSS ~).

Methods of **indirect measurement** use similar triangles and measurements to find distances that are difficult to measure directly.

14. If $\triangle MNP \sim \triangle RST$, which angles are congruent? Write an extended proportion to indicate the proportional corresponding sides of the triangles.

15. Art An artist is creating a stained glass window and wants it to be a golden rectangle. To the nearest inch, what should be the length if the width is 24 in.?

The triangles are similar. Find the similarity ratio of the first to the second.

16.
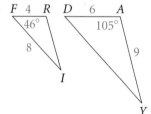

17.
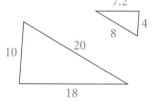

x^2 **Algebra** **The polygons are similar. Find the value of each variable.**

18.
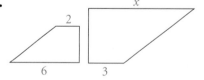

19.
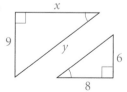

Are the triangles similar? If so, write the similarity statement and name the postulate or theorem you used. If not, explain.

20.
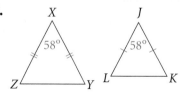

21.
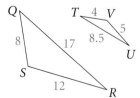

22. Two right triangles have an acute angle with the same measure. Name the theorem or postulate that is the most direct way to prove the triangles similar.

23. Indirect Measurement A crate is 1.5 ft high and casts a 2-ft shadow. At the same time, an apple tree casts an 18-ft shadow. How tall is the tree?

7-4 Objective

▼ To find and use relationships in similar right triangles

The **geometric mean** of two positive numbers a and b is the positive number x such that $\frac{a}{x} = \frac{x}{b}$.

When the altitude is drawn to the hypotenuse of a right triangle:

- the two triangles formed are similar to the original triangle and to each other;
- the length of the altitude is the geometric mean of the lengths of the segments of the hypotenuse; and
- the length of each leg is the geometric mean of the length of the adjacent hypotenuse segment and the length of the hypotenuse.

x^2 **Algebra** **Find the geometric mean of each pair of numbers.**

24. 4 and 25   **25.** 3 and 300   **26.** 5 and 12

27. $\frac{1}{9}$ and 28   **28.** 0.36 and 4   **29.** $2\sqrt{3}$ and $\sqrt{12}$

x^2 **Algebra** **Find the values of the variables. When an answer is not a whole number, leave it in simplest radical form.**

30.   **31.** 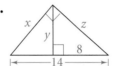   **32.**

7-5 Objectives

▼ To use the Side-Splitter Theorem

▼ To use the Triangle-Angle-Bisector Theorem

The **Side-Splitter Theorem** states that if a line is parallel to one side of a triangle and intersects the other two sides, then it divides those sides proportionally. If three parallel lines intersect two transversals, then the segments intercepted on the transversals are proportional.

The **Triangle-Angle-Bisector Theorem** states that if a ray bisects an angle of a triangle, then it divides the opposite side into two segments that are proportional to the other two sides of the triangle.

x^2 **Algebra** **Find the value of x.**

33.   **34.** 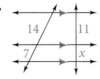   **35.**

36.   **37.** 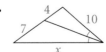   **38.**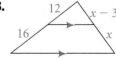

Chapter 7

Chapter Test

Go Online
PHSchool.com

For: Chapter Test
Web Code: aua-0752

x^2 **Algebra** Solve each proportion.

1. $\frac{4}{5} = \frac{x}{20}$ **2.** $\frac{6}{x} = \frac{10}{7}$ **3.** $\frac{x}{3} = \frac{8}{12}$

x^2 **Algebra** The figures in each pair are similar. Find the value of each variable.

4.
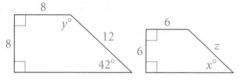

5.

6.
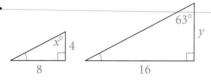

Are the triangles similar? If *yes*, write the similarity statement and name the postulate or theorem you can use to prove they are similar. If *no*, explain.

7.

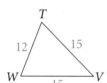

8.

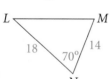

9.
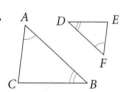

Use the information shown in the diagram and write a proportion for each triangle.

10.
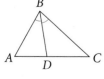

11.
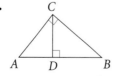

12. Indirect Measurement A meter stick is held perpendicular to the ground. It casts a shadow 1.5 m long. At the same time, a telephone pole casts a shadow that is 9 m long. How tall is the telephone pole?

13. Photography A photographic negative is 3 cm by 2 cm. If a similar print from the negative is 9 cm long on its shorter side, what is the length of its longer side?

Find the geometric mean of each pair of numbers. If the answer is not a whole number, write it in simplest radical form.

14. 10, 15   **15.** 4, 9   **16.** 6, 12

17. Open-Ended Draw an isosceles triangle, $\triangle ABC$. Then draw $\triangle DEF$ so that $\triangle ABC \sim \triangle DEF$. State the similarity ratio of $\triangle ABC$ to $\triangle DEF$.

x^2 **Algebra** Find the value of x.

18.

19.
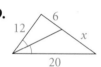

20.

21.
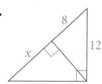

22. Writing Describe an object whose height or length would be difficult to measure directly. Then describe a method for measuring the object that involves using similar triangles.

Explain why the triangles are similar. Then find the value of x.

23.
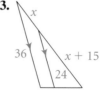

24.

Regents Test Prep

Reading Comprehension Read the passage below. Then answer the questions on the basis of what is *stated* or *implied* in the passage.

In Balance To suspend a mobile so that it balances, you need to know how to find the center of mass. If a figure has a "nice" geometric shape, you can find the center of mass, or centroid, using geometric methods. There are alternative ways to find centers of mass in figures without "nice" shapes.

Here is how one artist found the center of mass for the irregularly shaped flat sheet of metal shown at the right.

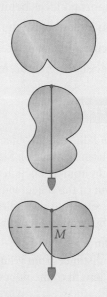

1. She suspended the shape from a point at its edge. She also suspended a *plumb line* (a weight on a string) from the same point. She traced the plumb line onto the shape.
2. She suspended the shape from another point, and again traced the plumb line. She found the center of mass of the shape at the intersection *M* of the two traced lines.

This method works because each traced plumb line passes through the center of mass. If you place the metal shape horizontally with a traced plumb line along the thin edge of a ruler, the shape will balance on the ruler. Since the shape balances along each traced plumb line, it will balance at the point of intersection of two such lines.

---

1. What should you learn how to do from the passage?
   (1) suspend a mobile
   (2) find a center of mass using geometric methods
   (3) find the center of mass of any flat shape
   (4) attach a plumb line to any flat shape

2. Why is a plumb line necessary?
   (1) to show a straight line
   (2) to find a vertical line
   (3) to split the shape into two equal halves
   (4) to suspend the shape at its center of mass

3. What geometric fact is applied in the passage?
   (1) Two nonparallel planes determine a line.
   (2) Two intersecting lines lie in one plane.
   (3) Two nonparallel lines determine a point.
   (4) Two points lie in one line.

4. Why does the shape balance when you place it on the edge of a ruler along a traced plumb line?
   (1) The center of mass of the shape rests on the edge of the ruler.
   (2) The traced plumb line is longer than the edge of the ruler.
   (3) The shape is placed horizontally on the edge of the ruler.
   (4) Half the mass of the shape is on each side of the edge of the ruler.

5. If the metal shape were an isosceles triangle, what would be other names for some traced plumb lines? Justify your answer.

6. If the metal shape were a circular ring with its inner circle cut out, where would you find the center of mass?

Data Analysis

Activity Lab

NY **G.PS.8:** Determine information required to solve a problem, choose methods for obtaining the information, and define parameters for acceptable solutions.

Accuracy and Precision in Measuring

Just how precise do you want—or need—a measurement to be?

A Global Positioning System (GPS) handheld device can locate you to within 10 feet of your true position. A surveyor's transit can locate you to within 1 inch.

Three important ideas behind any measurement you make are

• how you intend to use the measurement

• how accurate the measurement must be

• how precise the measurement must be

1 ACTIVITY

1. Pick a tall, vertical object, such as a tree or a flagpole, near your home or school. Stand far enough away from it so that you can hold a pencil at arm's length and line up the top and bottom of the object with the top and bottom of the pencil. Make sure that you hold the pencil so that it is parallel to the object and perpendicular to the ground.

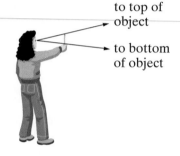

to top of object

to bottom of object

2. Measure:
   • the distance from your eye to the pencil at arm's length, and
   • the length of the pencil.

3. Walk to the base of the object, taking even strides. Count the number of strides. Measure the length of one stride. Then show how to use these data to find the distance to the object.

4. In the diagram at the right, BD is the length of the pencil. EG is the height of the object.
   a. Prove: $\triangle ABD \sim \triangle AEG$.
   b. You measured arm's length, AC (Exercise 2), and the distance to the object, AF (Exercise 3). These are heights of $\triangle ABD$ and $\triangle AEG$, respectively. Prove: Corresponding heights of similar triangles are proportional to the corresponding sides of the triangle.

5. Use a proportion to find the height of the object.

6. How confident are you in your result for Exercise 5? What kinds of measurement errors could you have made?

The *precision* of a measurement describes how repeatable the results are. The micrometer shown here has a precision of ±0.00005 in. This means that if you use this tool to make the same measurement repeatedly, the results would almost always be within 0.00005 in. of the mean of the measurements.

Precision is not the same as accuracy. For example, you may be very good at making precise measurements with a ruler. If, however, the ruler is warped or bent, your measurements will not likely be accurate.

2 ACTIVITY

In Activity 1, you found the height of a tall object. To help you understand how accurate this indirect measurement is, you can find the *greatest possible error* of each direct measurement you made. Copy the table below.

	Measurement	GPE	Minimum	Maximum
Pencil Length				
Arm's Length				
Object's Distance				
Object's Height				

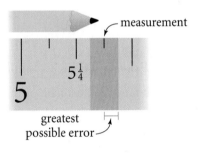

7. For your measurement of pencil length, look at the markings on the ruler. Your measurement was likely to the nearest mark. The pencil shown at the right would probably be measured as $5\frac{3}{8}$ in. So the actual length of the pencil is within ± half the distance between two adjacent marks on the ruler. If you measured your pencil to the nearest $\frac{1}{8}$ inch, then the greatest possible error of your measurement is $\pm\frac{1}{16}$ inch. In your table, record the greatest possible error (GPE) for your pencil measurement.

8. Use the greatest possible error. Record the minimum and maximum possible lengths of the pencil.

9. Find and record the greatest possible error in your arm's length measurement. Record the minimum and maximum possible arm's lengths.

10. Measure the length of your stride at least nine more times.
    a. Find the mean, minimum, and maximum of these measurements.
    b. Find the difference between the mean and the minimum; then between the maximum and the mean. Record the larger of the two differences as the greatest possible error.
    c. Use this greatest possible error to find the minimum and maximum values for the distance to the object.

11. Use the proportion from Exercise 5 and the minimum and maximum values for the first three measurements in the table to calculate the minimum and maximum possible heights of the object.

What You've Learned

NY Learning Standards for Mathematics

- **G.G.36:** In Chapter 3, you learned how to use the Triangle Angle-Sum Theorem and the Polygon Angle-Sum Theorem to find the measures of angles in polygons.

- **G.G.31:** In Chapter 4, you learned about the properties of isosceles and equilateral triangles.

- **G.G.44:** In Chapter 7, you learned how to prove triangles similar and that corresponding sides of similar triangles are in proportion.

 Check Your Readiness   **for Help** to the Lesson in green.

Solving Proportions (Lesson 7-1)

x^2 **Algebra** Solve for x. Round answers to the nearest thousandth.

1. $0.2734 = \frac{x}{17}$ **2.** $0.5858 = \frac{24}{x}$ **3.** $0.8572 = \frac{5271}{x}$ **4.** $0.5 = \frac{x}{3x + 5}$

Proving Triangles Similar (Lesson 7-3)

Name the postulate or theorem that proves each pair of triangles similar.

5. $\overline{CD} \parallel \overline{AB}$ **6.**   **7.** $\overline{JK} \perp \overline{ML}$

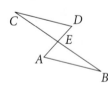   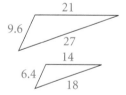   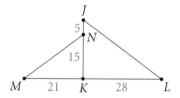

Similarity in Right Triangles (Lesson 7-4)

x^2 **Algebra** Find the unknown quantity in $\triangle ABC$ with right $\angle ACB$ and altitude $\overline{CD}$.

8.   **9.**   **10.**   **11.**

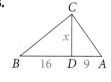   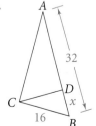   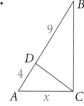   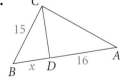

Right Triangles and Trigonometry

◀)) Key Vocabulary
- angle of depression (p. 445)
- angle of elevation (p. 445)
- cosine (p. 439)
- identity (p. 440)
- initial point (p. 452)
- magnitude (p. 452)
- Pythagorean triple (p. 417)
- resultant (p. 454)
- sine (p. 439)
- tangent (p. 432)
- terminal point (p. 452)
- vector (p. 452)

What You'll Learn Next

NY Learning Standards for Mathematics

- **G.G.48:** In this chapter, you will learn the Pythagorean Theorem and its converse.

- **G.G.48:** You will use the Pythagorean Theorem to find relationships in special right triangles.

- You will use similar right triangles to define the sine, cosine, and tangent ratios.

- With these ratios, you will solve height and distance problems using angles of elevation and angles of depression.

Activity Lab Applying what you have learned, you will do activities on pages 466 and 467 involving views from the tops of tall structures.

415

The Pythagorean Theorem

The Pythagorean Theorem is introduced in Lesson 8-1. The Activity and exercises below will help you understand why the theorem works.

NY G.G.48:
Investigate, justify, and apply the Pythagorean theorem and its converse.

ACTIVITY

Using graph paper, draw any rectangle. Label the sides a and b. Cut four rectangles with length a and width b from the graph paper. Then cut each rectangle on its diagonal, c, forming eight congruent triangles.

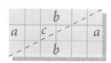

Cut three squares from colored paper, one with sides of length a, one with sides of length b, and one with sides of length c.

Separate the eleven pieces into groups.

    Group 1: four triangles and the two smaller squares

    Group 2: four triangles and the largest square

Arrange the pieces of each group to form a square.

1. a. How do the areas of the two squares you formed in the last step above compare?
  b. Write an algebraic expression for the area of each of these squares.
  c. What can you conclude about the areas of the three squares you cut from colored paper?

2. Repeat this investigation using a new rectangle with different a and b values. What do you notice?

3. Express your conclusion as an algebraic equation.

4. Use your ruler with any rectangle to find actual measures for $a, b,$ and c. Do these measures confirm that $a^2 + b^2 = c^2$?

EXERCISES

5. Explain how the diagram at the right represents your conclusion in Exercise 3.

6. Does the equation from Exercise 4 work for triangles other than right triangles? Explore and explain.

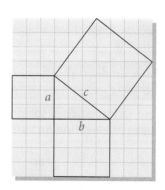

The Pythagorean Theorem and Its Converse

G.G.48 Investigate, justify, and apply the Pythagorean theorem and its converse.

✓ **Check Skills You'll Need**

 for Help Skills Handbook, p. 753

Square the lengths of the sides of each triangle. What do you notice?

1.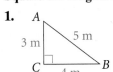

2.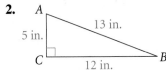

3.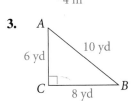

4.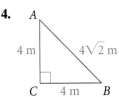

◀》 **New Vocabulary** • Pythagorean triple

---

1 The Pythagorean Theorem

The well-known right triangle relationship called the Pythagorean Theorem is named for Pythagoras, a Greek mathematician who lived in the sixth century B.C. We now know that the Babylonians, Egyptians, and Chinese were aware of this relationship before its discovery by Pythagoras.

There are many proofs of the Pythagorean Theorem. You will see one proof in Exercise 48 and others later in the book.

 Key Concepts

Theorem 8-1	**Pythagorean Theorem**

In a right triangle, the sum of the squares of the lengths of the legs is equal to the square of the length of the hypotenuse.

$$a^2 + b^2 = c^2$$

Vocabulary Tip

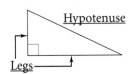

A **Pythagorean triple** is a set of nonzero whole numbers a, b, and c that satisfy the equation $a^2 + b^2 = c^2$. Here are some common Pythagorean triples.

3, 4, 5   5, 12, 13   8, 15, 17   7, 24, 25

If you multiply each number in a Pythagorean triple by the same whole number, the three numbers that result also form a Pythagorean triple.

1 EXAMPLE **Pythagorean Triples**

Find the length of the hypotenuse of $\triangle ABC$. Do the lengths of the sides of $\triangle ABC$ form a Pythagorean triple?

$a^2 + b^2 = c^2$ **Use the Pythagorean Theorem.**

$21^2 + 20^2 = c^2$ **Substitute 21 for a and 20 for b.**

$441 + 400 = c^2$ **Simplify.**

$841 = c^2$

$c = 29$ **Take the square root.**

The length of the hypotenuse is 29. The lengths of the sides, 20, 21, and 29, form a Pythagorean triple because they are whole numbers that satisfy $a^2 + b^2 = c^2$.

✓ Quick Check **1** A right triangle has a hypotenuse of length 25 and a leg of length 10. Find the length of the other leg. Do the lengths of the sides form a Pythagorean triple?

In some cases, you will write the length of a side in simplest radical form.

2 EXAMPLE **Using Simplest Radical Form**

Algebra Find the value of x. Leave your answer in simplest radical form (page 390).

$a^2 + b^2 = c^2$ **Pythagorean Theorem**

$8^2 + x^2 = 20^2$ **Substitute.**

$64 + x^2 = 400$ **Simplify.**

$x^2 = 336$ **Subtract 64 from each side.**

$x = \sqrt{336}$ **Take the square root.**

$x = \sqrt{16(21)}$ **Simplify.**

$x = 4\sqrt{21}$

✓ Quick Check **2** The hypotenuse of a right triangle has length 12. One leg has length 6. Find the length of the other leg. Leave your answer in simplest radical form.

3 EXAMPLE **Real-World** 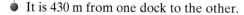 **Connection**

Gridded Response The Parks Department rents paddle boats at docks near each entrance to the park. To the nearest meter, how far is it to paddle from one dock to the other?

$a^2 + b^2 = c^2$ **Pythagorean Theorem**

$250^2 + 350^2 = c^2$ **Substitute.**

$185,000 = c^2$ **Simplify.**

$c = \sqrt{185,000}$ **Take the square root.**

$c = 430.11626$ **Use a calculator.**

It is 430 m from one dock to the other.

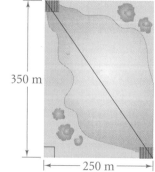

✓ Quick Check **3** **Critical Thinking** When you want to know how far you have to paddle a boat, why is an approximate answer more useful than an answer in simplest radical form?

You can use the Converse of the Pythagorean Theorem to determine whether a triangle is a right triangle. You will prove Theorem 8-2 in Exercise 58.

Key Concepts

Theorem 8-2	Converse of the Pythagorean Theorem

If the square of the length of one side of a triangle is equal to the sum of the squares of the lengths of the other two sides, then the triangle is a right triangle.

For: Pythagorean Activity
Use: Interactive Textbook, 8-1

4 EXAMPLE **Is It a Right Triangle?**

Is this triangle a right triangle?

$$c^2 \stackrel{?}{=} a^2 + b^2$$

$85^2 \stackrel{?}{=} 13^2 + 84^2$ **Substitute the greatest length for c.**

$7225 \stackrel{?}{=} 169 + 7056$ **Simplify.**

$7225 = 7225 \checkmark$

$c^2 = a^2 + b^2$, so the triangle is a right triangle.

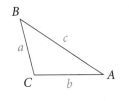

Quick Check **4** A triangle has sides of lengths 16, 48, and 50. Is the triangle a right triangle?

You can also use the squares of the lengths of the sides of a triangle to find whether the triangle is acute or obtuse. The following two theorems tell how.

Key Concepts

Theorem 8-3

If the square of the length of the longest side of a triangle is greater than the sum of the squares of the lengths of the other two sides, the triangle is obtuse.

If $c^2 > a^2 + b^2$, the triangle is obtuse.

Theorem 8-4

If the square of the length of the longest side of a triangle is less than the sum of the squares of the lengths of the other two sides, the triangle is acute.

If $c^2 < a^2 + b^2$, the triangle is acute.

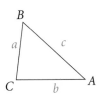

Real-World Connection

The length to the brace along each leg is 36 in. The brace is 26 in. long to guarantee that the triangle is acute.

5 EXAMPLE **Classifying Triangles as Acute, Obtuse, or Right**

Classify the triangle whose side lengths are 6, 11, and 14 as acute, obtuse, or right.

$14^2 \stackrel{?}{=} 6^2 + 11^2$ **Compare c² to a² + b². Substitute the greatest length for c.**

$196 \stackrel{?}{=} 36 + 121$

$196 > 157$

Since $c^2 > a^2 + b^2$, the triangle is obtuse.

Quick Check **5** A triangle has sides of lengths 7, 8, and 9. Classify the triangle by its angles.

EXERCISES

For more exercises, see *Extra Skill, Word Problem, and Proof Practice*.

Practice and Problem Solving

A Practice by Example x^2 Algebra **Find the value of x.**

Example 1
(page 418)

GO for Help

1.

2.

3.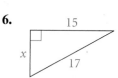

4.

5.

6.

Does each set of numbers form a Pythagorean triple? Explain.

7. $4, 5, 6$ **8.** $10, 24, 26$ **9.** $15, 20, 25$

Example 2
(page 418) x^2 Algebra **Find the value of x. Leave your answer in simplest radical form.**

10.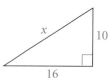

11.

12.

13.

14.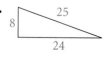

15.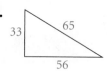

Example 3
(page 418)

16. Home Maintenance A painter leans a 15-ft ladder against a house. The base of the ladder is 5 ft from the house.
  a. To the nearest tenth of a foot, how high on the house does the ladder reach?
  b. The ladder in part (a) reaches too high on the house. By how much should the painter move the ladder's base away from the house to lower the top by 1 ft?

17. A walkway forms the diagonal of a square playground. The walkway is 24 m long. To the nearest tenth of a meter, how long is a side of the playground?

Example 4
(page 419)

Is each triangle a right triangle? Explain.

18.

19.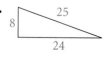

20.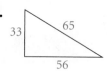

Example 5
(page 419)

The lengths of the sides of a triangle are given. Classify each triangle as acute, right, or obtuse.

21. $4, 5, 6$ **22.** $0.3, 0.4, 0.6$ **23.** $11, 12, 15$

24. $\sqrt{3}, 2, 3$ **25.** $30, 40, 50$ **26.** $\sqrt{11}, \sqrt{7}, 4$

27.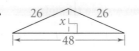

28.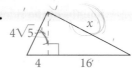

29.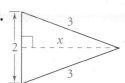

30. Writing Each year in an ancient land, a large river overflowed its banks, often destroying boundary markers. The royal surveyors used a rope with knots at 12 equal intervals to help reconstruct boundaries. Explain how a surveyor could use this rope to form a right angle. (*Hint:* Use the Pythagorean triple 3, 4, 5.)

31. Multiple Choice Which triangle is *not* a right triangle?

A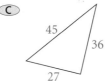

B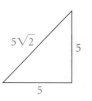

C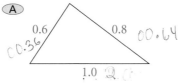

D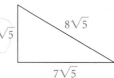

GO for Help

For a guide to solving Exercise 32, see p. 424.

32. Embroidery You want to embroider a square design. You have an embroidery hoop with a 6 in. diameter. Find the largest value of *x* so that the entire square will fit in the hoop. Round to the nearest tenth.

33. In parallelogram $RSTW$, $RS = 7$, $ST = 24$, and $RT = 25$. Is $RSTW$ a rectangle? Explain.

Proof **34. Coordinate Geometry** You can use the Pythagorean Theorem to prove the Distance Formula. Let points $P(x_1, y_1)$ and $Q(x_2, y_2)$ be the endpoints of the hypotenuse of a right triangle.
  a. Write an algebraic expression to complete each of the following:
  $PR = $ ■ and $QR = $ ■.
  b. By the Pythagorean Theorem, $PQ^2 = PR^2 + QR^2$. Rewrite this statement substituting the algebraic expressions you found for PR and QR in part (a).
  c. Complete the proof by taking the square root of each side of the equation that you wrote in part (b).

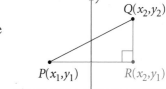

35. Constructions Explain how to construct a segment of length $\sqrt{n}$, where *n* is any positive integer, and you are given a segment of length 1. (*Hint:* See the diagram.)

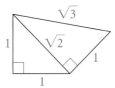

GO Online

Homework Video Tutor

Visit: PHSchool.com
Web Code: aue-0801

Find a third whole number so that the three numbers form a Pythagorean triple.

36. 20, 21   **37.** 14, 48   **38.** 13, 85   **39.** 12, 37

Find integers j and k so that (a) the two given integers and j represent the lengths of the sides of an acute triangle and (b) the two given integers and k represent the lengths of the sides of an obtuse triangle.

40. $4, 5$ **41.** $2, 4$ **42.** $6, 9$ **43.** $5, 10$

44. $6, 7$ **45.** $9, 12$ **46.** $8, 17$ **47.** $9, 40$

Proof **48.** Prove the Pythagorean Theorem.

Given: $\triangle ABC$ is a right triangle
Prove: $a^2 + b^2 = c^2$

(*Hint*: Begin with proportions suggested by Theorem 7-3 or its corollaries.)

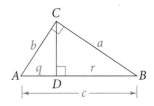

49. Astronomy The Hubble Space Telescope is orbiting Earth 600 km above Earth's surface. Earth's radius is about 6370 km. Use the Pythagorean Theorem to find the distance x from the telescope to Earth's horizon. Round your answer to the nearest ten kilometers.

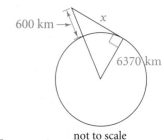

not to scale

The figures below are drawn on centimeter grid paper. Find the perimeter of each shaded figure to the nearest tenth.

50.   **51.** 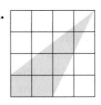   **52.**

53. a. The ancient Greek philosopher Plato used the expressions $2n, n^2 - 1$, and $n^2 + 1$ to produce Pythagorean triples. Choose any integer greater than 1. Substitute for n and evaluate the three expressions.
    b. Verify that your answers to part (a) form a Pythagorean triple.
    c. Show that, in general, $(2n)^2 + (n^2 - 1)^2 = (n^2 + 1)^2$ for any n.

C Challenge

54. Geometry in 3 Dimensions The box at the right is a rectangular solid.
    a. Use $\triangle ABC$ to find the length d_1 of the diagonal of the base.
    b. Use $\triangle ABD$ to find the length d_2 of the diagonal of the box.
    c. You can generalize the steps in parts (a) and (b). Use the facts that $AC^2 + BC^2 = d_1{}^2$ and $d_1{}^2 + BD^2 = d_2{}^2$ to write a one-step formula to find d_2.
    d. Use the formula you wrote to find the length of the longest fishing pole you can pack in a box with dimensions 18 in., 24 in., and 16 in.

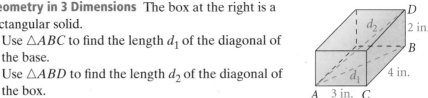

Geometry in 3 Dimensions Points $P(x_1, y_1, z_1)$ and $Q(x_2, y_2, z_2)$ at the left are points in a three-dimensional coordinate system. Use the following formula to find PQ. Leave your answer in simplest radical form.

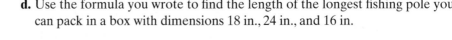

$$d = \sqrt{(x_2 - x_1)^2 + (y_2 - y_1)^2 + (z_2 - z_1)^2}$$

55. $P(0, 0, 0), Q(1, 2, 3)$ **56.** $P(0, 0, 0), Q(-3, 4, -6)$ **57.** $P(-1, 3, 5), Q(2, 1, 7)$

Proof **58.** Use the plan and write a proof of Theorem 8-2, the Converse of the Pythagorean Theorem.

Given: $\triangle ABC$ with sides of length $a, b,$ and c where $a^2 + b^2 = c^2$

Prove: $\triangle ABC$ is a right triangle.

Plan: Draw a right triangle (not $\triangle ABC$) with legs of lengths a and b. Label the hypotenuse x. By the Pythagorean Theorem, $a^2 + b^2 = x^2$. Use substitution to compare the lengths of the sides of your triangle and $\triangle ABC$. Then prove the triangles congruent.

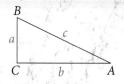

Test Prep

Gridded Response

59. The lengths of the legs of a right triangle are 17 m and 20 m. To the nearest tenth of a meter, what is the length of the hypotenuse?

60. The hypotenuse of a right triangle is 34 ft. One leg is 16 ft. Find the length of the other leg in feet.

61. What whole number forms a Pythagorean triple with 40 and 41?

62. The two shorter sides of an obtuse triangle are 20 and 30. What is the least whole number length possible for the third side?

63. Each leg of an isosceles right triangle has measure 10 cm. To the nearest tenth of a centimeter, what is the length of the hypotenuse?

64. The legs of a right triangle have lengths 3 and 4. What is the length, to the nearest tenth, of the altitude to the hypotenuse?

Mixed Review

Lesson 7-5

For the figure at the right, complete the proportion.

65. $\frac{10}{\blacksquare} = \frac{y}{18}$ **66.** $\frac{\blacksquare}{x} = \frac{y}{18}$

67. Find the values of x and y.

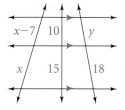

Lesson 5-2

In the second figure, $\overrightarrow{PS}$ bisects $\angle RPT$. Solve for each variable. Then find RS.

68. $RS = 2x + 19, ST = 7x - 16; x = \blacksquare, RS = \blacksquare$

69. $RS = 2(7y - 11), ST = 5y + 5; y = \blacksquare, RS = \blacksquare$

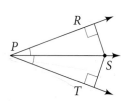

Lesson 4-1

$\triangle PQR \cong \triangle STV$. Solve for each variable.

70. $m\angle P = 4w + 5, m\angle S = 6w - 15$ **71.** $RQ = 10y - 6, VT = 5y + 9$

72. $m\angle T = 2x - 40, m\angle Q = x + 10$ **73.** $PR = 2z + 3, SV = 4z - 11$

Understanding Word Problems Read the problem below. Let Sharleen's thinking guide you through the solution. Check your understanding with the exercises at the bottom of the page.

Embroidery You want to embroider a square design. You have an embroidery hoop with a 6 in. diameter. Find the largest value of x so that the entire square will fit in the hoop. Round to the nearest tenth.

What Sharleen Thinks

I'll draw the diagram without the embroidery and mark in it the information I know. In particular, I'll show a 6-in. diameter that *also* happens to be a diagonal of the square.

What Sharleen Writes

Diameter = 6, so length of diagonal = 6

Now I have two right triangles. Each right triangle has two legs of length x in. and a hypotenuse of 6 in. I can apply the Pythagorean Theorem to find x.

$a^2 + b^2 = c^2$

$a = x$, $b = x$, and $c = 6$

$x^2 + x^2 = 6^2$

Combine like terms.

Divide each side by 2.

Take the square root of each side.

Use a calculator. Then round to the nearest tenth.

$2x^2 = 36$

$x^2 = 18$

$x = \sqrt{18}$

$x \approx 4.2426$

The largest square that you can embroider with this hoop has sides of 4.2 in. (Don't forget to include the unit of measure.)

$x \approx 4.2$ in.

EXERCISES

1. In the problem above, what is the value of x in simplest radical form?

Suppose the embroidery hoop had the given diameter. What is the largest value of x possible? Give your answer in simplest radical form.

2. 12 in.   **3.** 18 in.   **4.** 9 in.   **5.** 3 in.

6. Based on your results for Exercises 1–5, make a conjecture about a relationship between a side and a diagonal of a square.

8-2

Special Right Triangles

NY Learning Standards for Mathematics

G.G.48 Investigate, justify, and apply the Pythagorean theorem and its converse.

 Check Skills You'll Need

GO for **Help** Lesson 1-6

Use a protractor to find the measures of the angles of each triangle.

1. 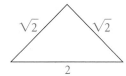 **2.** 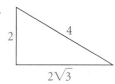 **3.**

---

1 Using 45°-45°-90° Triangles

The acute angles of an isosceles right triangle are both 45° angles. Another name for an isosceles right triangle is a 45°-45°-90° triangle. If each leg has length x and the hypotenuse has length y, you can solve for y in terms of x.

$x^2 + x^2 = y^2$ **Use the Pythagorean Theorem.**

$2x^2 = y^2$ **Simplify.**

$x\sqrt{2} = y$ **Take the square root of each side.**

You have just proved the following theorem.

 Key Concepts

Theorem 8-5	45°-45°-90° Triangle Theorem

In a 45°-45°-90° triangle, both legs are congruent and the length of the hypotenuse is $\sqrt{2}$ times the length of a leg.

$$\text{hypotenuse} = \sqrt{2} \cdot \text{leg}$$

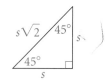

Test-Taking Tip

If you forget the formula for a 45°-45°-90° triangle, use the Pythagorean Theorem. The triangle is isosceles, so the legs have the same length.

1 **EXAMPLE** **Finding the Length of the Hypotenuse**

Find the value of each variable.

a. 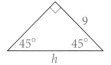   **b.**

$h = \sqrt{2} \cdot 9$ ← hypotenuse $= \sqrt{2} \cdot$ leg → $x = \sqrt{2} \cdot 2\sqrt{2}$

$h = 9\sqrt{2}$ ← Simplify. → $x = 4$

 **Quick Check** **1** Find the length of the hypotenuse of a 45°-45°-90° triangle with legs of length $5\sqrt{3}$.

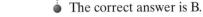 **EXAMPLE** Finding the Length of a Leg

Multiple Choice What is the value of x?

Ⓐ 3   Ⓑ $3\sqrt{2}$ Ⓒ 6   Ⓓ $6\sqrt{2}$

$6 = \sqrt{2} \cdot x$ **hypotenuse = $\sqrt{2}$ · leg**

$x = \dfrac{6}{\sqrt{2}}$ **Divide each side by $\sqrt{2}$.**

$x = \dfrac{6}{\sqrt{2}} \cdot \dfrac{\sqrt{2}}{\sqrt{2}} = \dfrac{6\sqrt{2}}{2}$ **Multiply by a form of 1.**

$x = 3\sqrt{2}$ **Simplify.**

● The correct answer is B.

 Quick Check ❷ Find the length of a leg of a 45°-45°-90° triangle with a hypotenuse of length 10.

When you apply the 45°-45°-90° Triangle Theorem to a real-life example, you can use a calculator to evaluate square roots.

③ EXAMPLE **Real-World** **Connection**

Softball A high school softball diamond is a square. The distance from base to base is 60 ft. To the nearest foot, how far does a catcher throw the ball from home plate to second base?

The distance d from home plate to second base is the length of the hypotenuse of a 45°-45°-90° triangle.

$d = 60\sqrt{2}$ **hypotenuse = $\sqrt{2}$ · leg**

$d = \mathbf{84.852814}$ **Use a calculator.**

On a high school softball diamond, the catcher throws the ball about 85 ft from home plate to second base.

Real-World 🌎 **Connection**

Careers Opportunities for coaching in women's sports have soared since the passage of Title IX in 1972.

 **Quick Check** ❸ A square garden has sides 100 ft long. You want to build a brick path along a diagonal of the square. How long will the path be? Round your answer to the nearest foot.

2 Using 30°-60°-90° Triangles

Another type of special right triangle is a 30°-60°-90° triangle.

 Key Concepts

Theorem 8-6	**30°-60°-90° Triangle Theorem**

In a 30°-60°-90° triangle, the length of the hypotenuse is twice the length of the shorter leg. The length of the longer leg is $\sqrt{3}$ times the length of the shorter leg.

     hypotenuse $= 2 \cdot$ shorter leg

     longer leg $= \sqrt{3} \cdot$ shorter leg

To prove Theorem 8-6, draw a 30°-60°-90° triangle using an equilateral triangle.

Proof

Proof of Theorem 8-6

For 30°-60°-90° $\triangle WXY$ in equilateral $\triangle WXZ$,
$\overline{WY}$ is the perpendicular bisector of $\overline{XZ}$.
Thus, $XY = \frac{1}{2}XZ = \frac{1}{2}XW$, or $XW = 2XY = 2s$.
Also,

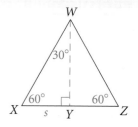

$XY^2 + YW^2 = XW^2$	**Use the Pythagorean Theorem.**
$s^2 + YW^2 = (2s)^2$	**Substitute s for XY and 2s for XW.**
$YW^2 = 4s^2 - s^2$	**Subtract s^2 from each side.**
$YW^2 = 3s^2$	**Simplify.**
$YW = s\sqrt{3}$	**Find the square root of each side.**

The 30°-60°-90° Triangle Theorem, like the 45°-45°-90° Triangle Theorem, lets you find two sides of a right triangle when you know the length of the third side.

4 EXAMPLE Using the Length of One Side

Algebra Find the value of each variable.

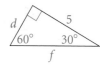

$5 = d\sqrt{3}$	**longer leg = $\sqrt{3}$ · shorter leg**
$d = \frac{5}{\sqrt{3}} \cdot \frac{\sqrt{3}}{\sqrt{3}} = \frac{5\sqrt{3}}{3}$	**Solve for d.**
$f = 2d$	**hypotenuse = 2 · shorter leg**
$f = 2 \cdot \frac{5\sqrt{3}}{3} = \frac{10\sqrt{3}}{3}$	**Substitute $\frac{5\sqrt{3}}{3}$ for d.**

✓ Quick Check **4** Find the value of each variable.

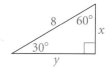

5 EXAMPLE Real-World  Connection

Road Signs The moose warning sign at the left is an equilateral triangle. The height of the sign is 1 m. Find the length *s* of each side of the sign to the nearest tenth of a meter.

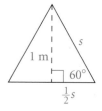

The triangle is equilateral, so the altitude divides the triangle into two 30°-60°-90° triangles as shown in the diagram. The altitude also bisects the base, so the shorter leg of each 30°-60°-90° triangle is $\frac{1}{2}s$.

$1 = \sqrt{3}\left(\frac{1}{2}s\right)$	**longer leg = $\sqrt{3}$ · shorter leg**
$\frac{2}{\sqrt{3}} = s$	**Solve for s.**
$s \approx 1.155$	**Simplify. Use a calculator.**

Each side of the sign is about 1.2 m long.

✓ Quick Check **5** If the sides of the sign are 1 m long, what is the height?

Practice by Example

Example 1
(page 425)

 GO for Help

Find the value of each variable. If your answer is not an integer, leave it in simplest radical form.

1.

2.

3.

Examples 2 and 3
(page 426)

4.

5.

6.

Exercise 7

7. Dinnerware Design You are designing dinnerware. What is the length of a side of the smallest square plate on which a 20-cm chopstick can fit along a diagonal without any overhang? Round your answer to the nearest tenth of a centimeter.

8. Helicopters The four blades of a helicopter meet at right angles and are all the same length. The distance between the tips of two adjacent blades is 36 ft. How long is each blade? Round your answer to the nearest tenth.

Examples 4 and 5
(page 427)

x^2 **Algebra** **Find the value of each variable. If your answer is not an integer, leave it in simplest radical form.**

9.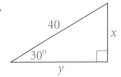

10.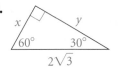

11.

12.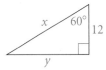

13.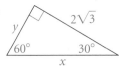

14.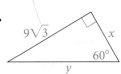

15. Architecture An escalator lifts people to the second floor, 25 ft above the first floor. The escalator rises at a 30° angle. How far does a person travel from the bottom to the top of the escalator?

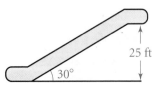

16. City Planning Jefferson Park sits on one square city block 300 ft on each side. Sidewalks join opposite corners. About how long is each diagonal sidewalk?

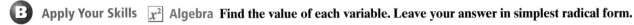

B **Apply Your Skills** x^2 **Algebra** **Find the value of each variable. Leave your answer in simplest radical form.**

17.

18.

19.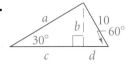

20.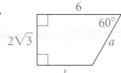

21.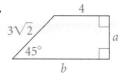

22.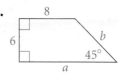

23. Error Analysis Sandra drew the triangle at the right. Rika said that the lengths couldn't be correct. With which student do you agree? Explain your answer.

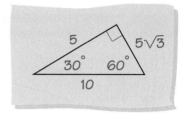

24. Open-Ended Write a real-life problem that you can solve using a 30°-60°-90° triangle with a 12-ft hypotenuse. Show your solution.

25. Farming A conveyor belt carries bales of hay from the ground to the barn loft 24 ft above the ground. The belt makes a 60° angle with the ground.
  a. How far does a bale of hay travel from one end of the conveyor belt to the other? Round your answer to the nearest foot.
  b. The conveyor belt moves at 100 ft/min. How long does it take for a bale of hay to go from the ground to the barn loft?

Exercise 25

26. House Repair After heavy winds damaged a farmhouse, workers placed a 6-m brace against its side at a 45° angle. Then, at the same spot on the ground, they placed a second, longer brace to make a 30° angle with the side of the house.
  a. How long is the longer brace? Round your answer to the nearest tenth of a meter.
  b. About how much higher on the house does the longer brace reach than the shorter brace?

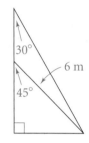

 Challenge

27. Geometry in 3 Dimensions Find the length d, in simplest radical form, of the diagonal of a cube with sides of the given length.
  a. 1 unit   **b.** 2 units   **c.** s units

28. Constructions Construct a 30°-60°-90° triangle given a segment that is
  a. the shorter leg.
  b. the hypotenuse.
  c. the longer leg.

 **REGENTS**

Test Prep

Multiple Choice

29. What is the length of a diagonal of a square with sides of length 4?
  A. 2   **B.** $\sqrt{2}$ **C.** $2\sqrt{2}$ **D.** $4\sqrt{2}$

30. The longer leg of a 30°-60°-90° triangle is 6. What is the length of the hypotenuse?
  F. $2\sqrt{3}$ **G.** $3\sqrt{2}$ **H.** $4\sqrt{3}$ **J.** 12

31. The hypotenuse of a 30°-60°-90° triangle is 30. What is the length of one of its legs?
  A. $3\sqrt{10}$ **B.** $10\sqrt{3}$ **C.** $15\sqrt{2}$ **D.** 15

32. The vertex angle of an isosceles triangle is 120°. The length of its base is 24 cm.
  a. Find the height of the triangle from the vertex angle to its base.
  b. Find the length of each leg of the isosceles triangle.

Mixed Review

GO for Help

Lesson 8-1

For Exercises 33 and 34, leave your answers in simplest radical form.

33. A right triangle has a 6-in. hypotenuse and a 5-in. leg. Find the length of the other leg.

34. An isosceles triangle has 20-cm legs and a 16-cm base. Find the length of the altitude to the base.

Lesson 6-4

Determine whether each quadrilateral must be a parallelogram. If not, provide a counterexample.

35. The diagonals are congruent and perpendicular to each other.

36. Two opposite angles are right angles and two opposite sides are 5 cm long.

37. One pair of sides is congruent and the other pair of sides is parallel.

Lesson 4-3

Can you conclude that $\triangle TRY \cong \triangle ANG$ from the given conditions? If so, name the postulate or theorem that justifies your conclusion.

38. $\angle A \cong \angle T, \angle Y \cong \angle G, \overline{TR} \cong \overline{AN}$ **39.** $\angle T \cong \angle A, \angle R \cong \angle N, \angle Y \cong \angle G$

40. $\angle R \cong \angle N, \overline{TR} \cong \overline{AN}, \overline{TY} \cong \overline{AG}$ **41.** $\angle G \cong \angle Y, \angle N \cong \angle R, \overline{RY} \cong \overline{NG}$

Checkpoint Quiz 1   Lessons 8-1 through 8-2

x^2 **Algebra** Find the value of each variable. Leave your answer in simplest radical form.

1.

2.

3.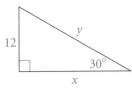

4.

5.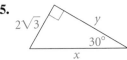

6.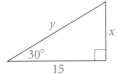

The lengths of the sides of a triangle are given. Classify each triangle as *acute*, *obtuse*, or *right*.

7. 7, 8, 9   **8.** 15, 36, 39   **9.** 10, 12, 16

10. A square has a 40-cm diagonal. How long is each side of the square? Round your answer to the nearest tenth of a centimeter.

The Staff and Stadiascope

Ancient peoples used measuring sticks of various kinds for direct measurements. Such a device was not practical, however, for finding the height of a tree or the width of a river. Two early instruments for indirect measurements were the staff—to measure horizontal distances—and the stadiascope—to measure heights. Each used similar right triangles.

NY G.CN.8: Develop an appreciation for the historical development of mathematics.

1 ACTIVITY

The *staff* is made of two pieces of wood, fastened together to form right angles. To measure the distance across a river, you move the crosspiece ($\overline{CP}$) vertically until the line of sight from the top of the staff (S) through the end of the crosspiece (P) lines up with the opposite river bank (R).

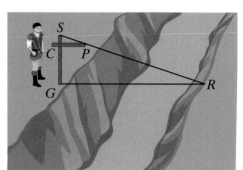

1. What two similar triangles are formed? Why are they similar?

2. What measurements do you know for the staff? Explain how to use these measurements to find the distance across the river.

3. Build your own staff and use it to find a distance in your schoolyard. Also, measure the width directly. Compare results. How confident are you in each method of measurement? Explain.

2 ACTIVITY

The development of the *stadia method* of indirect measurement is usually credited to James Watt, a Scottish inventor who lived in the mid-1700s.

A *stadiascope* is a long tube with a series of parallel lines—equally spaced by a known amount—on a clear end cap. To measure a distant object, you keep the stadiascope horizontal and as close to the ground as possible. Look through the plain end and line up the bottom of the grid end with the bottom of the object.

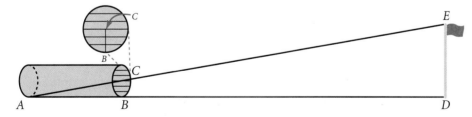

4. Explain how to proceed to find the height of the object.

5. Why is there a grid on the end of the stadiascope rather than a simple vertical ruler?

6. Could you use a stadiascope to find horizontal distance? A staff to find height? Explain.

7. Build your own stadiascope and use it to find the height of a tall object at your school. Compare all the results from your class. Find the mean of the results that you consider reasonable. How confident are you in your result versus the mean result? Explain.

The Tangent Ratio

What You'll Learn

• To use tangent ratios to determine side lengths in triangles

. . . And Why

To use the tangent ratio to estimate distance to a distant object, as in Example 2

✓ **Check Skills You'll Need**

GO **for Help** Lessons 7-1 and 8-2

Find the ratios $\frac{BC}{AB}$, $\frac{AC}{AB}$, and $\frac{BC}{AC}$. Round answers to the nearest hundredth.

1.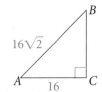

2.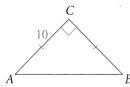

3.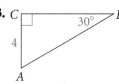

x^2 **Algebra** Solve each proportion.

4. $\frac{x}{3} = \frac{4}{7}$
5. $\frac{6}{11} = \frac{x}{9}$
6. $\frac{8}{15} = \frac{4}{x}$
7. $\frac{5}{x} = \frac{7}{12}$

🔊 **New Vocabulary** • tangent

---

1 Using Tangents in Triangles

Activity: Tangent Ratios

Work in groups of three or four.

• Have your group select one angle measure from {10°, 20°, . . . , 80°}. Then have each member of your group draw a right triangle, △ABC, where ∠A has the selected measure. Make the triangles different sizes.

• Measure the legs of each △ABC to the nearest millimeter.

1. Compute the ratio $\frac{\text{leg opposite } \angle A}{\text{leg adjacent to } \angle A}$ and round to two decimal places.

2. Compare the ratios in your group. Make a conjecture.

Vocabulary Tip

Trigonometry comes from the Greek words *trigonon* and *metria* meaning "triangle measurement."

For each pair of complementary angles, ∠A and ∠B, there is a family of similar right triangles. In each such family, the ratio

$$\frac{\text{length of leg opposite } \angle A}{\text{length of leg adjacent to } \angle A}$$

is constant no matter the size of △ABC. This trigonometric ratio is called the tangent ratio.

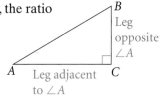

tangent of ∠A = $\dfrac{\text{length of leg opposite } \angle A}{\text{length of leg adjacent to } \angle A}$

You can abbreviate this equation as tan A = $\dfrac{\text{opposite}}{\text{adjacent}}$.

1 EXAMPLE Writing Tangent Ratios

Write the tangent ratios for $\angle T$ and $\angle U$.

$$\tan T = \frac{\text{opposite}}{\text{adjacent}} = \frac{UV}{TV} = \frac{3}{4}$$

$$\tan U = \frac{\text{opposite}}{\text{adjacent}} = \frac{TV}{UV} = \frac{4}{3}$$

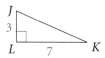

✓ Quick Check **1** **a.** Write the tangent ratios for $\angle K$ and $\angle J$.
b. How is $\tan K$ related to $\tan J$?

As stated on the facing page, the tangent ratio
for an acute angle does not depend on leg
lengths of a right triangle. To see why this is
so, consider the congruent angles, $\angle T$ and
$\angle T'$, in the two right triangles shown here.

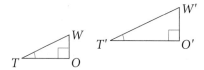

$\triangle TOW \sim T'O'W'$ **AA Similarity Postulate**

$\dfrac{OW}{TO} = \dfrac{O'W'}{T'O'}$ **Corresponding sides of ~ triangles are proportional.**

$\tan T = \tan T'$ **Substitute.**

You can use the tangent ratio to measure distances that would be difficult to
measure directly.

2 EXAMPLE Real-World Connection

Cross-Country Skiing Your goal in Bryce
Canyon National Park is the distant cliff.
About how far away is the cliff?

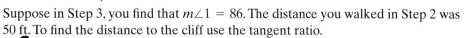

Step 1 Point your compass at a
distinctive feature of the cliff
and note the reading.

Step 2 Turn 90° and stride 50 ft
in a straight path.

Step 3 Turn and point the compass
again at the same feature seen in Step 1. Take a reading.

Suppose in Step 3, you find that $m\angle 1 = 86$. The distance you walked in Step 2 was
50 ft. To find the distance to the cliff use the tangent ratio.

$\tan 86° = \dfrac{x}{50}$ **Use the tangent ratio.**

$x = 50(\tan 86°)$ **Solve for x.**

50 `TAN` 86 `ENTER` *715.03331*   **Use a calculator.**

The cliff is about 715 ft away.

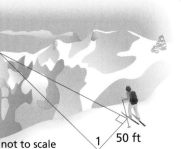

not to scale

✓ Quick Check **2** Find the value of w to the nearest tenth.

a.
b.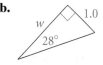
c.

If you know leg lengths for a right triangle, you can find the tangent ratio for each acute angle. Conversely, if you know the tangent ratio for an angle, you can use inverse of tangent, $\tan^{-1}$, to find the measure of the angle.

3 EXAMPLE Using the Inverse of Tangent

The lengths of the sides of $\triangle BHX$ are given. Find $m\angle X$ to the nearest degree.

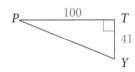

$\tan X = \frac{6}{8} = 0.75$ **Find the tangent ratio.**

$m\angle X = \tan^{-1}(0.75)$ **Use the inverse of tangent.**

$\text{TAN}^{-1}\ 0.75\ \boxed{\text{ENTER}}\ 36.869898$ **Use a calculator.**

● So $m\angle X \approx 37$.

✓ **Quick Check** ❸ Find $m\angle Y$ to the nearest degree.

EXERCISES

For more exercises, see *Extra Skill, Word Problem, and Proof Practice.*

Practice and Problem Solving

Ⓐ **Practice by Example**   **Write the tangent ratios for $\angle A$ and $\angle B$.**

Example 1
(page 433)

for Help

1.

2.

3.
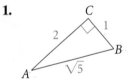

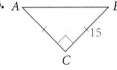

Example 2
(page 433)

Find the value of x to the nearest tenth.

4.

5.

6.

7.

8.

9.
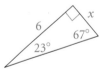

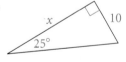

10. Surveying To find the distance from the boathouse on shore to the cabin on the island, a surveyor measures from the boathouse to point X as shown. He then finds $m\angle X$ with an instrument called a transit. Use the surveyor's measurements to find the distance from the boathouse to the cabin.

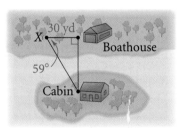

Example 3
(page 434)

Find the value of *x* to the nearest degree.

11.

12.

13.

14.

15.

16.

Find each missing value to the nearest tenth.

17. $\tan \blacksquare° = 3.5$

18. $\tan 34° = \dfrac{\blacksquare}{20}$

19. $\tan 2° = \dfrac{4}{\blacksquare}$

20. $\tan \blacksquare° = 90$

B Apply Your Skills

21. The lengths of the diagonals of a rhombus are 2 in. and 5 in. Find the measures of the angles of the rhombus to the nearest degree.

22. Pyramids All but two of the pyramids built by the ancient Egyptians have faces inclined at 52° angles. Suppose an archaeologist discovers the ruins of a pyramid. Most of the pyramid has eroded, but she is able to determine that the length of a side of the square base is 82 m. How tall was the pyramid, assuming its faces were inclined at 52°? Round your answer to the nearest meter.

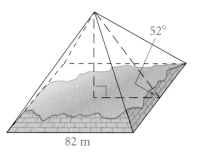

23. Multiple Choice The legs of a right triangle have lengths 5 and 12. What is the tangent of the angle opposite the leg with length 12?

Ⓐ $\dfrac{5}{13}$ Ⓑ $\dfrac{12}{13}$ Ⓒ $\dfrac{13}{12}$ Ⓓ $\dfrac{12}{5}$

24. Writing Explain why $\tan 60° = \sqrt{3}$. Include a diagram with your explanation.

25. Explain why $\tan^{-1} \dfrac{\sqrt{2}}{\sqrt{2}} = 45°$.

26. A rectangle is 80 cm long and 20 cm wide. To the nearest degree, find the measures of the angles formed by the diagonals at the center of the rectangle.

Find the value of *w*, then *x*. Round lengths of segments to the nearest tenth. Round angle measures to the nearest degree.

27.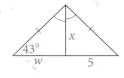

28.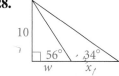

29.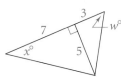

30. a. Coordinate Geometry Complete the table of values at the right. Give table entries to the nearest tenth.
b. Plot the points $(x, \tan x°)$ on the coordinate plane. Connect the points with a smooth curve.
c. What happens to the tangent ratio as the angle measure *x* approaches 0? Approaches 90?
d. Use your graph to estimate each value.
   $\tan \blacksquare° = 7$ $\tan 68° = \blacksquare$ $\tan \blacksquare° = 3.5$

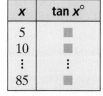

x	tan *x*°
5	▦
10	▦
⋮	⋮
85	▦

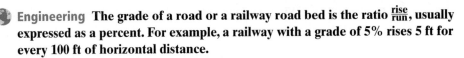

Engineering The grade of a road or a railway road bed is the ratio $\frac{rise}{run}$, usually expressed as a percent. For example, a railway with a grade of 5% rises 5 ft for every 100 ft of horizontal distance.

31. The Katoomba Railway, pictured at left, has a grade of 122%. What angle does its roadbed make with the horizontal?

32. The Johnstown, Pennsylvania, inclined railway was built as a "lifesaver" after the Johnstown flood of 1889. It has a 987-ft run at a 71% grade. How high does this railway lift its passengers?

33. The Fenelon Place Elevator railway in Dubuque, Iowa, lifts passengers 189 ft to the top of a bluff. It has an 83% grade. How long is this railway?

34. The Duquesne Incline Plane Company's roadway in Pittsburgh, Pennsylvania, climbs Mt. Washington, located above the mouth of the Monongahela River. It reaches a height of 400 ft with a 793-ft incline. What is its grade?

Real-World  **Connection**

The world's steepest railway is the Katoomba Scenic Railway in Australia's Blue Mountains.

Find the missing value to the nearest tenth.

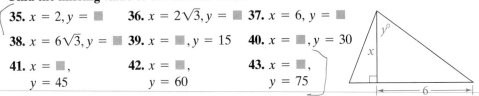

35. $x = 2, y = \blacksquare$ 36. $x = 2\sqrt{3}, y = \blacksquare$ 37. $x = 6, y = \blacksquare$

38. $x = 6\sqrt{3}, y = \blacksquare$ 39. $x = \blacksquare, y = 15$ 40. $x = \blacksquare, y = 30$

41. $x = \blacksquare,$ 42. $x = \blacksquare,$ 43. $x = \blacksquare,$
    $y = 45$ $y = 60$ $y = 75$

44. **a. Critical Thinking** Does $\tan A + \tan B = \tan(A + B)$ when $A + B < 90$? Explain.

   b. Reasoning Does $\tan A - \tan B = \tan(A - B)$ when $A - B > 0$? Use part (a) and indirect reasoning to explain.

45. **Graphing Calculator** Use the **TABLE** feature of your graphing calculator to study tan X as X gets close to 90. In the Y= screen, enter Y1 = tan X.

   a. Use the **TBLSET** feature so that X starts at 80 and changes by 1. Access the **TABLE**. From the table, what is tan X for X = 89?

   b. Perform a "numerical zoom in." Use the **TBLSET** feature, so that X starts with 89 and changes by 0.1. What is tan X for X = 89.9?

   c. Continue to numerically zoom in on values close to 90. What is the greatest value you can get for tan X on your calculator? How close is X to 90?

   d. Writing Use right triangles to explain the behavior of tan X found above.

Go Online
PHSchool.com
For: Graphing calculator procedures
Web Code: aue-2111

Challenge

46. **Graphing Calculator** Use the **TABLE** and graphing features of your graphing calculator to study the product tan X · tan (90 − X). In the Y= screen, enter Y1 = tan X · tan (90 − X).

   a. Use the **TBLSET** feature so that X starts at 1 and changes by 1. Access the **TABLE**. What do you notice?

   b. Press GRAPH. What do you notice?

   Proof **c.** Make a conjecture about tan X · tan (90 − X) based on parts (a) and (b). Write a paragraph proof of your conjecture.

Use the given information and $\tan^{-1}$ to find $m\angle A$ to the nearest whole number.

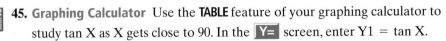

47. $\tan 2A = 9.5144$ 48. $\tan \frac{A}{3} = 0.4663$

49. $(\tan 5A)^2 = 0.3333$ 50. $\frac{\tan A}{1 + \tan A} = 0.5437$

Problem Solving Hint

Think of $\tan^{-1}(x)$ as "the angle whose tangent is x."

Simplify each expression.

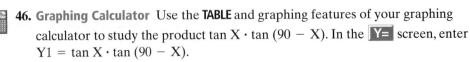

51. $\tan(\tan^{-1} x)$ 52. $\tan^{-1}(\tan X)$

Coordinate Geometry You can use the slope of a line to find the measure of the acute angle that the line forms with any horizontal line.

$$\text{slope} = \frac{\text{rise}}{\text{run}} = 3$$

$$\tan A = \frac{\text{opposite}}{\text{adjacent}} = 3$$

$$m\angle A = \tan^{-1}(3) \approx 71.6$$

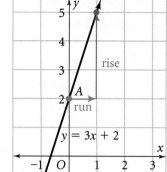

To the nearest tenth, find the measure of the acute angle that the line forms with a horizontal line.

53. $y = \frac{1}{2}x + 6$ **54.** $y = 6x - 1$

55. $y = 5x - 7$ **56.** $y = \frac{4}{3}x - 1$

57. $3x - 4y = 8$ **58.** $-2x + 3y = 6$

Test Prep

Gridded Response

59. What is tan 84° to the nearest tenth?

60. What is the whole number value of $\tan^{-1}\sqrt{3}$?

In Exercises 61–64 what is the value of x to the nearest tenth?

61.

62.

63.
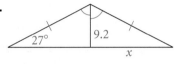

64.

65. The tangent of an angle is 7.5. What is the measure of the angle to the nearest tenth?

Mixed Review

Lesson 8-2

66. A diagonal of a square is 10 units. Find the length of a side of the square. Leave your answer in simplest radical form.

Lesson 8-1

The lengths of the sides of a triangle are given. Classify each triangle as *acute*, *right*, or *obtuse*.

67. 5, 8, 4 **68.** 15, 15, 20 **69.** 0.5, 1.2, 1.3

Lesson 6-7

70. For the kite pictured at the right, give the coordinates of the midpoints of its sides.

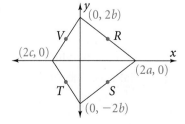

Exploring Trigonometric Ratios

FOR USE WITH LESSON 8-4

Construct

Use geometry software to construct $\overrightarrow{AB}$ and $\overrightarrow{AC}$ so that $\angle A$ is acute. Through a point D on $\overrightarrow{AB}$ construct a line perpendicular to $\overrightarrow{AB}$ that intersects $\overrightarrow{AC}$ in point E.

Moving point D enlarges or reduces $\triangle ADE$. Moving point C changes the size of $\angle A$.

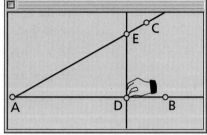

EXERCISES

1. • Measure $\angle A$.

   • Find the lengths of the sides of $\triangle ADE$.

   • Calculate the ratio $\dfrac{\text{leg opposite } \angle A}{\text{hypotenuse}}$, which is $\dfrac{ED}{AE}$.

   • Move point D to change the size of the right triangle without changing the size of $\angle A$.

   What do you observe about the ratio as the size of $\triangle ADE$ changes?

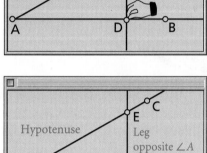

2. • Move point C to change the size of $\angle A$.
   a. What do you observe about the ratio as the size of $\angle A$ changes?
   b. What value does the ratio approach as $m\angle A$ approaches 0? As $m\angle A$ approaches 90?

3. • Make a table that shows values for $m\angle A$ and the ratio $\dfrac{\text{leg opposite } \angle A}{\text{hypotenuse}}$. In your table, include $10, 20, 30, \ldots, 80$ for $m\angle A$.

   • Compare your table with the table of trigonometric ratios on page 769.

   Do your values for $\dfrac{\text{leg opposite } \angle A}{\text{hypotenuse}}$ match the values in one of the columns of the table? What is the name of this ratio in the table?

Extend

4. Repeat Exercises 1–3 for the ratio $\dfrac{\text{leg adjacent to } \angle A}{\text{hypotenuse}}$, which is $\dfrac{AD}{AE}$.

5. Repeat Exercises 1–3 for the ratio $\dfrac{\text{leg opposite } \angle A}{\text{leg adjacent to } \angle A}$, which is $\dfrac{ED}{AD}$.

6. • Choose a measure for $\angle A$ and determine the ratio $r = \dfrac{\text{leg opposite } \angle A}{\text{hypotenuse}}$. Record $m\angle A$ and this ratio.

   • Manipulate the triangle so that $\dfrac{\text{leg adjacent to } \angle A}{\text{hypotenuse}}$ has the same value r. Record this $m\angle A$ and compare it with your first value of $m\angle A$.

   • Repeat this procedure several times.

   • Look for a pattern in the two measures of $\angle A$ that you found for the different values of r.

   Make a conjecture.

Sine and Cosine Ratios

What You'll Learn

- To use sine and cosine to determine side lengths in triangles

... And Why

To use the sine ratio to estimate astronomical distances indirectly, as in Example 2

 Check Skills You'll Need

 for Help Lesson 8-3

For each triangle, find (a) the length of the leg opposite ∠B and (b) the length of the leg adjacent to ∠B.

1.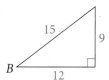

2.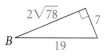

3.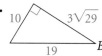

🔊 **New Vocabulary** • sine • cosine • identity

---

1 Using Sine and Cosine in Triangles

The tangent ratio, as you have seen, involves both legs of a right triangle. The sine and cosine ratios involve one leg and the hypotenuse.

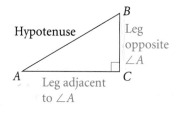

sine of $\angle A = \dfrac{\text{leg opposite } \angle A}{\text{hypotenuse}}$

cosine of $\angle A = \dfrac{\text{leg adjacent to } \angle A}{\text{hypotenuse}}$

These equations can be abbreviated:

$$\sin A = \frac{\text{opposite}}{\text{hypotenuse}} \qquad \cos A = \frac{\text{adjacent}}{\text{hypotenuse}}$$

Real-World 🌎 **Connection**

For an angle of a given size, the sine and cosine ratios are constant, no matter where the angle is located.

1 EXAMPLE **Writing Sine and Cosine Ratios**

Use the triangle to write each ratio.

a. sin T $\qquad \sin T = \dfrac{\text{opposite}}{\text{hypotenuse}} = \dfrac{8}{17}$

b. cos T $\qquad \cos T = \dfrac{\text{adjacent}}{\text{hypotenuse}} = \dfrac{15}{17}$

c. sin G $\qquad \sin G = \dfrac{\text{opposite}}{\text{hypotenuse}} = \dfrac{15}{17}$

d. cos G $\qquad \cos G = \dfrac{\text{adjacent}}{\text{hypotenuse}} = \dfrac{8}{17}$

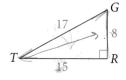

✓ **Quick Check** **1 a.** Write the sine and cosine ratios for ∠X and ∠Y.
b. Critical Thinking When does sin X = cos Y? Explain.

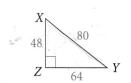

One way to describe the relationship of sine and cosine is to say that $\sin x° = \cos(90 - x)°$ for values of x between 0 and 90. This type of equation is called an **identity** because it is true for all the allowed values of the variable. You will discover other identities in the exercises.

2 EXAMPLE   Real-World Connection

Astronomy The trigonometric ratios have been known for centuries by peoples in many cultures. The Polish astronomer Nicolaus Copernicus (1473−1543) developed a method for determining the sizes of orbits of planets closer to the sun than Earth. The key to his method was determining when the planets were in the position shown in the diagram, and then measuring the angle to find a.

If $a = 22.3$ for Mercury, how far is Mercury from the sun in astronomical units (AU)? One astronomical unit is defined as the average distance from Earth to the center of the sun, about 93 million miles.

not to scale

$$\sin 22.3° = \frac{x}{1} \qquad \text{Use the sine ratio.}$$
$$x = \sin 22.3° \qquad \text{Solve for } x.$$

SIN 22.3 **ENTER** $.37945616$ **Use a calculator.**

● Mercury is about 0.38 AU from the sun.

✓ **Quick Check** ❷ **a.** If $a = 46$ for Venus, how far is Venus from the sun in AU?
b. About how many miles from the sun is Venus? Mercury?

When you know the leg and hypotenuse lengths of a right triangle, you can use inverse of sine and inverse of cosine to find the measures of the acute angles.

3 EXAMPLE   Using the Inverse of Cosine and Sine

Find $m\angle L$ to the nearest degree.

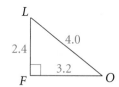

Method 1

$$\cos L = \frac{2.4}{4.0} \quad \longleftarrow \text{Find the trigonometric ratio.} \longrightarrow$$
$$m\angle L = \cos^{-1}\left(\frac{2.4}{4.0}\right) \quad \longleftarrow \text{Use the inverse.} \longrightarrow$$

COS⁻¹ 2.4 **÷** 4.0 **ENTER** ← Use a calculator. →
53.130102

● $m\angle L \approx 53$

Method 2

$$\sin L = \frac{3.2}{4.0}$$
$$m\angle L = \sin^{-1}\left(\frac{3.2}{4.0}\right)$$

SIN⁻¹ 3.2 **÷** 4.0 **ENTER**
53.130102

$m\angle L \approx 53$

✓ **Quick Check** ❸ Find the value of x. Round your answer to the nearest degree.

a.

b.

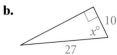

EXERCISES

For more exercises, see *Extra Skill, Word Problem, and Proof Practice.*

Practice and Problem Solving

Ⓐ Practice by Example

Example 1
(page 439)

GO for Help

Write the ratios for sin *M* and cos *M*.

1.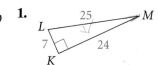

2.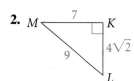

3.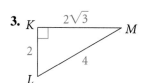

Example 2
(page 440)

Find the value of *x*. Round answers to the nearest tenth.

4.

5.

6.

7.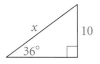

8.

9.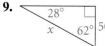

10. Escalators An escalator in the subway system of St. Petersburg, Russia, has a vertical rise of 195 ft 9.5 in., and rises at an angle of 10.4°. How long is the escalator? Round your answer to the nearest foot.

Example 3
(page 440)

Find the value of *x*. Round answers to the nearest degree.

11.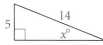

12.

13.

14.

15.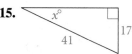

16.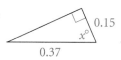

Ⓑ Apply Your Skills

17. Construction Carlos is planning to build a grain bin with a radius of 15 ft. He reads that the recommended slant of the roof is 25°. He wants the roof to overhang the edge of the bin by 1 ft. What should the length *x* be? Give your answer in feet and inches.

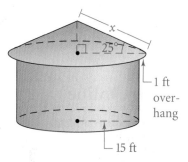

Use what you know about trigonometric ratios (and other identities) to show that each equation is an identity.

18. $\tan X = \dfrac{\sin X}{\cos X}$

19. $\sin X = \cos X \cdot \tan X$

20. $\cos X = \dfrac{\sin X}{\tan X}$

21. Error Analysis A student states that sin *A* > sin *X* because the lengths of the sides of △*ABC* are greater than the lengths of the sides of △*XYZ*. Is the student correct? Explain.

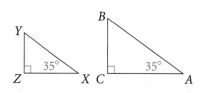

Real-World 🌎 Connection

Corn that fills the bin in Exercise 17 would make 28,500 gallons of ethanol.

Lesson 8-4 Sine and Cosine Ratios | **441**

Find the values of *w* and then *x*. Round lengths to the nearest tenth and angle measures to the nearest degree.

22.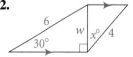

23.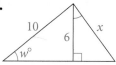

24.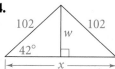

25. **a.** In $\triangle ABC$, how does sin *A* compare to cos *B*? Is this true for the acute angles of other right triangles?

    b. **Reading Math** The word cosine is derived from the words *complement's sine* (see page 694). Which angle in $\triangle ABC$ is the complement of $\angle A$? Of $\angle B$?

    c. Explain why the derivation of the word cosine makes sense.

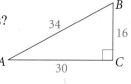

26. Find each ratio.

    a. sin *P*   **b.** cos *P*
    c. sin *R*   **d.** cos *R*
    e. Make a conjecture about how the sine and cosine of a 45° angle are related.

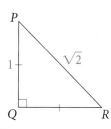

27. **Writing** Leona said that if she had a diagram that showed the measure of one acute angle and the length of one side of a right triangle, she could find the measure of the other acute angle and the lengths of the other sides. Is she correct? Explain.

28. Find each ratio.

    a. sin *S*   **b.** cos *S*
    c. sin *T*   **d.** cos *T*
    e. Make a conjecture about how the sine and cosine of a 30° angle are related.
    f. Make a conjecture about how the sine and cosine of a 60° angle are related.

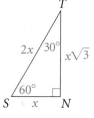

Proof 29. For right $\triangle ABC$ with right $\angle C$, prove each of the following.

    a. sin $A < 1$, no matter how large $\angle A$ is.
    b. cos $A < 1$, no matter how small $\angle A$ is.

30. **Graphing Calculator** Use the **TABLE** feature of your graphing calculator to study sin X as X gets close (but ≠) to 90. In the $\boxed{Y=}$ screen, enter Y1 = sin X.

    a. Use the **TBLSET** feature so that X starts at 80 and changes by 1. Access the **TABLE**. From the table, what is sin X for X = 89?

    b. Perform a "numerical zoom in." Use the **TBLSET** feature, so that X starts with 89 and changes by 0.1. What is sin X for X = 89.9?

    c. Continue to numerically zoom in on values close to 90. What is the greatest value you can get for sin X on your calculator? How close is X to 90? Does your result contradict what you are asked to prove in Exercise 29a?

    d. **Writing** Use right triangles to explain the behavior of sin X found above.

 Challenge

Show that each equation is an identity by showing that each expression on the left simplifies to 1.

31. $(\sin A)^2 + (\cos A)^2 = 1$ 32. $(\sin B)^2 + (\cos B)^2 = 1$

33. $\dfrac{1}{(\cos A)^2} - (\tan A)^2 = 1$ 34. $\dfrac{1}{(\sin A)^2} - \dfrac{1}{(\tan A)^2} = 1$

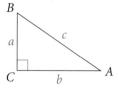

35. Show that $(\tan A)^2 - (\sin A)^2 = (\tan A)^2 (\sin A)^2$ is an identity.

Real-World Connection

Poland honored Copernicus with this 1000-zloty note, last used in 1995.

36. Astronomy Copernicus devised a method different from the one in Example 2 in order to find the sizes of the orbits of planets farther from the sun than Earth. His method involved noting the number of days between the times that a planet was in the positions labeled A and B in the diagram. Using this time and the number of days in each planet's year, he calculated c and d.

a. For Mars, $c = 55.2$ and $d = 103.8$. How far is Mars from the sun in astronomical units (AU)?

b. For Jupiter, $c = 21.9$ and $d = 100.8$. How far is Jupiter from the sun in astronomical units?

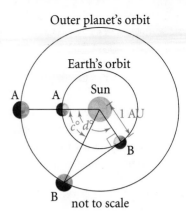

Outer planet's orbit

Earth's orbit

not to scale

Test Prep

Multiple Choice

37. What is the value of x to the nearest whole number?
   A. 2   **B.** 3
   C. 4   **D.** 6

38. What is the value of y to the nearest tenth?
   F. 5.4   **G.** 5.5
   H. 5.6   **J.** 5.7

39. What is the value of x to the nearest whole number?
   A. 53   **B.** 47
   C. 43   **D.** 37

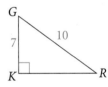

Short Response

40. Use the figure at the right.
   a. Find $m\angle G$. Show your work.
   b. Find $m\angle R$ by two different methods. Show your work.

Mixed Review

Lesson 8-3   Find the value of x. Round answers to the nearest tenth.

41.

42.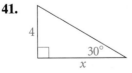

43.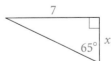

Lesson 7-2   **44.** The wall of a room is in the shape of a golden rectangle. If the height of the wall is 8 ft, what are the possible lengths of the wall to the nearest tenth?

Lesson 6-2   Find the value of x for each parallelogram.

45.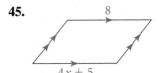

46.

47.

You have learned about two tools—the staff and the stadiascope (page 431)—that were used by people of earlier times to find distances and heights. In this activity, you will make another tool to find heights. All three tools require you to use similar right triangles, but this time you will use trigonometric ratios.

ACTIVITY

1. The device shown below is called an *inclinometer*. Make your own inclinometer using a protractor, a piece of string, and a washer.

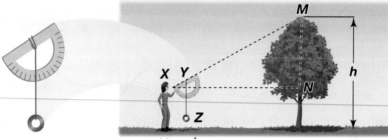

not to scale

2. The string on the inclinometer shows $m\angle XYZ$. Explain how you can find $m\angle X$ if you know $m\angle XYZ$.

3. Think about how you can calculate an approximate height of the tree using the inclinometer and a trigonometric ratio.
   a. What other measurement do you need besides the $m\angle X$?
   b. What trigonometric ratio do you need?
   c. Show how to find the height of the tree.

4. Use an inclinometer to find the height of a tall object at your school. If you found the height of the same object as part of the activity on page 431, compare your results. In which result do you have the greater confidence? Why?

EXERCISES

5. You are on a steep hillside directly across from the top of a tree. Explain how you could use an inclinometer, a trigonometric ratio, and the distance from the hilltop to the base of the tree to find the approximate height of the tree.

not to scale

6. Suppose you could climb up only to point P on the hillside. Explain how you could use an inclinometer and trigonometric ratios to find the height of the tree from that spot.

7. Use the diagram at the right. Show that the height of the tree can be found using the formula

$$h = d(\cos b)(\tan a + \tan b).$$

Angles of Elevation and Depression

What You'll Learn

• To use angles of elevation and depression to solve problems

. . . And Why

To use the angle of elevation to calculate the height of a natural wonder, as in Example 2

✓ Check Skills You'll Need

Refer to rectangle *ABCD* to complete the statements.

1. ∠1 ≅ ■
2. ∠5 ≅ ■
3. ∠3 ≅ ■
4. m∠1 + m∠5 = ■
5. m∠10 + m∠3 = ■
6. ∠10 ≅ ■

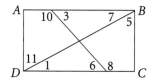

🔊 New Vocabulary   • angle of elevation   • angle of depression

GO for Help Lesson 6-1

---

1 Using Angles of Elevation and Depression

Suppose a person on the ground sees a hot-air balloon gondola at a 38° angle above a horizontal line.

This angle is the **angle of elevation.**

At the same time, a person in the hot-air balloon sees the person on the ground at a 38° angle below a horizontal line.

This angle is the **angle of depression.**

Examine the diagram. The angle of elevation is congruent to the angle of depression because they are alternate interior angles.

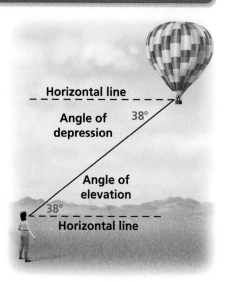

1 EXAMPLE   Identifying Angles of Elevation and Depression

Describe each angle as it relates to the situation shown.

a. ∠1   ∠1 is the angle of depression from the peak to the hiker.

b. ∠4   ∠4 is the angle of elevation from the hut to the hiker.

✓ **Quick Check** ① Describe each angle as it relates to the situation in Example 1.
   a. ∠2   **b.** ∠3

Surveyors use two instruments, the transit and the theodolite, to measure angles of elevation and depression. On both instruments, the surveyor sets the horizon line perpendicular to the direction of gravity. Using gravity to find the horizon line ensures accurate measures even on sloping surfaces.

2 EXAMPLE Real-World Connection

Surveying To find the height of Delicate Arch in Arches National Park in Utah, a surveyor levels a theodolite with the bottom of the arch. From there, she measures the angle of elevation to the top of the arch. She then measures the distance from where she stands to a point directly under the arch. Her results are shown in the diagram. What is the height of the arch?

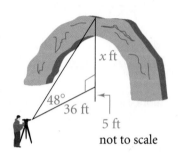
not to scale

$$\tan 48° = \frac{x}{36} \qquad \text{Use the tangent ratio.}$$
$$x = 36(\tan 48°) \qquad \text{Solve for } x.$$

36 **TAN** 48 **ENTER** *39.982051*   Use a calculator.

So $x \approx 40$. To find the height of the arch, add the height of the theodolite. Since $40 + 5 = 45$, Delicate Arch is about 45 feet high.

✓ **Quick Check** ❷ You sight a rock climber on a cliff at a 32° angle of elevation. The horizontal ground distance to the cliff is 1000 ft. Find the line-of-sight distance to the rock climber.

3 EXAMPLE Real-World Connection

Multiple Choice To approach runway 17 of the Ponca City Municipal Airport in Oklahoma, the pilot must begin a 3° descent starting from an altitude of 2714 ft. The airport altitude is 1007 ft. How many miles from the runway is the airplane at the start of this approach?

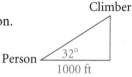

   Ⓐ 3.6 mi   Ⓑ 5.7 mi   Ⓒ 6.2 mi   Ⓓ 9.8 mi

The airplane is $2714 - 1007$, or 1707 ft above the level of the airport.

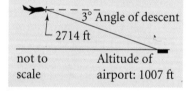

$$\sin 3° = \frac{1707}{x} \qquad \text{Use the sine ratio.}$$
$$x = \frac{1707}{\sin 3°} \qquad \text{Solve for } x.$$

1707 **÷** **SIN** 3 **ENTER** *32616.2*   Use a calculator.

÷ 5280 **ENTER** *6.1773105*   Divide by 5280 to convert feet to miles.

The airplane is about 6.2 mi from the runway. The correct answer is C.

✓ **Quick Check** ❸ An airplane pilot sights a life raft at a 26° angle of depression. The airplane's altitude is 3 km. What is the airplane's surface distance d from the raft?

EXERCISES

For more exercises, see *Extra Skill, Word Problem, and Proof Practice.*

Practice and Problem Solving

A Practice by Example

Example 1
(page 445)

GO for Help

Describe each angle as it relates to the situation in the diagram.

1. ∠1   **2.** ∠2   **3.** ∠3   **4.** ∠4   **5.** ∠5   **6.** ∠6   **7.** ∠7   **8.** ∠8

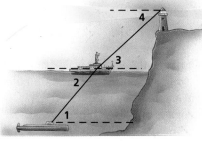

Jim   Kelley

Example 2
(page 446)

Find the value of *x*. Round the lengths to the nearest tenth of a unit.

9.
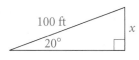
100 ft   *x*   20°

10.
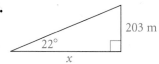
203 m   22°   *x*

11. Meteorology A meteorologist measures the angle of elevation of a weather balloon as 41°. A radio signal from the balloon indicates that it is 1503 m from his location. To the nearest meter, how high above the ground is the balloon?

Example 3
(page 446)

Find the value of *x*. Round the lengths to the nearest tenth of a unit.

12.
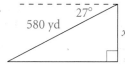
27°   580 yd   *x*

13.
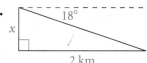
18°   *x*   2 km

14. Indirect Measurement Miguel looks out from the crown of the Statue of Liberty approximately 250 ft above ground. He sights a ship coming into New York harbor and measures the angle of depression as 18°. Find the distance from the base of the statue to the ship to the nearest foot.

B Apply Your Skills

15. Flagpole The world's tallest unsupported flagpole is a 282-ft-tall steel pole in Surrey, British Columbia. The shortest shadow cast by the pole during the year is 137 ft long. To the nearest degree, what is the angle of elevation of the sun when the shortest shadow is cast?

16. Engineering The Americans with Disabilities Act states that wheelchair ramps can have a slope no greater than $\frac{1}{12}$. Find the angle of elevation of a ramp with this slope. Round your answer to the nearest tenth.

17. Construction Two office buildings are 51 m apart. The height of the taller building is 207 m. The angle of depression from the top of the taller building to the top of the shorter building is 15°. Find the height of the shorter building to the nearest meter.

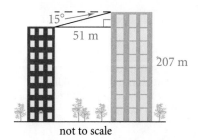

15°   51 m   207 m

not to scale

GO Online
Homework Video Tutor
Visit: PHSchool.com
Web Code: aue-0805

 **for Help**

For a guide to solving
Exercise 18, see p. 451.

18. Aerial Television A blimp is providing aerial television views of a football game. The television camera sights the stadium at a 7° angle of depression. The blimp's altitude is 400 m. What is the line-of-sight distance from the TV camera to the stadium, to the nearest hundred meters?

x^2 **Algebra** The angle of elevation e from A to B and the angle of depression d from B to A are shown below. Find the measure of each angle.

19. $e: (7x - 5)°, d: 4(x + 7)°$ **20.** $e: (3x + 1)°, d: 2(x + 8)°$

21. $e: (x + 21)°, d: 3(x + 3)°$ **22.** $e: 5(x - 2)°, d: (x + 14)°$

23. Multiple Choice An engineer is 980 ft from the base of a fountain at Fountain Hills, Arizona. The angle of elevation to the top of the column of water is 29.7°. The surveyor's angle measuring device is at the same level as the base of the fountain. Find the height of the column of water to the nearest 10 ft.

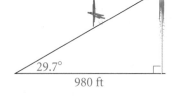

Ⓐ 490 ft   Ⓑ 560 ft   Ⓒ 850 ft   Ⓓ 1720 ft

24. Writing A communications tower is located on a plot of flat land. The tower is supported by several guy wires. Assume that you are able to measure distances along the ground, as well as angles formed by the guy wires and the ground. Explain how you could estimate each of the following measurements.
a. the length of any guy wire
b. how high on the tower each wire is attached

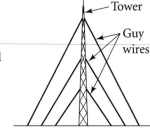

Flying An airplane at altitude a flies distance d towards you with velocity v. You watch for time t and measure its angles of elevation, $\angle E_1$ and $\angle E_2$, at the start and end of your watch. Find the missing information.

25. $a = \blacksquare$ mi, $v = 5$ mi/min, $t = 1$ min, $m\angle E_1 = 45$, $m\angle E_2 = 90$

26. $a = 2$ mi, $v = \blacksquare$ mi/min, $t = 15$ s, $m\angle E_1 = 40$, $m\angle E_2 = 50$

27. $a = 4$ mi, $d = 3$ mi, $v = 6$ mi/min, $t = \blacksquare$ min, $m\angle E_1 = 50$, $m\angle E_2 = \blacksquare$

28. Meteorology One method that meteorologists could use to find the height of a layer of clouds above the ground is to shine a bright spotlight directly up onto the cloud layer and measure the angle of elevation from a known distance away. Find the height of the cloud layer in the diagram to the nearest 10 m.

Real-World Connection

Careers Atmospheric scientists specialize by linking meteorology with another field such as agriculture.

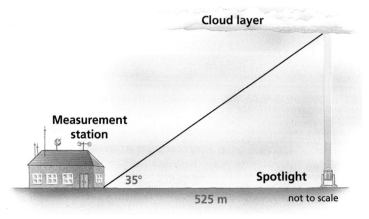

C Challenge 🌐 **29. Firefighting** A firefighter on the ground sees fire break through a window near the top of the building. There is voice contact between the ground and firefighters on the roof. The angle of elevation to the windowsill is 28°. The angle of elevation to the top of the building is 42°. The firefighter is 75 ft from the building and her eyes are 5 ft above the ground. What roof-to-windowsill distance can she report to the firefighters on the roof?

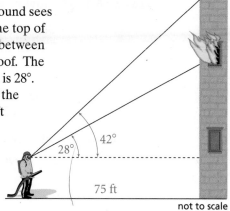

not to scale

🌐 **30. Geography** For locations in the United States, the relationship between the latitude ℓ and the greatest angle of elevation a of the sun at noon on the first day of summer is $a = 90° - \ell + 23\frac{1}{2}°$. Find the latitude of your town. Then determine the greatest angle of elevation of the sun for your town on the first day of summer.

Test Prep

Multiple Choice

31. A 107-ft-tall building casts a shadow of 90 ft. To the nearest whole degree, what is the angle of elevation to the sun?
   A. 33°   **B.** 40°   **C.** 50°   **D.** 57°

32. The angle of depression of a submarine from another Navy ship is 28°. The submarine is 791 ft from the ship. About how deep is the submarine?
   F. 371 ft   **G.** 421 ft   **H.** 563 ft   **J.** 698 ft

33. A kite on a 100-ft string has an angle of elevation of 18°. The hand holding the string is 4 ft from the ground. How high above the ground is the kite?
   A. 95 ft   **B.** 35 ft   **C.** 31 ft   **D.** 22 ft

Short Response

34. A 6-ft-tall man is viewing the top of a tree with an angle of elevation of 83°. He is standing 12 ft from the base of the tree.
   a. Draw a sketch of the situation. Show a stick figure for the man. Label the angle of elevation, the height of the man, and the distance the man is standing from the tree.
   b. Write and solve an equation to find the height of the tree. Round your answer to the nearest foot.

Mixed Review

GO for Help

Lesson 8-4

Find the value of *x*. Round answers to the nearest tenth.

35.

36.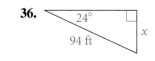

37.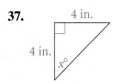

Lesson 6-1 $\boxed{x^2}$ **Algebra Find the value of each variable. Then find the length of each side.**

38.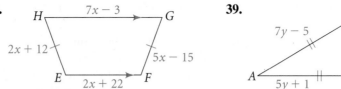

39.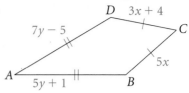

Lesson 4-4 **40. Given:** $\angle QPS \cong \angle RSP, \angle Q \cong \angle R$
Prove: $\overline{PQ} \cong \overline{SR}$

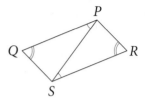

Checkpoint Quiz 2   Lessons 8-3 through 8-5

Write the tangent, sine, and cosine ratios for $\angle A$ and $\angle B$.

1.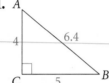
2.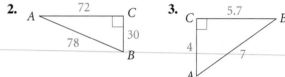
3.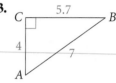

$\boxed{x^2}$ **Algebra Find the value of x. Round each segment length to the nearest tenth and each angle measure to the nearest whole number.**

4.
5.
6.

7. Landmarks The Leaning Tower of Pisa, shown at the right, reopened in 2001 after a 10-year project reduced its tilt from vertical by 0.5°. How far from the base of the tower will an object land if it is dropped the 150 ft shown in the photo?

8. Navigation A captain of a sailboat sights the top of a lighthouse at a 17° angle of elevation. A navigation chart shows the height of the lighthouse to be 120 m. How far is the sailboat from the lighthouse?

9. Writing How do you decide which trigonometric ratio to use to solve a problem?

10. Hang Gliding Students in a hang gliding class stand on the top of a cliff 70 m high. They watch a hang glider land on the beach below. The angle of depression to the hang glider is 72°. How far is the hang glider from the base of the cliff?

Understanding Word Problems Read the problem below. Then let Curtis's thinking guide you through the solution. Check your understanding with the exercises at the bottom of the page.

A blimp is providing aerial television views of a football game. The television camera sights the stadium at a 7° angle of depression. The blimp's altitude is 400 m. What is the line-of-sight distance from the TV camera to the stadium, to the nearest hundred meters?

What Curtis Thinks

To start, I'll draw and label a sketch.
  Angle of depression = 7°
  Altitude of blimp = 400 m

I'm supposed to find the line-of-sight distance from the TV camera to the stadium. That's the diagonal segment from the blimp's gondola to the 50-yd line. I'll label it x.

To find x, I'll probably need a trig ratio. I've labeled a leg and the hypotenuse of a right triangle 400 m and x. I can use a *sine* or *cosine* ratio if I can find one of the acute angles.

Because alternate interior angles must be congruent, the smaller acute angle is 7°.

Now I'll write an equation.

I'll solve the equation for x.

I'll use a calculator and round.

Now I can state the answer.

What Curtis Writes

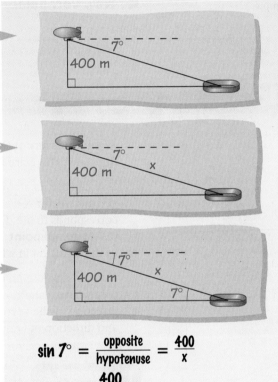

$$\sin 7° = \frac{\text{opposite}}{\text{hypotenuse}} = \frac{400}{x}$$

$$x = \frac{400}{\sin 7°}$$

$$x \approx 3282 \approx 3300$$

The line-of-sight distance is about 3300 m.

EXERCISES

1. The blimp moves straight up and the angle of depression for the TV camera increases by 1°. How high did the blimp rise?

2. A pedestrian sights the top of a building at an angle of elevation of 75°. She is standing 50 ft from the base of the building. How high above her eye level is the top of the building to the nearest foot?

8-6

Vectors

What You'll Learn

• To describe vectors

• To solve problems that involve vector addition

. . . And Why

To use vectors to describe the distance and direction of an airplane flight, as in Example 3

✓ **Check Skills You'll Need**

GO for Help Lesson 8-1

x^2 **Algebra** Find the value of x. Leave your answers in simplest radical form.

1.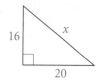

2.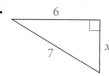

3.

🔊 **New Vocabulary** • vector • magnitude • initial point • terminal point • resultant

---

1 Describing Vectors

A **vector** is any quantity with magnitude (size) and direction. There are many models for a vector.

You can use an arrow for a vector as shown by the velocity vector $\overrightarrow{KW}$ in the photo. The **magnitude** corresponds to the distance from **initial point** K to the **terminal point** W. The direction corresponds to the direction in which the arrow points.

You can also use an ordered pair $\langle x, y \rangle$ in the coordinate plane for a vector. The magnitude and direction of the vector correspond to the distance and direction of $\langle x, y \rangle$ from the origin.

1 EXAMPLE **Describing a Vector**

Coordinate Geometry Describe $\overrightarrow{OL}$ as an ordered pair. Give the coordinates to the nearest tenth.

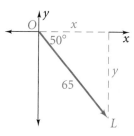

Use the sine and cosine ratios to find the values of x and y.

$\cos 50° = \dfrac{x}{65}$ $\sin 50° = \dfrac{y}{65}$ **Use sine and cosine.**

$x = 65(\cos 50°)$ $y = 65(\sin 50°)$ **Solve for the variable.**

≈ 41.78119463 ≈ 49.7928888 **Use a calculator.**

L is in the fourth quadrant so the y-coordinate is negative. $\overrightarrow{OL} \approx \langle 41.8, -49.8 \rangle$.

① Describe the vector at the right as an ordered pair. Give the coordinates to the nearest tenth.

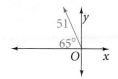

Real-World ⊕ Connection

A velocity vector for a "bullet train" can have magnitude 275 km/h paired with any direction point on a compass.

In many applications of vectors, you use the compass directions north, south, east, and west to describe the direction of a vector.

② **EXAMPLE** **Describing a Vector Direction**

Use compass directions to describe the direction of each vector.

a.

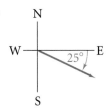

25° south of east

b.

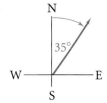

35° east of north

② **a.** Sketch a vector that has the direction 30° west of north.
b. Critical Thinking Give a second description for the direction of this vector.

Example 3 shows how to describe a vector's magnitude and direction when you are given its description as an ordered pair.

③ **EXAMPLE** **Real-World ⊕ Connection**

Aviation An airplane lands 40 km west and 25 km south from where it took off. The result of the trip can be described by the vector $\langle -40, -25 \rangle$. Use distance (for magnitude) and direction to describe this vector a second way.

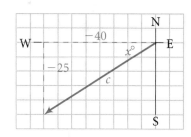

To find the distance, use the Distance Formula:

$d = \sqrt{(-40 - 0)^2 + (-25 - 0)^2}$

$d = \sqrt{1600 + 625}$ **Simplify.**

$d = \sqrt{2225}$

$d \approx 47.169906$ **Use a calculator to find the square root.**

To find the direction of the flight, find the angle of the vector south of west.

$\tan x° = \frac{25}{40}$ **Find the tangent ratio.**

$x = \tan^{-1}\left(\frac{25}{40}\right)$ **Use the inverse of tangent.**

TAN⁻¹ 25 ⊟ 40 **ENTER** 32.005383 **Use a calculator.**

The airplane flew about 47 km at 32° south of west.

③ A small airplane lands at a point 246 mi east and 76 mi north of the point from which it took off. Describe the magnitude and the direction of its flight vector.

You can also use a single lowercase letter, such as $\vec{u}$, to name a vector.

This map shows vectors representing a flight from Houston to Memphis with a stopover in New Orleans. The vector from Houston to Memphis is called the sum, or **resultant,** of the other two vectors. You write this as

$$\vec{w} = \vec{u} + \vec{v}.$$

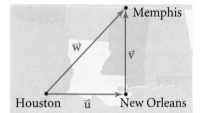

You can add vectors by adding their coordinates. You can also show the sum geometrically.

 Key Concepts

Property	**Adding Vectors**

For $\vec{a} = \langle x_1, y_1 \rangle$ and $\vec{c} = \langle x_2, y_2 \rangle$, $\vec{a} + \vec{c} = \langle x_1 + x_2, y_1 + y_2 \rangle$.

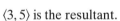

4 **EXAMPLE** **Adding Vectors**

Online
active math

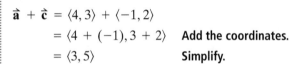

For: Adding Vectors Activity
Use: Interactive Textbook, 8-6

Vectors $\vec{a}$ $\langle 4, 3 \rangle$ and $\vec{c}$ $\langle -1, 2 \rangle$ are shown in the diagram. Write the sum of the two vectors as an ordered pair. Then draw $\vec{e}$, the sum of $\vec{a}$ and $\vec{c}$.

$$\begin{aligned}
\vec{a} + \vec{c} &= \langle 4, 3 \rangle + \langle -1, 2 \rangle \\
&= \langle 4 + (-1), 3 + 2 \rangle \quad \textbf{Add the coordinates.} \\
&= \langle 3, 5 \rangle \quad \textbf{Simplify.}
\end{aligned}$$

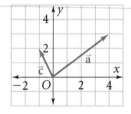

$\langle 3, 5 \rangle$ is the resultant.

Draw $\vec{a}$ with its initial point at the origin. Then draw $\vec{c}$ with its initial point at the terminal point of $\vec{a}$. Finally, draw the resultant $\vec{e}$ from the initial point of $\vec{a}$ to the terminal point of $\vec{c}$.

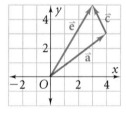

✓ **Quick Check** **4** Write the sum of the two vectors $\langle 2, 3 \rangle$ and $\langle -4, -2 \rangle$ as an ordered pair.

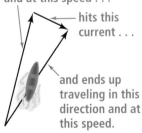

A canoe traveling in this direction and at this speed . . . hits this current . . . and ends up traveling in this direction and at this speed.

A vector sum can show the result of vectors that occur in sequence, such as in the airplane flight described above.

A vector sum can also show the result of vectors that act at the same time, such as when you row in a direction different from that of the current. See diagram at left.

The velocity of the canoe is the vector sum of the velocities of the paddlers and the stream.

5 EXAMPLE Real-World 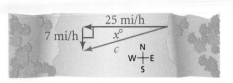 Connection

Real-World Connection

Ferry service is essential in remote regions such as on the Mackenzie River in Canada's Northwest Territories.

Navigation A ferry shuttles people from one side of a river to the other. The speed of the ferry in still water is 25 mi/h. The river flows directly south at 7 mi/h. If the ferry heads directly west, what are the ferry's resultant speed and direction?

The diagram shows the sum of the two vectors. To find the ferry's resultant speed, use the Pythagorean Theorem.

$c^2 = 25^2 + 7^2$ **The lengths of the legs are 25 and 7.**

$c^2 = 674$ **Simplify.**

$c \approx 25.961510$ **Use a calculator.**

To find the ferry's resultant direction, use trigonometry.

$\tan x° = \frac{7}{25}$ **Use the tangent ratio.**

$x = \tan^{-1}\left(\frac{7}{25}\right)$ **Use the inverse of the tangent.**

$x \approx 15.642246$ **Use a calculator.**

● The ferry's speed is about 26 mi/h. Its direction is about 16° south of west.

✓ Quick Check ⑤ **Critical Thinking** Use the diagram to find the angle at which the ferry must head upriver in order to travel directly across the river.

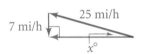

EXERCISES

For more exercises, see *Extra Skill, Word Problem, and Proof Practice.*

Practice and Problem Solving

Ⓐ Practice by Example

Example 1
(page 452)

GO for Help

Describe each vector as an ordered pair. Give the coordinates to the nearest tenth.

1.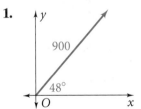

2.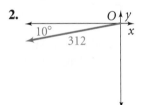

3.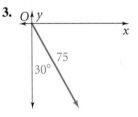

Example 2
(page 453)

Use compass directions to describe the direction of each vector.

4.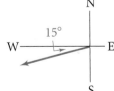

5.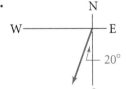

6.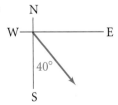

Sketch a vector that has the given direction.

7. 50° south of east   **8.** 20° north of west   **9.** 45° northeast

10. 70° west of north   **11.** 45° southwest   **12.** 10° east of south

Example 3
(page 453)

13. History Homing pigeons have the ability or instinct to find their way home when released hundreds of miles away from home. Homing pigeons carried news of Olympic victories to various cities in ancient Greece. Suppose one such pigeon took off from Athens and landed in Sparta, which is 73 mi west and 64 mi south of Athens. Find the distance and direction of its flight.

Find the magnitude and direction of each vector.

14.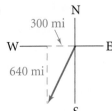

15.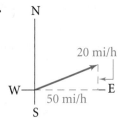

16.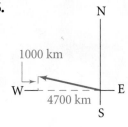

Example 4
(page 454)

In Exercises 17–22, (a) write the resultant as an ordered pair and (b) draw the resultant.

17.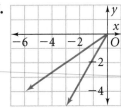

18.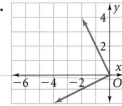

19.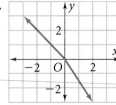

20.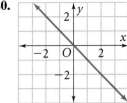

21.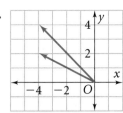

22.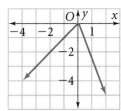

Example 5
(page 455)

Write the sum of the two vectors as an ordered pair.

23. $\langle 2, 1 \rangle$ and $\langle -3, 2 \rangle$ **24.** $\langle 0, 0 \rangle$ and $\langle 4, -6 \rangle$ **25.** $\langle -1, 1 \rangle$ and $\langle -1, 2 \rangle$

Navigation **The speed of a powerboat in still water is 35 mi/h. It is traveling on a river that flows directly south at 8 mi/h.**

26. The boat heads directly west across the river. What are the resulting speed and direction of the boat? Round answers to the nearest tenth.

27. At what angle should the boat head upriver in order to travel directly west?

28. Aviation A twin-engine airplane has a speed of 300 mi/h in still air. Suppose this airplane heads directly south and encounters a 50 mi/h wind blowing due east. Find the resulting speed and direction of the plane. Round your answers to the nearest unit.

B **Apply Your Skills**

29. Critical Thinking Valerie described the direction of a vector as 35° south of east. Pablo described it as 55° east of south. Could the two be describing the same vector? Explain.

30. Error Analysis Ely says that the magnitude of vector $\langle 6, 1 \rangle$ is 3 times that of vector $\langle 2, 1 \rangle$ since 6 is 3 times 2. Explain why Ely's statement is incorrect.

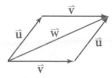
$\vec{v}$ $\vec{v}$
$\vec{u}$ $\vec{w}$ $\vec{u}$
$\vec{v}$

31. The diagram at the left shows that you can add vectors in any order. That is, $\vec{u} + \vec{v} = \vec{v} + \vec{u}$. Notice also that the four vectors shown in red form a parallelogram. The resultant $\vec{w}$ is the diagonal of the parallelogram. This representation of vector addition is called *The Parallelogram Rule*.

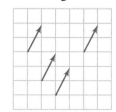

a. Copy the diagram at the right. Draw a parallelogram that has the given vectors as adjacent sides.

b. Find the magnitude and direction of the resultant.

Problem Solving Hint

You can also model vector addition with the *Triangle Rule* as shown in Example 4, and by either triangular half of the diagram above.

32. Use the diagrams below to write a definition of *equal vectors*.

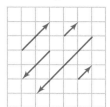

These vectors are equal.

No two of these vectors are equal.

33. Use the diagrams below to write a definition of *parallel vectors*.

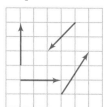

These vectors are parallel.

No two of these vectors are parallel.

34. **Multiple Choice** A Red Cross helicopter takes off and flies 75 km at 20° south of west. There, it drops off some relief supplies. It then flies 130 km at 20° west of north to pick up three medics. What is the helicopter's distance from its point of origin?

   Ⓐ 75 mi   Ⓑ 130 mi   Ⓒ 150 mi   Ⓓ 205 mi

35. a. Find the sum of $\vec{a}$ and $\vec{c}$, where $\vec{a} = \langle 45, -60 \rangle$ and $\vec{c} = \langle -45, 60 \rangle$.

   b. **Writing** Based on your answer to part (a), how can you describe $\vec{a}$ and $\vec{c}$?

36. **Aviation** In still air, the WP-3D (see below) flies at 374 mi/h. Suppose that a WP-3D flies due west and meets a hurricane wind blowing due south at 95 mi/h. What are the resultant speed and direction of the airplane to the nearest unit?

Flying into a Hurricane

When most pilots hear a forecast for gale force winds, they don't think, "Time to fly." Then again, most pilots don't work for the National Oceanic and Atmospheric Administration. NOAA fly their four-engine WP-3D turboprops directly into hurricanes. These aircraft carry eight crew members, up to ten scientists, and a load of data-collection equipment. Some of this equipment is in the WP-3D's long "snout," which also serves as a lightning rod. In a routine flight, the WP-3D is struck by lightning three or four times. Surprisingly, small burn holes are the only damage from these strikes. To help overcome temporary blindness caused by lightning flashes, the pilot sets the cockpit lights at the brightest level.

The vector $\langle -5, 5 \rangle$ can be written as the *column matrix* $\begin{bmatrix} -5 \\ 5 \end{bmatrix}$. **Find the sum of the vectors in column matrix form.**

37. $\begin{bmatrix} 2 \\ -4 \end{bmatrix} + \begin{bmatrix} -3 \\ 2 \end{bmatrix}$ **38.** $\begin{bmatrix} 8 \\ -1 \end{bmatrix} + \begin{bmatrix} 3 \\ -4 \end{bmatrix}$ **39.** $\begin{bmatrix} 4 \\ -5 \end{bmatrix} + \begin{bmatrix} -5 \\ 5 \end{bmatrix}$

40. Aviation An airplane takes off from a runway in the direction 10° east of south. When it reaches 5000 ft, it turns right 45°. It cruises at this altitude for 60 mi. Then it turns left 160°, descends, and lands. Match each vector with the appropriate portion of the flight.

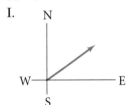
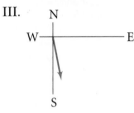

I. II. III.

A. The plane takes off.   B. The plane cruises.   C. The plane lands.

> **Problem Solving Hint**
> In Exercise 40, remember that any vector is equal to one whose initial point is the origin.

41. Aviation The cruising speed of a Boeing 767 in still air is 530 mi/h. Suppose that a 767 is cruising directly east when it encounters an 80 mi/h wind blowing 40° south of west.
  a. Sketch the vectors for the velocities of the airplane and the wind.
  b. Express both vectors from part (a) in ordered pair notation.
  c. Find the sum of the vectors from part (b).
  d. Find the magnitude and direction of the vector from part (c).

Give the sum of $\vec{a}$ and $\vec{b}$. Show $\vec{a}$ and $\vec{b}$ and their sum in the coordinate plane.

42. $\vec{a} \langle -5, -2 \rangle, \vec{b} \langle 2, -5 \rangle$ **43.** $\vec{a} \langle 5, -2 \rangle, \vec{b} \langle -5, -2 \rangle$ **44.** $\vec{a} \langle 5, -5 \rangle, \vec{b} \langle -2, 2 \rangle$

45. Writing How are vectors $\overrightarrow{AB}$ and $\overrightarrow{BA}$ alike? How are they different?

46. Open-Ended Name four other vectors with the same magnitude as $\langle -7, -24 \rangle$.

47. Navigation A fishing boat leaves its home port and travels 150 mi directly east. It then changes course and travels 40 mi due north.
  a. In what direction should the boat head to return to home port?
  b. How long will the return trip take if the boat averages 23 mi/h?

THE FAR SIDE®   **BY GARY LARSON**

POINT
A

POINT
B

"Well, lemme think. ... You've stumped me, son. Most folks only wanna know how to go the other way."

Exercise 45

48. Navigation A boat left dock A, traveled north for 10 miles, then 45° east of north for 20 miles, and docked at B.
  a. How far north did the boat travel? How far east did it travel?
  b. Find the magnitude and direction of the direct-path vector $\overrightarrow{AB}$.

C Challenge

49. Geometry in 3 Dimensions A hot-air balloon traveled 2000 ft north and 900 ft east, while rising 400 ft. This trip can be described with the three-coordinate vector $\langle 2000, 900, 400 \rangle$. What is the magnitude of the vector? What is the angle of elevation of the balloon from its starting point?

50. a. Probability You choose two of the vectors at the right at random. Find the probability that the magnitude of their resultant vector is greater than that of the third vector.

b. Open-Ended Draw three vectors of your own. Then do part (a) for your vectors.

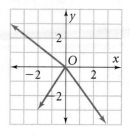

51. Writing Think of the number zero and its properties. Define a *zero vector* and justify your definition.

52. Aviation A helicopter starts at $(0, 0)$ and makes three parts of a flight represented by the vectors $\langle 10, 10 \rangle$, $\langle 5, -4 \rangle$, and $\langle -3, 5 \rangle$, in that order.

a. If another helicopter starts at $(0, 0)$ and flies the same three parts in a different order, would it end in the same place? Justify your answer.

b. If yet another helicopter flew the three parts of the flight in a different order from the original trip, could the second part of the flight end at the same place as the second part of the original trip? Justify your answer.

Test Prep

Multiple Choice

53. $\vec{\mathbf{c}}$, $\vec{\mathbf{s}}$, and $\vec{\mathbf{u}}$ are vectors. $\vec{\mathbf{c}} = \langle -8, 10 \rangle$, $\vec{\mathbf{s}} = \langle 0, -3 \rangle$, and $\vec{\mathbf{u}} = \vec{\mathbf{c}} + \vec{\mathbf{s}}$. What are the coordinates of $\vec{\mathbf{u}}$?

A. $\langle 7, -8 \rangle$ **B.** $\langle -7, 8 \rangle$ **C.** $\langle 8, -7 \rangle$ **D.** $\langle -8, 7 \rangle$

Short Response

54. A boat heads due south directly across a river at 30 ft/min. The river is flowing east at 20 ft/min.
a. What is the resultant speed of the boat?
b. What is the resultant direction of the boat?

Extended Response

55. A small aircraft is traveling east at 400 mi/h. It encounters a 50 mi/h wind blowing 30° west of south.
a. Sketch and label vectors for the velocities of the aircraft and the wind.
b. Express both vectors in ordered pair notation.
c. Find the sum of the vectors.
d. Find the magnitude and direction of the vector from part (c).

Mixed Review

Lesson 8-5

56. Indirect Measurement A hot-air balloon pilot sights the landing field from a height of 2000 ft. The angle of depression is 24°. To the nearest foot, what is the ground distance from the hot-air balloon to the landing field?

Lesson 8-2

Find the value of each variable.

57.

58.

59.

Lesson 6-1

60. Classify the quadrilateral with vertices $A(-1, -5)$, $B(6, -5)$, $C(9, 3)$, and $D(2, 3)$.

Before you begin working a problem in earnest, or if you do not know how to do a problem, you usually can eliminate some answer choices. Cross out the answers you eliminate. But do this in the test booklet, not on the answer sheet.

1 EXAMPLE

The length of a diagonal of a square is 12 cm. What is the area of the square?

  Ⓐ 49 cm^2 Ⓑ 72 cm^2 Ⓒ 100 cm^2 Ⓓ 145 cm^2 Ⓔ 225 cm^2

The area of a square with side length s is s^2. Since the length of the side of the square is less than the length of the diagonal, you can eliminate answers D and E. Those areas are both larger than $12^2 = 144$. Answer A is the area of a square whose side is 7 cm, and this square does not have a diagonal whose length is 12 cm.
● One of B or C must be the correct answer.

2 EXAMPLE

A kite at the end of a 100-ft string has an angle of elevation of 28°. The end of the string is staked to the ground. Which is the best estimate of the kite's height?

  Ⓐ 38 ft   Ⓑ 47 ft   Ⓒ 49 ft   Ⓓ 52 ft   Ⓔ 58 ft

Draw a diagram. Since 28° is a little less than 30°, you can estimate the answer by using a 30°-60°-90° triangle. If the angle of elevation were 30°, then the height of the kite would be 50 ft. Since the actual angle is less than 30°, you know the actual height is less than 50 ft. You can eliminate answers D and E since they are greater than 50 ft.
● The correct answer must be A, B, or C.

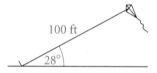

EXERCISES

1. If d is the length of a diagonal of a square, what is the area of the square in terms of d? What is the correct answer in Example 1?

2. Use the sine ratio to find the height of the kite above the ground in Example 2.

Use the following question for Exercises 3–5.

  The lengths of the diagonals of a rhombus are 6 cm and 10 cm. What is the measure of each acute angle of the rhombus?

  Ⓐ 31.0   Ⓑ 45   Ⓒ 61.9   Ⓓ 75   Ⓔ 118.1

3. Explain why you can immediately eliminate answer E.

4. The diagonals of a rhombus are perpendicular and bisect each other. Draw a diagram of the rhombus and its diagonals. Use the three sides of a right triangle to explain why you can eliminate answer A.

5. Explain how you now choose between B, C, and D.

6. Explain how it is possible to use an accurately drawn diagram, such as the one in Example 2, to help you eliminate answer choices.

Chapter Review

Vocabulary Review

🔊 angle of depression (p. 445)
angle of elevation (p. 445)
cosine (p. 439)
identity (p. 440)

initial point (p. 452)
magnitude (p. 452)
Pythagorean triple (p. 417)
resultant (p. 454)

sine (p. 439)
tangent (p. 432)
terminal point (p. 452)
vector (p. 452)

Choose the correct term to complete each sentence.

1. Any quantity that has magnitude and direction is called a(n) __?__.

2. The __?__ is the angle formed by a horizontal line and the line of sight to an object above that horizontal line.

3. The __?__ of $\angle A$ is $\dfrac{\text{leg adjacent } \angle A}{\text{hypotenuse}}$.

4. The __?__ of $\angle A$ is $\dfrac{\text{leg opposite } \angle A}{\text{hypotenuse}}$.

5. An equation that is true for all allowed values of the variable is called a(n) __?__.

6. The sum of two vectors is the __?__.

7. A set of three nonzero whole numbers that satisfy the equation $a^2 + b^2 = c^2$ form a(n) __?__.

Go Online
PHSchool.com
For: Vocabulary quiz
Web Code: auj-0851

Skills and Concepts

8-1 and 8-2 Objectives

▼ To use the Pythagorean Theorem

▼ To use the Converse of the Pythagorean Theorem

▼ To use properties of 45°-45°-90° triangles

▼ To use properties of 30°-60°-90° triangles

The **Pythagorean Theorem** states that in a right triangle, the sum of the squares of the lengths of the legs equals the square of the length of the hypotenuse, or $a^2 + b^2 = c^2$.

Positive integers a, b, and c form a **Pythagorean triple** if $a^2 + b^2 = c^2$.

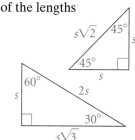

The **Pythagorean Theorem Converse** states that if the square of the length of one side of a triangle is equal to the sum of the squares of the lengths of the other two sides, then the triangle is a right triangle.

In a 45°-45°-90° triangle, the legs are congruent and the length of the hypotenuse is $\sqrt{2}$ times the length of a leg.

In a 30°-60°-90° triangle, the length of the hypotenuse is twice the length of the shorter leg. The length of the longer leg is $\sqrt{3}$ times the length of the shorter leg.

Find the value of each variable. If your answer is not an integer, leave it in simplest radical form.

8.

9.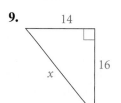

10.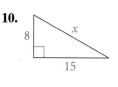

Find the value of each variable. If your answer is not an integer, leave it in simplest radical form.

11.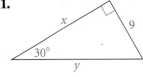

12.

13.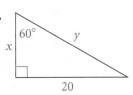

8-3 and 8-4 Objectives

▼ To use tangent ratios to determine side lengths in triangles

▼ To use sine and cosine to determine side lengths in triangles

In right $\triangle ABC$,

sine of $\angle A = \sin A = \dfrac{\text{leg opposite } \angle A}{\text{hypotenuse}}$

cosine of $\angle A = \cos A = \dfrac{\text{leg adjacent to } \angle A}{\text{hypotenuse}}$

and **tangent** of $\angle A = \tan A = \dfrac{\text{leg opposite } \angle A}{\text{leg adjacent to } \angle A}$

You can use the inverses of sine, cosine, and tangent to find the measures of the acute angles of a right triangle.

A trigonometric **identity** is an equation that is always true for all the allowed values of the variable.

Find each missing value to the nearest whole number.

14. $\tan \blacksquare° = 0.9$ **15.** $\sin 17° = \dfrac{\blacksquare}{7}$ **16.** $\tan 27° = \dfrac{1}{\blacksquare}$ **17.** $\cos \blacksquare° = 0.39$

18. $\sin \blacksquare° = 0.39$ **19.** $\sin \blacksquare° = \dfrac{2}{3}$ **20.** $\tan 76° = \dfrac{\blacksquare}{3}$ **21.** $\cos 83° = \dfrac{1}{\blacksquare}$

Express $\sin A$, $\cos A$, and $\tan A$ as ratios.

22.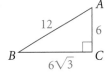

23.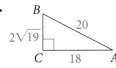

24.

Find the value of x. Round lengths of segments to the nearest tenth. Round angle measures to the nearest degree.

25.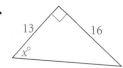

26.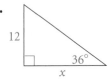

27.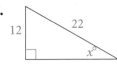

8-5 Objective

▼ To use angles of elevation and depression to solve problems

A horizontal line and the line of sight to an object above the horizontal line form an **angle of elevation**. A horizontal line and the line of sight to an object below that horizontal line form an **angle of depression**.

Solve each problem.

🌐 **28. Elevation** Two hills are 2 mi apart. The taller hill is 2707 ft high. The angle of depression from the top of the taller hill to the top of the shorter hill is 7°. Find the height of the shorter hill to the nearest foot. (1 mi = 5280 ft)

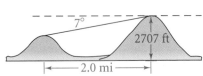

29. Surveying A surveyor is 305 ft from the base of the new courthouse. Her angle measuring device is 5 ft above the ground. The angle of elevation to the top of the courthouse is 42°. Find the height of the courthouse to the nearest foot.

30. Indirect Measurement Linda is flying a kite. She lets out 45 yd of string and anchors it to the ground. She determines that the angle of elevation of the kite is 58°. What is the height of the kite from the ground?

8-6 Objective

▼ To describe vectors

▼ To solve problems that involve vector addition

A **vector** is any quantity that has magnitude and direction. You can describe a vector by its horizontal and vertical change (ordered pair) or by its size and direction. Direction is often described in relation to north, south, east, and west.

The sum of two vectors is the **resultant.** Vector sums can show the result of vector actions that take place one after the other. Additionally, vector sums can show the result of two vector actions that occur at the same time. You can add vectors by adding their coordinates. You can also show the sum geometrically.

Describe each vector using ordered pair notation.

31.

32.

33.

Find the magnitude and direction of each vector.

34.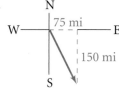

35.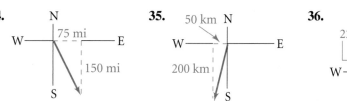

36.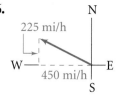

37. Open-Ended Write three vectors with the same direction as ⟨3, 4⟩. Explain how you found your answers.

Sketch a vector that has the given direction.

38. 25° east of north

39. 45° north of east

40. 60° west of south

Find the sum of each pair of vectors. Give your answers in ordered pair notation.

41.

42.

43.

 44. Navigation A whale-watching tour leaves port and travels 12 mi directly north. The tour then travels 5 mi due east.
a. In what direction should the boat head to return directly to port?
b. How long will the return trip take if the boat averages 20 mi/h?

 Algebra Find the value of each variable. Leave your answer in simplest radical form.

1.

2.
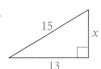

3.

4.
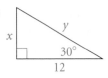

The lengths of three sides of a triangle are given. Describe each triangle as *acute, right,* or *obtuse.*

5. 9 cm, 10 cm, 12 cm

6. 8 m, 15 m, 17 m

7. 5 in., 6 in., 10 in.

Express sin *B*, cos *B*, and tan *B* as ratios.

8.
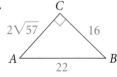

9.

Find each missing value to the nearest tenth.

10. $\tan \blacksquare° = 1.11$

11. $\tan 18° = \dfrac{\blacksquare}{87}$

12. $\sin 34° = \dfrac{5}{\blacksquare}$

13. $\sin \blacksquare° = 0.996$

14. $\cos \blacksquare° = \dfrac{12}{15}$

15. $\cos 76° = \dfrac{\blacksquare}{24}$

Find the value of *x*. Round lengths to the nearest tenth and angle measures to the nearest degree.

16.

17.

18.
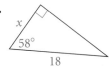

19.
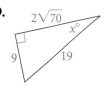

Indirect Measurement Solve each problem.

20. A surveyor measuring the tallest tree in a park is 100 ft from the tree. His angle-measuring device is 5 ft above the ground. The angle of elevation to the top of the tree is 48°. How tall is the tree?

21. Twenty minutes after being launched, a hot-air balloon has risen to an altitude of 300 m. The pilot can still see the starting point on the ground at a 25° angle of depression. How many meters is the balloon from the starting point?

22. A 5-foot-tall woman stands 15 ft from a statue. She must look up at an angle of 60° to see the top of the statue. How tall is the statue?

23. Writing Explain why $\sin x° = \cos (90 - x)°$. Include a diagram with your explanation.

Describe each vector using ordered pair notation. Round the coordinates to the nearest unit.

24.
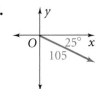

25.
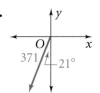

Find the magnitude and direction of each vector.

26.
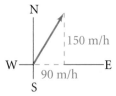

27.
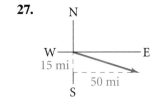

Describe the resultant as an ordered pair.

28.

29.
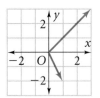

30. A canoe heading 30° west of north is being paddled at a rate of 7 mi/h. The current is pushing the canoe 20° south of west at a rate of 3 mi/h. Find the resulting speed and direction of the canoe.

Regents Test Prep

Multiple Choice

For Exercises 1–12, choose the correct letter.

1. What is the center of the circle that circumscribes △*OMN*?

   (1) (0, 0)
   (2) (0, −1)
   (3) (−1, 0)
   (4) (−1, −1)

2. Which quadrilateral cannot contain four right angles?
   (1) square   **(2)** trapezoid
   (3) rectangle   **(4)** rhombus

3. What is the distance between points *W*(−2, −10) and *R*(6, −1) to the nearest tenth?
   (1) 12.0   **(2)** 11.7
   (3) 9.8   **(4)** 6.8

4. **Algebra** For which value of *x* are lines *g* and *h* parallel?
   (1) 12
   (2) 15
   (3) 18
   (4) 25

5. Which is an equation of the line that has slope 3 and contains point *P*(2, 5)?
   (1) $y = x + 3$ **(2)** $y = x - 3$
   (3) $y = 3x + 1$ **(4)** $y = 3x - 1$

6. **Algebra** What is the value of *x* for this kite?
   (1) 48
   (2) 52
   (3) 62
   (4) 68

7. **Algebra** What is the value of *x*?
   (1) 3.75
   (2) 3.9
   (3) 4
   (4) 4.25

   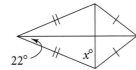

8. △*ABC* has *AB* = 7, *BC* = 24, and *CA* = 24. Which statement is true?
   (1) △*ABC* is an equilateral triangle.
   (2) △*ABC* is an isosceles triangle.
   (3) ∠*C* is the largest angle.
   (4) ∠*B* is the smallest angle.

9. Identify the converse of the statement: If you study in front of the TV, then you do not score well on exams.
   (1) If you do not study in front of the TV, then you score well on exams.
   (2) If you score well on exams, then you do not study in front of the TV.
   (3) If you do not score well on exams, then you study in front of the TV.
   (4) If you study in front of the TV, then you score well on exams.

10. In △*HTQ*, if *m*∠*H* = 72 and *m*∠*Q* = 55, what is the correct order of the lengths of the sides from least to greatest?
    (1) *TQ, HQ, HT*   **(2)** *TQ, HT, HQ*
    (3) *HQ, HT, TQ*   **(4)** *HQ, TQ, HT*

11. Which equation is NOT equivalent to $\frac{x}{y} = \frac{a}{c}$? Assume no denominators are 0.
    (1) $\frac{x}{a} = \frac{y}{c}$ **(2)** $cx = ay$
    (3) $\frac{x + y}{y} = \frac{a + c}{c}$ **(4)** $\frac{x + k}{y} = \frac{a + k}{c}$

12. Which quadrilateral has congruent diagonals?
    (1) kite   **(2)** parallelogram
    (3) rectangle   **(4)** rhombus

Gridded Response

13. The measure of the vertex angle of an isosceles triangle is 112. Find the measure of a base angle.

Short Response

14. **Constructions** Draw line *m* with point *A* on it. Construct a line perpendicular to *m* at *A*.

Extended Response

15. In the diagram, *AB* = *FE*, *BC* = *ED*, and *AE* = *FB*.
    a. Is there enough information to prove △*BCG* ≅ △*EDG*?
    b. What one additional piece of information would allow you to prove △*BCD* ≅ △*EDC*?
    c. What can you conclude from the diagram that would help you prove △*BAF* ≅ △*EFA*?
    d. In part (c), is △*BAF* ≅ △*EFA* by SAS or SSS?

Activity Lab

How Far Can You See?

Applying the Pythagorean Theorem Imagine that you're standing on an ocean beach looking out across the water. The deep blue sky is clearer than you've ever seen it, and it seems as though you can see forever! Well, you know that isn't really possible on Earth. The extent of your vision is limited by Earth's curvature. You can see to the horizon—and perhaps slightly beyond.

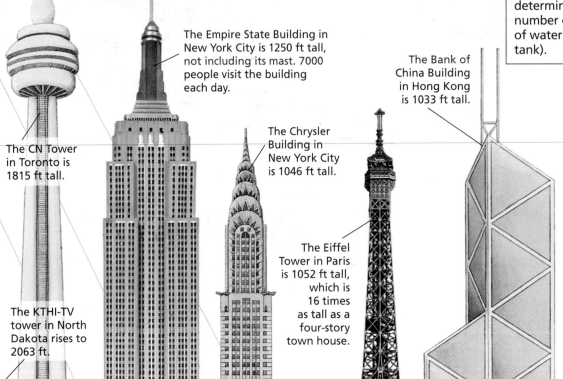

The CN Tower in Toronto is 1815 ft tall.

The KTHI-TV tower in North Dakota rises to 2063 ft.

The Empire State Building in New York City is 1250 ft tall, not including its mast. 7000 people visit the building each day.

The Chrysler Building in New York City is 1046 ft tall.

The Eiffel Tower in Paris is 1052 ft tall, which is 16 times as tall as a four-story town house.

Four-story town house, 66 ft tall

The Bank of China Building in Hong Kong is 1033 ft tall.

A chain of 8000 paper clips dangled from the top floor of 1 Canada Square in London would reach the ground, 797 ft below.

The *Saturn V* rocket is 364 ft tall.

Activity 1

Choose one of the structures on these pages. Imagine climbing to the very top to get a good view of the horizon. Assume that Earth is spherical and has a radius of 3963 mi. Also assume that you see a smooth horizon such as that of an ocean or desert. Find the distance from the top of the structure to the horizon. (*Hint:* Use a calculator and the Pythagorean Theorem.)

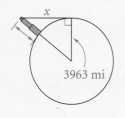

Go Online
PHSchool.com

For: Information about buildings
Web Code: aue-0853

Crow's nest, lookout post for land and ships

Flag indicates ship's origin.

Galleon
Galleons were fighting ships, with 40 to 50 cannons on board.

Activity 2

Sailors used to climb into the crow's nest on a ship's mast so they could spot land and other ships at a greater distance than was possible on deck. Imagine that you are on watch in a crow's nest so that your eyes are 40 ft above the water.

a. Determine the farthest distance from which you could spot the top of a 50-ft tree on a sea-level island.

b. Determine the farthest distance from which you could spot the tree if you were standing on deck with your eyes 15 ft above the water.

c. Compare your answers from parts (a) and (b).

With a spyglass, distant objects appear closer.

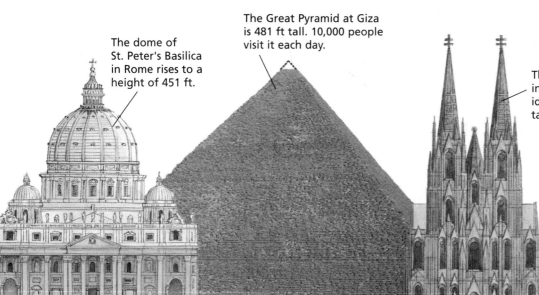

The dome of St. Peter's Basilica in Rome rises to a height of 451 ft.

The Great Pyramid at Giza is 481 ft tall. 10,000 people visit it each day.

The Cologne Cathedral in Germany has two identical spires that taper to 513-ft heights.

The Leaning Tower in Pisa, Italy, is about 185 ft high.

What You've Learned

NY Learning Standards for Mathematics

- **G.G.57:** In Chapter 4, you learned that you can move one of two congruent figures so that it can fit exactly on the other one.

- **G.G.45:** In Chapter 7, you learned about similar figures and scale factors.

- In Chapter 8, you learned how vectors can be used to represent distance and direction.

 Check Your Readiness

 for Help to the Lesson in green.

Regular Polygons (Lesson 3-5)

Determine the measure of an angle of the given regular polygon.

1. pentagon   **2.** octagon   **3.** decagon   **4.** 18-gon

Congruent Figures (Lesson 4-1)

The triangles are congruent. Complete the congruence statement, $\triangle ABC \cong \underline{\ ?\ }$.

5. 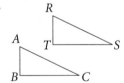   **6.** 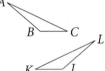   **7.** 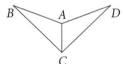   **8.**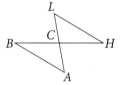

Quadrilaterals (Lessons 6-2, 6-4, and 6-5)

Determine whether a diagonal of the given quadrilateral *always, sometimes,* or *never* produces congruent triangles.

9. rectangle   **10.** isosceles trapezoid   **11.** kite   **12.** parallelogram

Scale Drawing (Lesson 7-1)

The scale of a blueprint is 1 in. = 20 ft.

13. What is the actual length of a wall that measures 2.5 in. in the blueprint?

14. What is the blueprint length of an entrance that is 5 ft wide?

Vectors (Lesson 8-6)

Write the sum of the two vectors as an ordered pair.

15. $\langle 3, 1 \rangle$ and $\langle -2, 4 \rangle$ **16.** $\langle -1, 2 \rangle$ and $\langle -5, 0 \rangle$ **17.** $\langle 0, 0 \rangle$ and $\langle -6, 3 \rangle$

Transformations

What You'll Learn Next

NY Learning Standards for Mathematics

- **G.G.54:** In this chapter, you will learn how to use transformations known as reflections, translations, and rotations to create a congruent image of a given shape.

- **G.G.57:** You will learn to use transformations for relating two given congruent shapes to each other.

- You will learn the effects of applying two transformations, one after the other.

- By learning about transformations, you will understand such terms as *symmetry* and *tessellation*.

Activity Lab Applying what you learn, you will do activities on pages 528 and 529 involving transformations in three dimensions.

◀)) Key Vocabulary

- center of a regular polygon (p. 484)
- composition (p. 472)
- dilation (p. 498) –
- enlargement (p. 498)
- glide reflection (p. 508)
- glide reflectional symmetry (p. 516)
- image (p. 470)
- isometry (p. 470)
- line symmetry (p. 492)
- point symmetry (p. 493)
- preimage (p. 470)
- reduction (p. 498)
- reflection (p. 478)
- reflectional symmetry (p. 492)
- rotation (p. 484)
- rotational symmetry (p. 493)
- symmetry (p. 492)
- tessellation (p. 515)
- tiling (p. 515)
- transformation (p. 470)
- translation (p. 471)
- translational symmetry (p. 516)

Translations

 Learning Standards for Mathematics

G.G.61 Investigate, justify, and apply the analytical representations for translations, rotations about the origin of 90° and 180°, reflections over the lines $x = 0$, $y = 0$, and $y = x$, and dilations centered at the origin.

✓ **Check Skills You'll Need**

GO for Help Lesson 4-1

$\triangle ABC \cong \triangle EFG$.

Complete the congruence statements.

1. $\overline{AB} \cong$ ___?___ 2. $\overline{EG} \cong$ ___?___
3. $\overline{FG} \cong$ ___?___ 4. $\angle C \cong$ ___?___
5. $\angle E \cong$ ___?___ 6. $\angle B \cong$ ___?___

7. Complete: If $\triangle KTQ \cong \triangle LGR$, then $\overline{TK} \cong$ ___?___ and $\angle TQK \cong$ ___?___ .

◀)) **New Vocabulary** • transformation • preimage • image
• isometry • translation • composition

---

1 Identifying Isometries

Vocabulary Tip

In general, the word <u>transformation</u> can refer to any kind of change in appearance.

A **transformation** of a geometric figure is a change in its position, shape, or size. When you assemble a jigsaw puzzle, you often move the puzzle pieces by flipping them, sliding them, or turning them. Each move is a type of transformation. The photos below illustrate some basic transformations that you will study.

The figure flips.

The figure slides.

The figure turns.

The original figure is the **preimage.** The resulting figure is an **image.** An **isometry** is a transformation in which the preimage and image are congruent. Each transformation above is an isometry.

1 EXAMPLE **Identifying Isometries**

Does the transformation appear to be an isometry? Explain.

No, this transformation involves a change in size. The sides of the preimage square and the sides of its image are not congruent.

Preimage   Image

✓ **Quick Check** ❶ Does the transformation appear to be an isometry? Explain.

a.

b. Preimage

Image

A transformation maps a figure onto its image and may be described with arrow (→) notation. Prime (') notation is sometimes used to identify image points. In the diagram at the right, K' is the image of K ($K \to K'$).

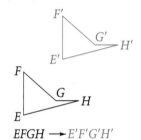

$\triangle JKQ \to \triangle J'K'Q'$
$\triangle JKQ$ maps onto $\triangle J'K'Q'$.

Notice that you list corresponding points of the preimage and image in the same order, as you do for corresponding points of congruent or similar figures.

2 EXAMPLE Naming Images and Corresponding Parts

In the diagram, $E'F'G'H'$ is an image of $EFGH$.

a. Name the images of $\angle F$ and $\angle H$.

$\angle F'$ is the image of $\angle F$.
$\angle H'$ is the image of $\angle H$.

b. List all pairs of corresponding sides.

$\overline{EF}$ and $\overline{E'F'}$; $\overline{FG}$ and $\overline{F'G'}$;
$\overline{EH}$ and $\overline{E'H'}$; $\overline{GH}$ and $\overline{G'H'}$

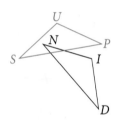

$EFGH \to E'F'G'H'$

✓ Quick Check ❷ In the diagram, $NID \to SUP$.
a. Name the images of $\angle I$ and point D.
b. List all pairs of corresponding sides.

A **translation** (or *slide*) is an isometry that maps all points of a figure the same distance in the same direction. The diagram at the right shows a translation of the black square by 4 units right and 2 units down. Using variables, you can say that each (x, y) pair in the original figure is mapped to $(x + 4, y - 2)$.

3 EXAMPLE Finding a Translation Image

Find the image of $\triangle XYZ$ under the translation $(x, y) \to (x - 2, y - 5)$.

Use the rule to find each vertex in the translated image.
$X(2, 1)$ translates to $(2 - 2, 1 - 5)$, or $X'(0, -4)$.
$Y(3, 3)$ translates to $(3 - 2, 3 - 5)$, or $Y'(1, -2)$.
$Z(-1, 3)$ translates to $(-1 - 2, 3 - 5)$, or $Z'(-3, -2)$.

The image of $\triangle XYZ$ is $\triangle X'Y'Z'$ with $X'(0, -4)$, $Y'(1, -2)$, and $Z'(-3, -2)$.

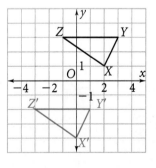

✓ Quick Check ❸ Find the image of $\triangle XYZ$ for the translation $(x, y) \to (x - 4, y + 1)$.

EXAMPLE Writing a Rule to Describe a Translation

Write a rule to describe the translation
$PQRS \rightarrow P'Q'R'S'$.

Use $P(-1, -2)$ and its image $P'(-5, -1)$.
Horizontal change: $-5 - (-1) = -4$
Vertical change: $-1 - (-2) = 1$

The rule is $(x, y) \rightarrow (x - 4, y + 1)$.

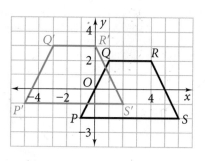

Quick Check ④ The translation image of $\triangle LMN$ is
$\triangle L'M'N'$ with $L'(1, -2), M'(3, -4), N'(6, -2)$.
Write a rule to describe the translation.

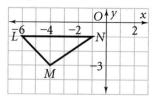

A **composition** of transformations is a combination
of two or more transformations. In a composition,
each transformation is performed on the image of
the preceding transformation.

In a knight's move on a chessboard, the translation
indicated in blue is the composition of two translations
indicated in red.

In general, the composition of any two translations
is a translation.

5 **EXAMPLE** Real-World  Connection

Tourism Yolanda Perez is visiting San Francisco. From her hotel near Union
Square, she walks 4 blocks east and 4 blocks north to the Wells Fargo History
Museum to see a stagecoach and relics of the Gold Rush. Then she walks 5 blocks
west and 3 blocks north to the Cable Car Barn Museum. Now how many blocks
is she from her hotel?

Use $(0, 0)$ to represent Yolanda's hotel.

$(x, y) \rightarrow (x + 4, y + 4)$ represents a walk of 4 blocks east and 4 blocks north.
$(x, y) \rightarrow (x - 5, y + 3)$ represents a walk of 5 blocks west and 3 blocks north.

Yolanda's current position is the composition of the two translations. First,

$(0, 0)$ translates to $(0 + 4, 0 + 4)$ or $(4, 4)$. Then,
$(4, 4)$ translates to $(4 - 5, 4 + 3)$ or $(-1, 7)$.

Yolanda is 1 block west and 7 blocks north of her hotel.

Quick Check ⑤ Yolanda next walks to a restaurant 2 blocks east and 4 blocks south of the
Cable Car Barn Museum. Now how many blocks is she from her hotel?

EXERCISES

For more exercises, see *Extra Skill, Word Problem, and Proof Practice.*

Practice and Problem Solving

  Practice by Example

State whether the transformation appears to be an isometry. Explain.

Example 1
(page 470)

for Help

1. Preimage Image

2. 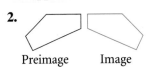 Preimage Image

3. Preimage Image

Example 2
(page 471)

In each diagram, the blue figure is an image of the black figure.
(a) Choose an angle or point from the preimage and name its image.
(b) List all pairs of corresponding sides.

4.

5.

6.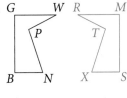

Example 3
(page 471)

Find the image of each figure under the given translation.

7. $(x, y) \rightarrow (x + 3, y + 2)$

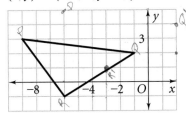

8. $(x, y) \rightarrow (x + 5, y - 1)$

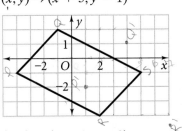

9. $(x, y) \rightarrow (x - 2, y + 5)$

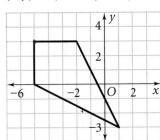

10. $(x, y) \rightarrow (x - 4, y + 3)$

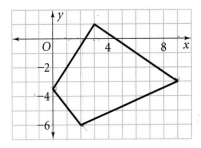

Example 4
(page 472)

In Exercises 11–14, the blue figure is a translation image of the red figure. Write a rule to describe each translation.

11.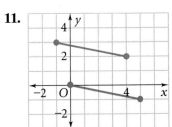

12.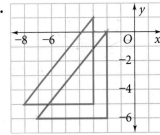

13.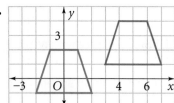

14.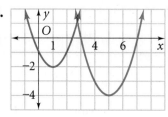

Example 5
(page 472)

15. Emily left Galveston Bay at the east jetty and sailed 4 km north to an oil rig. She then sailed 5 km west to Redfish Island. Finally, she sailed 3 km southwest to Spinnaker Restaurant.
  a. Draw a diagram on graph paper that shows her journey.
  b. Describe where Spinnaker Restaurant is from where Emily started.

16. Nakesha and her parents are visiting colleges. They leave their home in Enid, Oklahoma, and drive to Tulsa, which is 107 mi east and 18 mi south of Enid. From Tulsa, they go to Norman, 83 mi west and 63 mi south of Tulsa. Draw a diagram to show their trip. Then, tell where Norman is in relation to Enid.

B **Apply Your Skills**

The orange figure is a translation image of the red figure. Write a rule to describe each translation.

17.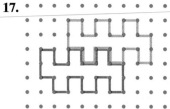

18.

19. Multiple Choice For which translation does the image of $\triangle JKL$ have a vertex at the origin?

  Ⓐ $(x, y) \rightarrow (x - 2, y - 2)$
  Ⓑ $(x, y) \rightarrow (x + 4, y + 1)$
  Ⓒ $(x, y) \rightarrow (x + 2, y - 3)$
  Ⓓ $(x, y) \rightarrow (x + 4, y - 4)$

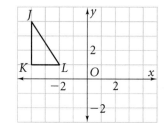

Graph each polygon and its image for the given translation.

20. $\triangle ACE$ with vertices $A(7, 2), C(-8, 5), E(0, -6)$
  translation $(x, y) \rightarrow (x - 9, y + 4)$

21. $\square PLAT$ with vertices $P(-2, 0), L(-1, 1), A(0, 1), T(-1, 0)$
  translation $(x, y) \rightarrow (x + 1, y)$

22. $\square NILE$ with vertices $N(2, -5), I(2, 2), L(-3, 4), E(-3, -3)$
  translation $(x, y) \rightarrow (x - 3, y - 4)$

23. Landscaping The Michelsons want to build a storage shed in their backyard. The diagram at the right shows the original site plan. Local law, however, requires the shed to set back at least 15 feet from property lines.
  a. Describe in words how the Michelsons should move the shed.
  b. Write a translation rule for moving the shed.

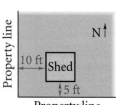

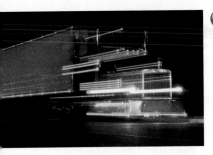

Real-World 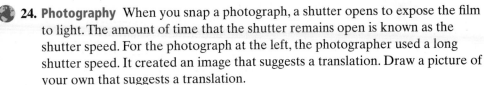 **Connection**

A time-lapse photo captures translations of different points of the truck.

24. Photography When you snap a photograph, a shutter opens to expose the film to light. The amount of time that the shutter remains open is known as the shutter speed. For the photograph at the left, the photographer used a long shutter speed. It created an image that suggests a translation. Draw a picture of your own that suggests a translation.

25. Coordinate Geometry $\triangle MUG$ has coordinates $M(2, -4)$, $U(6, 6)$, and $G(7, 2)$. A translation maps point M to $M'(-3, 6)$. Find the coordinates of U' and G' under this translation.

26. Coordinate Geometry $\square ABCD$ has vertices $A(3, 6)$, $B(5, 5)$, $C(4, 2)$, and $D(2, 3)$. The figure is translated so that the image of point C is the origin.
  a. Find the rule that describes the translation.
  b. Graph $\square ABCD$ and its image.

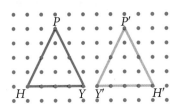

27. Writing Is the transformation at the right, $\triangle HYP \rightarrow \triangle H'Y'P'$, a translation? Explain.

Find a single translation that has the same effect as each composition of translations.

28. $(x, y) \rightarrow (x + 2, y + 5)$ followed by $(x, y) \rightarrow (x - 4, y + 9)$

29. $(x, y) \rightarrow (x + 12, y + 0.5)$ followed by $(x, y) \rightarrow (x + 1, y - 3)$

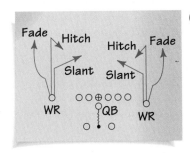

30. Football The play chart at the left shows routes that a wide receiver (WR) can choose to run when the team is in the "red zone" (within 20 yards of the goal line). The quarterback (QB) drops back two steps to make the pass to the wide receiver.
  a. Suppose a wide receiver runs a slant. Describe the two translations involved and the composition of those two translations.
  b. Describe the two intended translations of the football during the play, and the composition of the translations.
  c. What is the intended outcome of the two compositions in parts (a) and (b)?

Geometry in 3 Dimensions Use each figure, graph paper, and the given translation to draw a three-dimensional figure.

  SAMPLE Use the rectangle and $(x, y) \rightarrow (x + 3, y + 1)$ to draw a box.

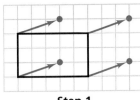

Step 1

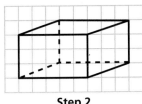

Step 2

31.

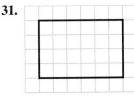

$(x, y) \rightarrow (x + 2, y - 1)$

32.

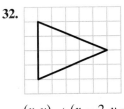

$(x, y) \rightarrow (x - 2, y + 2)$

33.

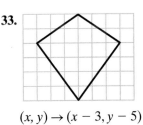

$(x, y) \rightarrow (x - 3, y - 5)$

34. Open-Ended You work for a company that specializes in creating unique, artistic designs for business stationery. One of your clients is Totter Toy Co. You have been assigned to create a border design for the top of their stationery. Create a design that involves translations to present to your client.

35. a. △*ABC* has vertices *A*(−2, 5), *B*(−4, −1), and *C*(2, −3). Find the image of △*ABC* under the translation (*x*, *y*) → (*x* + 4, *y* + 2).

b. Show that the images of the midpoints of the sides of △*ABC* are the midpoints of △*A′B′C′*.

36. Writing Explain how a parallelogram could be defined in terms of translations.

NY REGENTS

Test Prep

Multiple Choice

37. What is the image of (6, −2) under the translation (*x*, *y*) → (*x* − 5, *y* − 8)?
   A. (14, 3)   **B.** (−2, −7)   **C.** (11, 6)   **D.** (1, −10)

38. The point (5, −9) is the image under the translation (*x*, *y*) → (*x* + 3, *y* + 2). What is the preimage?
   F. (2, −11)   **G.** (8, −7)   **H.** (2, −7)   **J.** (8, −11)

39. What rule describes the translation 4 units up and 12 units left?
   A. (*x*, *y*) → (*x* + 4, *y* + 12)   **B.** (*x*, *y*) → (*x* − 12, *y* + 4)
   C. (*x*, *y*) → (*x* + 4, *y* − 12)   **D.** (*x*, *y*) → (*x* + 12, *y* + 4)

40. △*XYZ* has vertices *X*(−5, 2), *Y*(0, −4), and *Z*(3, 3). What are the vertices of the image of △*XYZ* under the translation (*x* + *y*) → (*x* + 7, *y* − 5)?
   F. *X′*(2, −3), *Y′*(7, −9), *Z′*(10, −2)   **G.** *X′*(−12, 7), *Y′*(−7, 1), *Z′*(−4, 8)
   H. *X′*(−12, −3), *Y′*(−7, −9), *Z′*(−4, −2)   **J.** *X′*(2, −3), *Y′*(10, −2), *Z′*(7, −9)

Short Response

41. △*ABC* has coordinates *A*(0, −3), *B*(−4, −2), and *C*(2, 1). A translation maps point *B* to (10, −3).
   a. What rule describes the translation?
   b. What are the images of *A* and *C* under this translation?

Mixed Review

 for Help

Lesson 8-6

42. Navigation An airplane lands at a point 100 km east and 420 km south from where it took off. Describe the magnitude and the direction of its flight vector.

Lesson 7-5   **Solve for *x*.**

43.
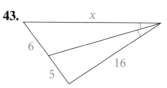

44.
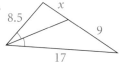

Lesson 4-4   **45. Given:** $\overline{BC} \cong \overline{EF}, \overline{BC} \parallel \overline{EF}$
                         $\overline{AD} \cong \overline{DC} \cong \overline{CF}$
                 Prove: $\overline{AB} \cong \overline{DE}$

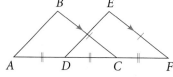

46. Given: $\overline{AB} \cong \overline{CB}, \overline{BD} \perp \overline{AC}$
         Prove: $\angle ABD \cong \angle CBD$

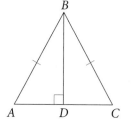

Paper Folding and Reflections

Activity 1 will help you understand how a figure and its reflection image are related. Then you will use the properties you've discovered to construct a reflection image using a compass and a straightedge.

NY G.G.55: Investigate, justify, and apply the properties that remain invariant under translations, rotations, reflections, and glide reflections.

1 ACTIVITY

- Use a page of tracing paper and a straightedge. Using less than half the page, draw a large, scalene triangle. Label its vertices A, B, and C.

- Fold the paper so that your triangle is covered. Trace △ABC carefully using a straightedge.

- Unfold the paper. Label the traced points corresponding to A, B, and C as A′, B′, and C′, respectively. △A′B′C′ is a reflection image of △ABC. The fold is the reflection line.

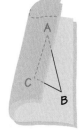

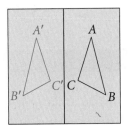 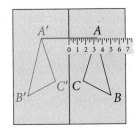

1. Use a ruler to draw $\overline{AA'}$. Measure the perpendicular distances from A to the fold and from A′ to the fold. What do you notice?

2. Measure the angles formed by the fold and $\overline{AA'}$.

3. Repeat Exercises 1 and 2 for B and B′ and for C and C′.

4. Make a conjecture: How is the reflection line related to the segment joining a point and its reflected image?

2 ACTIVITY

- On regular paper, draw a simple shape or design made of segments (on less than half the page).

- Draw a reflection line near your figure.

- Use a compass and straightedge to construct a perpendicular to the reflection line through one point of your drawing.

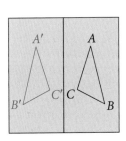

5. Explain how you can use a compass and the perpendicular you drew to find the reflection image of the point you chose.

6. Find the reflection images for several points of your shape. Connect the image points corresponding to the connections in your original figure and complete the image. Check the accuracy of your reflection image by folding the paper along the reflection line and holding the paper to the light.

Reflections

Learning Standards for Mathematics

G.G.57 Justify geometric relationships (perpendicularity, parallelism, congruence) using transformational techniques (translations, rotations, reflections).

G.G.61 Investigate, justify, and apply the analytical representations for translations, rotations about the origin of 90° and 180°, reflections over the lines $x = 0$, $y = 0$, and $y = x$, and dilations centered at the origin.

✓ **Check Skills You'll Need**

GO for Help Lesson 3-7

Write an equation for the line through point A that is perpendicular to the given line.

1.
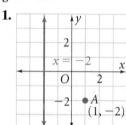

2.
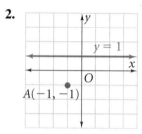

3.
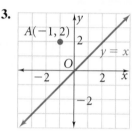

🔊 **New Vocabulary** • reflection

---

1 Finding Reflection Images

Real-World Connection

In your rear-view mirror you see the reflection "AMBULANCE."

A **reflection** (or *flip*) is an isometry in which a figure and its image have opposite orientations. Thus, a reflected image in a mirror appears "backwards." In the diagram below, $\triangle BUG$ is reflected across a line to produce $\triangle B'U'G'$. Since the reflection is an isometry, $\triangle BUG \cong \triangle B'U'G'$.

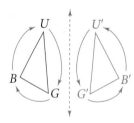

You can use the following two rules to reflect a figure across a line r.

- If a point A is on line r, then the image of A is A itself (that is, $A' = A$).
- If a point B is not on line r, then r is the perpendicular bisector of $\overline{BB'}$.

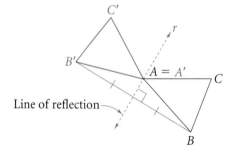

Line of reflection

As the diagram shows, a point on the line of reflection is mapped onto itself. Otherwise, a point and its image are equidistant from the line of reflection.

1 EXAMPLE Finding Reflection Images

Multiple Choice If point $P(2, -1)$ is reflected across the line $y = 1$, what are the coordinates of its reflection image?

Ⓐ $(2, 1)$ Ⓑ $(0, -1)$ Ⓒ $(2, 3)$ Ⓓ $(-2, 3)$

P is 2 units below the reflection line, so its image P' is 2 units above the reflection line. The reflection line is the perpendicular bisector of $\overline{PP'}$ if P' is at $(2, 3)$. The correct answer is C.

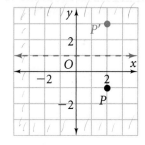

✔️ **Quick Check** ❶ What are the coordinates of the images of P if the reflection line is $x = 1$?

2 EXAMPLE Drawing Reflection Images

For: Reflection Activity
Use: Interactive Textbook, 9-2

Coordinate Geometry Given points $A(-3, 4)$, $B(0, 1)$, and $C(2, 3)$, draw $\triangle ABC$ and its reflection image across each line.

a. the x-axis

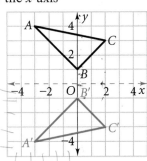

Locate points A', B', and C' such that the line of reflection is the perpendicular bisector of $\overline{AA'}$, $\overline{BB'}$, and $\overline{CC'}$.

b. the y-axis

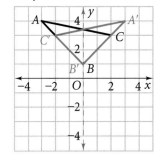

✔️ **Quick Check** ❷ Draw $\triangle ABC$ of Example 2. Then draw its reflection image across the line $x = 3$.

You can use the properties of reflections to solve real-world problems.

3 EXAMPLE Real-World 🌐 Connection

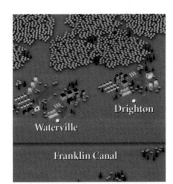

Engineering Town officials in Waterville and Drighton (see aerial view at left) are planning to construct a water pumping station along the Franklin Canal. The station will provide both towns with water. Where along the canal should the officials build the pumping station to minimize the amount of pipe needed?

You need to find the point P on ℓ such that $WP + PD$ is as small as possible. Locate D', the reflection image of D in ℓ. Because a reflection is an isometry, $PD = PD'$, and $WP + PD = WP + PD'$. By the Triangle Inequality Theorem, the sum $WP + PD'$ is smallest when W, P, and D' are collinear. So, the pump should be located at the point P where $\overline{WD'}$ intersects ℓ.

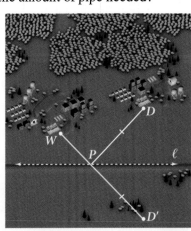

✔️ **Quick Check** ❸ **Critical Thinking** Angela began to solve the problem above by reflecting point W in line ℓ. Will her method work? Explain.

EXERCISES

For more exercises, see *Extra Skill, Word Problem,* and *Proof Practice.*

Practice and Problem Solving

A Practice by Example

Example 1
(page 479)

GO for Help

Each point is reflected across the line indicated. Find the coordinates of the images.

1. *Q* across $x = 1$

2. *R* across $y = -1$

3. *S* across the *y*-axis

4. *T* across $y = 0.5$

5. *U* across $x = -3$

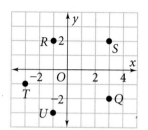

Example 2
(page 479)

Coordinate Geometry Given points $J(1, 4)$, $A(3, 5)$, and $R(2, 1)$, draw $\triangle JAR$ and its reflection image across each line.

6. the *x*-axis   **7.** the *y*-axis   **8.** $y = 2$ **9.** $y = 5$

10. $x = -1$ **11.** $x = 2$ **12.** $y = -x$ **13.** $y = x - 3$

Example 3
(page 479)

14. Trail Building A hiking club is building a new trail system. It wants to build trails to the Overlook and Balance Rock that will connect at a point on Summit Trail. Working under a tight budget, it wants to minimize the total length of these trails. If the trails cover similar terrain, at what point should they meet on Summit Trail?

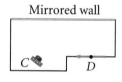

15. Security You are installing a security camera. At what point on the mirrored wall should you aim camera *C* in order to videotape door *D*?

Mirrored wall

B Apply Your Skills

Copy each pair of figures. Then draw the line of reflection you can use to map one figure onto the other.

16.

17.

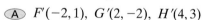

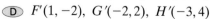

18.

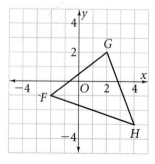

GO **O**nline
Homework Video Tutor
Visit: PHSchool.com
Web Code: aue-0902

19. Multiple Choice If you reflect $\triangle FGH$ across the *x*-axis, what will be the coordinates of the vertices of the image $\triangle F'G'H'$?

  Ⓐ $F'(-2, 1)$, $G'(2, -2)$, $H'(4, 3)$

  Ⓑ $F'(2, -1)$, $G'(-2, 2)$, $H'(-4, 3)$

  Ⓒ $F'(-1, 2)$, $G'(2, -2)$, $H'(3, -4)$

  Ⓓ $F'(1, -2)$, $G'(-2, 2)$, $H'(-3, 4)$

Copy each figure and line ℓ. Then draw each figure's reflection image across line ℓ.

20.

21.

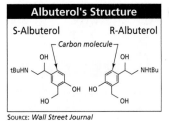

Albuterol's Structure

SOURCE: *Wall Street Journal*

The *R-isomer* of the drug albuterol relieves asthma. Its twin has been shown to increase the chances of having future attacks.

22. **Pharmaceuticals** Most drugs are made of two versions of the same molecule, each a mirror image of the other. One version is known as an *R-isomer* and the other as an *S-isomer*. While one isomer can help with what ails you, the other can create unwanted side effects. Models of two isomers are shown above. For this drug to cure an illness, it needs to fit into the "receptor molecule." Which isomer will cure the illness?

S-Isomer   R-Isomer   Receptor Molecule

23. **Open-Ended** Give three examples from everyday life of objects that come in a left-handed version and a right-handed version.

Coordinate Geometry **A point is reflected across the given line. How are the coordinates of the point and its image related? Explain.**

24. x-axis   25. y-axis   26. the line $y = x$

27. **History** The work of artist and scientist Leonardo da Vinci (1452–1519) has an unusual characteristic. His handwriting is a mirror image of normal handwriting.
   a. Write the mirror image of the sentence, "Leonardo da Vinci was left-handed." Use a mirror to check how well you did.
   b. Explain why the fact about da Vinci in part (a) might have made mirror writing seem natural to him.

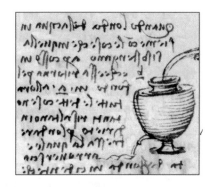

Find the image of $O(0, 0)$ after two reflections, first across ℓ_1 and then across ℓ_2.

28. $\ell_1: y = 3, \ell_2: x$-axis   29. $\ell_1: x = -2, \ell_2: y$-axis   30. $\ell_1: x$-axis, $\ell_2: y$-axis

31. $\ell_1: x = -2, \ell_2: y = 3$ 32. $\ell_1: y = 3, \ell_2: x = -2$ 33. $\ell_1: x = -2, \ell_2: y = x$

34. $\ell_1: x = a, \ell_2: y = b$ 35. $\ell_1: x = a, \ell_2: y = x$ 36. $\ell_1: y = b, \ell_2: y = x$

37. **Critical Thinking** Given that the transformation $\triangle ABC \rightarrow \triangle A'B'C'$ is an isometry, list everything you know about the two figures.

C Challenge   **Writing** **Can you form the given type of quadrilateral by drawing a triangle and then reflecting one or more times? Explain.**

38. parallelogram   39. isosceles trapezoid   40. kite

41. rhombus   42. rectangle   43. square

44. **Coordinate Geometry** Show that $B(b, a)$ is the reflection image of $A(a, b)$ across the line $y = x$. (*Hint*: Show that $y = x$ is the perpendicular bisector of $\overleftrightarrow{AB}$.)

45. **Coordinate Geometry** Find the line of reflection that maps $A(a, b)$ to $C(c, d)$.

46. Use the diagram at the right. Find the coordinates of the given point across the given line.
   a. A', the reflection image of A across $y = x$
   b. A'', the reflection image of A' across $y = -x$
   c. A''', the reflection image of A'' across $y = x$
   d. A'''', the reflection image of A''' across $y = -x$
   e. How are A and A'''' related?

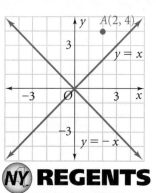

Test Prep

Multiple Choice

47. What is the reflection image of $(5, -3)$ across the y-axis?
   A. $(5, 3)$ **B.** $(-5, 3)$ **C.** $(-5, -3)$ **D.** $(-3, 5)$

48. What is the reflection image of $(5, -3)$ across the line $y = -x$?
   F. $(-3, 5)$ **G.** $(-3, -5)$ **H.** $(3, -5)$ **J.** $(3, 5)$

49. What is the reflection image of (a, b) across the line $y = -6$?
   A. $(a - 6, b)$ **B.** $(a, b - 6)$ **C.** $(-12 - a, b)$ **D.** $(a, -12 - b)$

50. The preimage of a reflection lies in Quadrant II and the image is in Quadrant III. Which line could be the line of reflection?
   F. $y = x$ **G.** $y = -x$ **H.** x-axis   **J.** y-axis

Short Response

51. A point is reflected across the line $y = x$. Its image is in Quadrant III. In which quadrant is the preimage? Explain.

52. A segment has endpoints $A(4, 0)$ and $B(-2, 5)$. Find the endpoints of the images when $\overline{AB}$ is reflected across the line $y = 2$ and then the line $x = 2$.

Mixed Review

Lesson 9-1

In each diagram, the blue figure is the translation image of the red figure. Write a rule to describe each translation.

53.

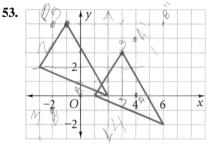

54.

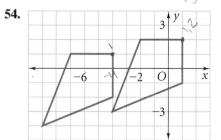

Lesson 7-1 **55. Maps** A map of Alberta, Canada, is drawn to the scale 1 cm = 25 km. On the map, the distance from Calgary to Edmonton is about 11.1 cm.
   a. About how far apart are the two cities?
   b. If 1 km = 0.62 mi, about how many miles apart are the cities?

Lesson 3-6   Write an equation in point-slope form of the line that contains the given points.

56. $M(2, 3), N(-2, 5)$ **57.** $A(-1, -1), B(0, 4)$ **58.** $X(1, -3), Y(7, 6)$

Rotations

 Learning Standards for Mathematics

G.G.54 Define, investigate, justify, and apply isometries in the plane (rotations, reflections, translations, glide reflections).

✓ **Check Skills You'll Need**   **GO** **for Help** Skills Handbook, p. 746, Lesson 1-7

Use a protractor to draw an angle with the given measure.

1. 120   **2.** 90   **3.** 72   **4.** 60   **5.** 45   **6.** 36

7. Draw a segment, $\overline{AB}$. Then construct $\overline{A'B'}$ congruent to $\overline{AB}$.

 New Vocabulary • rotation • center of a regular polygon

---

1 **Drawing and Identifying Rotation Images**

Hands-On Activity: Making Designs With Rotations

You will need at least three pieces of tracing paper for this Investigation.

Step 1 Place a piece of paper over the figure at the right. Trace the six points on the circle, the center of the circle, and the triangle.

Step 2 Place the point of your pencil on the center of the circle and then rotate the paper until the six points align again. Trace the triangle in its new location.

Step 3 Repeat Step 2 until there are six triangles on your paper. Compare your drawings to others to be sure that your results look the same.

Step 4 Now it's your turn to be creative. Place a piece of paper over the figure above, trace the six points on the circle and the center of the circle, and then draw your own triangle on the paper.

Step 5 Place the paper from Step 4 on your desk. On a blank piece of paper, repeat Steps 1–3. Color your drawing to make a design.

Turn the above diagram 60° about the center of the circle and each dot on the circle maps to the next dot on the circle. This is an example of a type of transformation known as a rotation (or *turn*).

To describe a **rotation**, you need to know the center of rotation (a point), the angle of rotation (a positive number of degrees), and whether the rotation is clockwise or counterclockwise. Unless stated otherwise, rotations in this book are counterclockwise.

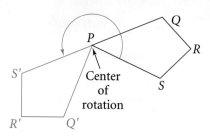

You can use the following two rules to rotate a figure through $x°$ about a point R:

- The image of R is itself (that is, $R' = R$).
- For any point V, $RV' = RV$ and $m\angle VRV' = x$.

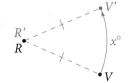

1 EXAMPLE Drawing a Rotation Image

Draw the image of $\triangle LOB$ for a $100°$ rotation about C.

Step 1
Use a protractor to draw a $100°$ angle with vertex C and side $\overline{CO}$.

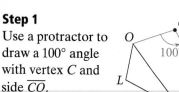

Step 2
Use a compass to construct $\overline{CO'} \cong \overline{CO}$.

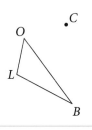

Step 3
Locate B' and L' in a similar manner. Draw $\triangle L'O'B'$.

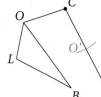

Quick Check ❶ Draw the image of $\triangle LOB$ from Example 1 for a $50°$ rotation about B. Label the vertices of the image.

A regular polygon has a **center** that is equidistant from its vertices. Segments that connect the center to the vertices divide the polygon into congruent triangles. You can use this fact to find rotation images of regular polygons.

2 EXAMPLE Identifying a Rotation Image

Regular pentagon $PENTA$ is divided into five congruent triangles.

a. Name the image of E for a $72°$ rotation about X.

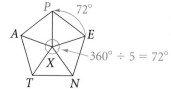

P is the image of E.

b. Name the image of P for a $216°$ rotation about X.

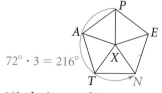

N is the image of P.

Quick Check ❷ Name the image of T for a $144°$ rotation about X.

Figures that are rotation images of themselves have a special beauty that you will study more closely in Lesson 9-4.

3 EXAMPLE Real-World Connection

Gridded Response You can find designs with rotation images in some types of Native American art. In the design pictured, how many degrees are in the angle of rotation about C that maps Q to X?

The eight-pointed star in the design divides the circle into eight congruent parts. $360 \div 8 = 45$, so each part has a $45°$ angle at the center. The angle of rotation that maps Q to X is $5 \cdot 45$, or $225°$.

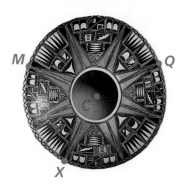

✓ **Quick Check** ❸ In the design above, find the angle of rotation about C that maps Q to M.

A composition of rotations about the same point is itself a rotation about that point. To sketch the image, add the angles of rotation to find the total rotation.

4 EXAMPLE Compositions of Rotations

Draw the image of the kite at the left for a composition of a $30°$ rotation and a $60°$ rotation, both about point K.

The total rotation is $90°$. Draw the kite. Locate image points of the vertices for a $90°$ rotation. Use the image points to sketch the entire image.

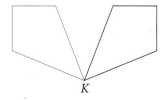

✓ **Quick Check** ❹ Draw the kite at the left. Then draw its image for a composition of two $90°$ rotations about point K.

EXERCISES

For more exercises, see *Extra Skill, Word Problem, and Proof Practice.*

Practice and Problem Solving

A Practice by Example

Example 1
(page 484)

Copy each figure and point P. Draw the image of each figure for the given rotation about P. Label the vertices of the image.

1. $60°$

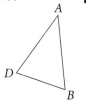

2. $90°$

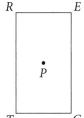

3. $90°$

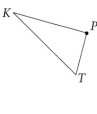

4. $180°$

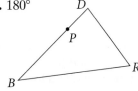

5. $140°$

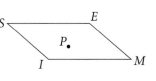

Copy each figure. Then draw the image of $\overline{JK}$ for a 180° rotation about P.

6.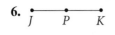

7. K J P

8. J K P

9. K J = P P

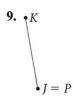

Example 2
(page 484)

The large triangle, quadrilateral, and hexagon are regular. Find the image of each point or segment for the given rotation. (*Hint:* Green segments form 30° angles.)

10. 120° rotation of B about O

11. 270° rotation of L about O

12. 60° rotation of E about O

13. 300° rotation of $\overline{IB}$ about O

14. 240° rotation of G about O

15. 180° rotation of $\overline{JK}$ about O

16. 120° rotation of F about H

17. 270° rotation of M about L

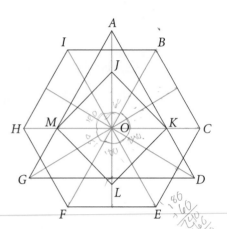

Example 3
(page 485)

Native American Art Find the angle of rotation about C that (a) maps Q to X and (b) maps X to Q.

18.

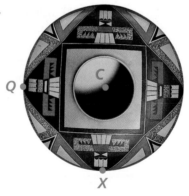

19.

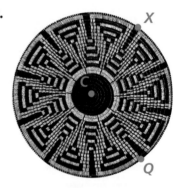

Example 4
(page 485)

For each of Exercises 20–25, copy △XYZ. Draw the image of △XYZ for the given composition of rotations about the given point.

20. 45°, then 45°; X

21. 45°, then 45°; Y

22. 30°, then 30°; Z

23. 20°, then 160°; Z

24. 135°, then 135°; Y

25. 180°, then 180°; X

B **Apply Your Skills**

26. $\overline{M'N'}$ is the rotation image of $\overline{MN}$ about point E. Name all pairs of congruent angles and all pairs of congruent segments in the diagram.

🌐 **27. Language Arts** The symbol ə is called a *schwa*. It is used in dictionaries to represent neutral vowel sounds such as *a* in *ago*, *i* in *sanity*, and *u* in *focus*. What transformation maps a ə to a lowercase e?

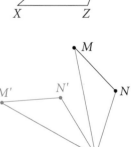

Find the angle of rotation about *C* that maps the black figure onto the blue figure.

28.
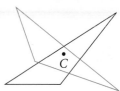

29.

• *C*

30.

• *C*

31. **a. Coordinate Geometry** Graph $A(5, 2)$. Then graph B, the image of A for a 90° rotation about the origin O. (*Hint:* Consider the slope of $\overline{OA}$.)
   b. Graph C, the image of A for a 180° rotation about O.
   c. Graph D, the image of A for a 270° rotation about O.
   d. What type of quadrilateral is $ABCD$? Explain.

GO for Help

For a guide to solving
Exercise 32, see p. 489.

32. **Reasoning** If you are given a figure and a rotation image of the figure, how can you find the center and angle of rotation?

33. **Writing** Describe compositions of rotations that have the same effect as a 360° rotation about a point X.

C Challenge

34. **Coordinate Geometry** Draw $\triangle LMN$ with vertices $L(2, -1)$, $M(6, -2)$, and $N(4, 2)$. Find the coordinates of the vertices after a 90° rotation about the origin and about each of the points L, M, and N.

NY REGENTS

Test Prep

Multiple Choice

35. Name the image of X for a 240° counterclockwise rotation about the center of the regular hexagon.
   A. A **B.** G **C.** O **D.** H

36. What is the image of $(1, -6)$ for a 90° counterclockwise rotation about the origin?
   F. $(6, 1)$ **G.** $(-1, 6)$ **H.** $(-6, -1)$ **J.** $(-1, -6)$

Read the passage below. Then answer the questions on the basis of what is *stated* or *implied* in the passage.

The same hemisphere of the moon always faces Earth. Thus, the motion of the moon about Earth for a given time interval can be modeled by a rotation. The center of the rotation is the center of Earth. The angle of rotation is determined by the time interval, given that one journey of the moon around Earth takes about $27\frac{1}{3}$ days.

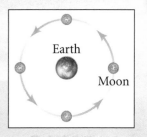

37. What rotation is modeled by the motion of the moon?
   A. a circle rotating about its center   **B.** a circle rotating about a point
   C. 2 circles rotating around each other   **D.** a circle rotating around a circle

38. In how many days does the moon complete a 90° angle of rotation?
   F. about 4   **G.** about 7   **H.** about 14   **J.** about 27

Short Response

39. $\triangle XYZ$ has vertices $X(1, 2)$, $Y(0, 5)$, and $Z(-8, 0)$.
   a. Graph $\triangle XYZ$ and its image after a 270° rotation about the origin.
   b. Name the coordinates of each vertex of the image.

Lesson 9-2

△*BIG* has vertices *B*(−4, 2), *I*(0, −3), and *G*(1, 0). Draw △*BIG* and then its reflection image across the given line.

40. the *y*-axis   **41.** the *x*-axis   **42.** *x* = 4

Lesson 8-5

Find the value of *x*. Round answers to the nearest tenth.

43.

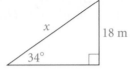

44.

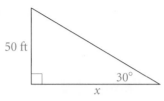

Lesson 3-7

Are the given lines *parallel, perpendicular,* or *neither*? Explain.

45. $y - \frac{2}{3}x = 2$ **46.** $x = -3$ **47.** $-2x + 4y = -8$

  $3x - 4 = 2y$ $x = 3$ $y = -2x + 13$

Checkpoint Quiz 1   Lessons 9-1 through 9-3

State whether the transformation appears to be an isometry. Explain.

1. 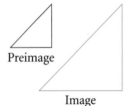   **2.** 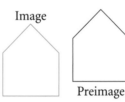   **3.**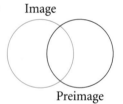

4. Describe in words the translation represented by $(x, y) \rightarrow (x - 3, y + 5)$.

5. Write a rule to describe a translation 5 units left and 10 units up.

6. Describe in words the result of the translation $(x, y) \rightarrow (x + 7, y - 2)$ followed by the translation $(x, y) \rightarrow (x - 3, y + 2)$.

Sketch each figure and point *A*. Draw the image of each figure for the given angle of rotation about *A*. Label the vertices of the image.

7. 40°   **8.** 90°   **9.** 180°

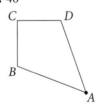   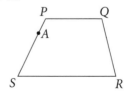   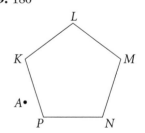

10. △*WXY* has vertices *W*(−4, 1), *X*(2, −7), and *Y*(0, −3). Find its image for the translation $(x, y) \rightarrow (x - 2, y + 5)$.

Understanding Math Problems Read the problem below. Then let Kate's thinking guide you through the solution. Check your understanding with the exercises at the bottom of the page.

Reasoning If you are given a figure and a rotation image of the figure, how can you find the center and angle of rotation?

What Kate Thinks   ### What Kate Draws

I'll use tracing paper and draw two congruent figures that have the same orientation.
I like the letter K, so I think I'll use it.
I'll tilt one K so that it's not a translation image.

I think I see a rotation. It will rotate the left K counterclockwise onto the right K. It will rotate point A to point A' and point B to point B'.

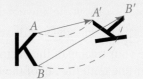

That means that A and A' are on a circle whose center is the center of rotation. The same is true for B and B'. The perpendicular bisectors of both $\overline{AA'}$ and $\overline{BB'}$ go through the center. I'll use blue to construct the perpendicular bisectors.

Point C must be a center of rotation.

I can check this using tracing paper.

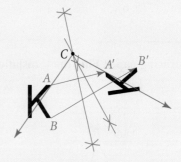

To find the angle of rotation, I just have to draw $\angle ACA'$ and measure it.

The center of rotation is C.
The angle of rotation is 98°.

EXERCISES

1. In a newspaper or magazine that you can cut, find two large identical letters that are the same size and type. Cut them out and place them at random on a large piece of paper. Then find the center and angle of rotation that maps one onto the other.

2. **Writing** Can a triangle be both a translation image and a rotation image of another triangle? Explain.

Tracing Paper Transformations

FOR USE WITH LESSON 9-3

In Lesson 9-1, you learned how to describe a translation using variables. In the Activity on page 477, you drew a reflection in a plane by using tracing paper. Now, you will use tracing paper to perform translations and rotations. You will also discover ways to describe certain rotations and reflections using variables.

NY G.G.54: Define, investigate, justify, and apply isometries in the plane (rotations, reflections, translations, glide reflections). *Note: Use proper function notation.*

1 ACTIVITY

You can use the vector arrow shown in the diagram to represent the translation $(x, y) \rightarrow (x + 4, y + 2)$. The translation shifts $\triangle ABC$ with $A(-3, 3)$, $B(-1, 1)$, and $C(1, 4)$ to $\triangle A'B'C'$ with $A'(1, 5)$, $B'(3, 3)$, and $C'(5, 6)$. You can see this translation using tracing paper as follows:

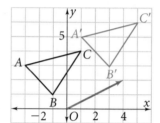

- Draw $\triangle ABC$ and the vector arrow on graph paper. Also, show the line containing the arrow.

- Trace $\triangle ABC$ and the vector arrow.

- Move your tracing of the vector arrow along the vector line until the tail of the tracing is on the head of the original vector arrow. Your tracing of $\triangle ABC$ should now be at $A'(1, 5)$, $B'(3, 3)$, and $C'(5, 6)$.

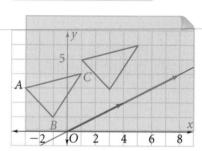

Use tracing paper. Find the translation image of each triangle for the given vector.

1.

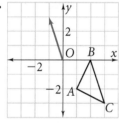

2.

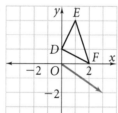

3.

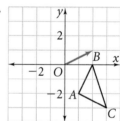

4. Show that the composition of the translation in Exercise 1 followed by the translation in Exercise 2 gives you the translation in Exercise 3.

2 ACTIVITY

To rotate a figure 90° about the origin, trace the figure, one axis, and the origin. Then turn your tracing paper counterclockwise, keeping the origin in place and aligning the traced axis with the other original axis.

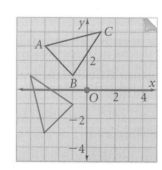

5. Use $\triangle ABC$ from Activity 1 with $A(-3, 3)$, $B(-1, 1)$, and $C(1, 4)$. What is the image of $\triangle ABC$ for a 90° turn about the origin?

6. Copy and complete the table. Use tracing paper to find point P', the image of point P, for a turn of 90° about the origin.

7. Study the pattern in your table. Complete this rule for a turn of 90° about the origin.

$(x, y) \rightarrow (\blacksquare, \blacksquare)$

8. Test your rule with tracing paper. For $\triangle TRN$ with $T(0, 2)$, $R(3, 0)$, and $N(4, 5)$, a 90° turn about the origin should result in $\triangle T'R'N'$ with $T'(-2, 0)$, $R'(0, 3)$, and $N'(-5, 4)$. Does it?

P	P'
(3, 4)	
(−3, 4)	
(−3, −4)	
(3, −4)	
(3, 0)	
(0, 4)	

9. In parts (a) and (b) below, what should be the result of each composition?

  a. a 90° turn about the origin followed by a 90° turn about the origin

  b. $(x, y) \rightarrow (-y, x)$ followed by $(x, y) \rightarrow (-y, x)$

  c. Use tracing paper. Test your conjectures from parts (a) and (b) on $\triangle ABC$ from Activity 1.

3 ACTIVITY

To reflect a figure across an axis using tracing paper, trace the figure, the axis, and the origin. Then turn over your tracing paper, keeping the origin in place and aligning the traced axis with the original axis.

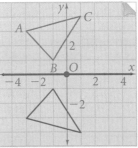

10. What is the reflection image of $\triangle ABC$ across the x-axis? Across the y-axis?

11. Copy and complete the table. Use tracing paper to find points P_x and P_y, the reflection images of point P across the x-axis and y-axis, respectively.

Study the patterns in the table. Complete these rules for reflections across the axes.

12. x-axis: $(x, y) \rightarrow (\blacksquare, \blacksquare)$ **13.** y-axis: $(x, y) \rightarrow (\blacksquare, \blacksquare)$

P	Pₓ	P_y
(3, 4)		
(−3, 4)		
(−3, −4)		
(3, −4)		
(3, 0)		
(0, 4)		

14. Test your rules on $\triangle KLM$ with $K(-2, 3)$, $L(3, 1)$, and $M(1, -2)$.

  a. Across the x-axis, $\triangle KLM$ should map to __?__ . Does it?

  b. Across the y-axis, $\triangle KLM$ should map to __?__ . Does it?

15. In parts (a) and (b) below, what should be the result of each composition?

  a. $(x, y) \rightarrow (x, -y)$ followed by $(x, y) \rightarrow (-x, y)$

  b. a reflection across the x-axis followed by a reflection across the y-axis

  c. Use tracing paper. Test your conjectures from parts (a) and (b) on $\triangle ABC$ from Activity 1.

EXERCISES

16. Use tracing paper to find a rule for a reflection across the line $y = x$. Test your rule on $\triangle ABC$ from Activity 1.

17. Compare the results of Exercises 9 and 15. Make a conjecture about the compositions suggested by each.

18. What other transformation is equivalent to a reflection across the line $y = x$ followed by a reflection across the line $y = -x$? Explain.

9-4

Symmetry

Check Skills You'll Need

GO for Help Lessons 9-2 and 9-3

The regular octagon at the right is divided into eight congruent triangles. Find the image of the given point or segment for the given rotation or reflection.

1. point A; a 90° rotation about the center
2. point H; a 180° rotation about the center
3. $\overrightarrow{AB}$; a reflection in $\overleftrightarrow{AE}$
4. $\overline{GH}$; a reflection in $\overleftrightarrow{AE}$

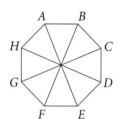

New Vocabulary • symmetry • reflectional symmetry • line symmetry • rotational symmetry • point symmetry

1 Identifying Types of Symmetry in Figures

A figure has **symmetry** if there is an isometry that maps the figure onto itself. If the isometry is the reflection of a plane figure, the figure has **reflectional symmetry** or **line symmetry.** One half of the figure is a mirror image of its other half. Fold the figure along the line of symmetry and the halves match exactly.

The image of the Inuit sculpture at the right has reflectional symmetry about a vertical line down the middle of the face.

It is possible for a figure to have more than one line of symmetry.

1 EXAMPLE Identifying Lines of Symmetry

Draw all lines of symmetry for a regular hexagon.

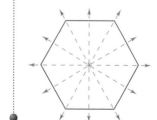

Draw a regular hexagon.

Then draw lines on the hexagon that make mirror-image congruent halves.

There are 6 lines of symmetry.

Quick Check ① Draw a rectangle and all of its lines of symmetry.

A figure that has **rotational symmetry** is its own image for some rotation of 180°
or less. A figure that has **point symmetry** has 180° rotational symmetry. A square
has 90° and 180° rotational symmetry with the center of rotation at the center of
the square. Thus, a square also has point symmetry.

2 EXAMPLE Identifying Rotational Symmetry

Judging from appearance, tell whether each triangle has rotational symmetry. If so,
give the angle of rotation.

a.

b.

This isosceles triangle does not
have rotational symmetry.

120°

The equilateral
triangle has rotational
symmetry. The angle
of rotation is 120°.

✓ **Quick Check** **2 a.** Judging from appearance, tell whether the
figure at the right has rotational symmetry.
If so, give the angle of rotation.
b. Does the figure have point symmetry?

Three-dimensional objects can have various types of symmetry, including
rotational symmetry about a line and reflectional symmetry in a plane.

3 EXAMPLE Real-World 🌐 Connection

Symmetric Design Tell whether each object has rotational symmetry about a line
and/or reflectional symmetry in a plane.

a.

b.

The paddle has both rotational
and reflectional symmetry.

The cup has reflectional symmetry.

✓ **Quick Check** **3** Tell whether the umbrella has
rotational symmetry about a
line and/or reflectional
symmetry in a plane.

Practice by Example

Examples 1, 2
(pages 492, 493)

 for Help

Tell what type(s) of symmetry each figure has. If it has line symmetry, sketch the figure and the line(s) of symmetry. If it has rotational symmetry, state the angle of rotation.

1.
2.
3.
4.

5.
6.
7.
8.

9.
10.
11.
12.

Draw each quadrilateral. Then draw all of its lines of symmetry.

13. rhombus   **14.** kite   **15.** square   **16.** parallelogram

Example 3
(page 493)

Tell whether each three-dimensional object has rotational symmetry about a line and/or reflectional symmetry in a plane.

17.
18.

B **Apply Your Skills**

19. Open-Ended The word C H E C K B O O K has a horizontal line of symmetry. Find two other words for which this is true.

20. Open-Ended Stack the letters of M A T H vertically and upright, and you can find a vertical line of symmetry. Find two other words for which this is true.

21. a. Alphabets Copy the chart. Use it to classify the letters of the English and Greek alphabets below. You will list some letters in more than one category.
   b. Which alphabet can you say is more symmetrical? Explain.

Type of Symmetry

Language	Horizontal Line	Vertical Line	Point
English			
Greek			

English: ABCDEFGHIJKLMNOPQRSTUVWXYZ

Greek: ΑΒΓΔEZHΘIKΛMNΞOΠPΣTYΦXΨΩ

Tell what type(s) of symmetry each image has. For line symmetry, sketch the image and the line(s) of symmetry. For rotational symmetry, state the angle of rotation.

22.

23.

24. Open-Ended The equation $\frac{10}{10} - 1 = 0 \div \frac{83}{83}$ is not only true, but also symmetrical (horizontally). Write four other equations or inequalities that are both true and symmetrical.

🌐 **Logos** **Describe the types of symmetry, if any, of each automobile logo.**

25.

26.

27.

28.

29.

30.

31.

32.

Problem Solving Hint

In Exercises 33–35, you can conclude a statement is false by finding a counterexample.

33. Is the line that contains the bisector of an angle a line of symmetry of the angle? Explain.

34. Is the line that contains the bisector of an angle of a triangle a line of symmetry of the triangle? Explain.

35. Is a bisector of a segment a line of symmetry of the segment? Explain.

36. Multiple Choice Which statement is true about the figure at the right?
  Ⓐ It has no rotational symmetry.
  Ⓑ It has no reflectional symmetry.
  Ⓒ It has rotational symmetry with an angle of rotation of 45°.
  Ⓓ It has reflectional symmetry with six lines of symmetry.

Coordinate Geometry **A figure has a vertex at (3, 4). If the figure has the given type of symmetry, state the coordinates of another vertex of the figure.**

37. line symmetry about the *y*-axis

38. line symmetry about the *x*-axis

39. point symmetry about the origin

40. line symmetry about the line $y = x$

Coordinate Geometry **Graph each equation. Describe the symmetry of each graph.**

41. $y = x$

42. $y = x^2$

43. $x = y^2$

44. $x^2 + y^2 = 9$

🅒 **Challenge**

45. $y = (x + 2)^2$

46. $y = x^3$

47. $y = |x|$

48. $x = |y|$

For each three-dimensional figure, draw a net that has rotational symmetry and a net that has 1, 2, or 4 lines of symmetry.

49.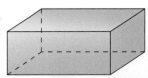

50.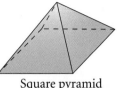
Square pyramid

Test Prep

Multiple Choice

51. Which figure does NOT have rotational symmetry?

    A.   B.   C. N   D.

52. Which figure, in general, has exactly two lines of symmetry?
    F. pentagon   G. circle   H. square   J. rectangle

53. Which quadrilateral has rotational symmetry but not reflectional symmetry?
    A. nonisosceles trapezoid   B. kite
    C. rhombus   D. parallelogram

54. What is the smallest angle through which you can rotate a regular hexagon onto itself?
    F. 30°   G. 60°   H. 90°   J. 120°

Short Response

55. Does a regular octagon have
    a. line symmetry? If so, how many lines of symmetry does it have?
    b. rotational symmetry? If so, what is the angle of rotation?

56. Use the figure at the right to answer the questions below.
    a. Does the figure have rotational symmetry?
        If so, identify the angle of rotation.
    b. Does the figure have reflectional symmetry?
        If so, how many lines of symmetry does it have?

Mixed Review

GO for Help

Lesson 9-3

57. Which capital letters of the alphabet are rotation images of themselves? Draw each letter and give an angle of rotation ($< 360°$).

Lesson 9-1

The blue figure is a transformation image of the black figure. Does the transformation appear to be an isometry? Explain.

58.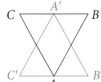

59.

60.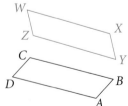

Lesson 6-6

61. Three vertices of an isosceles trapezoid are $(-2, 1)$, $(1, 4)$, and $(4, 4)$. Find possible pairs of coordinates for the fourth vertex.

To *extend* an idea means to enlarge the scope of the idea. One way is to substitute a new word for a key word. Another is to change the frame of reference. In each case, you then decide whether the new statement is true.

1 EXAMPLE

Is the following extension of the Pythagorean Theorem true?

The *semicircles* on the legs of a right triangle are equal in area to the *semicircle* on the hypotenuse.

$(2a)^2 + (2b)^2 = (2c)^2$ by the Pythagorean Theorem.

Using the Multiplication Property of Equality, it follows that $\frac{1}{2}\pi a^2 + \frac{1}{2}\pi b^2 = \frac{1}{2}\pi c^2$.

This extension of the Pythagorean Theorem to semicircles is true.

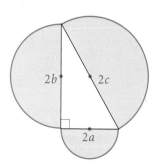

2 EXAMPLE

Extend the definition of *circle* to spherical geometry (see pp. 154–155). Give an unusual result.

Let a circle on a sphere be the set of all points equidistant from a given point. For this definition, one circle will have two distinct centers.

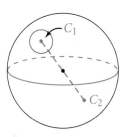

EXERCISES

1. Suppose you have similar triangles on the three sides of a right triangle. Are the triangles on the legs equal in area to the triangle on the hypotenuse? Explain.

2. Determine whether the Pythagorean Theorem extends as follows: The squares on the two shortest sides of an acute triangle are equal in area to the square on the longest side. Justify your response.

3. Example 2 extends the idea of a circle to spherical geometry and gives an unusual result. Find another unusual fact that follows from the definition in Example 2.

You *generalize* an idea if you can extend it to all possible cases (of a particular type).

4. The midpoint of the segment from a to b on a number line is $\frac{a+b}{2}$.

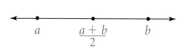

   a. Explain how you can extend this idea to both two and three dimensions.
   b. Explain how you can generalize the Midpoint Formula to be a definition of a midpoint in n dimensions, for any whole number n.

5. Explain how to extend the Midpoint Formula to points of trisection and points of quadrisection. Can you generalize?

6. Explain how to extend and generalize the Triangle Angle-Sum Theorem.

9-5

Dilations

Learning Standards for Mathematics

G.G.61 Investigate, justify, and apply the analytical representations for translations, rotations about the origin of 90° and 180°, reflections over the lines $x = 0$, $y = 0$, and $y = x$, and dilations centered at the origin.

✓ Check Skills You'll Need

GO for Help Lesson 7-1

Determine the scale drawing dimensions of a room using a scale of $\frac{1}{4}$ in. = 1 ft.

1. kitchen: 12 ft by 16 ft
2. bedroom: 8 ft by 10 ft
3. laundry room: 6 ft by 9 ft
4. bathroom: 5 ft by 7 ft

🔊 **New Vocabulary** • dilation • enlargement • reduction

---

1 Locating Dilation Images

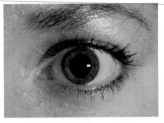

Real-World 🌐 Connection

Look closely at your pupil in a mirror, and you can watch it dilate.

A **dilation** is a transformation whose preimage and image are similar. Thus, a dilation is a similarity transformation. It is *not*, in general, an isometry.

Every dilation has a center and a scale factor n, $n > 0$. The scale factor describes the size change from the original figure to the image.

To find a dilation with center C and scale factor n, you can use the following two rules.

• The image of C is itself (that is, $C' = C$).
• For any point R, R' is on $\overrightarrow{CR}$ and $CR' = n \cdot CR$.

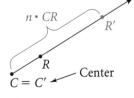

The dilation is an **enlargement** if the scale factor is greater than 1. The dilation is a **reduction** if the scale factor is between 0 and 1.

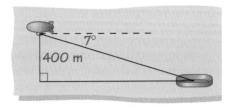

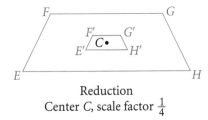

Reduction
Center C, scale factor $\frac{1}{4}$

Test-Taking Tip

When finding a scale factor, note first whether the dilation is an enlargement or a reduction.

1 EXAMPLE Finding a Scale Factor

The blue triangle is a dilation image of the red triangle. Describe the dilation.

The center is X. The image is larger than the preimage, so the dilation is an enlargement.

$$\frac{X'T'}{XT} = \frac{4 + 8}{4} = 3$$

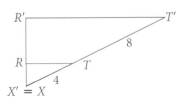

● The dilation has center X and scale factor 3.

✓ Quick Check ❶ The blue quadrilateral is a dilation image of the red quadrilateral. Describe the dilation.

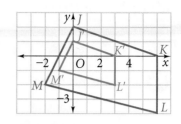

Scale factors help you understand scale models, both large and small.

2 EXAMPLE **Real-World 🌎 Connection**

Scale Models The packaging lists a model car's length as 7.6 cm. It also gives the scale as 1 : 63. What is the length of the actual car?

To "enlarge" the model car to the actual car, use the scale factor 63. Multiply 7.6 cm
● by 63 to get 478.8 cm, or about 4.8 m, for the length of the actual car.

✓ Quick Check ❷ The height of a tractor-trailer truck is 4.2 m. The scale factor for a model of the truck is $\frac{1}{54}$. Find the height of the model to the nearest centimeter.

Suppose a dilation is centered at the origin. You can find the dilation image of a point by multiplying its coordinates by the scale factor.

Scale factor 4, $(x, y) \longrightarrow (4x, 4y)$

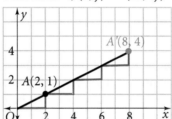

Scale factor $\frac{1}{3}$, $(x, y) \longrightarrow \left(\frac{1}{3}x, \frac{1}{3}y\right)$

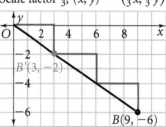

To dilate a triangle from the origin, find the dilation images of its vertices.

3 EXAMPLE **Graphing Dilation Images**

Multiple Choice $\triangle PZG$ has vertices $P(2, 0)$, $Z\left(-1, \frac{1}{2}\right)$, and $G(1, -2)$. What are the coordinates of the image of P for a dilation with center $(0, 0)$ and scale factor 3?

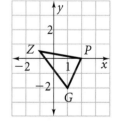

Ⓐ $(5, 3)$ Ⓑ $(6, 0)$

Ⓒ $\left(\frac{2}{3}, 0\right)$ Ⓓ $(3, -6)$

The scale factor is 3, so use the rule $(x, y) \rightarrow (3x, 3y)$.

$P(2, 0) \rightarrow P'(3 \cdot 2, 3 \cdot 0)$, or $P'(6, 0)$. The correct answer is B.

Z' is $\left(3 \cdot (-1), 3 \cdot \frac{1}{2}\right)$ and G' is $(3 \cdot 1, 3 \cdot (-2))$, so the vertices of the enlargement
● at the left are $P'(6, 0)$, $Z'\left(-3, \frac{3}{2}\right)$, and $G'(3, -6)$.

✓ Quick Check ❸ Find the image of $\triangle PZG$ for a dilation with center $(0, 0)$ and scale factor $\frac{1}{2}$. Draw the reduction.

EXERCISES

For more exercises, see *Extra Skill, Word Problem, and Proof Practice*.

Practice and Problem Solving

 Practice by Example

Example 1
(page 498)

GO for Help

The blue figure is a dilation image of the red figure. Describe the dilation.

1.

2.
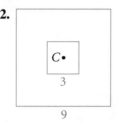

3.
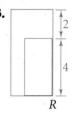

4.
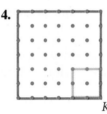

5.
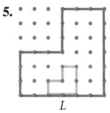

6.
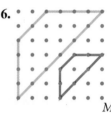

7.
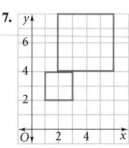

8.
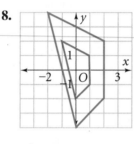

9.
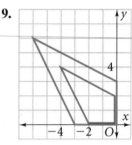

Example 2
(page 499)

Model Railroads The table shows scales for different types of model railroads. For each model in Exercises 10–12, what would be the actual measurement?

10. An HO-scale tank car is 1.4 in. high.

11. An S-scale boxcar has length 8 in.

12. A model of an engineer in a G-scale model train layout is 3 in. tall.

13. A diesel engine is 60 feet long. How long is its O-scale model?

14. Actual railroad tracks are 4 ft 8.5 in. apart. How far apart are N-scale tracks?

Model Railroad Scales

Scale Name	Scale Ratio
N	1 : 160
HO	1 : 87.1
S	1 : 64
O	1 : 48
G	1 : 22.5

Example 3
(page 499)

Find the image of △*PQR* for a dilation with center (0, 0) and the scale factor given. Draw the image.

15.
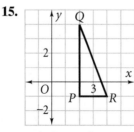
scale factor 3

16.
scale factor 10

17.
scale factor $\frac{3}{4}$

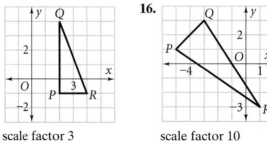

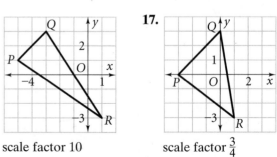

500 Chapter 9 Transformations

A dilation has center (0, 0). Find the image of each point for the scale factor given.

18. $D(1, -5); 2$

19. $L(-3, 0); 5$

20. $A(-6, 2); 1.5$

21. $T(0, 6); 3$

22. $M(0, 0); 10$

23. $N(-4, -7); 0.1$

24. $F(-3, -2); \frac{1}{3}$

25. $B(\frac{5}{4}, -\frac{2}{3}); \frac{1}{10}$

26. $Q(6, \frac{\sqrt{3}}{2}); \sqrt{6}$

B Apply Your Skills

Find the image of $QRTW$ for a dilation with center (0, 0) and the scale factor given.

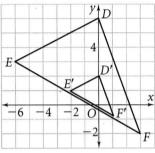

27. 3

28. 2

29. $\frac{1}{2}$

30. $\frac{1}{4}$

31. 0.6

32. 0.9

33. 10

34. 100

35. Writing An equilateral triangle has 4-in. sides. Describe its image for a dilation with scale factor 2.5. Explain.

36. Multiple Choice $\triangle D'E'F'$ is a dilation of $\triangle DEF$ with center $(0, 0)$. Which of the following statements is *not* true?

Ⓐ $\triangle DEF$ and $\triangle D'E'F'$ are similar.

Ⓑ $m\angle F'D'E' = \frac{1}{3} \cdot m\angle FDE$

Ⓒ The scale factor of the dilaton is $\frac{1}{3}$.

Ⓓ $\angle EFD \cong \angle E'F'D'$

Coordinate Geometry Graph *MNPQ* and its image *M'N'P'Q'* for a dilation with center (0, 0) and the scale factor given.

37. $M(-1, -1), N(1, -2), P(1, 2), Q(-1, 3)$; scale factor 2

38. $M(1, 3), N(-3, 3), P(-5, -3), Q(-1, -3)$; scale factor 3

39. $M(0, 0), N(4, 0), P(6, -2), Q(-2, -2)$; scale factor $\frac{1}{2}$

40. $M(2, 6), N(-4, 10), P(-4, -8), Q(-2, -12)$; scale factor $\frac{1}{4}$

41. Open-Ended Use the dilation command in geometry software or drawing software to create a design that involves repeated dilations. The software will prompt you to specify a center of dilation and a scale factor. Print your design and color it. Feel free to use other transformations along with dilations.

42. Copy Reduction Your copy of your family crest is 4.5 in. wide. You need a reduced copy for the front page of the family newsletter. The copy must fit in a space 1.8 in. wide. What scale factor should you use on the copy machine?

A dilation maps $\triangle HIJ$ to $\triangle H'I'J'$. Find the missing values.

43. $HI = 8$ in.
$IJ = 5$ in.
$HJ = 6$ in.
$H'I' = 16$ in.
$I'J' = \blacksquare$ in.
$H'J' = \blacksquare$ in.

44. $HI = 7$ cm
$IJ = 7$ cm
$HJ = \blacksquare$ cm
$H'I' = 5.25$ cm
$I'J' = \blacksquare$ cm
$H'J' = 9$ cm

45. $HI = \blacksquare$ ft
$IJ = 30$ ft
$HJ = 24$ ft
$H'I' = 8$ ft
$I'J' = \blacksquare$ ft
$H'J' = 6$ ft

46. Error Analysis Brendan says that when a rectangle with length 6 cm and width 4 cm is dilated by a scale factor of 2, the perimeter and area of the rectangle are doubled. Explain what is incorrect about Brendan's statement.

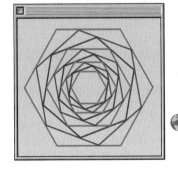

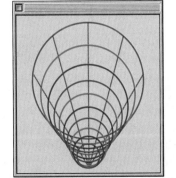

Exercise 41

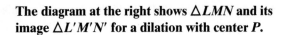
GO **Online**
Homework Video Tutor
Visit: PHSchool.com
Web Code: aue-0905

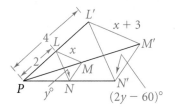

The diagram at the right shows △*LMN* and its image △*L'M'N'* for a dilation with center *P*.

x^2 **47. Algebra** Find the values of *x* and *y*.

48. Evaluate *y* and $2y - 60$. Explain why the two values must be equal.

Copy △*TBA* and point *O* for each of Exercises 49–52. Draw the dilation image △*T'B'A'* for the given center and scale factor.

49. center *O*, scale factor $\frac{1}{2}$

50. center *B*, scale factor 3

51. center *T*, scale factor $\frac{1}{3}$

52. center *O*, scale factor 2

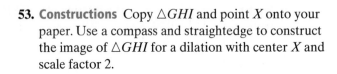

Real-World 🌐 **Connection**

An overhead projection is a dilation only when the mirror in the head is tilted at a 45° angle. Turning the head distorts the images.

53. Constructions Copy △*GHI* and point *X* onto your paper. Use a compass and straightedge to construct the image of △*GHI* for a dilation with center *X* and scale factor 2.

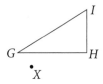

🌐 **Overhead Projection** An overhead projector can dilate figures on transparencies.

54. A segment on a transparency is 2 in. long. Its image on the screen is 2 ft long. What is the scale factor of the dilation?

55. The height of a parallelogram on the transparency is 4 cm. The scale factor is 15. What is the height of the parallelogram on the screen?

56. The area of a square on the screen is 3 ft². The scale factor is 16. What is the area of the square on the transparency?

Write *true* **or** *false* **for Exercises 57–61. Explain your answers.**

57. A dilation is an isometry.

58. A dilation changes orientation.

59. A dilation with a scale factor greater than 1 is a reduction.

60. For a dilation, corresponding angles of the image and preimage are congruent.

61. A dilation image cannot have any points in common with its preimage.

C **Challenge**

62. A flashlight projects an image of rectangle *ABCD* on a wall so that each vertex of *ABCD* is 3 ft away from the corresponding vertex of *A'B'C'D'*. The length of $\overline{AB}$ is 3 in. The length of $\overline{A'B'}$ is 1 ft. How far from each vertex of *ABCD* is the light?

not to scale

63. Critical Thinking You are given $\overline{AB}$ and its dilation image $\overline{A'B'}$ with *A, B, A',* and *B'* noncollinear. Explain how to find the center of dilation and scale factor.

Coordinate Geometry In the coordinate plane you can extend dilations to include scale factors that are negative numbers.

64. a. Graph △*PQR* with vertices $P(1, 2)$, $Q(3, 4)$, and $R(4, 1)$.
   b. For a dilation centered at the origin with a scale factor of -3, multiply the coordinates in part (a) by -3. List the results as $P', Q',$ and R'.
   c. Graph △*P'Q'R'* on the same set of axes.

65. a. A dilation with center at the origin and scale factor −1 (see Exercise 64) may be called a *reflection in a point.* For △*PQR* of Exercise 64, find the image △*P′Q′R′* for such a dilation.

   b. Writing Explain why the dilation described in part (a) may be called a *reflection in a point.* Extend your explanation to a new definition of point symmetry. Compare your new definition with the definition given on page 493.

66. Constructions Draw acute △*ABC.* Construct square *DEFG* so that $\overline{DG}$ is on $\overline{AC}$, and *E* and *F* are on the other two sides of △*ABC.* (*Hint:* First, try the special case with a right angle at *A* and use a dilation.)

Test Prep

Gridded Response

67. A dilation maps △*ABC* onto △*A′B′C′* with a scale factor of 0.3. If *A′B′* = 312 m, what is *AB* in meters?

68. A dilation maps △*CDE* onto △*C′D′E′.* If *CD* = 7.5 ft, *CE* = 15 ft, *D′E′* = 3.25 ft, and *C′D′* = 2.5 ft, what is *DE* in feet?

69. A dilation maps △*XYZ* onto △*X′Y′Z′.* If *XY* = 4 m, *YZ* = 29 m, *X′Z′* = 28.7 m, and *Y′Z′* = 29.145 m, what is *X′Y′* in meters?

70. The center of dilation of quadrilateral *ABCD* is point *X,* as shown at the right. The length of a side of quadrilateral *A′B′C′D′* is what percent of the length of the corresponding side of quadrilateral *ABCD*?

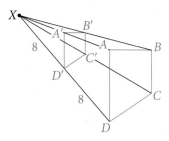

71. A dilation maps △*JKL* onto △*J′K′L′.* If *JK* = 28 cm, *KL* = 52 cm, *JL* = 40.2 cm, and *J′K′* = 616 cm, what is the scale factor?

Mixed Review

Lesson 9-4

Coordinate Geometry A figure has a vertex at (−2, 7). If the figure has the given type of symmetry, state the coordinates of another vertex of the figure.

72. line symmetry about the *x*-axis   **73.** line symmetry about the *y*-axis

74. point symmetry about the origin   **75.** line symmetry about the line *y* = *x*

Lesson 9-1

Write a rule for each translation.

76. 0 units to the right, 4 units up   **77.** 2 units to the left, 1 unit down

78. 3 units to the left, 6 units down   **79.** 8 units to the right, 10 units up

Lesson 8-5

For the angle of depression shown, find the value of *x* to the nearest tenth of a unit.

80.

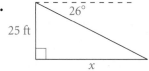

81.

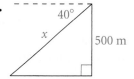

Transformations Using Vectors and Matrices

FOR USE WITH LESSON 9-5

In Lesson 8-6, you learned that vectors are quantities that have both direction and magnitude. A translation moves all the points of a figure in the same direction and the same distance, so you can use a vector to describe a translation.

1 EXAMPLE Using Vectors to Find Translation Images

a. Find the image of $D(1, -2)$ under the translation described by the vector $\langle 1, 4 \rangle$.

$\langle 1, 4 \rangle$ is a translation right 1 unit and up 4 units, the same as $(x, y) \rightarrow (x + 1, y + 4)$.
$D(1, -2)$ translates to $(1 + 1, -2 + 4)$, or $(2, 2)$.

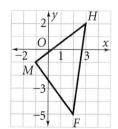

b. Find the image of $D(1, -2)$ under the composition of translations: $\langle 1, 4 \rangle$ followed by $\langle -4, -1 \rangle$.

Use vector addition: $\langle 1, 4 \rangle + \langle -4, -1 \rangle = \langle 1 + (-4), 4 + (-1) \rangle = \langle -3, 3 \rangle$
$D(1, -2)$ translates to $(1 - 3, -2 + 3)$, or $(-2, 1)$.

You can use a matrix to find the transformation images of several points at once. A *matrix* is a rectangular arrangement of numbers in rows and columns.

The matrix at the right is called a 2 × 3 (two-by-three) matrix because it has 2 rows and 3 columns. Each entry in a matrix is an *element*. This matrix has 6 elements.

$$\begin{bmatrix} -1 & 4 & 9 \\ 3 & -5 & 7 \end{bmatrix}$$

To find the translation image of a triangle, use a matrix made up of the three vertices and a translation matrix that shows the translation vector three times.

2 EXAMPLE Using Matrices to Find Translation Images

Use matrices to find the image of $\triangle MFH$ under the translation $\langle 4, -5 \rangle$.

$$\begin{array}{c} \\ x\text{-coordinate} \\ y\text{-coordinate} \end{array} \begin{array}{ccc} M & F & H \\ \begin{bmatrix} -1 & 2 & 3 \\ -1 & -5 & 2 \end{bmatrix} \end{array}$$ **Write a matrix for $\triangle MFH$.**

$$\begin{bmatrix} 4 & 4 & 4 \\ -5 & -5 & -5 \end{bmatrix}$$ **Write the translation matrix.**

$$\begin{bmatrix} -1 & 2 & 3 \\ -1 & -5 & 2 \end{bmatrix} + \begin{bmatrix} 4 & 4 & 4 \\ -5 & -5 & -5 \end{bmatrix} = \begin{bmatrix} 3 & 6 & 7 \\ -6 & -10 & -3 \end{bmatrix}$$ **Add the matrices.**

The image of $\triangle MFH$ is $\triangle M'F'H'$ with $M'(3, -6)$, $F'(6, -10)$, and $H'(7, -3)$.

To find the image of a triangle for a dilation from the origin, multiply each element in the vertex matrix for the triangle by the scale factor. Multiplying a matrix by a number in this way is called *scalar multiplication*.

3 EXAMPLE Using Scalar Multiplication

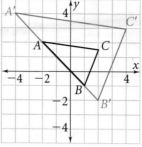

$\triangle ABC$ has vertices $A(-2, 2)$, $B(1, -1)$, and $C\left(2, 1\frac{1}{2}\right)$. Use scalar multiplication to find the image of $\triangle ABC$ for a dilation with center $(0, 0)$ and scale factor 2. Draw the enlargement.

$$\begin{array}{ccc} & A & B & C \end{array}$$

x-coordinate $\begin{bmatrix} -2 & 1 & 2 \\ 2 & -1 & 1\frac{1}{2} \end{bmatrix}$ **Write a matrix for $\triangle ABC$.**
y-coordinate

$2 \cdot \begin{bmatrix} -2 & 1 & 2 \\ 2 & -1 & 1\frac{1}{2} \end{bmatrix} = \begin{bmatrix} -4 & 2 & 4 \\ 4 & -2 & 3 \end{bmatrix}$ **Multiply each element of the matrix by the scale factor, or scalar, 2.**

● The vertices of the enlargement are $A'(-4, 4)$, $B'(2, -2)$, and $C'(4, 3)$.

EXERCISES

Find the image of each figure under the translation described by the given vector.

1. $\langle 4, -1 \rangle$

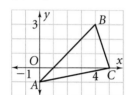

2. $\langle 2, 0 \rangle$

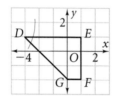

3. $\langle -3, -2 \rangle$

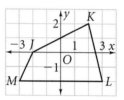

Find a single translation that has the same effect as each composition of translations.

4. $\langle -2, 3 \rangle$ followed by $\langle 1, 5 \rangle$

5. $\langle -6, 0 \rangle$ followed by $\langle -4, 7 \rangle$

6. $\langle 3, -4 \rangle$ followed by $\langle -4, 3 \rangle$

7. $\langle 8, 10 \rangle$ followed by $\langle -7, 6 \rangle$

Use matrices to find the image of each figure under the given translation.

8. $\langle 2, -1 \rangle$

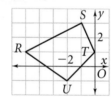

9. $\langle 4, -3 \rangle$

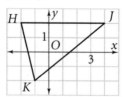

10. $\langle 0, 2 \rangle$

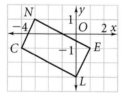

Use a matrix and scalar multiplication. Find the image of each figure for a dilation with center $(0, 0)$ and the scale factor given. Use graph paper. Draw the figure and its image.

11. 3

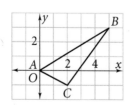

12. $\frac{1}{3}$

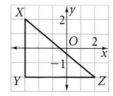

13. $1\frac{1}{2}$

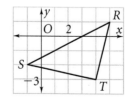

Compositions of Reflections

 Learning Standards for Mathematics

G.G.57 Justify geometric relationships (perpendicularity, parallelism, congruence) using transformational techniques (translations, rotations, reflections).

G.G.60 Identify specific similarities by observing orientation, numbers of invariant points, and/or parallelism.

✓ **Check Skills You'll Need**

GO for Help   Lessons 9-1 and 9-2

Given points $R(-1, 1)$, $S(-4, 3)$, and $T(-2, 5)$, draw $\triangle RST$ and its reflection image in each line.

1. the y-axis   **2.** the x-axis   **3.** $y = 1$

Draw $\triangle RST$ described above and its translation image for each translation.

4. $(x, y) \rightarrow (x, y - 3)$ **5.** $(x, y) \rightarrow (x + 4, y)$

6. $(x, y) \rightarrow (x + 2, y - 5)$

7. Copy the figure at the right. Draw the image of the figure for a reflection across $\overleftrightarrow{DG}$.

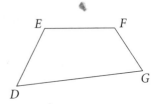

◀)) **New Vocabulary**   • glide reflection

---

1   **Compositions of Reflections**

If two figures are congruent, there is a transformation that maps one onto the other. If no reflection is involved, then the figures are either translation or rotation images of each other.

1 EXAMPLE   **Recognizing the Transformation**

The two figures are congruent. Is one figure a translation image of the other, a rotation image, or neither? Explain.

The orientations of these congruent figures do not appear to be opposite, so one is a translation image or a rotation image of the other. Clearly, it's not a translation image, so it must be a rotation image.

✓ **Quick Check** ❶ The two figures are congruent. Is one figure a translation image of the other, a rotation image, or neither? Explain.

Any translation or rotation can be expressed as the composition of two reflections.

 **Key Concepts**

Theorem 9-1

A translation or rotation is a composition of two reflections.

The examples that illustrate Theorems 9-2 and 9-3 suggest a proof of Theorem 9-1 (how to find two reflections for a given translation or rotation).

Theorems 9-2 and 9-3 together form the converse of Theorem 9-1.

 **Key Concepts**

| **Theorem 9-2** |
| A composition of reflections across two parallel lines is a translation. |

| **Theorem 9-3** |
| A composition of reflections across two intersecting lines is a rotation. |

2 EXAMPLE **Composition of Reflections Across Parallel Lines**

Find the image of R for a reflection across line ℓ followed by a reflection across line m. Describe the resulting translation.

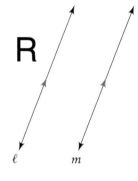
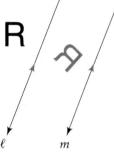
Reflect in ℓ.
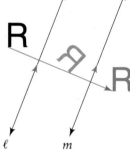
Reflect in m.

Real-World Connection

Each mirror shows a reverse image. But bend the mirrors like this ⌒ and you get compositions of reflections.

R is translated the distance and direction shown by the green arrow. The arrow is perpendicular to lines ℓ and m with length equal to twice the distance from ℓ to m.

 **Quick Check** ❷ Draw lines ℓ and m as shown above. Draw R between ℓ and m. Find the image of R for a reflection across line ℓ and then across line m. Describe the resulting translation.

3 EXAMPLE **Composition of Reflections in Intersecting Lines**

Lines a and b intersect in point C and form acute $\angle 1$ with measure 35. Find the image of R for a reflection across line a and then a reflection across line b. Describe the resulting rotation.

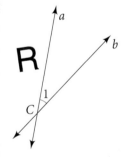
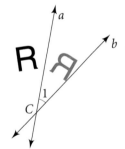
Reflect in a.
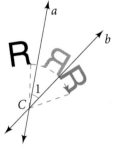
Reflect in b.

R rotates clockwise through the angle shown by the green arrow. The center of rotation is C and the measure of the angle is twice $m\angle 1$, or 70.

 **Quick Check** ❸ Repeat Example 3, but begin with R in a different position.

Two plane figures A and B can be congruent with opposite orientations. Reflect A and you get a figure A′ that has the same orientation as B. Thus, B is a translation or rotation image of A′. By Theorem 9-1, two reflections map A′ to B. The net result is that three reflections map A to B.

This is summarized in what is sometimes called the Fundamental Theorem of Isometries.

 Key Concepts

Theorem 9-4	Fundamental Theorem of Isometries

In a plane, one of two congruent figures can be mapped onto the other by a composition of at most three reflections.

Real-World ⊕ Connection

A computer can translate an image and then reflect it, or vice versa. The two rabbit images are glide reflection images of each other.

If two figures are congruent and have opposite orientations (but are not simply reflections of each other), then there is a slide and a reflection that will map one onto the other. A **glide reflection** is the composition of a glide (translation) and a reflection across a line parallel to the direction of translation.

4 **EXAMPLE** **Finding a Glide Reflection Image**

Coordinate Geometry Find the image of $\triangle TEX$ for a glide reflection where the translation is $(x, y) \rightarrow (x, y - 5)$ and the reflection line is $x = 0$.

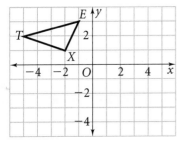

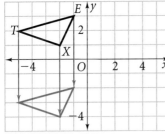

Translate $\triangle TEX$.

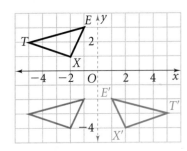

Reflect the image in $x = 0$.

✓ **Quick Check** **4** Use $\triangle TEX$ from Example 4 above.
a. Find the image of $\triangle TEX$ under a glide reflection where the translation is $(x, y) \rightarrow (x + 1, y)$ and the reflection line is $y = -2$.
b. **Critical Thinking** Would the result of part (a) be the same if you reflected $\triangle TEX$ first, and then translated it? Explain.

You can map one of any two congruent figures onto the other by a single reflection, translation, rotation, or glide reflection. Thus, you are able to classify any isometry.

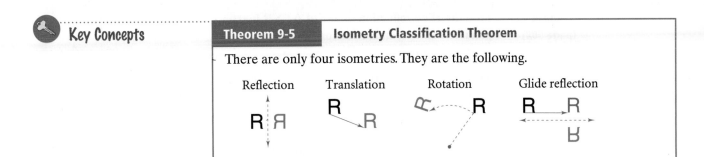

Key Concepts

Theorem 9-5   Isometry Classification Theorem

There are only four isometries. They are the following.

Reflection   Translation   Rotation   Glide reflection

5 EXAMPLE Classifying Isometries

Each figure is an isometry image of the figure at the left. Tell whether their orientations are the same or opposite. Then classify the isometry.

a.   **b.**   **c.**   **d.**

opposite;   opposite;   same;   same;
a reflection   a glide reflection   a translation   a rotation

✓ Quick Check **5** Classify the isometry.

EXERCISES

For more exercises, see *Extra Skill, Word Problem, and Proof Practice.*

Practice and Problem Solving

A Practice by Example

Example 1
(page 506)

GO for Help

The two figures in each pair are congruent. Is one figure a translation image of the other, a rotation image, or neither? Explain.

1.   **2.** bye bye   **3.**

Example 2
(page 507)

Find the image of each letter for a reflection across line ℓ and then a reflection across line m. Describe the resulting translation or rotation.

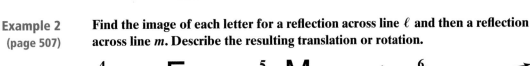

4.   **5.**   **6.**

Example 3
(page 507)

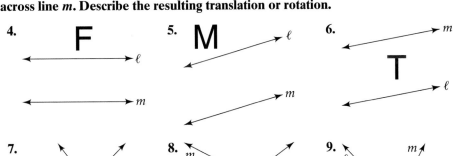

7.   **8.**   **9.**

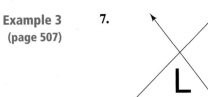
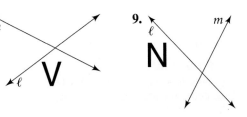

Example 4
(page 508)

Find the glide reflection image of △PNB for the given translation and reflection line.

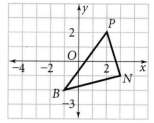

10. $(x, y) \rightarrow (x + 2, y); y = 3$

11. $(x, y) \rightarrow (x, y - 3); x = 0$

12. $(x, y) \rightarrow (x + 2, y + 2); y = x$

13. $(x, y) \rightarrow (x - 1, y + 1); y = -x$

14. $(x, y) \rightarrow (x, y - 1); x = 2$ **15.** $(x, y) \rightarrow (x - 2, y - 2); y = x$

Example 5
(page 509)

Each figure is an isometry image of the figure at the left. Tell whether their orientations are the same or opposite. Then classify the isometry.

16.   **17.**   **18.**   **19.**

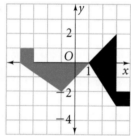

20. 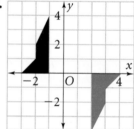   **21.**   **22.**   **23.**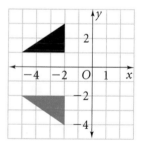

B **Apply Your Skills**

The two figures are congruent. Name the isometry that maps one onto the other.

24.

25.

26. Multiple Choice Which transformation maps the black triangle onto the blue triangle?

Ⓐ a translation $(x, y) \rightarrow (x, y - 3)$ followed by a reflection across $x = -2$

Ⓑ a rotation of 180° about the origin

Ⓒ a reflection across $y = -\frac{1}{2}$

Ⓓ a reflection across the y-axis followed by a 180° rotation about the origin

27. Writing Reflections and glide reflections are *odd isometries*, while translations and rotations are *even isometries*. Use what you learned in this lesson to explain why these categories make sense.

28. Open-Ended Draw △ABC. Then, describe a reflection, a translation, a rotation, and a glide reflection, and draw the image of △ABC for each transformation.

29. For center of rotation P, does an $x°$ rotation followed by a $y°$ rotation give the same image as a $y°$ rotation followed by an $x°$ rotation? Explain.

30. Does an $x°$ rotation about a point P followed by a reflection in a line ℓ give the same image as a reflection in ℓ followed by an $x°$ rotation about P? Explain.

GO Online
Homework Video Tutor
Visit: PHSchool.com
Web Code: aue-0906

To learn more about
kaleidoscopes, see p. 513.

Kaleidoscopes The vibrant images of a kaleidoscope are produced by compositions of reflections in intersecting mirrors. Determine the angle between the mirrors in each kaleidoscope image.

31.

32.

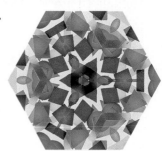

33.

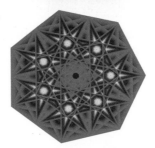

34.

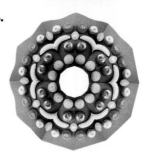

Identify each mapping as a reflection, translation, rotation, or glide reflection. Find the reflection line, translation rule, center and angle of rotation, or glide translation and reflection line.

35. $\triangle ABC \rightarrow \triangle EDC$

36. $\triangle EDC \rightarrow \triangle PQM$

37. $\triangle MNJ \rightarrow \triangle EDC$

38. $\triangle HIF \rightarrow \triangle HGF$

39. $\triangle PQM \rightarrow \triangle JLM$

40. $\triangle MNP \rightarrow \triangle EDC$

41. $\triangle JLM \rightarrow \triangle MNJ$

42. $\triangle PQM \rightarrow \triangle KJN$　　**43.** $\triangle KJN \rightarrow \triangle ABC$　　**44.** $\triangle HGF \rightarrow \triangle KJN$

 Challenge

45. Describe a glide and a reflection that maps the red R to the blue R.

For the given transformation mapping $\overline{XY}$ to $\overline{X'Y'}$, give a convincing argument why $\overline{XY} \cong \overline{X'Y'}$.

46. a reflection　　　**47.** a translation　　　**48.** a rotation

49. The definition states that a glide reflection is the composition of a translation and a reflection. Explain why these can occur in either order.

50. For lines of reflection r and s, does a reflection in r followed by a reflection in s give the same image as a reflection in s followed by a reflection in r? Explain.

$P \rightarrow P'(3, -1)$ for the given translation and reflection line. Find the coordinates of P.

51. $(x, y) \rightarrow (x - 3, y); y = 2$　　　　　**52.** $(x, y) \rightarrow (x, y - 3); y = 2$

53. $(x, y) \rightarrow (x - 3, y - 3); y = x$　　　　**54.** $(x, y) \rightarrow (x + 4, y - 4); y = -x$

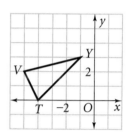
Multiple Choice

55. Find the image of P(11, −5) for the translation (x, y) → (x − 12, y − 6) followed by a reflection in x = 0.
  A. (1, −11)   **B.** (−1, 11)   **C.** (1, 11)   **D.** (−1, −11)

56. A reflection in the y-axis followed by a reflection in the x-axis does NOT give the same result as which of the following transformations?
  F. a reflection in the x-axis followed by a reflection in the y-axis
  G. a rotation of 180°
  H. a rotation of 90° followed by a reflection in the x-axis
  J. a reflection in the line y = x followed by a reflection in the line y = −x

Short Response

57. Find the image of △VTY for the given glide reflection. Show all your steps.
    translation: (x, y) → (x + 3, y − 3)
    reflection line: y = −x

Extended Response

58. Copy the diagram with s ∥ t.
  a. Draw the image of F for a composition of two reflections. Reflect first in line s and then in line t.
  b. Explain why the resulting image is the same image as found by translating F in a direction parallel to $\overline{PQ}$ through a distance 2 · PQ.

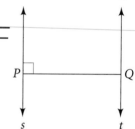

Mixed Review

GO for Help

Lesson 9-5

Coordinate Geometry Find the image of △ABC for a dilation with center (0, 0) and the scale factor given.

59. A(0, 4), B(0, 0), C(3, 0); 3

60. A(4, 2), B(0, 0), C(6, 0); 1.5

61. A(2, 3), B(−4, −2), C(5, −3); 2

62. A(7, 8), B(5, 4), C(9, 6); 0.5

Lesson 7-4 x^2 **Algebra** Find the value of x.

63. 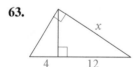   **64.**    **65.**

Lesson 5-4

Identify the two statements that contradict each other.

66. I. △ABC is right.
    II. △ABC is equiangular.
    III. △ABC is isosceles.

67. I. In right △ABC, m∠B = 90.
    II. In right △ABC, m∠A = 80.
    III. In right △ABC, m∠C = 90.

Kaleidoscopes

FOR USE WITH LESSON 9-6

The mirrors in a kaleidoscope provide compositions of reflections to create a *symmetrical* design. You can create your own kaleidoscope.

Construct

- Use geometry software. Draw a line and construct a point on the line. Rotate the line 60° about the point and repeat to get a third line.

- Construct a polygon in the interior of an angle, as shown. Reflect the polygon in the closest line, then in the next line, and so on until the kaleidoscope is filled and you have a symmetrical design. Then hide the lines of reflection.

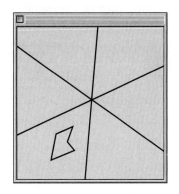
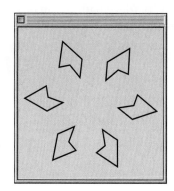

Investigate

- Manipulate the original figure by dragging any of its vertices or selecting and moving it. As you manipulate the figure, what happens to the images? Does the design remain symmetrical? Continue manipulating the original figure until you are satisfied with your design. Print the design and color it.

- Now add other figures beside the original polygon. Reflect these figures to create a more interesting design, as shown at the right. (You may need to temporarily show the lines of reflection.) Print your design and color it.

EXERCISES

1. Create a kaleidoscope with four lines of reflection. Draw a line and construct a point on the line. Rotate the line 45° about the point and repeat two more times to get four lines. Add a figure to the interior of an angle and reflect it as described above.

A *tessellation* is another type of interesting design that can be constructed using geometry software. An example is shown in Exercise 2 and tessellations are explained more fully in Lesson 9-7.

2. Construct a regular hexagon. Translate the hexagon several times to create the tessellation shown. Can you make a similar tessellation with other regular polygons? Can you find two or more polygons that, when placed together repeatedly, form a tessellation?

A frieze pattern is a design that repeats itself along a straight line. Frieze patterns are popular forms of design trim, often appearing along the edges of buildings or rooms.

Every frieze pattern can be mapped onto itself by a translation. Some frieze patterns can be mapped onto themselves by reflections.

A.

B.

Some frieze patterns have repeated rotational symmetry.

C.

D.

Some frieze patterns show glide reflections.

E.

F.

EXERCISES

For Exercises 1–4, refer to the frieze patterns above. You may find tracing paper helpful.

1. For each frieze pattern, find a portion (as small as possible) that you could translate repeatedly to form the entire pattern.

2. Which patterns show reflectional symmetry? Find their reflection lines.

3. Which patterns show rotational symmetry? Find their centers of rotation.

4. Which patterns show glide reflections? Find a glide translation and reflection line.

For each frieze pattern below, describe all the transformations that map the pattern onto itself.

5.

6.

7.

8.

9.

10.

9-7

Tessellations

What You'll Learn

- To identify transformations in tessellations, and figures that will tessellate
- To identify symmetries in tessellations

. . . And Why

To identify a tessellation in art, as in Example 1

✓ **Check Skills You'll Need**

GO> for Help Lesson 3-5

Classify the polygon with the given number of sides.

1. five   **2.** eight   **3.** twelve

Find the measure of an angle of each regular polygon.

4. triangle   **5.** quadrilateral   **6.** hexagon

7. octagon   **8.** decagon   **9.** 14-gon

🔊 **New Vocabulary**   • tessellation   • tiling   • translational symmetry
• glide reflectional symmetry

▼1 Identifying Transformations in Tessellations

Vocabulary Tip

A figure that creates a tessellation is said to <u>tessellate</u>.

A **tessellation,** or **tiling,** is a repeating pattern of figures that completely covers a plane, without gaps or overlaps. You can create tessellations with translations, rotations, and reflections. You can find tessellations in art (see below), nature (cells in a honeycomb), and everyday life (tiled floors).

1 EXAMPLE **Identifying the Transformation in a Tessellation**

Art Identify a transformation and the repeating figures in this tessellation.

Repeating figures

The arrow shows a translation.

✓ **Quick Check** **1** Identify a transformation and the repeating figures in each tessellation below.

a.

b.

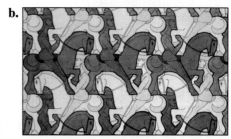

Because the figures in a tessellation do not overlap or leave gaps, the sum of the measures of the angles around any vertex must be 360°. If the angles around a vertex are all congruent, then the measure of each angle must be a factor of 360.

2 EXAMPLE Determining Figures That Will Tessellate

Determine whether a regular 18-gon tessellates a plane.

$a = \dfrac{180(n - 2)}{n}$ **Use the formula for the measure of an angle of a regular polygon.**

$a = \dfrac{180(18 - 2)}{18}$ **Substitute 18 for *n*.**

$a = 160$ **Simplify.**

● Since 160 is not a factor of 360, the 18-gon will not tessellate.

✓ **Quick Check** ② Explain why you can tessellate a plane with an equilateral triangle.

A figure does not have to be a regular polygon to tessellate.

🔑 **Key Concepts**

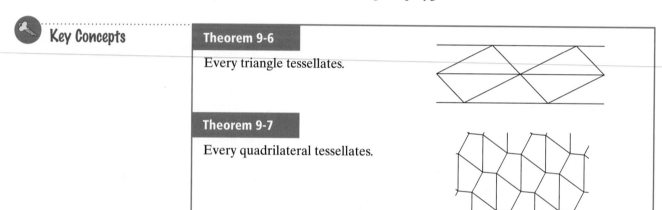

Theorem 9-6
Every triangle tessellates.

Theorem 9-7
Every quadrilateral tessellates.

2 Identifying Symmetries in Tessellations

The tessellation with regular hexagons at the right has reflectional symmetry in each of the blue lines. It has rotational symmetry centered at each of the red points. The tessellation also has translational symmetry and glide reflectional symmetry, as shown below.

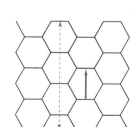

Translational Symmetry

A translation maps the tessellation onto itself.

Glide Reflectional Symmetry

A glide reflection maps the tessellation onto itself.

3 EXAMPLE **Identifying Symmetries in Tessellations**

List the symmetries in the tessellation.

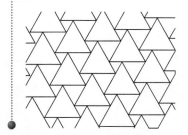

Rotational symmetry centered at each red point
Translational symmetry (blue arrow)

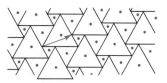

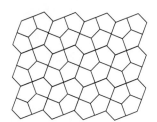

Quick Check ❸ List the symmetries in the tessellation at the right.

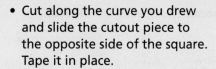

The following Activity shows the steps for making creative tessellations.

Hands-On Activity: Creating Tessellations

• Draw a 1.5-inch square on a blank piece of paper and cut it out.

• Draw a curve joining two consecutive vertices.

• Cut along the curve you drew and slide the cutout piece to the opposite side of the square. Tape it in place.

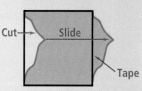

• Repeat this process using the other two opposite sides of the square.

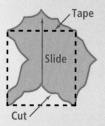

• Rotate the resulting figure. What does your imagination suggest it looks like? Is it a penguin wearing a hat or a knight on horseback? Could it be a dog with floppy ears? Draw the image on your figure.

• Create a tessellation using your figure.

EXERCISES

For more exercises, see *Extra Skill, Word Problem, and Proof Practice.*

Practice and Problem Solving

A **Practice by Example**

Example 1
(page 515)

Does the picture show a tessellation of repeating figures? If so, identify a transformation and the repeating figure.

1.

2.

3.

4.
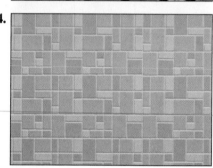

Example 2
(page 516)

Determine whether each figure will tessellate a plane.

5. equilateral triangle   6. square   7. regular pentagon

8. regular heptagon   9. regular octagon   10. regular nonagon

Example 3
(page 517)

List the symmetries in each tessellation.

11.
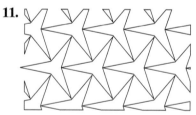

12.
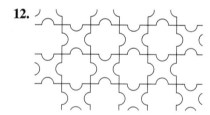

13.
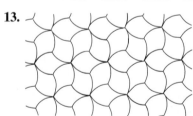

14.
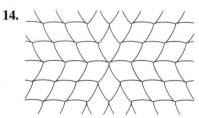

B **Apply Your Skills**

Use each figure to create a tessellation on dot paper.

15.
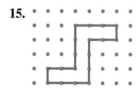

16.
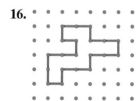

17.
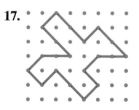

18. Multiple Choice Which jigsaw puzzle piece can tessellate a plane using *only* translation images of itself?

Show how to tessellate with each figure described below. Try to draw two different tessellations. If you think that two are not possible, explain.

19. a scalene triangle

20. the pentagon at the right

21. a quadrilateral with no sides parallel or congruent

22. Writing A *pure tessellation* is a tessellation made up of congruent copies of one figure. Explain why there are three, and only three, pure tessellations that use regular polygons. (*Hint:* See Exercises 5–10.)

Decide whether a semiregular tessellation (see photo) is possible using the given pair of regular polygons. If so, draw a sketch.

23.

24.

A *semiregular tessellation* is made from two or more regular polygons.

Can each set of polygons be used to create a tessellation? If so, draw a sketch.

25.

26.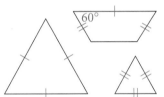

List the symmetries of each tessellation.

27.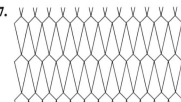

28.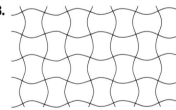

⊙ Challenge

Copy the Venn diagram. Write each exercise number in the correct region of the diagram.

Regular figures   Polygons

29. scalene triangle   **30.** obtuse triangle

31. equilateral △   **32.** isosceles △

33. kite   **34.** rhombus

35. square   **36.** regular pentagon

37. regular hexagon   **38.** regular octagon   **39.** regular decagon

Figures that tessellate

40. On graph paper, draw quadrilateral $ABCD$ with no two sides congruent. Locate M, the midpoint of $\overline{AB}$, and N, the midpoint of $\overline{BC}$.
   a. Draw the image of $ABCD$ under a 180° rotation about M.
   b. Draw the image of $ABCD$ under a 180° rotation about N.
   c. Draw the image of $ABCD$ under the translation that maps D to B.
   d. Make a conjecture about whether your quadrilateral tessellates, using the pattern in parts (a)–(c). Justify your answer.

41. List steps (like those in Exercise 40) that suggest a way to tessellate with any scalene triangle. Then list a second set of steps that suggest another way.

Test Prep

NY REGENTS

Multiple Choice

42. Which figure will NOT tessellate a plane?
   A.   **B.**   **C.**   **D.**

43. You can tessellate a plane using a regular octagon together with which other type of regular polygon?
   F. triangle   **G.** square   **H.** pentagon   **J.** hexagon

44. Which is NOT a symmetry for the tessellation?
   A. line symmetry
   B. translational symmetry
   C. rotational symmetry
   D. glide reflectional symmetry

Short Response

45. Is it possible to tile a plane with regular pentagons? Justify your answer.

Extended Response

46. Unit squares form this tessellation. Tell whether this tessellation has each type of symmetry (line, point, rotational, translational, or glide reflectional). Explain.

Mixed Review

Lesson 9-6

47. A triangle has vertices $A(3, 2)$, $B(4, 1)$, and $C(4, 3)$. Find the coordinates of the images of A, B, and C for a glide reflection with translation $(x, y) \rightarrow (x, y + 1)$ and reflection line $x = 0$.

Lesson 8-1

Algebra Find the value of x.

$\boxed{x^2}$ **48.**

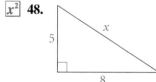

49.

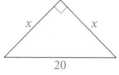

50.

Lesson 5-5

The lengths of two sides of a triangle are given. What are the possible lengths for the third side?

51. 16 in., 26 in.   **52.** 19.5 ft, 20.5 ft   **53.** 9 m, 9 m

54. 2 cm, 7 cm   **55.** $4\frac{1}{2}$ yd, 8 yd   **56.** 1 km, 2 km

The two figures in each pair are congruent. Is one figure a translation image, a rotation image, or a reflection image of the other? Explain.

1.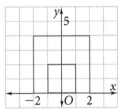

2.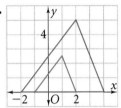

The blue figure is a dilation image of the red figure. Describe the dilation.

3.

4.

Tell what type(s) of symmetry (line, rotational, or point) each figure has. For line symmetry, sketch the figure and the line(s) of symmetry. For rotational symmetry, state the angle of rotation.

5.   6.   7.   8.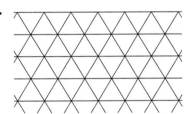

List the symmetries of each tessellation.

9.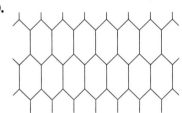

10.

.... **A Point in Time**

1500   1600   1700   1800   1900   2000

A mosaic is a picture or design made by setting tiny pieces of glass, stone, or other materials in clay or plaster. A mosaic may be a tessellation. Most mosaics, however, do not have a repeating pattern of figures. Mosaics go back at least 6000 years to the Sumerians, who used tiles to both decorate and reinforce walls.

During 100 and 200 A.D., Roman architects used two million tiles to create the magnificent mosaic of Dionysus in Germany. In the years 1941–1951, Mexican artist Juan O'Gorman covered all four sides of a 10-story library in Mexico with 7.5 million stones—the largest mosaic ever. It depicts Mexico's cultural history.

Go **O**nline   **For:** Information about mosaics
PHSchool.com **Web Code:** aue-2032

Answering the Question Asked

When answering a question, be sure to answer the question that is asked.
Read the question carefully and identify the quantity that you are asked to find.
Some answer choices are answers to related questions, so you have to be careful.

1 EXAMPLE

The point $L(a, b)$ in Quadrant I is rotated 90° about the origin and is then reflected across the y-axis. What is the x-coordinate of the image?

A a B $-b$ C b D $-a$ E 0

The question asks for the x-coordinate of the image. The x-coordinate of the preimage is a. After the 90° rotation about the origin, the x-coordinate is $-b$. After a reflection across the y-axis, the x-coordinate of the image is b.

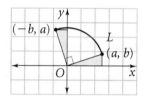 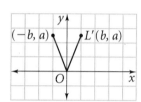

The correct answer is C. Choice A is the y-coordinate of the image, but that is not what is asked for.

2 EXAMPLE

What is the image of $M(4, 3)$ for the translation $(x, y) \rightarrow (x - 2, y + 1)$ followed by a reflection across the line $x = -1$?

A $(2, -6)$ B $(-4, 4)$ C $(0, 4)$ D $(-8, 4)$ E $(2, 4)$

The translation puts the point at $(2, 4)$. A reflection across $x = -1$ then puts the point at $(-4, 4)$. The correct answer is B.

Answers A and C result from the translation followed by a reflection across a line other than $x = -1$. In fact, if you reflect $(2, 4)$ across $x = 1$, which is one unit in the negative direction from $(2, 4)$, you get $(0, 4)$, or choice C, as your answer. If you do the steps in the wrong order, you will get answer D.

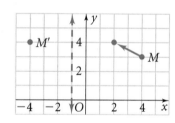

EXERCISES

Answer the question. Then ask a related question that would lead to an incorrect choice.

1. The point $K(4, 3)$ is rotated 90° about the origin. What is the sum of the x- and y-coordinates of the image?
   A -7 B -1 C 0 D 1 E 7

2. The point $P(1, 3)$ is rotated 180° about the origin, and then reflected across the line $y = 0$. What is the y-coordinate of the image?
   A -3 B -1 C 0 D 1 E 3

3. The point $T(5, -1)$ is reflected in the y-axis. What is the distance between the image and the preimage?
   A 1 B 2 C $\sqrt{26}$ D 10 E 12

Chapter Review

Vocabulary Review

center of a regular polygon (p. 484)
composition (p. 472)
dilation (p. 498)
enlargement (p. 498)
glide reflection (p. 508)
glide reflectional symmetry (p. 516)
image (p. 470)
isometry (p. 470)

line symmetry (p. 492)
point symmetry (p. 493)
preimage (p. 470)
reduction (p. 498)
reflection (p. 478)
reflectional symmetry (p. 492)
rotation (p. 484)

rotational symmetry (p. 493)
symmetry (p. 492)
tessellation (p. 515)
tiling (p. 515)
transformation (p. 470)
translation (p. 471)
translational symmetry (p. 516)

To complete each definition, find the appropriate word in the second column.

1. A(n) __?__ is a change in position, shape, or size of a figure.

2. A(n) __?__ is a transformation in which the preimage and its image are congruent.

3. A __?__ is an isometry in which a figure and its image have opposite orientations.

4. A __?__ is an isometry in which all points of a figure move the same distance in the same direction.

5. A(n) __?__ is a translation followed by a reflection in a line parallel to the translation vector.

6. A(n) __?__ is a repeating pattern of figures that completely covers a plane, without gaps or overlaps.

7. A(n) __?__ is a transformation that proportionally reduces or enlarges a figure.

A. dilation

B. glide reflection

C. tessellation

D. isometry

E. reflection

F. transformation

G. translation

Go Online
PHSchool.com

For: Vocabulary quiz
Web Code: auj-0951

Skills and Concepts

9-1 and 9-2 Objectives

▼ To identify isometries

▼ To find translation images of figures

▼ To find reflection images of figures

A **transformation** of a geometric figure is a change in its position, shape, or size. An **isometry** is a transformation in which the **preimage** and **image** are congruent. A transformation maps a figure onto its image.

A **translation** is an isometry that maps all points of a figure the same distance in the same direction. A translation is an isometry that does not change orientation.

The second diagram shows a **reflection** of B to B' across line r. A reflection is an isometry in which a figure and its image have opposite orientations.

A **composition** of transformations is a combination of two or more transformations. Each transformation is performed on the image of the preceding transformation.

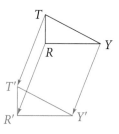

Use matrices to find the image of each triangle for the given translation.

8. △ABC with vertices $A(5, 9), B(6, 3), C(1, 2)$;
translation: $(x, y) \rightarrow (x + 2, y + 3)$

9. △RST with vertices $R(0, -4), S(-2, -1), T(-6, 1)$;
translation: $(x, y) \rightarrow (x - 4, y + 7)$

Find a single translation that has the same effect as each composition.

10. $(x, y) \rightarrow (x - 5, y + 7)$ followed by $(x, y) \rightarrow (x + 3, y - 6)$

11. $(x, y) \rightarrow (x + 10, y - 9)$ followed by $(x, y) \rightarrow (x + 1, y + 5)$

Given points $A(6, 4), B(-2, 1),$ and $C(5, 0)$, draw △ABC and its reflection image in each line.

12. the x-axis   13. $x = 4$ 14. $y = x$

9-3 Objective

▼ To draw and identify rotation images of figures

The diagram shows a rotation of point V about point R through $x°$. A **rotation** is an isometry that does not change orientation.

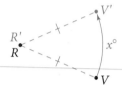

Copy each figure and point P. Draw the image of each figure for the given rotation about P. Label the vertices of the image.

15. 180°   16. 60°   17. 90°

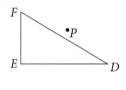   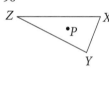

Find the image of each point for a 90° rotation about the origin.

18. $(5, 2)$ 19. $(0, 3)$ 20. $(-4, 1)$ 21. $(7, 0)$ 22. $(-2, -8)$

9-4 Objective

▼ To identify the type of symmetry in a figure

A figure has **symmetry** if there is an isometry that maps the figure onto itself. A plane figure has **reflectional symmetry,** or **line symmetry,** if one half of the figure is a mirror image of its other half. A figure that has **rotational symmetry** is its own image for some rotation of 180° or less. A figure that has **point symmetry** has 180° rotational symmetry.

Point Symmetry

Tell what type(s) of symmetry each figure has. If it has rotational symmetry, state the angle of rotation.

23.   24.

9-5 Objective

▼ To locate dilation images of figures

The diagram shows a **dilation** with center C and scale factor n. A dilation is a similarity transformation because its preimage and image are similar figures. When the scale factor is greater than 1, the dilation is an **enlargement.** When the scale factor is between 0 and 1, the dilation is a **reduction.** In the coordinate plane, you can use **scalar multiplication** to find the image of a figure under a dilation centered at the origin.

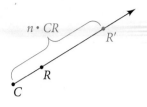

A dilation has center (0, 0). Find the image of each point for the scale factor given.

25. $A(0, 3); 4$ **26.** $B(-2, 6); 0.5$ **27.** $C(1.5, -2); 10$

Find the image of each set of points for a dilation with center at the origin and the scale factor given.

28. $M(-3, 4), A(-6, -1), T(0, 0), H(3, 2)$; scale factor 5

29. $F(-4, 0), U(5, 0), N(-2, -5)$; scale factor $\frac{1}{2}$

9-6 Objectives

▼ To use a composition of reflections

▼ To identify glide reflections

A composition of reflections in two parallel lines is a translation. A composition of reflections in two intersecting lines is a rotation. A **glide reflection** is the composition of a glide (translation) and a reflection in a line parallel to the translation vector. The only four isometries are reflection, translation, rotation, and glide reflection.

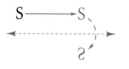

For the figure at the left below, four isometry images are shown. Tell whether orientations are the same or opposite. Then classify the isometry.

30. **31.** **32.** **33.**

34. $\triangle TAM$ has vertices $T(0, 5), A(4, 1),$ and $M(3, 6)$. Find the image of $\triangle TAM$ where the translation is $(x, y) \rightarrow (x - 4, y)$ and the reflection is in the line $y = -2$.

9-7 Objectives

▼ To identify transformations in tessellations and figures that will tessellate

▼ To identify symmetries in tessellations

A **tessellation,** or **tiling,** is a repeating pattern of figures that completely covers a plane, without gaps or overlaps. A tessellation can have **translational symmetry** if there is a translation that maps the tessellation onto itself. If a tessellation can be mapped onto itself by a glide reflection, then the tessellation has **glide reflectional symmetry.**

For each tessellation, (a) identify a transformation and the repeating figures, and (b) list the symmetries.

35. **36.**

Chapter 9

Chapter Test

Go Online
PHSchool.com
For: Chapter Test
Web Code: aua-0952

Find the coordinates of the vertices of the image of *ABCD* for each transformation.

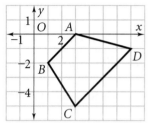

1. reflection across the line $x = -4$

2. translation $(x, y) \rightarrow (x - 6, y + 8)$

3. rotation of 90° about the point $(0, 0)$

4. dilation centered at $(0, 0)$ with scale factor $\frac{2}{3}$

5. glide reflection with translation $(x, y) \rightarrow (x, y + 3)$ and reflection across the line $x = 0$

6. reflection across the line $y = x$

7. rotation of 270° about $(0, 0)$

8. dilation centered at the origin with scale factor 5

9. glide reflection with translation $(x, y) \rightarrow (x - 2, y)$ and reflection across the line $y = 5$

10. translation 3 units right and 1 unit down

What type of transformation has the same effect as each composition of transformations?

11. translation $(x, y) \rightarrow (x + 4, y)$ followed by a reflection across the line $y = -4$

12. translation $(x, y) \rightarrow (x + 4, y + 8)$ followed by $(x, y) \rightarrow (x - 2, y + 9)$

13. reflection across the line $y = 7$, and then across the line $y = 3$

14. reflection across the line $y = x$, and then across the line $y = 2x + 5$

Draw a figure that has each type of symmetry.

15. reflectional   16. rotational   17. point

What type(s) of symmetry does each figure have?

18.    19.

20. **Writing** Line *m* intersects $\overline{UH}$ at *N*, and $UN = NH$. Must *H* be the reflection image of *U* across line *m*? Explain.

21. Describe the symmetries of this tessellation. Copy a portion of the tessellation and draw any centers of rotational symmetry or lines of symmetry.

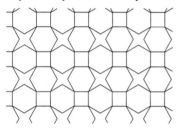

Does each letter tessellate? If so, sketch a tessellation. If not, explain why it cannot tessellate.

22.   23.   24. K

Find the image of $\triangle ABC$ for a dilation with center $(0, 0)$ and the scale factor given.

25. $A(-2, 2), B(2, -2), C(3, 4)$; scale factor 3

26. $A(0, 0), B(-3, 2), C(1, 7)$; scale factor $\frac{1}{2}$

27. The blue figure is a translation image of the red figure. Write a rule to describe the translation.

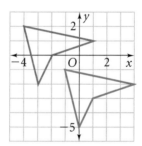

28. A dilation with center $(0, 0)$ and scale factor 2.5 maps $(4, -10)$ to (a, b). Find the values of *a* and *b*.

29. A dilation maps $\triangle LMN$ to $\triangle L'M'N'$. Find the missing values.

$LM = 36$ ft, $LN = 26$ ft, and $MN = 45$ ft; $L'M' = 9$ ft, $L'N' = \blacksquare$ ft, and $M'N' = \blacksquare$ ft; scale factor = $\blacksquare$

30. A dilation with scale factor 4 maps square A onto square B. The area of square B is 25. Find the area of square A.

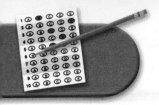

Regents Test Prep

Reading Comprehension Read the passage below. Then answer the questions on the basis of what is *stated* or *implied* in the passage.

Hanging a Picture A picture has a wire from side to side across its back. Hang the picture from one hook and it can easily swing, or slide into a tilt. Use two hooks and the picture will hang level. Here is how to place two hooks on the wall to hang the picture level and precisely where you want it.

On your wall, mark two level points *A* and *B* where you want the top corners of the picture. For example, assume the back of a 21-in. wide picture is rigged as shown at the left and you want the two hooks 12 in. apart.

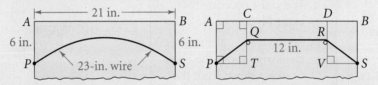

At the right, the matching diagrams in the upper corners locate points *Q* and *R* where hooks would hold the wire. Determine measurements as follows:

Calculate to find that $QP = 5.5$ in. $= RS$.
Calculate to find that $TP = 4.5$ in. $= VS$.
Then $QT = \sqrt{5.5^2 - 4.5^2} \approx 3.2$ in. $= RV$, so $CQ = 2.8$ in. $= DR$.

From points *A* and *B* on the wall, measure 4.5 in. towards each other and 2.8 in. down to find points *Q* and *R*, respectively. Attach picture hangers to support the wire at *Q* and *R*, and hang your picture perfectly!

---

1. From the passage, what should you learn about hanging a picture?

   (1) How to use one hook so that you can easily slide the picture to hang straight.

   (2) How to use two hooks so that the picture hangs straight and where you want it.

   (3) How to use one hook so that the picture can easily swing, or slide into a tilt.

   (4) How to use two hooks at the top corners.

2. How do you calculate *QP*?

   (1) $\dfrac{\text{wire length} - 12}{2}$ **(2)** $\dfrac{PQ + RS}{2}$

   (3) $\dfrac{AB - 10}{2}$ **(4)** $\dfrac{PT + SV + 2}{2}$

3. How do you calculate *TP*?

   (1) $\dfrac{\text{wire length} - 14}{2}$ **(2)** $\dfrac{PQ + RS - 2}{2}$

   (3) $\dfrac{AB - 12}{2}$ **(4)** $\dfrac{PT + SV}{2}$

4. Which theorem do you use to calculate *QT*?

   (1) 30°-60°-90° Triangle   **(2)** Pythagorean

   (3) 45°-45°-90° Triangle   **(4)** Triangle Midsegment

5. How do you calculate *CQ*?

   (1) $CQ = DR$ **(2)** $CQ + QP = 8.3$

   (3) $CQ = 6 - CQ$ **(4)** $CQ + QT = CT$

6. What kind of quadrilateral is *DBSV*? Justify your answer.

Describe how to locate the hooks for hanging.

7. A picture is 30 in. wide. The hanging wire is 34 in. long, attached at the sides of the picture, 9 in. from the top. The hooks are 14 in. apart.

8. A circular mirror has diameter 22 in. The hanging wire is 28 in. long, attached at the endpoints of a diameter. The hooks are 10 in. apart.

Activity Lab

How'd They Do That?

Applying Translations and Rotations To create computer-generated characters and objects, computer animators and designers first define every point of a wire-frame model within a three-dimensional coordinate system. To make the model move and rotate, the animators must move and rotate its points.

Wire-frame wings stretch out in flight or fold in close to the dragon's body.

 G.CN.4:
Understand how concepts, procedures, and mathematical results in one area of mathematics can be used to solve problems in other areas of mathematics.

Dragon Lore

Unlike European dragons, which breathe fire and wreak havoc across the countryside, Asian dragons are peace-loving protectors of the heavens.

Tail shape changes as points move and lines stretch.

Activity

Use the diagram at the right.

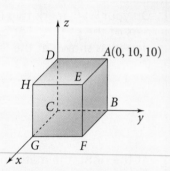

a. Write the coordinates of all eight vertices of the cube.

b. Suppose the cube rotates 90° clockwise about the z-axis (looking down from the positive z-axis). Find the new coordinates of vertices A–H.

c. Starting from the cube's position at the end of part (b), rotate the cube 90° clockwise about the x-axis (looking toward the origin from the positive x-axis). Find the new coordinates of vertices A–H.

d. Describe a composition of rotations that will move point E from its original location to $(-10, -10, -10)$.

e. Open-Ended You can also find a composition of translations, each parallel to an axis, to move point E from its original location to the original location of point C, $(0, 0, 0)$. Describe compositions of rotations (about axes) and translations (parallel to axes) that move point E to the locations of two vertices of the cube. Let each composition include at least one translation and one rotation.

f. Suppose the cube returns to its original position and then rotates 30° clockwise about the y-axis (looking toward the origin from the positive y-axis). Find the new coordinates of the eight vertices. (*Hint:* Use trigonometric ratios.)

Muscle structure added to frame

Exterior skin detail added and placed within live-action footage

Mouth and snout digitally animated

Eyes are modeled on those of a lizard— they move, blink, open, and close.

Go Online
PHSchool.com

For: Information about computer animation
Web Code: aue-0953

What You've Learned

NY Learning Standards for Mathematics

- **G.PS.6:** In Chapter 1, you learned two postulates about area and how to find the areas of rectangles and circles.

- **G.CM.11:** In Chapters 3 and 6, you learned how to classify polygons, including special quadrilaterals.

- **G.G.28:** In Chapter 4, you learned conditions necessary for two polygons, particularly triangles, to be congruent.

 Check Your Readiness   **for Help** to the Lesson in green.

Squaring Numbers and Finding Square Roots (Skills Handbook page 753)

Simplify.

1. 3^2 **2.** 8^2 **3.** 12^2 **4.** 15^2

5. $\sqrt{16}$ **6.** $\sqrt{64}$ **7.** $\sqrt{100}$ **8.** $\sqrt{169}$

Solve each quadratic equation. Round to the nearest tenth or whole number.

9. $x^2 = 36$ **10.** $a^2 = 104$ **11.** $x^2 - 48 = 0$ **12.** $b^2 - 65 = 0$

Simplifying Radicals (Skills Handbook page 755)

Simplify. Leave your answer in simplest radical form.

13. $\sqrt{8}$ **14.** $\sqrt{27}$ **15.** $\sqrt{48}$ **16.** $6\sqrt{72}$

Probability (Skills Handbook page 762)

A jar contains 3 red balls and 2 green balls. You draw one ball at random. Determine the probability of selecting a ball of the given color.

17. red   **18.** green   **19.** blue   **20.** red or green

Classifying Quadrilaterals (Lesson 6-1)

Classify each quadrilateral as specifically as possible.

21.   **22.**    **23.**

Area

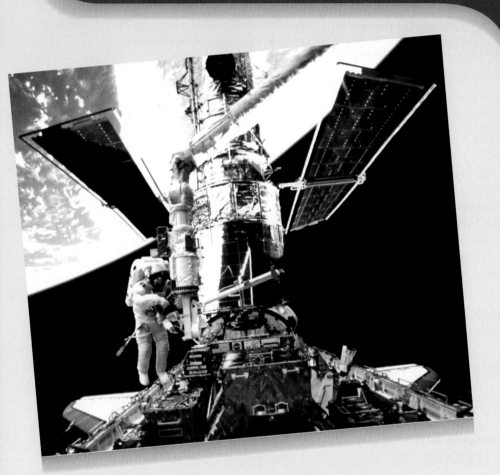

◀)) **Key Vocabulary**

What You'll Learn Next

NY Learning Standards for Mathematics

In this chapter, you will learn how finding the area of a rectangle can help you find the areas of parallelograms and triangles.

- You will learn how to find the areas of special quadrilaterals and regular polygons.

- You will learn how perimeters and areas of similar figures are related.

- You will also learn how to find measurements of parts of circles.

Data Analysis **Activity Lab** Applying what you learn, you will investigate on pages 594 and 595 a relationship between area and the graphical representation of data.

531

Transforming to Find Area

You can use transformations to find area formulas.

In these activities, you will cut polygons into pieces. You will use isometry transformations on the pieces to form different polygons. Because a preimage and its image are congruent and congruent figures have the same area (Postulate 1-9), you can find area formulas for the new polygons.

You will need several pieces of grid paper. Let the side of each grid square represent one unit of length.

1 ACTIVITY

- For the parallelogram shown here, count and record the number of units in the base of the parallelogram. Do the same for the height.
- Copy the parallelogram onto grid paper.
- Cut out the parallelogram. Then cut it into two pieces as shown.
- Translate the triangle to the right through a distance equal to the base of the parallelogram.

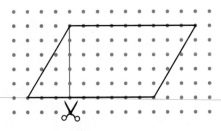

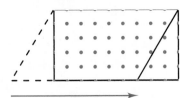

The translation image is a rectangle. The parallelogram and the rectangle have the same area.

1. For the rectangle, how many units are in the base? The height?

2. How do the base and the height of the rectangle compare to the base and height of the parallelogram?

3. Write the formula for the area of the rectangle. Explain why you can use this formula to find the area of a parallelogram.

2 ACTIVITY

Now you have the formula for the area of a parallelogram. You can use transformations and this formula to find an area formula for a triangle.

- Count and record the base and height of the triangle.
- Copy the triangle. Mark the midpoints A and B and draw the midsegment $\overline{AB}$.
- Cut out the triangle. Then cut it along $\overline{AB}$.
- Rotate the small triangle 180° about point B.

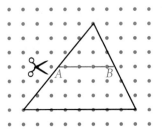

The bottom part of the original triangle and the image of the top part form a parallelogram.

4. For the parallelogram, how many units are in the base? The height?

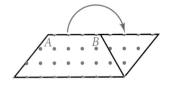

5. How do the base and height of the parallelogram compare to the base and height of the original triangle? Write an expression for the height of the parallelogram in terms of the height, h, of the triangle.

6. Write your formula for the area of a parallelogram from Exercise 3. To find an area formula for a triangle, substitute into the formula the expression you wrote in Exercise 5 for the height of the parallelogram.

3 ACTIVITY

You can find an area formula for a trapezoid using a transformation similar to the one you used for a triangle.

- Draw a trapezoid like the one shown here.
- Count and record its bases and height.
- Find the midpoints of the legs and the midsegment of the trapezoid.
- Cut out the trapezoid. Then cut it along the midsegment.

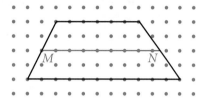

7. What transformation can you apply to the top piece of the trapezoid to form a parallelogram?

8. a. Write an expression for the base of the parallelogram in terms of the two bases, b_1 and b_2, of the trapezoid.
b. Write an expression for the height of the parallelogram in terms of h, the height of the trapezoid.
c. To find an area formula for a trapezoid, substitute your answers for parts (a) and (b) into your area formula for a parallelogram.

EXERCISES

9. In Activity 2, is there another rotation of the small triangle that will form a parallelogram? Explain.

10. Make another copy of the Activity 2 triangle. Mark the midpoints A and B. Find a rotation of the entire triangle so that the preimage and image together form a parallelogram. Show how to use it and your formula for the area of a parallelogram to find the formula for the area of a triangle.

11. For Exercise 10, there are in fact three rotations of the entire triangle that you can use. Find all three and describe them.

12. Here is the trapezoid with a different cut. What transformation can you apply to the top piece to form a triangle from the trapezoid? Use your formula for the area of a triangle to find a formula for the area of a trapezoid.

13. Show how you can find an area formula for a kite using a reflection. (*Hint:* Reflect half of the kite across its line of symmetry d_1 by folding the kite along d_1. How is the area of the triangle formed related to the area of the kite?)

10-1

Areas of Parallelograms and Triangles

What You'll Learn

- To find the area of a parallelogram
- To find the area of a triangle

. . . And Why

To find the force of wind against the side of a building, as in Example 4

 Check Skills You'll Need

 for Help Lesson 1-9

Find the area of each figure.

1. a square with 5-cm sides

2. a rectangle with base 4 in. and height 7 in.

3. a 4.6 m-by-2.5 m rectangle

4. a rectangle with length 3 ft and width $\frac{1}{2}$ ft

Each rectangle is divided into two congruent triangles. Find the area of each triangle.

5.

6.

7.

◀)) **New Vocabulary**
- base of a parallelogram
- height of a parallelogram
- height of a triangle
- altitude of a parallelogram
- base of a triangle

---

1   Area of a Parallelogram

The diagrams at the top of page 532 show that a parallelogram with the same base and height as a rectangle has the same area as the rectangle.

 Key Concepts

Theorem 10-1   **Area of a Rectangle**
The area of a rectangle is the product of its base and height. $A = bh$

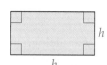

Theorem 10-2   **Area of a Parallelogram**
The area of a parallelogram is the product of a base and the corresponding height. $A = bh$

 Vocabulary Tip

The term <u>base</u> is used to represent both a segment and its length.

A **base of a parallelogram** is any of its sides. The corresponding **altitude** is a segment perpendicular to the line containing that base, drawn from the side opposite the base. The **height** is the length of an altitude.

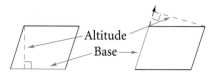

534   Chapter 10   Area

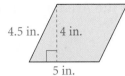 **Finding the Area of a Parallelogram**

Find the area of each parallelogram.

a.

4.5 in.   4 in.

5 in.

b.

4.6 cm   3.5 cm

2 cm

You are given each height. Choose the corresponding side to use as the base.

$$A = bh$$
$$= 5(4) = 20$$ ←**Substitute.**→

$$A = bh$$
$$= 2(3.5) = 7$$

The area is 20 in.^{2}.   The area is 7 cm^2.

 **Quick Check** ❶ Find the area of a parallelogram with base 12 m and height 9 m.

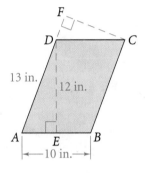
❷ **EXAMPLE** **Finding a Missing Dimension**

For □$ABCD$, find CF to the nearest tenth.

First, find the area of □$ABCD$. Then use the area formula a second time to find CF.

$$A = bh$$
$$= 10(12) = 120$$ **Use base *AB* and height *DE*.**

The area of □$ABCD$ is 120 in.^{2}.

$$A = bh$$
$$120 = 13(CF)$$ **Use base *AD* and height *CF*.**
$$CF = \frac{120}{13} \approx 9.2$$

CF is about 9.2 in.

F

D   C

13 in.   12 in.

A   E   B

←—10 in.—→

 **Quick Check** ❷ A parallelogram has sides 15 cm and 18 cm. The height corresponding to a 15-cm base is 9 cm. Find the height corresponding to an 18-cm base.

---

2 Area of a Triangle

You can rotate a triangle about the midpoint of a side to form a parallelogram.

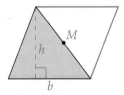

The area of the triangle is half the area of the parallelogram.

Key Concepts

Theorem 10-3	Area of a Triangle

The area of a triangle is half the product
of a base and the corresponding height.

$$A = \tfrac{1}{2}bh$$

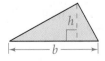

A **base of a triangle** is any of its sides. The corresponding **height** is the length of the altitude to the line containing that base.

3 EXAMPLE Finding the Area of a Triangle

Find the area of the triangle.

$$A = \tfrac{1}{2}bh$$
$$= \tfrac{1}{2}(10)(6.4) = 32 \quad \textbf{Substitute and simplify.}$$

● The area of the triangle is 32 ft^2.

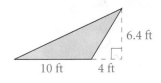

✓ **Quick Check** ③ Find the area of the triangle.

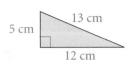

4 EXAMPLE Real-World  Connection

Structural Design When designing a building, you must be sure that the building can withstand hurricane-force winds, which have a velocity of 73 mi/h or more. The formula $F = 0.004Av^2$ gives the force F in pounds exerted by a wind blowing against a flat surface. A is the area of the surface in square feet, and v is the wind velocity in miles per hour.

How much force is exerted by a 73 mi/h wind blowing directly against the side of the building shown here?

Find the area of the side of the building.
  triangle area $= \tfrac{1}{2}bh = \tfrac{1}{2}(20)6 = 60$ ft^2
  rectangle area $= bh = 20(12) = 240$ ft^2
  area of the side $= 60 + 240 = 300$ ft^2

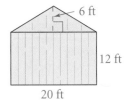

Use the area of the side of the building and the velocity of the wind to find the force.

$$F = 0.004Av^2 \qquad \textbf{Use the formula for force.}$$
$$= 0.004(300)(73)^2 \quad \textbf{Substitute 300 for } A \textbf{ and 73 for } v.$$
$$= 6394.8$$

The force is about 6400 lb, or 3.2 tons.

✓ **Quick Check** ④ **Critical Thinking** Suppose the bases of the rectangle and triangle in the building above are doubled to 40 ft, but the height of each figure remains the same. How is the force of the wind against the side of the building affected?

EXERCISES
For more exercises, see *Extra Skill, Word Problem, and Proof Practice.*

Practice and Problem Solving

A **Practice by Example**

GO for Help

Example 1
(page 535)

Find the area of each parallelogram.

1.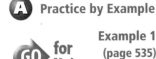

2.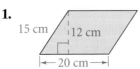

3.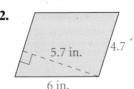

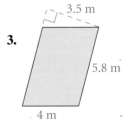

Example 2
(page 535)

Find the value of *h* for each parallelogram.

4.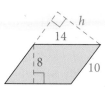

5.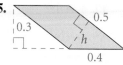

6.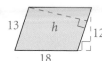

Example 3
(page 536)

Find the area of each triangle.

7.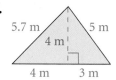

8.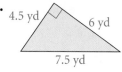

9.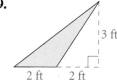

Example 4
(page 536)

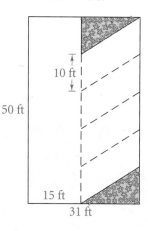

10. **Landscaping** Taisha's Bakery has a plan for a 50 ft-by-31 ft parking lot. The four parking spaces are congruent parallelograms, the driving region is a rectangle, and the two unpaved areas for flowers are congruent triangles.
 a. Find the area of the surface to be paved by adding the areas of the driving region and the four parking spaces.
 b. Describe another method for finding the area of the surface to be paved.
 c. Use your method from part (b) to find the area. Then compare answers from parts (a) and (b) to check your work.

 **Apply Your Skills**

11. The area of a parallelogram is 24 in.^{2} and the height is 6 in. Find the corresponding base.

12. **Multiple Choice** What is the area of the figure at the right?
 Ⓐ 64 cm^2 Ⓑ 88 cm^2
 Ⓒ 96 cm^2 Ⓓ 112 cm^2

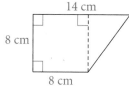

13. An isosceles right triangle has area of 98 cm^2. Find the length of each leg.

x^2 14. **Algebra** In a triangle, a base and a corresponding height are in the ratio 3 : 2. The area is 108 in.^{2}. Find the base and the corresponding height.

15. **Technology** Ki used geometry software to create the figure at the right. She constructed $\overleftrightarrow{AB}$ and a point *C* not on $\overleftrightarrow{AB}$. Then she constructed line *k* parallel to $\overleftrightarrow{AB}$ through point *C*. Next, Ki constructed point *D* on line *k* as well as $\overline{AD}$

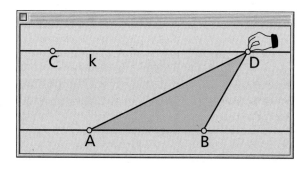

and $\overline{BD}$. She dragged point *D* along line *k* to manipulate $\triangle ABD$. How does the area of $\triangle ABD$ change? Explain.

16. **Open-Ended** Using graph paper, draw an acute triangle, an obtuse triangle, and a right triangle, each with area 12 units^2.

GO Online
Homework Video Tutor
Visit: PHSchool.com
Web Code: aue-1001

Find the area of each figure.

17. $\square ABJF$ **18.** $\triangle BDJ$

19. $\triangle DKJ$ **20.** $\square BDKJ$

21. $\square ADKF$ **22.** $\triangle BCJ$

23. $ADJF$

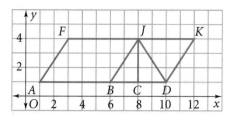

In Exercises 24–27, (a) graph the lines and (b) find the area of the triangle enclosed by the lines.

24. $y = x, x = 0, y = 7$ **25.** $y = x + 2, y = 2, x = 6$

26. $y = -\frac{1}{2}x + 3, y = 0, x = -2$ **27.** $y = \frac{3}{4}x - 2, y = -2, x = 4$

28. Find the area of the yellow triangular patch in the large field in the photo at the left. It has a base of 60 yd and a height of 140 yd.

29. Probability Ann drew these three figures on a grid. A fly lands at random at a point on the grid.

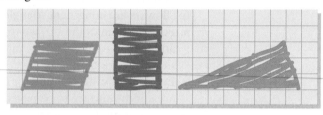

 a. Writing Is the fly more likely to land on one of the figures or on the blank grid? Explain.
     b. Suppose you know the fly lands on one of the figures. Is the fly more likely to land on one figure than on another? Explain.

Coordinate Geometry **Find the area of a polygon with the given vertices.**

30. $A(3, 9), B(8, 9), C(2, -3), D(-3, -3)$ **31.** $E(1, 1), F(4, 5), G(11, 5), H(8, 1)$

32. $D(0, 0), E(2, 4), F(6, 4), G(6, 0)$ **33.** $K(-7, -2), L(-7, 6), M(1, 6), N(7, -2)$

Find the area of each figure.

34.
     25 ft
     25 ft
     25 ft

35.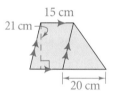
     15 cm
     21 cm
     20 cm

36.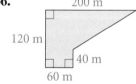
     200 m
     120 m
     40 m
     60 m

C **Challenge** 🌐 **History** **The ancient Greek mathematician Heron is most famous for this formula for the area of a triangle in terms of the lengths of its sides a, b, and c.**

$$A = \sqrt{s(s - a)(s - b)(s - c)}, \text{ where } s = \frac{1}{2}(a + b + c)$$

Use Heron's Formula and a calculator to find the area of each triangle. Round your answer to the nearest whole number.

37. $a = 8$ in., $b = 9$ in., $c = 10$ in.   **38.** $a = 15$ m, $b = 17$ m, $c = 21$ m

39. a. Use Heron's Formula to find the area of this triangle.
     b. Verify your answer to part (a) by using the formula $A = \frac{1}{2}bh$.

15 in.   9 in.
12 in.

Multiple Choice

40. The lengths of the sides of a right triangle are 10 in., 24 in., and 26 in. What is the area of the triangle?
A. 116 in.²   B. 120 in.²   C. 130 in.²   D. 156 in.²

41. What is the area of ▱*ABCD* at the right?
F. 32 in.²   G. 64 in.²
H. 91.2 in.²   J. 45.6 in.²

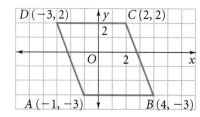

42. A parallelogram has adjacent sides of 176 ft and 312 ft. The altitude to the shorter side is 290 ft. What is the area of the parallelogram?
A. 51,040 ft²   B. 51,352 ft²   C. 54,912 ft²   D. 55,202 ft²

43. The perimeter of an equilateral triangle is 60 m. Its height is 17.3 m. What is its area?
F. 173 m²   G. 200 m²   H. 348 m²   J. 1044 m²

Short Response

44. a. For ▱*ABCD*, explain how to determine the length of an altitude drawn to base $\overline{AB}$.
b. Find the area of ▱*ABCD*.

Lesson 9-7

List the symmetries in each tessellation.

45.

46.

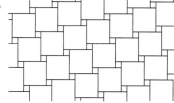

Lesson 4-5

The base of the isosceles triangle is a side of a regular pentagon *PENTA*. Find the measure of each angle.

47. ∠*APE*   **48.** ∠*APN*

49. ∠*PAN*   **50.** ∠*PNA*

51. ∠*EPN*   **52.** ∠*ANT*

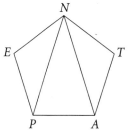

Lesson 3-8

Use a compass and straightedge for the following constructions.

53. Draw a segment and label it $\overline{AB}$. Construct $\overleftrightarrow{AD}$ so that $\overleftrightarrow{AD} \perp \overline{AB}$ at point *A*.

54. Draw a segment. Label it $\overline{EF}$. Construct a line $\overleftrightarrow{GH}$ so that $\overleftrightarrow{GH} \parallel \overline{EF}$.

55. Draw a segment and label it $\overline{KL}$. Draw a point *X* not on $\overleftrightarrow{KL}$. Construct a perpendicular from point *X* to $\overline{KL}$ (or to $\overleftrightarrow{KL}$).

Areas of Trapezoids, Rhombuses, and Kites

What You'll Learn

• To find the area of a trapezoid

• To find the area of a rhombus or a kite

. . . And Why

To use a map and the trapezoid area formula to approximate the area of Arkansas, as in Example 1

✓ **Check Skills You'll Need**

🔵 **for Help** Lesson 10-1

Write the formula for the area of each type of figure.

1. a rectangle

2. a triangle

Find the area of each trapezoid by using the formulas for area of a rectangle and area of a triangle.

3.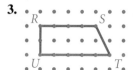

4.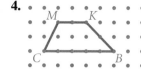

5.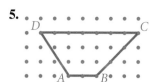

🔊 **New Vocabulary** • height of a trapezoid

---

1 Area of a Trapezoid

Vocabulary Tip

The term <u>base</u> can refer to either a line segment or its length.

In Lesson 6-5, you learned that the bases of a trapezoid are the parallel sides and the legs are the nonparallel sides. The **height of a trapezoid** is the perpendicular distance h between the bases.

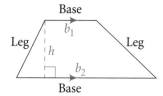

🔑 **Key Concepts**

Theorem 10-4	**Area of a Trapezoid**
The area of a trapezoid is half the product of the height and the sum of the bases. $$A = \tfrac{1}{2}h(b_1 + b_2)$$	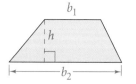

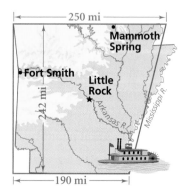

1 EXAMPLE **Real-World** 🌐 **Connection**

Geography Approximate the area of Arkansas by finding the area of the trapezoid shown.

$$A = \tfrac{1}{2}h(b_1 + b_2)$$ **Use the formula for area of a trapezoid.**

$$= \tfrac{1}{2}(242)(190 + 250)$$ **Substitute 242 for h, 190 for b_1, and 250 for b_2.**

$$= 53{,}240$$ **Simplify.**

● The area of Arkansas is about 53,240 mi².

✓ **Quick Check** **①** Find the area of a trapezoid with height 7 cm and bases 12 cm and 15 cm.

Properties of special right triangles can help you find the area of a trapezoid.

2 EXAMPLE **Finding Area Using a Right Triangle**

Multiple Choice What is the area of trapezoid $PQRS$?

Ⓐ 6 m^2 Ⓑ $10\sqrt{3} \text{ m}^2$
Ⓒ $12\sqrt{3} \text{ m}^2$ Ⓓ 35 m^2

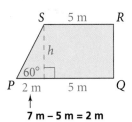

You can draw an altitude that divides the trapezoid into a rectangle and a 30°-60°-90° triangle. Since the opposite sides of a rectangle are congruent, the longer base of the trapezoid is divided into segments of lengths 2 m and 5 m.

Find h.

Test-Taking Tip

You can also find the area by adding the areas of the triangle and rectangle.

$h = 2\sqrt{3}$ longer leg = shorter leg · $\sqrt{3}$

$A = \frac{1}{2}h(b_1 + b_2)$ Use the trapezoid area formula.

$\quad = \frac{1}{2}(2\sqrt{3})(7 + 5)$ Substitute.

$\quad = 12\sqrt{3}$ Simplify.

● The area of trapezoid $PQRS$ is $12\sqrt{3} \text{ m}^2$. The answer is C.

✓ Quick Check ❷ In Example 2, suppose h is made smaller so that $m\angle P = 45$ while bases and angles R and Q are unchanged. Find the area of trapezoid $PQRS$.

---

2   Finding Areas of Rhombuses and Kites

Rhombuses and kites have perpendicular diagonals. This property allows you to find areas using the following theorem.

Key Concepts

Theorem 10-5	Area of a Rhombus or a Kite

The area of a rhombus or a kite is half the product of the lengths of its diagonals.

$$A = \frac{1}{2}d_1d_2$$

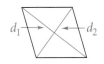

3 EXAMPLE **Finding the Area of a Kite**

Find the area of kite $KLMN$.

For the two diagonals, $KM = 2 + 5 = 7 \text{ m}$ and $LN = 3 + 3 = 6 \text{ m}$.

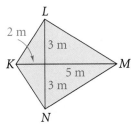

$A = \frac{1}{2}d_1d_2$ Use the formula for area of a kite.

$A = \frac{1}{2}(7)(6)$ Substitute 7 for d_1 and 6 for d_2.

$A = 21$ Simplify.

● The area of kite $KLMN$ is 21 m^2.

✓ Quick Check ❸ Find the area of a kite with diagonals that are 12 in. and 9 in. long.

The fact that the diagonals of a rhombus bisect each other can help you find the area.

4 EXAMPLE **Finding the Area of a Rhombus**

Find the area of rhombus $ABCD$.

$\triangle BEC$ is a right triangle. Using a Pythagorean triple, $BE = 9$. Since the diagonals of a rhombus bisect each other, $AC = 24$ and $BD = 18$.

$A = \frac{1}{2}d_1d_2$ **Use the formula for area of a rhombus.**

$A = \frac{1}{2}(24)(18)$ **Substitute 24 for d_1 and 18 for d_2.**

$A = 216$ **Simplify.**

The area is 216 m².

✓ **Quick Check** **4 Critical Thinking** In Example 4, explain how you can use a Pythagorean triple to conclude that $BE = 9$.

EXERCISES

For more exercises, see *Extra Skill, Word Problem, and Proof Practice.*

Practice and Problem Solving

A Practice by Example

Example 1
(page 540)

GO for Help

GO nline

Video Tutor Help
Visit: PhSchool.com
Web Code: aue-0775

Find the area of each trapezoid.

1.
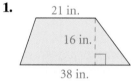
21 in.
16 in.
38 in.

2.
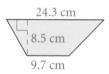
24.3 cm
8.5 cm
9.7 cm

3.

9 ft
6 ft
18 ft

4. Geography Approximate the area of Nevada by finding the area of the trapezoid shown.

5. Find the area of a trapezoid with bases 12 cm and 18 cm and height 10 cm.

6. Find the area of a trapezoid with bases 2 ft and 3 ft and height $\frac{1}{3}$ ft.

7. Geography The border of Tennessee resembles a trapezoid with bases 342 mi and 438 mi, and height 111 mi. Approximate the area of Tennessee by finding the area of this trapezoid.

205 mi
Humboldt R.
309 mi
Reno
★ Carson City
511 mi
Las Vegas

Example 2
(page 541)

Find the area of each trapezoid. If your answer is not an integer, leave it in simplest radical form.

8.

5 ft
3 ft
6 ft

9.
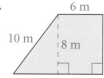
6 m
10 m
8 m

10.

8 ft
60°
15 ft

Example 3
(page 541)

Find the area of each kite.

11.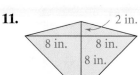
2 in.
8 in. 8 in.
8 in.

12.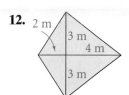
2 m
3 m
4 m
3 m

13.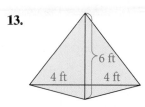
6 ft
4 ft 4 ft

Example 4
(page 542)

Find the area of each rhombus.

14.
20 ft
30 ft

15.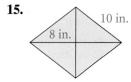
10 in.
8 in.

16.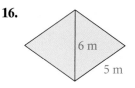
6 m
5 m

17. The end of the rain gutter has the shape of a trapezoid with the measurements shown. Find the area of this end.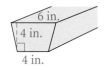
6 in.
4 in.
4 in.

18. A trapezoid has two right angles, 12-m and 18-m bases, and 8-m height.
  a. Sketch the trapezoid.   **b.** Find the perimeter.   **c.** Find the area.

19. Open-Ended Draw a kite. Measure the lengths of its diagonals. Find its area.

Gold Bars **Find the area of each trapezoidal face of the gold bars.**

20. End face: bases 4 cm and 2 cm, height 3 cm.

21. Side face: bases 8 cm and 5 cm, height 3 cm.

Find the area of each trapezoid to the nearest tenth.

22.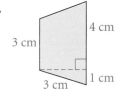
4 cm
3 cm
3 cm
1 cm

23.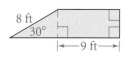
8 ft
30°
9 ft

24.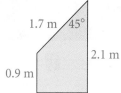
1.7 m 45°
2.1 m
0.9 m

Real-World **Connection**

On each gold bar the four trapezoidal faces tip inwards. This simplifies the molding process.

Coordinate Geometry **In Exercises 25–27, find the area of quadrilateral QRST.**

25.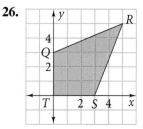

26.

27.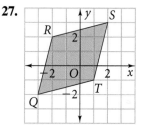

28. Multiple Choice What is the area of the kite at the right?
  Ⓐ 90 m²   Ⓑ 108 m²
  Ⓒ 135 m²   Ⓓ 216 m²

9√2 m
45°
6 m

29. a. Coordinate Geometry Graph the lines $x = 0$, $x = 6$, $y = 0$, and $y = x + 4$.
  b. What type of quadrilateral do the lines form?
  c. Find the area of the quadrilateral.

Find the area of each rhombus. Leave your answer in simplest radical form.

30.

45°
3 cm

31.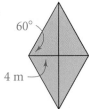

60°
4 m

32.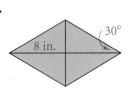

8 in.
30°

33. Draw a trapezoid. Label its bases and height b_1, b_2, and h, respectively. Then draw a diagonal of the trapezoid.
  a. Write equations for the area of each of the two triangles formed.
  b. Writing Explain how you can justify the trapezoid area formula using the areas of the two triangles.

34. Visualization The kite has diagonals d_1 and d_2 congruent to the sides of the rectangle. Explain why the area of the kite is $\frac{1}{2}d_1d_2$.

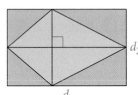

d_2
d_1

C **Challenge** x^2 **35. Algebra** One base of a trapezoid is twice the other. The height is the average of the two bases. The area is 324 cm^2. Find the height and the bases. (*Hint:* Let the smaller base be $2x$.)

36. Gravity Sports Ty wants to paint one end of his homemade skateboarding ramp. The ramp is 4 m wide. Its surface is modeled by the equation $y = 0.25x^2$. Use the trapezoids and triangles shown to estimate the area to be painted.

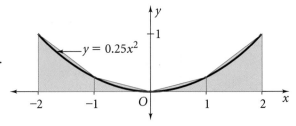

$y = 0.25x^2$

37. In trapezoid $ABCD$, $\overline{AB} \parallel \overline{DC}$. Find the area of $ABCD$.

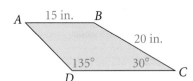

A 15 in. B
20 in.
135° 30°
D C

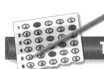

Test Prep

Multiple Choice

38. The area of a kite is 120 cm^2. The length of one diagonal is 20 cm. What is the length of the other diagonal?
  A. 12 cm   **B.** 20 cm
  C. 24 cm   **D.** 48 cm

39. What is the area of the trapezoid at the right?
  F. 39 m²   **G.** 60 m²
  H. 78 m²   **J.** 96 m²

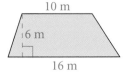

10 m
6 m
16 m

40. The lengths of the sides of a rhombus and one of its diagonals are each 10 m. What is the area of the rhombus?
  A. $100\sqrt{3}$ m²   **B.** $50\sqrt{3}$ m²   **C.** $25\sqrt{3}$ m²   **D.** $12.5\sqrt{3}$ m²

41. The area of a trapezoid is 100 ft². The sum of the two bases is 25 ft. What is the height of the trapezoid?

　　F. 2 ft 　　　　**G.** 4 ft 　　　　**H.** 8 ft 　　　　**J.** 10 ft

Short Response　**42.** The area of an isosceles trapezoid is 160 cm². Its height is 8 cm and the length of one leg is 10 cm.

　　a. Draw and label a diagram representing the given information.

　　b. Find the length of each base. Show your work.

Mixed Review

Lesson 10-1　**43.** Find the area of an isosceles right triangle that has one leg of length 12 cm.

44. An isosceles right triangle has area of 112.5 ft². Find the length of each leg.

Lesson 5-3　**Fill in the blank with *always*, *sometimes*, or *never* to form a true statement.**

45. The incenter of a triangle __?__ lies inside the triangle.

46. The orthocenter of a triangle __?__ lies outside the triangle.

47. The centroid of a triangle __?__ lies on the triangle.

Lesson 3-5　**48.** Find the measure of an interior angle of a regular 9-gon.

A Point in Time

1500　1600　1700　1800　1900　2000

Presidents are known more often for their foreign policy than for their mathematical creativity. James Garfield, the 20th President of the United States, is an exception. In 1876, Garfield demonstrated this proof of the Pythagorean Theorem.

In the diagram, $\triangle NRM$ and $\triangle RPQ$ are congruent right triangles with sides of lengths a, b, and c. The legs of isosceles right triangle NRP have length c. The three triangles form trapezoid $MNPQ$. The sum of the areas of the three triangles equals the area of trapezoid $MNPQ$.

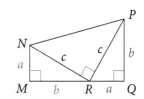

Areas of Triangles = Area of Trapezoid

$$\frac{1}{2}ab + \frac{1}{2}ab + \frac{1}{2}c^2 = \frac{1}{2}(a+b)(a+b)$$

$$ab + \frac{1}{2}c^2 = \frac{1}{2}a^2 + ab + \frac{1}{2}b^2$$

$$\frac{1}{2}c^2 = \frac{1}{2}a^2 + \frac{1}{2}b^2$$

$$c^2 = a^2 + b^2$$

For: Information about Pythagorean Theorem proofs
Web Code: aue-2032

Areas of Regular Polygons

What You'll Learn

• To find the area of a regular polygon

...And Why

To find the area of pieces of honeycomb material used to build boats, as in Example 3

Find the area of each regular polygon. If your answer involves a radical, leave it in simplest radical form.

1. 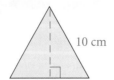 10 cm

2. 10 ft

3. 10 m

Find the perimeter of the regular polygon.

4. a hexagon with sides of 4 in.

5. an octagon with sides of $2\sqrt{3}$ cm

🔊 **New Vocabulary** • radius of a regular polygon • apothem

---

1 Areas of Regular Polygons

Vocabulary Tip

The terms radius and apothem (AP uh them) can each refer to either a segment or its length.

You can circumscribe a circle about any regular polygon. The center of a regular polygon is the center of the circumscribed circle. The **radius** is the distance from the center to a vertex. The **apothem** is the perpendicular distance from the center to a side.

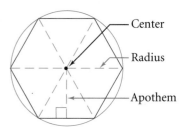

— Center
— Radius
— Apothem

1 EXAMPLE Finding Angle Measures

The figure at the right is a regular pentagon with radii and an apothem drawn. Find the measure of each numbered angle.

$m\angle 1 = \frac{360}{5} = 72$ Divide 360 by the number of sides.

$m\angle 2 = \frac{1}{2}m\angle 1$ The apothem bisects the vertex angle of the isosceles triangle formed by the radii.

$= \frac{1}{2}(72) = 36$

$90 + 36 + m\angle 3 = 180$ The sum of the measures of the angles of a triangle is 180.

$m\angle 3 = 54$

$m\angle 1 = 72, m\angle 2 = 36,$ and $m\angle 3 = 54$

✓ **Quick Check** ❶ At the right, a portion of a regular octagon has radii and an apothem drawn. Find the measure of each numbered angle.

Suppose you have a regular *n*-gon with side *s*. The radii divide the figure into *n* congruent isosceles triangles. Each isosceles triangle has area equal to $\frac{1}{2}as$.

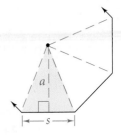

Since there are *n* congruent triangles, the area of the *n*-gon is $A = n \cdot \frac{1}{2}as$. The perimeter *p* of the *n*-gon is *ns*. Substituting *p* for *ns* results in a formula for the area in terms of *a* and *p*: $A = \frac{1}{2}ap$.

Key Concepts

Theorem 10-6	Area of a Regular Polygon

The area of a regular polygon is half the product of the apothem and the perimeter.

$$A = \tfrac{1}{2}ap$$

---

2 EXAMPLE **Finding the Area of a Regular Polygon**

Find the area of a regular decagon with a 12.3-in. apothem and 8-in. sides.

$p = ns$ **Find the perimeter.**

$\quad = 10(8) = 80$ in.   **A decagon has 10 sides, so *n* = 10.**

$A = \frac{1}{2}ap$ **Use the formula for the area of a regular polygon.**

$\quad = \frac{1}{2}(12.3)(80) = 492$

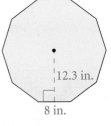
12.3 in.

8 in.

● The regular decagon has area 492 in.^{2}.

✓ **Quick Check** **2** Find the area of a regular pentagon with 11.6-cm sides and an 8-cm apothem.

---

3 EXAMPLE **Real-World 🌐 Connection**

Boat Racing Some boats used for racing have bodies made of a honeycomb of regular hexagonal prisms sandwiched between two layers of outer material. At the right is an end of one hexagonal cell. Find its area.

30°   60°

a

10 mm

5 mm

The radii form six 60° angles at the center. You can use a 30°-60°-90° triangle to find the apothem *a*.

$a = 5\sqrt{3}$ **longer leg = $\sqrt{3} \cdot$ shorter leg**

$p = ns$ **Find the perimeter of the hexagon.**

$\quad = 6(10) = 60$ **Substitute 6 for *n* and 10 for *s*.**

$A = \frac{1}{2}ap$ **Find the area.**

$\quad = \frac{1}{2}(5\sqrt{3})(60)$ **Substitute $5\sqrt{3}$ for *a* and 60 for *p*.**

$\quad \approx 259.80762$ **Use a calculator.**

● The area is about 260 mm^2.

✓ **Quick Check** **3** The side of a regular hexagon is 16 ft. Find the area of the hexagon.

EXERCISES

For more exercises, see *Extra Skill, Word Problem, and Proof Practice.*

Practice and Problem Solving

A Practice by Example

Example 1
(page 546)

GO for Help

Each regular polygon has radii and apothem as shown. Find the measure of each numbered angle.

1.

2.

3.
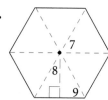

Example 2
(page 547)

Find the area of each regular polygon with the given apothem a and side length s.

4. pentagon, $a = 24.3$ cm, $s = 35.3$ cm

5. 7-gon, $a = 29.1$ ft, $s = 28$ ft

6. octagon, $a = 60.4$ in., $s = 50$ in.

7. nonagon, $a = 27.5$ in., $s = 20$ in.

8. decagon, $a = 19$ m, $s = 12.3$ m

9. dodecagon, $a = 26.1$ cm, $s = 14$ cm

Example 3
(page 547)

Find the area of each regular polygon. Round your answer to the nearest tenth.

10.

18 ft

11.

8 in.

12.
6 m

13. Art The smaller triangles in the Minneapolis sculpture at the left are equilateral. Each has a 12.7-in. radius. What is the area of each to the nearest square inch?

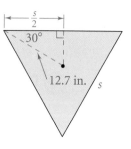

Find the area of each regular polygon with the given radius or apothem. If your answer is not an integer, leave it in simplest radical form.

14.

6 cm

15.

$8\sqrt{3}$ in.

16.

$6\sqrt{3}$ m

17.

5 m

18.

4 in.

Exercise 13

B Apply Your Skills

Find the measures of the angles formed by (a) two consecutive radii and (b) a radius and a side of the given regular polygon.

19. pentagon

20. octagon

21. nonagon

22. dodecagon

GO for Help

For a guide to solving Exercise 23, see p. 552.

23. Satellites One of the smallest space satellites ever developed has the shape of a pyramid. Each of the four faces of the pyramid is an equilateral triangle with sides about 13 cm long. What is the area of one equilateral triangular face of the satellite? Round your answer to the nearest whole number.

24. Multiple Choice The gazebo in the photo is built in the shape of a regular octagon. Each side is 8 ft long, and its apothem is 9.7 ft. What is the area enclosed by the gazebo?

Ⓐ 38.8 ft²   Ⓑ 77.6 ft²

Ⓒ 232.8 ft²   Ⓓ 310.4 ft²

25. The area of a regular polygon is 36 in.². Find the length of a side if the polygon has the given number of sides. Round your answer to the nearest tenth.

a. 3   **b.** 4   **c.** 6

d. Estimation Suppose the polygon is a pentagon. What would you expect the length of its side to be? Explain.

26. A portion of a regular decagon has radii and an apothem drawn. Find the measure of each numbered angle.

27. Writing Explain why the radius of a regular polygon is greater than the apothem.

28. Constructions Use a compass to construct a circle.
a. Construct two perpendicular diameters of the circle.
b. Construct diameters that bisect each of the four right angles.
c. Connect the consecutive points where the diameters intersect the circle. What regular polygon have you constructed?
d. Critical Thinking How can a circle help you construct a regular hexagon?

29. A regular hexagon has perimeter 120 m. Find its area.

30. Open-Ended Create a design using equilateral triangles and regular hexagons that have sides of the same length. Find the area of the completed design.

Find the area of each regular polygon. Show your answers in simplest radical form and rounded to the nearest tenth.

31.

8 cm

32.   4 cm

33.
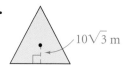
10√3 m

34. To find the area of an equilateral triangle, you can use the formula $A = \frac{1}{2}bh$ or $A = \frac{1}{2}ap$. A third way to find the area of an equilateral triangle is to use the formula $A = \frac{1}{4}s^2\sqrt{3}$. Verify the formula $A = \frac{1}{4}s^2\sqrt{3}$ in two ways as follows:

a. Find the area of Figure 1 using the formula $A = \frac{1}{2}bh$.

b. Find the area of Figure 2 using the formula $A = \frac{1}{2}ap$.

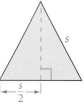

Figure 1   Figure 2

<u>Proof</u> **35.** For Example 1 on page 546, write a proof that the apothem bisects the vertex angle of the isosceles triangle formed by the radii.

C Challenge <u>Proof</u> **36.** Prove that the bisectors of the angles of a regular polygon (given congruent sides and angles) are concurrent and that they are, in fact, radii of the polygon. (*Hint:* For regular *n*-gon *ABCDE* . . ., let *P* be the intersection of the bisectors of $\angle ABC$ and $\angle BCD$. Show that $\overrightarrow{DP}$ must be the bisector of $\angle CDE$.)

37. Coordinate Geometry A regular octagon with center at the origin and radius 4 is graphed in the coordinate plane.

a. Since V_2 lies on the line $y = x$, its *x*- and *y*-coordinates are equal. Use the Distance Formula to find the coordinates of V_2 to the nearest tenth.

b. Use the coordinates of V_2 and the formula $A = \frac{1}{2}bh$ to find the area of $\triangle V_1 O V_2$ to the nearest tenth.

c. Use your answer to part (b) to find the area of the octagon to the nearest whole number.

38. In $\triangle ABC$, $\angle C$ is acute.

a. Show that the area of $\triangle ABC = \frac{1}{2}ab \sin C$.

b. Complete: The area of a triangle is half the product of __?__ and the sine of the __?__ angle.

c. Show that the area of a regular *n*-gon with radius *r* is $\frac{nr^2}{2} \sin\left(\frac{360}{n}\right)$.

Real-World **Connection**

Horizontal cross sections of the Wenfeng Pagoda in Yangzhou, China, are regular octagons.

Test Prep

Multiple Choice **39.** What is the area of a regular pentagon whose apothem is 25.1 mm and perimeter is 182 mm?
   A. 913.6 mm²   **B.** 2284.1 mm²   **C.** 3654.6 mm²   **D.** 4568.2 mm²

40. The area of a regular octagonal garden is 1235.2 yd². The apothem is 19.3 yd. What is the perimeter of the garden?
   F. 128 yd   **G.** 154.4 yd   **H.** 186.6 yd   **J.** 192 yd

41. The radius of a regular hexagonal sandbox is 5 ft. What is the area to the nearest square foot?
   A. 30 ft²   **B.** 65 ft²   **C.** 75 ft²   **D.** 130 ft²

Short Response **42.** The perimeter of a regular decagon is 220 in. Its radius is 35.6 in.
   a. Explain how to use the given information to find its area.
   b. Find the area.

Extended Response **43.** In regular hexagon *ABCDEF*, $BC = 8\sqrt{3}$ ft.
   a. Find the area of $\triangle BCG$.
   b. Find the area of hexagon *ABCDEF*.
   c. Describe two different methods for finding the area of hexagon *ABCDEF*.

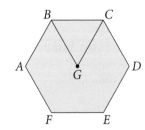

GO for Help

Lesson 10-2

44. Find the area of a kite with diagonals 8 m and 11.5 m.

45. The area of a kite is 150 in.². The length of one diagonal is 10 in. Find the length of the other diagonal.

46. The area of a trapezoid is 42 m². The trapezoid has a height of 7 m and one base of 4 m. Find the length of the other base.

Lesson 4-4 **Name the pairs of triangles you would have to prove congruent so that the indicated congruences are true by CPCTC.**

Given: $\angle DAB \cong \angle CBA$,
$\overline{AD} \cong \overline{BC}$,
$\overline{DF}$ bisects $\angle ADB$,
$\overline{CG}$ bisects $\angle BCA$.

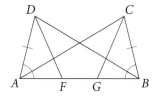

47. $\overline{AC} \cong \overline{BD}$ **48.** $\overline{AG} \cong \overline{BF}$ **49.** $\angle DFA \cong \angle CGB$

Lesson 1-9 **50. a. Biology** The size of a jaguar's territory depends on how much food is available. Where there is a lot of food, such as in a forest, jaguars have circular territories about 3 mi in diameter. Use 3.14 for π to estimate the area of such a region to the nearest tenth.

 b. Where food is less available, a jaguar may need up to 200 mi². Estimate the radius of this circular territory.

✓ Checkpoint Quiz 1 Lessons 10-1 through 10-3

Find the area of each figure.

1.
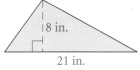

2.
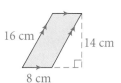

3.
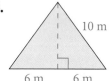

Find the area of each trapezoid, rhombus, or regular polygon. You may leave answers in simplest radical form.

4.

5.

6.

7.

8.

9.

10. A regular hexagon has a radius of $3\sqrt{3}$ m. Find the area. Show your answer in simplest radical form and rounded to the nearest tenth.

Understanding Word Problems Read the problem below. Let Jake's thinking guide you through the solution. Check your understanding with the exercises at the bottom of the page.

Satellites One of the smallest space satellites ever developed has the shape of a pyramid. Each of the four faces of the pyramid is an equilateral triangle with sides about 13 cm long. What is the area of one equilateral triangular face of the satellite? Round your answer to the nearest whole number.

What Jake Thinks	What Jake Writes
I'll make a sketch. First, the pyramid. Then a face. Each equilateral-triangle face has 13-cm sides.	
I have to find the area. I'll use red for an altitude. That gives me half of an equilateral triangle, or a 30°-60°-90° triangle. The short leg of the triangle is 6.5 cm. The altitude is $\sqrt{3}$ times the short leg.	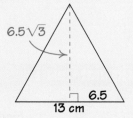 $6.5\sqrt{3}$ 6.5 13 cm
The area is $\frac{1}{2}$ base times height.	$A = \frac{1}{2}bh$
I'll use a calculator.	$A = \frac{1}{2} \cdot 13 \cdot 6.5\sqrt{3}$
I have to round to the nearest whole number.	$A \approx 73.18 \text{ cm}^2$ $A \approx 73 \text{ cm}^2$

EXERCISES

What would be the area of a face of the satellite for each side length given?

1. 6.5 cm   **2.** 26 cm   **3.** 1.3 m   **4.** 2.6 m

5. If you change the length of each side of the satellite by a factor k to $13k$ cm, how does the area of a face change?

Perimeters and Areas of Similar Figures

What You'll Learn

• To find the perimeters and areas of similar figures

...And Why

To find the expected yield of a garden, as in Example 3

✓ **Check Skills You'll Need**

GO for Help Lesson 1-9

Find the perimeter and area of each figure.

1.
7 in.

2.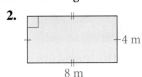
4 m
8 m

3.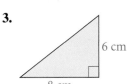
6 cm
8 cm

Find the perimeter and area of each rectangle with the given base and height.

4. $b = 1$ cm, $h = 3$ cm   **5.** $b = 2$ cm, $h = 6$ cm   **6.** $b = 3$ cm, $h = 9$ cm

---

1 Finding Perimeters and Areas of Similar Figures

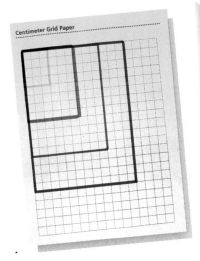
Centimeter Grid Paper

Hands-On Activity: Perimeters and Areas of Similar Rectangles

• On a piece of grid paper, draw a 3-unit by 4-unit rectangle.

• Draw three different rectangles, each similar to the original rectangle. Label them I, II, and III.

1. Use your drawings to complete a chart like this.

Rectangle	Perimeter	Area
Original		
I		
II		
III		

2. Use the information from the first chart to complete a chart like this.

Rectangle	Similarity Ratio	Ratio of Perimeters	Ratio of Areas
I to Original			
II to Original			
III to Original			

3. How do the ratios of perimeters and the ratios of areas compare with the similarity ratios?

To compare areas of similar figures, you can square the similarity ratio.

Theorem 10-7	Perimeters and Areas of Similar Figures

If the similarity ratio of two similar figures is $\frac{a}{b}$, then
(1) the ratio of their perimeters is $\frac{a}{b}$ and
(2) the ratio of their areas is $\frac{a^2}{b^2}$.

1 EXAMPLE **Finding Ratios in Similar Figures**

The trapezoids at the right are similar. The ratio of the lengths of corresponding sides is $\frac{6}{9}$, or $\frac{2}{3}$.

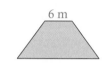

a. Find the ratio (smaller to larger) of the perimeters.

The ratio of the perimeters is the same as the ratio of corresponding sides, which is $\frac{2}{3}$.

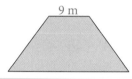

b. Find the ratio (smaller to larger) of the areas.

The ratio of the areas is the square of the ratio of corresponding sides, which is $\frac{2^2}{3^2}$, or $\frac{4}{9}$.

Quick Check **1** Two similar polygons have corresponding sides in the ratio 5 : 7.
a. Find the ratio of their perimeters.
b. Find the ratio of their areas.

When you know the area of one of two similar polygons, you can use a proportion to find the area of the other polygon.

2 EXAMPLE **Finding Areas Using Similar Figures**

Multiple Choice The area of the smaller regular pentagon is about 27.5 cm². What is the best approximation for the area of the larger regular pentagon?

(A) 11 cm²   (B) 69 cm²
(C) 172 cm²   (D) 275 cm²

4 cm   10 cm

Regular pentagons are similar because all angles measure 108 and all sides in each are congruent. Here the ratio of corresponding-side lengths is $\frac{4}{10}$, or $\frac{2}{5}$. The ratio of the areas is $\frac{2^2}{5^2}$, or $\frac{4}{25}$.

$\frac{4}{25} = \frac{27.5}{A}$ Write a proportion.

$4A = 687.5$ Cross-Product Property

$A = \frac{687.5}{4} = 171.875$ Solve for A.

The area of the larger pentagon is about 172 cm². The answer is C.

Quick Check **2** The corresponding sides of two similar parallelograms are in the ratio $\frac{3}{4}$. The area of the larger parallelogram is 96 in.². Find the area of the smaller parallelogram.

3 EXAMPLE Real-World Connection

Community Service During the summer, a group of high school students used a plot of city land and harvested 13 bushels of vegetables that they gave to a food pantry. Their project was so successful that next summer the city will let them use a larger, similar plot of land.

In the new plot, each dimension is 2.5 times the corresponding dimension of the original plot. How many bushels can they expect to harvest next year?

The ratio of the dimensions is 2.5 : 1. So, the ratio of the areas is $(2.5)^2 : 1^2$, or 6.25 : 1. With 6.25 times as much land next year, the students can expect to harvest 6.25(13), or about 81 bushels.

Real-World Connection

Many cities make city land available to the community for gardening.

✓ **Quick Check** **③** The similarity ratio of the dimensions of two similar pieces of window glass is 3 : 5. The smaller piece costs $2.50. What should be the cost of the larger piece?

When you know the ratio of the areas of two similar figures, you can work backward to find the ratio of their perimeters.

4 EXAMPLE Finding Similarity and Perimeter Ratios

The areas of two similar triangles are 50 cm^2 and 98 cm^2. What is the similarity ratio? What is the ratio of their perimeters?

Find the similarity ratio $a : b$.

$\dfrac{a^2}{b^2} = \dfrac{50}{98}$ **The ratio of the areas is $a^2 : b^2$.**

$\dfrac{a^2}{b^2} = \dfrac{25}{49}$ **Simplify.**

$\dfrac{a}{b} = \dfrac{5}{7}$ **Take square roots.**

The ratio of the perimeters equals the similarity ratio 5 : 7.

**Online
active math**

For: Perimeter and Area Activity
Use: Interactive Textbook, 10-4

✓ **Quick Check** **④** The areas of two similar rectangles are 1875 ft^2 and 135 ft^2. Find the ratio of their perimeters.

EXERCISES

For more exercises, see *Extra Skill, Word Problem, and Proof Practice.*

Practice and Problem Solving

A Practice by Example

Example 1
(page 554)

GO for Help

The figures in each pair are similar. Compare the first figure to the second. Give the ratio of the perimeters and the ratio of the areas.

1. 2 in.   4 in.

2.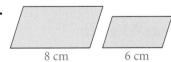
   8 cm   6 cm

3. 14 m   21 m

4.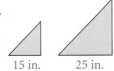
   15 in.   25 in.

Example 2
(page 554)

The figures in each pair are similar. The area of one figure is given. Find the area of the other figure to the nearest whole number.

5.

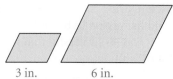

3 in.   6 in.

Area of smaller parallelogram = 6 in.^{2}

6.

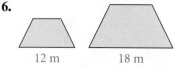

12 m   18 m

Area of larger trapezoid = 121 m^2

7.

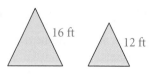

16 ft   12 ft

Area of larger triangle = 105 ft^2

8.

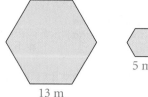

13 m   5 m

Area of smaller hexagon = 65 m^2

Example 3
(page 555)

9. Remodeling It costs a family $216 to have a 9 ft-by-12 ft wooden floor refinished. At that rate, how much would it cost them to have a 12 ft-by-16 ft wooden floor refinished?

10. Decorating An embroidered placemat costs $2.95. An embroidered tablecloth is similar to the placemat, but four times as long and four times as wide. How much would you expect to pay for the tablecloth?

Example 4
(page 555)

Find the similarity ratio and the ratio of perimeters for each pair of similar figures.

11. two regular octagons with areas 4 ft^2 and 16 ft^2

12. two triangles with areas 75 m^2 and 12 m^2

13. two trapezoids with areas 49 cm^2 and 9 cm^2

14. two parallelograms with areas 18 in.^{2} and 32 in.^{2}

15. two equilateral triangles with areas $16\sqrt{3}$ ft^2 and $\sqrt{3}$ ft^2

16. two circles with areas 2π cm^2 and 200π cm^2

B Apply Your Skills

The similarity ratio of two similar polygons is given. Find the ratio of their perimeters and the ratio of their areas.

17. 3 : 1   **18.** 2 : 5   **19.** $\frac{2}{3}$ **20.** $\frac{7}{4}$ **21.** 6 : 1

22. Multiple Choice The area of a regular decagon is 50 cm^2. What is the area of a regular decagon with sides four times the sides of the smaller decagon?

Ⓐ 200 cm^2 Ⓑ 500 cm^2 Ⓒ 800 cm^2 Ⓓ 2000 cm^2

23. Error Analysis A reporter used the graphic below to show that the number of houses with more than two televisions had doubled in the past few years. Explain why this graphic is misleading.

Homework Video Tutor
Visit: PHSchool.com
Web Code: aue-1004

Now

Then

Real-World **Connection**

Careers Doctors use enlarged images to aid in certain medical procedures.

Problem Solving Hint

For Exercise 34, recall the length of a diagonal of a square with 2-in. sides.

---

 24. Medicine For some medical imaging, the scale of the image is 3 : 1. That means that if an image is 3 cm long, the corresponding length on the person's body is 1 cm. Find the actual area of a lesion if its image has area 2.7 cm².

25. The longer sides of a parallelogram are 5 m. The longer sides of a similar parallelogram are 15 m. The area of the smaller parallelogram is 28 m². What is the area of the larger parallelogram?

x^2 **Algebra** **Find the values of x and y when the smaller triangle shown here has the given area.**

26. 3 cm²   **27.** 6 cm²   **28.** 12 cm²

29. 16 cm²   **30.** 24 cm²   **31.** 48 cm²

32. Two similar rectangles have areas 27 in.² and 48 in.². The length of one side of the larger rectangle is 16 in. What are the dimensions of both rectangles?

33. In △RST, RS = 20 m, ST = 25 m, and RT = 40 m.
   a. Open-Ended Choose a convenient scale. Then use a ruler and compass to draw △R'S'T' ~ △RST.
   b. Constructions Construct an altitude of △R'S'T' and measure its length. Find the area of △R'S'T'.
   c. Estimation Estimate the area of △RST.

34. Drawing Draw a square with an area of 8 in.². Draw a second square with an area that is four times as large. What is the ratio of their perimeters?

Compare the blue figure to the red figure. Find the ratios of (a) their perimeters and (b) their areas.

35.

36.
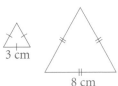

37.

 38. Writing The enrollment at an elementary school is going to increase from 200 students to 395 students. A parents' group is planning to increase the 100 ft-by-200 ft playground area to a larger area that is 200 ft by 400 ft. What would you tell the parents' group when they ask your opinion about whether the new playground will be large enough?

 39. a. Surveying A surveyor measured one side and two angles of a field as shown in the diagram. Use a ruler and a protractor to draw a similar triangle.
   b. Measure the sides and altitude of your triangle and find its perimeter and area.
   c. Estimation Estimate the perimeter and area of the field.

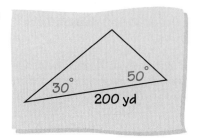

40. a. Find the area of a regular hexagon with sides 2 cm long. Leave your answer in simplest radical form.
   b. Use your answer to part (a) and Theorem 10-7 to find the areas of the regular polygons shown at the right.

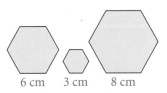

Complete each statement with *sometimes*, *always*, or *never*. Justify your answers.

41. Two similar rectangles with the same perimeter are __?__ congruent.

42. Two rectangles with the same area are __?__ similar.

43. Two rectangles with the same area and different perimeters are __?__ similar.

44. Similar figures __?__ have the same area.

Test Prep

Gridded Response

45. Two regular hexagons have sides in the ratio 3 : 5. The area of the smaller hexagon is 81 m². In square meters, what is the area of the larger hexagon?

46. Two similar polygons have areas in the ratio 9 : 16. The perimeter of the larger polygon is 900. What is the perimeter of the smaller polygon?

47. The two triangles are similar. Their perimeters have the ratio 1 : 3. In square feet, what is the area of the smaller triangle?

486 ft²

48. A rectangle has a perimeter of 104 in. A similar rectangle has a perimeter of 286 in. The area of the smaller rectangle is 672 in.². In square inches, what is the area of the larger rectangle?

49. The area of a polygon is 3267 cm². The area of a similar polygon is 9075 cm². The perimeter of the larger polygon is 270 cm. In centimeters, what is the perimeter of the smaller one?

Mixed Review

Lesson 10-3

Find the area of each regular polygon.

50. a square with a 10-cm diagonal

51. a pentagon with apothem 13.8 and side length 20

52. an octagon with apothem 12 and side length 10

53. a 12-sided polygon with apothem 3.7 and side length 2.0

Lesson 7-5

54. Solve for x and y in the diagram at the right.

55. An angle bisector divides the opposite side of a triangle into segments 4 cm and 6 cm long. A second side of the triangle is 8 cm long. Find all possible lengths for the third side of the triangle.

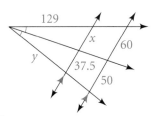

129

x

60

y

37.5

50

Lesson 3-6 x^2 **Algebra** Write the equation in slope-intercept form, and graph the line.

56. $6x - 2y = 8$ **57.** $x + y = -2$ **58.** $3x = 4y$ **59.** $2x + \frac{1}{2}y = \frac{5}{2}$

x^2 **Algebra** Write an equation of the line described.

60. has slope -3 and contains point $(1, -2)$

61. contains points $(3, 7)$ and $(0, -1)$

Trigonometry and Area

What You'll Learn

- To find the area of a regular polygon using trigonometry
- To find the area of a triangle using trigonometry

. . . And Why

To find the area of a courtyard, as in Example 2

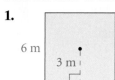 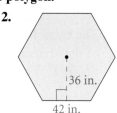 

✓ **Check Skills You'll Need**

GO **for Help** Lesson 10-3

Find the area of each regular polygon.

1. 6 m, 3 m

2. 36 in., 42 in.

3. 7 ft, 8 ft

1 Finding the Area of a Regular Polygon

In Lesson 10-3, you learned to find the area of a regular polygon by using the formula $A = \frac{1}{2}ap$, where a is the apothem and p is the perimeter. By using this formula and trigonometric ratios, you can solve other types of problems.

1 EXAMPLE Finding Area

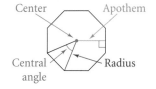

Vocabulary Tip

Center / Apothem

Central angle / Radius

Find the area of a regular pentagon with 8-cm sides.

To use the formula $A = \frac{1}{2}ap$, you need the apothem and perimeter. The perimeter is $5 \cdot 8$, or 40 cm.

To find the apothem, use trigonometry.
The measure of the central angle $\angle XCZ$ is $\frac{360}{5}$, or 72.
$m\angle XCY = \frac{1}{2}m\angle XCZ = 36$.
$XY = \frac{1}{2}XZ$, so $XY = 4$.

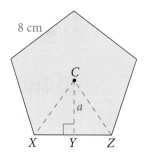

$\tan 36° = \frac{4}{a}$ **Use the tangent ratio.**

$a = \frac{4}{\tan 36°}$ **Solve for a.**

Now substitute into the area formula.

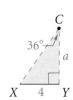

$A = \frac{1}{2}ap$

$= \frac{1}{2} \cdot \frac{4}{\tan 36°} \cdot 40$ **Substitute for a and p.**

$= \frac{80}{\tan 36°}$ **Simplify.**

80 ÷ TAN 36 ENTER *110.11055*   **Use a calculator.**

● The area of the regular pentagon is about 110 cm².

✓ **Quick Check** ❶ Find the area of a regular octagon with a perimeter of 80 in. Give the area to the nearest tenth.

Sometimes you can use trigonometry to find both apothem and perimeter.

2 EXAMPLE **Real-World**  **Connection**

Architecture The Castel del Monte, built on a hill in southern Italy circa 1240, makes extraordinary use of regular octagons. One regular octagon, the inner courtyard, has radius 16 m. Find the area of the courtyard.

The measure of a central angle of the octagon is $\frac{360}{8}$, or 45. So $m\angle C = \frac{1}{2}(45) = 22.5$.

Use the cosine ratio to find the apothem.

$$\cos 22.5° = \frac{a}{16}$$
$$a = 16(\cos 22.5°)$$

Use the sine ratio to find the perimeter.

$$\sin 22.5° = \frac{x}{16}$$
$$x = 16(\sin 22.5°)$$

$$p = 8 \cdot \text{length of a side}$$
$$= 8 \cdot 2x \qquad \textbf{The length of each side is 2x.}$$
$$= 8 \cdot 2 \cdot 16(\sin 22.5°) \qquad \textbf{Substitute for x.}$$
$$= 256(\sin 22.5°) \qquad \textbf{Simplify.}$$

Substitute into the area formula, $A = \frac{1}{2}ap$.

$$A = \frac{1}{2} \cdot 16(\cos 22.5°) \cdot 256(\sin 22.5°) \qquad \textbf{Substitute for a and p.}$$
$$\approx 724.07734 \qquad \textbf{Use a calculator.}$$

● The area of the courtyard is about 724 m².

✓ Quick Check **2 Critical Thinking** If the radius of the main structure is twice the radius of the inner courtyard, how does the area it covers compare to the area of the courtyard?

2 Finding the Area of a Triangle

Suppose you want to find the area of $\triangle ABC$, but you know only $m\angle A$ and the lengths b and c. To use the formula Area $= \frac{1}{2}bh$, you need to know the height. You can find the height by using the sine ratio.

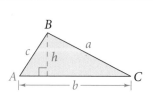

$$\sin A = \frac{h}{c} \qquad \textbf{Use the sine ratio.}$$
$$h = c(\sin A) \qquad \textbf{Solve for h.}$$

Now substitute for h in the formula Area $= \frac{1}{2}bh$.

$$\text{Area} = \frac{1}{2}bc(\sin A)$$

Your work at the bottom of page 560 completes a proof of the following theorem for the case in which $\angle A$ is acute.

Key Concepts

Theorem 10-8 **Area of a Triangle Given SAS**

The area of a triangle is one half the product of the lengths of two sides and the sine of the included angle.

$$\text{Area of } \triangle ABC = \tfrac{1}{2}bc(\sin A)$$

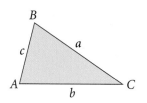

3 EXAMPLE **Real-World Connection**

Surveying The surveyed lengths of two adjacent sides of a triangular plot of land are 412 ft and 386 ft. The angle between the sides is 71°. Find the area of the plot.

Area $= \tfrac{1}{2} \cdot$ side length $\cdot$ side length $\cdot$ sine of included angle

$\quad = \tfrac{1}{2} \cdot 412 \cdot 386 \cdot \sin 71°$ **Substitute.**

$\quad \approx 75183.855$ **Use a calculator.**

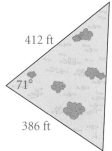

- The area of the plot is approximately 75,200 ft^2.

Quick Check ③ Two sides of a triangular building plot are 120 ft and 85 ft long. They include an angle of 85°. Find the area of the building plot to the nearest square foot.

EXERCISES

For more exercises, see *Extra Skill, Word Problem, and Proof Practice.*

Practice and Problem Solving

A Practice by Example

Example 1
(page 559)

for Help

Find the area of each regular polygon. Give answers to the nearest tenth.

1. octagon with side length 6 cm

2. pentagon with side length 7 in.

3. hexagon with perimeter 60 m

4. 15-gon with perimeter 180 yd

5. *PQRST* is a regular pentagon with center *O* and radius 10 in.
  a. Find $m\angle POQ$.
  b. Find $m\angle POX$.
  c. Find *OX*.
  d. Find *PQ*.
  e. Find the perimeter.
  f. Find the area.

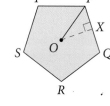

Example 2
(page 560)

Find the area of each regular polygon. Give answers to the nearest tenth.

6. hexagon with radius 10 ft

7. decagon with radius 4 in.

8. octagon with radius 20 cm

9. square with radius 2 ft

10. Architecture Each of the eight small towers around Castel del Monte (page 560) is a regular octagon. The radius is 7.3 m. Find the area each tower covers to the nearest square meter.

Example 3
(page 561)

Find the area of each triangle. Give answers to the nearest tenth.

11.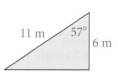

12.

13.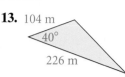

14.

15.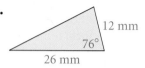

16.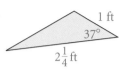

17. Surveying A surveyor marks off a triangular parcel of land. One side of the triangle extends 80 yd. A second side of 150 yd forms an angle of 67° with the first side. Determine the area of the parcel of land to the nearest square yard.

B Apply Your Skills

 18. Industrial Design Refer to the diagram of the regular hexagonal nut. Round each answer to the nearest unit.
  a. Find the area of the circular hole in the hexagonal nut.
  b. Find the area of the hexagonal face.

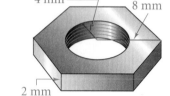

19. Writing Describe two ways to find the area of an equilateral triangle that has a 1-in. radius.

Real-World Connection

The length of each side of the Pentagon courtyard is 356 ft. A football field is 360 ft.

 20. Architecture The Pentagon, in Arlington, Virginia, is one of the world's largest office buildings. It is a regular pentagon, and the length of each of its sides is 921 ft. Find the area of this pentagon to the nearest thousand square feet.

Find the perimeter and area of each regular polygon to the nearest tenth.

21.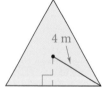

22.

23.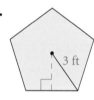

24.

25.

26.

27. Multiple Choice Replacement glass for more energy efficient windows costs $5/ft². Approximately how much will you pay for replacement glass for a regular hexagonal window with a radius of 2 ft?
  Ⓐ $10.39   Ⓑ $27.78   Ⓒ $45.98   Ⓓ $51.96

Regular polygons A and B are similar. Compare their areas.

28. The radius of square A is twice the radius of square B.

29. The apothem of pentagon A equals the radius of pentagon B.

30. The length of a side of hexagon A equals the radius of hexagon B.

31. The radius of octagon A equals the apothem of octagon B.

32. The perimeter of decagon A equals the length of a side of decagon B.

Homework Video Tutor

Visit: PHSchool.com
Web Code: aue-1005

 33. Road Signs The length of a side of the standard stop sign shown at left is 1 ft $\frac{1}{4}$ in. Find the area of the stop sign to the nearest tenth of a square foot.

34. Suppose a circle is inscribed in a regular n-gon with center C and apothem 1, as shown in the diagram. In parts (a)–(d), explain why each statement is true.

   a. $m\angle C = \frac{1}{2}\left(\frac{360}{n}\right) = \frac{180}{n}$
   b. $s = \tan C$
   c. The perimeter of the n-gon is $2n(\tan C)$.
   d. The area of the n-gon is $n\left(\tan \frac{180}{n}\right)$.

   **e. Graphing Calculator** Use the TABLE feature of your graphing calculator to study the areas of regular n-gons of apothem 1 as n increases. What should you enter as Y1?

   f. Use the **TBLSET** feature so that X starts at 3 and changes by 1. Access the **TABLE**. Tell what happens in the X and Y1 columns as you scroll down.

   g. For what value of X does Y1 take on a new value for the last time? Explain why this is so. Also, in terms of the circle and the circumscribing regular n-gons, interpret what you observe.

C Challenge

35. Suppose a circle is circumscribed about a regular n-gon with center C and radius 1. Proceed with steps similar to those in Exercise 34 to study the areas of regular n-gons inscribed in a circle as n increases.

36. Surveying A surveyor wants to mark off a triangular parcel with an area of 1 acre (1 acre $= 43,560$ ft^2). One side of the triangle extends 300 ft along a straight road. A second side extends at an angle of 65° from one end of the first side. Draw a triangle to represent the piece of land. Determine the length of the second boundary line to the nearest foot.

37. Segments are drawn between the midpoints of consecutive sides of a regular pentagon to form another regular pentagon. Find, to the nearest hundredth, the ratio of the area of the smaller pentagon to the area of the larger pentagon.

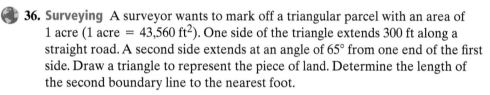

Test Prep

Multiple Choice   In Exercises 38–40, the polygons are regular.

38. The perimeter is 54 m. The apothem is $3\sqrt{3}$ m. To the nearest tenth, what is the area?
   A. 46.8 m^2 **B.** 140.3 m^2 **C.** 243.0 m^2 **D.** 280.6 m^2

39. The area is 1623.8 yd^2. The perimeter is 150 yd. What is the apothem?
   F. 10.8 yd   **G.** 21.7 yd   **H.** 32.5 yd   **J.** 43.3 yd

40. The area is 100 cm^2. The apothem is 5 m. What is the perimeter?
   A. 20 cm   **B.** 40 cm   **C.** 50 cm   **D.** 100 cm

Short Response   **41.** Sketch all possible right triangles ABC with $m\angle A = 40$ and $AC = 10$. Find the area of each.

Extended Response   **42. a.** In the regular pentagon, find x. Then use x to find a.
   b. Explain how you can use the perimeter of a regular polygon to find the area of the polygon.
   c. Use the perimeter of the pentagon shown here to find the area to the nearest tenth. Show your work.

GO for Help

Lesson 10-4

43. Two regular octagons are pictured.
  a. What is the similarity ratio of the smaller octagon to the larger octagon?
  b. The area of the larger octagon is 391.1 in.^{2}. What is the area of the smaller octagon to the nearest tenth of a square inch?

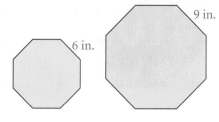

9 in.

6 in.

Lesson 8-6

44. Find the sum $\vec{\mathbf{a}} + \vec{\mathbf{c}}$. Give your answer as an ordered pair.

45. Describe a vector $\vec{\mathbf{d}}$ such that $\vec{\mathbf{a}} + \vec{\mathbf{d}} = \vec{\mathbf{c}}$.

46. Describe a vector $\vec{\mathbf{e}}$ such that $\vec{\mathbf{c}} + \vec{\mathbf{e}} = \vec{\mathbf{a}}$.

47. Which two vectors have $\vec{\mathbf{c}}$ as their sum?
  $\langle -2, -2 \rangle$ $\langle 2, -4 \rangle$ $\langle 4, -2 \rangle$

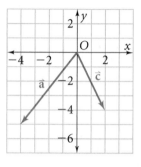

Lesson 6-6

Coordinate Geometry Find the coordinates of the midpoint of $\overline{WZ}$. Then find WZ.

48. rectangle

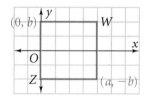

49. kite

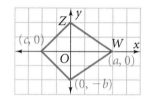

Geometry at Work

·······Surveyor

Surveyors calculate the locations, shapes, and areas of plots of land. A survey begins with a benchmark, a reference point whose latitude, longitude, and elevation are known. The surveyor uses a device called a transit to measure the boundary angles for the plot of land and the distances of key points from the benchmark. Using trigonometry, the surveyor finds the latitude, longitude, and elevation of each key point in the survey. These points are located on a map and an accurate sketch of the plot is drawn. Finally, the area is calculated.

One method involves dividing the plot into triangles and measuring the lengths of two sides and the included angle of each. The formula $A = \frac{1}{2}ab(\sin C)$ gives the area of each triangle. The area of the entire plot is the sum of the areas of the triangles.

Go Online
PHSchool.com   **For:** Information about surveying
Web Code: aub-2031

Laws of Sines and Cosines

You have seen *sine* and *cosine* defined in terms of acute angles in right triangles. In math courses after Algebra 2, you will learn how to extend the trigonometric relationships to angle measures in general—obtuse angle measures, right angle measures, even angle measures that are 180° or more, or 0° or less!

One result of this will be two special laws that can be applied to any triangle.

Law of Sines

$$\frac{\sin A}{a} = \frac{\sin B}{b} = \frac{\sin C}{c}$$

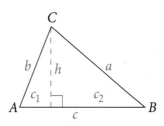

Law of Cosines

$$a^2 = b^2 + c^2 - 2bc \cos A$$
$$b^2 = a^2 + c^2 - 2ac \cos B$$
$$c^2 = a^2 + b^2 - 2ab \cos C$$

EXERCISES

1. Here is the first step of a proof that $\frac{\sin A}{a} = \frac{\sin B}{b}$:

1 $\frac{h}{ab} = \frac{h}{ab}$

Complete the proof.

2. Here are the first three steps of a proof that $a^2 = b^2 + c^2 - 2bc \cos A$:

1 $\quad b^2 \qquad\qquad = \quad c_1^2 + \qquad h^2$

2 $\quad c^2 \quad = \quad (c_1 + c_2)^2 = \quad c_1^2 + 2c_1c_2 + c_2^2$

3 $\quad -2bc \cos A = -2b(c_1 + c_2)\frac{c_1}{b} = -2c_1^2 - 2c_1c_2$

Complete the proof.

The Law of Sines lets you find missing measures in a triangle when you know the measures of two angles and a side, or two sides and a nonincluded angle. Use it to find the missing measures in each triangle.

3.

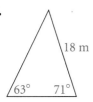

4.

5.

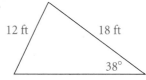

The Law of Cosines lets you find missing measures in a triangle when you know the measures of two sides and the included angle, or three sides. Use it to find the missing measures in each triangle.

6.

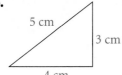

7.

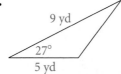

8.

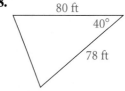

Circles and Arcs

What You'll Learn

- To find the measures of central angles and arcs
- To find circumference and arc length

. . . And Why

To use the turning radius of a car to compare the distances that its tires travel, as in Example 4

 Check Skills You'll Need   **GO for Help** Lesson 1-9 and Skills Handbook, p. 761

Find the diameter or radius of each circle.

1. $r = 7$ cm, $d = \blacksquare$ 2. $r = 1.6$ m, $d = \blacksquare$
3. $d = 10$ ft, $r = \blacksquare$ 4. $d = 5$ in., $r = \blacksquare$

Round to the nearest whole number.

5. 9% of 360   6. 38% of 360   7. 50% of 360   8. 21% of 360

 **New Vocabulary**   • circle • center • radius • congruent circles
• diameter • central angle • semicircle • minor arc
• major arc • adjacent arcs • circumference • pi
• concentric circles • arc length • congruent arcs

---

1 Central Angles and Arcs

Vocabulary Tip

<u>Diameter</u> comes from the classical Greek words *dia*, meaning through, and *meter*, meaning measure.

In a plane, a **circle** is the set of all points equidistant from a given point called the **center.** You name a circle by its center. Circle P ($\odot P$) is shown at the right.

A **radius** is a segment that has one endpoint at the center and the other endpoint on the circle. $\overline{PC}$ is a radius. $\overline{PA}$ and $\overline{PB}$ are also radii. **Congruent circles** have congruent radii.

A **diameter** is a segment that contains the center of a circle and has both endpoints on the circle. $\overline{AB}$ is a diameter.

A **central angle** is an angle whose vertex is the center of the circle. $\angle CPA$ is a central angle.

1 EXAMPLE   Real-World Connection

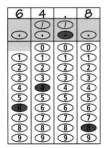

Gridded Response To learn how people really spend their time, a research firm studied the hour-by-hour activities of 3600 people. The participants were between 18 and 90 years old. Each participant was sent a 24-hour recording sheet every March for three years from 2000 to 2002.

Some information from the study is shown in this circle graph. What is the measure, in degrees, of the central angle used for the Entertainment part?

There are 360 degrees in a circle. To find the measure of a central angle in the circle graph, find the corresponding percent of 360. Entertainment is 18%, and 18% of 360 = 0.18 · 360, or 64.8.

You can use the same method to find the measures of the other central angles.

Sleep: 31% of 360 = 111.6 Other: 15% of 360 = 54

Food: 9% of 360 = 32.4 Must Do: 7% of 360 = 25.2

Work: 20% of 360 = 72

✓ **Quick Check** **1** **a.** **Critical Thinking** Each section of the circle graph represents a measurable quantity. What is that quantity?

b. Each section of the circle graph represents an average. Explain.

Test-Taking Tip

You can also find the measure a of a central angle by using a proportion. For Entertainment (18%) in Example 1: $\frac{18}{100} = \frac{a}{360}$.

An arc is a part of a circle. One type of arc, a **semicircle,** is half of a circle. A **minor arc** is smaller than a semicircle. A **major arc** is greater than a semicircle.

$\widehat{TRS}$ is a semicircle.
$m\widehat{TRS} = 180$

$\widehat{RS}$ is a minor arc.
$m\widehat{RS} = m\angle RPS$

$\widehat{RTS}$ is a major arc.
$m\widehat{RTS} = 360 - m\widehat{RS}$

The measure of a semicircle is 180. The measure of a minor arc is the measure of its corresponding central angle. The measure of a major arc is 360 minus the measure of its related minor arc.

2 **EXAMPLE** **Identifying Arcs**

Identify the following in ⊙O.

a. the minor arcs

$\widehat{AD}, \widehat{CE}, \widehat{AC}$, and $\widehat{DE}$ are minor arcs.

b. the semicircles

$\widehat{ACE}, \widehat{CED}, \widehat{EDA}$, and $\widehat{DAC}$ are semicircles.

c. the major arcs that contain point A

$\widehat{ACD}, \widehat{CEA}, \widehat{EDC}$, and $\widehat{DAE}$ are major arcs that contain point A.

The water line separates a circle into a major arc and a minor arc.

✓ **Quick Check** **2** Identify the four major arcs of ⊙O that contain point E.

Adjacent arcs are arcs of the same circle that have exactly one point in common. You can add the measures of adjacent arcs just as you can add the measures of adjacent angles.

 Key Concepts

Postulate 10-1	**Arc Addition Postulate**

The measure of the arc formed by two adjacent arcs is the sum of the measures of the two arcs.

$$m\widehat{ABC} = m\widehat{AB} + m\widehat{BC}$$

3 EXAMPLE Finding the Measures of Arcs

Find the measure of each arc.

a. $\overset{\frown}{BC}$ $m\overset{\frown}{BC} = m\angle BOC = 32$

b. $\overset{\frown}{BD}$ $m\overset{\frown}{BD} = m\overset{\frown}{BC} + m\overset{\frown}{CD}$
             $m\overset{\frown}{BD} = 32 + 58 = 90$

c. $\overset{\frown}{ABC}$ $\overset{\frown}{ABC}$ is a semicircle.
             $m\overset{\frown}{ABC} = 180$

d. $\overset{\frown}{AB}$ $m\overset{\frown}{AB} = 180 - 32 = 148$

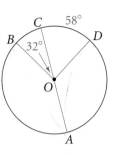

✓ Quick Check ③ Find $m\angle COD$, $m\overset{\frown}{CDA}$, $m\overset{\frown}{AD}$ and $m\overset{\frown}{BAD}$.

---

2   Circumference and Arc Length

The **circumference** of a circle is the distance around the circle. The number **pi** (π) is the ratio of the circumference of a circle to its diameter.

🔑 Key Concepts

> **Theorem 10-9**   **Circumference of a Circle**
>
> The circumference of a circle is π times the diameter.
>
> $$C = \pi d \text{ or } C = 2\pi r$$
>
>

Since the number π is irrational, you cannot write it as a terminating or repeating decimal. To approximate π, you can use 3.14, $\frac{22}{7}$, or the $\boxed{\pi}$ key on your calculator.

Circles that lie in the same plane and have the same center are **concentric circles.**

4 EXAMPLE Real-World 🌐 Connection

Automobiles A car has a turning radius of 16.1 ft. The distance between the two front tires is 4.7 ft. In completing the (outer) turning circle, how much farther does a tire travel than a tire on the concentric inner circle?

To find the radius of the inner circle, subtract 4.7 ft from the turning radius.

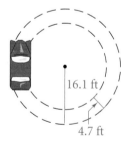

circumference of outer circle $= C = 2\pi r = 2\pi(16.1) = 32.2\pi$

radius of the inner circle $= 16.1 - 4.7 = 11.4$

circumference of inner circle $= C = 2\pi r = 2\pi(11.4) = 22.8\pi$

The difference in the two distances is $32.2\pi - 22.8\pi$, or 9.4π.

$9.4\pi \approx$ **29.530971**   **Use a calculator.**

A tire on the turning circle travels about 29.5 ft farther than a tire on the inner circle.

✓ Quick Check ④ The diameter of a bicycle wheel is 22 in. To the nearest whole number, how many revolutions does the wheel make when the bicycle travels 100 ft?

The measure of an arc is in degrees while the **arc length** is a fraction of a circle's circumference. An arc of 60° represents $\frac{60}{360}$ or $\frac{1}{6}$ of the circle. Its arc length is $\frac{1}{6}$ the circumference of the circle. This observation suggests the following theorem.

 **Key Concepts**

| Theorem 10-10 | Arc Length |

The length of an arc of a circle is the product of the ratio $\frac{\text{measure of the arc}}{360}$ and the circumference of the circle.

$$\text{length of } \overarc{AB} = \frac{m\overarc{AB}}{360} \cdot 2\pi r$$

5 EXAMPLE Finding Arc Length

Find the length of each arc shown in red. Leave your answer in terms of π.

a.

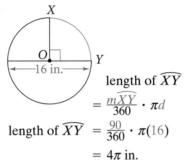

length of $\overarc{XY}$
$= \frac{m\overarc{XY}}{360} \cdot \pi d$

length of $\overarc{XY} = \frac{90}{360} \cdot \pi(16)$

$= 4\pi$ in.

b.

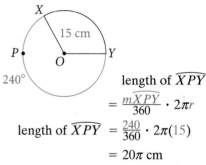

length of $\overarc{XPY}$
$= \frac{m\overarc{XPY}}{360} \cdot 2\pi r$

length of $\overarc{XPY} = \frac{240}{360} \cdot 2\pi(15)$

$= 20\pi$ cm

 **Quick Check** **5** Find the length of a semicircle with radius 1.3 m. Leave your answer in terms of π.

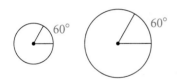

It is possible for two arcs of different circles to have the same measure but different lengths, as shown at the left. It is also possible for two arcs of different circles to have the same length but different measures. **Congruent arcs** are arcs that have the same measure *and* are in the same circle or in congruent circles.

EXERCISES

For more exercises, see *Extra Skill, Word Problem, and Proof Practice.*

Practice and Problem Solving

A Practice by Example

Example 1
(page 566)

GO for Help

Trash **The graph shows types of trash in a typical American city. Find the measure of each central angle to the nearest whole number.**

1. Glass   2. Metals

3. Plastics   4. Wood

5. Food Waste   6. Yard Waste

7. Other   8. Paper and Paperboard

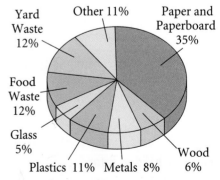

Source: Environmental Protection Agency, 2003.
Go to www.PHSchool.com for a data update.
Web Code: aug-9041

Example 2
(page 567)

Identify the following in ⊙O.

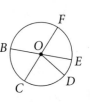

9. a minor arc   10. a major arc

11. a semicircle   12. a pair of adjacent arcs

13. an acute central angle   14. a pair of congruent angles

Example 3
(page 568)

Find the measure of each arc in ⊙P.

15. $\overset{\frown}{TC}$ 16. $\overset{\frown}{TBD}$ 17. $\overset{\frown}{BTC}$ 18. $\overset{\frown}{TCB}$

19. $\overset{\frown}{CD}$ 20. $\overset{\frown}{CBD}$ 21. $\overset{\frown}{TCD}$ 22. $\overset{\frown}{DB}$

23. $\overset{\frown}{TDC}$ 24. $\overset{\frown}{TB}$ 25. $\overset{\frown}{BC}$ 26. $\overset{\frown}{BCD}$

Example 4
(page 568)

Find the circumference of each circle. Leave your answer in terms of π.

27.
20 cm

28.
3 ft

29.
4.2 m

30.
14 in.

31.
$\frac{1}{2}$ m

32.
29 cm

33. The wheel of an adult's bicycle has diameter 26 in. The wheel of a child's bicycle has diameter 18 in. To the nearest inch, how much farther does the larger bicycle wheel travel in one revolution than the smaller bicycle wheel?

Example 5
(page 569)

Find the length of each arc shown in red. Leave your answer in terms of π.

34.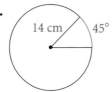
14 cm   45°

35.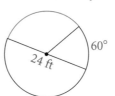
24 ft   60°

36.
18 m

37.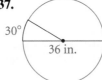
30°   36 in.

38.
23 m

39.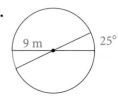
9 m   25°

Ⓑ Apply Your Skills

40. Use a compass to draw ⊙A and ⊙B with different radii. Then use a protractor to draw $\overset{\frown}{XY}$ on ⊙A and $\overset{\frown}{ZW}$ on ⊙B so that $m\overset{\frown}{XY} = m\overset{\frown}{ZW}$. Is $\overset{\frown}{XY} \cong \overset{\frown}{ZW}$?

41. **Surveys** Use the data in the table to construct a circle graph.

Interest in Languages by Students at McClellan High School

German	24%
Japanese	13%
Chinese	12%
French	25%
Spanish	26%

GO Online
Homework Video Tutor
Visit: PHSchool.com
Web Code: aue-1006

Real-World 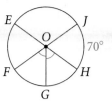 Connection

In 5 minutes, the tip of the minute hand of Boston's Custom House Tower travels 6 ft 10 in.

Find each indicated measure for ⊙O.

42. $m\angle EOF$ **43.** $m\widehat{EJH}$ **44.** $m\widehat{FH}$

45. $m\angle FOG$ **46.** $m\widehat{JEG}$ **47.** $m\widehat{HFJ}$

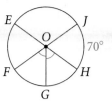

48. Open-Ended Make a circle graph showing how you spend a 24-hour weekday.

🌐 **Time Hands of a clock suggest an angle whose measure is continually changing.**

49. Through how many degrees does a minute hand move in each time interval?
 a. 1 minute   **b.** 5 minutes   **c.** 20 minutes

50. Through how many degrees does an hour hand move in each time interval?
 a. 1 minute   **b.** 5 minutes   **c.** 20 minutes

51. What is the measure of the angle formed by the hands of a clock at 7:20?

x^2 **Algebra Find the value of each variable.**

52.

53.

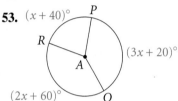

The circumference of a circle is 100π in. Find each of the following.

54. the diameter   **55.** the radius

56. the length of an arc of 120°

57. Multiple Choice Five streets come together at a traffic circle. The diameter of the circle is 200 ft. If traffic travels counterclockwise, what is the approximate distance from East St. to Neponset St.?
 Ⓐ 227 ft   Ⓑ 244 ft
 Ⓒ 454 ft   Ⓓ 488 ft

Problem Solving Hint
For Exercise 58, draw ⊙A and ⊙B concentric. Draw 60° and 45° angles that share a side. To have equal arc lengths, which circle must be larger?

58. A 60° arc of ⊙A has the same length as a 45° arc of ⊙B. Find the ratio of the radius of ⊙A to the radius of ⊙B.

🌐 **59. Metalworking** Nina designed an arch made of wrought iron for the top of a mall entrance. The 11 segments between the two concentric semicircles are each 3 ft long. Find the total length of wrought iron used to make this structure. Round your answer to the nearest foot.

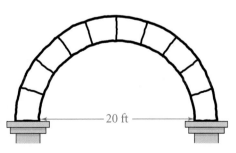

60. History In Exercise 31 on page 133, you learned that in 220 B.C., Eratosthenes estimated the circumference of Earth. He did so by finding that on a great circle of Earth, an arc of approximately 500 mi has a central angle of 7.2°.
 a. Use Eratosthenes's measurements to estimate the circumference of Earth.
 b. Compare your answer in part (a) to the actual circumference of Earth (at the equator) of 24,902 mi.

The Distance and Midpoint Formulas are on pages 53 and 55.

Coordinate Geometry A diameter of a circle has endpoints $A(1, 3)$ and $B(4, 7)$. Find each of the following.

61. the coordinates of the center

62. the circumference

Find the length of each arc shown in red. Leave your answer in terms of π.

63.

4.1 ft
45°

64.

50°
7.2 in.

65.

6 m

Use what you learn from Calvin's father to answer Exercises 66 and 67.

Calvin and Hobbes   by Bill Watterson

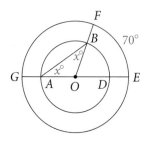

66. In one revolution, how much farther does a point 10 cm from the center of the record travel than a point 3 cm from the center? Round your answer to the nearest tenth.

67. Writing Kendra and her mother plan to ride the carousel. Two horses on the carousel are side by side. For a more exciting ride, should Kendra sit on the inside or the outside? Explain your reasoning.

68. In $\odot O$, the length of $\overset{\frown}{AB}$ is 6π cm and $m\overset{\frown}{AB}$ is 120. What is the diameter of $\odot O$?

69. Coordinate Geometry Find the length of a semicircle with endpoints $(3, 7)$ and $(3, -1)$. Round your answer to the nearest tenth.

Challenge

70. The two circles shown below are concentric.
  a. Name two arcs that have the same measure.
  b. Find the value of x.

71. Find the perimeter of the shaded portion of the figure below. Leave your answer in terms of π. Explain your reasoning and state what assumptions you make.

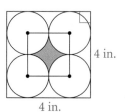
4 in.
4 in.

72. Sports An athletic field is a rectangle, 100 yd by 40 yd, with a semicircle at each of the short sides. A running track 10 yd wide surrounds the field. If the track is divided into eight lanes of equal width, find the distance around the track along the inside edge of each lane.

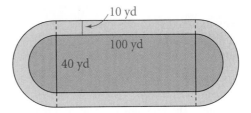

10 yd

100 yd

40 yd

Real-World Connection

The track is longer for runners on the outside, so the start of the race is staggered.

Test Prep

Multiple Choice

73. The radius of a circle is 12 cm. What is the length of a 60° arc?
   A. 3π cm   **B.** 4π cm   **C.** 5π cm   **D.** 6π cm

74. A 240° arc has length 16π ft. What is the radius of the circle?
   F. 6 ft   **G.** 12 ft   **H.** 15 ft   **J.** 24 ft

Short Response

75. Amy is constructing a curved path through a rectangular yard. She will edge the two sides of the curved path with plastic edging. Find the total length, in meters, of plastic edging she will need. Show your work or explain how you found the total.

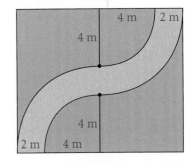

Mixed Review

Lesson 10-5

Part of a regular 12-gon is shown at the right.

76. Find the measure of each numbered angle.

77. The radius is 19.3 mm. Find the apothem.

78. Find the area of the 12-gon to the nearest square millimeter.

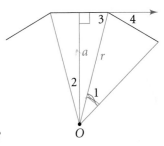

Lesson 6-3

Can you conclude that the figure is a parallelogram? Explain.

79.

80.

81.

Lesson 3-7

Indicate whether each statement is *always, sometimes,* or *never* true.

82. Two nonvertical parallel lines have the same slope.

83. Two perpendicular lines have slopes that are reciprocals.

Dimensional Analysis

FOR USE WITH LESSON 10-7

You can use conversion factors to change from one unit of measure to another. The process of analyzing units to decide which conversion factors to use is called *dimensional analysis*.

Since 60 min = 1 h, $\frac{60 \text{ min}}{1 \text{ h}}$ equals 1. You can use $\frac{60 \text{ min}}{1 \text{ h}}$ to convert hours to minutes.

$7 \text{ h} \cdot \frac{60 \text{ min}}{1 \text{ h}} = 420 \text{ min}$ **The hour units cancel, and the result is minutes.**

Sometimes you need to use a conversion factor more than once.

EXAMPLE

The area of the top of a circular table is 8 ft². Convert the area to square inches.

You need to convert feet to inches.

$$\begin{array}{cc} \text{feet to} & \text{feet to} \\ \text{inches} & \text{inches} \\ \downarrow & \downarrow \end{array}$$

$8 \text{ ft} \cdot \text{ft} \cdot \frac{12 \text{ in.}}{1 \text{ ft}} \cdot \frac{12 \text{ in.}}{1 \text{ ft}} = 8 \text{ ft} \cdot \text{ft} \cdot \frac{12 \text{ in.}}{1 \text{ ft}} \cdot \frac{12 \text{ in.}}{1 \text{ ft}}$ **The feet units cancel. The result is square inches.**

$\qquad\qquad = 8 \cdot 12 \cdot 12 \text{ in.}^2$ **Simplify.**

$\qquad\qquad = 1152 \text{ in.}^2$

The area of the table top is 1152 in.².

EXERCISES

Choose the correct conversion factor for changing the units.

1. centimeters to meters

 A. $\frac{100 \text{ cm}}{1 \text{ m}}$ B. $\frac{1 \text{ m}}{100 \text{ cm}}$

2. yards to feet

 A. $\frac{3 \text{ ft}}{1 \text{ yd}}$ B. $\frac{1 \text{ yd}}{3 \text{ ft}}$

3. inches to yards

 A. $\frac{36 \text{ in.}}{1 \text{ yd}}$ B. $\frac{1 \text{ yd}}{36 \text{ in.}}$

4. square feet to square inches

 A. $\frac{144 \text{ in.}^2}{1 \text{ ft}^2}$ B. $\frac{1 \text{ ft}^2}{144 \text{ in.}^2}$

5. cubic meters to cubic kilometers

 A. $\frac{1{,}000{,}000{,}000 \text{ m}^3}{1 \text{ km}^3}$ B. $\frac{1 \text{ km}^3}{1{,}000{,}000{,}000 \text{ m}^3}$

Write each quantity in the given unit.

6. 4 m = ■ cm

7. 360 in. = ■ yd

8. 9 mm = ■ cm

9. 17 yd = ■ ft

10. 2.5 ft = ■ in.

11. 35 m = ■ km

12. 2 yd = ■ in.

13. 23 cm = ■ mm

14. 2 ft² = ■ in.²

15. 500 mm² = ■ cm²

16. 840 in.² = ■ ft²

17. 3 km² = ■ cm²

18. 7 m² = ■ km²

19. 360 in.² = ■ yd²

20. 900 cm³ = ■ m³

21. 4 yd³ = ■ ft³

22. 75 m² = ■ km²

23. 250 in.² = ■ yd²

24. 3 km³ = ■ m³

25. 2 km² = ■ mm²

26. 60 in.² = ■ yd²

10-7

Areas of Circles and Sectors

 Check Skills You'll Need

GO **for Help** Lesson 10-6

1. What is the radius of a circle with diameter 9 cm?
2. What is the diameter of a circle with radius 8 ft?
3. Find the circumference of a circle with diameter 12 in.
4. Find the circumference of a circle with radius 3 m.

 New Vocabulary • sector of a circle • segment of a circle

---

1   **Finding Areas of Circles and Parts of Circles**

Hands-On Activity: Exploring the Area of a Circle

- Use a compass to draw a large circle. Fold the circle horizontally and vertically. Cut the circle into four wedges on the fold lines.

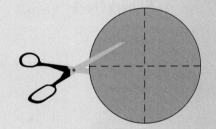

- Fold each wedge into quarters. Cut each wedge on the fold lines. You will have 16 wedges.

- Tape the wedges to a piece of paper to form the figure shown here.

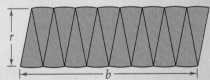

1. How does the area of the figure compare with area of the circle?

2. The base of the figure is formed by arcs of the circle. Explain how the length b relates to the circumference C of the circle.

3. Explain how the length b relates to the radius r of the circle.

4. If you increase the number of wedges, the figure you create becomes more and more like a rectangle with base b and height r. Write an expression for the area of the rectangle in terms of r.

In the diagrams on the preceding page,

area of a circle = area of a "parallelogram" ≈ $b \cdot r \approx \frac{1}{2}C \cdot r = \pi r^2$,

and the approximations improve as the circle is cut into more pieces.

 Key Concepts

Theorem 10-11   **Area of a Circle**

The area of a circle is the product of π and the square of the radius.

$$A = \pi r^2$$

---

1 EXAMPLE   **Real-World Connection**

Food How much more pizza is in a 12-in.-diameter pizza than in a 10-in. pizza?

radius of small pizza = $\frac{10}{2}$ = 5

radius of medium pizza = $\frac{12}{2}$ = 6

Find the radii.

area of small pizza = $\pi(5)^2 = 25\pi$

area of medium pizza = $\pi(6)^2 = 36\pi$

Use the formula for area of a circle.

difference in area = $36\pi - 25\pi = 11\pi$

≈ 34.557519 **Use a calculator.**

- There is about 35 in.^{2} more pizza in the medium pizza.

 Quick Check ❶ How much more pizza is in a 14-in.-diameter pizza than in a 12-in. pizza?

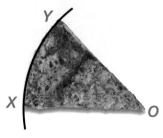

A **sector of a circle** is a region bounded by an arc of the circle and the two radii to the arc's endpoints. You name a sector using one arc endpoint, the center of the circle, and the other arc endpoint. The slice of pizza at the left is sector XOY of a circle O.

The area of a sector is a fractional part of the area of a circle. The ratio of a sector's area to a circle's area is $\frac{\text{measure of the arc}}{360}$.

 Key Concepts

Theorem 10-12   **Area of a Sector of a Circle**

The area of a sector of a circle is the product of the ratio $\frac{\text{measure of the arc}}{360}$ and the area of the circle.

Area of sector $AOB = \frac{m\widehat{AB}}{360} \cdot \pi r^2$

---

2 EXAMPLE   **Finding the Area of a Sector of a Circle**

Find the area of sector ZOM. Leave your answer in terms of π.

area of sector $ZOM = \frac{m\widehat{ZM}}{360} \cdot \pi r^2$

$= \frac{72}{360} \cdot \pi(20)^2$

$= 80\pi$

- The area of sector ZOM is 80π cm^2.

 Quick Check ❷ **Critical Thinking** A circle has a diameter of 20 cm. What is the area of a sector bounded by a 208° major arc? Round your answer to the nearest tenth.

This pizza is cut into two segments.

A part of a circle bounded by an arc and the segment joining its endpoints is a **segment of a circle.** To find the area of a segment for a minor arc, draw radii to form a sector. The area of the segment equals the area of the sector minus the area of the triangle formed.

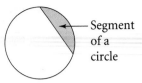

Area of sector − Area of triangle = Area of segment

3 EXAMPLE **Finding the Area of a Segment of a Circle**

Find the area of the shaded segment. Round your answer to the nearest tenth.

$\text{area of sector } AOB = \frac{m\overarc{AB}}{360} \cdot \pi r^2$ **Use the formula for area of a sector.**

$= \frac{90}{360} \cdot \pi(10)^2$ **Substitute.**

$= \frac{1}{4} \cdot 100\pi$

$= 25\pi$

$\text{area of } \triangle AOB = \frac{1}{2}bh$ **Use the formula for area of a triangle.**

$= \frac{1}{2}(10)(10)$ **Substitute.**

$= 50$

$\text{area of segment} = 25\pi - 50$

≈ 28.539816 **Use a calculator.**

The area of the segment is about 28.5 in.^{2}

✓ **Quick Check** ❸ A circle has a radius of 12 cm. Find the area of the smaller segment of the circle determined by a 60° arc. Round your answer to the nearest tenth.

EXERCISES
For more exercises, see *Extra Skill, Word Problem, and Proof Practice.*

Practice and Problem Solving

A Practice by Example

Example 1
(page 576)

GO for Help

Find the area of each circle. Leave your answer in terms of π.

1. 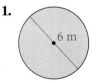 6 m

2. 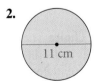 11 cm

3. 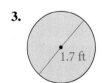 1.7 ft

4. 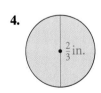 $\frac{2}{3}$ in.

5. **Agriculture** Some farmers use a circular irrigation method. An irrigation arm acts as the radius of an irrigation circle. How much more land is covered with an irrigation arm of 300 ft than by an irrigation arm of 250 ft?

6. What is the difference in the areas of a circular table with diameter 6 ft and a circular table with diameter 8 ft?

Example 2
(page 576)

Find the area of each shaded sector of a circle. Leave your answer in terms of π.

7.

8.

9.

10.

11.

12.

Find the area of sector *TOP* in ⊙*O* using the given information. Leave your answer in terms of π.

13. $r = 5$ m, $m\widehat{TP} = 90°$

14. $r = 6$ ft, $m\widehat{TP} = 15°$

15. $d = 16$ in., $m\widehat{PT} = 135°$

16. $d = 15$ cm, $m\widehat{PT} = 180°$

Example 3
(page 577)

Find the area of each shaded segment. Round your answer to the nearest tenth.

17.

18.

19.

A circle has the given radius. Find the area of the smaller segment of the circle determined by an arc with the given measure. Round to the nearest tenth.

20. radius 15 m, arc 60°

21. radius 14 cm, arc 120°

B **Apply Your Skills**

Find the area of the shaded region. Leave your answer in terms of π and in simplest radical form.

22.

23.

24.

25.

26.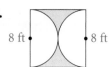

27.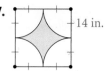

28. Multiple Choice The diver at the right is collecting samples from the ocean floor. The line to the diver is 100 ft long, and the diver is working at a depth of 80 ft. What is the approximate area of the circle that the diver can cover?

Ⓐ 11,300 ft²   Ⓑ 25,400 ft²

Ⓒ 31,400 ft²   Ⓓ 51,400 ft²

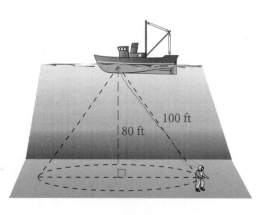

29. **Writing** The American Institute of Baking suggests a technique for cutting and serving a tiered cake. The tiers of a cake have the same height and have radii 8 in. and 13 in. The top tier and the cake directly under it are each cut into 8 wedges as shown. The outer ring of the 13-inch tier is cut into 12 pieces. Which would be larger, a piece from the top or a piece from the outer ring? Explain.

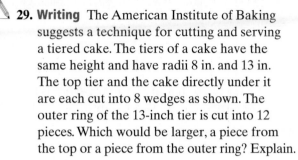
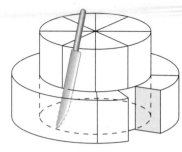

30. How many circles with radius 4 in. will have the same total area as a circle with radius 12 in.?

31. **Games** A dart board has diameter 20 in. and is divided into 20 congruent sectors. Find the area of one sector. Round your answer to the nearest tenth.

32. In a circle, a 90° sector has area 36π in.². What is the circle's radius?

33. **Open-Ended** Draw a circle and a sector so that the area of the sector is 16π cm². Give the radius of the circle and the measure of the arc of the sector.

34. A method for finding the area of a segment determined by a minor arc is described on page 577.
   a. Describe two ways to find the area of a segment determined by a major arc.
   b. If $m\overarc{AB} = 90$ in a circle of radius 10, find the areas of the two segments determined by $\overarc{AB}$.

🔴 **Challenge**

Find the area of the shaded region. Leave your answer in terms of π.

35.

7 m

36.

10 m

37.
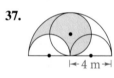
4 m

38. Circle *O* at the right is inscribed in square *ABCD* and circumscribed about square *PQRS*. Which is smaller, the blue region or the yellow region? Explain.

39. Circles *T* and *U* each have a radius 10 and *TU* = 10. Find the area of the region that is contained inside both circles.

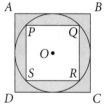

🌐 40. **Recreation** An 8 ft-by-10 ft floating dock is anchored in the middle of a pond. The bow of a canoe is tied to a corner of the dock with a 10-ft rope as shown in the picture below.
   a. Sketch a diagram of the region in which the bow of the canoe can travel.
   b. Write a plan for finding the area.
   c. Find the area. Round your answer to the nearest square foot.

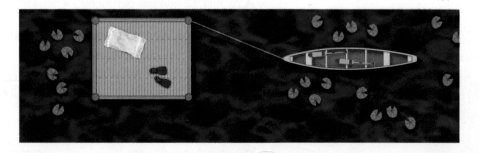

Multiple Choice

41. A circle has area 72π yd². What is the area of a 10° sector of the circle?
  A. 2π yd²   **B.** 3π yd²   **C.** 4π yd²   **D.** 6π yd²

42. A sector of 90° has area π mm². What is the area of the circle?
  F. 2π mm²   **G.** 4π mm²   **H.** 8π mm²   **J.** 4 mm²

Short Response

43. Each of three water sprinklers covers a semicircle of radius 2 m. The shaded region remains dry. Find the area of the shaded region to the nearest square meter. Show your work or explain how you found the area.

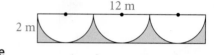

Mixed Review

GO for Help

Lesson 10-6

Find the length of $\overarc{AB}$ in each circle. Leave your answers in terms of π.

44.

45.

46.

Lesson 6-5

47. Three sides of a trapezoid are congruent. The fourth side is 4 in. longer than each of the other three. The perimeter is 49 in. Find the length of each side.

Lesson 4-7

48. Write a proof. (*Hint:* First prove overlapping triangles are congruent.)

  Given: $\overline{AC} \cong \overline{BC}, \angle A \cong \angle B$
  Prove: $\triangle BDF \cong \triangle AEF$

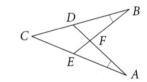

✓ Checkpoint Quiz 2   **Lessons 10-4 through 10-7**

The similarity ratio of △*ABC* to △*DEF* is 3 : 5. Fill in the missing information.

1. The perimeter of $\triangle ABC$ is 36 in.
The perimeter of $\triangle DEF$ is _?_.

2. The area of $\triangle ABC$ is _?_.
The area of $\triangle DEF$ is 125 in.².

3. The areas of two similar triangles are 1.44 and 1.00. Find their similarity ratio.

4. Find the area of $\triangle ABC$ if $a = 4$ ft, $b = 7$ ft and $m\angle C = 20$.

5. A regular 12-gon has perimeter 24 cm. Find its area to the nearest tenth.

Find the area of each shaded region. Leave answers in terms of π.

6.

7.

8.

9.

10. In a circle of radius 18 mm, $m\overarc{AB} = 45$. Find the length of $\overarc{AB}$ in terms of π.

Exploring Area and Circumference

A polygon that is *inscribed* in a circle has all its vertices on the circle. Work in pairs or small groups. Investigate the ratios of the perimeters and areas of inscribed regular polygons to the circumference and area of the circle in which they are inscribed.

> **NY** **Prepares for G.G.12:** Know and apply that the volume of a prism is the product of the area of the base and the altitude.

Begin by making a table like this.

Regular Polygon			Circle		Ratios	
Sides	Perimeter	Area	Circumference	Area	Perimeter / Circumference	Polygon Area / Circle Area
3						

Construct

Use geometry software to construct a circle. Find its circumference and area and record them in your table. Inscribe an equilateral triangle in the circle. Your software may be able to do this for you automatically, or you can construct three points on the circle and move them so they are approximately evenly spaced on the circle. Then draw a triangle.

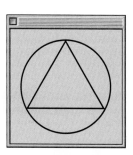

Investigate

Use your geometry software to measure the perimeter and area of the triangle and to calculate the ratios $\frac{\text{triangle perimeter}}{\text{circle circumference}}$ and $\frac{\text{triangle area}}{\text{circle area}}$. Record the results.

Manipulate the circle to change its size. Do the ratios you calculate stay the same or change?

Now inscribe a square in a circle and fill in your table for a polygon of four sides. Do the same for a regular pentagon.

EXERCISES

1. Make a Conjecture What will happen to the ratios
$\frac{\text{perimeter}}{\text{circumference}}$ and $\frac{\text{polygon area}}{\text{circle area}}$
as you increase the number of sides of the polygon?

Extend

2. Extend your table to include polygons of 12 sides.
   a. Does your conjecture still hold?
   b. Compare the two columns of ratios in your table. How do they differ?

3. Estimate the perimeter and area of a polygon of 100 sides that is inscribed in a circle with a radius of 10 cm.

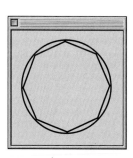

regular octagon

Geometric Probability

What You'll Learn

- To use segment and area models to find the probabilities of events

. . . And Why

To find the probability of winning a carnival game, as in Example 4

✓ **Check Skills You'll Need**

 **for Help** Skills Handbook pages 756 and 762

Find and simplify each ratio.

1. $\dfrac{BD}{AE}$ **2.** $\dfrac{CE}{AF}$ **3.** $\dfrac{AB}{BC}$

A	B		C	D			E	F

```
0  1  2  3  4  5  6  7  8  9  10
```

4. Two circles have radii 1 m and 2 m, respectively. What is the simplest form of the fraction with numerator equal to the area of the smaller circle and denominator equal to the area of the larger circle?

You roll a number cube. Find the probability of rolling each of the following.

5. 4

6. an odd number

7. 2 or 5

8. a prime number

◀ᴗ))) **New Vocabulary** • geometric probability

---

1 Using Segment and Area Models

Vocabulary Tip

<u>P(event)</u> is read "the probability of an event."

You may recall that the probability of an event is the ratio of the number of favorable outcomes to the number of possible outcomes.

$$P(\text{event}) = \frac{\text{favorable outcomes}}{\text{possible outcomes}}$$

Sometimes you can use a **geometric probability** model in which you let points represent outcomes. You find probabilities by comparing measurements of sets of points. For example, if points of segments represent outcomes, then

$$P(\text{event}) = \frac{\text{length of favorable segment}}{\text{length of entire segment}}.$$

1 EXAMPLE Finding Probability Using Segments

A gnat lands at a random point on the ruler's edge. Find the probability that the point is between 3 and 7.

$P(\text{landing between 3 and 7}) = \dfrac{\text{length of favorable segment}}{\text{length of entire segment}} = \dfrac{4}{12}$, or $\dfrac{1}{3}$

✓ **Quick Check** ❶ A point on $\overline{AB}$ is selected at random. What is the probability that it is a point on $\overline{CD}$?

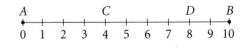

You can use a segment model to find the probability of how long you will wait for a bus.

2 EXAMPLE **Real-World Connection**

Commuting Elena's bus runs every 25 minutes. If she arrives at her bus stop at a random time, what is the probability that she will have to wait at least 10 minutes for the bus?

Assume that a stop takes very little time, and let $\overline{AB}$ represent the 25 minutes between buses.

If Elena arrives at any time between A and C, she has to wait at least 10 minutes until B.

$$P(\text{waiting at least 10 min}) = \frac{\text{length of } \overline{AC}}{\text{length of } \overline{AB}} = \frac{15}{25}, \text{ or } \frac{3}{5}$$

The probability that Elena will have to wait at least 10 minutes for the bus is $\frac{3}{5}$ or 60%.

✓ Quick Check ❷ What is the probability that Elena will have to wait no more than 10 minutes for the bus?

If the points of a region represent equally-likely outcomes, then you can find probabilities by comparing areas.

$$P(\text{event}) = \frac{\text{area of favorable region}}{\text{area of entire region}}$$

3 EXAMPLE **Finding Probability Using Area**

Target Game Assume that a dart you throw will land on the 1-ft square dartboard and is equally likely to land at any point on the board. Find the probability of hitting each of the blue, yellow, and red regions. The radii of the concentric circles are 1, 2, and 3 inches, respectively.

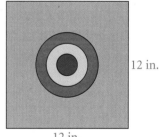

12 in.

12 in.

$$P(\text{blue}) = \frac{\text{area of blue region}}{\text{area of square}} = \frac{\pi(1)^2}{12^2} = \frac{\pi}{144} \approx 0.022, \text{ or } 2.2\% \quad \textbf{Use a calculator.}$$

$$P(\text{yellow}) = \frac{\text{area of yellow region}}{\text{area of square}} = \frac{\pi(2)^2 - \pi(1)^2}{12^2} = \frac{3\pi}{144} \approx 0.065, \text{ or } 6.5\%$$

$$P(\text{red}) = \frac{\text{area of red region}}{\text{area of square}} = \frac{\pi(3^2) - \pi(2)^2}{12^2} = \frac{5\pi}{144} \approx 0.109, \text{ or } 10.9\%$$

The probabilities of hitting the blue, yellow, and red regions are about 2.2%, 6.5%, and 10.9%, respectively.

✓ Quick Check ❸ If you change the blue circle as indicated, how does the probability of hitting the blue circle change? Explain.
a. Double the radius.　　　　　　　　　　**b.** Triple the radius.

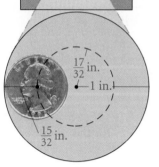

As Example 3 suggests, you can apply geometric probability to some games. This can help you decide how easy or difficult it may be to win such games.

4 EXAMPLE **Real-World Connection**

Coin Toss To win a prize in a carnival game, you must toss a quarter so that it lands entirely within the circle as shown at the left. Find the probability of this happening on one toss. Assume that the center of a tossed quarter is equally likely to land at any point within the 8-in. square.

The radius of the circle is 1 in. The radius of a quarter is $\frac{15}{32}$ in. The favorable points are those that are less than $\frac{17}{32}$ in. from the center of the circle. They are the points within the dashed circle.

$$P(\text{quarter landing in circle}) = \frac{\text{area of dashed circle}}{\text{area of square}}$$

$$= \frac{\pi\left(\frac{17}{32}\right)^2}{8^2} \approx 0.014, \text{ or } 1.4\%$$

● The probability of a quarter landing in the circle is about 1.4%.

✓ Quick Check ❹ **Critical Thinking** Suppose you toss 100 quarters. Would you expect to win a prize? Explain.

EXERCISES

For more exercises, see *Extra Skill, Word Problem, and Proof Practice.*

Practice and Problem Solving

A **Practice by Example**

Example 1
(page 582)

GO for Help

Find the probability that a point chosen at random from $\overline{AK}$ is on the given segment.

```
A  B  C  D  E  F  G  H  I  J  K
+--+--+--+--+--+--+--+--+--+--+
0  1  2  3  4  5  6  7  8  9  10
```

1. $\overline{CH}$ **2.** $\overline{FG}$ **3.** $\overline{DJ}$ **4.** $\overline{EI}$ **5.** $\overline{AK}$

6. Points M and N are on $\overline{ZB}$ with $ZM = 5, NB = 9,$ and $ZB = 20.$ A point is chosen at random from $\overline{ZB}.$ What is the probability that the point is on $\overline{MN}$?

Example 2
(page 583)

7. Transportation A rapid transit line runs trains every 10 minutes. Draw a geometric model and find the probability that randomly arriving passengers will not have to wait more than 4 minutes.

Traffic Patterns **Main Street intersects each street below. The traffic lights on Main follow the cycles shown. As you travel along Main and approach the intersection, what is the probability that the first color you see is green?**

8. Durham Avenue: green 30 s, yellow 5 s, red 25 s

9. Martin Luther King Boulevard: green 20 s, yellow 5 s, red 50 s

10. Yonge Street: green 40 s, yellow 5 s, red 25 s

11. International Drive: green 25 s, yellow 5 s, red 45 s

12. Tamiami Trail: green 35 s, yellow 8 s, red 32 s

13. Flutie Pass: green 50 s, yellow 4 s, red 26 s

14. During May, a certain drawbridge over the Intracoastal Waterway is raised every half hour to allow boats to pass. It remains open for 5 min. What is the probability that a motorist arriving at the bridge in May will find it raised?

Examples 3, 4
(pages 583 and 584)

Target Games Darts are thrown at each of the boards shown below. A dart hits the board at a random point. Judging by appearances, find the probability that it will land in the shaded region.

15.

16.

17.

18.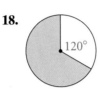

19.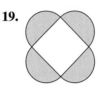

20.

Real-World 🌐 **Connection**

An archer receives from 1 to 10 points for an arrow that hits the target. A hit in the center zone is worth 10 points.

B Apply Your Skills

21. Archery An archery target with a radius of 61 cm has 5 scoring zones formed by concentric circles. The colors of the zones are yellow, red, blue, black, and white. The radius of the yellow circle is 12.2 cm. The width of each ring is also 12.2 cm. If an arrow hits the target at a random point, what is the probability that it hits the center yellow zone?

22. $\overline{BZ}$ contains $\overline{MN}$ and $BZ = 20$. A point is chosen at random from $\overline{BZ}$. The probability that the point is also on $\overline{MN}$ is 0.3, or 30%. Find MN.

🌐 **Target Games** A dart hits each square dartboard at a random point. Find the probability that the dart lands inside a circle. Leave your answer in terms of π.

23.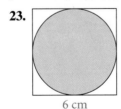
6 cm

24.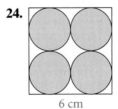
6 cm

25.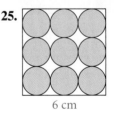
6 cm

26. A dartboard is a square of radius 10 in. You throw a dart and hit the target. Find the probability that the dart lies within $\sqrt{10}$ in. of the center of the square.

27. Critical Thinking Use the information given in Example 4.
 a. For each 1000 quarters tossed, about how many prizes would be won?
 b. Suppose the game prize costs the carnival $10. About how much profit would the carnival expect for every 1000 quarters tossed?

🌐 **28. Commuting** Suppose a bus arrives at a bus stop every 25 min and waits 5 min before leaving. Sketch a geometric model. Use it to find the probability that a person has to wait more than 10 min for a bus to leave.

🌐 **29. Traffic Patterns** The traffic lights at Fourth and Commercial Streets repeat themselves in 60-second cycles. Ms. Li regularly has students drive on Fourth Street through the Commercial Street intersection. By experience, she knows that they will face a red light 60% of the time. Use this information to estimate how long the Fourth Street light is red during each 1-min cycle.

GO 🌐 **nline**
Homework Video Tutor
Visit: PHSchool.com
Web Code: aue-1008

For Exercises 30 and 31, sketch a geometric model and solve.

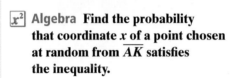

30. Astronomy Meteoroids (mostly dust-particle size) are continually bombarding Earth. The surface area of Earth is about 65.7 million square miles. The area of the United States is about 3.7 million square miles. What is the probability that a meteoroid landing on Earth will land in the United States?

31. Tape Recording Amy made a tape recording of a chorus rehearsal. The recording began 21 min into the 60-min tape and lasted 8 min. Later she accidentally erased a 15-min segment somewhere on the tape.
   a. In your model show the possible starting times of the erasure. Explain how you know that the erasure did not start after the 45-min mark.
   b. In your model show the starting times of the erasures that would erase the entire rehearsal. Find the probability that the entire rehearsal was erased.

Problem Solving Hint
$0 \leq P(\text{event}) \leq 1$
$P(\text{event}) = 0$ means the event will not occur.
$P(\text{event}) = 1$ means the event will occur.

x^2 **Algebra** Find the probability that coordinate x of a point chosen at random from $\overline{AK}$ satisfies the inequality.

A B C D E F G H I J K
0 1 2 3 4 5 6 7 8 9 10

32. $2 \leq x \leq 8$ **33.** $x \geq 7$ **34.** $2x \leq 9$ **35.** $\frac{1}{2}x - 5 \geq 0$

36. $2 \leq 4x \leq 3$ **37.** $0 \leq \frac{1}{3}x + 1 \leq 5$ **38.** $|x - 6| \leq 1.5$ **39.** $\sqrt{2} \leq \pi x \leq \sqrt{10}$

Dunk Tank At a fund-raiser, a volunteer sits on a platform above a tank of water. She gets dunked when you throw a ball and hit the red target. The radius of the ball is 3.6 cm. What is the probability that a ball heading randomly for the given background shape would hit the given target shape?

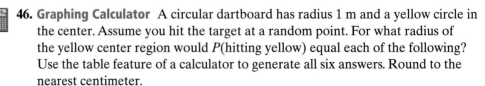

|← 40 cm →|

10 cm

40. Background (at right): a circle 40 cm across
   Target: a circle with 10-cm radius

41. Background: a square with 40-cm sides
   Target: a circle with 10-cm radius

42. Background: a circle 40-cm across   **43.** Background: a square with 40-cm sides
   Target: a square with 20-cm sides   Target: a square with 20-cm sides

Real-World Connection

A mere touch of the target by the ball triggers the dunk.

44. Kimi has a 4-in. straw and a 6-in. straw. She wants to cut the 6-in. straw into two pieces so that the three pieces form a triangle.
   a. If she cuts the straw to get two 3-in. pieces, can she form a triangle?
   b. If the two pieces are 1 in. and 5 in., can she form a triangle?
   c. If Kimi cuts the straw at a random point, what is the probability that she can form a triangle?

45. a. Open-Ended Design a dartboard game to be used at a charity fair. Specify the size and shape of the regions of the board.
   b. Writing Describe the rules for using your dartboard and the prizes that winners receive. Explain how much money you would expect to raise if the game were played 100 times.

ⓒ Challenge

46. Graphing Calculator A circular dartboard has radius 1 m and a yellow circle in the center. Assume you hit the target at a random point. For what radius of the yellow center region would P(hitting yellow) equal each of the following? Use the table feature of a calculator to generate all six answers. Round to the nearest centimeter.
   a. 0.2   **b.** 0.4   **c.** 0.5
   d. 0.6   **e.** 0.8   **f.** 1.0

47. Target Game A target has a central circle and three concentric rings. The diameters of the circles are 2 cm, 6 cm, 10 cm, and 14 cm. Find the probability of landing in the gray region. Compare it with the probability of landing in *either* the blue or red region.

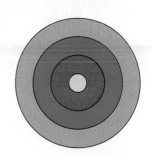

Test Prep

Multiple Choice

48. A dart hits the dartboard shown. Find the probability that it lands in the shaded region.
  A. 21%　　**B.** 25%　　**C.** 50%　　**D.** 79%

49. A dart hits the dartboard shown. Find the probability that it lands in a circle.
  F. 21%　　**G.** 25%　　**H.** 50%　　**J.** 79%

4 m

Short Response

50. On this dartboard, the circle with 1-m radius is inscribed in an equilateral triangle. Find the probability that a dart that hits the board lands in the circular region. Justify your answer.

1 m

Extended Response

51. The radius of a circle is 28 m. The measure of the central angle is 120.
  a. Find the area of the sector in terms of π. Justify your answer.
  b. Find the area of the shaded segment to the nearest tenth. Justify your answer.

120°　28 m

Mixed Review

Lesson 10-7

52. A circle has circumference 20π ft. What is its area?

53. A circle has radius 12 cm. What is the area of a sector of the circle with a 30° central angle?

54. What is the area of a semicircle with diameter 20 ft?

Lesson 6-2

x^2 **Algebra** **Find the values of the variables in each parallelogram.**

55.

$y°$　$x°$　$4x°$

56.

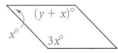

$(y + x)°$　$x°$　$3x°$

Lesson 5-1

57. The coordinates of the vertices of a triangle are $A(1, -4)$, $B(5, 6)$, and $C(-3, 2)$.
  a. Find the coordinates of D, the midpoint of $\overline{AB}$, and E, the midpoint of $\overline{BC}$.
  b. Find the slope of $\overline{DE}$ and the slope of $\overline{AC}$.
  c. Verify that $\overline{DE} \parallel \overline{AC}$.
  d. Find DE and AC.
  e. Verify that $DE = \frac{1}{2}AC$.

Finding Multiple Correct Answers

In Multiple-Correct-Answer questions, you have to determine whether a number of statements are true or false. As you test each statement, mark it as true or false. You then choose those that are true.

EXAMPLE

The perimeter of a rectangle is 20 cm.
Which of the following could be true?
    I. One side of the rectangle is 2 cm.
    II. The area of the rectangle is 24 cm².
    III. The diagonal of the rectangle is 5 cm.

  (A) I only   **(B)** II only   **(C)** I and II only   **(D)** II and III only   **(E)** I, II, and III

Test each statement to see if it could be true in a rectangle with perimeter 20 cm.

Statement I is possible in a rectangle with dimensions 2 cm by 8 cm.

Statement II is possible in a rectangle with dimensions 4 cm by 6 cm.

Statement III is not possible. A diagonal is longer than each side. In any rectangle with perimeter 20 cm, at least one side has to be 5 cm or more.

Only statements I and II could be true. The correct choice is B.

EXERCISES

1. Explain why it is possible for the rectangle in the Example to have a side of length 0.1 cm. What would its area be?

2. In $\triangle ABC$, $AB = \sqrt{1}$, $BC = \sqrt{2}$, and $AC = \sqrt{3}$.
    Which of the following are true?
        I. The largest angle is $\angle B$.
        II. The triangle is a right triangle.
        III. One angle of the triangle has measure 30.

  (A) I only   **(B)** II only   **(C)** I and II only   **(D)** II and III only   **(E)** I, II, and III

3. A square and an equilateral triangle each have apothem of length 1 cm.
    Which of the following are always true?
        I. The perimeter of the square is greater than the perimeter of the triangle.
        II. The area of the square is less than the area of the triangle.
        III. The radius of the square is less than the radius of the triangle.

  (A) I only   **(B)** II only   **(C)** I and II only   **(D)** II and III only   **(E)** I, II, and III

4. A polygon is equilateral.
    Which of the following are always true?
        I. The sum of the measures of its exterior angles, one at each vertex, is 360.
        II. The sum of the measures of its interior angles is divisible by 180.
        III. The polygon is equiangular.

  (A) I only   **(B)** II only   **(C)** I and II only   **(D)** II and III only   **(E)** I, II, and III

Chapter Review

Vocabulary Review

🔊 adjacent arcs (p. 567)
altitude of a parallelogram (p. 534)
apothem (p. 546)
arc length (p. 569)
base of a parallelogram (p. 534)
base of a triangle (p. 536)
center of a circle (p. 566)
central angle (p. 566)
circle (p. 566)

circumference (p. 568)
concentric circles (p. 568)
congruent arcs (p. 569)
congruent circles (p. 566)
diameter (p. 566)
geometric probability (p. 582)
height of a parallelogram (p. 534)
height of a trapezoid (p. 540)
height of a triangle (p. 536)

major arc (p. 567)
minor arc (p. 567)
pi (p. 568)
radius (p. 566)
radius of a regular polygon (p. 546)
sector of a circle (p. 576)
segment of a circle (p. 577)
semicircle (p. 567)

Go Online
PHSchool.com
For: Vocabulary quiz
Web Code: auj-1051

Choose the correct term to complete each sentence.

1. You can use any side as the (*altitude, base*) of a triangle.

2. A (*sector, segment*) of a circle is a region bounded by two radii and the intercepted arc.

3. A segment that contains the center of a circle and has both endpoints on the circle is the (*diameter, circumference*) of a circle.

4. In a regular polygon, the perpendicular distance from the center to a side is the (*apothem, radius*) of the parallelogram.

5. Two arcs of a circle with exactly one point in common are (*congruent arcs, adjacent arcs*).

Skills and Concepts

10-1, 10-2, and 10-3 Objectives

▼ To find the area of a parallelogram

▼ To find the area of a triangle

▼ To find the area of a trapezoid

▼ To find the area of a rhombus or a kite

▼ To find the area of a regular polygon

You can find the area of a rectangle, a parallelogram, or a triangle if you know the **base,** b, and **height,** h. The area of a rectangle or a parallelogram is $A = bh$.

The area of a triangle is $A = \frac{1}{2}bh$.

The **height of a trapezoid,** h, is the perpendicular distance between the bases, b_1 and b_2. The area of a trapezoid is $A = \frac{1}{2}h(b_1 + b_2)$.

The **center of a regular polygon,** C, is the center of its circumscribed circle. The **radius,** r, is the distance from the center to a vertex. The **apothem,** a, is the perpendicular distance from the center to a side. The area of a regular polygon with apothem a and perimeter p is $A = \frac{1}{2}ap$.

The area of a rhombus or kite is $A = \frac{1}{2}d_1d_2$.

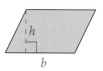

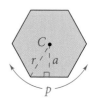

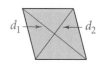

Find the area of each figure. If your answer is not an integer, leave it in simplest radical form.

6.

5 m

4 m

7.

10 in.

9 in.

8.
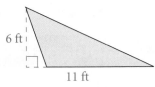
6 ft

11 ft

9.
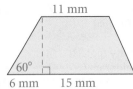
11 mm

60°

6 mm   15 mm

10.

8 ft

8 ft

10 ft

11.
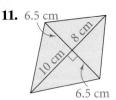
6.5 cm

10 cm   8 cm

6.5 cm

Sketch each regular polygon with the given radius. Then find its area. Round your answers to the nearest tenth.

12. triangle; radius 4 in.   **13.** square; radius 8 mm   **14.** hexagon; radius 7 cm

---

10-4 Objective

▼ To find the perimeters and areas of similar figures

If the similarity ratio of two similar figures is $\frac{a}{b}$, then the ratio of their perimeters is $\frac{a}{b}$, and the ratio of their areas is $\frac{a^2}{b^2}$.

For each pair of similar figures, find the ratio of the area of the first figure to the area of the second.

15.
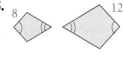
8   12

16.
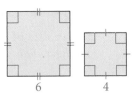
6   4

17.
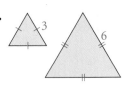
3

6

18. If the ratio of areas of two similar hexagons is 8 : 25, what is the ratio of their perimeters?

10-5 Objectives

▼ To find the area of a regular polygon using trigonometry

▼ To find the area of a triangle using trigonometry

You can use trigonometry to find the areas of regular polygons.

You can also use trigonometry to find the area of a triangle when you know the lengths of two sides and the measure of the included angle.

Area of triangle $= \frac{1}{2} \cdot$ side length $\cdot$ side length $\cdot$ sine of included angle

Find the area of each polygon. Round your answers to the nearest tenth.

19. regular decagon with radius 5 ft

20. regular pentagon with apothem 8 cm

21. regular hexagon with apothem 6 in.

22. regular quadrilateral with radius 2 m

23.
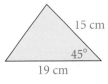
15 cm

45°

19 cm

24.

65°
15 ft   13 ft

25.
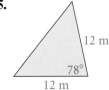
12 m

78°

12 m

10-6 and 10-7 Objectives

▼ To find the measures of central angles and arcs

▼ To find circumference and arc length

▼ To find the areas of circles, sectors, and segments of circles

A **circle** is the set of all points in a plane equidistant from one point called the **center.** The measure of a **minor arc** is the measure of its corresponding central angle. The measure of a **major arc** is 360 minus the measure of its related minor arc. **Adjacent arcs** have exactly one point in common.

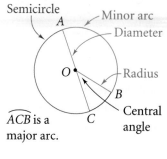

$\overset{\frown}{ACB}$ is a major arc.

The **circumference** of a circle is $C = \pi d$ or $C = 2\pi r$. The area of a circle is $A = \pi r^2$.

Arc length is a fraction of a circle's circumference. The length of $\overset{\frown}{AB} = \dfrac{m\overset{\frown}{AB}}{360} 2\pi r$.

A **sector of a circle** is a region bounded by two radii and their intercepted arc.
The area of sector $APB = \dfrac{m\overset{\frown}{AB}}{360} \pi r^2$.

A **segment of a circle** is the part of a circle bounded by an arc and the segment joining its endpoints. The area of a segment of a circle is the difference between the areas of the related sector and the related triangle.

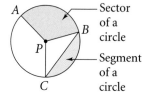

Find each measure.

26. $m\angle APD$ **27.** $m\overset{\frown}{AC}$

28. $m\overset{\frown}{ABD}$ **29.** $m\angle CPA$

Find the length of each arc shown in red. Leave your answer in terms of π.

30.

31.

Find the area of each shaded region. Round your answer to the nearest tenth.

32.

33.

10-8 Objective

▼ To use segment and area models to find the probabilities of events

Geometric probability uses geometric figures to represent occurrences of events. You can use a segment model or an area model. Compare the part that represents favorable outcomes to the whole, which represents all outcomes.

A dart hits each dartboard at a random point. Find the probability that it lands in the shaded area.

34.

35.

36.

Chapter Test

For: Chapter Test
Web Code: aua-1052

Find the area of each figure. If your answer is not an integer, round to the nearest tenth.

1.

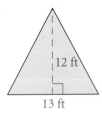

12 ft

13 ft

2.

6 mm

3.

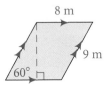

8 m

9 m

60°

4.

3 in.

3 in.

6 in.

Find the area of each regular polygon. Round to the nearest tenth.

5.

4 ft

6.

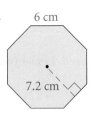

6 cm

7.2 cm

For each pair of similar figures, find the ratio of the area of the first figure to the area of the second.

7.

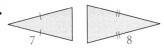

7   8

8.

12   8

Find the area of each polygon to the nearest tenth.

9.

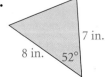

7 in.

8 in.   52°

10.

23 mm

69°

26 mm

11. a regular hexagon with apothem 5 ft

12. a regular pentagon with radius 3 cm

Find each measure for ⊙P.

13. $m\angle BPC$

14. $m\widehat{AB}$

15. $m\widehat{ADC}$

16. $m\widehat{ADB}$

A   B
50°   C
P
D

Find the length of each arc shown in red. Leave your answer in terms of π.

17.

5 in.

120°

18.

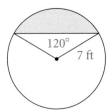

3 cm

Find the area of each shaded region to the nearest hundredth.

19.

6 m

80°

20.

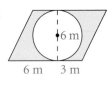

120°

7 ft

21. Open-Ended Use a compass to draw a circle. Shade a sector of the circle and find its area.

Find the area of each shaded region. Leave your answers in terms of π.

22.

12 cm

12 cm

23.

6 m

6 m   3 m

24. Probability Every 20 minutes from 4:00 P.M. to 7:00 P.M., a commuter train crosses Main Street. For three minutes a gate stops cars from passing as the train goes by. What is the probability that a motorist approaching the train crossing during this time interval will have to stop for the train?

Regents Test Prep

Multiple Choice

For Exercises 1–9, choose the correct letter.

1. One leg of an isosceles right triangle is 3 in. long. What is the length of the hypotenuse?

   (1) 3 in.   (2) $3\sqrt{2}$ in.

   (3) $3\sqrt{3}$ in.   (4) 6 in.

2. What is the ratio of the areas of similar triangles whose similarity ratio is 4 : 9?

   (1) 2 : 3   (2) 8 : 27

   (3) 16 : 81   (4) 64 : 729

3. What is the most precise name of the figure?

   (1) quadrilateral   (2) parallelogram
   (3) rectangle   (4) square

4. For what value of x will the two triangles be similar?

   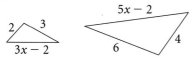

   (1) 10   (2) 8   (3) 4   (4) 2

5. What is the area of the square?

   (1) 50 ft^2 (2) 100 ft^2
   (3) $100\sqrt{2}$ ft^2 (4) 150 ft^2

6. Find the area of a circle with radius 6 in.
   (1) 9π in.^{2} (2) 36π in.^{2}
   (3) 81π in.^{2} (4) 133π in.^{2}

7. Which of the following could be the side lengths of a right triangle?
   (1) 4.1, 6.2, 7.3
   (2) 3.2, 5.4, 6.2
   (3) 40, 60, 72
   (4) 33, 56, 65

8. What is the area of the trapezoid?

   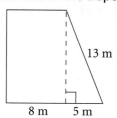

   (1) 60.5 m^2 (2) 126 m^2
   (3) 136.5 m^2 (4) 169 m^2

9. Evaluate $\frac{y}{x}$.

   (1) $\frac{16}{27}$

   (2) $\frac{27}{16}$

   (3) $\frac{\sqrt{985}}{27}$

   (4) $\frac{\sqrt{985}}{16}$

   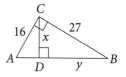

Gridded Response

10. Find the area in square centimeters of a regular pentagon with side length 4 cm. Round your answer to the nearest tenth.

Short Response

11. Find the circumference of the circle shown below and then the length of $\overset{\frown}{AB}$. Leave your answers in terms of π.

   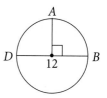

Extended Response

12. Find the area of the shaded region. Leave your answer in terms of π. Explain your work.

   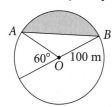

Activity Lab

Misleading Graphs

Eye-catching data displays make you want to look at them. They also try to give you information. Unfortunately, poor designs sometimes give information that is different from what was intended.

NY G.R.4: Select appropriate representations to solve problem situations.

1 ACTIVITY

A real-estate company made the graph at the right. It wanted to show that the company's annual income from rental properties doubled from 2005 to 2006.

1. Sketch an outline of a building like the one shown in the graph. Assign integer values to the width and height of the rectangle and to the height of the isosceles triangle.

2. Find the area of your building sketch.

3. Now sketch a second building similar to the first but with dimensions that are twice what you found for your first sketch. (Multiply the dimensions of the first building by the scale factor 2.)

4. **a.** Find the area of your second building sketch. Then find and simplify the ratio $\frac{\text{area of larger sketch}}{\text{area of first sketch}}$. Explain why this ratio is not 2.

   b. Why is the Rental Property Income graph misleading?

5. Copy the table below. Use the scale factors shown to sketch more buildings. Find the area of each sketch. Then find the $\frac{\text{greater area}}{\text{first area}}$ ratio to complete the table.

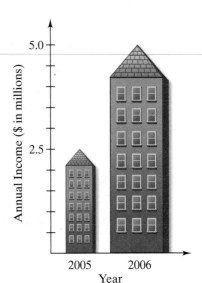

Rental Property Income

Scale factor	3	4	5	6
Area of larger building				
Area of first building				
Greater area / First area				

6. What scale factor should you use to get the following?
   a. $\frac{\text{greater area}}{\text{lesser area}} = 2$ **b.** $\frac{\text{greater area}}{\text{lesser area}} = k$

7. Suppose the real estate company's annual income in 2006 was 150% of its 2005 income. What scale factor should you use in a graph like the one above to accurately represent this?

The graph at the right shows the results of a state election for senator.

Senate Race Election Results

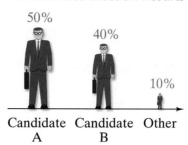

50% 40% 10%

Candidate   Candidate   Other
A   B

8. The dimensions of the silhouette for Candidate B are 0.8 those of the silhouette for Candidate A. Explain why someone might have chosen this scale factor.

9. Calculate the scale factors for the two smaller silhouettes so that their areas—not their heights—will be in proportion to the percents they represent. (*Hint:* Use your results from Activity 1.)

10. Use the corrected scale factors to redraw the graph.

11. Describe how the original graph may have been misleading.

Draw a bounding rectangle. Scale the dimensions of the rectangle to form a bounding rectangle for the scaled silhouette.

EXERCISES

12. The graph at the right shows the results of the mayor's race.
    a. Find the ratio of the shaded areas.
    b. Find the ratio of the election results.
    c. Explain why these ratios are not the same.
    d. Why is this graph misleading?
    e. Redraw the graph correctly.

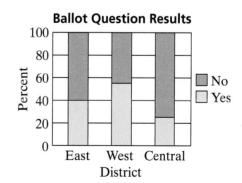

Mayor's Race Election Results

60%   40%

Candidate A   Candidate B

13. A newspaper showed ballot question results with the graph at the right.
    a. The East District has 8400 voters. The West District has 12,900 voters. The Central District has 10,500 voters. Explain how the graph is misleading.
    b. Redraw the graph so the heights of the bars more accurately reflect the ballot question results.

Ballot Question Results

Percent

100
80
60
40
20
0

East   West   Central
District

■ No
□ Yes

What You've Learned

NY Learning Standards for Mathematics

- In Chapter 1, you learned that geometric figures can lie in a plane or be three-dimensional.

- In Chapter 10, you learned how to find the areas of certain plane figures such as triangles, special quadrilaterals, and regular polygons.

- In Chapter 10, you also learned how perimeters and areas of similar figures are related.

Check Your Readiness

GO **for Help** to the Lesson in green.

The Pythagorean Theorem (Lesson 8-1)

x^2 **Algebra** Solve for a, b, or c in right $\triangle ABC$ where a and b are the lengths of the legs and c is the length of the hypotenuse.

1. $a = 8; b = 15; c = \underline{\ ?\ }$ **2.** $a = \underline{\ ?\ }; b = 4; c = 12$ **3.** $a = 2\sqrt{3}; b = 2\sqrt{6}; c = \underline{\ ?\ }$

Special Right Triangles (Lesson 8-2)

4. Find the length of the shorter leg of a 30°-60°-90° right triangle with hypotenuse $8\sqrt{5}$.

5. Find the length of the diagonal of a square whose perimeter is 24.

6. Find the height of an equilateral triangle with sides of length 8.

Area (Lessons 10-1, 10-2, 10-3)

Find the area of each figure. Leave your answers in simplest radical form.

7.   **8.** 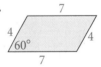   **9.**    **10.**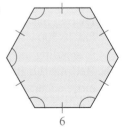

Perimeters and Areas of Similar Figures (Lesson 10-4)

11. Two similar triangles have corresponding sides in a ratio of 3:5. Find the perimeter of the smaller triangle if the larger triangle has perimeter 40.

12. Two regular hexagons have areas of 8 and 25. Find the ratio of corresponding sides.

Surface Area and Volume

🔊 **Key Vocabulary**

- altitude
  (pp. 608, 610, 617, 619)
- base(s)
  (pp. 608, 610, 617, 619)
- cone (p. 619)
- cross section (p. 600)
- cylinder (p. 610)
- edge (p. 598)
- face (p. 598)
- height (pp. 608, 610, 617, 619)
- polyhedron (p. 598)
- prism (p. 608)
- pyramid (p. 617)
- similar solids (p. 646)
- sphere (p. 638)
- surface area
  (pp. 608, 610, 618, 619)
- volume (p. 624)

What You'll Learn Next

(NY) Learning Standards for Mathematics

- In this chapter, you will learn about special three-dimensional figures built from two-dimensional figures such as triangles and rectangles.

- **G.G.13:** You will use what you know about finding perimeter and area to help you find surface area and volume.

- **G.G.13:** You will also learn how lengths, surface areas, and volumes for similar solids are related.

 Activity Lab Applying what you learn, you will solve problems on pages 658 and 659 involving shapes of colossal size.

11-1

Space Figures and Cross Sections

What You'll Learn

- To recognize polyhedra and their parts
- To visualize cross sections of space figures

. . . And Why

To learn about medical techniques, as in Exercise 44.

✓ **Check Skills You'll Need**

For each exercise, make a copy of the cube at the right. Shade the plane that contains the indicated points.

1. $A, B,$ and C
2. $A, B,$ and G
3. $A, C,$ and G
4. $A, D,$ and G
5. $F, D,$ and G
6. $B, D,$ and G
7. the midpoints of $\overline{AD}, \overline{CD}, \overline{EH},$ and $\overline{GH}$

GO for Help Lesson 1-3

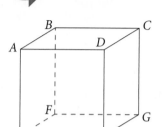

◀)) **New Vocabulary** • **polyhedron** • **face** • **edge** • **vertex** • **cross section**

---

1 Identifying Parts of a Polyhedron

Vocabulary Tip

Polyhedron comes from the Greek *poly* for "many" and *hedron* for "side." A cube is a polyhedron with six sides, or faces, each of which is a square.

A **polyhedron** is a three-dimensional figure whose surfaces are polygons. Each polygon is a **face** of the polyhedron. An **edge** is a segment that is formed by the intersection of two faces. A **vertex** is a point where three or more edges intersect.

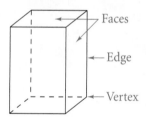

Faces
Edge
Vertex

1 EXAMPLE Identifying Vertices, Edges, and Faces

a. How many vertices are there in the polyhedron at the right? List them.

There are five vertices: $D, E, F, G,$ and $H.$

b. How many edges are there? List them.

There are eight edges: $\overline{DE}, \overline{EF}, \overline{FG}, \overline{GD},$ $\overline{DH}, \overline{EH}, \overline{FH},$ and $\overline{GH}.$

c. How many faces are there? List them.

There are five faces: $\triangle DEH, \triangle EFH, \triangle FGH, \triangle GDH,$ and the quadrilateral $DEFG.$

✓ **Quick Check** ❶ List the vertices, edges, and faces of the polyhedron.

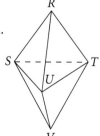

Leonhard Euler, a Swiss mathematician, discovered a relationship among the numbers of faces, vertices, and edges of any polyhedron. The result is known as Euler's Formula.

Key Concepts

Formula	Euler's Formula

The numbers of faces (F), vertices (V), and edges (E) of a polyhedron are related by the formula $F + V = E + 2$.

2 EXAMPLE Using Euler's Formula

Count faces and edges. Then use Euler's Formula to find the number of vertices in the polyhedron at the right.

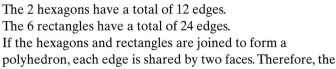

The polyhedron has 2 hexagons and 6 rectangles for a total of 8 faces.

The 2 hexagons have a total of 12 edges.
The 6 rectangles have a total of 24 edges.
If the hexagons and rectangles are joined to form a polyhedron, each edge is shared by two faces. Therefore, the number of edges in the polyhedron is one half of the total of 36, or 18.

$F + V = E + 2$ **Euler's Formula**
$8 + V = 18 + 2$ **Substitute.**
$V = 12$ **Simplify.**

● Count the number of vertices in the figure to verify the result.

 **Quick Check** ❷ Use Euler's Formula to find the number of edges on a polyhedron with eight triangular faces.

In two dimensions, Euler's Formula reduces to
$$F + V = E + 1$$
where F is the number of regions formed by V vertices linked by E segments.

3 EXAMPLE Verifying Euler's Formula

Verify Euler's Formula for a two-dimensional net of the solid in Example 2.

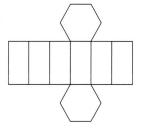

Draw a net:
Count the regions: $F = 8$
Count the vertices: $V = 22$
Count the segments: $E = 29$

$8 + 22 = 29 + 1$

 **Quick Check** ❸ The figure at the right is a trapezoidal prism.
a. Verify Euler's formula $F + V = E + 2$ for the prism.
b. Draw a net for the prism.
c. Verify Euler's formula $F + V = E + 1$ for your two-dimensional net.

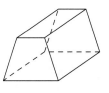

Real-World 🌎 **Connection**

Euler's Formula applies to the polyhedron suggested by the panels on a volleyball.

A **cross section** is the intersection of a solid and a plane. You can think of a cross section as a very thin slice of the solid.

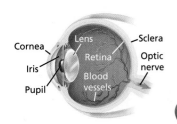

4 EXAMPLE Describing a Cross Section

Describe each cross section.

a.

The cross section is a square.

b.

The cross section is a triangle.

Quick Check ④ For the funnel shown, sketch each of the following.
   a. a horizontal cross section
   b. a vertical cross section that contains the axis of symmetry

To draw a cross section, you can sometimes use the idea from Postulate 1-3 that the intersection of two planes is exactly one line.

5 EXAMPLE Drawing a Cross Section

Visualization Draw and describe a cross section formed by a vertical plane intersecting the front and right faces of the cube.

A vertical plane cuts the vertical faces of the cube in parallel segments.

Draw the parallel segments.

Join their endpoints. Shade the cross section.

The cross section is a rectangle.

Quick Check ⑤ Draw and describe the cross section formed by a horizontal plane intersecting the left and right faces of the cube.

EXERCISES

For more exercises, see *Extra Skill, Word Problem, and Proof Practice*.

Practice and Problem Solving

 Practice by Example

Example 1
(page 598)

For each polyhedron, how many vertices, edges, and faces are there? List them.

1.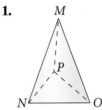

2.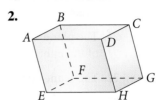

3.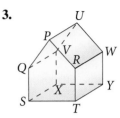

Example 2
(page 599)

Use Euler's Formula to find the missing number.

4. Faces: ▪
Edges: 15
Vertices: 9

5. Faces: 8
Edges: ▪
Vertices: 6

6. Faces: 20
Edges: 30
Vertices: ▪

Use Euler's Formula to find the number of vertices in each polyhedron described below.

7. 6 square faces

8. 5 faces: 1 rectangle and 4 triangles

9. 9 faces: 1 octagon and 8 triangles

Example 3
(page 599)

Verify Euler's Formula for each polyhedron. Then draw a net for the figure and verify Euler's Formula for the two-dimensional figure.

10.

11.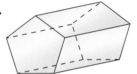

12.

Example 4
(page 600)

Describe each cross section.

13.

14.

15.

16. For the nut shown, sketch each of following.
a. a horizontal cross section
b. a vertical cross section that contains the vertical line of symmetry

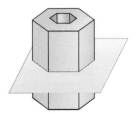

Example 5
(page 600)

Visualization Draw and describe a cross section formed by a vertical plane intersecting the cube as follows.

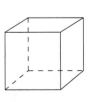

17. The vertical plane intersects the front and left faces of the cube.

18. The vertical plane intersects opposite faces of the cube.

19. The vertical plane contains opposite edges of the cube.

20. **a. Open-Ended** Sketch a polyhedron whose faces are all rectangles. Label the lengths of its edges.
    b. Use graph paper to draw two different nets for the polyhedron.

Visualization **Draw and describe a cross section formed by a plane intersecting the cube as follows.**

21. The plane is tilted and intersects the left and right faces of the cube.

22. The plane contains opposite horizontal edges of the cube.

23. The plane cuts off a corner of the cube.

Describe the cross section shown.

24.

25.

26.

Visualization **A plane region that revolves completely about a line sweeps out a *solid of revolution*. Use the sample to help you describe the solid of revolution you get by revolving each region about line ℓ.**

Sample: Revolve the rectangular region about the line ℓ and you get a cylinder as a solid of revolution.

 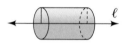

27.

28.

29.

Sports Equipment **Some balls are made from panels that suggest polygons. The ball then suggests a polyhedron to which Euler's Formula, $F + V = E + 2$, applies.**

30. A soccer ball suggests a polyhedron with 20 regular hexagons and 12 regular pentagons. How many vertices does this polyhedron have?

31. Show how Euler's Formula applies to the polyhedron suggested by the volleyball pictured on page 599. (*Hint:* It has 6 sets of 3 panels.)

Euler's Formula $F + V = E + 1$ applies to any two-dimensional network where F is the number of regions formed by V vertices linked by E edges (or paths). Verify Euler's Formula for each network shown.

32.

33.

34.

35. Draw a network of your own. Verify Euler's Formula for it.

36. There are five regular polyhedrons. They are called *regular* because all their faces are congruent regular polygons, and the same number of faces meet at each vertex. They are also called Platonic Solids after the Greek philosopher Plato (427–347 B.C.).

Tetrahedron

Hexahedron

Octahedron

Dodecahedron

Icosahedron

a. Match each net below with a Platonic Solid.

A.　　　B.　　　C.　　　D.　　　E.

b. The first two Platonic solids have more familiar names. What are they?
c. Verify that Euler's Formula is true for the first three Platonic solids.

37. Multiple Choice A cube has a net with area 216 in.². How long is an edge of the cube?

Ⓐ 6 in.　　　Ⓑ 15 in.　　　Ⓒ 36 in.　　　Ⓓ 54 in.

Draw each object. Then draw a horizontal and a vertical cross section.

38. a golf tee

39. a football

40. a baseball bat

41. a banana

42. a pear

43. a bagel

44. Writing Cross sections are used in medical training and research. Research and write a paragraph on how magnetic resonance imaging (MRI) is used to study cross sections of the brain.

C Challenge

45. Draw a solid that has the following cross sections.

horizontal

vertical

Visualization Draw a plane intersecting a cube to get the cross section indicated.

46. scalene triangle

47. isosceles triangle

48. equilateral triangle

49. trapezoid

50. isosceles trapezoid

51. parallelogram

52. rhombus

53. pentagon

54. hexagon

Multiple Choice   For Exercises 55–56, you may need Euler's Formula, $F + V = E + 2$.

55. A polyhedron has four vertices and six edges. How many faces does it have?
    A. 2   **B.** 4   **C.** 5   **D.** 10

56. A polyhedron has three rectangular faces and two triangular faces. How many vertices does it have?
    F. 5   **G.** 6   **H.** 10   **J.** 12

57. The plane is horizontal. What best describes the shape of the cross section?
    A. rhombus   **B.** trapezoid
    C. parallelogram   **D.** square

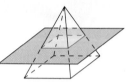

58. The plane is vertical. What best describes the shape of the cross section?
    F. pentagon   **G.** square
    H. rectangle   **J.** triangle

Short Response   **59.** Draw and describe a cross section formed by a plane intersecting a cube as follows.
    a. The plane is parallel to a horizontal face of the cube.
    b. The plane cuts off two corners of the cube.

Mixed Review

Lesson 10-8   **60. Probability** A shuttle bus to an airport terminal leaves every 20 min from a remote parking lot. Draw a geometric model and find the probability that a traveler who arrives at a random time will have to wait at least 8 min for the bus to leave the parking lot.

 61. Games A dartboard is a circle with a 12-in. radius. You throw a dart that hits the dartboard. What is the probability that the dart lands within 6 in. of the center of the dartboard?

Lesson 10-3   **Find the area of each equilateral triangle with the given measure. Leave answers in simplest radical form.**

62. side 2 ft   **63.** apothem 8 cm   **64.** radius 100 in.

Lesson 8-3   **Find the value of x to the nearest tenth.**

65.
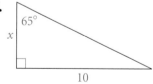

66.

67. The lengths of the diagonals of a rhombus are 4 cm and 6 cm. Find the measures of the angles of the rhombus to the nearest degree.

Perspective Drawing

You can create a three-dimensional space figure with a two-dimensional *perspective drawing*. Suppose two lines are parallel in three dimensions but recede from the viewer. You draw them—and create perspective—so that they meet at a *vanishing point* on a *horizon line*.

1 EXAMPLE

Draw a cube in one-point perspective.

Step 1: Draw a square. Then draw a horizon line and a vanishing point on the line.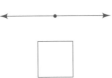

Step 2: Lightly draw segments from the vertices of the square to the vanishing point.

Step 3: Draw a square for the back of the cube. Each vertex should lie on a segment you drew in Step 2.

Step 4: Complete the figure by using dashes for hidden edges of the cube. Erase unneeded lines.

Two-point perspective involves the use of two vanishing points.

2 EXAMPLE

Draw a box in two-point perspective.

Step 1: Draw a vertical segment. Then draw a horizon line and two vanishing points on the line.

Step 2: Lightly draw segments from the endpoints of the vertical segment to each vanishing point.

Step 3: Draw two vertical segments between the segments of Step 2.

Step 4: Draw segments from the endpoints of the segments you drew in Step 3 to the vanishing points.

Step 5: Complete the figure by using dashes for hidden edges of the figure. Erase unneeded lines.

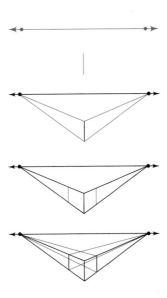

In one-point perspective, the front of the cube is parallel to the drawing surface. A two-point perspective drawing generally looks like a corner view. For either type of drawing, you should be able to envision each vanishing point.

EXERCISES

Is each object drawn in one- or two-point perspective?

1.

2.

3.

4.

5.

6.

Draw each object in one-point perspective and then in two-point perspective.

7. a shoe box

8. a building in your town that sits on a street corner

Draw each container using one-point perspective. Show a base at the front.

9. triangular carton

10. hexagonal box

Copy each figure and locate the vanishing point(s).

11.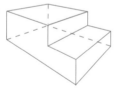

12.

13.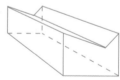

Optical Illusions **What is the optical illusion? Explain the role of perspective in each illusion.**

14.

15.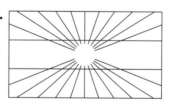

16. Open Ended You can draw block letters in either one-point perspective or two-point perspective. Write your initals in block letters using one-point perspective and two-point perspective.

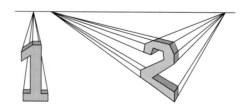

Literal Equations

A *literal equation* is an equation involving two or more variables. A formula is a special type of literal equation. You can transform a formula by solving for one variable in terms of the others.

1 EXAMPLE

Algebra The formula for the volume of a cylinder is $V = \pi r^2 h$. Find a formula for the height in terms of the radius and volume.

$$V = \pi r^2 h$$

$$\frac{V}{\pi r^2} = \frac{\pi r^2 h}{\pi r^2} \quad \textbf{Divide each side by } \pi r^2, r \neq 0.$$

$$\frac{V}{\pi r^2} = h \quad \textbf{Simplify.}$$

● The formula for the height is $h = \dfrac{V}{\pi r^2}$.

Solving literal equations also allows you to build other formulas.

2 EXAMPLE

Algebra Find a formula for the area of a square in terms of its perimeter.

$$P = 4s \quad \textbf{Use the formula for perimeter.}$$

$$\frac{P}{4} = s \quad \textbf{Solve for } s \textbf{ in terms of } P.$$

$$A = s^2 \quad \textbf{Use the formula for area.}$$

$$A = \left(\frac{P}{4}\right)^2 \quad \textbf{Substitute.}$$

$$A = \frac{P^2}{16} \quad \textbf{Simplify.}$$

● The formula for the area is $A = \dfrac{P^2}{16}$.

EXERCISES

$\boxed{x^2}$ **Algebra** Solve each equation for the variable in red.

1. $C = 2\pi r$ **2.** $A = \frac{1}{2}bh$ **3.** $A = \pi r^2$

$\boxed{x^2}$ **Algebra** Solve for the variable in red. Then solve for the variable in blue.

4. $P = 2w + 2\ell$ **5.** $\tan A = \frac{y}{x}$ **6.** $A = \frac{1}{2}(b_1 + b_2)h$

Find a formula as stated.

7. the circumference of a circle in terms of its area

8. the area of an isosceles right triangle in terms of the hypotenuse

9. the apothem of a regular hexagon in terms of the area of the hexagon

10. Solve $A = \frac{1}{2}ab \sin C$ for $m\angle C$.

Surface Areas of Prisms and Cylinders

Learning Standards for Mathematics

G.G.10 Know and apply that the lateral edges of a prism are congruent and parallel.

G.G.14 Apply the properties of a cylinder, including:
• bases are congruent.
• volume equals the product of the area of the base and the altitude.

✓ **Check Skills You'll Need**

GO **for Help**

Lessons 1-9 and 10-3

Find the area of each net.

1.
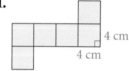
4 cm
4 cm

2.
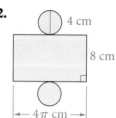
4 cm
8 cm
← 4π cm →

3.
6 m

🔊 **New Vocabulary** • prism • bases, lateral faces, altitude, height, lateral area, surface area (of a prism) • right prism • oblique prism • cylinder • bases, altitude, height, lateral area, surface area (of a cylinder) • right cylinder • oblique cylinder

---

1 Finding Surface Area of a Prism

Real-World 🌐 Connection

A triangular prism breaks white light into rainbow colors.

A **prism** is a polyhedron with exactly two congruent, parallel faces, called **bases.** Other faces are **lateral faces.** You name a prism by the shape of its bases.

An **altitude** of a prism is a perpendicular segment that joins the planes of the bases. The **height** h of the prism is the length of an altitude.

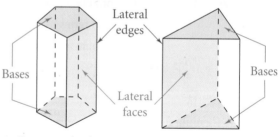

Lateral edges

Bases

Lateral faces

Bases

Pentagonal prism   Triangular prism

A prism may either be right or oblique.

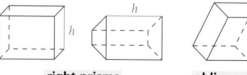

right prisms   **oblique prism**

In a right prism the lateral faces are rectangles and a lateral edge is an altitude. In this book you may assume that a prism is a right prism unless stated or pictured otherwise.

The **lateral area** of a prism is the sum of the areas of the lateral faces. The **surface area** is the sum of the lateral area and the area of the two bases.

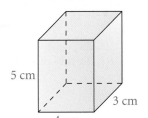

5 cm
3 cm
4 cm

1 EXAMPLE Finding Surface Area of a Prism

Use a net to find the surface area of the prism at the left.

Surface Area = Lateral Area + area of bases

$= \text{sum of areas of lateral faces} + \text{area of bases}$

$= (5 \cdot 4 + 5 \cdot 3 + 5 \cdot 4 + 5 \cdot 3) + 2(3)(4)$

$= 70 + 24$

$= 94$

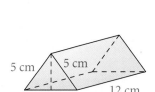

The surface area of the prism is 94 cm².

 **Quick Check** 1 Use a net to find the surface area of the triangular prism.

5 cm   5 cm
6 cm   12 cm

You can find formulas for lateral and surface areas by looking at a net for a prism.

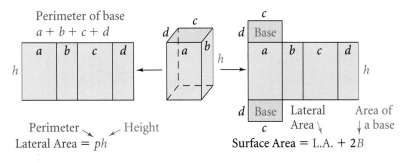

Perimeter of base
$a + b + c + d$

Perimeter   Height
Lateral Area = ph

Surface Area = L.A. + 2B

You can use the formulas with any right prism.

2 EXAMPLE Using Formulas to Find Surface Area

Multiple Choice What is the surface area of the prism?

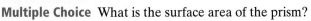

A 72 cm²   B 78 cm²   C 84 cm²   D 96 cm²

By the Pythagorean Theorem, the hypotenuse of the triangular base is 5 cm.

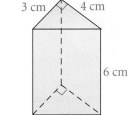

3 cm   4 cm
6 cm

L.A. $= ph$ **Use the formula for lateral area.**

$= 12 \cdot 6$ **p = 3 + 4 + 5 = 12 cm**

$= 72$

The lateral area of the prism is 72 cm².

Now use the formula for surface area.

S.A. $=$ L.A. $+ 2B$

$= 72 + 2(6) = 84$ **B = $\frac{1}{2}$(3 · 4) = 6 cm²**

The surface area of the prism is 84 cm². Choice C is the answer.

 **Quick Check** 2 Use formulas to find the lateral area and surface area of the prism.

12 m
6 m

Key Concepts

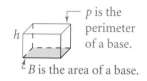

| **Theorem 11-1** | **Lateral and Surface Areas of a Prism** |

The lateral area of a right prism is the product of the perimeter of the base and the height.

$$\text{L.A.} = ph$$

The surface area of a right prism is the sum of the lateral area and the areas of the two bases.

$$\text{S.A.} = \text{L.A.} + 2B$$

p is the perimeter of a base.

B is the area of a base.

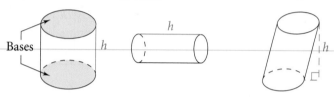

2 Finding Surface Area of a Cylinder

Like a prism, a **cylinder** has two congruent parallel **bases.** However, the bases of a cylinder are circles. An **altitude** of a cylinder is a perpendicular segment that joins the planes of the bases. The **height** *h* of a cylinder is the length of an altitude.

Bases   *h*   *h*   *h*

right cylinders   **oblique cylinder**

In this book you may assume that a cylinder is a right cylinder unless stated or pictured otherwise.

To find the area of the curved surface of a cylinder, visualize "unrolling" it. The area of the resulting rectangle is the **lateral area** of the cylinder. The **surface area** of a cylinder is the sum of the lateral area and the areas of the two circular bases. You can find formulas for these areas by looking at a net for a cylinder.

Real-World Connection

A full turn of the roller inks a rectangle with area equal to the roller's lateral area.

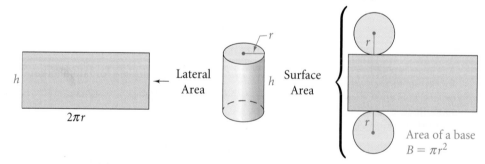

Lateral Area   *h*   Surface Area

$2\pi r$

Area of a base
$B = \pi r^2$

Key Concepts

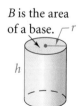

| **Theorem 11-2** | **Lateral and Surface Areas of a Cylinder** |

The lateral area of a right cylinder is the product of the circumference of the base and the height of the cylinder.

$$\text{L.A.} = 2\pi rh, \text{ or L.A.} = \pi dh$$

The surface area of a right cylinder is the sum of the lateral area and the areas of the two bases.

$$\text{S.A.} = \text{L.A.} + 2B, \text{ or S.A.} = 2\pi rh + 2\pi r^2$$

B is the area of a base.

3 EXAMPLE Finding Surface Area of a Cylinder

The radius of the base of a cylinder is 4 in. and its height is 6 in. Find the surface area of the cylinder in terms of π.

S.A. = L.A. + 2*B*	Use the formula for surface area of a cylinder.
$= 2\pi rh + 2(\pi r^2)$	Substitute the formulas for lateral area and area of a circle.
$= 2\pi(4)(6) + 2\pi(4^2)$	Substitute 4 for *r* and 6 for *h*.
$= 48\pi + 32\pi$	Simplify.
$= 80\pi$	

● The surface area of the cylinder is 80π in.^{2}.

✓ **Quick Check** ❸ Find the surface area of a cylinder with height 10 cm and radius 10 cm in terms of π.

4 EXAMPLE Real-World 🌐 Connection

Machinery The drums of the roller at the left are cylinders of length 3.5 ft. The diameter of the large drum is 4.2 ft. What area does the large drum cover in one full turn? Round your answer to the nearest square foot.

The area covered is the lateral area of a cylinder that has a diameter of 4.2 ft and a height of 3.5 ft.

L.A. = πdh	Use the formula for lateral area of a cylinder.
$= \pi(4.2)(3.5)$	Substitute.
$= 46.181412$	Use a calculator.

● In one full turn, the large drum covers about 46 ft^2.

✓ **Quick Check** ❹ The small drum has diameter 3 ft.
  a. To the nearest square foot, what area does the small drum cover in one turn?
  b. Critical Thinking What area does the small drum cover in one turn of the large drum?

EXERCISES
For more exercises, see *Extra Skill, Word Problem, and Proof Practice*.

Practice and Problem Solving

A Practice by Example

Example 1
(page 609)

GO for Help

Use a net to find the surface area of each prism.

1.
29 cm
6.5 cm
19 cm

2.
6 ft
6 ft
6 ft

3.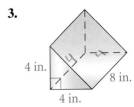
4 in.
8 in.
4 in.

4. a. Classify the prism.
  b. Find the lateral area of the prism.
  c. The bases are regular hexagons. Find the sum of their areas.
  d. Find the surface area of the prism.

10 cm
4 cm

Example 2
(page 609)

Use formulas to find the lateral area and surface area of each prism. Show your answer to the nearest whole number.

5. 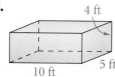 4 ft, 10 ft, 5 ft

6. 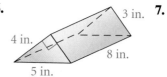 3 in., 4 in., 8 in., 5 in.

7. 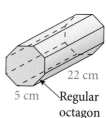 22 cm, 5 cm, Regular octagon

Example 3
(page 611)

Find the surface area of each cylinder in terms of π.

8. 2 cm, 8 cm

9. 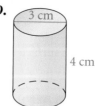 3 cm, 4 cm

10. 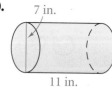 7 in., 11 in.

11. A standard drinking straw is 19.5 cm long and has a diameter of 0.6 cm. How many square centimeters of plastic are used in one straw? Round your answer to the nearest tenth.

Example 4
(page 611)

12. Packaging A cylindrical carton of oatmeal with radius 3.5 in. is 9 in. tall. If all surfaces except the top are made of cardboard, how much cardboard is used to make the oatmeal carton? Round your answer to the nearest square inch.

Find the surface area of each cylinder to the nearest whole number.

13. 4 in., $6\frac{1}{2}$ in.

14. 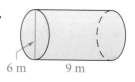 6 m, 9 m

15. 8 cm, 20 cm

B **Apply Your Skills**

16. A triangular prism has base edges 4 cm, 5 cm, and 6 cm long. Its lateral area is 300 cm². What is the height of the prism?

17. Estimation Estimate the surface area of a cube with edges 4.95 cm long.

18. Writing Explain how a cylinder and a prism are alike and how they are different.

19. A hexagonal pencil is a hexagonal prism. A base edge of the pencil has length 4 mm. The pencil (without eraser) has height 170 mm. How much surface area of a hexagonal pencil gets painted?

20. Open-Ended Draw a net for a rectangular prism with a surface area of 220 cm².

21. Consider a box with dimensions 3, 4, and 5.
   a. Find its surface area.
   b. Double each dimension and then find the new surface area.
   c. Find the ratio of the new surface area to the original surface area.
   d. Repeat parts (a)–(c) for a box with dimensions 6, 9, and 11.
   e. Make a Conjecture How does doubling the dimensions of a rectangular prism affect the surface area?

22. Multiple Choice A cylindrical can of cocoa has the dimensions shown at the right. What is the approximate surface area available for the label?

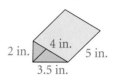
7 in.
5 in.
COCOA

Ⓐ 110 in.^{2} Ⓑ 148 in.^{2}
Ⓒ 179 in.^{2} Ⓓ 219 in.^{2}

23. Pest Control A flour moth trap has the shape of a triangular prism that is open on both ends. An environmentally safe chemical draws the moth inside the prism, which is lined with an adhesive. Find the surface area of the trap.

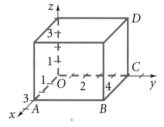
2 in. 4 in. 5 in.
3.5 in.

24. Packaging A typical box for a videocassette tape is open on one side as pictured at the left. How many square inches of cardboard are in a typical box for a videocassette tape?

1 in.

SUPER VHS
VIDEOCASSETTE

$7\frac{1}{2}$ in.

2 HOURS
ST-120

4 in.
Exercise 24

25. Suppose that a cylinder has a radius of r units, and that the height of the cylinder is also r units. The lateral area of the cylinder is 98π square units.

x^2 **a. Algebra** Find the value of r.
  b. Find the surface area of the cylinder.

26. a. Geometry in 3 Dimensions Find the three coordinates of each vertex A, B, C, and D of the rectangular prism.
  b. Find AB.
  c. Find BC.
  d. Find CD.
  e. Find the surface area of the prism.

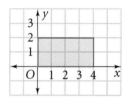
z
D
3
1
C
1 O 2 4 y
3
x A B

Visualization The plane region is revolved completely about the given line to sweep out a solid of revolution. Describe the solid and find its surface area in terms of π.

27. the y-axis **28.** the x-axis
29. the line $y = 2$ **30.** the line $x = 4$

y
3
2
1
O 1 2 3 4 x

31. a. Critical Thinking Suppose you double the radius of a right cylinder. How does that affect the lateral area?
  b. How does that affect the surface area?
  c. Use the formula for surface area of a right cylinder to explain why the surface area in part (b) was not doubled.

GO for Help

For a guide to solving Exercise 32, see p. 615.

32. a. Packaging Some cylinders have wrappers with a spiral seam. Peeled off, the wrapper has the shape of a parallelogram. The wrapper for a biscuit container has base 7.5 in. and height 6 in. Find the radius and height of the container.
  b. Find the surface area of the container.

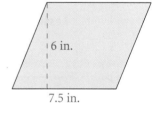
6 in.
7.5 in.

ⓒ Challenge

Judging by appearances, what is the surface area of each solid?

33.

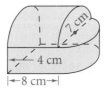
7 cm
4 cm
8 cm

34.
4 m
6 m 3 m

35.

8 in.
3 in.
10 in.

x^2 **36. Algebra** The sum of the height and radius of a cylinder is 9 m. The surface area of the cylinder is 54π m^2. Find the height and the radius.

37. Each edge of the large cube at the right is 12 inches long. The cube is painted on the outside, and then cut into 27 smaller cubes. Answer these questions about the 27 cubes.
a. How many are painted on 4, 3, 2, 1, and 0 faces?
b. What is the total surface area that is unpainted?

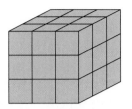

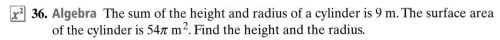
NY REGENTS

Test Prep

Multiple Choice

38. What is the surface area of the figure to the nearest tenth?
A. 335.7 m^2 **B.** 411.6 m^2
C. 671.5 m^2 **D.** 721.2 m^2

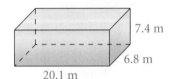
7.4 m
6.8 m
20.1 m

39. If the radius and height of a cylinder are both doubled, then the surface area is   ? .
F. the same   **G.** doubled   **H.** tripled   **J.** quadrupled

40. A cylinder of radius r sits snugly inside a cube. Which expression represents the difference of their lateral areas?
A. $2r^2(8 - \pi)$ **B.** $2r(\pi - 2)$ **C.** $2r(4 - \pi)$ **D.** $4r^2(4 - \pi)$

Short Response

41. The sides of a base of a right triangular prism are 6 in., 8 in., and 10 in. The lateral area of the prism is 48 in.^{2}.
a. Find the height of the prism. Explain your reasoning.
b. What is the surface area of the prism?

Mixed Review

GO for Help

Lesson 11-1

Sketch each space figure and then draw a net for it. Label the net with its dimensions.

42. a rectangular box with height 5 cm and a base 3 cm by 4 cm

43. a cube with 2-in. sides

Lesson 10-7

Find the area of each part of the circle to the nearest tenth.

44. sector QOP

45. the segment of the circle bounded by $\overline{QP}$ and $\overparen{QP}$

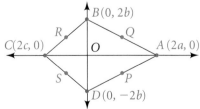
P   120°   Q
O   6 cm

Lesson 6-7

46. In the kite at the right $AB = AD$ and $CB = CD$. Points $P, Q, R,$ and S are midpoints.
a. Determine the coordinates of the midpoints.
b. $RQ = \blacksquare$; $SP = \blacksquare$; $PQ = \blacksquare$; $SR = \blacksquare$
c. Use your answers to part (b) to explain why $PQRS$ must be a parallelogram.

$B(0, 2b)$
R
Q
$C(2c, 0)$
O
$A(2a, 0)$
S
P
$D(0, -2b)$

Guided Problem Solving

Understanding Word Problems Read the problem below. Then let Ray's thinking guide you through the solution. Check your understanding with the exercises at the bottom of the page.

a. Packaging Some cylinders have wrappers with a spiral seam. Peeled off, the wrapper has the shape of a parallelogram. The wrapper for a biscuit container has base 7.5 in. and height 6 in. Find the radius and height of the container.

b. Find the surface area of the container.

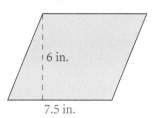

6 in.

7.5 in.

What Ray Thinks

A cutout model I made suggests that the diameter of the cylinder is about 2.3 in. When I roll the parallelogram into a cylinder, the height stays the same. The circumference, $2\pi r$, is the base of the parallelogram, 7.5.

I can solve for r. I'll use a calculator.

So, $r \approx 1.2$ in. and $h = 6$ in. The 1.2 in. radius agrees with what I found with my cutout model!

Surface area = lateral area + base areas. I can replace $2\pi r$ with 7.5 from above.

What Ray Writes

a. $2\pi r = 7.5$

$r = \dfrac{7.5}{2\pi}$

≈ 1.19

$r \approx 1.2$ in. and $h = 6$ in.

b. S.A. = L.A. + 2B

$= 2\pi rh + 2\pi r^2$

$\approx (7.5) \cdot 6 + 2\pi(1.2)^2$

≈ 54.0

The surface area is about 54 in.^{2}.

EXERCISES

For Exercises 1–3, refer to the exercise above.

1. For the given base and height, how does the slant of the parallelogram affect the surface area of the biscuit cylinder?

2. The slanted edge has length 6.4 in. If it were used for the base, what would be the height, lateral area, and surface area of the cylinder formed?

3. How big should the small angle of the parallelogram be so that the two endpoints of the seam will be in line vertically when the can is wrapped?

Exploring Surface Area

At room temperature, 1 L, 1000 mL, and 1000 cm^3 all represent the same amount of water. Thus, one type of model for a liter is any square prism that holds 1000 cm^3. The best model, perhaps, is a 10-cm cube as shown here. But there are many others.

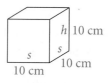

Go Online
PHSchool.com
For: Graphing calculator procedures
Web Code: aue-2121

You can use graphing calculator lists to study how height (h) and surface area (S.A.) of a 1-L square prism change as the length (s) of each side of a base changes.

The volume of a prism equals the area of a base times its height ($V = Bh$ or $V = s^2h$). You can solve for h in each equation to find $h = \frac{V}{B} = \frac{V}{s^2}$. The surface area equals two times the area of a base plus four times the area of a face, or

$$\begin{aligned}
\text{S.A.} = 2B + 4sh &= 2s^2 + 4sh \\
&= 2s^2 + 4s\frac{V}{s^2} \quad \textbf{Substitute.} \\
&= 2s^2 + \frac{4V}{s} \quad \textbf{Simplify.}
\end{aligned}$$

Use the commands shown on the screens below to generate lists L_1, L_2, and L_3, for s, h, and S.A., respectively. The fourth screen shows what the lists should look like.

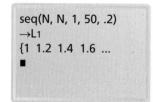

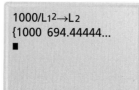

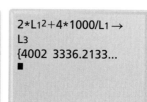

L_1	L_2	L_3	1
1	1000	4002	
1.2	694.44	3336.2	
1.4	510.2	2861.1	
1.6	390.63	2505.1	
1.8	308.64	2228.7	
2	250	2008	
2.2	206.61	1827.9	

$L_1(1)=1$

EXERCISES

Generate the lists (shown above) on your graphing calculator. Scroll down to study them.

1. How small can the surface area be? How large can it be?

2. a. Which dimensions give a very large surface area?
  b. Which dimensions give the smallest surface area?
  c. How do s and h compare in the prism with the smallest surface area?
  d. What is the shape of the prism that has the smallest surface area?

Extend

3. If a square prism must have a volume of 100 cm^3, what dimensions would give the smallest surface area?

4. A cereal manufacturer is designing a cereal box that has a capacity of 3000 cm^3. Surface area should be large to provide space for advertising. What else should be considered for the box design? Use a graphing calculator as needed to support your conclusions.

11-3

Surface Areas of Pyramids and Cones

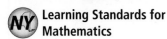
Learning Standards for Mathematics

G.G.13 Apply the properties of a regular pyramid, such as the volume of a pyramid equals one-third the product of the area of the base and the altitude.

G.G.15 Apply the properties of a right circular cone, such as the volume is one-third the product of the area of the base and its altitude.

✓ **Check Skills You'll Need**

Find the length of the hypotenuse in simplest radical form.

1.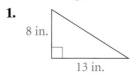
8 in.
13 in.

2.
9 m
7 m

3.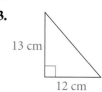
13 cm
12 cm

🔊 **New Vocabulary**
• pyramid • base, lateral faces, vertex, altitude, height, slant height, lateral area, surface area (of a pyramid)
• regular pyramid • cone • base, altitude, vertex, height, slant height, lateral area, surface area (of a cone) • right cone

---

1 **Finding Surface Area of a Pyramid**

A **pyramid** is a polyhedron in which one face (the **base**) can be any polygon and the other faces (the **lateral faces**) are triangles that meet at a common vertex (called the **vertex** of the pyramid).

Vocabulary Tip

If the base is a hexagon, the pyramid is a hexagonal pyramid.

You can name a pyramid by the shape of its base. The **altitude** of a pyramid is the perpendicular segment from the vertex to the plane of the base. The length of the altitude is the **height** h of the pyramid.

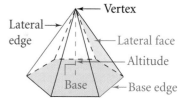
Vertex
Lateral edge
Lateral face
Altitude
Base
Base edge

A **regular pyramid** is a pyramid whose base is a regular polygon and whose lateral faces are congruent isosceles triangles. The **slant height** ℓ is the length of the altitude of a lateral face of the pyramid.

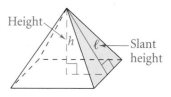
Height
h ℓ
Slant height

In this book, you can assume that a pyramid is regular unless stated otherwise.

The **lateral area** of a pyramid is the sum of the areas of the congruent lateral faces. You can find a formula for the lateral area of a pyramid by looking at its net.

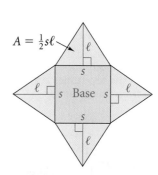
$A = \frac{1}{2}s\ell$
ℓ
s
ℓ s Base s ℓ
s
ℓ

L.A. $= 4\left(\frac{1}{2}s\ell\right)$ The area of each lateral face is $\frac{1}{2}s\ell$.

      $= \frac{1}{2}(4s)\ell$ **Commutative and Associative Properties of Multiplication**

      $= \frac{1}{2}p\ell$ The perimeter p of the base is $4s$.

To find the **surface area** of a pyramid, add the area of its base to its lateral area.

 Key Concepts

Theorem 11-3	**Lateral and Surface Areas of a Regular Pyramid**

The lateral area of a regular pyramid is half the product of the perimeter of the base and the slant height.

$$\text{L.A.} = \tfrac{1}{2}p\ell$$

The surface area of a regular pyramid is the sum of the lateral area and the area of the base.

$$\text{S.A.} = \text{L.A.} + B$$

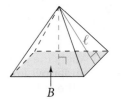

---

1 EXAMPLE Finding Surface Area of a Pyramid

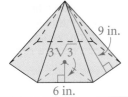

Find the surface area of the hexagonal pyramid at the left.

S.A. = L.A. + B	**Use the formula for surface area.**
$= \tfrac{1}{2}p\ell + \tfrac{1}{2}ap$	**Substitute the formulas for L.A. and B.**
$= \tfrac{1}{2}(36)(9) + \tfrac{1}{2}(3\sqrt{3})(36)$	**Substitute.**
≈ 255.53074	**Use a calculator.**

● The surface area of the pyramid is about 256 in.^{2}.

✓ **Quick Check** ❶ Find the surface area of a square pyramid with base edges 5 m and slant height 3 m.

Sometimes the slant height of a pyramid is not given. You must calculate it before you can find the lateral or surface area.

2 EXAMPLE Real-World Connection

Social Studies The Great Pyramid at Giza, Egypt, pictured at the left, was built about 2580 B.C. as a final resting place for Pharaoh Khufu. At the time it was built, its height was about 481 ft. Each edge of the square base was about 756 ft long. What was the lateral area of the Great Pyramid?

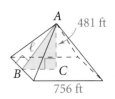

The legs of right $\triangle ABC$ are the height of the pyramid and the apothem of the base. The height of the pyramid was 481 ft. The apothem of the base was $\frac{756}{2}$, or 378 ft. You can use the Pythagorean Theorem to find the slant height ℓ.

Real-World Connection

Today, most casing stones (used to smooth the sides) and some of the top stones are gone from this pyramid.

L.A. $= \tfrac{1}{2}p\ell$	**Use the formula for lateral area.**
$= \tfrac{1}{2}(4s)\sqrt{a^2 + b^2}$	**Substitute the formulas for p and ℓ.**
$= \tfrac{1}{2}(4 \cdot 756)\sqrt{378^2 + 481^2}$	**Substitute.**
≈ 924974.57	**Use a calculator.**

● The lateral area of the Great Pyramid was about 925,000 ft^2.

✓ **Quick Check** ❷ Find the surface area of the Great Pyramid to the nearest square foot.

A **cone** is "pointed" like a pyramid, but its **base** is a circle. In a **right cone,** the **altitude** is a perpendicular segment from the **vertex** to the center of the base. The **height** h is the length of the altitude. The **slant height** ℓ is the distance from the vertex to a point on the edge of the base.

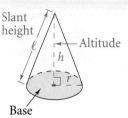

As with a pyramid, the **lateral area** is $\frac{1}{2}$ the perimeter (circumference) of the base times the slant height. The formulas for the lateral area and **surface area** of a cone are similar to those for a pyramid.

Key Concepts

Theorem 11-4	Lateral and Surface Areas of a Cone

The lateral area of a right cone is half the product of the circumference of the base and the slant height.

$$\text{L.A.} = \frac{1}{2} \cdot 2\pi r \cdot \ell, \text{ or L.A.} = \pi r \ell$$

The surface area of a right cone is the sum of the lateral area and the area of the base.

$$\text{S.A.} = \text{L.A.} + B$$

In this book, you can assume that a cone is a right cone unless stated or pictured otherwise.

3 EXAMPLE Finding Surface Area of a Cone

Find the surface area of the cone in terms of π.

S.A. = L.A. + B	**Use the surface area formula.**
$= \pi r \ell + \pi r^2$	**Substitute the formulas for L.A. and B.**
$= \pi(15)(25) + \pi(15)^2$	**Substitute.**
$= 375\pi + 225\pi$	**Simplify.**
$= 600\pi$	

The surface area of the cone is 600π cm^2.

Quick Check ❸ The radius of the base of a cone is 16 m. Its slant height is 28 m. Find the surface area in terms of π.

By cutting a cone and laying it out flat, you can see how the formula for lateral area of a cone $\left(\text{L.A.} = \frac{1}{2} \cdot C_{\text{base}} \cdot \ell\right)$ resembles that for the area of a triangle $\left(A = \frac{1}{2} bh\right)$.

Careers Successful chemists attend to detail, persevere, and work independently.

 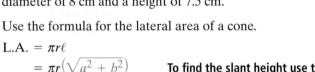
Chemistry The funnel is in the shape of a cone. How much filter paper do you need to line the funnel?

The area covered is the lateral area of a cone that has a diameter of 8 cm and a height of 7.5 cm.

Use the formula for the lateral area of a cone.

$$L.A. = \pi r \ell$$
$$= \pi r\left(\sqrt{a^2 + b^2}\right) \quad \text{To find the slant height use the Pythagorean Theorem.}$$
$$= \pi(4)\left(\sqrt{4^2 + 7.5^2}\right) \quad \text{Substitute. If } d = 8, \text{ then } r = 4.$$
$$= 106.81415 \quad \text{Use a calculator.}$$

You need about 107 cm² of filter paper to line the funnel.

✓ **Quick Check** ④ Find the lateral area of a cone with radius 15 in. and height 20 in.

EXERCISES

For more exercises, see *Extra Skill, Word Problem, and Proof Practice.*

Practice and Problem Solving

Ⓐ **Practice by Example**

Example 1
(page 618)

Find the surface area of each pyramid to the nearest whole number.

1.  11 in.
12 in.

2. 8 m
$2\sqrt{3}$ m   4 m

3. 7.2 in.
8 in.

Example 2
(page 618)

Find the lateral area of each pyramid to the nearest whole number.

4. 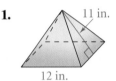 6 m
12 m

5. 8 cm — ℓ
$5\sqrt{3}$ cm   10 cm

6. 6 m
4 m
4 m

7. **Social Studies** The original height of the pyramid built for Khafre, next to the Great Pyramid, was about 471 ft. Each side of its square base was about 708 ft. What is the lateral area to the nearest square foot of a pyramid with those dimensions?

Example 3
(page 619)

Find the surface area of each cone in terms of π.

8.  18 cm
12 cm

9. 8 ft
6 ft

10. 10 cm
7 cm

Example 4
(page 620)

Find the lateral area of each cone to the nearest whole number.

11.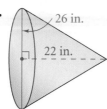
26 in.
22 in.

12.
4.5 m
4 m

13.
3 cm
4 cm

 Apply Your Skills

Problem Solving Hint

In Exercise 14, explain why $\overline{PT}$ is shorter than $\overline{PR}$, and then why $\overline{PR}$ is shorter than $\overline{PC}$.

14. Writing Explain why the altitude $\overline{PT}$ in the pyramid at the right must be shorter than each edge $\overline{PA}, \overline{PB}, \overline{PC},$ and $\overline{PD}$.

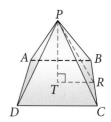

15. Reasoning Suppose you could climb to the top of the Great Pyramid in Egypt. Which route would be shorter, a route along a lateral edge or along the altitude of a side? Which of these routes is steeper? Explain your answers.

16. A square pyramid has base edges 10 in. long and height 4 in. Sketch the pyramid and find its surface area. Round your answer to the nearest tenth.

x^2 **17. Algebra** The lateral area of a pyramid with a square base is 240 ft². Its base edges are 12 ft long. Find the height of the pyramid.

Find the surface area to the nearest whole number.

18.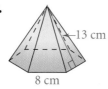
13 cm
8 cm

19.
6 cm
6 cm

20.
4 in.
4 in.

21. The lateral area of a cone is 4.8π in.². The radius is 1.2 in. Find the slant height.

22. Open-Ended Draw a square pyramid with a lateral area of 48 cm². Label its dimensions. Then find its surface area.

23. Architecture The roof of a tower in a castle is shaped like a cone. The height of the roof is 30 ft and the radius of the base is 15 ft. What is the area of the roof? Round your answer to the nearest tenth.

24. The hourglass shown at the right is made by connecting two glass cones inside a glass cylinder. Which has more glass, the two cones or the cylinder? Explain.

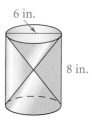
6 in.
8 in.

25. You can use the formula S.A. $= (\ell + r)\, r\pi$ to find the surface area of a cone. Explain why this formula works. Also, explain why you may prefer to use this formula when finding surface area with a calculator.

26. Find a formula for each of the following.
a. the slant height of a cone in terms of the surface area and radius
b. the radius of a cone in terms of the surface area and slant height

27. Multiple Choice The roof of a tower is a square pyramid with base edges 10 ft long. The height of the pyramid is 6 ft. What is the approximate area of the roofing material needed to cover the roof?
Ⓐ 156 ft²　　Ⓑ 233 ft²　　Ⓒ 256 ft²　　Ⓓ 333 ft²

Exercise 23

Find the surface area to the nearest whole number.

28.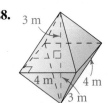

29.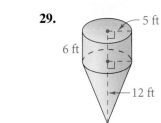

30.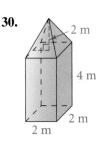

The length of a side of the base (s), slant height, height, lateral area, and surface area are measurements of a square pyramid. Given two of the measurements, find the other three to the nearest tenth.

31. $s = 3$ in., S.A. $= 39$ in.^{2}

32. $h = 8$ m, $\ell = 10$ m

33. $\ell = 5$ ft, L.A. $= 20$ ft^2

34. L.A. $= 118$ cm^2, S.A. $= 182$ cm^2

35. A cone with radius 9 cm has the same surface area as a cylinder with radius 6 cm and height 18 cm. What is the height of the cone to the nearest tenth?

36. Find the surface area of the hexagonal pyramid at the right.

Visualization The plane region is revolved completely about the given line to sweep out a solid of revolution. Describe the solid. Then find its surface area in terms of π.

37. about the y-axis

38. about the x-axis

39. about the line $x = 4$

40. about the line $y = 3$

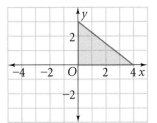

The given figure fits inside a 10-cm cube. The figure's base is in one face of the cube and is as large as possible. The figure's vertex is in the opposite face of the cube. Draw a sketch and find the lateral and surface areas of the figure.

41. a square pyramid

42. a cone

43. A sector has been cut out of the disk. The radii of the part that remains are taped together, without overlapping, to form the cone. The cone has a lateral area of 64π cm^2. Find the measure of the central angle of the cut-out sector.

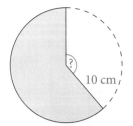 

NY REGENTS

Test Prep

Multiple Choice

44. To the nearest whole number, what is the surface area of a cone with diameter 27 m and slant height 19 m?

A. 1378 m^2 **B.** 1951 m^2 **C.** 2757 m^2 **D.** 3902 m^2

45. To the nearest whole number, what is the surface area of a cone with radius 14 cm and slant height 18 cm?

F. 448 cm^2 **G.** 836 cm^2 **H.** 1012 cm^2 **J.** 1407 cm^2

46. To the nearest whole number, what is the surface area of a square pyramid with each side of the base 30 yd and slant height 42 yd?
A. 900 yd²   **B.** 2520 yd²   **C.** 3420 yd²   **D.** 3600 yd²

47. A cylinder and a cone each have height 1 and radius $\sqrt{3}$. How does the cylinder's lateral area x compare with the cone's lateral area y?
F. $x = y$ **G.** $x = 2y$ **H.** $x > 2y$ **J.** $x < 2y$

Short Response

48. A square pyramid is 8 m on each side. Its surface area is 240 m². What is its slant height? Show your work and explain your reasoning.

Extended Response

49. The lateral area of a cone is twice the area of its base.
     a. What is its slant height in terms of the radius r? Show your work.
     b. What is the lateral area to the nearest tenth if the radius is 6 centimeters? Show your work.

Mixed Review

Lesson 11-2

50. How much cardboard do you need to make a closed box that is 4 ft by 5 ft by 2 ft?

51. How much posterboard do you need to make a cylinder, open at each end, with height 9 in. and diameter $4\frac{1}{2}$ in.? Round your answer to the nearest square inch.

Lesson 10-2

52. The area of a rhombus is 714 cm². One diagonal is 42 cm long. Find the length of the other diagonal.

53. A kite with area 195 in.² has a 15-in. diagonal. How long is the other diagonal?

Lesson 8-5

54. A TV camera views a tall building 400 m away with a 35° angle of elevation to the top. How tall is the building if the camera lens is 160 cm off the ground?

✓ Checkpoint Quiz 1   Lessons 11-1 through 11-3

Draw a net for each figure. Label the net with its dimensions.

1.
11 cm
4 cm

2.
12 in.
4 in.
6.3 in.

3.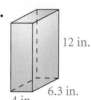
40 m
ℓ
60 m
60 m

4. Find the surface area of the cylinder in Exercise 1.

5. Find the surface area of the prism in Exercise 2.

6. Find the surface area of the pyramid in Exercise 3.

7. A cone has a radius of 8 in. and a slant height of 13 in. Find the lateral area of the cone. Leave your answer in terms of π.

8. Open-Ended Draw a net for a regular hexagonal prism.

Draw a cube. Shade the cube to show each cross section.

9. a rectangle

10. a trapezoid

11-4 Volumes of Prisms and Cylinders

 Learning Standards for Mathematics

G.G.11 Know and apply that two prisms have equal volumes if their bases have equal areas and their altitudes are equal.

G.G.14 Apply the properties of a cylinder, including:
• bases are congruent.
• volume equals the product of the area of the base and the altitude.

Check Skills You'll Need

 for Help Lessons 1-9 and 10-1

Find the area of each figure. For answers that are not whole numbers, round to the nearest tenth.

1. a square with side length 7 cm

2. a circle with diameter 15 in.

3. a circle with radius 10 mm

4. a rectangle with length 3 ft and width 1 ft

5. a rectangle with base 14 in. and height 11 in.

6. a triangle with base 11 cm and height 5 cm

7. an equilateral triangle that is 8 in. on each side

New Vocabulary • volume • composite space figure

---

1 Finding Volume of a Prism

Hands-On Activity: Finding Volume

Explore the volume of a prism with unit cubes.

• Make a one-layer rectangular prism that is 4 cubes long and 2 cubes wide. The prism will be 4 units by 2 units by 1 unit.

1. How many cubes are in the prism?

2. Add a second layer to your prism to make a prism 4 units by 2 units by 2 units. How many cubes are in this prism?

3. Add a third layer to your prism to make a prism 4 units by 2 units by 3 units. How many cubes are in this prism?

4. How many cubes would be in the prism if you added two additional layers of cubes for a total of 5 layers?

5. How many cubes would be in the prism if there were 10 layers?

Volume is the space that a figure occupies. It is measured in cubic units such as cubic inches (in.^{3}), cubic feet (ft^3), or cubic centimeters (cm^3). The volume of a cube is the cube of the length of its edge, or $V = e^3$.

Both stacks of paper below contain the same number of sheets.

The first stack forms a right prism. The second forms an oblique prism. The stacks have the same height. The area of every cross section parallel to a base is the area of one sheet of paper. The stacks have the same volume. These stacks illustrate the following principle.

 Key Concepts

| **Theorem 11-5** | **Cavalieri's Principle** |

If two space figures have the same height and the same cross-sectional area at every level, then they have the same volume.

The area of each shaded cross section below is 6 cm^2. Since the prisms have the same height, their volumes must be the same by Cavalieri's Principle.

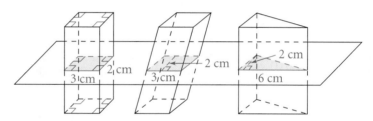

You can find the volume of a right prism by multiplying the area of the base by the height. Cavalieri's Principle lets you extend this idea to any prism.

 Key Concepts

| **Theorem 11-6** | **Volume of a Prism** |

The volume of a prism is the product of the area of a base and the height of the prism.

$$V = Bh$$

Online
active math

For: Prism, Cylinder Activity
Use: Interactive Textbook, 11-4

1 EXAMPLE Finding Volume of a Rectangular Prism

Find the volume of the prism at the right.

$V = Bh$ **Use the formula for volume.**

$\quad = 480 \cdot 10$ $B = 24 \cdot 20 = 480$ cm^2

$\quad = 4800$ **Simplify.**

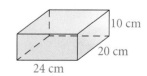

● The volume of the rectangular prism is 4800 cm^3.

✓ Quick Check   ❶ **Critical Thinking** Suppose the prism in Example 1 is turned so that the base is 20 cm by 10 cm and the height is 24 cm. Explain why the volume does not change.

2 EXAMPLE Finding Volume of a Triangular Prism

Multiple Choice Find the approximate volume of
the triangular prism at the right.

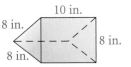

- A 188 in.^{3}
- B 277 in.^{3}
- C 295 in.^{3}
- D 554 in.^{3}

Each base of the triangular prism is an equilateral
triangle. An altitude of the triangle divides it into
two 30°-60°-90° triangles. The area of the base is
$\frac{1}{2} \cdot 8 \cdot 4\sqrt{3}$, or $16\sqrt{3}$ in.^{2}.

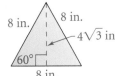

$V = Bh$ **Use the formula for the volume of a prism.**

$\quad = 16\sqrt{3} \cdot 10$ **Substitute.**

$\quad = 160\sqrt{3}$ **Simplify.**

$\quad = 277.12813$ **Use a calculator.**

● The volume of the triangular prism is about 277 in.^{3}. The answer is B.

Quick Check ❷ Find the volume of the triangular prism at the right.

---

2 Finding Volume of a Cylinder

To find the volume of a cylinder, you use the same formula $V = Bh$ that you use to
find the volume of a prism. Now, however, B is the area of the circle, so you use the
formula $B = \pi r^2$ to find its value.

Key Concepts

Theorem 11-7	Volume of a Cylinder

The volume of a cylinder is the product of the
area of the base and the height of the cylinder.

$$V = Bh, \text{ or } V = \pi r^2 h$$

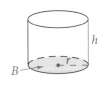

GO Online

Video Tutor Help

Visit: PHSchool.com
Web Code: aue-0775

3 EXAMPLE Finding Volume of a Cylinder

Find the volume of the cylinder at the right. Leave your
answer in terms of π.

$V = \pi r^2 h$ **Use the formula for the volume of a cylinder.**

$\quad = \pi(3)^2(8)$ **Substitute.**

$\quad = \pi(72)$ **Simplify.**

● The volume of the cylinder is 72π cm^3.

Quick Check ❸ The cylinder at the right is oblique.
  a. Find its volume in terms of π.
  b. Find its volume to the nearest tenth of a cubic meter.

A **composite space figure** is a three-dimensional figure that is the combination of two or more simpler figures. A space probe, for example, might begin as a composite figure—a cylindrical rocket engine in combination with a nose cone.

You can find the volume of a composite space figure by adding the volumes of the figures that are combined.

4 EXAMPLE Finding Volume of a Composite Figure

Estimation Use a composite space figure to estimate the volume of the backpack shown at the left.

Step 1: You can use a prism and half of a cylinder to approximate the shape, and therefore the volume, of the backpack.

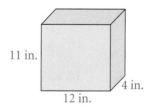
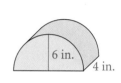

Step 2: Volume of the prism $= Bh = (12 \cdot 4)11 = 528$

Step 3: Volume of the half cylinder $= \frac{1}{2}(\pi r^2 h) = \frac{1}{2}\pi(6)^2(4)$
$$= \frac{1}{2}\pi(36)(4) \approx 226$$

Step 4: Sum of the two volumes $= 528 + 226 = 754$

● The approximate volume of the backpack is 754 in.^{3}.

✓ **Quick Check** ④ Find the volume of the composite space figure.

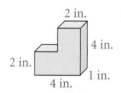

EXERCISES

For more exercises, see *Extra Skill, Word Problem, and Proof Practice*.

Practice and Problem Solving

A Practice by Example

Example 1
(page 625)

GO for Help

In Exercises 1–8, find the volume of each prism.

1.

2.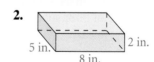

3.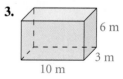

4. The base is a square, 2 cm on a side. The height is 3.5 cm.

Example 2
(page 626)

5.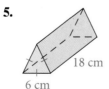

6.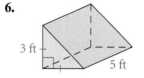

7.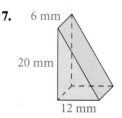

8. The base is a 45°-45°-90° triangle with a leg of 5 in. The height is 1.8 in.

Example 3
(page 626)

Find the volume of each cylinder in terms of π and to the nearest tenth.

9.
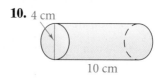
8 in.
6 in.

10. 4 cm

10 cm

11.
5 m

6 m

Example 4
(page 627)

Find the volume of each composite space figure to the nearest whole number.

12.
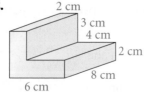
2 cm
3 cm
4 cm
2 cm
8 cm
6 cm

13.
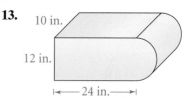
10 in.
12 in.
24 in.

B **Apply Your Skills**

14. a. What is the volume of a waterbed mattress that is 7 ft by 4 ft by 1 ft?
   b. To the nearest pound, what is the weight of the water in a full mattress? (Water weighs 62.4 lb/ft³.)

15. Find the volume of the lunch box shown at the right to the nearest cubic inch.

3 in.
6 in.
6 in.
10 in.

16. Open-Ended Give the dimensions of two rectangular prisms that have volumes of 80 cm³ each but also have different surface areas.

Find the height of each figure with the given volume.

17.

h
9 cm
$V = 234\pi$ cm³

18.

h
5 in.
5 in.
$V = 125$ in.³

19.
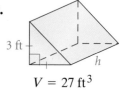
3 ft
h
$V = 27$ ft³

20. Ecology The isolation cube at the left measures 27 in. on each side. What is its volume in cubic feet?

21. Environmental Engineering A scientist suggests keeping indoor air relatively clean as follows: Provide two or three pots of flowers for every 100 square feet of floor space under a ceiling of 8 feet. If your classroom has an 8-ft ceiling and measures 35 ft by 40 ft, how many pots of flowers should it have?

22. Find the volume of the oblique prism pictured at the right.

23. Tank Capacity The main tank at an aquarium is a cylinder with diameter 203 ft and height 25 ft.
   a. Find the volume of the tank to the nearest cubic foot.
   b. Convert your answer to part (a) to cubic inches.
   c. If 1 gallon ≈ 231 in.³, about how many gallons does the tank hold?

6 ft
4 ft

24. Writing The figures at the right can be covered by equal numbers of straws that are the same length. Describe how Cavalieri's Principle could be adapted to compare the areas of these figures.

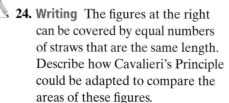

25. Coordinate Geometry Find the volume of the rectangular prism at the right.

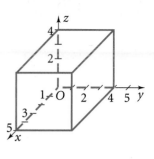

26. The volume of a cylinder is 600π cm³. The radius of a base of the cylinder is 5 cm. What is the height of the cylinder?

27. The volume of a cylinder is 135π cm³. The height of the cylinder is 15 cm. What is the radius of a base of the cylinder?

28. Multiple Choice A cylindrical water tank has a diameter of 9 inches and a height of 12 inches. The water surface is 2.5 inches from the top. About how much water is in the tank?

- Ⓐ 604 in.³
- Ⓑ 636 in.³
- Ⓒ 668 in.³
- Ⓓ 763 in.³

29. Landscaping To landscape her 70 ft-by-60 ft rectangular backyard, Joy is planning first to put down a 4-in. layer of topsoil. She can buy bags of topsoil at $2.50 per 3-ft³ bag, with free delivery. Or, she can buy bulk topsoil for $22.00/yd³, plus a $20 delivery fee. Which option is less expensive? Explain.

Visualization The plane region is revolved completely about the given line to sweep out a solid of revolution. Describe the solid and find its volume in terms of π.

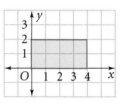

30. the x-axis

31. the y-axis

32. the line $y = 2$

33. the line $x = 5$

A cylinder has been cut out of each solid. Find the volume of the remaining solid. Round your answer to the nearest tenth.

34.

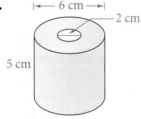

35.

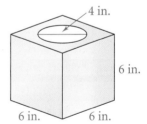

36. A closed box is 9 in. by 14 in. by 6 in. on the inside and 11 in. by 16 in. by 7 in. on the outside. Find each measurement.
- **a.** the outside surface area
- **b.** the inside surface area
- **c.** the inside volume
- **d.** the volume of the material needed to make the box

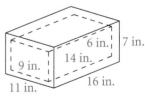

Ⓒ Challenge

37. Any rectangular sheet of paper can be rolled into a right cylinder in two ways.
- **a.** Use ordinary sheets of paper to model the two cylinders. Compute the volume of each cylinder. How do they compare?
- **b.** Of all sheets of paper with perimeter 39 in., which size can be rolled into a right cylinder with greatest volume? (*Hint:* See Activity Lab, page 616.)

Problem Solving Hint

In Exercise 25, find the length, width, and height along the axes.

38. The outside diameter of a pipe is 5 cm. The inside diameter is 4 cm. The pipe is 4 m long. What is the volume of the material used for this length of pipe? Round your answer to the nearest cubic centimeter.

39. A cube has a volume of $2M$ cubic units and a total surface area of $3M$ square units. Find the length of an edge of the cube.

40. The radius of cylinder B is twice the radius of cylinder A. The height of cylinder B is half the height of cylinder A. Compare their volumes.

Test Prep

Multiple Choice

41. What is the volume of a rectangular prism whose edges measure 2 ft, 2 ft, and 3 ft?
  A. 7 ft³   **B.** 12 ft³   **C.** 14 ft³   **D.** 16 ft³

42. One gallon fills about 231 in.³. A right cylindrical carton is 12 in. tall and holds 9 gal when full. Find the radius of the carton to the nearest tenth of an inch.
  F. 0.5 in.   **G.** 7.4 in.   **H.** 37.7 in.   **J.** 55.1 in.

43. The height of a triangular prism is 8 feet. One side of the base measures 6 feet. What additional information do you need to find the volume?
  A. the perimeter of the base
  B. the length of a second side of the base
  C. the altitude of the base to the 6-foot side
  D. the area of each rectangular face of the prism

44. A rectangular prism has a volume of 100 ft³. If the base measures 5 ft by 8 ft, what is the height of the prism?
  F. 2.5 ft   **G.** 12.5 ft   **H.** 20 ft   **J.** 40 ft

Short Response

45. How is the formula for finding the lateral area of a cylinder like the formula for finding the area of a rectangle?

Mixed Review

Lesson 11-3

Find the lateral area of each figure to the nearest tenth.

46. a right circular cone with height 12 mm and radius 5 mm

47. a regular hexagonal pyramid with base edges 9.2 ft long and slant height 17 ft

Lesson 10-6 $\boxed{x^2}$ **Algebra** **Find the value of each variable and the measure of each labeled angle.**

48.
49.
50.

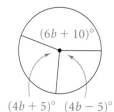

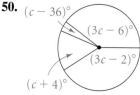

Lesson 7-3

51. You want to find the height of a tree near your school. Your shadow is three-fourths of your height. The tree's shadow is 57 feet. How tall is the tree?

11-5

Volumes of Pyramids and Cones

 Learning Standards for Mathematics

G.G.13 Apply the properties of a regular pyramid.

G.G.15 Apply the properties of a right circular cone.

✓ **Check Skills You'll Need** **GO for Help** Lesson 8-1

Use the Pythagorean Theorem to find the value of the variable.

1.
15 cm
h
9 cm

2.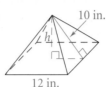
10 in.
h
12 in.

3.
ℓ
2 m
1.5 m

---

1 Finding Volume of a Pyramid

Hands-On Activity: Finding Volume

You know how to find the volume of a prism. Use the following to explore finding the volume of a pyramid.

- Draw the nets shown at the right on cardboard.

- Cut out the nets and tape them together to make a cube and a regular square pyramid. Each model will have one open face.

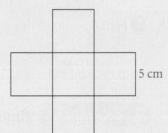
5 cm

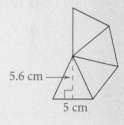
5.6 cm
5 cm

1. How do the areas of the bases of the cube and the pyramid compare?

2. How do the heights of the cube and pyramid compare?

3. Fill the pyramid with rice or other material. Then pour the rice from the pyramid into the cube. How many pyramids full of rice does the cube hold?

4. The volume of the pyramid is what fractional part of the volume of the cube?

The volume of a pyramid is a particular fraction of the volume of a prism that has the same base and height as the pyramid. The fraction is shown and this fact is stated as Theorem 11-8 at the top of the next page.

Key Concepts

Theorem 11-8	Volume of a Pyramid

The volume of a pyramid is one third the product of the area of the base and the height of the pyramid.

$$V = \frac{1}{3}Bh$$

GO for Help

You can find Cavalieri's Principle on page 625.

Because of Cavalieri's Principle, the volume formula is true for all pyramids, including oblique pyramids. The height h of an oblique pyramid is the length of the perpendicular segment from the vertex to the plane of the base.

Oblique Pyramid

1 EXAMPLE **Real-World Connection**

Architecture The Pyramid is an arena in Memphis, Tennessee. The area of the base of The Pyramid is about 300,000 ft². Its height is 321 ft. What is the volume of The Pyramid?

$V = \frac{1}{3}Bh$ **Use the formula for volume of a pyramid.**

$= \frac{1}{3}(300,000)(321)$ **Substitute.**

$= 32,100,000$ **Simplify.**

● The volume is about 32,100,000 ft³.

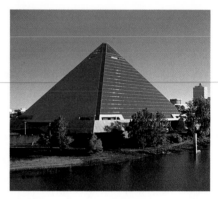

✓ Quick Check **1** Find the volume of a square pyramid with base edges 12 in. and height 8 in.

To find the volume of a pyramid you may first need to find its height.

2 EXAMPLE **Finding Volume of a Pyramid**

Gridded Response Find the volume in cubic feet of a square pyramid with base edges 40 ft and slant height 25 ft.

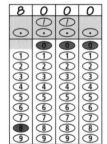

Step 1: Find the height of the pyramid.

$25^2 = h^2 + 20^2$ **Use the Pythagorean Theorem.**

$625 = h^2 + 400$ **Simplify.**

$h^2 = 225$ **Solve for h^2.**

$h = 15$ **Take square roots.**

Step 2: Find the volume of the pyramid.

$V = \frac{1}{3}Bh$ **Use the formula for volume of a pyramid.**

$= \frac{1}{3}(40 \cdot 40)15$ **Substitute.**

$= 8000$ **Simplify.**

● The volume of the pyramid is 8000 ft³.

✓ Quick Check **2** Find the volume of a square pyramid with base edges 24 m and slant height 13 m.

For: Pyramid, Cone Activity
Use: Interactive Textbook, 11-5

You have seen that the volume of a pyramid is one third the volume of a prism with the same base and height. Similarly, the volume of a cone is one third the volume of a cylinder with the same base and height.

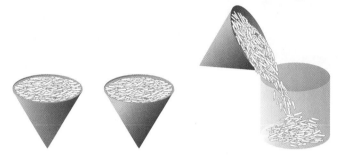

The cones and the cylinder have the same base and height.
It takes three cones full of rice to fill the cylinder.

 Key Concepts

Theorem 11-9	Volume of a Cone

The volume of a cone is one third the product of the area of the base and the height of the cone.

$$V = \frac{1}{3}Bh, \text{ or } V = \frac{1}{3}\pi r^2 h$$

This volume formula applies to all cones, including oblique cones.

3 EXAMPLE   **Finding Volume of an Oblique Cone**

Find the volume of an oblique cone with diameter 30 ft and height 25 ft. Give your answer in terms of π and also rounded to the nearest cubic foot.

$V = \frac{1}{3}\pi r^2 h$ **Use the formula for volume of a cone.**

$= \frac{1}{3}\pi(15)^2 25$ **Substitute 15 for r and 25 for h.**

$= 1875\pi$ **Simplify.**

≈ 5890.4862 **Use a calculator.**

● The volume of the cone is 1875π ft^3, or about 5890 ft^3.

Test-Taking Tip

The formula $V = \frac{1}{3}Bh$ can be used to find the volume of a pyramid or a cone. The only difference is how you calculate B, the area of the base.

✓ **Quick Check**   ❸ Find the volume of each cone in terms of π and also rounded as indicated.

a. to the nearest cubic meter   **b.** to the nearest cubic millimeter

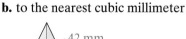

A cone-shaped structure can be particularly strong, as downward forces at the vertex are distributed to all points in its circular base.

4 EXAMPLE <u>Real-World</u> Connection

Dwelling The covering on a teepee rests on poles that come together like concurrent lines. The resulting structure approximates a cone. If the teepee pictured is 12 ft. high with a base diameter 14 ft, estimate its volume.

$V = \frac{1}{3}\pi r^2 h$ Use the formula for the volume of a cone.

$V = \frac{1}{3}\pi(7)^2(12)$ Substitute; $r = \frac{1}{2}(14) = 7$.

$= 615.75216$ Use a calculator.

● The volume of the teepee is approximately 616 ft³.

✓ **Quick Check** ④ A small child's teepee is 6 ft tall and 7 ft in diameter. Find the volume of the teepee to the nearest cubic foot.

EXERCISES

For more exercises, see *Extra Skill, Word Problem, and Proof Practice.*

Practice and Problem Solving

A Practice by Example

Example 1
(page 632)

GO for Help

1. The entrance to the Louvre Museum in Paris, France is a pyramid. The height of the pyramid is about 70 ft and the area of its base is about 10,000 ft². What is the volume of the pyramid?

Find the volume of a square pyramid with the following dimensions.

2. base edges 10 cm, height 6 cm

3. base edges 18 in., height 12 in.

4. base edges 5 m, height 6 m

Find the volume of each square pyramid.

5.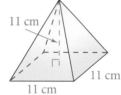
11 cm   11 cm   11 cm

6.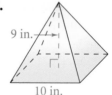
9 in.   10 in.

7.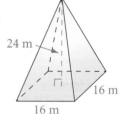
24 m   16 m   16 m

Example 2
(page 632)

8.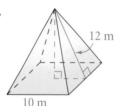
12 m   10 m

9.
24 mm   23 mm

10.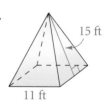
15 ft   11 ft

Example 3
(page 633)

Find the volume of each cone in terms of π and also rounded as indicated.

11. nearest cubic foot

4 ft   4 ft

12. nearest cubic inch

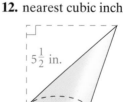
$5\frac{1}{2}$ in.   4 in.

13. nearest cubic inch

9 in.   7 in.

Example 4
(page 634)

14. Chemistry In a chemistry lab you use a filter paper cone to filter a liquid. The diameter of the cone is 6.5 cm and its height is 6 cm. How much liquid will the cone hold when it is full?

15. Chemistry This funnel has a filter that was being used to remove impurities from a solution but became clogged and stopped draining. The remaining solution is represented by the shaded region. How many cubic centimeters of the solution remain in the funnel?

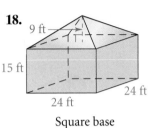

3 cm

2 cm

B **Apply Your Skills** Find the volume to the nearest whole number.

16.

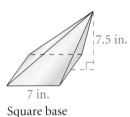

7.5 in.

7 in.

Square base

17.

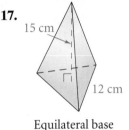

15 cm

12 cm

Equilateral base

18.

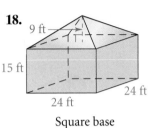

9 ft

15 ft

24 ft

24 ft

Square base

19. Writing The two cylinders pictured at the right are congruent. How does the volume of the larger cone compare to the total volume of the two smaller cones? Explain.

20. Architecture The Transamerica Pyramid in San Francisco (see photo at left) is 853 ft tall with a square base that is 149 ft on each side.
a. What is its volume to the nearest thousand cubic feet?
b. How tall would a prism-shaped building with the same square base as the Pyramid have to be to have the same volume as the Pyramid?

x^2 **Algebra** Find the value of the variable in each figure. Leave answers in simplest radical form. The diagrams are not to scale.

21.

6

x

x x

Volume = $18\sqrt{3}$

22.

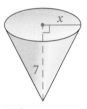

x

7

Volume = 21π

23.

4

r

Volume = 24π

24. Hardware Builders use a plumb bob to find a vertical line. The plumb bob shown combines a regular hexagonal prism with a pyramid. Find its volume to the nearest cubic centimeter.

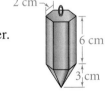

2 cm

6 cm

3 cm

25. Multiple Choice A cone has a volume of 600π in.^{3} and a height of 50 in. What is the radius of the cone?
  Ⓐ 3.5 in.   Ⓑ 6.0 in.   Ⓒ 10.6 in.   Ⓓ 36.0 in.

26. A cone with radius 1 fits snugly inside a square pyramid which fits snugly inside a cube. What are the volumes of the three figures?

27. A cone with radius 3 ft and height 10 ft has a volume of 30π ft^3. What is the volume of the cone formed when the following happens to the original cone?
a. The radius is doubled.   **b.** The height is doubled.
c. The radius and the height are both doubled.

Exercise 20

GO Online
Homework Video Tutor
Visit: PHSchool.com
Web Code: aue-1105

28. List the volumes of the cone, prism, and pyramid in order from least to greatest.

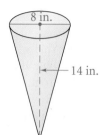
8 in.
14 in.

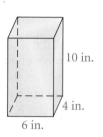
10 in.
4 in.
6 in.

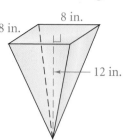
8 in.
8 in.
8 in.
12 in.

Visualization The plane region is revolved completely about the given line to sweep out a solid of revolution. Describe the solid. Then find its volume in terms of π.

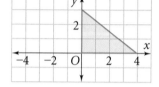

29. the y-axis

30. the x-axis

Ⓒ Challenge

31. the line $x = 4$

32. the line $y = -1$

33. A *frustum* of a cone is the part that remains when the vertex is cut off by a plane parallel to the base.
  a. Explain how to use the formula for the volume of a cone to find the volume of a frustum of a cone.
  b. Containers A 9-in. tall popcorn container is the frustum of a cone. Its small radius is 4.5 in. and its large radius is 6 in. What is its volume?

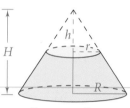
Frustum of cone

34. A disk has radius 10 m. A 90° sector is cut away, and a cone is formed.
  a. What is the circumference of the base of the cone?
  b. What is the area of the base of the cone?
  c. What is the volume of the cone? (*Hint:* Use the slant height and the radius of the base to find the height.)

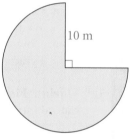
10 m

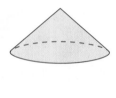

📟 **Graphing Calculator** In Exercises 35 and 36, the volume of the solid is 1000 cm³. Use the Activity Lab on page 616 to help you complete each exercise.

35. For a square pyramid, find the length of a side of the base for which the lateral area is as small as possible.

36. For a cone, find the radius for which the lateral area is as small as possible.

NY REGENTS

Test Prep

Multiple Choice

37. What is the volume of a 6-ft high square pyramid with base edges 8 ft?
  A. 128 ft³   **B.** 192 ft³   **C.** 256 ft³   **D.** 384 ft³

38. What is the volume of a cone with diameter 21 m and height 4 m?
  F. 147π m³   **G.** 220.5π m³   **H.** 294π m³   **J.** 441π m³

39. What is the volume of an oblique cone with radius 9 cm and height 12 cm?
  A. 324π cm³   **B.** 486π cm³   **C.** 648π cm³   **D.** 972π cm³

40. What is the volume of the square pyramid at the right?
   F. 1568 m³   **G.** 1633 m³
   H. 2352 m³   **J.** 2450 m³

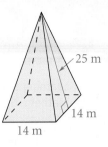

25 m

14 m

14 m

41. What is the volume of an oblique square pyramid
with base edges 25 in. and height 24 in.?
   A. 5000 in.³   **B.** 7500 in.³
   C. 10,000 in.³   **D.** 15,000 in.³

Short Response   **42.** The volume of a cone is 82,418π cm³. Its diameter is 203 cm. What is its height? Show all your work, including any formulas that you use.

Mixed Review

Lesson 11-4

43. Sports A cylindrical hockey puck is 1 in. high and 3 in. in diameter. What is its volume in cubic inches? Round your answer to the nearest tenth.

44. A triangular prism has height 30 cm. Its base is a right triangle with legs 10 cm and 24 cm. Find the volume of the prism.

Lesson 10-5

45. Find the area of a regular pentagon with a radius 5 in. Give your answer to the nearest tenth of a square inch.

Lesson 8-2

Find the area of each equilateral triangle to the nearest tenth of a square unit.

46. The triangle has 12 cm sides.   **47.** The triangle has 10-in. altitudes.

48. Find the area of a 30°-60°-90° triangle with shorter leg of length 4 cm.

Geometry at Work

Package Designer

Each year, more than one trillion dollars in manufactured goods are packaged in containers. To create each new box, bag, or carton, package designers must balance such factors as safety, environmental impact, and attractiveness against cost of production.

Consider the three boxes of dishwasher detergent. All three boxes have standard volumes of 108 in.³. The boxes have different shapes, however, and different surface areas. The box on the left has

9 in.

6 in.

6 in.

6 in.

2 in.

6 in.

3 in.

4½ in.

4 in.

the greatest surface area and therefore costs the most to produce. Despite the higher cost, the box on the left has become standard. In this case, the least expensive package on the right is too difficult for a consumer to pick up and pour.

Go Online
PHSchool.com

For: Information about package design
Web Code: aub-2031

11-6

Surface Areas and Volumes of Spheres

 Learning Standards for Mathematics

G.G.16 Apply the properties of a sphere, including:
- the intersection of a plane and a sphere is a circle.
- a great circle is the largest circle that can be drawn on a sphere.
- surface area is $4\pi r^2$.
- volume is $\frac{4}{3}\pi r^3$.

✓ **Check Skills You'll Need**

GO **for Help** Lesson 1-9

Find the area and circumference of a circle with the given radius. Round your answers to the nearest tenth.

1. 6 in.   **2.** 5 cm   **3.** 2.5 ft
4. 1.2 m   **5.** 15 yd   **6.** 12 mm

◀)) **New Vocabulary**   • sphere   • center, radius, diameter, circumference (of a sphere)   • great circle   • hemisphere

---

1   Finding Surface Area and Volume of a Sphere

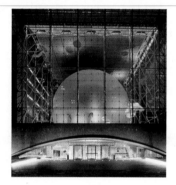

Real-World 🌐 **Connection**

The diameter of the Hayden Sphere in New York City is 87 ft.

A **sphere** is the set of all points in space equidistant from a given point called the **center**. A **radius** is a segment that has one endpoint at the center and the other endpoint on the sphere. A **diameter** is a segment passing through the center with endpoints on the sphere.

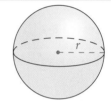

The *length r* is *the* radius of the sphere. 2*r* is *the* diameter.

When a plane and a sphere intersect in more than one point, the intersection is a circle. If the center of the circle is also the center of the sphere, the circle is called a **great circle** of the sphere. The circumference of a great circle is the **circumference** of the sphere. A great circle divides a sphere into two **hemispheres.**

A baseball is a model of a sphere. To approximate its surface area, you can take apart its covering. Each of the two sections suggests a pair of circles with radius *r* approximately the radius of the ball. The area of the four circles, $4\pi r^2$, suggests the surface area of the ball.

 Key Concepts

Theorem 11-10	**Surface Area of a Sphere**

The surface area of a sphere is four times the product of π and the square of the radius of the sphere.

$$\text{S.A.} = 4\pi r^2$$

---

1 EXAMPLE Finding Surface Area

Find the surface area of this sphere. Leave your answer in terms of π.

S.A. $= 4\pi r^2$ **Use the formula for surface area.**

$\quad = 4\pi 4^2$ **Substitute $r = \frac{8}{2} = 4$.**

$\quad = 64\pi$ **Simplify.**

● The surface area is 64π m².

 **1** Find the surface area of a sphere with $d = 14$ in. Give your answer two ways, in terms of π and rounded to the nearest square inch.

You can use spheres to approximate the surface areas of real-world objects.

2 EXAMPLE Real-World Connection

Geography Earth's equator is about 24,902 mi long. Approximate the surface area of Earth by finding the surface area of a sphere with circumference 24,902 mi.

Step 1 Find the radius.

$\quad C = 2\pi r$ **Use the formula for circumference.**

$\quad 24{,}902 = 2\pi r$ **Substitute.**

$\quad \dfrac{24{,}902}{2\pi} = r$ **Solve for r.**

$\quad r = 3963.2764$ **Use a calculator.**

Step 2 Use the radius to find the surface area.

S.A. $= 4\pi r^2$ **Use the formula for surface area.**

$\quad = 4\pi$ ANS $\boxed{x^2}$ $\boxed{\text{ENTER}}$ **Use a calculator.**

$\quad = 197387020$

● The surface area of Earth is about 197,400,000 mi².

Go Online
PHSchool.com

For: Graphing calculator procedures
Web Code: aue-2101

 **2** Find the surface area of a melon with circumference 18 in. Round your answer to the nearest ten square inches.

The following model suggests a formula for the volume of a sphere.

Fill a sphere with a large number n of small pyramids. The vertex of each pyramid is the center of the sphere. The height of each pyramid is approximately the radius r of the sphere. The sum of the areas of the bases of the n pyramids approximates the surface area of the sphere. The sum of the volumes of the n pyramids should approximate the volume of the sphere.

Volume of each pyramid $= \frac{1}{3}Bh$

Sum of the volumes of n pyramids $\approx n \cdot \frac{1}{3}Br$ **Substitute r for h.**

$\quad = \frac{1}{3} \cdot (nB) \cdot r$

$\quad \approx \frac{1}{3} \cdot (4\pi r^2) \cdot r$ **Replace nB with the surface area of a sphere.**

$\quad = \frac{4}{3}\pi r^3$

It is reasonable to conjecture that the volume of a sphere is $\frac{4}{3}\pi r^3$.

 **Key Concepts**

Theorem 11-11	**Volume of a Sphere**

The volume of a sphere is four thirds the product
of π and the cube of the radius of the sphere.

$$V = \tfrac{4}{3}\pi r^3$$

3 EXAMPLE **Finding Volume**

Find the volume of the sphere. Leave your answer in terms of π.

$V = \tfrac{4}{3}\pi r^3$ **Use the formula for volume.**

$= \tfrac{4}{3}\pi 6^3$ **Substitute.**

$= 288\pi$

6 m

● The volume of the sphere is 288π m^3.

 **Quick Check** ❸ Find the volume to the nearest cubic inch of a sphere with diameter 60 in.

If you know the volume of a sphere, you can find its surface area.

4 EXAMPLE **Using Volume to Find Surface Area**

The volume of a sphere is 5000 m^3. What is the surface area of the sphere?

Step 1 Find the radius r.

$V = \tfrac{4}{3}\pi r^3$ **Use the formula for volume of a sphere.**

$5000 = \tfrac{4}{3}\pi r^3$ **Substitute.**

$5000\left(\tfrac{3}{4\pi}\right) = r^3$ **Solve for r^3.**

$\sqrt[3]{5000\left(\tfrac{3}{4\pi}\right)} = r$ **Take cube roots.**

$r = 10.607844$ **Use a calculator.**

Step 2 Find the surface area of the sphere.

S.A. $= 4\pi r^2$ **Use the formula for the surface area of a sphere.**

$= 4\pi$ ANS $\boxed{x^2}$ $\boxed{\text{ENTER}}$ **Use a calculator.**

$= 1414.0479$

● The surface area of the sphere is about 1414 m^2.

> **Problem Solving Hint**
>
> The cube root of x, $\sqrt[3]{x}$, is the number whose third power is x.

 **Quick Check** ❹ The volume of a sphere is 4200 ft^3. Find the surface area to the nearest tenth.

EXERCISES

For more exercises, see *Extra Skill, Word Problem, and Proof Practice.*

Practice and Problem Solving

 **A** **Practice by Example**

Example 1
(page 639)

 GO for Help

Find the surface area of the sphere with the given diameter or radius. Leave your answer in terms of π.

1. $d = 30$ m   **2.** $r = 10$ in.   **3.** $d = 32$ mm   **4.** $r = 100$ yd

Find the surface area of each ball. Leave each answer in terms of π.

5.

$d = 68$ mm

6.

$d = 24$ cm

7.

$d = 2\frac{3}{4}$ in.

Example 2
(page 639)

Use the given circumference to find the surface area of each spherical object. Round your answer to the nearest whole number.

8. a grapefruit with $C = 14$ cm

9. a bowling ball with $C = 27$ in.

10. a pincushion with $C = 8$ cm

11. a head of lettuce with $C = 22$ in.

Example 3
(page 640)

Find the volume of each sphere. Give each answer in terms of π and rounded to the nearest cubic unit.

12.

5 ft

13.

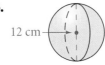

12 cm

14.

15 in.

15.

8 cm

16.

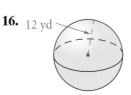

12 yd

17.

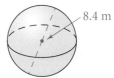

8.4 m

Example 4
(page 640)

A sphere has the volume given. Find its surface area to the nearest whole number.

18. $V = 900$ in.^{3}

19. $V = 3000$ m^3

20. $V = 140$ cm^3

B Apply Your Skills

21. Mental Math Use $\pi \approx 3$ to estimate the surface area and volume of a sphere with radius 3 cm.

22. Visualization The region enclosed by the semicircle at the right is revolved completely about the x-axis.
  a. Describe the solid of revolution that is formed.
  b. Find its volume in terms of π.
  c. Find its surface area in terms of π.

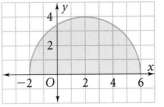

23. Food A sphere of frozen yogurt was pressed into the cone as shown at the left. If the yogurt melts into the cone, would the cone overflow? Explain.

24. Multiple Choice The sphere at the right fits snugly inside a cube with 6-in. edges. What is the approximate volume of the space between the sphere and cube?

6 in.

  Ⓐ 28.3 in.^{3} Ⓑ 76.5 in.^{3} Ⓒ 102.9 in.^{3} Ⓓ 113.1 in.^{3}

Geometry in 3 Dimensions **A sphere has center $(0, 0, 0)$ and radius 5.**

25. Name the coordinates of six points on the sphere.

26. Tell whether each of the following points is inside, outside, or on the sphere.
$A(0, -3, 4), B(1, -1, -1), C(4, -6, -10)$

4 cm

4 cm   4 cm

12 cm

Exercise 23

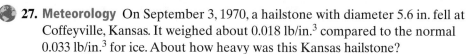

27. Meteorology On September 3, 1970, a hailstone with diameter 5.6 in. fell at Coffeyville, Kansas. It weighed about 0.018 lb/in.³ compared to the normal 0.033 lb/in.³ for ice. About how heavy was this Kansas hailstone?

28. Critical Thinking Which is greater, the total volume of three spheres, each of which has diameter 3 in., or the volume of one sphere that has diameter 8 in.?

Find the volume in terms of π of each sphere with the given surface area.

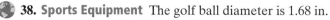

29. 4π m²   **30.** 36π in.²   **31.** 9π ft²   **32.** 100π mm²

33. 25π yd²   **34.** 144π cm²   **35.** 49π m²   **36.** 225π mi²

37. A balloon has a 14-in. diameter when it is fully inflated. Half of the air is let out of the balloon. Assume that the balloon is a sphere.
  a. Find the volume of the fully-inflated balloon in terms of π.
  b. Find the volume of the half-inflated balloon in terms of π.
  c. What is the diameter of the half-inflated balloon to the nearest inch?

38. Sports Equipment The golf ball diameter is 1.68 in.
  a. Approximate the surface area of the golf ball.
  b. Critical Thinking Do you think that the value you found in part (a) is greater or less than the actual surface area of the golf ball? Explain.

39. Open-Ended Give the dimensions of a cylinder and a sphere that have the same volume.

Find the surface area and volume of each figure.

40.   **41.** 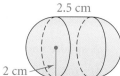   **42.**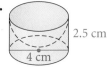

43. Science The density of steel is about 0.28 lb/in.³. Could you lift a solid steel ball with radius 4 in.? With radius 6 in.? Explain.

44. A cube with edges 6 in. long fits snugly inside the sphere. The diagonal of the cube is the diameter of the sphere.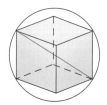
  a. Find the length of the diagonal and the radius of the sphere. Leave your answers in simplest radical form.
  b. What is the volume of the space between the sphere and the cube to the nearest tenth?

Graphing Calculator The sphere has a 10-cm radius. Of all the cylinders that fit snugly inside the sphere, such as the one shown here, find the dimensions of the one with the greatest measure indicated. (*Hint:* Use the Exploration on page 616.)

45. lateral area   **46.** volume

47. A plane intersects a sphere to form a circular cross section. The radius of the sphere is 17 cm and the plane comes to within 8 cm of the center. Draw a sketch and find the area of the cross section, to the nearest whole number.

48. Suppose a cube and a sphere have the same volume.
  a. Which has the greater surface area? Explain.
  b. Writing Explain why spheres are rarely used for packaging.

GO Online
Homework Video Tutor
Visit: PHSchool.com
Web Code: aue-1106

Challenge

Find the radius of a sphere with the given property.

49. The number of square meters of surface area equals the number of cubic meters of volume.

50. The ratio of surface area in square meters to volume in cubic meters is 1 : 5.

51. **History** The sphere fits snugly inside the cylinder. Archimedes (c. 287–212 B.C.) asked that such a figure be put on his gravestone along with the ratio of their volumes, a finding that he regarded as his greatest. What is that ratio?

Multiple Choice

Read the passage below, then answer Exercises 52–54 based on what is stated in the passage.

Believe It Or Not

J.C. Payne, a Texas farmer, is the world champion string collector. The ball of string he wound over a three-year period has a circumference of 41.5 ft. It weighs 13,000 lb.

Listed in the Guinness Book of World Records, the ball of string is now in a museum devoted to oddities. It took almost a dozen men with fork-lift trucks to load the ball onto a truck to move it there.

52. What is the radius of the ball of string to the nearest tenth of a foot?
   A. 3.3 ft   B. 6.6 ft   C. 13.2 ft   D. 20.75 ft

53. Which is the best approximation of the volume of the ball of string?
   F. 300 ft³   G. 600 ft³   H. 1200 ft³   J. 2400 ft³

54. If Mr. Payne wound the same amount of string each year, which is the best estimate of the radius after one year?
   A. 2.2 ft   B. 2.3 ft   C. 4.4 ft   D. 4.6 ft

55. What is the surface area of a sphere whose radius is 7.5 m?
   F. 75π m²   G. 112.5π m²   H. 225π m²   J. 562.5π m²

56. What is the volume of a sphere whose radius is 6 ft?
   A. 48π ft³   B. 144π ft³   C. 288π ft³   D. 324π ft³

57. The volume of a sphere is 26,244π cm³. What is its surface area?
   F. 1070π cm²   G. 1402π cm²   H. 2448π cm²   J. 2916π cm²

58. The surface area of a sphere is 576π in.². What is its diameter?
   A. 1 ft   B. 1.7 ft   C. 2 ft   D. 3.5 ft

Short Response

59. The surface area of a sphere is 36π ft². Find the radius and volume. Show your work.

GO for Help

Lesson 11-5

Find the volume of each figure to the nearest cubic unit.

60.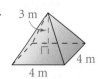

61.

62.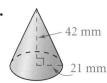

Lesson 10-4

The similarity ratio of a pair of similar isosceles trapezoids is 2 : 3. A diagonal of the smaller figure has length 7 cm.

63. Find the length of a diagonal in the larger trapezoid.

64. The perimeter of the larger trapezoid is 40.5 cm. What is the perimeter of the smaller trapezoid?

65. The area of the smaller trapezoid is 30 cm². What is the area of the larger trapezoid?

Lesson 8-4

66. A leg of a right triangle measures 4 cm and the hypotenuse measures 7 cm. Find the measure of each acute angle of the triangle to the nearest degree.

67. The length of each side of a rhombus is 16. The longer diagonal has length 26. Find the measures of the angles of the rhombus to the nearest degree.

✓ Checkpoint Quiz 2   **Lessons 11-4 through 11-6**

Find the surface area and volume of each figure to the nearest tenth.

1.

2.

3.

4.

5.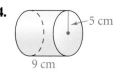

6.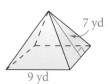

7.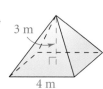

8.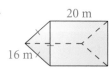

9.

10. Critical Thinking Tennis balls fit snugly inside a cylinder as shown. Which is greater, the volume of the three tennis balls or the volume of the space around the balls? Explain.

Exploring Similar Solids

To explore the surface areas and volumes of similar rectangular prisms, you can set up a spreadsheet like the one below. You choose the numbers for the length, width, height, and similarity ratio. The computer will use formulas to calculate all the other numbers.

	A	B	C	D	E	F	G	H	I
1					Surface		Similarity		
2		Length	Width	Height	Area	Volume	Ratio (II : I)	Ratio of	Ratio of
3	Rectangular Prism I	6	4	23	508	552	2	Surface	Volumes
4								Areas (II : I)	(II : I)
5	Similar Prism II	12	8	46	2032	4416		4	8

In cell E3 enter the formula $=2*(B3*C3+B3*D3+C3*D3)$.
This will calculate the sum of the areas of the six faces of Prism I.
In cell F3 enter the formula $=B3*C3*D3$. This will calculate the volume of Prism I.

In cells B5, C5, and D5 enter the formulas $=G3*B3$, $=G3*C3$, and $=G3*D3$, respectively. These will calculate the dimensions of similar Prism II. Copy the formulas from E3 and F3 into E5 and F5 to calculate the surface area and volume of Prism II.

In cell H5 enter the formula $=E5/E3$ and in cell I5 enter the formula $=F5/F3$. These will calculate the ratios of the surface areas and volumes.

Investigate

In row 3, enter numbers for the length, width, height, and similarity ratio. Change those numbers to investigate how the ratio of the surface areas and the ratio of the volumes are each related to the similarity ratio.

EXERCISES

Make a Conjecture **State a relationship that seems to be true about the similarity ratio and each given ratio.**

1. the ratio of volumes

2. the ratio of surface areas

Extend

Set up spreadsheets that allow you to investigate the following ratios. State a conclusion from each investigation.

3. the volumes of similar cylinders

4. the lateral areas of similar cylinders

5. the surface areas of similar cylinders

6. the volumes of similar square pyramids

7. the lateral areas of similar square pyramids

8. the surface areas of similar square pyramids

11-7

Areas and Volumes of Similar Solids

What You'll Learn

• To find relationships between the ratios of the areas and volumes of similar solids

. . . And Why

To use similarity ratios to find the weight of an object, as in Example 4

✓ **Check Skills You'll Need**

GO for Help   Lessons 7-2 and 11-4

Are the figures similar? Explain. Include the similarity ratio as appropriate.

1. two squares, one with 3-in. sides and the other with 1-in. sides

2. two isosceles right triangles, one with a 3-cm hypotenuse and the other with a 1-cm leg

Find the volume of each space figure.

3. a cube with a 3-in. edge

4. a 3 m-by-5 m-by-9 m rectangular prism

5. a cylinder with radius 4 cm and height 8 cm

◀)) **New Vocabulary** • similar solids • similarity ratio

---

1 Finding Relationships in Area and Volume

Similar solids have the same shape, and all their corresponding dimensions are proportional. The ratio of corresponding linear dimensions of two similar solids is the **similarity ratio.** Any two cubes are similar, as are any two spheres.

Real-World Connection

These Russian nesting dolls suggest similar solids.

1 EXAMPLE Identifying Similar Solids

Are the two rectangular prisms similar? If so, give the similarity ratio.

a.

$$\frac{3}{6} = \frac{3}{6} = \frac{2}{4}$$

The rectangular prisms are similar because the ratios of the corresponding linear dimensions are equal.

The similarity ratio is $\frac{1}{2}$.

b.

 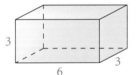

$$\frac{2}{3} \neq \frac{3}{6}$$

The rectangular prisms are not similar because the ratios of corresponding linear dimensions are not equal.

✓ **Quick Check** **1** Are the two cylinders similar? If so, give the similarity ratio.

The two similar prisms shown here suggest two important relationships for similar solids.

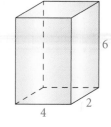

The ratio of the side lengths is 1 : 2.
The ratio of the surface areas is 22 : 88, or 1 : 4.
The ratio of the volumes is 6 : 48, or 1 : 8.

The ratio of the surface areas equals the square of the similarity ratio. The ratio of the volumes equals the cube of the similarity ratio. These two facts apply to all similar solids.

S.A. = 22 m^2 S.A. = 88 m^2
V = 6 m^3 V = 48 m^3

 Key Concepts

Theorem 11-12   **Areas and Volumes of Similar Solids**
If the similarity ratio of two similar solids is $a : b$, then (1) the ratio of their corresponding areas is $a^2 : b^2$, and (2) the ratio of their volumes is $a^3 : b^3$.

2 EXAMPLE   Finding the Similarity Ratio

Find the similarity ratio of two cubes with volumes of 729 cm^3 and 1331 cm^3.

$\dfrac{a^3}{b^3} = \dfrac{729}{1331}$ **The ratio of the volumes is $a^3 : b^3$.**

$\dfrac{a}{b} = \dfrac{9}{11}$ **Take cube roots.**

● The similarity ratio is 9 : 11.

✓ Quick Check ❷ Find the similarity ratio of two similar prisms with surface areas 144 m^2 and 324 m^2.

3 EXAMPLE   Using a Similarity Ratio

Paint Cans The lateral areas of two similar paint cans are 1019 cm^2 and 425 cm^2. The volume of the small can is 1157 cm^3. Find the volume of the large can.

First find the similarity ratio $a : b$.

$\dfrac{a^2}{b^2} = \dfrac{1019}{425}$ **The ratio of the surface areas is $a^2 : b^2$.**

$\dfrac{a}{b} = \dfrac{\sqrt{1019}}{\sqrt{425}}$ **Take square roots.**

Use the similarity ratio to find the volume.

$\dfrac{V_{large}}{V_{small}} = \dfrac{\sqrt{1019}^3}{\sqrt{425}^3}$ **The ratio of the volumes is $a^3 : b^3$.**

$\dfrac{V_{large}}{1157} = \dfrac{\sqrt{1019}^3}{\sqrt{425}^3}$ **Substitute 1157 for V_{small}.**

$V_{large} = 1157 \cdot \dfrac{\sqrt{1019}^3}{\sqrt{425}^3}$ **Solve for V_{large}.**

$V_{large} \approx 4295$ **Use a calculator.**

● The volume of the large paint can is about 4295 cm^3.

Vocabulary Tip

The underlined subscripts "large" and "small" let you use the letter V for the volumes of both the large can (V_{large}) and small can (V_{small}).

✓ Quick Check ❸ The volumes of two similar solids are 128 m^3 and 250 m^3. The surface area of the larger solid is 250 m^2. What is the surface area of the smaller solid?

The weights of solid objects made of the same material are proportional to their volumes.

4 EXAMPLE Real-World  Connection

Paperweights A marble paperweight shaped like a pyramid weighs 0.15 lb. How much does a similarly shaped marble paperweight weigh if each dimension is three times as large?

The similarity ratio is 1 : 3. The ratio of the volumes, and hence the ratio of the weights, is $1^3 : 3^3$, or 1 : 27.

$\frac{1}{27} = \frac{0.15}{x}$ **Let x = the weight of the larger paperweight.**

$x = 27(0.15)$ **Use the Cross Product Property.**

$x = 4.05$

● The larger paperweight weighs about 4 lb.

✓ **Quick Check** ❹ Find the weight of a marble bead that is similar to the paperweight in Example 4 but has dimensions half as large.

EXERCISES

For more exercises, see *Extra Skill, Word Problem, and Proof Practice.*

Practice and Problem Solving

A Practice by Example

Example 1
(page 646)

 for Help

Are the two figures similar? If so, give the similarity ratio.

1.
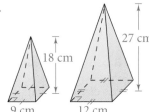

2.

3.
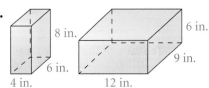

4.
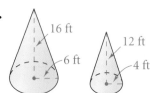

5. two cubes, one with 3-cm edges, the other with 4.5-cm edges

6. a cylinder and a square prism each with 3-in. radii and 1-in. heights

Example 2
(page 647)

Each pair of figures is similar. Use the given information to find the similarity ratio of the smaller figure to the larger figure.

7.
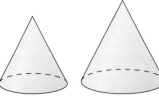

$V = 250\pi$ ft^3 $V = 432\pi$ ft^3

8.

$V = 216$ in.^{3} $V = 343$ in.^{3}

9.

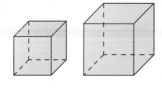

S.A. = 18 m²   S.A. = 32 m²

10.

S.A. = 20π yd²   S.A. = 125π yd²

Example 3
(page 647)

The surface areas of two similar figures are given. The volume of the larger figure is given. Find the volume of the smaller figure.

11. S.A. = 248 in.²
S.A. = 558 in.²
V = 810 in.³

12. S.A. = 192 m²
S.A. = 1728 m²
V = 4860 m³

13. S.A. = 52 ft²
S.A. = 208 ft²
V = 192 ft³

The volumes of two similar figures are given. The surface area of the smaller figure is given. Find the surface area of the larger figure.

14. V = 27 in.³
V = 125 in.³
S.A. = 63 in.²

15. V = 27 m³
V = 64 m³
S.A. = 63 m²

16. V = 2 yd³
V = 250 yd³
S.A. = 13 yd²

Example 4
(page 648)

17. Packaging There are 750 toothpicks in a regular-sized box. If a jumbo box is made by doubling all the dimensions of the regular-sized box, how many toothpicks will the jumbo box hold?

18. Packaging A cylinder 4 in. in diameter and 6 in. high holds 1 lb of oatmeal. To the nearest ounce, how much oatmeal will a similar 10-in.-high cylinder hold? (*Hint:* 1 lb = 16 oz)

19. A regular pentagonal solid prism has 9-cm base edges. A larger, similar solid prism of the same material has 36-cm base edges. How does each indicated measurement for the larger prism compare to the same measurement for the smaller prism?
a. the volume   **b.** the weight

20. Two similar prisms have heights 4 cm and 10 cm.
a. What is their similarity ratio?
b. What is the ratio of their surface areas?
c. What is the ratio of their volumes?

21. Atomic Clock A company announced that it had developed the technology to reduce the size of its atomic clock, which is used in electronic devices that transmit data. The company claims that the smaller clock will be similar to the existing clock made of the same material. It will be $\frac{1}{10}$ the size of its existing atomic clocks and $\frac{1}{100}$ the weight. Do these ratios make sense? Explain.

22. Is there a value of x for which the rectangular prisms below are similar? Explain.

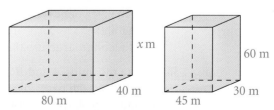

80 m   40 m x m   45 m   30 m   60 m

23. The volume of a spherical balloon with radius 3.1 cm is about 125 cm³. Estimate the volume of a similar balloon with radius 6.2 cm.

Real-World Connection

The time scale of an atomic clock is based on vibrations of atoms and molecules.

24. Critical Thinking A carpenter is making a blanket chest based on an antique chest. Both chests have the shape of a rectangular prism. The length, width, and height of the new chest will all be 4 in. greater than the respective dimensions of the antique. Will the chests be similar? Explain.

25. Writing Explain why all spheres are similar.

26. Two similar pyramids have lateral area 20 ft^2 and 45 ft^2. The volume of the smaller pyramid is 8 ft^3. Find the volume of the larger pyramid.

27. The volumes of two spheres are 729 in.^{3} and 27 in.^{3}.
   a. Find the ratio of their radii.
   b. Find the ratio of their surface areas.

28. The volumes of two similar pyramids are 1331 cm^3 and 2744 cm^3.
   a. Find the ratio of their heights.
   b. Find the ratio of their surface areas.

29. A clown's face on a balloon is 4 in. high when the balloon holds 108 in.^{3} of air. How much air must the balloon hold for the face to be 8 in. high?

Copy and complete the table for the similar solids.

	Similarity Ratio	Ratio of Surface Areas	Ratio of Volumes
30.	1 : 2	▓ : ▓	▓ : ▓
31.	3 : 5	▓ : ▓	▓ : ▓
32.	▓ : ▓	49 : 81	▓ : ▓
33.	▓ : ▓	▓ : ▓	125 : 512

34. Literature In *Gulliver's Travels* by Jonathan Swift, Gulliver first traveled to Lilliput. The Lilliputian average height was one twelfth of Gulliver's height.
   a. How many Lilliputian coats could be made from the material in Gulliver's coat? (*Hint:* Use the ratio of surface areas.)
   b. How many Lilliputian meals would be needed to make a meal for Gulliver? (*Hint:* Use the ratio of volumes.)

C Challenge

35. Indirect Reasoning Some stories say that Paul Bunyan was ten times as tall as the average human. Assume that Paul Bunyan's bone structure was proportional to that of ordinary people.
   a. Strength of bones is proportional to the area of their cross section. How many times as strong as the average person's bones would Paul Bunyan's bones be?
   b. Weights of objects made of like material are proportional to their volumes. How many times the average person's weight would Paul Bunyan's weight be?
   c. Human leg bones can support about 6 times the average person's weight. Use your answers to parts (a) and (b) to explain why Paul Bunyan could not exist with a bone structure that was proportional to that of ordinary people.

36. Square pyramids *A* and *B* are similar. In pyramid *A*, each base edge is 12 cm. In pyramid *B*, each base edge is 3 cm and the volume is 6 cm^3.
   a. Find the volume of pyramid *A*.
   b. Find the ratio of the surface area of *A* to the surface area of *B*.
   c. Find the surface area of each pyramid.

37. The cone is cut by a plane parallel to its base. The small cone on top is similar to the large cone. The ratio of the slant heights of the cones is 1 : 2. Find the ratio indicated.

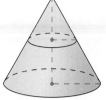

  a. the surface area of the large cone to that of the small cone; the volume of the large cone to that of the small cone

  b. the surface area of the frustum to that of the large cone; to that of the small cone

  c. the volume of the frustum to that of the large cone; to that of the small cone

GO for Help

A frustum of a cone is defined on page 636.

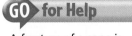

Test Prep

Gridded Response

38. The slant heights of two similar pyramids are in the ratio 1 : 5. The volume of the smaller pyramid is 60 m³. What is the volume in cubic meters of the larger pyramid?

39. A lawn chair weighs 8 lb. A child's lawn chair has dimensions exactly one half those of the larger chair. How many pounds does the child's chair weigh?

40. A model of a historical home has dimensions that are one fifteenth the dimensions of the actual home. The area of a window in the model is 2 cm². What is the area in square centimeters of the corresponding window in the actual home?

41. The volumes of two similar rectangular prisms are 64 cm³ and 1000 cm³. The surface area of the smaller figure is 112 cm². What is the surface area in square centimeters of the larger figure?

42. The surface areas of two similar cylinders are 6π ft² and 54π ft². The volume of the smaller cylinder is 2π ft³. What is the volume in cubic feet of the larger cylinder?

Mixed Review

GO for Help

Lesson 11-6

43. Sports Equipment The circumference of a regulation basketball is between 75 cm and 78 cm. What are the smallest and the largest surface areas that a basketball can have? Give your answers to the nearest whole unit.

Find the volume and surface area of each sphere to the nearest tenth.

44. diameter = 6 in.   **45.** circumference = 2.5π m   **46.** radius = 6 in.

Lesson 7-4

47. The altitude to the hypotenuse of a right triangle ABC divides the hypotenuse into 12-mm and 16-mm segments. Find the length of each of the following.
  a. the altitude to the hypotenuse
  b. the shorter leg of $\triangle ABC$
  c. the longer leg of $\triangle ABC$

Lesson 7-1 x^2 **Algebra** Solve each proportion.

48. $\dfrac{25}{16} = \dfrac{x}{16}$ **49.** $\dfrac{21}{x} = \dfrac{8}{5}$ **50.** $\dfrac{3}{8} = \dfrac{n}{n+4}$

Choosing "Cannot Be Determined"

Some multiple-choice questions do not contain enough information. In such cases, you will not be able to find an answer, and one of the answer choices should be "cannot be determined." If, however, you see "cannot be determined" before reading the question, don't assume that you cannot answer the question. You still have to try to answer it on its own merits.

EXAMPLE

The rectangular prism is inscribed in the cylinder. The diameter and height of the cylinder are both s. What is the ratio of the volume of the prism to the volume of the cylinder?

   (A) $\frac{2}{\pi}$ (B) $\frac{1}{\pi}$ (C) $\frac{\pi}{2}$

   (D) $\frac{1}{2}$ (E) cannot be determined

To find the ratio of the volumes, you need to find the volume of each solid. The volume of a cylinder is $V = \pi r^2 h$. You know that the height of this cylinder is s and its radius is one half the diameter, or $\frac{s}{2}$. The volume of this cylinder is $V = \pi r^2 h = \pi\left(\frac{s}{2}\right)^2 \cdot s = \frac{\pi s^3}{4}$.

The volume of the prism is $V = Bh$ where B is the area of the base and h is the height. The height of the prism is s, but what is the area of the base? The only thing you know about the base is that it is a rectangle whose diagonal is s. Since the length of the diagonal of a rectangle does not fix the dimensions of the rectangle, there is not enough information about the base to find its area.

There is not enough information to determine the volume of the prism. The correct answer choice is E, "cannot be determined."

EXERCISES

Can the answer to each exercise be determined? If *yes*, state the answer. If *no*, explain.

1. The sum of the circumference and height of a cylinder is 20 cm. What is the surface area of the cylinder?

2. **a.** The surface area of a cylinder is 44 cm^2. What is the volume of the cylinder?
   b. The surface area of a sphere is 44 cm^2. What is the volume of the sphere?

3. A polyhedron has 10 faces and the number of edges is 8 more than the number of vertices. What is the number of edges of the polyhedron?

4. An equilateral triangle has an area of 100 in.^{2}. What is the length of a side?

Chapter Review

Vocabulary Review

🔊))
altitude (pp. 608, 610, 617, 619)
base(s) (pp. 608, 610, 617, 619)
center of a sphere (p. 638)
circumference of a sphere (p. 638)
composite space figure (p. 627)
cone (p. 619)
cross section (p. 600)
cylinder (p. 610)
diameter of a sphere (p. 638)
edge (p. 598)
face (p. 598)

great circle (p. 638)
height (pp. 608, 610, 617, 619)
hemisphere (p. 638)
lateral area (pp. 608, 610, 617, 619)
lateral faces (pp. 608, 617)
oblique cylinder (p. 610)
oblique prism (p. 608)
polyhedron (p. 598)
prism (p. 608)
pyramid (p. 617)
radius of a sphere (p. 638)

regular pyramid (p. 617)
right cone (p. 619)
right cylinder (p. 610)
right prism (p. 608)
similar solids (p. 646)
similarity ratio (p. 646)
slant height (p. 617, 619)
sphere (p. 638)
surface area (pp. 608, 610, 618, 619)
vertex (pp. 598, 617, 619)
volume (p. 624)

Choose the correct term to complete each sentence.

1. A set of points in space equidistant from a given point is called a (*circle, sphere*).

2. The (*height, altitude*) of a cylinder is a segment that joins the bases and is perpendicular to each base.

3. A (*pyramid, prism*) is a polyhedron in which one face can be any polygon and the lateral faces are triangles that meet at a common vertex.

4. If you slice a prism with a plane, the intersection of the prism and the plane is a (*lateral area, cross section*) of the prism.

5. In a(n) (*right, oblique*) prism, the lateral faces are rectangles and a lateral edge is an altitude.

Go **O**nline
PHSchool.com
For: Vocabulary quiz
Web Code: auj-1151

Skills and Concepts

11-1 Objectives

▼ To recognize polyhedra and their parts

▼ To visualize cross sections of space figures

A **polyhedron** is a three-dimensional figure whose surfaces are polygons. The polygons are **faces** of the polyhedron. An **edge** is a segment that is the intersection of two faces. A **vertex** is a point where three or more edges intersect. A **net** is a two-dimensional pattern that folds to form a three-dimensional figure.

Draw a net for each three-dimensional figure.

6.    7.   8.

The number of faces (F), vertices (V), and edges (E) of a polyhedron are related by Euler's Formula $F + V = E + 2$.

Use Euler's Formula to find the missing number.

9. $F = 5, V = 5, E = $ ▦   10. $F = 6, V = $ ▦ $, E = 12$

A **cross section** is the intersection of a solid and a plane.

11. Sketch a cube with an equilateral triangle cross section.

11-2 and 11-4 Objectives

▼ To find the surface area of a prism

▼ To find the surface area of a cylinder

▼ To find the volume of a prism

▼ To find the volume of a cylinder

The **lateral area** of a **right prism** is the product of the perimeter of the base and the height. The **lateral area** of a **right cylinder** is the product of the circumference of the base and the height of the cylinder. The **surface area** of each solid is the sum of the lateral area and the areas of the bases.

The **volume** of a space figure is the space that the figure occupies. Volume is measured in cubic units. The **volume** of a **prism** and the **volume** of a **cylinder** are the product of the area of a base and the height of the solid.

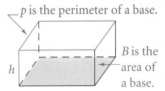
p is the perimeter of a base.
B is the area of a base.

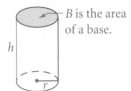
B is the area of a base.

L.A. (prism) = ph
S.A. = L.A. + 2B
V = Bh

L.A. (cylinder) = $2\pi rh$ or πdh
S.A. = L.A. + 2B
V = Bh

Find the surface area and volume of each figure. Leave your answers in terms of π.

12.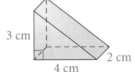
3 cm
4 cm
2 cm

13.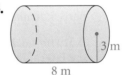
3 m
8 m

14.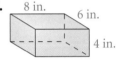
8 in.
6 in.
4 in.

11-3 and 11-5 Objectives

▼ To find the surface area of a pyramid

▼ To find the surface area of a cone

▼ To find the volume of a pyramid

▼ To find the volume of a cone

The **lateral area** of a **regular pyramid** is half the product of the perimeter of the base and the slant height.

The **surface area** of a pyramid is the sum of the lateral area and the area of the base.

The **volume** of a pyramid is one third the product of the area of the base and the height of the solid.

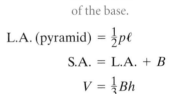
B is the area of the base.

L.A. (pyramid) = $\frac{1}{2}p\ell$
S.A. = L.A. + B
V = $\frac{1}{3}Bh$

The **lateral area** of a **right cone** is half the product of the circumference of the base and the slant height.

The **surface area** of a cone is the sum of the lateral area and the area of the base.

The **volume** of a cone is one third the product of the area of the base and the height of the solid.

B is the area of the base.

L.A. (cone) = $\pi r\ell$
S.A. = L.A. + B
V = $\frac{1}{3}Bh$

Find the surface area and volume of each figure. Round to the nearest tenth.

15.
10 ft
10.8 ft
4 ft

16.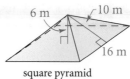
6 m
10 m
16 m
square pyramid

17.
4 in.
6 in.

18.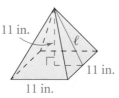
11 in.
ℓ
11 in.
11 in.

19.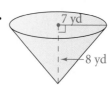
7 yd
8 yd

20.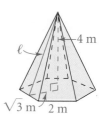
4 m
ℓ
√3 m   2 m

Find each formula as stated.

21. the base area of a prism in terms of the surface area and lateral area

22. the radius of a cylinder in terms of the surface area and height (*Hint:* Use the quadratic formula.)

11-6 Objectives

▼ To find the surface area and volume of a sphere

The **surface area of a sphere** is four times the product of π and the square of the radius of the sphere. The **volume of a sphere** is $\frac{4}{3}$ the product of π and the cube of the radius of the sphere.

Find the surface area and volume of a sphere with the given radius or diameter. Round answers to the nearest tenth.

23. $r = 5$ in.   **24.** $d = 7$ cm   **25.** $d = 4$ ft   **26.** $r = 0.8$ ft

 27. Sports Equipment The circumference of a lacrosse ball is 8 in. Find its volume to the nearest tenth of a cubic inch.

11-7 Objectives

▼ To find relationships between the ratios of the areas and volumes of similar solids

Similar solids have the same shape and all their corresponding dimensions are proportional.

If the **similarity ratio** of two similar solids is $a : b$, then the ratio of their corresponding surface areas is $a^2 : b^2$, and the ratio of their volumes is $a^3 : b^3$.

28. Open-Ended Sketch two similar solids whose surface areas are in the ratio 16 : 25. Include dimensions.

For each pair of similar solids, find the ratio of the volume of the first figure to the volume of the second.

29.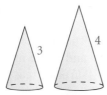
3
4

30.
12
9

Chapter
11

Chapter Test

Go Online
PHSchool.com
For: Chapter Test
Web Code: aua-1152

Draw a net for each figure. Label the net with appropriate dimensions.

1.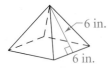
6 in.
6 in.

2.
4 cm
10 cm

Use the polyhedron at the right for Exercises 3 and 4.

3. Verify Euler's Formula for the polyhedron.

4. Draw a net for the polyhedron. Verify $F + V = E + 1$ for the net.

5. Find the number of edges in a pyramid with seven faces.

Describe the cross section formed in each diagram.

6.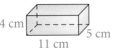

7.

🌐 **Aviation** The "black box" data recorders on commercial airliners are rectangular prisms.

8. The base of a recorder is 15 in. by 8 in. Its height ranges from 15 in. to 22 in. What are the largest and smallest possible volumes for the recorder?

9. New flight data recorders are smaller and record more data. A new recorder might be 8 in. by 8 in. by 13 in. What is its volume?

Find the volume and surface area of each figure to the nearest tenth.

10.
4 cm
5 cm
11 cm

11.
4 ft

12.
6 m   5 m

13.
8 cm
3 cm

14.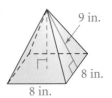
9 in.
8 in.
8 in.

15.
1 in.   12 in.
6 in.

16. **Open-Ended** Draw two different space figures that have a volume of 100 in.^{3}. Label the dimensions of each figure.

17. **Visualization** The triangle is revolved completely about the y-axis.
    a. Describe the solid of revolution that is formed.
    b. Find its lateral area and volume in terms of π.

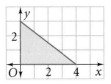

18. **Painting** The floor of a bedroom is 12 ft by 15 ft and the walls are 7 ft high. One gallon of paint covers about 450 ft^2. How many gallons of paint do you need to paint the walls of the bedroom?

19. List these space figures in order from the one with least volume to the one with greatest volume.
    A. cube with an edge of 5 cm
    B. cylinder with radius 4 cm and height 4 cm
    C. square pyramid with base sides of 6 cm and height 6 cm
    D. cone with radius 4 cm and height 9 cm
    E. rectangular prism with a 5 cm-by-5 cm base and height 6 cm

20. **Writing** Describe a real-world situation in which you would need to know the volume of an object. Then describe another situation in which you would need to know the lateral area of an object.

21. The two solids are similar. Find the ratio of the volume of the first figure to the volume of the second.

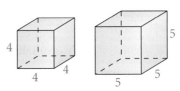
4   5
4   4   5   5

22. The volumes of two spheres are 327π mm^3 and 8829π mm^3. What is the ratio of their surface areas?

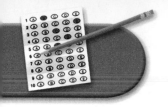
Regents Test Prep

Reading Comprehension Read the passage below. Then answer the questions on the basis of what is *stated* or *implied* in the passage.

Packaging In the cosmetics section of a department store you can see many unusual and eye-catching geometric shapes. There are star-shaped perfume sprayers, gourd-shaped bath-oil bottles, and hexagonal jars of skin cream. In general, the easiest way to package and ship such items is to use rectangular boxes.

Suppose a bar of soap has a "footprint" that is a parallelogram with sides of lengths 3 and 4. There are two ways to "box" the parallelogram (shown in red and blue at the right) so that the box aligns with one side of the parallelogram and otherwise wastes no space.

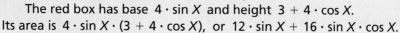

The red box has base $4 \cdot \sin X$ and height $3 + 4 \cdot \cos X$.
Its area is $4 \cdot \sin X \cdot (3 + 4 \cdot \cos X)$, or $12 \cdot \sin X + 16 \cdot \sin X \cdot \cos X$.
The blue box has base $4 + 3 \cdot \cos X$ and height $3 \cdot \sin X$.
Its area is $3 \cdot \sin X \cdot (4 + 3 \cdot \cos X)$, or $12 \cdot \sin X + 9 \cdot \sin X \cdot \cos X$.
Thus the area of the red box is $7 \cdot \sin X \cdot \cos X$ greater than the area of the blue box. This value is greatest when $X = 45$.
It can also be shown that the perimeter of the red box exceeds the perimeter of the blue box by $2 \cdot \sin X + 2 \cdot \cos X - 2$, which is also greatest when $X = 45$.
Clearly the blue box is a better design for packaging the parallelogram.

1. Why is it important to design rectangular boxes for packages with unusual shapes?

   (1) for easy arrangement on store shelves
   (2) to disguise the contents
   (3) for easy shipment
   (4) so the packages don't rattle around inside

2. In the passage, why is the blue box a better design for packaging the parallelogram?

   (1) In the blue box the parallelogram has base 4.
   (2) The blue box is horizontal.
   (3) The parallelogram fits perfectly inside the blue box.
   (4) The blue box requires less package materials.

3. With how many different rectangles can you box an equilateral triangle so that the box contains one side of the triangle and otherwise wastes no space?

   (1) 1   **(2)** 2   **(3)** 3   **(4)** infinitely many

With how many different rectangles can you box the given shape so that the box contains at least one side of the shape and otherwise wastes no space? Justify each answer.

4. a scalene triangle   5. a rhombus

6. an isosceles trapezoid   7. a regular hexagon

8. Which fact below allows you to express the dimensions of both rectangles in terms of X?
   (1) The red and blue triangles are congruent.
   (2) Sine equals cosine of the complement.
   (3) Vertical angles are congruent.
   (4) All right angles are congruent.

9. Explain how you can check that $7 \cdot \sin X \cdot \cos X$ is greatest when $X = 45$.

10. For the boxes above, the red box has the greater perimeter. Show that it exceeds the perimeter of the blue box by $2 \cdot \sin X + 2 \cdot \cos X - 2$.

Activity Lab

A Colossal Task

Applying Volume The Statue of Liberty stands in New York harbor, welcoming people to the United States. A similar giant statue once stood at the entrance to the harbor of the Greek island of Rhodes. A stone base and an iron framework supported the bronze statue, which took 12 years to construct. The statue remained standing for only about 66 years, falling to the ground after a violent earthquake weakened its knees.

The Colossus of Rhodes

The Colossus is believed to have included 12.9 tons of bronze and 7.7 tons of iron. Some historians believe that the sculptor modeled the head of the Colossus on that of Alexander the Great.

Activity 1

Materials: ruler, paper and pencil

a. Measure the height of one person attaching bronze plates to the shin of the Colossus. Estimate the probable height of the finished statue.

b. Using the painting of the Colossus (above left) as a guide, sketch the statue. Use proportions to determine at least eight dimensions on the statue. Add these dimensions to your sketch.

The Sculptor

Chares of Lindos (in red robes) probably made small models and scaled them up to get correct proportions.

Artist's rendition

Activity 2

a. The dimensions of the larger block are twice the dimensions of the smaller block. Calculate the ratio of the larger block's volume to the smaller block's volume.

3 cm
4 cm
2 cm

b. Determine what the volume ratio would be if the larger block's dimensions were 10 times the smaller block's dimensions.

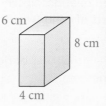

6 cm
8 cm
4 cm

The Statue of Liberty

French sculptor Frederic-Auguste Bartholdi used his mother as the model for the Statue of Liberty. The statue was built in France and shipped to New York in 350 pieces. The seven spikes on the crown symbolize the seven seas and seven continents.

Light Lady Liberty

From heel to top of head, the Statue of Liberty is about 1.3 times as long as a mature blue whale but only about 1.6 times as heavy. The statue weighs less than you'd think because, rather than being solid, it is a thin layer of copper over an iron framework.

The Statue of Liberty weighs 225 tons.

A fully grown blue whale weighs about 143 tons.

Activity 3

Materials: newspaper or Internet

Suppose the residents in your area have decided to honor you, a friend, or a local hero by building a 100-ft solid-gold statue in the center of town. Draw a sketch of the statue. Estimate the weight and cost.

Hints:

- Use what you learned in Activity 2 to estimate the ratio of the statue's volume to its subject's volume.

- Think about how the weight ratio relates to the volume ratio.

- You'll need to find the density of gold. You'll also need to compare the density of gold to that of a person. Keep in mind that our bodies are mostly water.

- Research the current price of gold.

Go Online
PHSchool.com

For: Information about statues
Web Code: aue-1153

659

What You've Learned

NY Learning Standards for Mathematics

- **G.PS.6:** In Chapters 1 and 10, you learned how to find the circumference and area of a circle.

- **G.G.29:** In Chapters 4 and 5, you learned relationships involving congruent triangles, corresponding parts of congruent triangles, and special segments within a triangle.

- **G.G.48:** In Chapter 8, you learned special properties of right triangles, including the Pythagorean Theorem.

 Check Your Readiness

 for Help to the Lesson in green.

Solving Equations (Skills Handbook Page 758)

x^2 **Algebra** Solve for x.

1. $\frac{1}{2}(x + 42) = 62$ **2.** $(5 + 3)8 = (4 + x)6$ **3.** $(9 + x)2 = (12 + 4)3$

Distance Formula (Lesson 1-8)

Find the distance between each pair of points.

4. $(13, 7), (6, 31)$ **5.** $(-4, 4), (2, -4)$ **6.** $(-3, -1), (0, 3)$ **7.** $\left(2\sqrt{3}, 5\right), \left(-\sqrt{3}, 2\right)$

Isosceles and Equilateral Triangles (Lesson 4-5)

x^2 **Algebra** Find the value of x.

8.   **9.**   **10.**   **11.**

The Pythagorean Theorem (Lesson 8-1)

x^2 **Algebra** Find the value of x. Leave your answer in simplest radical form.

12.   **13.** 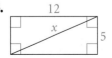   **14.** 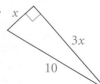   **15.**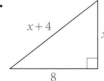

Circles

🔊 **Key Vocabulary**

- chord (p. 670)
- circumscribed about (p. 664)
- inscribed angle (p. 678)
- inscribed in (p. 664)
- intercepted arc (p. 678)
- locus (p. 701)
- point of tangency (p. 662)
- secant (p. 687)
- standard form of an equation of a circle (p. 695)
- tangent to a circle (p. 662)

What You'll Learn Next

(NY) Learning Standards for Mathematics

- **G.G.49:** In this chapter, you will learn the many properties of circles and of lines and segments that intersect circles.

- **G.G.51:** When these lines and segments meet to form angles, you will learn how the angles are related to the arcs they intercept on a circle.

- **G.G.22:** You will also learn how to describe a set of points as a locus.

Data Analysis **Activity Lab** Applying what you learn, you will design packaging on pages 714 and 715.

661

Tangent Lines

 Learning Standards for Mathematics

G.G.50 Investigate, justify, and apply theorems about tangent lines to a circle:
- a perpendicular to the tangent at the point of tangency.
- two tangents to a circle from the same external point.
- common tangents of two non-intersecting or tangent circles.

 Check Skills You'll Need **GO for Help** Skills Handbook page 754 and Lesson 8-1

Find each product.

1. $(p + 3)^2$

2. $(w + 10)^2$

3. $(m - 2)^2$

x^2 **Algebra** **Find the value of x. Leave your answer in simplest radical form.**

4.

5.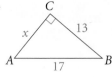

6.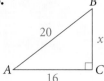

New Vocabulary • tangent to a circle • point of tangency • inscribed in
• circumscribed about

1 Using the Radius-Tangent Relationship

In Chapter 8, you studied the tangent ratio in right triangles. The tangents you will study here relate to circles.

Vocabulary Tip

The word <u>tangent</u> may refer to a line, ray, or segment.

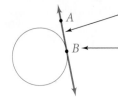

A **tangent to a circle** is a line in the plane of the circle that intersects the circle in exactly one point.

The point where a circle and a tangent intersect is the **point of tangency.**

$\overrightarrow{BA}$ is a tangent ray and $\overline{BA}$ is a tangent segment.

Theorem 12-1 relates a tangent and a radius in a given circle. You will write an indirect proof for Theorem 12-1 in Exercise 29.

 Key Concepts

Theorem 12-1

If a line is tangent to a circle, then the line is perpendicular to the radius drawn to the point of tangency.

$$\overleftrightarrow{AB} \perp \overline{OP}$$

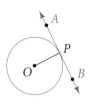

You can use Theorem 12-1 to solve problems involving tangents to circles.

1 EXAMPLE Finding Angle Measures

Multiple Choice $\overline{ML}$ and $\overline{MN}$ are tangent to $\odot O$. Find the value of x.

(A) 58   (B) 63   (C) 90   (D) 117

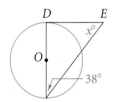

Since $\overline{ML}$ and $\overline{MN}$ are tangent to $\odot O$, $\angle L$ and $\angle N$ are right angles. $LMNO$ is a quadrilateral whose angle measures have a sum of 360.

$$m\angle L + m\angle M + m\angle N + m\angle O = 360$$

$$90 + x + 90 + 117 = 360 \quad \textbf{Substitute.}$$

$$297 + x = 360 \quad \textbf{Simplify.}$$

$$x = 63 \quad \textbf{Solve.}$$

The correct answer is B.

 **Quick Check** **1** $\overline{ED}$ is tangent to $\odot O$. Find the value of x.

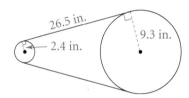

2 EXAMPLE Real-World 🌐 Connection

Dirt Bikes A dirt bike chain fits tightly around two gears. The chain and gears form a figure like the one at the right. Find the distance between the centers of the gears.

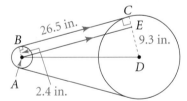

Label the diagram.

Draw $\overline{AE}$ parallel to $\overline{BC}$.

Real-World 🌐 Connection

This motorcycle and many other two-wheeled vehicles have chain-drive systems like the one shown in Example 2.

$ABCE$ is a rectangle. $\triangle AED$ is a right triangle with $AE = 26.5$ in. and $ED = 9.3 - 2.4 = 6.9$ in.

$$AD^2 = AE^2 + ED^2 \quad \textbf{Pythagorean Theorem}$$

$$AD^2 = 26.5^2 + 6.9^2 \quad \textbf{Substitute.}$$

$$AD^2 = 749.86 \quad \textbf{Simplify.}$$

$$AD \approx 27.383572 \quad \textbf{Use a calculator to find the square root.}$$

The distance between the centers is about 27.4 in.

 **Quick Check** **2** A belt fits tightly around two circular pulleys, as shown at the right. Find the distance between the centers of the pulleys.

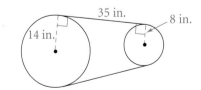

Theorem 12-2 (next page) is the converse of Theorem 12-1. You can use it to prove that a line or segment is tangent to a circle. You can also use it to construct a tangent to a circle (see Exercise 24). You will prove this theorem in Exercise 33.

Key Concepts

Theorem 12-2

If a line in the plane of a circle is perpendicular to a radius at its endpoint on the circle, then the line is tangent to the circle.

$\overleftrightarrow{AB}$ is tangent to $\odot O$.

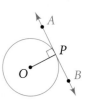

---

3 EXAMPLE · Finding a Tangent

Is $\overline{ML}$ tangent to $\odot N$ at L? Explain.

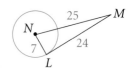

$NL^2 + LM^2 \stackrel{?}{=} NM^2$ **Is $\triangle MLN$ a right triangle?**

$7^2 + 24^2 \stackrel{?}{=} 25^2$ **Substitute.**

$625 = 625$ ✓ **Simplify.**

By the Converse of the Pythagorean Theorem, $\triangle MLN$ is a right triangle with right $\angle L$. Therefore $\overline{ML} \perp \overline{NL}$, and $\overline{ML}$ is tangent to $\odot N$ at L by Theorem 12-2.

✓ Quick Check ③ If $NL = 4$, $LM = 7$, and $NM = 8$, is $\overline{ML}$ tangent to $\odot N$ at L? Explain.

---

2 Using Multiple Tangents

If a circle is circumscribed about a triangle (Chapter 5), the triangle is inscribed in the circle. Similarly, when a circle is **inscribed in** a triangle, as in the diagram, the triangle is **circumscribed about** the circle. Each side of the triangle is tangent to the circle. The tangent segments from each vertex are congruent. You will prove this theorem in Exercise 30.

Key Concepts

Theorem 12-3

The two segments tangent to a circle from a point outside the circle are congruent.

$\overline{AB} \cong \overline{CB}$

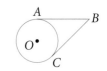

---

4 EXAMPLE · Using Theorem 12-3

The diagram represents a chain drive system on a bicycle. Give a convincing argument that $BC = GF$.

Extend $\overline{BC}$ and $\overline{GF}$ to intersect in point H. By Theorem 12-3, $HC = HF$, or $HB + BC = HG + GF$. By Theorem 12-3 again, $HB = HG$, so by the Subtraction Property of Equality, $BC = GF$.

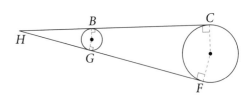

✓ Quick Check ④ **Critical Thinking** Give a convincing argument that $BC = GF$ above if you know that $\overleftrightarrow{BC}$ and $\overleftrightarrow{GF}$ never intersect.

5 **EXAMPLE** **Circles Inscribed in Polygons**

$\odot O$ is inscribed in $\triangle ABC$. Find the perimeter of $\triangle ABC$.

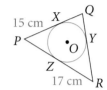

$AD = AF = 10$ cm   **The two segments tangent to a**
$BD = BE = 15$ cm   **circle from a point outside the**
$CF = CE = 8$ cm   **circle are congruent.**

$p = AB + BC + CA$ **Definition of perimeter p**

   $= AD + DB + BE + EC + CF + FA$ **Segment Addition Postulate**

   $= 10\ + 15\ + 15\ + 8\ \ + 8\ \ + 10$ **Substitute.**

   $= 66$

● The perimeter is 66 cm.

✓ **Quick Check** **5** $\odot O$ is inscribed in $\triangle PQR$. $\triangle PQR$ has a perimeter of 88 cm. Find QY.

EXERCISES

For more exercises, see *Extra Skill, Word Problem, and Proof Practice.*

Practice and Problem Solving

A **Practice by Example**

Example 1
(page 663)

GO for Help

$\boxed{x^2}$ **Algebra** **Assume that lines that appear to be tangent are tangent. O is the center of each circle. Find the value of x.**

1.

2.
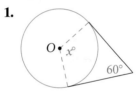

3.

Examples 2, 4
(pages 663, 664)

A belt fits snugly around the two circular pulleys shown.

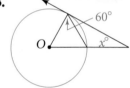

4. Find the distance between the centers of the pulleys. Round to the nearest hundredth.

5. Give a convincing argument why the belt lengths RS and QP are equal.

Vocabulary Tip

$\overline{RS}$ and $\overline{QP}$ are common tangents.

For the pulley system shown, use the lengths given below. Find the missing length to the nearest tenth.

Exercises 4–7

6. $MQ = 10$ cm, $NP = 4$ cm, $QP = 14$ cm, $MN = $ ▇ cm

7. $MQ = 5$ in., $NP = 4$ in., $QP = 20$ in., $MN = $ ▇ in.

Example 3
(page 664)

Determine whether a tangent line is shown in each diagram. Explain.

8.
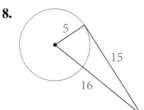

9.
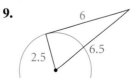

10.
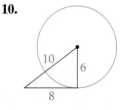

Example 5
(page 665)

Each polygon circumscribes a circle. Find the perimeter of the polygon.

11.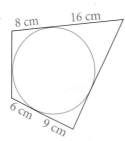

8 cm   16 cm

6 cm   9 cm

12.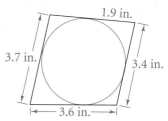

1.9 in.

3.7 in.   3.4 in.

3.6 in.

B **Apply Your Skills** x^2 **Algebra** **Assume that lines that appear to be tangent are tangent. O is the center of each circle. Find the value of x to the nearest tenth.**

13.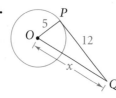

P
5
O
12
x
Q

14.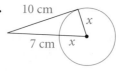

10 cm
x
7 cm   x

15.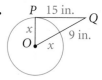

P   15 in.   Q
x
O   x   9 in.

16. Solar Eclipse Common tangents to two circles may be *internal* or *external*. If you draw a segment joining the centers of the circles, a common internal tangent will intersect the segment. A common external tangent will not.

For this cross-sectional diagram of the sun, moon, and Earth during a solar eclipse, use the terms above to describe the types of tangents of each color.

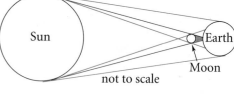

Sun

Earth
Moon
not to scale

   a. red   **b.** blue   **c.** green
   d. Which tangents show the extent on Earth's surface of total eclipse? Of partial eclipse?
   e. Reasoning In general, does every pair of circles have common tangents of both types? Explain.

Real-World Connection

This "diamond ring" effect in a solar eclipse may be seen by a person on Earth at the end of a common external tangent of the sun and moon. (See diagram at right.)

Earth The circle at the right represents Earth. The radius of Earth is about 6400 km. Find the distance d that a person can see on a clear day from each of the following heights h above Earth. Round your answer to the nearest tenth of a kilometer.

h
d
r
r

17. 100 m   **18.** 500 m   **19.** 1 km

20. $\overline{BD}$ and $\overline{CK}$ at the right are diameters of $\odot A$. $\overline{BP}$ and $\overline{QP}$ are tangents to $\odot A$. What is $m\angle CDA$?

B   25°
C   A   K   P
D   Q

21. History Leonardo da Vinci wrote, "When each of two squares touch the same circle at four points, one is double the other."

   a. Sketch a figure that illustrates this statement.
   b. Writing Explain why the statement is true.

22. Multiple Choice A regular hexagon is circumscribed about the ring surrounding the clock face. The diameter of the ring is 10 in. What is the perimeter of the hexagon?

   Ⓐ 30.0 in.   Ⓑ 34.6 in.
   Ⓒ 43.3 in.   Ⓓ 51.7 in.

GO Online
Homework Video Tutor

Visit: PHSchool.com
Web Code: aue-1201

23. Critical Thinking A nickel, a dime, and a quarter are touching as shown. Tangents are drawn from point *A* to both sides of each coin. What can you conclude about the four tangent segments? Explain.

24. Constructions Draw a circle. Label the center *T*. Locate a point on the circle and label it *R*. Construct a tangent to ⊙*T* at *R*.

$\overleftrightarrow{AC}$ **is tangent to ⊙*O* at *A*, and *m*∠1 = 70.**

25. Find *m*∠4.

26. Let *m*∠1 = *x*. Find *m*∠4 in terms of *x*. What is the relationship between ∠1 and ∠4?

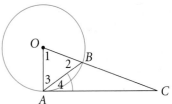

27. Coordinate Geometry Graph the equation $x^2 + y^2 = 9$. Then draw a segment from $(0, 5)$ tangent to the circle. Find the length of the segment.

28. Maintenance Mr. Gonzales is replacing a cylindrical air-conditioning duct. He estimates the radius of the duct by folding a ruler to form two 6-in. tangents to the duct. The tangents form an angle. Mr. Gonzales measures the angle bisector from the vertex to the duct. It is about $2\frac{3}{4}$ in. long. What is the radius of the duct?

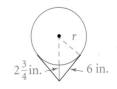

29. Complete the following indirect proof of Theorem 12-1.

Given: Line *n* is tangent to ⊙*O* at *P*.

Prove: line *n* ⊥ $\overline{OP}$

Step 1 Assume that line *n* is not perpendicular to $\overline{OP}$.

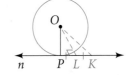

Step 2 If line *n* is not perpendicular to $\overline{OP}$, some other segment $\overline{OL}$ must be **a.** ? to line *n*. By the Ruler Postulate, there is a point *K* on $\overrightarrow{PL}$ such that *PK* = 2*PL*, so *PL* = **b.** ? . △*OPL* ≅ △*OKL* by **c.** ? , so *OP* = *OK* because **d.** ? . Since *P* and *K* are the same distance from *O*, both *K* and *P* are on ⊙*O*. This contradicts the given fact that line *n* is **e.** ? to ⊙*O* at *P*.

Step 3 The assumption that line *n* is not perpendicular to $\overline{OP}$ is **f.** ? , so line *n* ⊥ $\overline{OP}$.

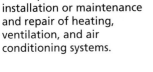

Proof **30.** Prove Theorem 12-3.

Given: $\overline{BA}$ and $\overline{BC}$ are tangent to ⊙*O* at *A* and *C*, respectively.

Prove: $\overline{BA} \cong \overline{BC}$

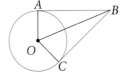

31. Given: $\overline{BC}$ is tangent to ⊙*A* at *D*.
$\overline{DB} \cong \overline{DC}$

Prove: $\overline{AB} \cong \overline{AC}$

32. Given: ⊙*A* and ⊙*B* with common tangents $\overline{DF}$ and $\overline{CE}$

Prove: △*GDC* ~ △*GFE*

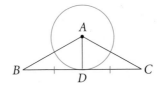

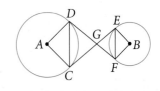

C Challenge *Proof* **33.** Write an indirect proof of Theorem 12-2.

> **Given:** $\overleftrightarrow{AB} \perp \overline{OP}$ at P.
>
> **Prove:** $\overleftrightarrow{AB}$ is tangent to $\odot O$.

34. Two circles that have one point in common are *tangent circles*. Given any triangle, explain how to draw three circles that are centered at each vertex of the triangle and are tangent to each other.

NY REGENTS

Multiple Choice Point O is the center of each circle. Assume the lines that appear tangent are tangent. What is the value of the variable?

35.

- **A.** 26
- **B.** 57
- **C.** 66
- **D.** 114

36.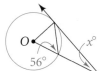

- **F.** 22
- **G.** 28
- **H.** 34
- **J.** 40

37.

- **A.** 8
- **B.** 9
- **C.** 15
- **D.** 17

38.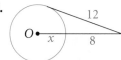

- **F.** 2
- **G.** 3
- **H.** 4
- **J.** 5

Short Response **39.** Find the value of x. Show your work.

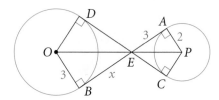

Mixed Review

Lesson 11-7 **Two cubes have heights 6 in. and 8 in. Find each ratio.**

40. similarity ratio **41.** ratio of surface areas **42.** ratio of volumes

Lesson 8-3 $\boxed{x^2}$ **Algebra** **Find the value of x. Round answers to the nearest tenth.**

43.

44. **45.**

Lesson 7-2 **The polygons are similar. (a) State the similarity ratio and (b) find the values of the variables.**

46.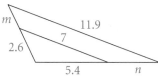

47.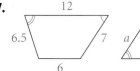

Paper Folding With Circles

A *chord* is a segment whose endpoints are on a circle. In these activities, you will explore some properties of chords.

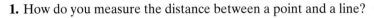

NY G.G.49:
Investigate, justify, and apply theorems regarding chords of a circle:

• perpendicular bisectors of chords.

• the relative lengths of chords as compared to their distance from the center of the circle.

1 ACTIVITY

- Use a compass. Draw a circle on tracing paper.
- Use a straightedge. Draw two radii.
- Set your compass to a distance shorter than the radii. Place the compass point at the center of the circle. Mark two congruent segments, one on each radius.
- At the point you marked on each radius, fold a line perpendicular to the radius.

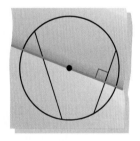

1. How do you measure the distance between a point and a line?

2. Each perpendicular contains a chord of the circle. Measure the length of each chord and compare. What do you notice?

3. Make a conjecture about the lengths of chords that are equidistant from the center of a circle.

2 ACTIVITY

- Use a compass. Draw a circle on tracing paper.
- Use a straightedge. Draw two chords of different lengths. Do not draw either one as a diameter.
- Fold the perpendicular bisector for each chord.

4. Where do the perpendicular bisectors appear to intersect?

5. Draw a third chord and fold its perpendicular bisector. Where does it appear to intersect the other two?

6. Make a conjecture about the perpendicular bisector of a chord of a circle.

EXERCISES

7. Write a proof of your conjecture from Exercise 3 or give a counterexample to prove it false.

8. What theorem provides a one-sentence proof of your conjecture from Exercise 6?

9. Suppose two chords have different lengths. Make a conjecture that compares their distances from the center of the circle.

10. You are building a circular table for a patio. For the umbrella, you have to drill a hole through the center of the tabletop. Explain how you can use a carpenter's square like the one shown to find the center.

12-2

Chords and Arcs

NY Learning Standards for Mathematics

G.G.49 Investigate, justify, and apply theorems regarding chords of a circle:
- perpendicular bisectors of chords.
- the relative lengths of chords as compared to their distance from the center of the circle.

✓ Check Skills You'll Need

GO for Help Lesson 8-2

Find the value of each variable. Leave your answer in simplest radical form.

1.

2.

3.

◀)) New Vocabulary • chord

---

1 Using Congruent Chords, Arcs, and Central Angles

A segment whose endpoints are on a circle is called a **chord.** The diagram shows the related chord and arc, $\overline{PQ}$ and $\overset{\frown}{PQ}$.

The following theorem is about related central angles, chords, and arcs. It says, for example, that if two central angles in a circle are congruent, then so are the two chords and two arcs that the angles intercept.

 Key Concepts

Online active math

For: Chords and Arcs Activity
Use: Interactive Textbook, 12-2

Theorem 12-4
Within a circle or in congruent circles
(1) Congruent central angles have congruent chords.
(2) Congruent chords have congruent arcs.
(3) Congruent arcs have congruent central angles.

You will prove Theorem 12-4 in Exercises 23, 24, and 35.

1 EXAMPLE Using Theorem 12-4

In the diagram, $\odot O \cong \odot P$. Given that $\overset{\frown}{BC} \cong \overset{\frown}{DF}$, what can you conclude?

By Theorem 12-4, $\angle O \cong \angle P$ and $\overline{BC} \cong \overline{DF}$.

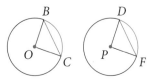

✓ Quick Check ❶ If you are instead given that $\overline{BC} \cong \overline{DF}$, what can you conclude?

Theorem 12-5 shows a relationship between two chords and their distances from the center of a circle. You will prove part (2) in Exercise 38.

 Key Concepts

> **Theorem 12-5**
>
> Within a circle or in congruent circles
>
> (1) Chords equidistant from the center are congruent.
>
> (2) Congruent chords are equidistant from the center.

Proof →

Proof of Theorem 12-5, Part (1)

Given: $\odot O, \overline{OE} \cong \overline{OF},$
$\overline{OE} \perp \overline{AB}, \overline{OF} \perp \overline{CD}$

Prove: $\overline{AB} \cong \overline{CD}$

Real-World Connection

Steel beams model congruent chords equidistant from the center to give the illusion of a circle.

Statements	Reasons
1. $\overline{OA} \cong \overline{OB} \cong \overline{OC} \cong \overline{OD}$	1. Radii of a circle are congruent.
2. $\overline{OE} \cong \overline{OF}, \overline{OE} \perp \overline{AB}, \overline{OF} \perp \overline{CD}$	2. Given
3. $\angle AEO$ and $\angle CFO$ are right angles.	3. Def. of perpendicular segments
4. $\triangle AEO \cong \triangle CFO$	4. HL Theorem
5. $\angle A \cong \angle C$	5. CPCTC
6. $\angle B \cong \angle A, \angle C \cong \angle D$	6. Isosceles Triangle Theorem
7. $\angle B \cong \angle D$	7. Transitive Property of Congruence
8. $\angle AOB \cong \angle COD$	8. If two ∠ of a △ are ≅ to two ∠ of another △, then the third ∠ are ≅.
9. $\overline{AB} \cong \overline{CD}$	9. ≅ central angles have ≅ chords.

You can use Theorem 12-5 to find missing lengths in circles.

② EXAMPLE **Using Theorem 12-5**

Multiple Choice What is the value of *a* in the circle at the right?

Ⓐ 9   Ⓑ 12.5   Ⓒ 18   Ⓓ 25

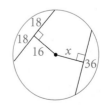

$PQ = QR = 12.5$ **Given**

$PQ + QR = PR$ **Segment Addition Postulate**

$25 = PR$ **Substitute.**

$a = PR$ **Chords equidistant from the center of a circle are congruent.**

$a = 25$ **Substitute.**

● The correct answer is D.

Test-Taking Tip

In a circle, the length of the perpendicular segment from the center to a chord is the distance from the center to the chord.

✓ Quick Check ② Find the value of *x* in the circle at the right.

The Converse of the Perpendicular Bisector Theorem from Lesson 5-2 has special applications to a circle and its diameters, chords, and arcs.

 Key Concepts

> **Theorem 12-6**
>
> In a circle, a diameter that is perpendicular to a chord bisects the chord and its arcs.
>
> **Theorem 12-7**
>
> In a circle, a diameter that bisects a chord (that is not a diameter) is perpendicular to the chord.
>
> **Theorem 12-8**
>
> In a circle, the perpendicular bisector of a chord contains the center of the circle.

Proof **Proof of Theorem 12-7**

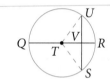

Given: $\odot T$ with diameter $\overline{QR}$ bisecting $\overline{SU}$ at V.

Prove: $\overline{QR} \perp \overline{SU}$

Proof: $TS = TU$ because the radii of a circle are congruent. $VS = VU$ by the definition of bisect. Thus, T and V are equidistant from S and U. By the Converse of the Perpendicular Bisector Theorem, T and V are on the perpendicular bisector of $\overline{SU}$. Since two points determine one line, $\overline{TV}$ is the perpendicular bisector of $\overline{SU}$. Another name for $\overline{TV}$ is $\overline{QR}$. Thus, $\overline{QR} \perp \overline{SU}$.

You will prove Theorems 12-6 and 12-8 in Exercises 25 and 36, respectively.

3 EXAMPLE **Using Diameters and Chords**

Algebra Find each missing length to the nearest tenth.

a.

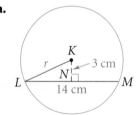

$LN = \frac{1}{2}(14) = 7$ A diameter $\perp$ to a chord bisects the chord.

$r^2 = 3^2 + 7^2$ Use the Pythagorean Theorem.

$r \approx 7.6$ Find the square root of each side.

b.

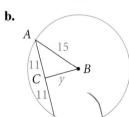

$\overline{BC} \perp \overline{AC}$ A diameter that bisects a chord that is not a diameter is $\perp$ to the chord.

$y^2 + 11^2 = 15^2$ Use the Pythagorean Theorem.

$y^2 = 104$ Solve for y^2.

$y \approx 10.2$ Find the square root of each side.

Real-World Connection

The center of the tire is located on the perpendicular bisector of the flat part.

3 Use the circle at the right.
  a. Find the length of the chord.
  b. Find the distance from the midpoint of the chord to the midpoint of its minor arc.

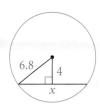

EXERCISES

For more exercises, see *Extra Skill, Word Problem, and Proof Practice.*

Practice and Problem Solving

A Practice by Example

Example 1
(page 670)

In Exercises 1 and 2, the circles are congruent. What can you conclude?

1.

2.

Example 2
(page 671)

Find the value of *x*.

3.

4.

5.

6.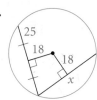

7.

8.

Example 3
(page 672)

Use the diagram at the right to complete Exercises 9 and 10.

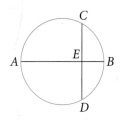

9. Given that $\overline{AB}$ is a diameter of the circle and $\overline{AB} \perp \overline{CD}$, then **a.** __?__ ≅ **b.** __?__ and **c.** __?__ ≅ **d.** __?__.

10. Given that $\overline{AB}$ is the perpendicular bisector of $\overline{CD}$, then $\overline{AB}$ contains __?__.

Algebra Find the value of *x* to the nearest tenth.

11.

12.

13.

14.

15.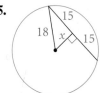

16.

Find $m\widehat{AB}$. (*Hint:* You will need to use trigonometry in Exercise 19.)

17.
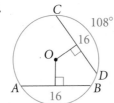

18.

19.
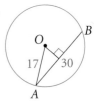

20. Archaeology An archaeologist found several jar fragments including a large piece of the circular rim. How can she find the center and radius of the rim to help her reconstruct the jar?

21. Geometry in 3 Dimensions In the figure at the right, sphere O with radius 13 cm is intersected by a plane 5 cm from center O. Find the radius of cross section $\odot A$.

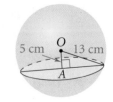

22. Geometry in 3 Dimensions A plane intersects a sphere that has radius 10 in. forming cross section $\odot B$ with radius 8 in. How far is the plane from the center of the sphere?

23. Complete this proof of Theorem 12-4, Part (1).

Given: $\odot P$ with $\angle KPM \cong \angle LPN$

Prove: $\overline{KM} \cong \overline{LN}$

Proof: $\overline{KP} \cong$ **a.** ? $\cong$ **b.** ? $\cong \overline{NP}$ because **c.** ? . $\triangle KPM \cong$ **d.** ? by **e.** ? . $\overline{KM} \cong \overline{LN}$ by **f.** ? .

24. Complete this proof of Theorem 12-4, Part (2).

Given: $\odot E$ with congruent chords $\overline{AB}$ and $\overline{CD}$

Prove: $\widehat{AB} \cong \widehat{CD}$

Problem Solving Hint

Recall that in a circle congruent central angles intercept congruent arcs.

$$\boxed{\overline{EA} \cong \overline{EC} \cong \overline{ED} \cong \overline{EB}}$$

a. ?

b. ?

c. ?

$\boxed{\triangle AEB \cong \triangle CED} \longrightarrow \boxed{\text{e.} \ ?}$ → $\boxed{\widehat{AB} \cong \widehat{CD}}$

d. ? CPCTC **f.** ?

Proof **25.** Prove Theorem 12-6.

Given: $\odot O$ with diameter $\overline{ED} \perp \overline{AB}$ at C

Prove: $\overline{AC} \cong \overline{BC}$ and $\widehat{AD} \cong \widehat{BD}$

(*Hint:* Begin by drawing $\overline{OA}$ and $\overline{OB}$.)

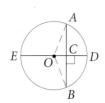

26. Two concentric circles have radii of 4 cm and 8 cm. A segment tangent to the smaller circle is a chord of the larger circle. What is the length of the segment?

For a guide to solving Exercise 26, see p. 677.

27. Error Analysis Scott looks at this figure and concludes that $\overline{ST} \cong \overline{PR}$. What is wrong with Scott's conclusion?

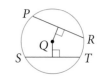

28. Open-Ended Use a circular object such as a can or a saucer to draw a circle. Construct the center of the circle.

29. Writing Theorems 12-4 and 12-5 both begin with the phrase "Within a circle or in congruent circles." Explain why "congruent" is essential for both theorems.

Real-World Connection

Careers Field archaeologists analyze artifacts to provide glimpses of life in the past.

GO for Help

30. $AB = 8$ in., $CD = 6$ in. How long is a radius?

31. $AB = 24$ cm, radius $= 13$ cm. How long is $\overline{CD}$?

32. radius $= 13$ ft, $CD = 24$ ft. How long is $\overline{AB}$?

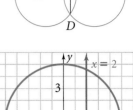

33. Multiple Choice In the diagram at the right, the endpoints of the chord are the points where the line $x = 2$ intersects the circle $x^2 + y^2 = 25$. What is the length of the chord?

  Ⓐ 4.6
  Ⓑ 8.0
  Ⓒ 9.2
  Ⓓ 10.0

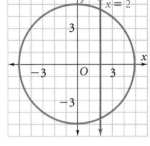

34. Critical Thinking The diameter of a circle is 20 cm. Two chords parallel to the diameter are 6 cm and 16 cm long. What are the possible distances between the chords to the nearest tenth of a centimeter?

Proof 35. Prove Theorem 12-4, Part (3).
  Given: ⊙*P* with $\overset{\frown}{QS} \cong \overset{\frown}{RT}$
  Prove: $\angle QPS \cong \angle RPT$

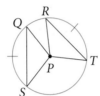

36. Prove Theorem 12-8.
  Given: ℓ is the ⊥ bisector of $\overline{WY}$.
  Prove: ℓ contains the center of ⊙*X*.

Proof 37. $\overline{PQ}$ and $\overline{PR}$ are chords of ⊙*C*. Prove that if *C* is on the bisector of $\angle QPR$, then $PQ = PR$.

 Challenge **Proof 38.** Prove Theorem 12-5, Part (2).
  Given: ⊙*O* with $\overline{AB} \cong \overline{CD}$
  Prove: $\overline{OE} \cong \overline{OF}$

39. Given: ⊙*A* with $\overline{CE} \perp \overline{BD}$
  Prove: $\overset{\frown}{BC} \cong \overset{\frown}{DC}$

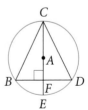

40. Dairy The diameter of the base of a cylindrical milk tank is 59 in. The length of the tank is 470 in. You estimate that the depth of the milk in the tank is 20 in. Find the number of gallons of milk in the tank to the nearest gallon. (1 gal $= 231$ in.^{3})

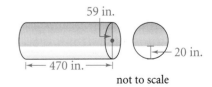

not to scale

Real-World 🌐 Connection

The cylinders used on milk tank trucks lie on the lateral surface and have a vertical base at each end.

Proof 41. If two circles are concentric and a chord of the larger circle is tangent to the smaller circle, prove that the point of tangency is the midpoint of the chord.

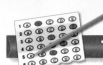

Multiple Choice

42. The diameter of a circle is 25 cm and a chord of the same circle is 16 cm. To the nearest tenth, what is the distance of the chord from the center of the circle?
   A. 9.0 cm   **B.** 9.6 cm   **C.** 18.0 cm   **D.** 19.2 cm

43. In the figure at the right, what is the value of x to the nearest tenth?
   F. 3.0   **G.** 6.2
   H. 6.8   **J.** 9.0

44. A 9-cm chord is 11 cm from the center of a circle. What is the radius of the circle?
   A. 9.0 cm   **B.** 11.9 cm   **C.** 13.0 cm   **D.** 14.2 cm

45. The radius of a circle is 10.8 ft. The length of a chord is 12 ft. What is the approximate distance of the chord from the center of the circle?
   F. 1.2 ft   **G.** 4.7 ft   **H.** 9.0 ft   **J.** 12.4 ft

46. What can you NOT conclude from the diagram at the right?
   A. $c = d$ **B.** $a = b$
   C. $c^2 + e^2 = b^2$ **D.** $e = d$

47. $\odot A$ and $\odot B$ intersect at points C and D. Each circle has radius 6 in. and $AB = 8$ in. What is CD?
   F. 4.5 in.   **G.** 6 in.   **H.** 8 in.   **J.** 8.9 in.

Short Response

48. Circles M and N are congruent with radii measuring 13 cm. $\overline{PQ}$ is a chord of both circles and $PQ = 18$ cm. To the nearest tenth, find MN. Justify your answer.

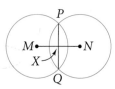

Mixed Review

Lesson 12-1

Assume that lines that appear to be tangent are tangent. O is the center of each circle. Find the value of x to the nearest tenth.

49.

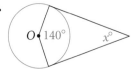

50.

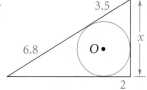

Lesson 8-5

51. From the top of a building you look down at an object on the ground. If your eyes are 50 feet above the ground and the angle of depression is 50°, how far is the object on the ground from the base of the building?

Lesson 7-5

52. The legs of a right triangle are 10 in. and 24 in. long. Find the lengths, to the nearest tenth, of the segments into which the bisector of the right angle divides the hypotenuse.

Understanding Math Problems Read the problem below. Then let Jamal's thinking guide you through the solution. Check your understanding with the exercises at the bottom of the page.

Two concentric circles have radii of 4 cm and 8 cm. A segment tangent to the smaller circle is a chord of the larger circle. What is the length of the segment?

What Jamal Thinks	What Jamal Writes

I need a picture to understand what this is about. I'll make a sketch.

Concentric circles have the same center, like circles in a bull's eye. The larger circle has radius 8 cm. The smaller circle has radius 4 cm.

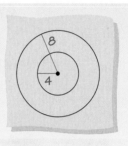

I need to draw a segment tangent to the smaller circle with both ends on the larger circle.

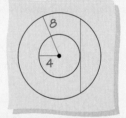

Now I need to find a relationship between these segments. I'll redo my sketch; move things around. Is a right triangle possible for my sketch so I can use the Pythagorean Theorem? Yes! Theorem 12-1 says a tangent to a circle and the radius to the point of tangency are perpendicular.

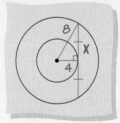

I'll use the Pythagorean Theorem . . .

. . . and a calculator to find $\sqrt{48}$.

$$4^2 + x^2 = 8^2$$
$$x^2 = 64 - 16$$
$$x = \sqrt{48} \approx 6.9$$

To find the length of the chord, I multiply 6.9 by 2 because Theorem 12-6 says that a diameter perpendicular to a chord bisects the chord.

The chord is about 13.8 cm long.

EXERCISES

1. Two parallel chords of a circle are each 24 cm long. The distance between them is 10 cm. Find the circumference of the circle.

2. A chord 1 in. from the center of a circle is 6 in. long. How long is a chord that is 3 in. from the center of the circle?

Inscribed Angles

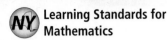
Learning Standards for Mathematics

G.G.51 Investigate, justify, and apply theorems about the arcs determined by the rays of angles formed by two lines intersecting a circle when the vertex is:
- inside the circle (two chords).
- on the circle (tangent and chord).
- outside the circle.

✓ Check Skills You'll Need

GO for Help   Lesson 10-6

Identify the following in ⊙*P* at the right.

1. a semicircle   **2.** a minor arc

3. a major arc   **4.** a central angle

Find the measure of each arc in ⊙*P*.

5. $\overarc{ST}$ **6.** $\overarc{STQ}$

7. $\overarc{RST}$ **8.** $\overarc{TQ}$

🔊 **New Vocabulary**   • inscribed angle   • intercepted arc

---

1   Finding the Measure of an Inscribed Angle

Hands-On Activity: Exploring Inscribed Angles

- Draw two large circles with a compass. Label the centers *X* and *Y*.

- On the circles, use a straightedge and copy the diagrams shown.

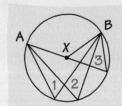 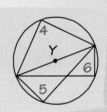

1. a. Patterns In ⊙*X*, use a protractor to measure ∠*AXB* and each numbered angle. Determine $m\overarc{AB}$. Record your results and look for patterns. Compare your results with others.

   b. Write a conjecture about the relationship between *m*∠1 and $m\overarc{AB}$.

   c. Write a conjecture about the measures of ∠1, ∠2, and ∠3.

2. a. Patterns Use a protractor to measure the numbered angles in ⊙*Y*. Record your results and look for patterns. Compare your results.

   b. Write a conjecture about an angle whose vertex is on a circle and whose sides intersect the endpoints of a diameter of the circle.

At the right, the vertex of ∠*C* is on ⊙*O*, and the sides of ∠*C* are chords of the circle. ∠*C* is an **inscribed angle.** $\overarc{AB}$ is the **intercepted arc** of ∠*C*.

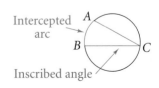

Intercepted arc

Inscribed angle

Theorem 12-9 describes the relationship between an inscribed angle and its intercepted arc.

> **Theorem 12-9** | **Inscribed Angle Theorem**
>
> The measure of an inscribed angle is half the measure of its intercepted arc.
>
> $$m\angle B = \tfrac{1}{2}m\widehat{AC}$$

Vocabulary Tip

When different conditions are possible, each possibility can be called a case. You can prove a theorem by proving it for all possible cases.

To prove Theorem 12-9, there are three cases to consider.

I: The center is on a side of the angle.   II: The center is inside the angle.   III: The center is outside the angle.

A proof of Case I is below. You will prove Cases II and III in Exercises 36 and 37.

Proof

Proof of Theorem 12-9, Case I

Given: $\odot O$ with inscribed $\angle B$ and diameter $\overline{BC}$
Prove: $m\angle B = \tfrac{1}{2}m\widehat{AC}$

Draw radius $\overline{OA}$ to form isosceles $\triangle AOB$ with $OA = OB$ and, hence, $m\angle A = m\angle B$.

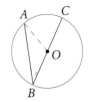

$m\widehat{AC} = m\angle AOC$	**Definition of $m\widehat{AC}$**
$= m\angle A + m\angle B$	**Triangle Exterior Angle Theorem**
$= 2m\angle B$	**Substitute and simplify.**
$\tfrac{1}{2}m\widehat{AC} = m\angle B$	**Solve for $m\angle B$.**

You can use the Inscribed Angle Theorem to find missing measures in circles.

1 EXAMPLE **Using the Inscribed Angle Theorem**

Find the values of a and b.

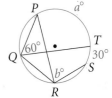

$m\angle PQT = \tfrac{1}{2}m\widehat{PT}$	**Inscribed Angle Theorem**
$60 = \tfrac{1}{2}a$	**Substitute.**
$120 = a$	**Solve for a.**

$m\angle PRS = \tfrac{1}{2}m\widehat{PS}$	**Inscribed Angle Theorem**
$b = \tfrac{1}{2}\left(m\widehat{PT} + m\widehat{TS}\right)$	**Arc Addition Postulate**
$= \tfrac{1}{2}(120 + 30)$	**Substitute.**
$b = 75$	**Simplify.**

 **Quick Check** 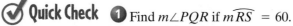 **1** Find $m\angle PQR$ if $m\widehat{RS} = 60$.

You will use three corollaries to the Inscribed Angle Theorem to find measures of angles in circles. You will justify these corollaries in Exercises 38, 39, and 40.

Key Concepts

Corollaries **Corollaries to the Inscribed Angle Theorem**

1. Two inscribed angles that intercept the same arc are congruent.

2. An angle inscribed in a semicircle is a right angle.

3. The opposite angles of a quadrilateral inscribed in a circle are supplementary.

2 EXAMPLE **Using Corollaries to Find Angle Measures**

Find the measure of the numbered angle.

a.

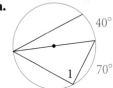

b.

For: Inscribed Angle Activity
Use: Interactive Textbook, 12-3

∠1 is inscribed in a semicircle.
By Corollary 2, ∠1 is a right angle.
$m\angle 1 = 90$

∠2 and the 38° angle intercept the same arc. By Corollary 1, the angles are congruent, so $m\angle 2 = 38$.

Quick Check **2** For the diagram at the right, find the measure of each numbered angle.

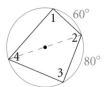

---

2 **The Angle Formed by a Tangent and a Chord**

In the diagram, B and C are fixed points, and point A moves along the circle. From the Inscribed Angle Theorem, you know that as A moves, $m\angle A$ remains the same and is $\frac{1}{2}m\widehat{BC}$. As the last diagram suggests, this is also true when A and C coincide.

Key Concepts

Theorem 12-10

The measure of an angle formed by a tangent and a chord is half the measure of the intercepted arc.

$$m\angle C = \frac{1}{2}m\widehat{BDC}$$

You will prove Theorem 12-10 in Exercise 41.

3 EXAMPLE Using Theorem 12-10

In the diagram at the right, $\overrightarrow{KJ}$ is tangent to the circle at J. Find the values of x and y.

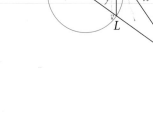

$x = \frac{1}{2}m\widehat{JL}$ **Theorem 12-10**

  $= m\angle Q$ **Inscribed Angle Theorem**

  $= 35$ **Substitution**

$y = \frac{1}{2}m\widehat{QJ}$ **Theorem 12-10**

  $= \frac{1}{2}\left(m\widehat{QL} - m\widehat{JL}\right)$ **Arc Addition Postulate**

  $= \frac{1}{2}(180 - 70)$ **Substitute.**

  $= 55$ **Simplify.**

✓ **Quick Check** ❸ Describe two ways to find $m\angle QJK$ using Theorem 12-10.

EXERCISES

For more exercises, see *Extra Skill, Word Problem, and Proof Practice.*

Practice and Problem Solving

A Practice by Example

Example 1
(page 679)

GO for Help

Identify the inscribed angle and its intercepted arc.

1.

2.
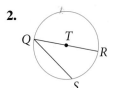

3.
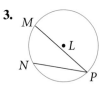

4. a. Name the four inscribed angles and their intercepted arcs.
   b. Which angles appear to intercept major arcs? What kind of angles do these appear to be?

Find the value of each variable.

5.
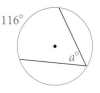

6.
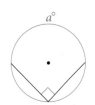

7.
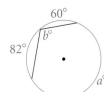

8.
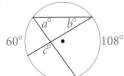

9.
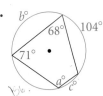

10.
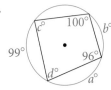

Example 2
(page 680)

11.
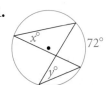

12.

13.

14.

Example 3
(page 681)

Find the value of each variable. You may assume that rays that appear to be tangent are tangent.

15.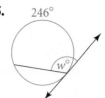

16.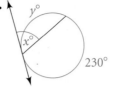

17.

18.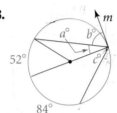

19.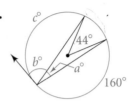

20.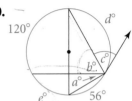

B **Apply Your Skills**

Find each indicated measure for ⊙O.

21. a. $m\widehat{BC}$
b. $m\angle B$
c. $m\angle C$
d. $m\widehat{AB}$

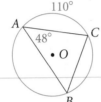

22. a. $m\angle A$
b. $m\widehat{CE}$
c. $m\angle C$
d. $m\angle D$
e. $m\angle ABE$

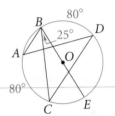

23. Multiple Choice What is the measure of $\widehat{EA}$ in the circle at the right?

Ⓐ 21　　Ⓑ 30　　Ⓒ 36　　Ⓓ 51

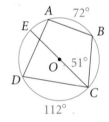

24. Find each indicated measure for the circle in Exercise 23.
a. $m\widehat{BC}$　　**b.** $m\angle A$　　**c.** $m\angle B$　　**d.** $m\angle BCD$

25. Writing Copy the diagram at the right on your paper. Draw chord $\overline{RQ}$. Explain why $m\widehat{PR} = m\widehat{QS}$.

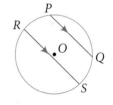

26. a. Open-Ended Sketch a trapezoid inscribed in a circle. Repeat several times using different circles.
b. Make a Conjecture What kind of trapezoid can be inscribed in a circle? Justify your response.

27. Landscape Architecture Some circular English gardens, like the one shown here, have paths in the shape of an inscribed regular star.
a. Find the measure of an inscribed angle formed by the star in the garden shown here.
b. What is the measure of an inscribed angle in a garden with a five-pointed star?

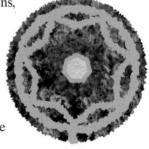

28. Critical Thinking A parallelogram inscribed in a circle must be what kind of parallelogram? Explain.

Graphing Calculator The diameter of a circle is 10 cm. Find the dimensions of the largest figure of each type that can be inscribed in the circle. (*Hint:* Use techniques demonstrated in the Activity Lab on page 616.)

29. a rectangle　　　　**30.** a triangle　　　　**31.** a right triangle

32. Television The director of a telecast wants the option of showing the same scene from three different views.
   a. Explain why cameras in the positions shown in the diagram will transmit the same scene.
   b. Critical Thinking Will the scenes look the same to the director when she views them on the control room monitors? Explain.

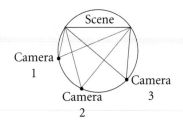

33. Constructions The diagrams below show the construction of a tangent to a circle from a point outside the circle. Explain why $\overleftrightarrow{BC}$ must be tangent to $\odot A$. (*Hint:* Copy the third diagram and draw $\overline{AC}$.)

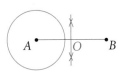

Given: $\odot A$ and point B. Construct the midpoint of $\overline{AB}$. Label the point O.

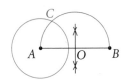

Construct a semicircle with radius OA and center O. Label its intersection with $\odot A$ as C.

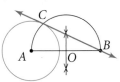

Draw $\overleftrightarrow{BC}$.

34. Technology Construct $\odot A$ and the chords shown with geometry software.
   a. As you move E on $\overarc{CED}$ between C and D, which inscribed angles remain congruent?
   b. Which inscribed angle remains a right angle?
   c. Which inscribed angles remain supplementary in quadrilateral $EFGD$?

35. Constructions Use Corollary 2 of Theorem 12-9 to construct a right triangle given one leg and the hypotenuse.

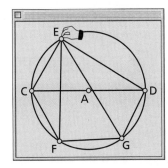

Proof → Write a proof for each of Exercises 36–41.

36. Inscribed Angle Theorem, Case II
   Given: $\odot O$ with inscribed $\angle ABC$
   Prove: $m\angle ABC = \frac{1}{2}m\overarc{AC}$

   Hint: Use the Inscribed Angle Theorem, Case I.

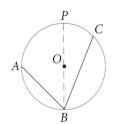

37. Inscribed Angle Theorem, Case III
   Given: $\odot S$ with inscribed $\angle PQR$
   Prove: $m\angle PQR = \frac{1}{2}m\overarc{PR}$

   Hint: Use the Inscribed Angle Theorem, Case I.

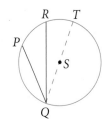

38. Inscribed Angle Theorem, Cor. 1
   Given: $\odot O$; $\angle A$ intercepts $\overarc{BC}$, and $\angle D$ intercepts $\overarc{BC}$.
   Prove: $\angle A \cong \angle D$

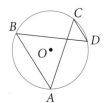

39. Inscribed Angle Theorem, Cor. 2
   Given: $\odot O$ with $\angle CAB$ inscribed in a semicircle
   Prove: $\angle CAB$ is a right angle.

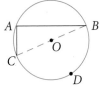

40. Inscribed Angle Theorem, Cor. 3

Given: quadrilateral *ABCD*
inscribed in ⊙*O*

Prove: ∠*A* and ∠*C* are supplementary.
∠*B* and ∠*D* are supplementary.

41. Theorem 12-10

Given: $\overline{GH}$ and tangent ℓ
intersecting at *H* on ⊙*E*

Prove: $m\angle GHI = \frac{1}{2}m\widehat{GFH}$

Problem Solving Hint

In Exercise 41, let $\overline{GH}$ first be a diameter.

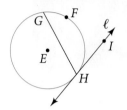

 **Challenge**

Critical Thinking Is the statement true or false? If true, give a convincing argument. If false, give a counterexample.

42. If two angles inscribed in a circle are congruent, then they intercept the same arc.

43. If an inscribed angle is a right angle, then it is inscribed in a semicircle.

44. A circle can always be circumscribed about a quadrilateral whose opposite angles are supplementary.

45. Constructions Draw two segments. Label their lengths *x* and *y*. Construct the geometric mean of *x* and *y*. (*Hint:* Recall a theorem about a geometric mean.)

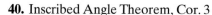

 **REGENTS**

Test Prep

Multiple Choice

In Exercises 46 and 47, what is the value of each variable?

46.

A. 25
B. 35
C. 45
D. 65

47.

F. 20
G. 30
H. 50
J. 60

48. In the figure at the right, a square is circumscribed about ⊙*A*. What is the area of the square?
A. 64 in.²   B. 192 in.²
C. 256 in.²   D. (256 + 16√3) in.²

Short Response

49. a. Explain how you can find *m*∠*XYZ*.
b. Find *m*∠*XYZ*.

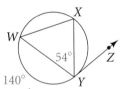

Extended Response

50. Use the figure at the right.
a. What is *m*∠*D*? Explain.
b. What is *m*∠*ACB*? Explain.
c. Use an equation to find the value of *x*.

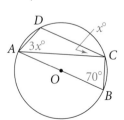

684   Chapter 12   Circles

GO for Help

Lesson 12-2 x^2 **Algebra** **Find the value of *x* to the nearest tenth.**

51.
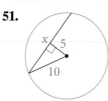

52.

53.

Lesson 10-5 **Find the area of each triangle. Give answers to the nearest tenth.**

54.
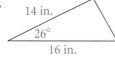

55.
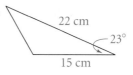

56.
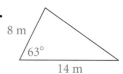

Lesson 7-3 **Indirect Measurement** To find the width of a river, you have made the measurements shown in the sketch.

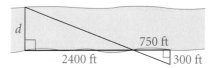

57. Explain why the triangles are similar.

58. a. Find the width of the river in feet.
   b. Find the width of the river in miles.

✓ **Checkpoint Quiz 1**   **Lessons 12-1 through 12-3**

Each polygon below circumscribes the circle. Find the perimeter of the polygon.

1.
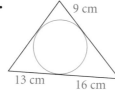

2.
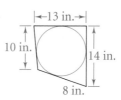

3.
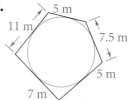

x^2 **Algebra** **Find the value of *x*.**

4.

5.

6.

x^2 **Algebra** **Find the value of each variable. Lines that appear to be tangent are tangent.**

7.
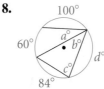

8.

9.

10.

Exploring Chords and Secants

Construct

Construct a circle with center A and two chords $\overline{BC}$ and $\overline{DE}$ that intersect each other at F.

Investigate

Measure $\overline{BF}$, $\overline{FC}$, $\overline{EF}$, and $\overline{FD}$. Use the calculator program of your software to find the products $BF \cdot FC$ and $EF \cdot FD$. Manipulate your construction and observe the products. What do you discover? Use your discovery to answer Exercise 1.

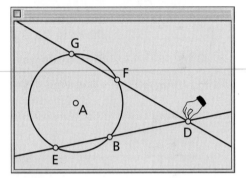

Construct

A *secant* is a line that intersects a circle in two points. Construct another circle and two secants that intersect in a point outside the circle. Label your construction as shown in the diagram.

Investigate

Measure $\overline{DG}$, $\overline{DF}$, $\overline{DE}$, and $\overline{DB}$. Calculate the products $DG \cdot DF$ and $DE \cdot DB$. Manipulate your construction and observe the products. What do you discover? Use your discovery to answer Exercise 2.

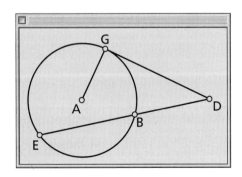

EXERCISES

Use results from your two investigations above. Make conjectures about the products of the lengths of the segments to a circle from the point of intersection of the figures given below.

1. two chords inside a circle

2. two secants outside a circle

Extend

3. Construct circle A with radius $\overline{AG}$ as shown. Construct a segment $\overline{DG}$ perpendicular to radius $\overline{AG}$. Then construct a secant $\overline{DE}$ that does not cross $\overline{AG}$. Hide $\overline{AG}$. Measure $\overline{DG}$, $\overline{DE}$, and $\overline{DB}$. Calculate DG^2 and the product $DE \cdot DB$. What is true about the products you calculated? Can you explain how this special case is related to the case of two secants?

4. Combine conjectures from Exercises 1–3 into a conjecture about any two intersecting lines, each of which also intersects a circle.

12-4

Angle Measures and Segment Lengths

 Learning Standards for Mathematics

G.G.53 Investigate, justify, and apply theorems regarding segments intersected by a circle:
- along two tangents or two secants from the same external point.
- along a tangent and a secant from the same external point.
- along two intersecting chords of a given circle.

 **Check Skills You'll Need**

 for Help Lessons 12-1 and 12-3

In the diagram at the right, $\overline{FE}$ and $\overline{FD}$ are tangents to $\odot C$. Find each arc measure, angle measure, or length.

1. $m\widehat{DE}$ 2. $m\widehat{AED}$ 3. $m\widehat{EBD}$

4. $m\angle EAD$ 5. $m\angle AEC$ 6. CE

7. DF 8. CF 9. $m\angle EFD$

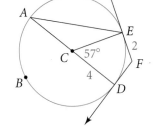

New Vocabulary • secant

---

1   Finding Angle Measures

Vocabulary Tip

The word secant may refer to a line, ray, or segment.

A **secant** is a line that intersects a circle at two points. $\overrightarrow{AB}$ is a secant ray, and $\overline{AB}$ is a secant segment.

Angles formed by secants, tangents, and chords intercept arcs on circles. The measures of the intercepted arcs can help you find the measures of the angles.

 Key Concepts

Theorem 12-11

The measure of an angle formed by two lines that

(1) intersect inside a circle is half the sum of the measures of the intercepted arcs.

$$m\angle 1 = \tfrac{1}{2}(x + y)$$

(2) intersect outside a circle is half the difference of the measures of the intercepted arcs.

$$m\angle 1 = \tfrac{1}{2}(x - y)$$

Part (1) is proved on the next page. The three cases of Part (2) are proved in Exercises 29 and 30.

Proof

Proof of Theorem 12-11, Part (1)

Given: $\odot O$ with intersecting chords $\overline{AC}$ and $\overline{BD}$

Prove: $m\angle 1 = \frac{1}{2}\left(m\widehat{AB} + m\widehat{CD}\right)$

Begin by drawing $\overline{AD}$ as shown in the diagram.

$m\angle BDA = \frac{1}{2}m\widehat{AB}$, and $m\angle CAD = \frac{1}{2}m\widehat{CD}$ $m\angle 1 = m\angle BDA + m\angle CAD$

Inscribed Angle Theorem   Exterior Angle Theorem

$m\angle 1 = \frac{1}{2}m\widehat{AB} + \frac{1}{2}m\widehat{CD}$

Substitute.

$m\angle 1 = \frac{1}{2}(m\widehat{AB} + m\widehat{CD})$

Distributive Property

You can use Theorem 12-11 to find the measures of angles and intercepted arcs.

1 EXAMPLE   Finding Angle Measures

Algebra Find the value of each variable.

a.

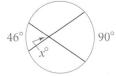

46°   90° $x°$

b. 95°

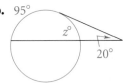

$z°$ 20°

$x = \frac{1}{2}(46 + 90)$ **Theorem 12-11 (1)** $20 = \frac{1}{2}(95 - z)$ **Theorem 12-11 (2)**

$x = 68$ **Simplify.** $40 = 95 - z$ **Solve for z.**

                                              $z = 55$

✓ Quick Check   **1** Find the value of each variable.

a.

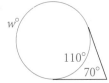

$w°$ 110°   70°

b.

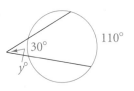

110°   30° $y°$

2 EXAMPLE   Real-World Connection

Gridded Response You focus your camera on a fountain. Your camera is at the vertex of the angle formed by tangents to the fountain. You estimate that this angle is 40°.

What is the measure, in degrees, of the arc of the circular basin of the fountain that will be in the photograph?

Let $m\widehat{AB} = x$.

Then $m\widehat{AEB} = 360 - x$.

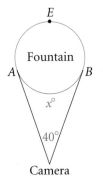

E   Fountain   A   B $x°$ 40°   Camera

Real-World Connection

Line-of-sight tangents to this fountain basin form a larger angle than do those to the distant basin.

Test-Taking Tip

For a gridded response answer, you grid only the numerical part. If there are units in the answer, omit them.

$40 = \frac{1}{2}\left(m\widehat{AEB} - m\widehat{AB}\right)$ **Theorem 12-11 (2)**

$40 = \frac{1}{2}[(360 - x) - x]$ **Substitute.**

$40 = \frac{1}{2}(360 - 2x)$ **Simplify.**

$40 = 180 - x$ **Distribute.**

$x = 140$ **Solve for x.**

● A 140° arc will be in the photograph.

✓ Quick Check ② **Critical Thinking** To photograph a 160° arc of the basin, should you move towards or away from the fountain? What angle should the tangents form?

2 Finding Segment Lengths

From a given point P, you can draw two segments to a circle along infinitely many lines. For example, $\overline{PA_1}$ and $\overline{PB_1}$ lie along one such line. Theorem 12-12 states the surprising result that, no matter which line you use, the product $PA \cdot PB$ remains constant.

$PA_i \cdot PB_i$ is constant.

 Key Concepts

Theorem 12-12

For a given point and circle, the product of the lengths of the two segments from the point to the circle is constant along any line through the point and circle.

I.

$a \cdot b = c \cdot d$

II.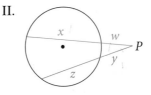

$(w + x)w = (y + z)y$

III.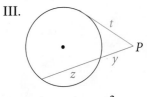

$(y + z)y = t^2$

Vocabulary Tip

The two segments to a circle along a secant are called <u>secant</u> <u>segments</u>.

Note in Case III that the tangent segment is used twice.

Here is a proof for Case I. You will prove II and III in Exercises 31 and 32.

Proof → **Proof of Theorem 12-12 (I)**

Given: a circle with chords $\overline{AB}$ and $\overline{CD}$ intersecting at P

Prove: $a \cdot b = c \cdot d$

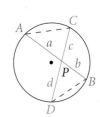

Draw $\overline{AC}$ and $\overline{BD}$. $\angle A \cong \angle D$ and $\angle C \cong \angle B$ because they are inscribed angles and each pair intercept the same arc. Thus, $\triangle APC \sim \triangle DPB$ by the Angle-Angle Similarity Postulate. The lengths of corresponding sides of similar triangles are proportional, so $\frac{a}{d} = \frac{c}{b}$. Therefore, $a \cdot b = c \cdot d$.

You can use Theorem 12-12 to find lengths of segments in circles.

3 EXAMPLE   Finding Segment Lengths

Algebra Find the value of the variable. If the answer is not a whole number, round to the nearest tenth.

a.

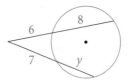

b.

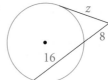

$(6 + 8)6 = (7 + y)7$ **Thm. 12-12 (II)**	$(8 + 16)8 = z^2$ **Thm. 12-12 (III)**
$84 = 49 + 7y$ **Solve for y.**	$192 = z^2$ **Solve for z.**
$35 = 7y$	$13.9 \approx z$
$5 = y$	

✓ Quick Check ❸ Find the value of the variable to the nearest tenth.

a.

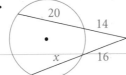

b.

4 EXAMPLE   Real-World 🌐 Connection

Real-World 🌐 Connection

The Taiko Bashi is a pedestrian bridge in the Japanese Tea Garden in San Francisco's Golden Gate Park.

Bridge Design The arch of the Taiko Bashi is an arc of a circle. A 14-ft chord is 4.8 ft from the edge of the circle. Find the radius of the circle.

Draw a diagram that shows a 14-ft chord 4.8 ft below the top of a circle. Let x represent the length of the part of the diameter from the chord to the bottom of the circle. Use x and Theorem 12-12 to find the radius.

$4.8x = 7 \cdot 7$ **Theorem 12-12 (I)**
$4.8x = 49$ **Solve for x.**
$x \approx 10.2$
diameter $\approx 10.2 + 4.8 = 15$ ft   **Add the segment lengths.**
radius ≈ 7.5 ft

● The radius is about 7.5 ft.

✓ Quick Check ❹ The basis of a design of a rotor for a Wankel engine is an equilateral triangle. Each side of the triangle is a chord to an arc of a circle. The opposite vertex of the triangle is the center of the arc. In the diagram at the right, each side of the equilateral triangle is 8 in. long.
a. Use what you know about equilateral triangles and find the value of x.
b. Critical Thinking Copy the diagram and complete the circle with the given center. Then use Theorem 12-12 to find the value of x. Show that your answers to parts (a) and (b) are equal.

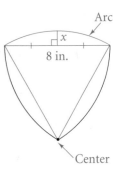

EXERCISES

For more exercises, see *Extra Skill, Word Problem, and Proof Practice.*

Practice and Problem Solving

A Practice by Example x^2 **Algebra** **Find the value of each variable.**

Example 1
(page 688)

GO for Help

1.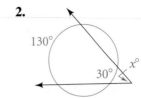

2.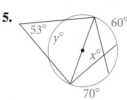

3.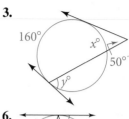

4.

5.

6.

Example 2
(pages 688–689)

7. Astroscience A departing space probe sends back a picture of Earth as it crosses the plane of Earth's equator. The angle formed by the two tangents to the equator is 20°. What arc of the equator is visible to the space probe?

8. At the left, the cross section of the ball is a circle. About how many degrees is the arc of the circle that is below the points of contact with the hands?

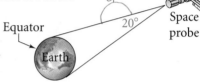

Example 3 x^2 **Algebra** **Find the value of each variable using the given chord, secant, and tangent**
(page 690) **lengths. If the answer is not a whole number, round to the nearest tenth.**

9.

10.

11.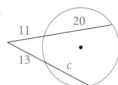

12.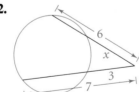

13.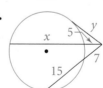

14.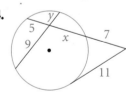

Example 4
(page 690)

Geology This natural arch, in Arches National Park, Utah, is an arc of a circle.

15. Find the diameter of the circle.

16. The chord length shown is rounded. It could range from 165 ft to 175 ft. Find the corresponding range for the diameter.

B Apply Your Skills x^2 **Algebra** $\overline{CA}$ and $\overline{CB}$ are tangents to $\odot O$. **Write an expression for each arc or angle in terms of the given variable.**

17. $m\widehat{ADB}$ using x **18.** $m\angle C$ using x **19.** $m\widehat{AB}$ using y

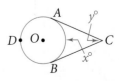

Find the diameter of ⊙O. If your answer is not a whole number, round it to the nearest tenth.

20.
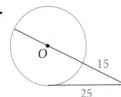

21.
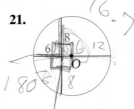

22.
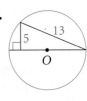

x^2 **Algebra** Find the values of x and y using the given chord, secant, and tangent lengths. If your answer is not a whole number, round it to the nearest tenth.

23.
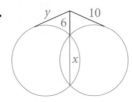

24.
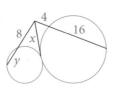

25.

26. Error Analysis To find the value of x, a student wrote the equation $(7.5)6 = x^2$. What error did the student make?

27. A circle is inscribed in a quadrilateral whose four angles have measures 85, 76, 94, and 105. Find the measures of the four arcs between consecutive points of tangency.

Exercise 26

28. Navigation The map at the left shows that the waters within $\overset{\frown}{AXB}$, a 300° arc, are unsafe. Here are what the letters represent.

A: a lighthouse B: a lighthouse X: locations of a ship on ⊙O
Y: locations of a ship inside ⊙O Z: locations of a ship outside ⊙O

  a. Critical Thinking What measures are possible for $\angle X$? For $\angle Y$? For $\angle Z$?
  b. Writing Using the angles a ship makes with the lighthouses (like angles X, Y, and Z), explain how a navigator can be sure the ship is in safe waters.

Proof **29.** Prove Theorem 12-11, Part (2), as it applies to two secants that intersect outside a circle.

  Given: ⊙O with secants $\overline{CA}$ and $\overline{CE}$ intersecting at C
  Prove: $m\angle ACE = \frac{1}{2}\left(m\overset{\frown}{AE} - m\overset{\frown}{BD}\right)$

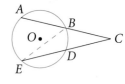

Proof **30.** Prove the other two cases of Theorem 12-11, Part (2). (See Exercise 29.)

Proof **For Exercises 31 and 32, write proofs that use similar triangles.**

  31. Prove Theorem 12-12 (II).   **32.** Prove Theorem 12-12 (III).

  33. Explain why Theorem 12-12 is true when the given point is on the circle.

C Challenge   Proof **In Exercises 34–37, prove each statement or theorem.**

  34. $m\angle 1 + m\overset{\frown}{PQ} = 180$ **35.** $m\angle 1 + m\angle 2 = m\overset{\frown}{QR}$

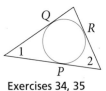

Exercises 34, 35

  36. the Pythagorean Theorem (Use the diagram at the right and the theorems of this lesson.)

  37. The tangents to a circle at the vertices of an inscribed equilateral triangle form an equilateral triangle.

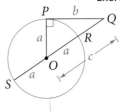

Gridded Response

38. If $m\widehat{AE} = 86$ and $m\widehat{BD} = 40$, find $m\angle BKD$.

39. If $AK = 14$, $EK = 17$, and $BK = 7$, find DK.

40. If $BC = 6$, $DC = 5$, and $CE = 12$, find AC.

41. If $m\angle C = 14$ and $m\widehat{AE} = 140$, find $m\widehat{BD}$.

42. If $m\widehat{AB} = 110$ and $m\widehat{DE} = 130$, find $m\angle AKE$.

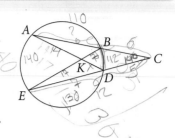

Mixed Review

Lesson 12-3

Find the value of each variable.

43.

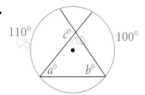

44.

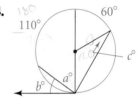

Lesson 10-4

45. The areas of two similar parallelograms are 20 cm^2 and 3.2 cm^2. Find the similarity ratio of the larger parallelogram to the smaller parallelogram.

Lesson 8-4

Find the value of x to the nearest degree.

46.

47.

48.

Geometry at Work

.... Aerospace Engineer

Aerospace engineers design and build all types of spacecraft, from the low-orbit space shuttle to interplanetary probes. Much of today's work involves communications satellites that relay television, telephone, computer, and other signals all over the world. The portion of Earth's surface that can communicate with a satellite increases as the height of the orbit increases.

Earth has a radius of about 3960 miles. The figure at the right shows a satellite 12,000 miles above Earth. $\widehat{AB}$ is the arc of Earth that is in the range of the satellite. You can find $m\widehat{AB}$ by finding $m\angle AEB$, which is twice $m\angle AES$.

$$m\widehat{AB} = m\angle AEB = 2m\angle AES = 2 \cdot \cos^{-1}\left(\frac{3960}{3960 + 12,000}\right) \approx 151.3$$

The measure of the arc of Earth in the range of the satellite is about 151.3.

Go Online

PHSchool.com

For: Information about aerospace careers
Web Code: aub-2031

Tangent Lines, Tangent Ratios

You learned about tangent ratios in Chapter 8 and tangent lines in this chapter. Are the two related? Yes, indeed! In fact, there are six trigonometric ratios that you can study in advanced mathematics.

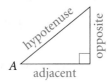

$$\text{sine } \angle A = \frac{\text{opposite}}{\text{hypotenuse}} \qquad \text{cosine } \angle A = \frac{\text{adjacent}}{\text{hypotenuse}} \qquad \text{tangent } \angle A = \frac{\text{opposite}}{\text{adjacent}}$$

$$\text{cosecant } \angle A = \frac{\text{hypotenuse}}{\text{opposite}} \qquad \text{secant } \angle A = \frac{\text{hypotenuse}}{\text{adjacent}} \qquad \text{cotangent } \angle A = \frac{\text{adjacent}}{\text{opposite}}$$

Each of these is related to the geometry of a unit circle, the circle shown in both figures below. Sine A, tangent A, and secant A are the segment lengths highlighted in Figure 1. Cosine A, cotangent A, and cosecant A are highlighted in Figure 2.

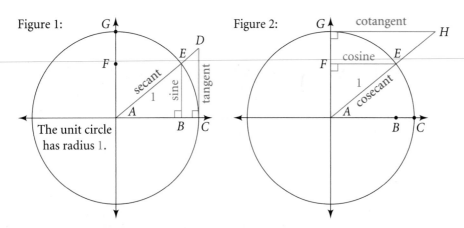

EXERCISES

1. For the four triangles shown above, complete these similarity statements:
$\triangle ABE \sim \triangle \underline{\ ?\ } \sim \triangle \underline{\ ?\ } \sim \triangle \underline{\ ?\ }$.

2. Explain why the sine A, tangent A, and secant A ratios have the same values as the segment lengths highlighted in Figure 1.

3. Explain why the cosine A, cotangent A, and cosecant A ratios have the same values as the segment lengths highlighted in Figure 2.

Describe the connection between each of the following.

4. the tangent A ratio and a tangent segment

5. the secant A ratio and a secant segment

6. $\angle EAB$ in Figure 1 and $\angle EAF$ in Figure 2;
what are the first two letters in a word that is commonly used to describe this connection?

Show that each equation is true.

7. $(\text{tangent } A)^2 = (\text{secant } A)^2 - 1$

8. $\text{tangent } A = \dfrac{\text{sine } A}{\text{cosine } A}$

9. $\text{cotangent } A = \dfrac{1}{\text{tangent } A}$

10. $(\text{sine } A)^2 + (\text{cosine } A)^2 = 1$

11. $\text{secant } A = \dfrac{1}{\text{cosine } A}$

12. $\text{cosecant } A = \dfrac{1}{\text{sine } A}$

13. For acute $\angle A, 0 < \text{sine } A < 1, 0 < \text{cosine } A < 1$, and $\text{tangent } A > 0$. What values are possible for cosecant A, secant A, and cotangent A? Explain.

Circles in the Coordinate Plane

NY Learning Standards for Mathematics

G.G.71 Write the equation of a circle, given its center and radius or given the endpoints of a diameter.

G.G.73 Find the center and radius of a circle, given the equation of the circle in center-radius form.

G.G.74 Graph circles of the form $(x - h)^2 + (j - k)^2 = r^2$.

✓ Check Skills You'll Need

GO for Help   Lesson 1-8

Find the length of each segment to the nearest tenth.

1.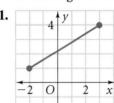

2.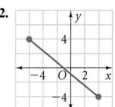

3.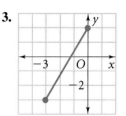

🔊 **New Vocabulary**   • standard form of an equation of a circle

---

1   Writing an Equation of a Circle

You can use the Distance Formula to find an equation of a circle with center (h, k) and radius r. Let (x, y) be any point on the circle. Then the radius r is the distance from (h, k) to (x, y).

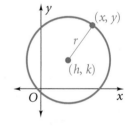

$$r = \sqrt{(x - h)^2 + (y - k)^2} \quad \textbf{Distance Formula}$$
$$r^2 = (x - h)^2 + (y - k)^2 \quad \textbf{Square both sides.}$$

This essentially proves the following theorem.

 Key Concepts

> **Theorem 12-13**
>
> An equation of a circle with center (h, k) and radius r is $(x - h)^2 + (y - k)^2 = r^2$.

The equation $(x - h)^2 + (y - k)^2 = r^2$ is in **standard form.** You may also call it the *standard equation* of a circle.

◯nline active math

For: Circle Activity
Use: Interactive Textbook, 12-5

1 EXAMPLE   Writing the Equation of a Circle

Write the standard equation of the circle with center $(5, -2)$ and radius 7.

$$(x - h)^2 + (y - k)^2 = r^2 \quad \textbf{Use standard form.}$$
$$(x - 5)^2 + [y - (-2)]^2 = 7^2 \quad \textbf{Substitute } (5, -2) \textbf{ for } (h, k), \textbf{ and 7 for } r.$$
$$(x - 5)^2 + (y + 2)^2 = 49 \quad \textbf{Simplify.}$$

 **Quick Check** **①** Write the standard equation of each circle.
   a. center $(3, 5)$; radius 6   **b.** center $(-2, -1)$; radius $\sqrt{2}$

If you know the center of a circle and a point on the circle, you can write the standard equation of the circle.

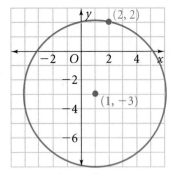

2 EXAMPLE Using the Center and a Point on a Circle

Write the standard equation of the circle with center $(1, -3)$ that passes through the point $(2, 2)$.

$r = \sqrt{(x - h)^2 + (y - k)^2}$ **Use the Distance Formula to find r.**

$= \sqrt{(2 - 1)^2 + (2 - (-3))^2}$ **Substitute (1, −3) for (h, k), and (2, 2) for (x, y).**

$= \sqrt{1 + 25} = \sqrt{26}$ **Simplify.**

$(x - h)^2 + (y - k)^2 = r^2$ **Use standard form.**

$(x - 1)^2 + [y - (-3)^2] = (\sqrt{26})^2$ **Substitute (1, −3) for (h, k), and $\sqrt{26}$ for r.**

$(x - 1)^2 + (y + 3)^2 = 26$ **Simplify.**

✓ Quick Check **2** Write the standard equation of the circle with center $(2, 3)$ that passes through the point $(-1, 1)$.

---

2 Finding the Center and Radius of a Circle

If you know the standard equation of a circle, you can describe the circle by naming its center and radius. Then you can use this information to graph the circle.

3 EXAMPLE Graphing a Circle Given its Equation

Find the center and radius of the circle with equation $(x - 7)^2 + (y + 2)^2 = 64$. Then graph the circle.

$(x - 7)^2 + (y + 2)^2 = 64$

$(x - 7)^2 + (y - (-2))^2 = 8^2$ **Use standard form.**

     ↑   ↑   ↑

     h k r

The center is $(7, -2)$ and the radius is 8.

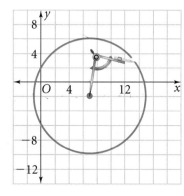

To graph the circle, place the compass point at the center $(7, -2)$ and draw a circle with radius 8.

✓ Quick Check **3** Find the center and radius of the circle with equation $(x - 2)^2 + (y - 3)^2 = 100$. Then graph the circle.

You can use equations of circles to model real-world situations.

4 EXAMPLE Real-World Connection

Communications When you make a call on a cellular phone, a tower receives the call. In the diagram, the centers of circles O and A are locations of cellular telephone towers.

a. The equation
$(x - 16)^2 + (y - 10)^2 = 100$
models the position and range of Tower A. Describe the position and range of Tower A.

$(x - 16)^2 + (y - 10)^2 = 10^2$
is in standard form. It shows that tower A is located at $(16, 10)$ and has a range of 10 units.

b. A new tower is to be built at B with range indicated in the graph. Write an equation that describes the position and range of this tower.

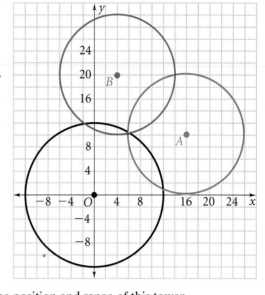

$\odot B$ has center $(4, 20)$ and radius 10. Substitute these into the standard equation.

$(x - h)^2 + (y - k)^2 = r^2$
$(x - 4)^2 + (y - 20)^2 = 10^2$ **Substitute.**
$(x - 4)^2 + (y - 20)^2 = 100$ **This is an equation for Tower B.**

Quick Check ④ Write an equation that describes the position and range of Tower O.

EXERCISES

For more exercises, see *Extra Skill, Word Problem, and Proof Practice.*

Practice and Problem Solving

A Practice by Example

Example 1
(page 695)

Write the standard equation of each circle.

1. center $(2, -8)$; $r = 9$ **2.** center $(0, 3)$; $r = 7$ **3.** center $(0.2, 1.1)$; $r = 0.4$

4. center $(5, -1)$; $r = 12$ **5.** center $(-6, 3)$; $r = 8$ **6.** center $(-9, -4)$; $r = \sqrt{5}$

7. center $(0, 0)$; $r = 4$ **8.** center $(-4, 0)$; $r = 3$ **9.** center $(-1, -1)$; $r = 1$

Example 2
(page 696)

Write the standard equation of the circle with the given center that passes through the given point.

10. center $(-2, 6)$; point $(-2, 10)$ **11.** center $(1, 2)$; point $(0, 6)$

12. center $(7, -2)$; point $(1, -6)$ **13.** center $(-10, -5)$; point $(-5, 5)$

14. center $(6, 5)$; point $(0, 0)$ **15.** center $(-1, -4)$; point $(-4, 0)$

Example 3
(page 696)

Find the center and radius of the circle with the given equation. Then graph the circle.

16. $(x + 7)^2 + (y - 5)^2 = 16$ **17.** $(x - 3)^2 + (y + 8)^2 = 100$

18. $(x + 4)^2 + (y - 1)^2 = 25$ **19.** $x^2 + y^2 = 36$

20. $(x - 0.3)^2 + y^2 = 0.04$ **21.** $(x + 5)^2 + (y + 2)^2 = 48$

Example 4
(page 697)

Use the diagram at the right. Write an equation that describes the position and radius of each circle.

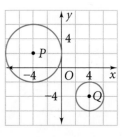

22. $\odot P$ **23.** $\odot Q$

24. Communications The plotted location of a cellular phone tower on a coordinate grid is $(-3, 2)$ and the range is 5 units. Write an equation that describes the position and range of the tower.

Each equation models the position and range of a tornado alert siren. Describe the position and range of each.

25. $(x - 5)^2 + (y - 7)^2 = 81$ **26.** $(x + 4)^2 + (y - 9)^2 = 144$

 Apply Your Skills

Write the standard equation of each circle.

27. 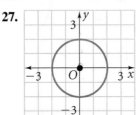   **28.** 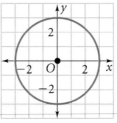   **29.**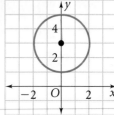

30. 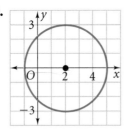   **31.** 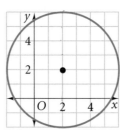   **32.**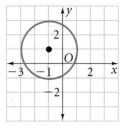

GO for Help

In Exercises 33–38, use the Midpoint Formula (p. 55) to find centers.

Write an equation of a circle with diameter $\overline{AB}$.

33. $A(0, 0), B(8, 6)$ **34.** $A(3, 0), B(7, 6)$ **35.** $A(1, 1), B(5, 5)$

36. $A(-1, 0), B(-5, -3)$ **37.** $A(-3, 1), B(0, 9)$ **38.** $A(-2, 3), B(6, -7)$

39. The *unit circle* has center $(0, 0)$ and radius 1. Write an equation for this circle.

40. Critical Thinking Describe the graph of $x^2 + y^2 = r^2$ when $r = 0$.

41. Open-Ended On graph paper, make a design that includes at least three circles. Write the standard equations of your circles.

Determine whether each equation is an equation of a circle. If not, explain.

42. $(x - 1)^2 + (y + 2)^2 = 9$ **43.** $x + y = 9$ **44.** $x + (y - 3)^2 = 9$

45. Find the circumference and area of the circle whose equation is $(x - 9)^2 + (y - 3)^2 = 64$. Leave your answers in terms of π.

46. Write an equation of a circle with area 36π and center $(4, 7)$.

47. What are the x- and y-intercepts of the line tangent to the circle $(x - 2)^2 + (y - 2)^2 = 5^2$ at the point $(5, 6)$?

48. For $(x - h)^2 + (y - k)^2 = r^2$, show that
$$y = \sqrt{r^2 - (x - h)^2} + k \text{ or } y = -\sqrt{r^2 - (x - h)^2} + k.$$

GO Online
Homework Video Tutor
Visit: PHSchool.com
Web Code: aue-1205

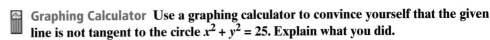

Graphing Calculator Use a graphing calculator to graph each circle. (*Hint:* See Exercise 48.) View the plotting in both sequential mode and simultaneous mode.

49. $(x - 3)^2 + (y - 2)^2 = 9$

50. $(x + 5)^2 + (y - 8)^2 = 1$

51. circle with center $(0, 0)$ and radius 7

52. circle with center $(-6, -3)$ and radius 2

Find all points of intersection of each pair of graphs. Make a sketch.

53. $x^2 + y^2 = 13$
$y = -x + 5$

54. $x^2 + y^2 = 17$
$y = -\frac{1}{4}x$

55. $x^2 + y^2 = 8$
$y = 2$

56. $x^2 + y^2 = 20$
$y = -\frac{1}{2}x + 5$

57. $(x + 1)^2 + (y - 1)^2 = 18$
$y = x + 8$

58. $(x - 2)^2 + (y - 2)^2 = 10$
$y = -\frac{1}{3}x + 6$

Graphing Calculator Use a graphing calculator to convince yourself that the given line is not tangent to the circle $x^2 + y^2 = 25$. Explain what you did.

59. $y = -5x + 26$

60. $3x + 5y = 29$

61. Writing Explain why it is not possible to conclude that a line and a circle are tangent by viewing their graphs.

C Challenge

62. Lines $y = \frac{2}{3}x + 3$ and $y = 5$ cut the ring formed by circles $(x - 3)^2 + (y - 5)^2 = 64$ and $(x - 3)^2 + (y - 5)^2 = 25$ into four parts. Find the area of each part.

63. Nautical Distance The radius of Earth's equator is about 3960 miles.
  a. Write the equation of the equator with the center of Earth as the origin.
  b. Find the length of a 1° arc on the equator to the nearest tenth of a mile.
  c. A 1° arc along the equator is 60 nautical miles long. How many miles are in a nautical mile? Round to the nearest tenth.
  d. History Columbus planned his trip to the East by going west. He thought each 1° arc was 45 miles long. He estimated that the trip would take 21 days. Use your answer to part (b) to find a better estimate.

Real-World 🌐 **Connection**

In 2005, Ellen MacArthur, age 28, sailed solo around the world in 71.6 days.

64. Geometry in 3 Dimensions The equation of a sphere is similar to the equation of a circle. The equation of a sphere with center (h, j, k) and radius r is
  $(x - h)^2 + (y - j)^2 + (z - k)^2 = r^2$.
  a. $M(-1, 3, 2)$ is the center of a sphere passing through $T(0, 5, 1)$. What is the radius of the sphere?
  b. Write an equation of the sphere.

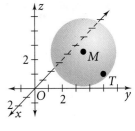

NY REGENTS

Test Prep

Multiple Choice

65. What is an equation of a circle with radius 16 and center $(2, -5)$?
  A. $(x - 2)^2 + (y + 5)^2 = 16$
  B. $(x + 2)^2 + (y - 5)^2 = 256$
  C. $(x + 2)^2 + (y - 5)^2 = 4$
  D. $(x - 2)^2 + (y + 5)^2 = 256$

66. What are the coordinates of the center of the circle whose equation is $(x - 9)^2 + (y + 4)^2 = 1$?
  F. $(3, -2)$
  G. $(-3, 2)$
  H. $(-9, 4)$
  J. $(9, -4)$

67. What is the diameter of the circle with equation $(x - 1)^2 + (y + 1)^2 = 4$?
  A. 1
  B. 2
  C. 4
  D. 16

68. Show how to find the radius of the circle whose equation is
$x^2 + (y + 8)^2 = 25$.

69. The line represented by the equation $y = -\frac{4}{3}x + 11$ is tangent to a circle at (6, 3). The center of the circle is on the *x*-axis. Write an equation of the circle. Show your work.

Mixed Review

Lesson 12-4

Find the value of each variable. Assume that lines that appear tangent are tangent.

70.

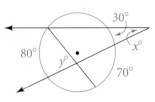

71.

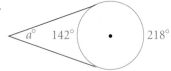

Lesson 8-6

For the given vectors $\vec{a}$ and $\vec{c}$, write the sum $\vec{a} + \vec{c}$ as an ordered pair.

72. $\vec{a} = \langle -2, 5 \rangle$ and $\vec{c} = \langle 8, 7 \rangle$

73. $\vec{a} = \langle -3, -4 \rangle$ and $\vec{c} = \langle -2, 6 \rangle$

74. $\vec{a} = \langle 3, 1 \rangle$ and $\vec{c} = \langle 1, 3 \rangle$

75. $\vec{a} = \langle 9, -6 \rangle$ and $\vec{c} = \langle 2, -1 \rangle$

Lesson 7-4

Find the geometric mean of each pair of numbers in simplest radical form.

76. 3 and 12

77. 9 and 27

78. 4 and 18

79. $\sqrt{3}$ and $\sqrt{27}$

80. $\sqrt{3}$ and $\sqrt{12}$

81. $\frac{3}{8}$ and $\frac{3}{2}$

Checkpoint Quiz 2   Lessons 12-4 through 12-5

Find the value of each variable. Assume that lines that appear tangent are tangent.

1.

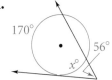

2.

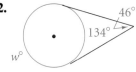

3.

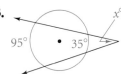

4.

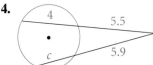

5.

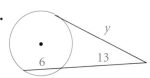

6. In the circle at the right, what is $m\widehat{BF}$?

7. Writing Explain the difference between a chord and a secant. Include a diagram.

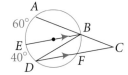

The endpoints of a diameter are given. Write an equation of the circle.

8. (3, 1) and (0, 0)

9. (−2, 5) and (9, −3)

10. (−4, −8) and (1, 0)

12-6

Locus: A Set of Points

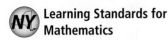
Learning Standards for Mathematics

G.G.22 Solve problems using compound loci.

 **Check Skills You'll Need**

 for Help Lessons 1-7 and 3-8

Sketch each of the following.

1. the perpendicular bisector of $\overline{CD}$

2. $\angle EFG$ bisected by $\overrightarrow{FH}$

3. line k parallel to line m and perpendicular to line w, all in plane N

🔊 **New Vocabulary** • locus

---

1 Drawing and Describing a Locus

A **locus** is a set of points, all of which meet a stated condition. To sketch a locus, draw points of the locus until you see a pattern.

1 EXAMPLE Describing a Locus in a Plane

a. Draw and describe the locus: In a plane, the points 1 cm from a given point C.

Draw a point C.
Sketch several points 1 cm from C.
Keep doing so until you see a pattern.
Draw the figure the pattern suggests.

The locus is a circle with center C and radius 1 cm.

b. Draw and describe the locus: In a plane, the points 1 cm from a segment $\overline{AB}$.

The locus is
• two segments parallel to $\overline{AB}$ and
• two semicircles centered at A and B

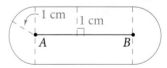

Real-World 🌐 Connection

The locus of footprints of children pushing the merry-go-round is a circle.

✓ **Quick Check** ❶ Draw and describe the locus: In a plane, the points 2 cm from a line $\overleftrightarrow{XY}$.

You can use locus descriptions for geometric terms.

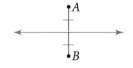

An angle bisector: The points in the interior of the angle that are equidistant from the sides of the angle.

A perpendicular bisector of a segment: In a plane, the points that are equidistant from the segment endpoints.

Vocabulary Tip

The word <u>locus</u> comes from the Latin word for "location." Its plural is *loci* (LOH sy).

Sometimes a locus is described by two conditions. You can draw the locus by first drawing the points that satisfy each condition. Then find their intersection.

2 EXAMPLE **Drawing a Locus for Two Conditions**

Draw the locus: In a plane, the points equidistant from two lines *k* and *m* and 5 cm from the point where *k* and *m* intersect.

The points in a plane equidistant from lines *k* and *m* are two lines that bisect the vertical angles formed by *k* and *m*.

The points in a plane 5 cm from the point where *k* and *m* intersect is a circle.

The locus that satisfies both conditions is the set of points *A*, *B*, *C*, and *D*.

✓ Quick Check ❷ Draw the locus: In a plane, the points equidistant from two points *X* and *Y* and 2 cm from the midpoint of $\overline{XY}$.

A locus in a plane and a locus in space can be quite different.

3 EXAMPLE **Describing a Locus in Space**

a. Draw and describe the locus: In space, the points that are *c* units from a point *D*.

The locus is a sphere with center at point *D* and radius *c*.

b. Draw and describe the locus: In space, the points that are 3 cm from a line ℓ.

The locus is an endless cylinder with radius 3 cm and center-line ℓ.

Real-World Connection

A soap bubble is a sphere, a locus of points in space that are a given distance from a given point.

✓ Quick Check ❸ Draw and describe each locus.
a. In a plane, the points that are equidistant from two parallel lines.
b. In space, the points that are equidistant from two parallel planes.

You can also think of a locus as a path. For example, the locus of the tip of a hand of a clock each day is the circle traced by the tip as it travels around the clock face. The locus of a point on the handle of a sliding-glass door when you enter a room is the line segment along which the point travels as the door slides back and forth.

EXERCISES

For more exercises, see *Extra Skill, Word Problem, and Proof Practice.*

Practice and Problem Solving

A Practice by Example

Example 1
(page 701)

GO for Help

Draw and describe each locus in a plane.

1. points 4 cm from a point X

2. points 2 in. from a segment $\overline{UV}$

3. points 3 mm from a line $\overleftrightarrow{LM}$

4. points 1 in. from a circle with radius 3 in.

5. points equidistant from the endpoints of $\overline{PQ}$

6. points in the interior of $\angle ABC$ and equidistant from the sides of $\angle ABC$

7. points equidistant from two perpendicular lines

8. midpoints of radii of a circle with radius 2 cm

Example 2
(page 702)

In a plane, draw the locus whose points satisfy the given conditions.

9. equidistant from points M and N and on a circle with center M and radius $= \frac{1}{2}MN$

10. 3 cm from $\overline{GH}$ and 5 cm from G, where $GH = 4.5$ cm

11. equidistant from the sides of $\angle PQR$ and on a circle with center P and radius PQ

12. equidistant from both points A and B and points C and D

13. equidistant from the sides of $\angle JKL$ and on $\odot C$

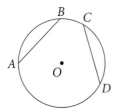

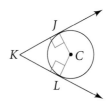

Example 3
(page 702)

Draw and describe each locus in space.

14. points 3 cm from a point F

15. points 4 cm from a line $\overleftrightarrow{DE}$

16. points 1 in. from plane M

17. points 5 mm from $\overrightarrow{PQ}$

B Apply Your Skills

Describe the locus that each blue figure represents.

18.

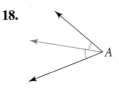

19.

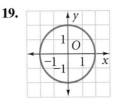

20.

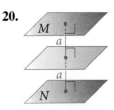

21. Open-Ended Give two examples of loci from everyday life, one in a plane and one in space.

22. Reasoning Rosie says that it is impossible to find a point equidistant from three collinear points. Is she correct? Explain.

Real-World Connection

A fingertip of the skater traces a locus as she twirls.

Coordinate Geometry Write an equation for the locus: In the plane, the points equidistant from the two given points.

23. $A(0, 2)$ and $B(2, 0)$

24. $P(1, 3)$ and $Q(5, 1)$

25. $T(2, -3)$ and $V(6, 1)$

26. Coordinate Geometry Complete the following locus description of the points highlighted in blue at the right: in the coordinate plane, the points 2 units from the <u>?</u> and 1 unit from the <u>?</u>-axis.

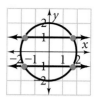

Make a drawing of each locus.

27. the path of a doorknob as a door opens

28. the path of a knot in the middle of a jump rope as it is being used

29. the path of the tip of your nose as you turn your head

30. the path of a fast-pitched softball

31. Jack and Julie Wilson take new jobs in Shrevetown and need a place to live. Jack says, "Let's try to move somewhere equidistant from both of our offices." Julie says, "Let's try to stay within three miles of downtown." Where on the map should the Wilsons look for a home?

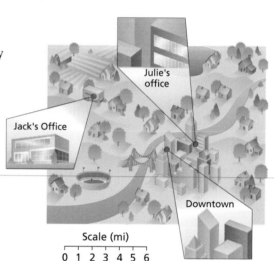

32. Critical Thinking Points *A* and *B* are 5 cm apart. Do the following loci in a plane have any points in common?

the points 3 cm from *A*

the points 4 cm from *B*

Illustrate with a sketch.

Coordinate Geometry **Draw each locus on the coordinate plane.**

33. all points 3 units from the origin

34. all points 2 units from $(-1, 3)$

35. all points 4 units from the *y*-axis

36. all points 5 units from $x = 2$

37. all points equidistant from $y = 3$ and $y = -1$

38. all points equidistant from $x = 4$ and $x = 5$

39. all points equidistant from the *x*- and *y*-axes

40. all points equidistant from $x = 3$ and $y = 2$

 Meteorology **In an anemometer, there are three cups mounted on an axis. Imagine a point on the edge of one of the cups.**

41. Describe the locus that this point traces as the cup spins in the wind.

42. Suppose the distance of the point from the axis of the anemometer is 2 in. Write an equation for the locus of part (a). Use the axis as the origin.

43. Robert draws a segment to use as the base of an isosceles triangle.
   a. Draw a segment to represent Robert's base. Locate three points that could be the vertex of the isosceles triangle.
   b. Describe the locus of possible vertices for Robert's isosceles triangle.
   c. Writing Explain why points in the locus you described are the only possibilities for the vertex of Robert's triangle.

Real-World Connection

An anemometer measures wind speed.

44. Describe the locus: The points in space equidistant from the points of a circle.

 **Challenge** 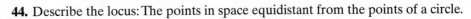 **Playground Equipment** **Think about the path of a child on each piece of playground equipment. Draw the path from (a) a top view, (b) a front view, and (c) a side view.**

45. a swing   **46.** a straight slide   **47.** a corkscrew slide

48. a merry-go-round   **49.** a firefighters' pole

50. In the diagram, three students are seated at uniform distances around a circular table. Copy the diagram. Shade the points on the table that are closer to Moesha than to Jan or Leandra.

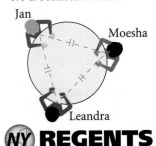

NY REGENTS

 Test Prep

Multiple Choice

51. Which graph shows the locus: The points in a plane 1 unit from the intersection of the lines $x + y = 2$ and $x - y = 4$?

A. 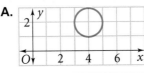   B.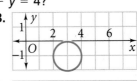

C. 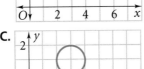   D.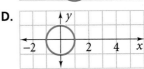

52. Which equation describes the locus: The points in the coordinate plane that are 5 units from the y-axis?

   F. $|y| = 5$ **G.** $|x| = 5$ **H.** $x + y = 5$ **J.** $x^2 + y^2 = 25$

Short Response

53. Margie's cordless telephone can transmit up to 0.5 mile from her home. Carol's cordless telephone can transmit up to 0.25 mile from her home. Carol and Margie live 0.25 mile from each other. Can Carol's telephone work in a region that Margie's cannot? Sketch and label a diagram.

Mixed Review

Lesson 12-5

Write an equation of the circle with center C and radius r.

54. $C(6, -10), r = 5$ **55.** $C(1, 7), r = 6$ **56.** $C(-8, 1), r = \sqrt{13}$

Lesson 11-2

Find the surface area of each figure to the nearest tenth.

57.  13 in. 15 in. 12 in.   **58.** 4 ft 12 ft

Lesson 10-7

In $\odot O$, find the area of sector AOB. Leave your answer in terms of π.

59. $OA = 4, m\widehat{AB} = 90$ **60.** $OA = 8, m\widehat{AB} = 72$ **61.** $OA = 10, m\widehat{AB} = 36$

Using Estimation

Estimation may help you find answers, check an answer, or eliminate one or more answer choices. Here are some decimal approximations that can be helpful. The symbol $\approx$ means "is approximately equal to."

$$\pi \approx 3 \qquad \frac{1}{\pi} \approx \frac{1}{3} \approx 0.3 \qquad \sqrt{2} \approx 1.4 \qquad \sqrt{3} \approx 1.7$$

EXAMPLE

Two circles have the same center O. The radius of the larger circle is twice the radius of the smaller circle, and chord $\overline{DB}$ is tangent to the smaller circle at C. Which of the following is the greatest?

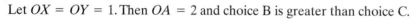

  (A) DB (B) OA (C) OX

  (D) length of $\overset{\frown}{XCY}$ (E) length of $\overset{\frown}{AB}$

Let $OX = OY = 1$. Then $OA = 2$ and choice B is greater than choice C.

The length of $\overset{\frown}{XCY}$ is one-half the circumference of the small circle.
The length of $\overset{\frown}{XCY} = \frac{1}{2}(2\pi r) = \pi \approx 3$. Thus, choice D is greater than choice B.

Draw $\overline{OC}$. $\triangle ODC$ has a right angle at C with $OD = 2$ and $OC = 1$.
By the Pythagorean Theorem, $DC = \sqrt{3}$. Therefore $DB = 2\sqrt{3} \approx 2(1.7) = 3.4$.
So, choice A is greater than choice D.

$\triangle OCD$ is a 30°-60°-90° triangle, so $m\overset{\frown}{AB} = 60$.
Thus, the length of $\overset{\frown}{AB}$ is $\frac{1}{6}$ the circumference of the larger circle.
The length of $\overset{\frown}{AB} = \frac{1}{6}(2\pi r) = \frac{1}{6}(4\pi) = \frac{2}{3}\pi \approx 2$, which is less than DB (choice A).

Therefore the greatest quantity is DB. The correct answer is A.

EXERCISES

1. Use the approximations above to estimate the value of each number.

  a. 2π **b.** $\frac{\pi}{2}$ **c.** $\frac{1}{\sqrt{2}}$ **d.** $\frac{1}{\sqrt{3}}$ **e.** $\sqrt{8}$

2. Which number is greatest? (*Hint:* Use estimation.)

  (A) π (B) $2\sqrt{3}$ (C) $3\sqrt{2}$ (D) $\sqrt{3} + \sqrt{2}$ (E) $\sqrt{3} + 2\sqrt{2}$

3. A student used a calculator to find the value of $\sqrt{8} + \sqrt{15}$ and got 4.796. Use estimation to explain why this answer is incorrect.

4. Which is the best estimate for $\sin A$?

  (A) 0.32   (B) 0.42   (C) 0.52

  (D) 0.62   (E) 0.78

5. Which is the best estimate for $\tan A$?

  (A) 0.14   (B) 0.24   (C) 0.34

  (D) 0.44   (E) 0.58

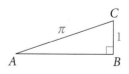

6. Use properties of special triangles to estimate the measures of $\angle X$ and $\angle Y$ in $\triangle XYZ$.

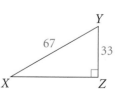

Chapter Review

Vocabulary Review

chord (p. 670)
circumscribed about (p. 664)
inscribed angle (p. 678)
inscribed in (p. 664)

intercepted arc (p. 678)
locus (p. 701)
point of tangency (p. 662)
secant (p. 687)

standard form of an equation of a
circle (p. 695)
tangent to a circle (p. 662)

Go Online
PHSchool.com

For: Vocabulary quiz
Web Code: auj-1251

Use the figure to choose the correct term to complete each sentence.

1. $\overline{CB}$ is *(a secant of, tangent to)* $\odot X$.

2. $\overline{DF}$ is a *(chord, locus)* of $\odot X$.

3. $\triangle DEF$ is *(inscribed in, circumscribed about)* $\odot X$.

4. $\angle DEF$ is an *(intercepted arc, inscribed angle)* of $\odot X$.

5. The set of "all points equidistant from the endpoints of $\overline{CB}$" is a *(locus, tangent)*.

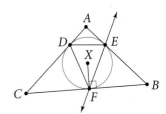

Skills and Concepts

12-1 Objectives

▼ To use the relationship between a radius and a tangent

▼ To use the relationship between two tangents from one point

A **tangent to a circle** is a line, ray, or segment in the plane of the circle that intersects the circle in exactly one point, the **point of tangency**. Two segments tangent to a circle from a point outside the circle are congruent. If a line is tangent to a circle, then the line is perpendicular to the radius drawn to the point of tangency. The converse is also true.

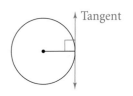

A triangle is **inscribed in** a circle if all of the vertices lie on the circle. When a triangle is **circumscribed about** a circle, each side is tangent to the circle.

Each polygon circumscribes a circle. Find the perimeter of the polygon.

6.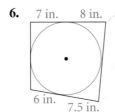

7.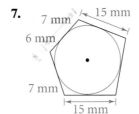

8.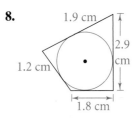

12-2 Objectives

▼ To use congruent chords, arcs, and central angles

▼ To recognize properties of lines through the center of a circle

Segments with endpoints on a circle are called **chords**. Within a circle or in congruent circles,

- congruent central angles have congruent chords.
- congruent chords have congruent arcs.
- congruent arcs have congruent central angles.

- chords equidistant from the center are congruent.
- congruent chords are equidistant from the center.

A diameter that is perpendicular to a chord bisects the chord and its arcs. A diameter that bisects a chord that is not a diameter is perpendicular to the chord. The perpendicular bisector of a chord contains the center of the circle.

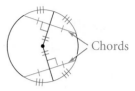
Chords

Find the value of x to the nearest tenth.

9.   10.   11.   12.

12-3 Objectives

▼ To find the measure of an inscribed angle

▼ To find the measure of an angle formed by a tangent and a chord

An angle is an **inscribed angle** if the vertex is on a circle and the sides of the angle are chords of the circle. Its **intercepted arc** is the arc whose endpoints are on the sides of the angle and whose remaining points lie in the interior of the angle.

The measure of an inscribed angle is half the measure of its intercepted arc. The measure of an angle formed by a tangent and a chord that intersect on a circle is half the measure of the intercepted arc.

Two inscribed angles that intercept the same arc are congruent. An angle inscribed in a semicircle is a right angle. The opposite angles of a quadrilateral inscribed in a circle are supplementary.

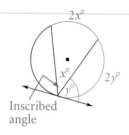
Inscribed angle

Assume that lines that appear tangent are tangent. Find the value of each variable.

13.   14.   15.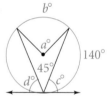

12-4 Objectives

▼ To find the measures of angles formed by chords, secants, and tangents

▼ To find the lengths of segments associated with circles

A **secant** is a line, ray, or segment that intersects a circle at two points.

The measure of an angle formed by two chords that intersect in a circle is half the sum of the measures of the intercepted arcs.

The measure of an angle formed by two secants, two tangents, or a secant and a tangent drawn from a point outside the circle is half the difference of the measures of the intercepted arcs.

$$m\angle 1 = \tfrac{1}{2}(x + y)$$

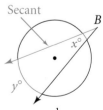
Secant

$$m\angle B = \tfrac{1}{2}(y - x)$$

For a given point and circle, the product of the lengths of the two segments from the point to the circle is constant along any line through the point and circle.

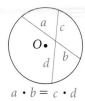

$a \cdot b = c \cdot d$

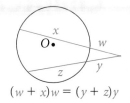

$(w + x)w = (y + z)y$

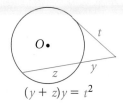

$(y + z)y = t^2$

Assume that lines that appear tangent are tangent. Find the value of each variable.

16.

17.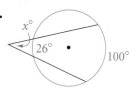

18.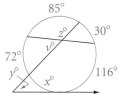

$\boxed{x^2}$ **Algebra** **Find the value of each variable using the given chords, secants, and tangents. If your answer is not an integer, round to the nearest tenth.**

19.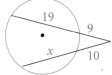

20.

21.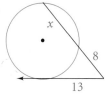

12-5 and 12-6 Objectives

▼ To write an equation of a circle

▼ To find the center and radius of a circle

▼ To draw and describe a locus

The equation $(x - h)^2 + (y - k)^2 = r^2$ is in **standard form.** You may also call it the *standard equation* of a circle.

If you know the center and a point on a circle, you can write the standard equation of the circle. Use the Distance Formula to find the radius. Then substitute the coordinates of the center for (h, k) and the radius for r in the equation of a circle. If you know the equation of a circle, you can identify the center and radius.

A set of points that meet a stated condition is a **locus.** Sometimes you can describe a figure as a locus.

Write the standard equation of the circle with center C and radius r.

22. $C(2, 5); r = 3.5$ **23.** $C(-3, 1); r = \sqrt{5}$ **24.** $C(9, -4); r = 4$

Write the standard equation of the circle with center C passing through point P.

25. $C(0, 1); P(4, 9)$ **26.** $C(-2, 3); P(4, -4)$ **27.** $C(10, 7); P(-8, -5)$

Describe the circle with the given equation.

28. $x^2 + (y - 8)^2 = 49$ **29.** $(x - 5)^2 + (y + 9)^2 = 40$ **30.** $(x + 1)^2 + y^2 = 9$

Sketch and label each locus.

31. all points in a plane 2 cm from a circle with radius 1 cm

32. all points in a plane equidistant from two points

33. all points in space a distance a from $\overline{DS}$

Go Online
PHSchool.com
For: Chapter Test
Web Code: aua-1252

 Algebra Assume that lines that appear tangent are tangent. Find the value of x.

1.

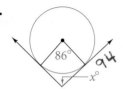

86°
$x°$ 94

2.

24
32
x 8

3. a. Open-Ended Draw a circle with two congruent chords that form an inscribed angle.

   b. Constructions Construct the bisector of the inscribed angle. What do you notice?

 Algebra Find the value of x. If your answer is not an integer, round to the nearest tenth.

4.

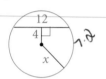

12
4
x 7.2

5.

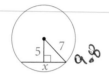

5 7
x 9.8

6. Coordinate Geometry An equation of a circle is $(x - 4)^2 + (y - 3)^2 = 25$. Find the center and radius of the circle. Then graph the circle.

Find $m\overarc{AB}$.

7.

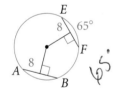

E
8 65°
F
8
A
B

65°

8.

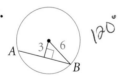

3 6
A
B

120°

9. Find the value of z.

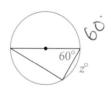

60°
$z°$

60°

10. Writing What is special about a rhombus inscribed in a circle? Justify your answer.

11. A chord of a circle has length 4.2 cm and is 8 cm from the center of the circle. What is the radius of the circle to the nearest hundredth? 8.27

Find the center and radius of each circle.

12. $(x + 3)^2 + (y - 2)^2 = 9$ (-3,2) 3

13. $x^2 + (y - 9)^2 = 225$ (0,9) 15

 Algebra For Exercises 14–19, lines that appear tangent are tangent. Find the value of each variable. If your answer is not an integer, round to the nearest tenth.

14.

55°
$a°$ $b°$
110° 70°
165°

15.

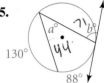

$a°$ 71 $b°$
130° 44
88°

16.

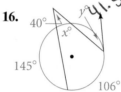

40° y
$x°$ 41.5
145°
106°

17.

3 2
10.5
x 7

18.

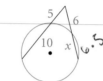

5 6
10 x 6.5

19.

x 8
4
12

Write the standard equation of each circle.

20.

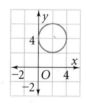

21.

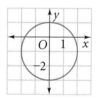

22. Write an equation of the circle with center $(3, 0)$ that passes through point $(-2, -4)$.

23. Find the circumference and area of the circle whose equation is $(x - 2)^2 + (y - 7)^2 = 81$. Round to the nearest tenth.

24. Write an equation for the locus: Points in the coordinate plane that are 4 units from $(-5, 2)$.

Coordinate Geometry Sketch each locus on a coordinate plane.

25. all points 6 units from the origin

26. all points 3 units from the line $y = -2$

27. all points equidistant from points $(2, 4)$ and $(0, 0)$

28. all points equidistant from the axes

CUMULATIVE REVIEW ▪ CHAPTERS 1–12

Regents Test Prep

Go Online
PHSchool.com
For: End-of-Course Test
Web Code: aua-1254

Multiple Choice

For Exercises 1–37, choose the correct letter.

1. Which word(s) best describes the triangle at the right?
(1) equilateral
(2) isosceles
(3) right
(4) isosceles right

2. What are the coordinates of the midpoint of $\overline{QS}$ with endpoints $Q(-2, -5)$ and $S(3, -8)$?
(1) $(-2.5, 1.5)$ (2) $(-2.5, 6.5)$
(3) $(0.5, -6.5)$ (4) $(0.5, 1.5)$

3. What is the volume of the figure?
(1) 72 ft^3
(2) 72π ft^3
(3) $(18 + 72\pi)$ ft^3
(4) $(72 + 18\pi)$ ft^3

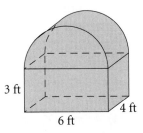
3 ft
4 ft
6 ft

4. Which of the following statements can be derived from the biconditional statement "The day is long if and only if it is summer"?
(1) If the day is long, then it is summer.
(2) If it is summer, then the day is long.
(3) If the day is not long, then it is not summer.
(4) all of the above

5. Which line or lines are perpendicular to the line $y = 4x - 1$?

I. $y = 4x + 7$ II. $y = \frac{1}{4}x + 3$
III. $y = -\frac{1}{4}x - 5$ IV. $x + 4y = 16$

(1) I only (2) II only
(3) III only (4) III and IV

6. Which solid has the least volume?
(1)
6 cm
6 cm

(2)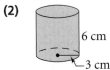
6 cm
3 cm

(3)
6 cm
2 cm

(4)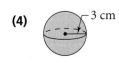
3 cm

7. Which figure below is the reflection across the x-axis of the figure at the right?

2
−2 O 2 x

(1)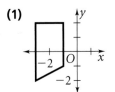
(2)
(3)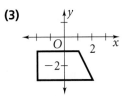
(4)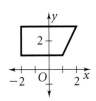

8. Which conditions allow you to conclude that a quadrilateral is a parallelogram?
(1) one pair of sides congruent, the other pair parallel
(2) perpendicular congruent diagonals
(3) congruent bisecting diagonals
(4) one diagonal bisecting opposite angles

9. Which triangle is drawn with its medians?
(1)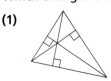
(2)
(3)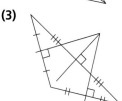
(4)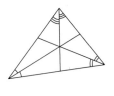

10. If quadrilateral *GFIH* is reflected across the x-axis to become quadrilateral *G'F'I'H'*, what will be the coordinates of *G'*?
(1) $(1, 2)$ (2) $(3, 2)$
(3) $(-3, -2)$ (4) $(3, -2)$

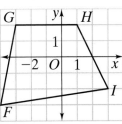
G y H
1
−2 O 1 x
F I

11. What is the surface area in square centimeters of a sphere with radius 7 cm?
(1) 196π (2) $\frac{196}{3}\pi$ (3) 49π (4) 14π

12. In $\triangle RST$, $RS = 4$, $ST = 5$, and $RT = 6$. Which angle is largest?

(1) $\angle R$ **(2)** $\angle S$ **(3)** $\angle T$

(4) cannot be determined

13. Which is true for both a rhombus and a kite?
(1) The diagonals are congruent.
(2) Opposite sides are congruent.
(3) The diagonals are perpendicular.
(4) Opposite sides are parallel.

14. To the nearest tenth, what is the value of x?

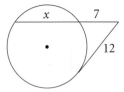

(1) 9.2   **(2)** 13.6
(3) 15.0   **(4)** 20.6

15. If $\frac{m}{n} = \frac{1}{3}$, which of the following must be true?

   I. $3m = n$ II. $m = 3n$

   III. $m + 1 = \frac{n + 3}{3}$ IV. $mn = 3$

(1) I only   **(2)** I and II only
(3) I and III only   **(4)** I and IV only

16. The length of the hypotenuse of an isosceles right triangle is 8 in. What is the length of one leg?

(1) $8\sqrt{2}$ in.   **(2)** $4\sqrt{2}$ in.   **(3)** 4 in.   **(4)** 2 in.

17. Which equation is that of a line that contains the point $P(5, 6)$ and has slope $-\frac{1}{3}$?

(1) $y = -\frac{1}{3}x + \frac{3}{23}$ **(2)** $y = -\frac{1}{3}x - \frac{3}{23}$
(3) $y = -\frac{1}{3}x - \frac{23}{3}$ **(4)** $y = -\frac{1}{3}x + \frac{23}{3}$

18. Which of the following must be true?

   I. $\angle BAC \cong \angle B$
   II. $\angle B \cong \angle C$
   III. $\overline{AD} \cong \overline{AB}$
   IV. $\overline{BD} \cong \overline{CD}$

(1) I and II only   **(2)** I and III only
(3) II and IV only   **(4)** III and IV only

19. What is the area of the trapezoid?
(1) 75 in.^{2}
(2) 43 in.^{2}
(3) 79.5 in.^{2}
(4) 159 in.^{2}

20. $\overleftrightarrow{AB}$ is tangent to $\odot C$ at point B. Which of the following can you NOT conclude is true?
(1) $m\angle CAB < m\angle ACB$
(2) $AB^2 + BC^2 = AC^2$
(3) $\angle CAB$ and $\angle ACB$ are complements.
(4) $\overleftrightarrow{AB} \perp \overleftrightarrow{BC}$

21. Sphere B has 4 times the surface area of sphere A. How many times the volume of sphere A is the volume of sphere B?

(1) 4   **(2)** 8   **(3)** 4π **(4)** 8π

22. What is the value of x?
(1) $\sqrt{30}$
(2) $\sqrt{39}$
(3) $3\sqrt{13}$
(4) $\sqrt{130}$

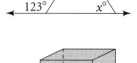

not to scale

23. What is the ratio of the surface areas of similar solids whose similarity ratio is 3 : 5?

(1) 6 : 10   **(2)** 9 : 15   **(3)** 9 : 25   **(4)** 27 : 125

24. What are the values of x and y?
(1) $x = 56$, $y = 68$
(2) $x = 68$, $y = 56$
(3) $x = 57$, $y = 66$
(4) $x = 66$, $y = 57$

25. What is the volume of the prism in cubic centimeters?
(1) 202   **(2)** 180
(3) 99   **(4)** 81

26. What is the circumference of a circle with radius 9?

(1) 4.5π **(2)** 9π **(3)** 18π **(4)** 81π

27. Which information CANNOT be used to prove that two triangles are congruent?

(1) SAS   **(2)** ASA   **(3)** AAS   **(4)** AAA

28. Which figure does NOT have an area of 15 ft^2?

(1)   **(2)**

 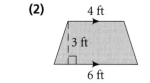

(3)   **(4)**

 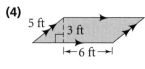

29. Which is greatest in $\triangle ABC$?
(1) $\sin A$ **(2)** $\cos A$
(3) $\tan A$ **(4)** $\tan B$

30. What is the surface area in square centimeters of a rectangular prism 9 cm by 8 cm by 10 cm?

(1) 240   **(2)** 242   **(3)** 484   **(4)** 720

31. What is the standard equation of a circle with center $(-2, 0)$ and radius 4?
(1) $(x - 2)^2 + y^2 = 4$ **(2)** $(x + 2)^2 + y^2 = 16$
(3) $x^2 + (y - 2)^2 = 2^2$ **(4)** $x^2 + (y + 2)^2 = 16$

32. An angle is five times as large as its supplement. What is the measure of the supplement?
(1) 20   (2) 30
(3) 100   (4) 150

33. Which quadrilateral CANNOT have one diagonal that bisects the other?
(1) square   (2) trapezoid
(3) kite   (4) parallelogram

34. Which is the contrapositive of the following conditional statement?

If Peg serves pudding, then it is not Laura's birthday.

(1) If Peg serves pudding, then it is Laura's birthday.
(2) If Peg does not serve pudding, then it is Laura's birthday.
(3) If it is not Laura's birthday, then Peg serves pudding.
(4) If it is Laura's birthday, then Peg does not serve pudding.

35. Two sides of a triangle measure 12 and 20. The measure of the included angle is 40. Find the area of the triangle to the nearest tenth.
(1) 77.1   (2) 91.9
(3) 120.0   (4) 154.3

36. A regular octagon has a side length of 8 m and an area of approximately 309 m^2. The side length of another regular octagon is 4 m. What is the approximate area of the smaller octagon?
(1) 39 m^2 (2) 77 m^2
(3) 98 m^2 (4) 154 m^2

37. How can you determine that a point lies on the perpendicular bisector of $\overline{PQ}$ with endpoints $P(-3, -6)$ and $Q(-3, 4)$?
(1) The point has x-coordinate -3.
(2) The point has y-coordinate -3.
(3) The point lies on the line $x = -1$.
(4) The point lies on the line $y = -1$.

Gridded Response

38. A diagonal of a rectangular garden makes a 70° angle with a side of the garden that is 60 ft long. To the nearest whole number of square feet, what is the area of the garden?

39. You are making a scale model of a building. The front of the actual building is 60 ft wide and 100 ft tall. The front of your model is 3 ft by 5 ft. What is the scale factor of the reduction?

40. You are 5 ft 6 in. tall. When your shadow is 6 ft long, the shadow of a sculpture is 30 ft long. How many feet tall is the sculpture?

41. What is the surface area in square centimeters of a right cone with slant height 5 cm and radius 3 cm? Use 3.14 for π and give your answer to the nearest tenth.

42. At 8 o'clock, what is the degree measure of the angle formed by the two hands of the clock?

Short Response

Show your work.

43. What is the area of an isosceles right triangle whose hypotenuse is $5\sqrt{2}$ m long?

44. The coordinates of the endpoints of $\overline{CD}$ are $C(5, 2.5)$ and $D(0, -9.5)$. Find the length of $\overline{CD}$ and the coordinates of the midpoint of $\overline{CD}$.

45. Write an equation of a line parallel to the line $y = 6x + 4$. Then write an equation of a line perpendicular to the line $y = 6x + 4$.

46. $\triangle DEB$ has vertices $D(3, 7)$, $E(1, 4)$, and $B(-1, 5)$. In which quadrant(s) is the image of $\triangle DEB$ for a 90° rotation about the origin?

Extended Response

Show your work.

47. Draw an angle. Then construct another angle congruent to the first.

48. Quadrilateral $ABCD$ has vertices $A(-1, -3)$, $B(4, -2)$, $C(3, 7)$, and $D(-7, 0)$. Find the vertices of the image of quadrilateral $ABCD$ under the translation $\langle -4, -1 \rangle$.

49. When airplane pilots make a visual sighting of an object outside the airplane, they often refer to the face of a dial clock to help locate the object. For example, an object at 12 o'clock is straight ahead, an object at 3 o'clock is 90° to the right, and so on.

Suppose that two pilots flying two airplanes in the same direction spot the same object. One pilot reports the object at 1 o'clock; the other pilot reports the object at 2 o'clock. At the same time the first pilot reports seeing the other airplane at 9 o'clock.

Draw a diagram showing the possible locations of the two planes and the object.

Data Analysis

Market Research

How a company packages a product can make a difference in sales. Despite the adage "Don't judge a book by its cover," the customer will do so anyway. He or she may choose one product over a similar product because the container looks like it is larger or because it has a more appealing appearance.

Market research is the study of consumer preference. A company will often use market research before it releases a product. It will try to find the buying public's reactions to both the product and its packaging.

1 ACTIVITY

You and a partner are to design the package for a new type of fruit-juice drink. The package must have a volume of exactly 1000 cm^3.

1. Begin by designing three different packages that are based on circles. One of your designs may be a cylinder. Another may be a cone. The third must be unique. (Some examples are shown at the right.)

2. Use posterboard and tape to make models of your package designs. Leave the models blank so that the only thing different about them is their shape.

3. Think of a name for your product. Also, list three ways that you might promote it (slogans, advertising themes, endorsements, and so on).

4. a. For each package, prepare a step-by-step explanation of how you found its volume.
   b. Trade designs and explanations with another pair of students. Verify their methods and calculations.

5. For three packages from your class, complete a survey like the one suggested by the table below .

	Strongly like	Like	Dislike	Strongly dislike
Package 1				
Package 2				
Package 3				

2 ACTIVITY

In this activity you will design and use a survey to do market research about your packaging. You want to gather information about the following.

A. Customers' reactions to your packages' different shapes
B. Customers' perceptions about which shape gives them the most for their money
C. The package that customers would most likely purchase

The types of questions you ask will determine how much you learn from the survey. For instance, if you ask customers only to pick their preferred shape, you won't learn anything about the strengths of, or reasons for, their preferences.

6. Write three survey questions for each of the following.
   a. What a customer likes about your packages
   b. What a customer dislikes about your packages

7. Work with your partner to write a survey that will give you good, easily summarized information about **A–C** above.

8. Survey at least 20 people outside of your classroom.

9. The Ice Cream Preferences graphs at the right (50 people surveyed) suggest two ways to show survey results. What can you learn from the stacked-bar graph that the bar graph does not show?

10. There are many other types of graphs including
    - line graphs
    - pie charts
    - stem-and-leaf plots
    - line plots
    - histograms
    - box-and-whisker plots

    Choose one or more types of graphs to display at least two results from your survey.

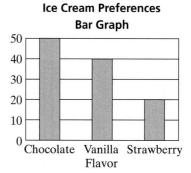

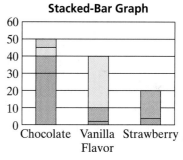

EXERCISES

11. Another interest of market research is the target audience. Describe three characteristics of the target audience for your product.

12. Market research can also test a product against its competition. Suppose you were to test your packaging side-by-side with a classmates' packaging. Write two survey questions that you would ask.

13. If you price your product too low, you won't make a profit. If you price it too high, you won't sell very many. Consider the following survey item.

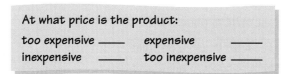

At what price is the product:

too expensive _____   expensive _____
inexpensive _____   too inexpensive _____

   a. How helpful do you feel that this survey item would be in deciding a price for your product? Explain.
   b. How could you improve on this item in order to find the best price?

Chapter 1

Extra Practice: Skills, Word Problems, and Proof

● **Lesson 1-1**

Find the next two terms in each sequence.

1. 12, 17, 22, 27, 32, . . .

2. 1, 1.1, 1.11, 1.111, 1.1111, . . .

3. 5000, 1000, 200, 40, . . .

4. 1, 12, 123, 1234, . . .

5. 3, 0.3, 0.03, 0.003, . . .

6. 1, 4, 9, 16, 25, . . .

Draw the next figure in each sequence.

7.

8.

● **Lesson 1-2**

Name the space figure that can be formed by folding each net.

9.

10.

11.

12.

Make (a) an isometric drawing and (b) an orthographic drawing for each foundation drawing.

13.

4	1
1	

Right
Front

14.

3	3
1	2

Right
Front

15.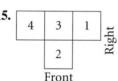

4	3	1
	2	

Right
Front

16.

	2
3	1

Right
Front

17. You can cut four of the lettered squares from the figure at the right and fold the remaining net to make a box that is open at one end. Write the letters of the squares you could remove to do this. List all the possibilities.

A	B	C
D	E	F
G	H	I

● **Lessons 1-3 and 1-4**

Write *true* or *false*.

18. A, D, F are coplanar.

19. $\overleftrightarrow{AC}$ and $\overleftrightarrow{FE}$ are coplanar.

20. A, B, E are coplanar.

21. D, A, B, E are coplanar.

22. $\overleftrightarrow{FC} \parallel \overleftrightarrow{EF}$

23. plane $ABC \parallel$ plane FDE

24. $\overleftrightarrow{BC}$ and $\overleftrightarrow{DF}$ are skew lines.

25. $\overrightarrow{AD}$ and $\overleftrightarrow{EB}$ are skew lines.

26. $\overleftrightarrow{DE} \parallel \overleftrightarrow{CF}$

27. $D, E,$ and B are collinear.

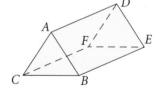

28. How many sets of four collinear points are there in a 4-by-4 geoboard as pictured at the right?

29. $\overline{AB}$ and $\overline{CD}$ do not intersect but $\overrightarrow{DC}$ intersects $\overline{AB}$ in one point. Make a sketch that shows this.

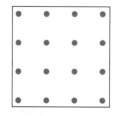

Lessons 1-5 and 1-6

Use the figure at the right for Exercises 30–35.

30. If $BC = 12$ and $CE = 15$, then $BE = $ ■.

31. ■ is the angle bisector of ■.

x^2 **32. Algebra** $BC = 3x + 2$ and $CD = 5x - 10$. Solve for x.

x^2 **33. Algebra** If $AC = 5x - 16$ and $CF = 2x - 4$, then $AF = $ ■.

34. $m\angle BCG = 60$, $m\angle GCA = $ ■, and $m\angle BCA = $ ■.

35. $m\angle ACD = 60$ and $m\angle DCH = 20$. Find $m\angle HCA$.

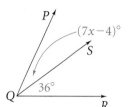

x^2 **36. Algebra** In the figure at the right, $m\angle PQR = 4x + 47$. Find $m\angle PQS$.

x^2 **37. Algebra** Points A, B, and C are collinear with B between A and C. $AB = 4x - 1$, $BC = 2x + 1$, and $AC = 8x - 4$. Find AB, BC, and AC.

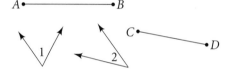

● **Lesson 1-7**

Make a diagram larger than the given one. Then do the construction.

38. Construct $\angle A$ so that $m\angle A = m\angle 1 + m\angle 2$.

39. Construct the perpendicular bisector of $\overline{AB}$.

40. Construct the angle bisector of $\angle 1$.

41. Construct $\overline{FG}$ so that $FG = AB + CD$.

● **Lesson 1-8**

(a) Find the distance between the points to the nearest tenth.
(b) Find the coordinates of the midpoint of the segments with the given endpoints.

42. $A(2, 1)$, $B(3, 0)$ **43.** $R(5, 2)$, $S(-2, 4)$

44. $Q(-7, -4)$, $T(6, 10)$ **45.** $C(-8, -1)$, $D(-5, -11)$

46. $J(0, -5)$, $N(3, 4)$ **47.** $Y(-2, 8)$, $Z(3, -5)$

48. A map of a city and suburbs shows an airport located at $A(25, 11)$.

An ambulance is on a straight expressway headed from the airport to Grant Hospital at $G(1, 1)$. The ambulance gets a flat tire at the midpoint M of $\overline{AG}$. As a result, the ambulance crew calls for helicopter assistance.
a. What are the coordinates of point M?
b. How far does the helicopter have to fly to get from M to G? Assume all coordinates are in miles.

● **Lesson 1-9**

Find the perimeter (or circumference) and area of each figure.

49.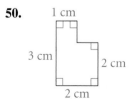

50.

51.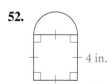

52.

● **Lessons 2-1 and 2-2**

For Exercises 1–3, identify the hypothesis and conclusion of each conditional.

1. If you can predict the future, then you can control the future.

2. If Dan is nearsighted, then Dan needs glasses.

3. If lines k and m are skew, then lines k and m are not perpendicular.

4. Write the converse of each statement in Exercises 1–3.

For each of the statements, write the conditional form and then the converse of the conditional. If the converse is true, combine the statements as a biconditional.

5. The number one is the smallest positive square.

6. Rectangles have four sides.

7. A square with area 100 m^2 has sides that measure 10 m.

8. Two numbers that add up to be less than 12 have a product less than 37.

9. Three points on the same line are collinear.

● **Lesson 2-2**

Is each statement a good definition? If not, find a counterexample.

10. A real number is an even number if its last digit is 0, 2, 4, 6, or 8.

11. A circle with center O and radius r is defined by the set of points in a plane a distance r from the point O.

12. A plane is defined by two lines.

13. Segments with the same length are congruent.

For Exercises 14 and 15, write the two statements that form each biconditional. Tell whether each statement is *true* or *false*.

14. Lines m and n are skew if and only if lines m and n do not intersect.

15. A person can be president of the United States if and only if the person is a citizen of the United States.

● **Lesson 2-3**

Using the statements below, apply the Law of Detachment or the Law of Syllogism to draw a conclusion.

16. If Jorge can't raise money, he can't buy a new car. Jorge can't raise money.

17. If Shauna is early for her meeting, she will gain a promotion. If Shauna wakes up early, she will be early for her meeting. Shauna wakes up early.

18. If Linda's band wins the contest, they will win $500. If Linda practices, her band will win the contest. Linda practices.

19. If Brendan learns the audition song, he will be selected for the chorus. If Brendan stays after school to practice, he will learn the audition song. Brendan stays after school to practice.

For Exercises 20–23, apply the Law of Detachment, the Law of Syllogism, or both to draw a conclusion. Tell which law(s) you used.

20. If you enjoy all foods, then you like cheese sandwiches. If you like cheese sandwiches, then you eat bread.

21. If you go to a monster movie, then you will have a nightmare. You go to a monster movie.

22. If Catherine is exceeding the speed limit, then she will get a speeding ticket. Catherine is driving at 80 mi/h. If Catherine is driving at 80 mi/h, then she is exceeding the speed limit.

23. If Carlos has more than $250, then he can afford the video game he wants. If Carlos worked more than 20 hours last week, then he has more than $250. If Carlos works 15 hours this week, then he worked more than 20 hours last week.

● **Lesson 2-4**

x^2 **24. Algebra** You are given that $2c^2 = 2bc + \frac{ac}{2}$ with $c \neq 0$. Show that $4b = 4c - a$ by filling in the blanks.

a. $2c^2 = 2bc + \frac{ac}{2}$ **a.** Given

b. $4c^2 = 4bc + ac$ **b.** __?__ and __?__

c. $4c = 4b + a$ **c.** __?__ and Distributive Property

d. __?__   **d.** Subtraction Property

e. $4b = 4c - a$ **e.** __?__

x^2 **25. Algebra** Solve for x. Show your work. Justify each step.

Given: $\overrightarrow{PF}$ bisects $\angle 1$.

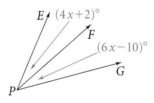

● **Lesson 2-5**

x^2 **Algebra Find the value of x.**

26.

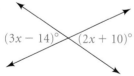

$(3x - 14)°$ $(2x + 10)°$

27.

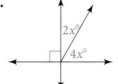

$2x°$
$4x°$

28.

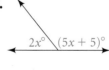

$2x°$ $(5x + 5)°$

29. Given: $\angle 1$ and $\angle 2$ are complementary.
         $\angle 3$ and $\angle 4$ are complementary.

    Prove: $\angle 5 \cong \angle 6$

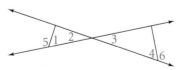

30. Prove or disprove the following statement.

If $\angle APB$ and $\angle CPD$ are vertical angles, $\angle APB$ and $\angle APE$ are complementary, and $\angle CPD$ and $\angle CPF$ are complementary, then $\angle APE$ and $\angle CPF$ are vertical angles.

Chapter 3 Extra Practice: Skills, Word Problems, and Proof

Lesson 3-1

Find $m\angle 1$ and then $m\angle 2$. State the theorems or postulates that justify your answers.

1.

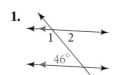

2.

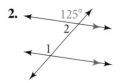

3.

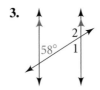

4.

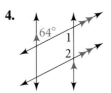

5. Complete the proof.

Given: $\ell \parallel m$, $a \parallel b$

Prove: $\angle 1 \cong \angle 5$

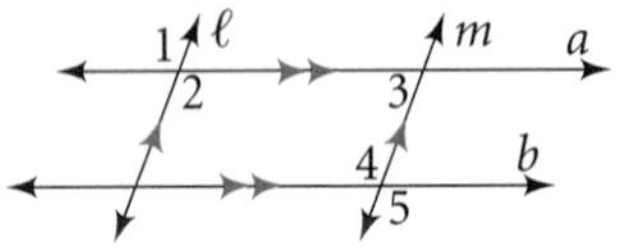

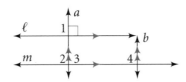

Statements	Reasons
1. $\ell \parallel m$, $a \parallel b$	**1.** Given
2. $\angle 1 \cong \angle 2$	**a.** __?__
3. $\angle 2$ and $\angle 3$ are supplementary.	**b.** __?__
4. $\angle 3$ and $\angle 4$ are supplementary.	**c.** __?__
5. $\angle 2 \cong \angle 4$	**d.** __?__
6. $\angle 1 \cong \angle 4$	**e.** __?__
7. $\angle 4 \cong \angle 5$	**f.** __?__
8. $\angle 1 \cong \angle 5$	**g.** __?__

Lessons 3-2 and 3-3

Refer to the diagram at the right. Use the given information to determine which lines, if any, must be parallel. If any lines are parallel, use a theorem or postulate to tell why.

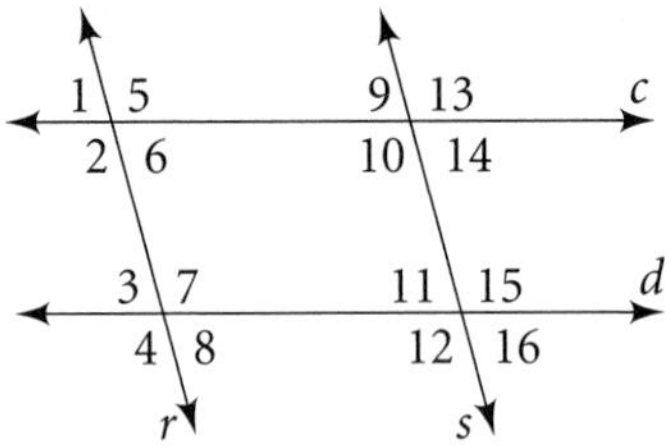

6. $\angle 9 \cong \angle 14$

7. $\angle 1 \cong \angle 9$

8. $\angle 2$ is supplementary to $\angle 3$.

9. $\angle 7 \cong \angle 10$

10. $m\angle 6 = 60, m\angle 13 = 120$

11. $\angle 4 \cong \angle 13$

12. $\angle 3$ is supplementary to $\angle 10$.

13. $\angle 10 \cong \angle 15$

14. Given: $\ell \parallel m$, $a \parallel b$, $a \perp \ell$

Prove: $b \perp m$

Lesson 3-4

Use a protractor and a centimeter ruler to measure the angles and the sides of each triangle. Classify each triangle by its angles and sides.

15.

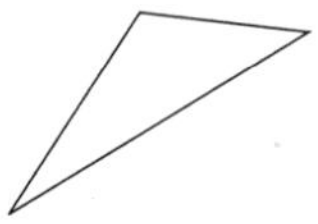

16.

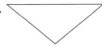

17.

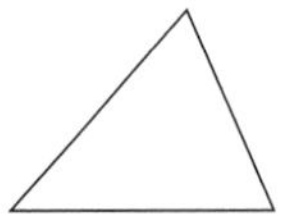

18.

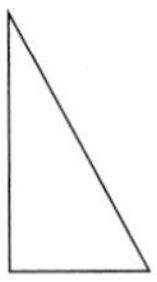

19. Use the figure at the right.
What is the relationship between $\overleftrightarrow{BC}$ and $\overleftrightarrow{DF}$?
Justify your answer.

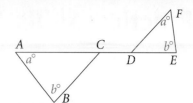

● **Lessons 3-4 and 3-5**

x^2 **Algebra** Find the value of each variable.

20.

21.

22.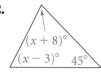

23.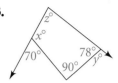

x^2 **Algebra** Find the missing angle measures.

24.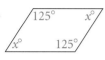

25.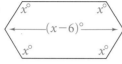

● **Lessons 3-6**

Write an equation in point-slope form of the line that contains the given points.

26. $A(4, 2)$, $B(6, -3)$ **27.** $C(-1, -1)$, $D(1, 1)$ **28.** $F(3, -5)$, $G(-5, 3)$ **29.** $K(5, 0)$, $L(-5, 2)$

Write an equation in slope-intercept form of the line through the given points.

30. $H(2, 7)$, $J(-3, 1)$ **31.** $M(-2, 4)$, $N(5, -8)$ **32.** $P(0, 2)$, $Q(6, 8)$ **33.** $K(5, 0)$, $L(-5, 2)$

● **Lessons 3-6 and 3-7**

Graph each pair of lines and state whether they are parallel, perpendicular, or neither. Explain.

34. $y = 4x - 8$
$y = 4x - 2$

35. $13y - x = 7$
$7 - \frac{y}{2} = x$

36. $y = \frac{-4}{3}x + 2$
$\frac{4}{3}y = x - 1$

37. $\frac{3}{5}y = -x + \frac{3}{2}$
$3x - \frac{15}{3}y = 0$

Without graphing, tell whether the lines are parallel, perpendicular, or neither. Explain.

38. $2x + 3y = 5$
$5x - 10y = 30$

39. $y = -2x + 7$
$x - 2y = 8$

40. $5x - 3y = 0$
$y = \frac{5}{3}x + 2$

41. $y = 3x + 8$
$x + 3y = 8$

42. On a city map, Washington Street is straight and passes through points at $(7, 13)$ and $(1, 5)$. Wellington Street is straight and passes through points at $(3, 24)$ and $(9, 32)$. Do Washington Street and Wellington Street intersect? How do you know?

● **Lesson 3-8**

Use the segments for each construction.

43. Construct a square with side length $2a$.

44. Construct a quadrilateral with one pair of parallel sides each of length $2b$.

45. Construct a rectangle with sides b and a.

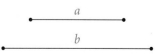

Extra Practice: Skills, Word Problems, and Proof

● **Lesson 4-1**

$\triangle SAT \cong \triangle GRE$. **Complete each congruence statement.**

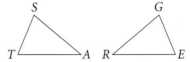

1. $\angle S \cong$ __?__

2. $\overline{GR} \cong$ __?__

3. $\angle E \cong$ __?__

4. $\overline{AT} \cong$ __?__

5. $\triangle ERG \cong$ __?__

6. $\overline{EG} \cong$ __?__

7. $\triangle REG \cong$ __?__

8. $\angle R \cong$ __?__

State whether the figures are congruent. Justify each answer.

9. $\triangle ABF$; $\triangle EDC$

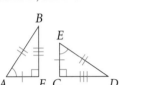

10. $\triangle TUV$; $\triangle UVW$

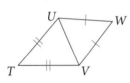

11. $\square XYZV$; $\square UTZV$

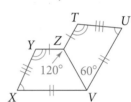

12. $\triangle ABD$; $\triangle EDB$

● **Lessons 4-2 and 4-3**

Where possible, explain how you would use SSS, SAS, ASA, or AAS to prove the triangles congruent. If not possible, write *not possible*.

13.

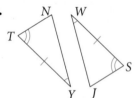

14.

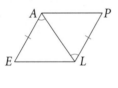

15.

16.

17. Given: $\overline{PX} \cong \overline{PY}$, $\overline{ZP}$ bisects $\overline{XY}$.

   Prove: $\triangle PXZ \cong \triangle PYZ$

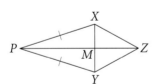

18. Given: $\angle 1 \cong \angle 2$, $\angle 3 \cong \angle 4$, $\overline{PD} \cong \overline{PC}$,
   P is the midpoint of $\overline{AB}$.

   Prove: $\triangle ADP \cong \triangle BCP$

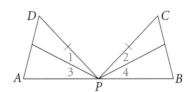

19. Given: $\angle 1 \cong \angle 2$, $\angle 3 \cong \angle 4$, $\overline{AP} \cong \overline{DP}$

   Prove: $\triangle ABP \cong \triangle DCP$

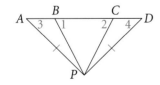

20. Given: $\overline{MP} \parallel \overline{NS}$, $\overline{RS} \parallel \overline{PQ}$, $\overline{MR} \cong \overline{NQ}$

   Prove: $\triangle MQP \cong \triangle NRS$

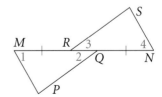

● **Lesson 4-4**

Explain how you would use SSS, SAS, ASA, or HL with CPCTC to prove each statement.

21. $\angle MLN \cong \angle ONL$

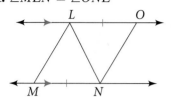

22. $\overline{TO} \cong \overline{ES}$

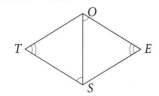

23. $\overline{MB} \cong \overline{RI}$

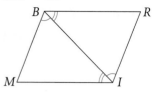

24. Given: $\angle 1 \cong \angle 2, \angle 3 \cong \angle 4$,
　　　　M is the midpoint of $\overline{PR}$
　　Prove: $\triangle PMQ \cong \triangle RMQ$

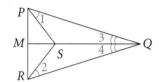

25. Given: $PO = QO, \angle 1 \cong \angle 2$,
　　Prove: $\angle A \cong \angle B$

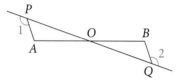

● **Lesson 4-5**

$\boxed{x^2}$ **Algebra** **Find the value of each variable.**

26.

27.

28.

29. Given: $\angle 5 \cong \angle 6, \overline{PX} \cong \overline{PY}$
　　Prove: $\triangle PAB$ is isosceles.

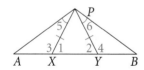

30. Given: $\overline{AP} \cong \overline{BP}, \overline{PC} \cong \overline{PD}$
　　Prove: $\triangle QCD$ is isosceles.

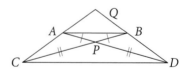

● **Lessons 4-6 and 4-7**

Name a pair of overlapping congruent triangles in each diagram. State whether the triangles are congruent by SSS, SAS, ASA, AAS, or HL.

31.

32.

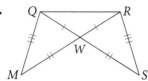

33.

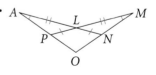

34. Given: M is the midpoint of $\overline{AB}$,
　　　　$\overline{MC} \perp \overline{AC}, \overline{MD} \perp \overline{BD}, \angle 1 \cong \angle 2$
　　Prove: $\triangle ACM \cong \triangle BDM$

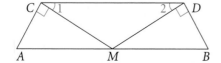

35. Given: $\triangle APQ \cong \triangle BQP$
　　　　$\overline{AP} \perp \overline{PQ}, \overline{BQ} \perp \overline{PQ}$
　　Prove: X is the midpoint of $\overline{AQ}$.

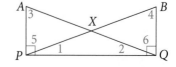

Extra Practice: Skills, Word Problems, and Proof

● **Lesson 5-1**

x^2 **Algebra Find the value of x.**

1.

$7x - 1$

48

2.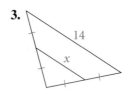

48

$3x$

3.

14

x

4.

5

x

5. A sinkhole caused the sudden collapse of a large section of highway. Highway safety investigators paced out the triangle shown in the figure to help them estimate the distance across the sinkhole. What is the distance across the sinkhole?

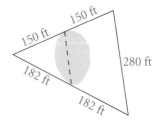

150 ft 150 ft 182 ft 280 ft 182 ft

● **Lessons 5-1 and 5-2**

x^2 **Algebra Use the figure at the right.**

6. Find the value of x.

7. Find the length of $\overline{AD}$.

8. Find the value of y.

9. Find the length of $\overline{EG}$.

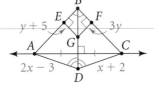

$y + 5$ $3y$

$2x - 3$ $x + 2$

10. **Given:** $\overline{AP} \cong \overline{AQ}, \overline{BP} \cong \overline{BQ},$
    $\overline{CP} \cong \overline{CQ}$
    Prove: A, B, and C are collinear.

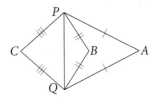

11. **Given:** $\overrightarrow{CX}$ bisects $\angle BCN$.
    $\overrightarrow{BX}$ bisects $\angle CBM$.
    Prove: X is on the bisector of $\angle A$

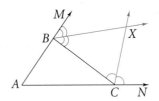

12. Find an equation in slope-intercept form for the perpendicular bisector of the segment with endpoints $H(-3, 2)$ and $K(7, -5)$.

● **Lesson 5-3**

Find the center of the circle that you can circumscribe about $\triangle ABC$.

13. $A(2, 8)$
    $B(0, 8)$
    $C(2, 2)$

14. $A(-3, 6)$
    $B(-3, -2)$
    $C(7, 6)$

15. $A(4, 3)$
    $B(-4, -3)$
    $C(4, -3)$

16. $A(-10, -2)$
    $B(-2, -2)$
    $C(-2, -10)$

Is $\overline{AB}$ an angle bisector, altitude, median, or perpendicular bisector?

17.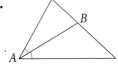

B

A

18.

A

B

19.

A

B

20.

B

A

21. Find the center of the circle that you can circumscribe about the triangle with vertices $A(1, 3)$, $B(5, 8)$, and $C(6, 3)$.

22. Tell which line contains each point for $\triangle ABC$.
   a. the circumcenter
   b. the orthocenter
   c. the centroid
   d. the incenter

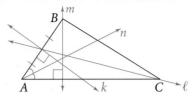

23. Draw an acute triangle and construct its inscribed circle.

● **Lesson 5-4**

Write (a) the inverse and (b) the contrapositive of each statement.

24. If two angles are vertical, then they are congruent.

25. If figures are similar, then their side lengths are proportional.

26. If a car is blue, then it has no doors.

27. If a triangle is scalene, then it is not equiangular.

28. Suppose you know that $\angle A$ is an obtuse angle in $\triangle ABC$. You want to prove that $\angle B$ is an acute angle. What assumption would you make to give an indirect proof?

Write the first step of an indirect proof of each statement.

29. $\triangle ABC$ is a right triangle.

30. Points J, K, and L are collinear.

31. Lines ℓ and m are not parallel.

32. $\square XYZV$ is a square.

● **Lesson 5-5**

List the sides of each triangle in order from shortest to longest.

33.

34.

35.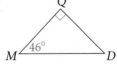

36.

Can a triangle have sides with the given lengths? Explain.

37. 2 in., 3 in., 5 in.   **38.** 9 cm, 11 cm, 15 cm   **39.** 8 ft, 9 ft, 18 ft

40. In $\triangle PQR$, $m\angle P = 55$, $m\angle Q = 82$, and $m\angle R = 43$. List the sides of the triangle in order from shortest to longest.

41. In $\triangle MNS$, $MN = 7$, $NS = 5$, and $MS = 9$. List the angles of the triangle in order from smallest to largest.

42. Two sides of a triangle have side lengths 8 units and 17 units. Describe the lengths x that are possible for the third side.

Extra Practice: Skills, Word Problems, and Proof

● **Lesson 6-1**

Graph the given points. Use slope and the Distance Formula to determine the most precise name for quadrilateral _ABCD_.

1. $A(3, 5), B(6, 5), C(2, 1), D(1, 3)$

2. $A(-1, 1), B(3, -1), C(-1, -3), D(-5, -1)$

3. $A(2, 1), B(5, -1), C(4, -4), D(1, -2)$

4. $A(-4, 5), B(-1, 3), C(-3, 0), D(-6, 2)$

● **Lesson 6-2**

x^2 **Algebra** **Find the values of the variables in each parallelogram.**

5.

6.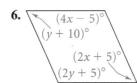

7.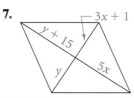

8.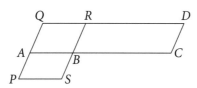

9. Given: $PQRS$ and $QDCA$ are parallelograms.

   Prove: $AP = BS$

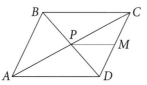

10. Given: $\square ABCD$
   M is the midpoint of $\overline{CD}$.

   Prove: $\overline{PM} \parallel \overline{AD}$

● **Lesson 6-3**

Based on the markings, decide whether each figure must be a parallelogram.

11.

12.

13.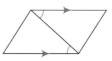

14.

15. Describe how you can use what you know about parallelograms to construct a point halfway between a given pair of parallel lines.

16. Given: $\square ABCD$
   $\overline{BX} \perp \overline{AC}$,
   $\overline{DY} \perp \overline{AC}$

   Prove: $BXDY$ is a parallelogram.

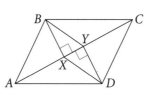

● **Lesson 6-4**

For each parallelogram, determine the most precise name and find the measures of the numbered angles.

17.

18.

19.

20.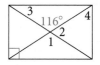

21. Use the information in the figure. Explain how you know that *ABCD* is a rectangle.

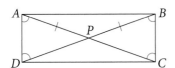

22. □*ABCD* is a rhombus. What is the relationship between ∠1 and ∠2? Explain.

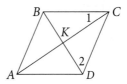

● **Lesson 6-5**

Find *m∠1* and *m∠2*.

23.

24.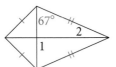

25.

26.

27. Suppose you manipulate the figure so that △*PAB*, △*PBC*, and △*PCD* are congruent isosceles triangles with their vertex angles at point *P*. What kind of figure is *ABCD*? Be sure to consider all the possibilities.

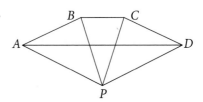

● **Lesson 6-6**

Give coordinates for points *D* and *S* without using any new variables.

28. rectangle

29. parallelogram

30. rhombus

31. square

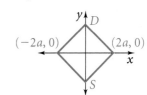

● **Lesson 6-7**

32. For the figure in Exercise 31, use coordinate geometry to prove that the midpoints of the sides of a square determine a square.

33. In the figure, △*PQR* is an isosceles triangle. Points *M* and *N* are the midpoints of $\overline{PQ}$ and $\overline{PR}$, respectively. Give a coordinate proof that the medians of isosceles triangle *PQR* intersect at $H\left(0, \frac{2b}{3}\right)$.

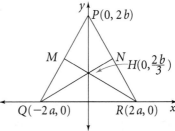

● **Lesson 7-1**

x^2 **Algebra** Solve each proportion.

1. $\frac{2}{3} = \frac{x}{15}$

2. $\frac{4}{9} = \frac{16}{x}$

3. $\frac{x}{4} = \frac{6}{12}$

4. $\frac{2}{x} = \frac{3}{9}$

5. $\frac{3}{4} = \frac{x}{6}$

6. $\frac{3}{7} = \frac{9}{x}$

● **Lesson 7-2**

x^2 **Algebra** The polygons are similar. Find the values of the variables.

7.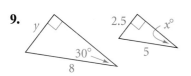

8.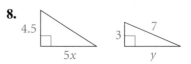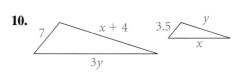

9.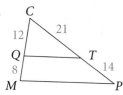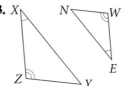

10.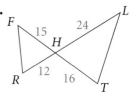

11. Are all equilateral quadrilaterals similar? Make a sketch to support your answer.

● **Lesson 7-3**

Can you prove that the triangles are similar? If so, write a similarity statement and tell whether you would use AA~, SAS~, or SSS~.

12.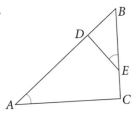

13.

14.

15.

16.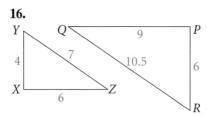

17.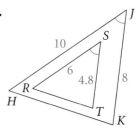

18. Refer to the figure at the right. Explain how you know that $\overline{AB} \parallel \overline{ED}$.

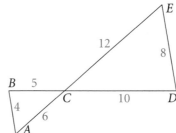

● **Lesson 7-4**

Algebra Find the value of each variable. If an answer is not a whole number, leave it in simplest radical form.

19.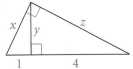

20.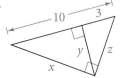

21.

22.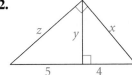

23. Give a coordinate proof of the converse of Corollary 1 to Theorem 7-3. That is, prove that if $\overline{CD}$ is the altitude from C to side $\overline{AB}$ of $\triangle ABC$, and if CD is the geometric mean of AD and DB, then $\triangle ABC$ is a right triangle with its right angle at C.

24. An artist is going to cut four similar right triangles from a rectangular piece of paper like the one shown below. What is the distance from B and D to the diagonal $\overline{AC}$?

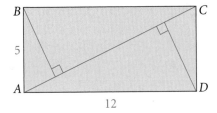

● **Lesson 7-5**

$\boxed{x^2}$ **Algebra** Find the value of x.

25.

26.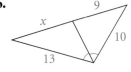

27.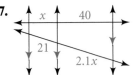

28.

29. Suppose you are given a segment $\overline{AB}$ of length 1 unit and a segment $\overline{CD}$ of length x units. Show how you can apply the Side-Splitter Theorem to construct a segment of length $\frac{1}{x}$.

30. The figure below shows the locations of a high school, a computer store, a library, and a convention center. The street along which the computer store and library are located bisects the obtuse angle formed by two of the other streets. Use the information in the figure to find the distance from the library to the convention center.

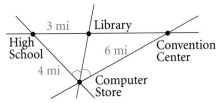

Extra Practice: Skills, Word Problems, and Proof

● **Lessons 8-1 and 8-2**

Find the value of *x*. If your answer is not a whole number, leave it in simplest radical form.

1.

2.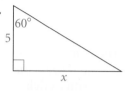

3.

4.

5. A rectangular lot is 165 feet long and 90 feet wide. How many feet of fencing are needed to make a diagonal fence for the lot? Round to the nearest foot.

Find the missing side lengths. Give answers in radical form if necessary.

6.

7.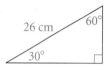

8.

9.

● **Lessons 8-3 and 8-4**

Find the value of *x*. Round lengths of segments to the nearest tenth and angle measures to the nearest degree.

10.

11.

12.

13.

14.

15.

16.

17.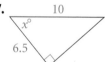

18. An architect includes wheelchair ramps in her plans for the entrance to a new museum. She wants the angle that the ramp makes with level ground to measure 4°. Will the dimensions shown in the figure work? If not, what change should she make?

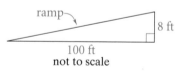

19. A 12-ft ladder is propped against a vertical wall. The top end is 11 ft above the ground. What is the measure of the angle formed by the ladder with the ground?

20. How long is the guy wire shown in the figure if it is attached to the top of a 50-ft antenna and makes a 70° angle with the ground? Round to the nearest tenth.

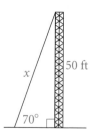

21. A 15-ft ladder is propped against a vertical wall and makes a 72° angle with the ground. How far is the foot of the ladder from the base of the wall? Round to the nearest tenth.

Lesson 8-5

Solve each problem. Round your answers to the nearest foot.

22. A couple is taking a balloon ride. After 25 minutes aloft, they measure the angle of depression from the balloon to its launch place as 16°. They are 180 ft above ground. Find the distance from the balloon to its launch place.

23. A surveyor is 300 ft from the base of an apartment building. The angle of elevation to the top of the building is 24°, and her angle-measuring device is 5 ft above the ground. Find the height of the building.

24. Oriana is flying a kite. She lets out 105 ft of string and anchors it to the ground. She determines that the angle of elevation of the kite is 48°. Find the height the kite is from the ground.

25. Two office buildings are 100 ft apart. From the edge of the shorter building, the angle of elevation to the top of the taller building is 28°, and the angle of depression to the bottom is 42°. How tall is each building? Round to the nearest foot.

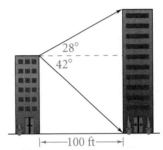

26. A plane flying at 10,000 ft spots a hot air balloon in the distance. The balloon is 9000 ft above ground. The angle of depression from the plane to the balloon is 30°. Find the distance from the plane to the balloon.

Lesson 8-6

(a) Describe each vector as an ordered pair. Give the coordinates to the nearest unit. (b) Write the resultant of each pair of vectors as an ordered pair.

27.

28.

29.

30.

Write the sum of the two vectors as an ordered pair.

31. $\langle 5, 9 \rangle$ and $\langle -3, 2 \rangle$

32. $\langle -1, 0 \rangle$ and $\langle 4, -6 \rangle$

33. $\langle 2, 4 \rangle$ and $\langle 0, 9 \rangle$

34. $\langle 4, -2 \rangle$ and $\langle -4, 2 \rangle$

35. A helicopter lands 55 km west and 14 km north of the airport from which it departed. It followed a straight flight path. Find the magnitude and direction of the resultant vector $\langle -55, 14 \rangle$.

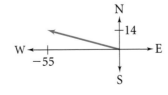

● **Lesson 9-1**

In Exercises 1–6, refer to the figure at the right.

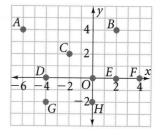

1. What is the image of C under $(x, y) \rightarrow (x + 4, y - 2)$?

2. What rule describes the translation $F \rightarrow B$?

3. What is the image of H under $(x, y) \rightarrow (x - 2, y + 4)$?

4. What rule describes the translation $D \rightarrow H$?

5. What is the image of C under $(x, y) \rightarrow (x - 2, y - 4)$?

6. What rule describes the translation $B \rightarrow A$?

Find the image of each figure under the given translation.

7. $\triangle ABC$ with vertices $A(-3, 4)$, $B(-1, -2)$, $C(1, 5)$; translation: $(x, y) \rightarrow (x - 2, y + 5)$

8. $\triangle EFG$ with vertices $E(0, 3)$, $F(6, -1)$, $G(4, 2)$; translation: $(x, y) \rightarrow (x + 1, y - 3)$

9. $\triangle PQR$ with vertices $P(-9, -4)$, $Q(-5, 1)$, $R(2, 8)$; translation: $(x, y) \rightarrow (x - 6, y - 7)$

10. Write two translation rules of the form $(x, y) \rightarrow (x + a, y + b)$ that map the line $y = x - 1$ to the line $y = x + 3$.

● **Lesson 9-2**

Given points $S(6, 1)$, $U(2, 5)$, and $B(-1, 2)$, draw $\triangle SUB$ and its reflection image across each line.

11. $y = 5$

12. $x = 7$

13. $y = -1$

14. the x-axis

15. $y = x$

16. $x = -1$

17. $y = 3$

18. the y-axis

19. What are the two shortest words in the English language that you can write with capital letters so that each word looks like its own reflection across a line?

20. The segments $\overline{AB}$ and $\overline{A'B'}$ are two different segments in the same plane. There is a translation such that $\overline{A'B'}$ is the translation image of $\overline{AB}$. There is also a line k in the plane such that $\overline{A'B'}$ is the reflection image of $\overline{AB}$ across line k. If $\overline{AB}$ and $\overline{A'B'}$ are opposite sides of a quadrilateral, what kind of quadrilateral is it?

● **Lesson 9-3**

Copy each figure and point P. Draw the image of each figure for the given rotation about P. Label the vertices of the image.

21. 90°

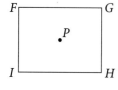

22. 60°

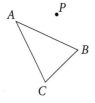

23. 45°

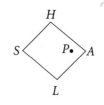

24. 180°

25. The right triangle ABC shown here has side lengths 3, 4, and 5. Point P is the incenter of the triangle. Copy the triangle and draw the image of the triangle for a 60° counterclockwise rotation about P.

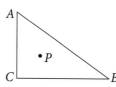

26. What is the smallest angle of rotation you can use to have the rotation image of the figure below exactly overlap the original figure?

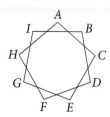

● **Lesson 9-4**

State what kind of symmetry each figure has.

27.

28.

29.

30.

31. Armando is going to draw a triangle that he will put on his backpack.
  a. If the triangle has a line of symmetry, what kind of triangle must it be?
  b. If the triangle has two lines of symmetry, what kind of triangle must it be?

● **Lessons 9-5 and 9-6**

The blue figure is the image of the gray figure. State whether the mapping is a reflection, rotation, translation, glide reflection, or dilation.

32.
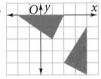

33.
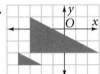

34.

35.
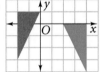

36. The vertices of trapezoid $ABCD$ are $A(-1, -1)$, $B(-1, 1)$, $C(2, 2)$, and $D(2, -1)$. Draw the trapezoid and its dilation image for a dilation with center $(0, 0)$ and scale factor 3.

37. Suppose you know the coordinates of the vertices of a polygon. Describe how you can use what you know about translations and dilations with respect to the origin to find the coordinates of the vertices of the image polygon if the center for the dilation is $(2, 5)$ and the scale factor is 3.

38. Find the image of the polygon for a reflection across line ℓ followed by a reflection across line m. Then use a separate diagram to repeat the process, but reflect across line m first and then across line ℓ. Each time, draw the intermediate image with dashed segments.

● **Lesson 9-7**

39. Which of the four figures in Exercises 27–30 will tessellate a plane?

40. Use a square and an equilateral triangle to make a tessellation. The square and equilateral triangle should have congruent sides.

● **Lesson 10-1**

If possible, find the perimeter and area of each figure. If not possible, state why.

1.

2.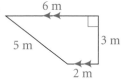

3.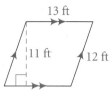

4.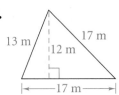

5.

6.

● **Lessons 10-2 and 10-3**

Find the area of each trapezoid or regular polygon. Leave your answer in simplest radical form.

7.

8.

9.

10.

11. The patio section of a restaurant is a trapezoid with the dimensions shown in the figure. What is the area of the patio section?

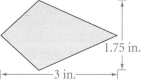

12. A mosaic design uses kite-shaped tiles with the dimensions shown in the figure. What is the area of each tile?

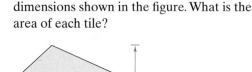

13. The tiles for a bathroom floor are regular hexagons that are $\frac{5}{8}$ in. on each side. Find the area of an individual tile. Express the answer in radical form.

14. The floor of a gazebo is a regular hexagon with sides that are 9 ft long. What is the area of the floor? Round to the nearest square foot.

● **Lesson 10-4**

Find the ratio of the perimeters and the ratio of the areas of the blue figure to the red figure.

15.

16.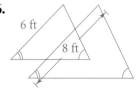

17.

18. A triangular banner has an area of 315 in.². A similar banner has sides $1\frac{1}{3}$ times as long as those of the smaller banner. What is the area of the larger banner?

19. You want to enlarge the picture on the front of a postcard by 10%. If the perimeter of the postcard is 44 cm, what will be the perimeter of the enlargement?

● Lesson 10-5

Find the area of each polygon. Round your answers to the nearest tenth.

20.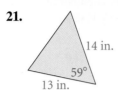

21.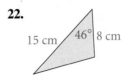

22.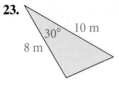

23.

24. a regular hexagon with an apothem of 3 ft

25. a regular octagon with radius 5 ft

● Lesson 10-6

(a) Find the circumference of each circle. (b) Find the length of the arc shown in red. Leave your answers in terms of π.

26.

27.

28.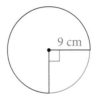

29.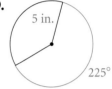

30. A bicycle wheel has a radius of 0.33 m. How many revolutions does the wheel make when the bicycle is ridden 1 km? Round to the nearest whole number.

● Lesson 10-7

Find the area of each shaded sector or segment. Leave your answers in terms of π.

31.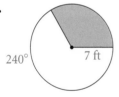

32.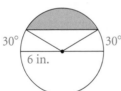

33.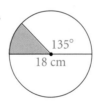

34.

35. A 14-in. diameter pizza is cut into 6 equal slices. About how many square inches of pizza are in each slice? Round to the nearest square inch.

● Lesson 10-8

Darts are thrown at random at each of the boards shown. If a dart hits the board, find the probability that it will land in the shaded area.

36.

37.

38.

39.

40. A square garden that is 80 ft on each side is surrounded by a cobblestone street that is 8 ft wide. If a child's balloon lands at random in the region formed by the garden and street, what is the probability that it lands on the street?

41. A dart hits the circular board shown in the figure at a random point. What is the probability that it does not hit the shaded square? Express your answer in terms of π.

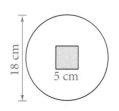

● **Lesson 11-1**

The diagrams in Exercises 1–4 each show a cube after part of it has been cut away. Identify the shape of the cross section formed by the cut. Also, verify Euler's Formula, $F + V = E + 2$, for the polyhedron that remains.

1.   **2.**   **3.**   **4.**

5. The bases of the prism shown at the right are equilateral triangles. Make a sketch that shows how you can have a plane intersect the prism to give a cross section that is an isosceles trapezoid.

● **Lessons 11-2 and 11-3**

Find the (a) lateral area and (b) surface area of each figure. Leave your answers in terms of π or in simplest radical form.

6.   **7.**   **8.**   **9.**

10. An optical instrument contains a triangular glass prism with the dimensions shown at the right. Find the lateral area and surface area of the prism. Round to the nearest tenth.

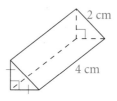

11. A company packages salt in a cylindrical box that has a diameter of 8 cm and a height of 13.5 cm. Find the lateral area and surface area of the box. Round to the nearest tenth.

Find the (a) lateral area and (b) surface area of each pyramid or cone. Assume that the base of each pyramid is a regular polygon. Round your answers to the nearest tenth.

12.   **13.** 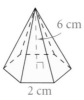 **14.**   **15.**

● **Lessons 11-4 and 11-5**

Find the volume of each figure. Round your answers to the nearest tenth.

16. 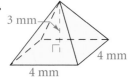 **17.** 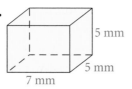 **18.**   **19.**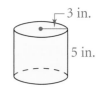

20.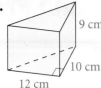

21.

22.

23.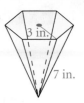

24. A greenhouse has the dimensions shown in the figure. What is the volume of the greenhouse? Round to the nearest cubic foot.

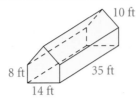

25. Find the volume of a can of chicken broth that has a diameter of 7.5 cm and a height of 11 cm. Round to the nearest tenth.

26. A paper drinking cup is a cone that has a diameter of $2\frac{1}{2}$ in. and a height of $3\frac{1}{2}$ in. How many cubic inches of water does the cup hold when it is full to the brim? Round to the nearest tenth.

● **Lesson 11-6**

Find the volume and surface area of a sphere with the given radius or diameter. Give each answer in terms of π and rounded to the nearest whole number.

27. $r = 5$ cm **28.** $r = 3$ ft **29.** $d = 8$ in.

30. $d = 2$ ft **31.** $r = 0.5$ in. **32.** $d = 9$ m

The surface area of each sphere is given. Find the volume of each sphere in terms of π.

33. 64π m^2 **34.** 16π in.^{2} **35.** 49π ft^2

36. A spherical beach ball has a diameter of 1.75 ft when it is full of air. What is the surface area of the beach ball, and how many cubic feet of air does it contain? Round to the nearest hundredth.

● **Lesson 11-7**

Copy and complete the table for three similar solids.

	Similarity Ratio	Ratio of Surface Areas	Ratio of Volumes
37.	2 : 3	▦ : ▦	▦ : ▦
38.	▦ : ▦	25 : 64	▦ : ▦
39.	▦ : ▦	▦ : ▦	27 : 64

40. How do the surface area and volume of a cylinder change if the radius and height are multiplied by $\frac{5}{4}$?

41. For two similar solids, how are the ratios of their volumes and surface areas related?

● **Lesson 12-1**

x^2 **Algebra** Assume that lines that appear to be tangent are tangent. *P* is the center of each circle. Find the value of *x*.

1.
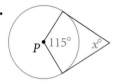

2.
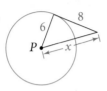

3.

4.
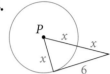

5. Given: Quadrilateral *ABCD* is circumscribed about ⊙*O*.

   Prove: $AB + DC = BC + AD$

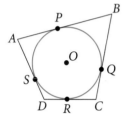

● **Lessons 12-2 and 12-3**

x^2 **Algebra** Find the value of each variable. If your answer is not a whole number, round it to the nearest tenth.

6.

7.

8.

9.

10.

11.
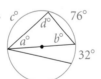

12.

13.
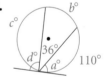

14. A polygon is inscribed in a circle. Are the perpendicular bisectors of the sides of the polygon concurrent? Explain.

15. A circle has a diameter of 4 units. A chord parallel to a diameter is 1.5 units from the center of the circle. The endpoints of the diameter and the chord are the vertices of an isosceles trapezoid. What is the distance from the center of the circle to each leg of the trapezoid? Round to the nearest hundredth.

16. Given: ∠*A* and ∠*D* are inscribed angles in ⊙*O* that
   intercept $\overparen{BC}$. $\overline{BD}$ and $\overline{AC}$ intersect at *P*.

   Prove: △*APB* ~ △*DPC*

● **Lesson 12-4**

x^2 **Algebra** Assume that lines that appear to be tangent are tangent. Find the value of each variable. If your answer is not a whole number, round it to the nearest tenth.

17.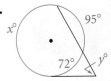

18.

19.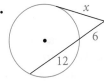

20.

21.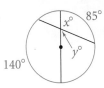

22.

23.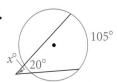

24.

25. The outer rim of a circular garden will be planted with three colors of tulips. The landscaper has stretched two strings from a point P to help workers see how much of the circular rim should be planted with each color. Use the information in the figure at the right to find x and y.

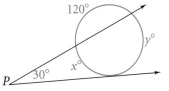

26. Planks are placed across the circular pool shown in the figure at the right. What is the length of the longer plank?

● **Lesson 12-5**

Write the standard equation for each circle with center P.

27. $P = (0,0); r = 4$

28. $P = (0,5); r = 3$

29. $P = (9,-3); r = 7$

30. $P = (-4,0)$; through $(2,1)$

31. $P = (-6,-2)$; through $(-8,1)$

32. $P = (-1,-3); r = 3$

33. When a coordinate grid is imposed over a map, the location of a radio station is given by $(113,215)$. A town located at $(149,138)$ is at the outermost edge of the circular region where clear reception is assured.
   a. Write an equation that describes the boundary of the clear reception region.
   b. If the radio station boosts power to increase the size of the clear-reception region by a factor of 4, what will be the equation for the new boundary for clear reception?

● **Lesson 12-6**

Draw and describe each locus.

34. all points in a plane 3 cm from a circle with $r = 2$ cm

35. all points in a plane 2 cm from $\overrightarrow{AB}$

36. all points in space 1.5 in. from a point Q

37. A dog is on a 20-ft leash. The leash is attached to a pipe at the midpoint of the back wall of a 30 ft-by-30 ft house, as shown in the diagram. Sketch and use shading to indicate the region in which the dog can play while attached to the leash. Include measurements to describe the region.

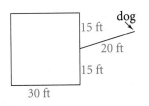

Using TI Technology for Selected Examples

This section includes calculator activities using the TI-83/84 Plus, TI-83/84 Plus Silver Edition, and TI-Nspire. All activities are tied to specific examples found in lessons throughout this course. These activities provide an alternate way to complete these examples using technology.

On pages 741 and 742, you will find two calculator activities that can be used as reference pages when using the GeoMaster application.

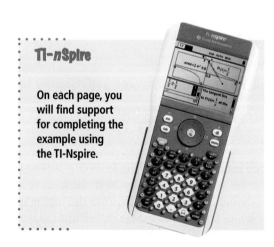

TI-*n*Spire

On each page, you will find support for completing the example using the TI-Nspire.

Drawing a Perpendicular Line

> **EXAMPLE** Graphing on the Coordinate Plane

Draw a line passing through (2, 3) and (22, 10).

The GeoMaster menu screen does not always appear at the bottom of the drawing screen. Press GRAPH to display the menu list. To turn the menu list off, press CLEAR . Press 2nd FORMAT and select AxesOn to graph with axes.

Press WINDOW to access the Draw menu. Select the Line tool to draw the line. Move the cursor to the point (2, 3) and press ENTER . Then move the cursor to (22, 10) and press ENTER to draw the line through these two points.

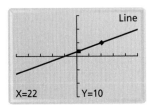

To change the coordinates of a point, move the cursor to the point and press ENTER . Input the new *x*-value and press ENTER . Repeat for the new *y*-value.

Calculate the slope of the line.

Access the Measurement menu by pressing ZOOM . Select the Slope tool. Move the cursor to a point on the line and ENTER to calculate its slope.

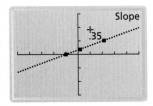

The slope is displayed near the cursor. Move the measurement to a clear section of the screen. Press ENTER to set it in on the screen.

Draw a perpendicular line going through the point (14, 2).

In the DRAW menu select the Perpendicular tool to draw a line perpendicular to the line on the screen. Select any point on the line and then move the cursor to the point (14, 2).

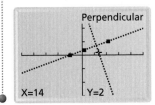

The lines are dashed because they have not been set. Press ENTER to set lines on the screen.

EXERCISES

Draw line *m* through (4, 23) and (6, 10) and complete each of the following.

1. Draw line *n* through (10, 10) and perpendicular to line *m*

2. Calculate the slope of both lines.

3. Draw line *l* parallel to line *m*.

Finding Perimeter and Area

● EXAMPLE Measuring on the Coordinate Plane

Draw a segment from (−6, 18) to (19, −9) and measure its length.

Press 2nd FORMAT . Select AxesOff. The coordinates of the cursor still appear.

In the DRAW menu select the Segment tool. Move the cursor to (−6, 18) and press ENTER . Move the cursor to (19, −9) and press ENTER . Select each endpoint.

Open the Measurement menu and select the Distance/Length tool. Select any point on the segment and press ENTER . Press ENTER to set the measurement on the screen.

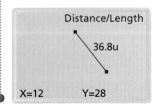

The length is measured in units, abbreviated as *u*.

Polygons can be drawn using the DRAW menu. Open the FILE menu and create a new file. To change the window settings, press CLEAR ZOOM and select ZDecimal.

● EXAMPLE Measuring Perimeter and Area

Draw a triangle with vertices (−1, −2), (3, 3) and (0, 3).

In the DRAW menu select the Triangle tool. Press ENTER at each coordinate, (−1, −2), (3, 3) and (0, 3), to draw the triangle.

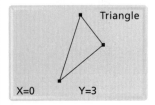

When using a tool in GeoMaster the name of the tool will appear in the top right corner. To exit out of a tool, press CLEAR .

Measure the perimeter and area of the triangle.

Select the Distance/Length tool to measure the perimeter. Press ENTER on any point on a side. Press ENTER again to place the measurement on the screen.

Now select the Area tool to measure the area of the triangle. Press ENTER on any point on a side. Press ENTER again to place the measurement.

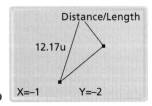
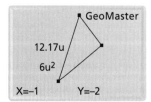

EXERCISES

1. Draw the triangle with vertices (−6, 8), (12, 2), and (14, −10)

2. Find the perimeter of the triangle.

3. Find the area of the triangle.

Using Measurement Tool to find Perimeter

See Lesson 1-9, Example 3

EXAMPLE **Finding Perimeter in the Coordinate Plane**

Find the perimeter of the triangle with vertices $A(-1, -2)$, $B(5, -2)$, and $C(5, 6)$.

Press **APPS** and select the GeoMaster application. Press **CLEAR** **WINDOW** to change the window settings so that x goes from -9.4 to 9.4 and y goes from -4.8 to 7.6.

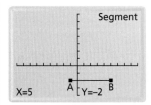

To display the axis, press **2nd** **ZOOM** and select AxesOn. Press **GRAPH** to return to the main screen.

Open the DRAW menu and select the Segment tool. Draw a segment with endpoints $(-1, -2)$ and $(5, -2)$.

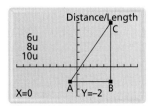

To label the endpoints A and B, press **ALPHA** followed by the letter after a point is set.

Draw $\overline{BC}$ and $\overline{AC}$ to complete the triangle.

Open the Measurement (MEAS) menu and select the Distance/Length tool to calculate the length of each side. Move the cursor to $\overline{AB}$ and press **ENTER**. Move the measurement to the left side of the screen and press **ENTER**. Repeat the steps to find the lengths of $\overline{BC}$ and $\overline{AC}$.

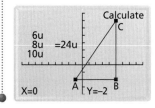

The u after each measurement represents units.

In the Measurement menu, select the Calculate tool. To add the side lengths, use the cursor to select a measurement then press ➕.

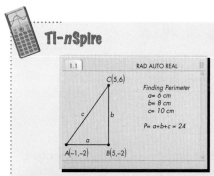

Press **STO▶** to display the perimeter of the triangle.

The Text tool allows you to write a general formula. You can use the Calculate tool to find the formula with the values on the screen.

EXERCISE

1. Find the perimeter of *KLMN* with vertices K $(-3, -3)$, L $(1, -3)$, M $(1, 4)$, and N $(-3, 1)$.

Drawing Perpendicular Lines

See Lesson 3-7, Example 5

EXAMPLE Writing Equations for Perpendicular Lines

Write an equation for the line through $(-3, 7)$ and perpendicular to $y = -3x - 5$.

Press **APPS** and select the GeoMaster application. Press **CLEAR** **WINDOW** and change the window settings so that x goes from -14.1 to 14.1 and y goes from -9.3 to 9.3.

Press **CLEAR** **Y=** to view the equation editor screen. Enter the equation $y = -3x - 5$ into Y1. Then press **GRAPH** to display the line.

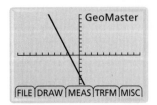

The default for GeoMaster is to not display the axis. To show them press **2nd** **ZOOM**, highlight AxesOn and press **ENTER**.

Next, draw a line segment of any length over the line $y = -3x - 5$. Start drawing the segment from the y-intercept.

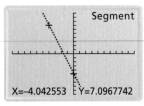

The segment tool is located in the DRAW menu.

Draw a perpendicular line to the segment by selecting any point on the line segment and press enter. The perpendicular tool is in the DRAW menu.

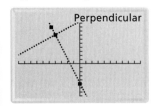

When the line appears, drag it as close to $(-3, 7)$ as possible.

Open the Measure (MEAS) menu and select the Eqn/Coords tool to display the equation of the perpendicular line. Select the line drawn through $(-3, 7)$. GeoMaster will display the equation. Move it to a location on the screen and press **ENTER** to set it.

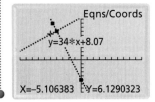

Exact values will not be shown.

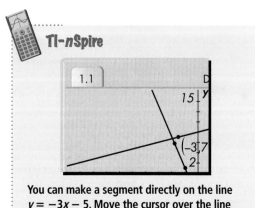

TI-nSpire

You can make a segment directly on the line $y = -3x - 5$. Move the cursor over the line and place the two points.

EXERCISES

Write the equations of the lines described.

1. Write the equation for the line through $(15, -4)$ perpendicular to $5y - x = 10$.

2. Write the equation of the line parallel to $y = \frac{1}{2}x + 3$ and passing through $(3, -4)$.

Graphing a System of Equations

See Lesson 4-6, Algebra 1 Review Example 2

EXAMPLE Solving Systems of Equations

Solve the system: $x + y = 3$
$\qquad\qquad\qquad 4x + 4y = 8$

The Polynomial Root Finder and Simultaneous Equation Solver will solve systems of equations. First, make sure the equations are in $Ax + By = C$ form.

Access PlySmlt2 from the **APPS** menu. Select **2: SIMULT EQN SOLVER.** Enter the number of equations in the system and the number of unknowns.

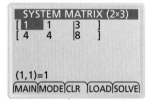

The Simultaneous Equation Solver is set to solve 2 equations and 2 unknowns.

Select **NEXT**. The calculator displays a 2×3 augmented matrix. Each row in the matrix represents one of the equations. Enter the coefficients of the variables and the constants into the system matrix. Enter each number and use the arrow keys to move.

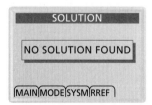

Remember, when there is no number in front of a variable, the coefficient is 1.

To solve the system, press select SOLVE.

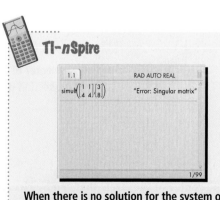

The solution to the system is: NO SOLUTION FOUND.
The lines represented by the equations are parallel.

The solution to the system is the point at which the equations intersect. If they do not intersect, they are parallel.

EXERCISES

Solve each system using the PlySmlt2 application.

1. $3x + 2y = -5$
$\quad -5x - 8y = -1$
$\quad x - 2y = 5$

2. $x - 2y = 5$
$\quad 2y = x + 5$

3. $-3x + 4y = -4$
$\quad x = -3y + 10$

TI-*n*Spire

When there is no solution for the system of equations, the display reads "Error: Singular matrix".

Using the Measurement Tool

See Lesson 5-3, Example 1

EXAMPLE Finding the Circumcenter

Find the center of the circle that you can circumscribe about $\triangle OPS$.

Press **APPS** and select the GeoMaster application. Press **CLEAR** **WINDOW** to change the settings so that x goes from -4 to 14.8 and y goes from -4 to 8.4.

Draw $\triangle OPS$ with $O(0, 0)$, $P(0, 6)$, and $S(4, 0)$. In the DRAW menu select the Triangle tool. Press **ENTER** on each of the three coordinates to set the vertices.

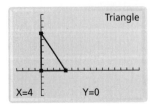

To display the *x*- and *y*-axis, press **2nd** **ZOOM** and select AxesOn.

Draw the perpendicular bisectors of each side of the triangle. Select the Perp Bisector tool from the DRAW menu. Select the bottom side and the left side of the triangle.

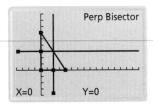

After using a tool, press **GRAPH** to display the menu options.

Use the Intersection tool in the DRAW menu. Press **ENTER** on each Perpendicular Bisector.

Use the Eqns/Coords tool from the Measurement (MEAS) menu to find the coordinate of the intersection point. Place the cursor at the point and press **ENTER**. Choose the Point option in the popup menu.

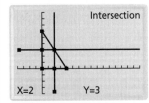
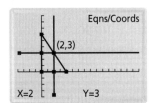

Use the Circle tool in the DRAW menu to circumscribe the triangle. Select the intersection point and then one of the vertices of the triangle.

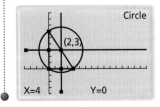

The center of the circle is the intersection point (2, 3).

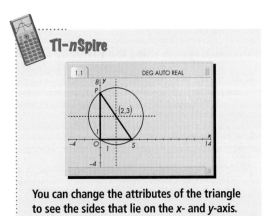

You can change the attributes of the triangle to see the sides that lie on the *x*- and *y*-axis.

EXERCISES

Find the center of the circle that can be circumscribed about each triangle.

1. $\triangle ABC$ with $A(3, 2)$, $B(-4, 5)$ and $C(-6, -4)$.

2. $\triangle XYZ$ with $X(10, 0)$, $Y(7, 8)$, and $Z(-3, 1)$

Drawing Polygons

See Lesson 6-1, Example 2

EXAMPLE Classifying by Coordinate Methods

Determine the most precise name for quadrilateral *LMNP* with vertices *L*(1, 2), *M*(3, 3), *N*(5, 2), and *P*(3, 1).

Press APPS and select the GeoMaster application. Press CLEAR WINDOW to change the window settings so that *x* goes from −1.4 to 8 and *y* goes from −1.4 to 4.8.

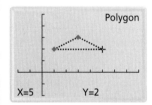

Press 2nd ZOOM and select AxesOn to display the axes on the graph.

Open the DRAW menu and select the Polygon tool to draw the quadrilateral. Move the cursor to (1, 2) to set the first point then press ENTER . Repeat for the other three vertices.

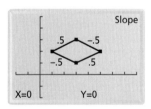

The sides of the polygon that have not yet been set will appear as dashed lines.

Use the Slope tool in the Measurement menu to calculate the slope of each side. Select one side of the polygon and then press ENTER to set the measurement.

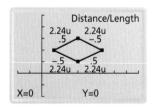

Both pairs of opposite sides are parallel, so *LMNP* is a parallelogram. No sides are perpendicular so *LMNP* is not a rectangle.

In the Measurement menu select the Distance/Length tool to find the length of each side of the quadrilateral. Select both endpoints of the side and then press ENTER to set the measurement.

When selecting an endpoint, if a dialog box pops up, press ENTER to select the point not the polygon.

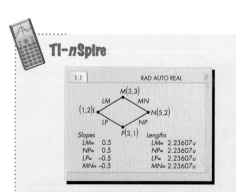

• All sides are congruent, so *LMNP* is a rhombus.

When the cursor hovers over a measurement, you can use the plus and minus keys to change the number of digits displayed.

EXERCISES

Determine the most precise name for the quadrilateral.

1. *ABCD* with vertices *A*(−3,3), *B*(2,4), *C*(3,1), and *D*(−2, −2).

2. *EFGH* with vertices *E*(0,3), *F*(5,4), *G*(4,−1), and *H*(−3, −4).

Find the Zeros of a Function

See Lesson 7-2, Algebra 1 Review Example 2

EXAMPLE Solving Quadratic Equations

Solve for x: $-3x^2 - 5x + 1 = 0$.

Graph the equation in Y= to estimate the solution(s) graphically. Make sure the equation is in standard form before entering it in the calculator. After entering the equation, press GRAPH. Change the window settings to zoom in on the graph.

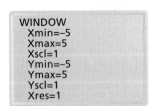

The solutions of the equation were located around the origin. The settings of −5 to 5 will show a better estimate of the solutions.

Press GRAPH to estimate the solutions.

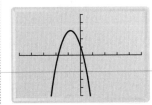

The solution to the left of zero is near −2. A closer approximation is not possible. The solution to the right of zero is a small decimal greater than zero.

Press 2nd TRACE and select **zero** to calculate the exact values for the solutions. The calculator will ask for the left bound value. Move the cursor to the left of the leftmost solution and press ENTER. Move the cursor to the right of the same solution to set the right bound value. Press ENTER. Press ENTER once more to find the calculated solution.

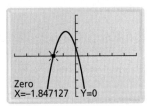

The calculator allows you to make a guess at the solution before giving the true solution.

Repeat the steps to find the second solution to the equation.

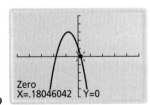

Notice that $y = 0$ since the solution represents an x-intercept.

EXERCISES

Solve each quadratic equation.

1. $y = -2x^2 + 4x - 1$

2. $0.8x^2 + 5x = 2$

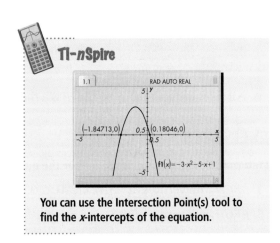

You can use the Intersection Point(s) tool to find the x-intercepts of the equation.

Measuring Angles

See Lesson 8-4, Example 3

EXAMPLE **Using the Inverse of Cosine and Sine**

Find $m\angle L$ to the nearest degree.

Use the GeoMaster™ Application. Press **APPS** and select GeoMastr. Press **CALC** **WINDOW** and change the window settings to show the first quadrant.

```
WINDOW
 Xmin=-1
 Xmax=5
 Xscl=1
 Ymin=-1
 Ymax=5
 Yscl=1
 Xres=1
```

Because the sides of the triangle have positive lengths, changing the graph to the first quadrant allows us to use the origin and only positive values.

Use the lengths to determine the coordinates of the vertices of the triangle. Select the Point tool from the DRAW menu to set F at $(0,0)$, O at $(3.2, 0)$ and L at $(0, 2.4)$. Use the triangle tool to connect the 3 points.

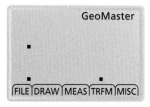
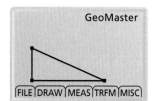

Select the Angle tool from the MEAS menu to measure the angle. Move the cursor to an endpoint of the angle. Press **ENTER**. Select the point representing the angle to be measured. Finally select the other endpoint of the angle and press **ENTER**.

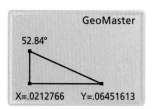

After pressing ENTER at the second endpoint of the angle the angle measurement is displayed. Press ENTER to set the measurement on the screen.

Confirm $m\angle L$ by using the side lengths and a trigonometric ratio. Since all three sides of the triangle are known, any of the trigonometric ratios will work. $\cos L$ represents the ratio of the side adjacent $\angle L$ to the hypotenuse of the triangle.

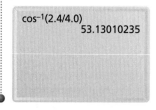

The inverse function is used to find an angle measurement. Be sure the calculator is in Degree mode.

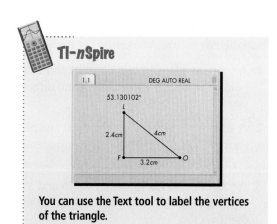

TI-*n*Spire

You can use the Text tool to label the vertices of the triangle.

EXERCISES

Find the measure of the indicated angle.

1. Find $m\angle C$ in $\triangle ABC$ with $A(2, 2)$ $B(7, 2)$ and $C(2, 9)$.

2. Find $m\angle X$ in $\triangle XYZ$ with $X(-3, -4)$, $Y(-3, -10)$ and $Z(0, -4)$.

Describing Vectors

See Lesson 8-6, Example 3

EXAMPLE · Real-World Connection

An airplane lands 40 km west and 25 km south from where it took off. The result of the trip can be described by the vector (−40, −25). Use the distance (for magnitude) and direction to describe this vector a second way.

Press APPS and select the SciTools application. Choose Vector Calculator. Then press Y= to access the *X/Y* entry tool.

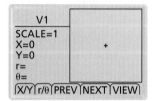

At the opening screen the distance of the vector *r*, and the angle of direction θ are blank.

Enter values for *x* and *y*. Since the plane lands west of its starting point, the *x*-value is −40. Since the plane lands south of its original starting point, *y* is −25.

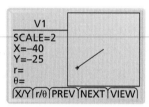

The scale has changed to 2 to allow the vector to fit the screen.

Press WINDOW to use the r/θ tool which calculates the distance of the vector and the measure of the angle, the direction.

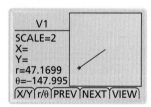

The angle is measured from the *x*-axis. Clockwise angles are considered negative angles. The measure of the plane south of west would be 180 − 148.

● The airplane is 47 miles from its starting point, the direction is 32° south of west.

EXERCISES

Find the distance and direction for each situation.

1. An airplane lands 312 miles east and 92 north of its starting point. Describe the magnitude and direction of its flight vector.

2. A small plane lands 42 miles west and 58 miles south of its starting point. Describe the magnitude and direction of its flight vector.

3. An airplane lands 135 miles south and 195 miles east of its starting point. Describe the magnitude and direction of its flight vector.

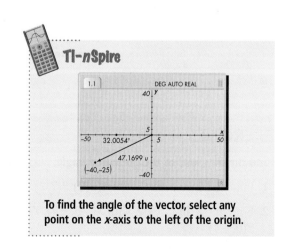

To find the angle of the vector, select any point on the *x*-axis to the left of the origin.

Finding Vector Sums

See Lesson 8-6, Example 4

> **EXAMPLE**   Adding Vectors

Find the sum of vectors $\vec{a}\langle 4, 3 \rangle$ and $\vec{c}\langle -1, 2 \rangle$ and draw the resultant vector.

Use the GeoMaster application to draw each vector. Press **APPS** and select GeoMastr. Press **CLEAR** **WINDOW** to change the window settings so that x goes from -1 to 8.4, y goes from -1 to 5.2 and the scale for x and y is 1.

Open the DRAW menu and select the Vector tool. Draw $\vec{a}$ from the origin to $(4, 3)$.

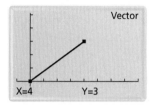

Draw $\vec{c}$ from the end of $\vec{a}$.

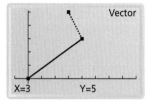

Vector $\vec{c}$ will end 1 unit to the left and 2 units up from $\vec{a}$.

In the Measurement (MEAS) menu select the Eqns/Coords tool. Select the endpoint of $\vec{c}$. Set the coordinate on the screen by pressing **ENTER**.

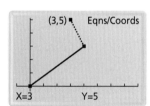

The coordinates of the endpoint of $\vec{c}$ is the sum of the two vectors.

Select the Vector Sum tool from the Measurement menu. Select each vector and press **ENTER**. Move the resultant vector to the origin and press **ENTER**.

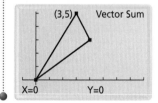

The resultant vector shows the path that can be completed in one vector.

EXERCISES

Find the measure of the indicated angle.

1. Sketch the resultant vector of $\vec{a}\langle -2, 5 \rangle$ and $\vec{c}\langle 4, -3 \rangle$.

2. Sketch the resultant vector of $\vec{a}\langle 7, -5 \rangle$ and $\vec{c}\langle -3, 0 \rangle$.

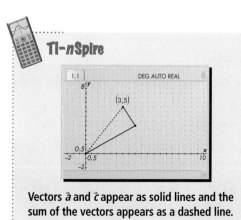

TI-*n*Spire

Vectors $\vec{a}$ and $\vec{c}$ appear as solid lines and the sum of the vectors appears as a dashed line.

Translating Polygons

See Lesson 9-1, Example 3

EXAMPLE Finding a Translation Image

Find the image of △XYZ under the translation $(x, y) \rightarrow (x - 2), (y - 5)$**. Triangle XYZ has vertices** $X(2, 1)$**,** $Y(3, 3)$**, and** $Z(21, 3)$**.**

Press APPS and select the GeoMaster application. Press CLEAR WINDOW and change the window setting so that x goes from -9.4 to 9.4 and y goes from -6.2 to 6.2.

Use the triangle tool in the DRAW menu to draw △XYZ.

Select the Vector tool from the DRAW menu. Select one vertex of the triangle; press ENTER. Move the cursor according to the translation, 2 units to the left and 5 units down.

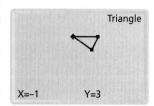

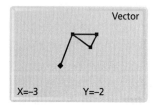

Select the Translation tool from the Transform(TRFM) menu to translate the triangle along the vector. Select the triangle by placing your cursor on one of its sides and pressing ENTER. Select the vector.

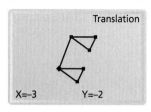

The triangle will appear as a dashed triangle when it is selected. Press ENTER to set the translated triangle on the screen.

Open the Measurement (MEAS) menu and select the Eqns/Coords tool to find the coordinates of the vertices of the translated triangle. Select each vertex and press ENTER to set the coordinate on the screen.

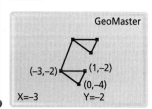

A popup will appear when the right-hand coordinate is selected. Select Point 5 to find the coordinate of the vertex of the triangle.

EXERCISES

Find the image of each triangle for the given translation.

1. △ABC with $A(2, 2)$ $B(9, -3)$ and $C(-2, -6)$, for the translation $(x, y) \rightarrow (x + 3, y - 2)$

2. △XYZ with $X(-6, 4)$, $Y(-3, -10)$ and $Z(0, 7)$, for the translation $(x, y) \rightarrow (x - 3, y + 5)$

3. △DEF with $D(-3, 2)$, $E(3, 2)$ and $F(4, -1)$, for the translation $(x, y) \rightarrow (x - 1, y + 4)$.

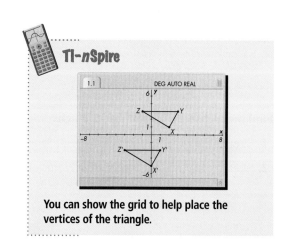

TI-nSpire

You can show the grid to help place the vertices of the triangle.

Measuring Area

See Lesson 10-1, Example 1

EXAMPLE Finding the Area of a Parallelogram

Find the area of the parallelogram.

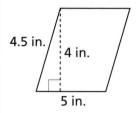

Press **APPS** and select the GeoMaster application. Press **CLEAR** **ZOOM** and choose ZDecimal to change the window settings.

```
WINDOW
  Xmin=-4.7
  Xmax=4.7
  Xscl=1
  Ymin=-3.1
  Ymax=3.1
  Yscl=1
  Xres=1
```

The parallelogram must be drawn with a base of 5 and a height of 4.

Use the Polygon tool from the DRAW menu to draw the parallelogram. Set the first vertex at $(-4, -2)$. Move the cursor to the right 5 units to $(1, -2)$ and set the second vertex.

Move the cursor 1 unit to the right and then up 4 units to $(2, 2)$. This will give the parallelogram a height of 4. Set the last vertex at $(-3, 2)$.

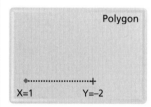

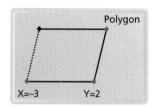

Select the Angle tool from the MEAS menu. Move the cursor to one of the sides of the parallelogram and select it by pressing **ENTER**. Press **ENTER** to set the displayed measurement on the screen.

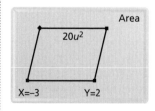

GeoMaster displays u^2 as the unit of area measure.

● The area of the parallelogram is 20 in.^{2}.

EXERCISES

Construct the parallelogram and calculate the area.

1. height 3 m and base 4 m

2. height 2.5 m and base 4.2 m

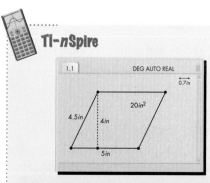

You can use the scale to change the units to match the units in the problem. In this case, inches are used.

Finding Centers and Radii

See Lesson 12-5, Example 3

EXAMPLE Graphing a Circle given its Equation

Find the center and radius of the circle with equation
$(x - 7)^2 + (y + 2)^2 = 64$.

Press **APPS** and select the Conics application. Choose Circle to graph the circle.

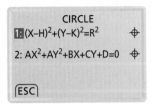

The Conics application can graph circles in different forms. Select option 1 as it matches the form of the equation given.

Enter the values of $h, k,$ and the radius. Press **ENTER** after each entry. Press **GRAPH** to graph the circle.

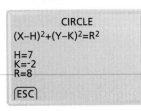

Be careful when determining h and k.
The standard form is
$(x - h)^2 + (y - k)^2 = r^2$.
So $(y + 2) = (y - (-2))$; k is -2.

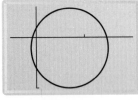

When you hit the **GRAPH** key, the Conics application will set the domain and range so the center of the circle is in the middle of the screen.

Press **Y=** to return to the entry screen. Press **ALPHA** **ENTER** to select SOLVE.

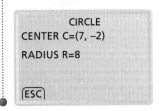

Compare these values to the original equation. Make sure they make sense for the equation.

EXERCISES

Find the radius and center for each circle.

1. $(x + 8)^2 + (y - 10)^2 = 121$

2. $(x - 5)^2 + (y - 8)^2 = 225$

3. $(x - 4)^2 + (y + 7)^2 = 49$

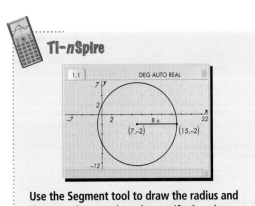

TI-*n*Spire

Use the Segment tool to draw the radius and then use the Length tool to verify that the length of the radius is correct.

Drawing a Locus

See Lesson 12-6, Example 1 part a

EXAMPLE Describing a Locus in a Plane

From the home screen of the TI-Nspire select Graphs & Geometry. Press `menu` and View > Plane Geometry View. To access the menus of tools press `menu`. To exit out of a tool press `esc`.

Go to the Points & Lines menu choose the Segment tool. Select a point on the screen and press `ENTER`. Move the cursor away and select a second point. A segment will be drawn.

Go to the Measurement menu and choose the Length tool. Select the segment and move the measurement to a clear part of the screen and press `ENTER`.

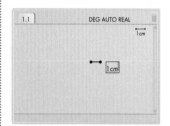

The segment measures greater than 1cm. Hover the cursor over the measurement and press `ENTER` twice. Enter 1 to change the length. The segment is now 1 cm long.

To keep the segment 1 cm in length, lock the measurement. In the Tools menu choose the Attributes tool. Select the measurement. A popup box will appear. Move the cursor down one space to the lock icon and move the cursor to the right to lock the measurement.

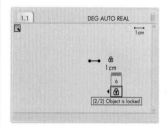

Press `ENTER` to exit the popup box. Because the length is locked the two points will stay 1 cm apart when either is moved.

Describe the locus by moving one point. In the Trace menu choose the Geometry Trace tool. Press `ENTER` on one of the endpoints. On the same point, hold down the Click key `click key` until the cursor changes to a hand. Use the arrow keys to move the point. Continue to move the point until the complete locus is created.

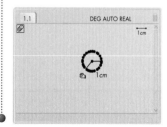

The locus is a circle. A circle is the set of points equidistant from a given point.

EXERCISES

Draw and describe each of the following.

1. the locus of all points 4 cm from point C

2. the locus of all points 2.3 cm from point C.

Skills Handbook

Problem Solving Strategies

You may find one or more of these strategies helpful in solving a word problem.

Strategy	When to Use it
Draw a Diagram	You need help in visualizing the problem.
Try, Check, Revise	Solving the problem directly is too complicated.
Make a Table	The problem has data that need organizing.
Look for a Pattern	The problem describes a relationship.
Solve a Simpler Problem	The problem is complex or has numbers that are too unmanageable to use at first.
Use Logical Reasoning	You need to reach a conclusion from some given information.
Work Backward	You undo various operations to arrive at the answer.

Problem Solving: Draw a Diagram

EXAMPLE

Antoine is 1.91 m tall. He measures his shadow and finds that it is 2.34 m long. He then measures the length of the shadow of a flagpole and finds that it is 13.2 m long. How tall is the flagpole?

Start by drawing a diagram showing the given information. The diagram shows that you can solve the problem by using a proportion.

$\frac{1.91}{2.34} = \frac{x}{13.2}$ **Write a proportion.**

$x \approx 10.77$ **Solve for x.**

● The flagpole is about 10.8 m tall.

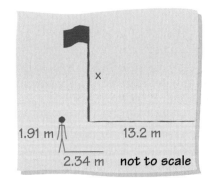

EXERCISES

1. Five people meet and shake hands with one another. How many handshakes are there in all?

2. Three tennis balls fit snugly in an ordinary, cylindrical tennis ball container. Which is greater, the circumference of a ball or the height of the container?

3. Three lines that all intersect a circle can determine at most 7 regions within the circle, as shown in the diagram. What is the greatest number of regions that can be determined by 5 lines?

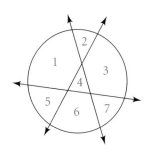

4. A triangle has vertices $(1, 3)$, $(2, 3)$, and $(7, 5)$. Find its area.

Problem Solving: Try, Check, Revise

Have you ever weighed yourself on a balance scale at a doctor's office? You start by guessing your weight, and then you see if the scale balances. If it doesn't, you slide the weights back and forth until the scale does balance. This is an example of the *Try, Check, Revise* strategy, a strategy helpful for solving many types of problems.

EXAMPLE

You have 100 ft of fencing and want to build a fence in the shape of a rectangle to enclose the largest possible area. What should be the dimensions of the rectangle?

Try 1:	10 ft wide by 40 ft long	Make an initial try with a perimeter
	$10 \cdot 40 = 400 \text{ ft}^2$	of 100 ft. Find the area.
Try 2:	20 ft wide by 30 ft long	Try again and find the area.
	$20 \cdot 30 = 600 \text{ ft}^2$	The area is larger than the initial try.
Try 3:	35 ft wide by 15 ft long	Continue trying and testing areas.
	$35 \cdot 15 = 525 \text{ ft}^2$	This area is smaller than the last try.
Try 4:	22 ft wide by 28 ft long	Notice that the areas are larger when the
	$22 \cdot 28 = 616 \text{ ft}^2$	width and length are closer together.
Try 5:	25 ft wide by 25 ft long	Choose dimensions that are as
	$25 \cdot 25 = 625 \text{ ft}^2$	close together as possible.

The dimensions of your rectangle should be 25 ft by 25 ft.

EXERCISES

1. The product of three consecutive even integers is 480. Find the integers.

2. The combined ages of a father and his twin daughters are 54 years. The father was 24 years old when the twins were born. How old is each of the three people?

3. What numbers can x represent in the rectangle?

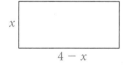

4. Alexandra has a collection of dimes and quarters. The number of dimes equals the number of quarters. She has a total of $2.80. How many of each coin does Alexandra have?

Use the Try, Check, Revise strategy to find the value of each variable.

5. $2a + 5 = 1$ 6. $10 - 3c = -2$ 7. $\frac{w}{-3} + 12 = -6$

8. $5y - 32 = 28$ 9. $12b + 11 = 14$ 10. $0.5x - 15 = -7$

11. Ruisa bought 7 rolls of film to take 192 pictures on a field trip. Some rolls had 36 exposures and the rest had 24 exposures. How many of each type did Ruisa buy?

12. The sum of five consecutive integers is 5. Find the integers.

13. Paul buys a coupon for $20 from a local theater that allows him to see movies for half price over the course of one year. The cost of seeing a movie is normally $7.50. What is the least number of movies Paul would have to see to pay less than the normal price per movie?

Problem Solving: Make a Table and Look for a Pattern

There are two important ways that making a table can help you solve a problem. First, a table is a handy method of organizing information. Second, once the information is in a table, it is easier for you to find patterns.

The squares below are made of toothpicks. How many toothpicks are in the square with 7 toothpicks on a side?

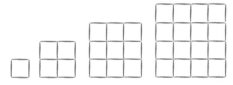

Use a table to organize the information.

Notice the pattern in the increases in the numbers of toothpicks in the squares. For each increase of 1 toothpick on a side, the increase increases by 4. The number of toothpicks in the 5th square is 40 + 20, or 60. The number in the 6th square is 60 + 24, or 84, and the number in the 7th square is 84 + 28, or 112.

Toothpicks on a side	1	2	3	4
Toothpicks in the square	4	12	24	40

$$+8 \quad +12 \quad +16$$

EXERCISES

1. The triangles are made of toothpicks. How many toothpicks are in Figure 10?

       Figure 1   Figure 2   Figure 3

2. In each figure, the vertices of the smallest square are midpoints of the sides of the next larger square. Find the area of the ninth shaded square.

1 in.

3. In each figure, the midpoints of the sides of the unshaded triangles are used as vertices of the shaded triangles. Find the total number of shaded triangles in Figure 8.

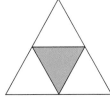 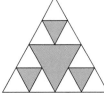 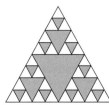

      Figure 1   Figure 2   Figure 3

Problem Solving: Solve a Simpler Problem

Looking at a simpler version of a problem can be helpful in suggesting a problem solving approach.

EXAMPLE

A fence along the highway is 570 m long. There is a fence post every 10 m. How many fence posts are there?

You may be tempted to divide 570 by 10, getting 57 as an answer, but looking at a simpler problem suggests that this answer isn't right. Suppose you have just 10 or 20 m of fencing.

10 m	20 m
two fence posts	three fence posts

These easier problems suggest that there is always *one more* fence post than one tenth the length. So for a 570-m fence, there are $\frac{570}{10} + 1$, or 58 fence posts.

EXERCISES

1. A farmer wishes to fence in a square lot 70 yards by 70 yards. He will install a fence post every 10 yards. How many fence posts will he need?

2. A snail is trying to escape from a well 10 ft deep. The snail can climb 2 ft each day, but each night it slides back 1 ft. How many days will the snail take to climb out of the well?

3. Janette is planning to walk from her house to her friend Barbara's house. How many different paths can she take to get there? Assume that she walks only east and south (along the grid lines).

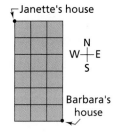

4. A square table can seat four people. For a banquet, a long rectangular table is formed by placing 14 such tables edge to edge in a straight line. How many people can sit at the long table?

5. Find the sum of the whole numbers from 1 to 999.

6. How many trapezoids are in the figure below? (*Hint:* Solve several simpler problems, and then look for a pattern.)

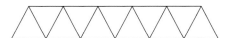

7. At a business luncheon, 424 handshakes took place. No two people shook hands with each other more than once. What is the least number of people in attendance at the luncheon?

8. On the occasion of his 50th birthday, the President was honored with a 21-gun salute. The sound of each gunshot lasted 1 second, and 4 seconds elapsed between shots. How long did the salute last?

9. In a tennis tournament, each athlete plays one match against each of the other athletes. There are 14 athletes scheduled to play in the tournament. How many matches will be played?

Problem Solving: Use Logical Reasoning

Some problems can be solved without the use of numbers. They can be solved by the use of logical reasoning, given some information.

EXAMPLE

Anna, Bill, Carla, and Doug are siblings. Each lives in a different state beginning with W. Use these clues to determine where each sibling lives:

(1) Neither sister lives in a state containing two words.

(2) Bill lives to the west of his sisters.

(3) Anna doesn't cross the Mississippi River when she visits Doug.

Make a table to organize what you know. Use an initial for each name.

State	A	B	C	D
West Virginia	✗	✗	✗	
Wisconsin		✗		
Wyoming		✗		
Washington	✗	✓	✗	✗

From clue 1, you know that neither Anna nor Carla lives in West Virginia.

Using clues 1 and 2, you know that Bill must live in Washington if he lives to the west of his sisters.

Use logical reasoning to complete the table.

State	A	B	C	D
West Virginia	✗	✗	✗	✓
Wisconsin	✓	✗	✗	✗
Wyoming	✗	✗	✓	✗
Washington	✗	✓	✗	✗

Doug lives in West Virginia because no other sibling does.

From clue 3, you know that Anna must live in Wisconsin.

Carla, therefore, lives in Wyoming.

EXERCISES

1. Harold has a dog, a parrot, a goldfish, and a hamster. Their names are J. T., Izzy, Arf, and Blinky. Izzy has neither feathers nor fins. Arf can't bark. J. T. weighs less than the four-legged pets. Neither the goldfish nor the dog has the longest name. Arf and Blinky don't get along well with the parrot. What is each pet's name?

2. At the state basketball championship tournament, 31 basketball games are played to determine the winner of the tournament. After each game, the loser is eliminated from the tournament. How many teams are in the tournament?

3. The sophomore class has 124 students. Of these students, 47 are involved in musical activities: 25 in band and 36 in choir. How many students are involved in both band and choir?

4. Tina's height is between Kimiko's and Ignacio's. Ignacio's height is between Jerome's and Kimiko's. Tina is taller than Jerome. List the people in order from shortest to tallest.

Problem Solving: Work Backward

In some situations it is easier to start with the end result and work backward to find the solution. You work backward in order to solve linear equations. The equation $2x + 3 = 11$ means "double x and add 3 to get 11." To find x, you "undo" those steps in reverse order.

$$2x + 3 = 11$$
$$2x = 8 \quad \text{Subtract 3 from each side.}$$
$$x = 4 \quad \text{Divide each side by 2.}$$

Another time it is convenient to work backward is when you want to "reverse" a set of directions.

 EXAMPLE

Algebra Sandy spent $\frac{1}{10}$ of the money in her purse for lunch. She then spent \$23.50 for a gift for her brother, then half of what she had left on a new CD. If Sandy has \$13 left in her purse, how much money did she have in it before lunch?

Start with the \$13 that Sandy has left in her purse.
She spent half of what she had before the \$13 on a new CD, so she must have had twice \$13, or \$26, before she bought the CD.

She spent \$23.50 on a gift for her brother, so add \$23.50 to \$26. Before buying the gift for her brother, she had \$49.50.

She spent $\frac{1}{10}$ of the money for lunch and was left with \$49.50. That means $\frac{9}{10}$ of what she had is \$49.50. Set up an equation.

$$\frac{9}{10}x = 49.50$$
$$x = 55$$

Sandy had \$55 in her purse before lunch.

EXERCISES

1. To go from Bedford to Worcester, take Route 4 south, then Route 128 south, and then Route 90 west. How do you get from Worcester to Bedford?

2. Algae are growing on a pond's surface. The area covered doubles each day. It takes 24 days to cover the pond completely. After how many days will the pond be half covered with algae?

3. Don sold $\frac{1}{5}$ as many raffle tickets as Carlita. Carlita sold 3 times as many as Ranesha. Ranesha sold 7 fewer than Russell. If Russell sold 12 tickets, how many did Don sell?

4. At 6% interest compounded annually, the balance in a bank account will double about every 12 years. If such an account has a balance of \$16,000 now, how much was deposited when the account was opened 36 years ago?

5. Solve the puzzle that Yuan gave to Inez: I am thinking of a number. If I triple the number and then halve the result, I get 12. What number am I thinking of?

6. Carlos paid a \$14.60 taxi fare from a hotel to the airport, including a \$2.00 tip. Green Cab Co. charges \$1.20 per passenger plus \$0.20 for each $\frac{1}{5}$ mile. How many miles is the hotel from the airport?

Using a Ruler and Protractor

Knowing how to use a ruler and protractor is crucial for success in geometry.

EXAMPLE

Draw a triangle that has sides of length 5.2 cm and 3.0 cm and a 68° angle between these two sides.

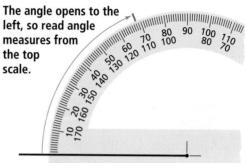

The angle opens to the left, so read angle measures from the top scale.

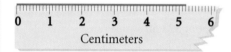

Step 1 Use a ruler to draw a segment 5.2 cm long.

Step 2 Place the crosshairs of a protractor at one endpoint of the segment. Make a small mark at the 68° position along the protractor.

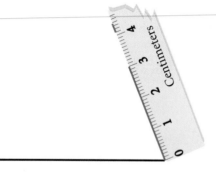

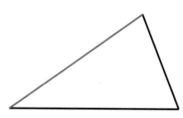

Step 3 Align the ruler along the small mark and the endpoint you used in Step 2. Place the zero point of the ruler at the endpoint. Draw a segment 3.0 cm long.

Step 4 Complete the triangle by connecting the endpoints of the first and second segments.

EXERCISES

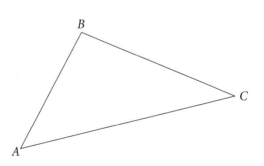

1. Measure sides $\overline{AB}$ and $\overline{BC}$ to the nearest millimeter.

2. Measure each angle of $\triangle ABC$ to the nearest degree.

3. Draw a triangle that has sides of length 4.8 cm and 3.7 cm and a 34° angle between these two sides.

4. Draw a triangle that has 43° and 102° angles and a side of length 5.4 cm between these two angles.

5. Draw a rhombus that has sides of length $2\frac{1}{4}$ in., and 68° and 112° angles.

6. Draw an isosceles trapezoid that has one pair of 48° base angles and a base of length 2 in. between these two base angles.

7. Draw an isosceles triangle that has two congruent sides $3\frac{1}{2}$ in. long and a 134° vertex angle.

Measurement Conversions

To convert from one unit of measure to another, you multiply by a conversion factor in the form of a fraction. The numerator and denominator are in different units, but they represent the same amount. So, you can think of this as multiplying by 1.

An example of a conversion factor is $\frac{1\ \text{ft}}{12\ \text{in.}}$. You can create other conversion factors using the table on page 766.

1 EXAMPLE

Complete each statement.

a. 88 in. = ▨ ft
$$88\ \text{in.} \cdot \frac{1\ \text{ft}}{12\ \text{in.}} = \frac{88}{12}\ \text{ft} = 7\frac{1}{3}\ \text{ft}$$

b. 5.3 m = ▨ cm
$$5.3\ \text{m} \cdot \frac{100\ \text{cm}}{1\ \text{m}} = 5.3(100)\ \text{cm} = 530\ \text{cm}$$

c. 3700 mm = ▨ cm
$$3700\ \text{mm} \cdot \frac{1\ \text{cm}}{10\ \text{mm}} = 370\ \text{cm}$$

d. 90 in. = ▨ yd
$$90\ \text{in.} \cdot \frac{1\ \text{ft}}{12\ \text{in.}} \cdot \frac{1\ \text{yd}}{3\ \text{ft}} = \frac{90}{36}\ \text{yd} = 2\frac{1}{2}\ \text{yd}$$

Area is always in square units, and volume is always in cubic units.

3 ft

1 yd = 3 ft

3 ft

3 ft

$1\ \text{yd}^2 = 9\ \text{ft}^2$

3 ft

3 ft

3 ft

$1\ \text{yd}^3 = 27\ \text{ft}^3$

2 EXAMPLE

Complete each statement.

a. 300 in.^{2} = ▨ ft^2
1 ft = 12 in., so 1 ft^2 = (12 in.)^2 = 144 in.^{2}
$$300\ \text{in.}^2 \cdot \frac{1\ \text{ft}^2}{144\ \text{in.}^2} = 2\frac{1}{12}\ \text{ft}^2$$

b. 200,000 cm^3 = ▨ m^3
1 m = 100 cm, so 1 m^3 = (100 cm)^3 = 1,000,000 cm^3
$$200,000\ \text{cm}^3 \cdot \frac{1\ \text{m}^3}{1,000,000\ \text{cm}^3} = 0.2\ \text{m}^3$$

EXERCISES

Complete each statement.

1. 40 cm = ▨ m

2. 1.5 kg = ▨ g

3. 60 cm = ▨ mm

4. 200 in. = ▨ ft

5. 28 yd = ▨ in.

6. 1.5 mi = ▨ ft

7. 42 fl oz = ▨ qt

8. 430 mg = ▨ g

9. 34 L = ▨ mL

10. 1.2 m = ▨ cm

11. 43 mm = ▨ cm

12. 3600 s = ▨ min

13. 15 g = ▨ mg

14. 12 qt = ▨ c

15. 0.03 kg = ▨ mg

16. 14 gal = ▨ qt

17. 4500 lb = ▨ t

18. 234 min = ▨ h

19. 12 mL = ▨ L

20. 2 pt = ▨ fl oz

21. 20 m/s = ▨ km/h

22. 3 ft^2 = ▨ in.^{2}

23. 108 m^2 = ▨ cm^2

24. 2100 mm^2 = ▨ cm^2

25. 1.4 yd^2 = ▨ ft^2

26. 0.45 km^2 = ▨ m^2

27. 1300 ft^2 = ▨ yd^2

28. 1030 in.^{2} = ▨ ft^2

29. 20,000,000 ft^2 = ▨ mi^2

30. 1000 cm^3 = ▨ m^3

Measurement, Rounding Error, and Reasonableness

There is no such thing as an *exact* measurement. Measurements are always approximate. No matter how precise it is, a measurement actually represents a range of values.

1 EXAMPLE

Chris's height, to the nearest inch, is 5 ft 8 in. Find the range of values this measurement represents.

The height is given to the nearest inch, so the error is $\frac{1}{2}$ in. Chris's height, then, is between 5 ft $7\frac{1}{2}$ in. and 5 ft $8\frac{1}{2}$ in., or 5 ft 8 in. $\pm \frac{1}{2}$ in. Within this range are all measures which, when rounded to the nearest inch, equal 5 ft 8 in.

As you calculate with measurements, errors can accumulate.

2 EXAMPLE

Jean drives 18 km to work each day. The distance is given to the nearest kilometer.
a. Find the range of values this measurement represents.

The driving distance is between 17.5 and 18.5 km, or 18 ± 0.5 km.

b. Find the error in the round-trip distance.

Double the lower limit, 17.5, and the upper limit, 18.5. Thus, the round trip can be anywhere between 35 and 37 km, or 36 ± 1 km. The error for the round trip is double the error of a single leg of the trip.

So that your answers will be reasonable, keep precision and error in mind as you calculate. For example, in finding AB, the length of the hypotenuse of $\triangle ABC$, it would be inappropriate to give the answer as 9.6566 if the sides are given to the nearest tenth. Round your answer to 9.7.

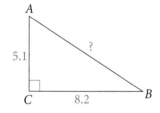

EXERCISES

Each measurement is followed by its unit of greatest precision. Find the range of values that each measurement represents.

1. 24 ft (ft)

2. 124 cm (cm)

3. 340 mL (mL)

4. $5\frac{1}{2}$ mi. $\left(\frac{1}{2} \text{ mi}\right)$

5. 73.2 mm (0.1 mm)

6. 34 yd^2 (yd^2)

7. 5.4 mi (0.1 mi)

8. 6 ft 5 in. (0.5 in.)

9. $15\frac{1}{2}$ yd $\left(\frac{1}{2} \text{ yd}\right)$

10. The lengths of the sides of *TJCM* are given to the nearest tenth of a centimeter. Find the range of values for the figure's perimeter.

11. To the nearest degree, two angles of a triangle are 49° and 73°. What is the range of values for the measure of the third angle?

12. The lengths of the legs of a right triangle are measured as 131 m and 162 m. You use a calculator to find the length of the hypotenuse. The calculator display reads *208.33867*. What should your answer be?

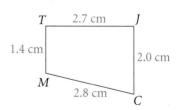

The Effect of Measurement Errors on Calculations

Measurements are always approximate, and calculations with these measurements produce error. Percent error is a measure of accuracy of a measurement or calculation. It is the ratio of the greatest possible error to the measurement.

$$\text{percent error} = \frac{\text{greatest possible error}}{\text{measurement}}$$

EXAMPLE

The dimensions of a box are measured as 18 in., 12 in., and 9 in. Find the percent error in calculating its volume.

The measurements are to the nearest inch, so the greatest possible length error is one half of one inch, or 0.5 in.

as measured

$V = \ell \cdot w \cdot h$

$\quad = 18 \cdot 12 \cdot 9$

$\quad = 1944$, or 1944 in.^{3}

maximum value

$V = \ell \cdot w \cdot h$

$\quad = 18.5 \cdot 12.5 \cdot 9.5$

$\quad \approx 2196.9$, or 2196.9 in.^{3}

minimum value

$V = \ell \cdot w \cdot h$

$\quad = 17.5 \cdot 11.5 \cdot 8.5$

$\quad \approx 1710.6$, or 1710.6 in.^{3}

Possible Error:

maximum − measured

2196.9 − 1944 = 252.9

measured − minimum

1944 − 1710.6 = 233.4

$$\text{percent error} = \frac{\text{greatest possible error}}{\text{measurement}}$$

$$= \frac{252.9}{1944}$$

$$\approx 0.1300926$$

$$\approx 13\%$$

● The percent error is about 13%.

EXERCISES

Find the percent error in calculating the volume of each box given its dimensions. Round to the nearest percent.

1. 10 cm by 5 cm by 20 cm

2. 12 in. by 6 in. by 2 in.

3. 1.2 mm by 5.7 mm by 2.0 mm

4. 7.5 m by 6.4 m by 2.7 m

5. 22.5 cm by 16.4 cm by 26.4 cm

6. 1.24 cm by 4.45 cm by 5.58 cm

7. $8\frac{1}{4}$ in. by $17\frac{1}{2}$ in. by 5 in.

8. $7\frac{3}{4}$ in. by $22\frac{1}{8}$ in. by $6\frac{1}{4}$ in.

Find the percent error in calculating the perimeter of each figure.

9.

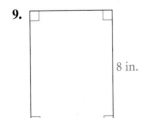

8 in.

6 in.

10.

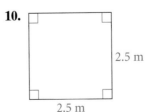

2.5 m

2.5 m

11.

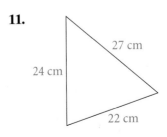

27 cm

24 cm

22 cm

Mean, Median, and Mode

Measures of central tendency, such as mean, median, and mode, are numbers that describe a set of data.

The **mean**, sometimes called the average, is the sum of the data items divided by the number of data items.

The **median** is the middle number when data items are placed in order and there are an odd number of data items. For an even number of data items, the median is the mean of the middle two numbers.

The **mode** is the data item that appears most frequently. A set of data may have more than one mode or no modes.

EXAMPLE

Eighteen students were asked to measure the angle formed by the three objects in the diagram. Their answers, in order from least to greatest, are as follows:

1 2 3 4 5 6 7 8 9 10 11 12 13 14 15 16 17
65, 66, 66, 66, 66, 66, 66, 67, 67, 67, 67, 67, 68, 68, 69, 70, 74, 113

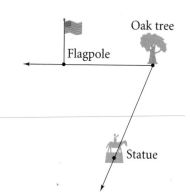

Oak tree

Flagpole

Statue

Find the mean, median, and mode of the data.

Mean: $\dfrac{\text{sum of the 18 measures}}{18} = \dfrac{1258}{18} = 69\dfrac{8}{9}$

Median:
This data list is already ordered. The two middle numbers—the ninth and the tenth numbers on the list—are both 67. So the median is 67.

Mode:
There are more 66's than any other number, so the mode is 66.

EXERCISES

Find the mean, median, and mode of each set of data.

1. Numbers of students per school in Newtown: 234, 341, 253, 313, 273, 301, 760

2. Lunch expenses: $4.50, $3.26, $5.02, $3.58, $1.25, $3.05, $4.24, $3.56, $3.31

3. Salaries at D. B. Widget & Co.: $15,000; 18,000; $18,000; $21,700; $26,500; $27,000; $29,300; $31,100; $43,000; $47,800; $69,000; $140,000

4. Population of towns in Brower County: 567, 632, 781, 902, 1034, 1100, 1598, 2164, 2193, 3062, 3074, 3108, 3721, 3800, 4104

5. In Exercise 3, which measure or measures of central tendency do you think best represent the data? Explain.

6. Find the mean, median, and mode of the exam scores at the right.

7. In the example, the student who reported the angle measure as 113 most likely made an error. If this measure is dropped from the list, what are the mean, median, and mode of the remaining 17 scores?

8. In the example, if the measurement 65 were instead 51, would the mean decrease? Would the median? Would the mode?

9. In the example, if the two students who measured the angle at 68 both reduced their measurements to 67, would the mode be affected? How?

Final Exam Scores
34, 47, 53, 56, 57, 62, 62, 64, 67, 70, 74, 74, 74, 78, 82, 85, 85, 85, 85, 86, 88, 92, 93, 93, 94, 95, 97

Bar Graphs and Line Graphs

Data displayed in a table can be very useful, but a table is not always as easy to interpret as a graph. Bar graphs and line graphs can show the same data, but sometimes one type of graph has advantages over the other.

EXAMPLE

Make a bar graph and a line graph showing the data in the table at the right.

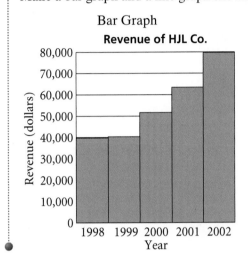

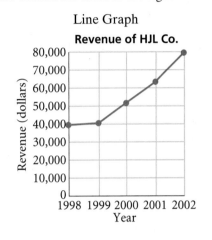

Revenue of HJL Co.

Year	Revenue
1998	$39,780
1999	$40,019
2000	$51,772
2001	$63,444
2002	$79,855

Bar graphs are useful when you wish to compare amounts. In the example above, the tallest bar is clearly twice the height of the shortest bar. At a glance, it is evident that in four years the revenue approximately doubled.

Line graphs allow you to see how a set of data changes over time. In the example, the slope of the line shows that revenue has increased at a steady rate since 1999.

Did revenue increase from 1998 to 1999? It is difficult to tell by looking at either graph; for that information, you should look back at the table.

EXERCISES

1. Create a bar graph and a line graph to display the data in the table below.

Sales of Rock Music (in millions of dollars)

Year	1995	1996	1997	1998	1999	2000
Sales	$4127	$4086	$3977	$3527	$3675	$3552

SOURCE: Recording Industry Association of America.
Go to **www.PHSchool.com** for a data update.
Web Code: afg-2041

For Exercises 2–6, refer to the line graph at the right.

2. What was the lowest temperature recorded between 6 A.M. and 6 P.M.?

3. During which time periods did the temperature appear to increase?

4. Estimate the temperature at 11 A.M. and at 5 P.M.

5. Can you tell from the graph what the actual maximum and minimum temperatures were between 6 A.M. and 6 P.M.? Explain.

6. The same data could be presented in a bar graph. Which presentation is better for these data, a line graph or a bar graph? Explain why.

Temperatures in Grand Island, Nebraska, on February 2

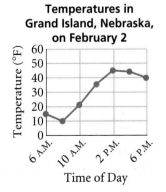

Box-and-Whisker Plots

A *box-and-whisker plot* is a way to display data on a number line. It provides a picture of how tightly the data cluster around the median and how wide a range the data have. The diagram below shows the various points associated with a box-and-whisker plot.

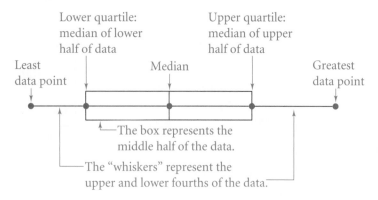

EXAMPLE

The heights, in inches, of 23 geometry students are as follows.

58, 61, 63, 63, 63, 64, 64, 65, 65, 65, 67, 68, 68, 68, 69, 70, 70, 70, 72, 72, 72, 74, 75

Draw a box-and-whisker plot.

The heights range from 58 in. to 75 in. Show 58 and 75 as endpoints on a line segment. The median is 68, so locate 68 in relation to 58 and 75. The lower quartile (the median of the lower eleven heights) is 64. The upper quartile (the median of the upper eleven heights) is 70. Locate 64 and 70 and draw a box enclosing them. Draw a vertical segment inside the box through the median.

EXERCISES

1. All of the physical education students at Martin Luther King, Jr., High School were timed sprinting the 100-meter dash. The box-and-whisker plot below summarizes the data. Use it to find the following.
   a. median   **b.** lower quartile   **c.** upper quartile

2. Make a box-and-whisker plot for the following data set, which lists the weights, in pounds, of the students trying out for the wrestling team at Benjamin Banneker High School.

   104, 121, 122, 130, 130, 131, 140, 144, 147, 147, 148, 155, 160, 163, 171

3. Make a box-and-whisker plot for the following set of data, which lists the numbers of pages in a set of books. (*Hint:* Order the data from smallest to largest.)

   205, 198, 312, 254, 185, 268, 297, 242, 356, 262

Squaring Numbers and Finding Square Roots

The square of a number is found by multiplying the number by itself. An exponent of 2 is used to indicate that a number is being squared.

1 EXAMPLE

Simplify.

a. 5^2

$5^2 = 5 \cdot 5$

$= 25$

b. $(-3.5)^2$

$(-3.5)^2 = (-3.5) \cdot (-3.5)$

$= 12.25$

c. $\left(\frac{2}{7}\right)^2$

$\left(\frac{2}{7}\right)^2 = \frac{2}{7} \cdot \frac{2}{7}$

$= \frac{4}{49}$

The square root of a number is itself a number that, when squared, results in the original number. A radical symbol ($\sqrt{}$) is used to represent the positive square root of a number.

2 EXAMPLE

Simplify. Round to the nearest tenth if necessary.

a. $\sqrt{36}$

$\sqrt{36} = 6$ since $6^2 = 36$

b. $\sqrt{174}$

$\sqrt{174} \approx 13.2$ since $13.2^2 \approx 174$

You can solve equations that include squared numbers.

3 EXAMPLE

Algebra Solve.

a. $x^2 = 144$

$x = 12$ or -12

b. $a^2 + 3^2 = 5^2$

$a^2 + 9 = 25$

$a^2 = 16$

$a = 4$ or -4

EXERCISES

Simplify.

1. 11^2

2. 16^2

3. $(-14)^2$

4. $(-21)^2$

5. 5.1^2

6. $\left(\frac{3}{7}\right)^2$

7. $\left(\frac{8}{5}\right)^2$

8. -6^2

$\boxed{x^2}$ **Simplify. Round to the nearest tenth if necessary.**

9. $\sqrt{100}$

10. $\sqrt{169}$

11. $\sqrt{74}$

12. $\sqrt{50}$

13. $\sqrt{400}$

14. $\sqrt{289}$

15. $\sqrt{\frac{4}{9}}$

16. $\sqrt{\frac{49}{81}}$

Algebra Solve. Round to the nearest tenth if necessary.

17. $x^2 = 49$

18. $a^2 = 9$

19. $y^2 + 7 = 8$

20. $5 + x^2 = 11$

21. $8^2 + b^2 = 10^2$

22. $5^2 + 4^2 = c^2$

23. $p^2 + 12^2 = 13^2$

24. $20^2 = 15^2 + a^2$

Evaluating and Simplifying Expressions

You evaluate an expression with variables by substituting a number for each variable. Then simplify the expression using the order of operations. Be especially careful with exponents and negative signs. For example, the expression $-x^2$ always yields a negative or zero value, and $(-x)^2$ is always positive or zero.

1 EXAMPLE

Algebra Evaluate each expression for $r = 4$.

a. $-r^2$

$$-r^2 = -(4^2) = -16$$

b. $-3r^2$

$$-3r^2 = -3(4^2) = -3(16) = -48$$

c. $(-3r)^2$

$$(-3r)^2 = (-3 \cdot 4)^2 = (-12)^2 = 144$$

To simplify an expression, you eliminate any parentheses and combine like terms.

2 EXAMPLE

Algebra Simplify each expression.

a. $5r - 2r + 1$

Combine like terms.
$$5r - 2r + 1 = 3r + 1$$

b. $\pi(3r - 1)$

Use the distributive property.
$$\pi(3r - 1) = 3\pi r - \pi$$

c. $(r + \pi)(r - \pi)$

Multiply polynomials.
$$(r + \pi)(r - \pi) = r^2 - \pi^2$$

EXERCISES

$\boxed{x^2}$ **Algebra** Evaluate each expression for $x = 5$ and $y = -3$.

1. $-2x^2$

2. $-y + x$

3. $-xy$

4. $(x + 5y) \div x$

5. $x + 5y \div x$

6. $(-2y)^2$

7. $(2y)^2$

8. $(x - y)^2$

9. $\dfrac{x + 1}{y}$

10. $y - (x - y)$

11. $-y^x$

12. $\dfrac{2(1 - x)}{y - x}$

13. $x \cdot y - x$

14. $x - y \cdot x$

15. $\dfrac{y^3 - x}{x - y}$

16. $-y(x - 3)^2$

17. Which expression gives the area of the shaded figure at the right?

A. $\pi(r - s)^2$
B. $\pi(r^2 - s^2)$
C. $\pi(s^2 - r^2)$
D. $\pi r^2 - 2\pi s$

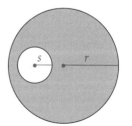

$\boxed{x^2}$ **Algebra** Simplify.

18. $6x - 4x + 8 - 5$

19. $2(\ell + w)$

20. $-(4x + 7)$

21. $-4x(x - 2)$

22. $3x - (5 + 2x)$

23. $2t^2 + 4t - 5t^2$

24. $(r - 1)^2$

25. $(1 - r)^2$

26. $(y + 1)(y - 3)$

27. $4h + 3h - 4 + 3$

28. $\pi r - (1 + \pi r)$

29. $(x + 4)(2x - 1)$

30. $2\pi h(1 - r)^2$

31. $3y^2 - (y^2 + 3y)$

32. $-(x + 4)^2$

Simplifying Radicals

A radical expression is in its simplest form when all three of the following statements are true.

1. The expression under the radical sign contains no perfect square factors (other than 1).
2. The expression under the radical sign does not contain a fraction.
3. The denominator does not contain a radical expression.

1 EXAMPLE

Simplify.

a. $\sqrt{\frac{4}{9}}$

$$\sqrt{\frac{4}{9}} = \frac{\sqrt{4}}{\sqrt{9}} = \frac{2}{3}$$

b. $\sqrt{12}$

$$\sqrt{12} = \sqrt{4} \cdot \sqrt{3} = 2\sqrt{3}$$

2 EXAMPLE

Find the length of the diagonal of rectangle $HJKL$.

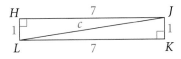

$c^2 = 7^2 + 1^2$	Use the Pythagorean Theorem.
$c^2 = 50$	Simplify the right side.
$c = \sqrt{50}$	Find the square root of each side.
$= \sqrt{25 \cdot 2}$	Find a perfect square factor of 50.
$= 5\sqrt{2}$	Simplify the radical.

3 EXAMPLE

Simplify $\frac{1}{\sqrt{3}}$.

$$\frac{1}{\sqrt{3}} \cdot \frac{\sqrt{3}}{\sqrt{3}} = \frac{\sqrt{3}}{3}$$ Multiply by $\frac{\sqrt{3}}{\sqrt{3}}$, or 1, to eliminate the radical in the denominator.

EXERCISES

Simplify each radical expression.

1. $\sqrt{27}$

2. $\sqrt{24}$

3. $\sqrt{150}$

4. $\sqrt{\frac{1}{9}}$

5. $\sqrt{\frac{72}{9}}$

6. $\frac{\sqrt{228}}{\sqrt{16}}$

7. $\sqrt{\frac{2}{5}}$

8. $\sqrt{\frac{27}{75}}$

9. $\frac{3}{\sqrt{8}}$

10. $\frac{6\sqrt{18}}{\sqrt{48}}$

$\boxed{x^2}$ **Algebra** Find the value of x. Leave your answer in simplest radical form.

11.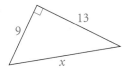

12.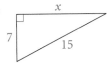

13.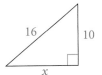

14.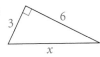

15.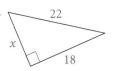

16.

17.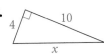

18.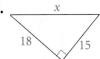

Simplifying Ratios

The ratio of the length of the shorter leg to the length of the longer leg for this right triangle is 4 to 6. This ratio can be written in several ways.

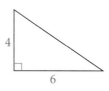

4 to 6 $\qquad \frac{4}{6} \qquad$ 4 : 6

EXAMPLE

Algebra Simplify each ratio.

a. 4 to 6

$$4 \text{ to } 6 = \frac{4}{6}$$

$$= \frac{2 \cdot 2}{2 \cdot 3} \quad \longleftarrow \text{Find and remove the common factor.} \longrightarrow$$

$$= \frac{2}{3}$$

b. $3ab : 27ab$

$$3ab : 27ab = \frac{3ab}{27ab}$$

$$= \frac{3ab}{9 \cdot 3ab}$$

$$= \frac{1}{9}$$

c. $\frac{4a + 4b}{a + b}$

$$\frac{4a + 4b}{a + b} = \frac{4(a + b)}{a + b} \quad \text{Factor the numerator. The denominator cannot be factored. Remove the common factor } (a + b).$$

$$= 4$$

EXERCISES

$\boxed{x^2}$ **Algebra** **Simplify each ratio.**

1. 25 to 15

2. 6 : 9

3. $\frac{36}{54}$

4. 0.8 to 2.4

5. $\frac{7}{14x}$

6. $\frac{12c}{14c}$

7. $22x^2$ to $35x$

8. $0.5ab : 8ab$

9. $\frac{4xy}{0.25x}$

10. $1\frac{1}{2}x$ to $5x$

11. $\frac{x^2 + x}{2x}$

12. $\frac{1}{4}r^2$ to $6r$

13. $0.72t : 7.2t^2$

14. $(2x - 6) : (6x - 4)$

15. $12xy : 8xy$

16. $(9x - 9y)$ to $(x - y)$

17. $\frac{\pi r}{r^2 + \pi r}$

18. $\frac{8ab}{32xy}$

Express each ratio in simplest form.

19. shorter leg : longer leg

20. hypotenuse to shorter leg

21. $\frac{\text{shorter leg}}{\text{hypotenuse}}$

22. hypotenuse : longer leg

23. longer leg to shorter leg

24. $\frac{\text{longer leg}}{\text{hypotenuse}}$

$\boxed{x^2}$ **Algebra** **Write an expression in simplest form for** $\frac{\text{area of shaded figure}}{\text{area of blue figure}}$.

25.

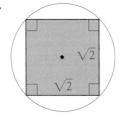

26.

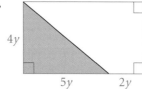

27.

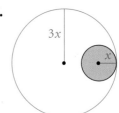

Absolute Value

Absolute value is used to represent the distance of a number from 0 on a number line. Since distance is always referred to as a nonnegative number, the absolute value of an expression is nonnegative.

On the number line at the right, both 4 and -4 are four units from zero. Therefore, $|4|$ and $|-4|$ are both equal to four.

Skills Handbook

1 EXAMPLE

Simplify each expression.

a. $|-7|$

$|-7| = 7$

b. $|15|$

$|15| = 15$

c. $|4| + |-19|$

$|4| + |-19| = 4 + 19 = 23$

When working with more complicated expressions, always remember to simplify within absolute value symbols first.

2 EXAMPLE

Simplify each expression.

a. $|4 - 8|$

$|4 - 8| = |-4|$

$\qquad = 4$

b. $-3|-7 - 4|$

$-3|-7 - 4| = -3|-11|$

$\qquad = -3 \cdot 11$

$\qquad = -33$

To solve an equation involving absolute value, remember that absolute value symbols cause both negative and positive values to become positive.

3 EXAMPLE

Algebra Solve.

a. $|x| = 7$

$x = 7 \text{ or } -7$

b. $|x| - 3 = 22$

$|x| - 3 = 22$

$|x| = 25$

$x = -25 \text{ or } 25$

EXERCISES

Simplify each expression.

1. $|-8|$

2. $|11|$

3. $|16|$

4. $|-23|$

5. $|-7| + |15|$

6. $|-12| - |-12|$

7. $|5| - |10|$

8. $|4| + |2|$

9. $10 - |-20|$

10. $|-9 - 11|$

11. $2|-21 + 16|$

12. $-8|-9 + 4|$

x^2 **Algebra Solve.**

13. $|x| = 16$

14. $2 = |x|$

15. $|x| + 7 = 27$

16. $|x| - 9 = 15$

Solving and Writing Linear Equations

To solve a linear equation, use the properties of equality and properties of real numbers to find the value of the variable that satisfies the equation.

1 EXAMPLE

Algebra Solve each equation.

a. $5x - 3 = 2$

$5x - 3 = 2$

$5x = 5$ **Add 3 to each side.**

$x = 1$ **Divide each side by 5.**

b. $1 - 2(x + 1) = x$

$1 - 2(x + 1) = x$

$1 - 2x - 2 = x$ **Use the Distributive Property.**

$-1 - 2x = x$ **Simplify the left side.**

$-1 = 3x$ **Add 2x to each side.**

$-\frac{1}{3} = x$ **Divide each side by 3.**

You will sometimes need to translate word problems into equations. Look for words that suggest a relationship or some type of mathematical operation.

2 EXAMPLE

Algebra A student has grades of 80, 65, 78, and 92 on four tests. What is the minimum grade she must earn on her next test to ensure an average of 80?

Relate average of 80, 65, 78, 92, and next test, is 80   **Pull out the key words and numbers.**

Define Let x = the grade on the next test.   **Let a variable represent what you are looking for.**

Write $\dfrac{80 + 65 + 78 + 92 + x}{5} = 80$ **Write an equation.**

$\dfrac{315 + x}{5} = 80$ **Combine like terms.**

$315 + x = 400$ **Multiply each side by 5.**

$x = 85$ **Subtract 315 from each side.**

The student must earn 85 on the next test for an average of 80.

EXERCISES

$\boxed{x^2}$ **Algebra** Solve each equation.

1. $3n + 2 = 17$

2. $5a - 2 = -12$

3. $2x + 4 = 10$

4. $3(n - 4) = 15$

5. $4 - 2y = 8$

6. $-6z + 1 = 13$

7. $6 - (3t + 4) = -17$

8. $7 = -2(4n - 4.5)$

9. $(w + 5) - (2w + 5) = 5$

10. $\frac{5}{7}p - 10 = 30$

11. $\frac{m}{-2} - 3 = 1$

12. $5k + 2(k + 1) = 23$

$\boxed{x^2}$ **Algebra** Write an equation and solve the problem.

13. Twice a number subtracted from 35 is 9. What is the number?

14. A new tenant pays the landlord the amount of the first month's rent, the same amount for the last month's rent, and half a month's rent for a security deposit. The total is $2437.50. How much is the monthly rent?

15. The Johnsons pay $9.95 a month plus $0.035 per minute for local phone service. Last month, they paid $12.75. How many minutes of local calls did they make?

Solving Literal Equations

An equation with two or more variables is called a literal equation. It is often necessary to solve a literal equation for a particular variable.

1 EXAMPLE

Algebra The formula $P = 2(\ell + w)$ gives the perimeter P of a rectangle with length ℓ and width w. Solve the equation for ℓ.

$$P = 2(\ell + w)$$
$$P = 2\ell + 2w \quad \textbf{Use the Distributive Property.}$$
$$P - 2w = 2\ell \quad \textbf{Subtract 2\textit{w} from each side.}$$
$$\frac{P - 2w}{2} = \ell \quad \textbf{Divide each side by 2.}$$

2 EXAMPLE

Algebra The formula $A = \frac{1}{2}(b_1 + b_2)h$ gives the area A of a trapezoid with bases b_1 and b_2 and height h. Solve for h.

$$A = \frac{1}{2}(b_1 + b_2)h$$
$$2A = h(b_1 + b_2) \quad \textbf{Multiply each side by 2.}$$
$$\frac{2A}{b_1 + b_2} = h \quad \textbf{Divide each side by } (\boldsymbol{b_1 + b_2}).$$

3 EXAMPLE

Algebra The formula for converting from degrees Celsius C to degrees Fahrenheit F is $F = \frac{9}{5}C + 32$. Solve for C.

$$F = \frac{9}{5}C + 32$$
$$F - 32 = \frac{9}{5}C \quad \textbf{Subtract 32 from each side.}$$
$$\frac{5}{9}(F - 32) = C \quad \textbf{Multiply each side by } \tfrac{5}{9}.$$

EXERCISES

x^2 **Algebra** **Solve each equation for the variable in red.**

1. Perimeter of rectangle: $P = 2w + 2\ell$

2. Volume of prism: $V = \ell wh$

3. Surface area of sphere: $S = 4\pi r^2$

4. Lateral area of cylinder: $A = 2\pi rh$

5. Area of kite or rhombus: $A = \frac{1}{2}d_1 d_2$

6. Area of circle: $A = \pi r^2$

7. Area of regular polygon: $A = \frac{1}{2}ap$

8. Volume of cylinder: $V = \pi r^2 h$

9. Area of triangle: $A = \frac{1}{2}bh$

10. Tangent of $\angle A$: $\tan A = \frac{y}{x}$

11. Euler's Formula: $F + V = E + 2$

12. Circumference of circle: $C = 2\pi r$

13. Cosine of $\angle A$: $\cos A = \frac{b}{c}$

14. Volume of cone: $V = \frac{1}{3}\pi r^2 h$

15. Surface area of right cone: $S = \pi r^2 + \pi r\ell$

16. Area of trapezoid: $A = \frac{1}{2}(b_1 + b_2)h$

17. Volume of pyramid: $V = \frac{1}{3}Bh$

18. Pythagorean Theorem: $a^2 + b^2 = c^2$

19. Surface area of regular pyramid: $S = B + \frac{1}{2}p\ell$

20. Surface area of right cylinder: $S = 2\pi r^2 + 2\pi rh$

Systems of Linear Equations

Normally, there are many ordered pairs that satisfy a given equation. For example, $(3, 4)$, $(4, 5)$, $(5, 6)$, and infinitely many other pairs all satisfy the equation $y = x + 1$. In solving a system of two linear equations, however, you need to find ordered pairs that satisfy both equations at once. Ordinarily, there is just one such ordered pair; it is the point where the graphs of the two lines intersect.

One method you can always use to solve a system of linear equations is the substitution method.

EXAMPLE

Algebra Solve the system. $\quad 2x - y = -10$
$$-3x - 2y = 1$$

Solve one of the equations for a variable. Looking at the two equations, it seems easiest to solve the first equation for y.

$2x - y = -10$

$\quad -y = -2x - 10 \quad$ **Subtract 2x from each side.**

$\quad\ \ y = 2x + 10 \quad$ **Multiply each side by −1.**

Now substitute $2x + 10$ for y in the other equation.

$\quad\quad\quad -3x - 2y = 1 \quad\quad$ **Write the other equation.**

$\quad -3x - 2(2x + 10) = 1 \quad\quad$ **Substitute (2x + 10) for y.**

$\quad\quad -3x - 4x - 20 = 1 \quad\quad$ **Use the Distributive Property.**

$\quad\quad\quad\quad\quad\quad -7x = 21 \quad\quad$ **Simplify and add 20 to each side.**

$\quad\quad\quad\quad\quad\quad\quad\ x = -3 \quad\quad$ **Divide each side by −7.**

So $x = -3$. To find y, substitute -3 for x in either equation.

$\quad\quad 2x - y = -10 \quad\quad$ **Write one of the equations.**

$\quad 2(-3) - y = -10 \quad\quad$ **Substitute −3 for x.**

$\quad\quad -6 - y = -10 \quad\quad$ **Simplify.**

$\quad\quad\quad\quad -y = -4 \quad\quad$ **Add 6 to each side.**

$\quad\quad\quad\quad\ \ y = 4 \quad\quad$ **Multiply each side by −1.**

So the solution is $x = -3$ and $y = 4$, or $(-3, 4)$. If you graph $2x - y = -10$ and $-3x - 2y = 1$, you'll find that the lines intersect at $(-3, 4)$.

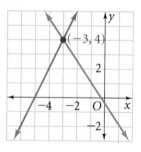

EXERCISES

$\boxed{x^2}$ **Algebra** Solve each system.

1. $x + y = 3$
$\quad x - y = 5$

2. $y - x = 4$
$\quad x + 3 = y$

3. $y = 1$
$\quad 5x - 2y = 18$

4. $3x - y = 5$
$\quad x = -2$

5. $4y - x = -3$
$\quad 2x - 6 = 8y$

6. $8x - 1 = 4y$
$\quad 3x = y + 1$

7. $2x + 2y = -4$
$\quad -x + 3y = 6$

8. $12y - 3x = 11$
$\quad x - 2y = -2$

9. $5x + 7y = 1$
$\quad 4x - 2y = 16$

10. Give an example of a system of linear equations with no solution. What do you know about the slopes of the lines of such a system?

Percents

A percent is a ratio in which a number is compared to 100. For example, the expression *60 percent* means "60 out of 100." The symbol % stands for "percent."

A percent can be written in decimal form by first writing it in ratio form, and then writing the ratio as a decimal. For example, 25% is equal to the ratio $\frac{25}{100}$ or $\frac{1}{4}$. As a decimal, $\frac{1}{4}$ is equal to 0.25. Note that 25% can also be written directly as a decimal by moving the decimal point two places to the left.

1 EXAMPLE

Convert each percent to a decimal.

a. 42%   **b.** 157%   **c.** 12.4%   **d.** 4%

42% = 0.42   157% = 1.57   12.4% = 0.124   4% = 0.04

To calculate a percent of a number, write the percent as a decimal and multiply.

2 EXAMPLE

Simplify. Where necessary, round to the nearest tenth.

a. 30% of 242   **b.** 7% of 38

30% of 242 = 0.3 · 242   7% of 38 = 0.07 · 38

 = 72.6 = 2.66 ≈ 2.7

For a percent problem, it is a good idea to check that your answer is reasonable by estimating it.

3 EXAMPLE

Estimate 23% of 96.

23% ≈ 25% and 96 ≈ 100. 25% $\left(\text{or } \frac{1}{4}\right)$ of 100 = 25. A reasonable estimate is 25.

EXERCISES

Convert each percent to a decimal.

1. 50%   **2.** 75%   **3.** 27%   **4.** 6%

5. 32.5%   **6.** 84.6%   **7.** 9%   **8.** 2.5%

Simplify. Where necessary, round to the nearest tenth.

9. 21% of 40   **10.** 45% of 200   **11.** 6% of 120   **12.** 2% of 54

13. 80.4% of 52   **14.** 23.8% of 176   **15.** 7.5% of 32   **16.** 9.25% of 89

Estimate.

17. 12% of 70   **18.** 48% of 87   **19.** 73% of 64   **20.** 77% of 42

Probability

Probability is a measure of the likelihood of an event occurring. All probabilities range from 0 to 1 where 0 is the probability of an event that cannot happen and 1 is the probability of an event that is certain to happen. An event with probability 0.5 or 50% has an equal chance of happening or not happening.

The formula $P(E) = \frac{\text{number of favorable outcomes}}{\text{number of possible outcomes}}$ is used to calculate the probability of event E.

1 EXAMPLE

The numbers 2 through 21 are written on pieces of paper and placed in a hat. One piece of paper is drawn at random. Determine the probability of selecting a perfect square.

The total number of outcomes, $2, 3, 4, \ldots, 21$, for this event is 20.

There are 3 favorable outcomes: 4, 9, 16.

$P(\text{selecting a perfect square}) = \frac{3}{20}$

2 EXAMPLE

Determine the probability of getting exactly two heads when two coins are tossed.

The total number of outcomes, $(H, H), (H, T), (T, H), (T, T)$, for this event is 4.

There is 1 favorable outcome, (H, H).

$P(\text{two heads}) = \frac{1}{4}$

EXERCISES

A jar contains 3 white balls, 7 red balls, and 4 green balls. A ball is selected at random from the jar. Determine the probability of selecting a ball with the given color.

1. red

2. white

3. green

4. green or white

5. A red ball is removed from the jar. Determine the probability that the next ball selected will be green.

6. Two green balls are removed from the jar. Determine the probability that the next ball selected will be green.

You roll a 12-sided polyhedron with the numbers 1–12 on its congruent faces. Determine the probability of each outcome.

7. rolling a 2

8. rolling a 4 or a 5

9. rolling an even number

10. rolling an odd number

11. rolling a prime number

12. rolling a factor of 8

A coin is flipped three times. Determine the probability of each outcome.

13. exactly two tails

14. two heads and one tail

15. no more than two tails

16. no more than one head

17. at least one tail

18. all tails or all heads

Introduction

Both the SAT and the ACT are standardized tests used nationally for college admissions. This appendix provides you with test-taking strategies and sample problems to help you with the Mathematics portions of both tests. You can find SAT-specific strategies on pages 798 and 799, and ACT-specific strategies on pages 800 and 801.

The SAT Test

One of the three parts of the SAT is Mathematics. You will be tested on numbers and operations, algebra and functions, geometry, statistics, probability, and data analysis.

The Mathematics portion contains these three sections (with their timing):

- 20 multiple-choice (25 minutes)

- 8 multiple-choice and 10 student-produced response (25 minutes)

- 16 multiple-choice (20 minutes)

On the 10 student-produced response questions, also called "grid-ins", you bubble in your answer on a four-column grid. The questions of the same type are ordered by difficulty, starting with the easiest.

Scoring The multiple-choice questions count the same—you are awarded 1 point for each correct answer and deducted $\frac{1}{4}$ point for each incorrect answer. No points are awarded or deducted for omitted questions and you can only be awarded points for a correct grid-in answer. Your raw score is rounded to the nearest whole number and then a conversion chart is used to find your official score from 0 to 800.

The multiple-choice answers are provided in A, B, C, D, E format. Be careful when omitting answers that you do not incorrectly complete your answer sheet.

The ACT Test

The ACT test has four subject tests, one of which is Mathematics. You will be tested on pre-algebra, elementary algebra, intermediate algebra, coordinate geometry, geometry, and elementary trigonometry. The Mathematics section consists of 60 multiple-choice questions to be answered in 60 minutes. All of the questions count the same, and there is no penalty for guessing.

Scoring The total number of correct answers is counted to get your raw score. No points are deducted for an incorrect answer. This raw score is then converted to the test score ranging from 1 (low) to 36 (high).

The multiple-choice answers alternate between A, B, C, D, E and F, G, H, J, K. When omitting answers, use the alternating answer choices to make sure you are answering the correct item number.

SAT and ACT Strategy: Time Management

Time management plays a very important role in your success. It is important that you do not spend extended periods of time on any one question. You may skip questions that are too difficult or time consuming. You can mark your booklet, so circle questions that you have skipped.

On the SAT *only*, the questions are arranged from easy to hard. To manage your time properly, you may want to omit the last 2 or 3 questions at the end of a section. They will be the most difficult and time consuming to complete. Use this time to look back over the section and redo any skipped questions. If there is time remaining once you are finished looking over the section, try to answer the last few questions.

On the ACT *only*, when you hear the announcement of five minutes remaining, go back and answer any skipped questions, even if it is only a guess. Put an answer for *every* question on the ACT, because there is no penalty for answering incorrectly.

EXAMPLE

You have 5 minutes to complete the questions below. Next to each problem, write the time you started.

1. What is the diameter of a circle with circumference 4?

   A. π B. 1   C. 4   D. $4 + \pi$ E. $\frac{4}{\pi}$

2. What is the slope of the line shown?

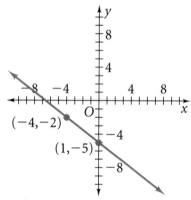

(−4,−2)

(1,−5)

   A. $-\frac{5}{3}$ B. $-\frac{3}{5}$ C. $\frac{3}{5}$ D. $\frac{5}{7}$ E. $\frac{5}{3}$

3. Which set of numbers could *not* be the measures of the sides of a right triangle?

   A. 5, 12, 13   B. 7, 24, 25   C. 9, 12, 15   D. 16, 30, 34   E. 20, 25, 30

4. What is the measure of an angle if twice the measure of its complement is 15 less than three times the angle?

   A. 16   B. 47   C. 39   D. 51   E. 141

5. The area of square A is 9 and the area of square B is 16. Which of the following is the difference of the square's perimeters?

   A. 1   B. 4   C. 5   D. 7   E. 8

After you've completed the exercises, answer the following questions.

6. Which problem was the most difficult for you?

7. Which problem took you the longest?

8. Was there a problem you did not answer because you ran out of time?

SAT and ACT Strategy: Eliminating Answers

You may solve a multiple-choice question by eliminating some of the answer choices. One way to eliminate answers is to use an "educated" guess. You may also use the answer choices to work backward through the problem. Each method requires less work and increases your chances of answering the question correctly.

EXAMPLE

In the figure at the right, the 4 shaded areas are triangles. What is the area of the unshaded region?

(A) $lw - 4x^2$

(B) $lw - 2x^2$

(C) $(l + x^2)(w + x^2)$

(D) $(l + 2x)(w + 2x)$

(E) $(l + 4x)(w + 4x)$

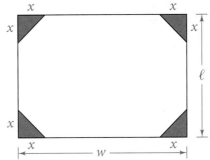

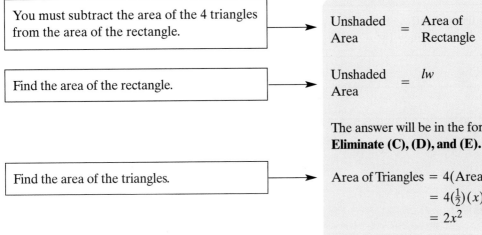

You must subtract the area of the 4 triangles from the area of the rectangle.

$$\text{Unshaded Area} = \text{Area of Rectangle} - \text{Area of Triangles}$$

Find the area of the rectangle.

$$\text{Unshaded Area} = lw - \text{Area of Triangles}$$

The answer will be in the form $lw - [\ \ ]$.
Eliminate (C), (D), and (E).

Find the area of the triangles.

$$\text{Area of Triangles} = 4(\text{Area of 1 triangle})$$
$$= 4(\tfrac{1}{2})(x)(x)$$
$$= 2x^2$$

The correct answer is (B).

$$\text{Unshaded Area} = lw - 2x^2$$

EXERCISES

1. What is the surface area of a cube with side length 3.1 in.?

   (A) 18.6 in.^{2} (B) 29.71 in.^{2} (C) 57.66 in.^{2} (D) 63.6 in.^{2} (E) 81.6 in.^{2}

2. The area of a rectangle is 36 in.^{2}. If the area is increased by 25%, which of the following could be the dimensions of the new rectangle?

   (A) 3 in. by 9 in.   (B) 5 in. by 6 in.   (C) 4 in. by 8 in.   (D) 6 in. by 8 in.   (E) 3 in. by 15 in.

3. If the perimeter of square A is double that of square B, then the area of A is how many times the area of B?

   (A) $\frac{1}{4}$ (B) $\frac{1}{2}$ (C) 1   (D) 4   (E) 8

SAT and ACT Strategy: Using a Variable

When you are solving a problem with unknown values, you may need to use variables. Choose letters that characterize the unknown value, for example d for distance or a for age.

You should know what the variables in common formulas represent. For example, the variables in the formula $V = \pi r^2 h$ stand for Volume, radius, and height. Often, the value of one or more of the variables needed for a formula is given in the problem. You can substitute given values into the formula to find the unknown variable.

EXAMPLE

What is the measure of an angle if twice the measure of its supplement is 24 more than the angle?

A. 90   B. 102   C. 110   D. 112   E. 136

Use x to represent the unknown angle. Use y to represent the supplement of the unknown angle.

$x =$ Unknown angle

$y =$ Supplement of x
$y = 180 - x$

Write an equation using x and y.
Substitute $180 - x$ for the value of y.

$$2y = 24 + x$$
$$2(180 - x) = 24 + x$$

Solve for x.

$$360 - 2x = 24 + x$$
$$360 - 24 - 2x = 24 - 24 + x$$
$$336 - 2x = x$$
$$336 - 2x + 2x = x + 2x$$
$$336 = 3x$$
$$\frac{336}{3} = \frac{3x}{3}$$
$$112 = x$$

● The correct answer is (D).

EXERCISES

1. Points A, B, and C are collinear. $\overline{AB}$ is 1 more than three times $\overline{BC}$, and $\overline{AC} = 17$. What is the length of $\overline{BC}$?

   (A) 3   (B) 4   (C) 13   (D) 15   (E) 17

2. A 20-foot ladder is placed against a wall. The bottom of the ladder is standing 12 feet from the base of the wall. If the top of the ladder slips down 4 feet, how many feet would the bottom of the ladder slip?

   (A) 18   (B) 16   (C) 12   (D) 4   (E) 2

3. What is the volume of a cube with the surface area of $54x^2$?

   (A) $3x$ (B) $9x^2$ (C) $81x^2$ (D) $27x^3$ (E) $81x^3$

SAT Strategy: Drawing a Diagram

Sometimes a problem will provide a diagram. These diagrams may have a note stating "Figure not drawn to scale". If the note is present, the diagram may mislead you toward an incorrect answer. It is best to redraw the diagram to solve the problem, matching the criteria provided.

If the diagram does not have a note, you can assume it is drawn to scale. A helpful strategy is to label the diagram with the information given in the problem. If a problem does not provide a diagram at all, it may be helpful to draw your own.

 EXAMPLE

In the figure shown at the right $\Delta = 15°$. If segments $\overline{CD}, \overline{CE}, \overline{CF}$, and $\overline{AB}$ were extended infinitely downward, how many points of intersection would there be in addition to B and C?

(A) 0

(B) 1

(C) 2

(D) 3

(E) 4

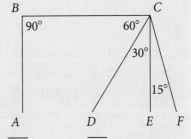

Note: Figure not drawn to scale.

If you use the current diagram to solve the problem, you might answer (C). But, you were told that the diagram was not drawn to scale.

Use the given information, $\Delta = 15°$, to redraw the diagram. Find the measurement of all marked angles.

$\angle ABC = 6\Delta° = 6 \times 15° = 90°$

$\angle BCD = 4\Delta° = 4 \times 15° = 60°$

$\angle DCE = 2\Delta° = 2 \times 15° = 30°$

$\angle ECF = \Delta° = 1 \times 15° = 15°$

$\overline{CD}$ intersects $\overline{AB}$ at 1 point.
The correct answer is (B).

EXERCISES

1. If $AB = 6BC$, what is the value of $\dfrac{BC}{AC}$?

(A) 1   (B) $\dfrac{1}{2}$ (C) $\dfrac{1}{3}$ (D) $\dfrac{1}{7}$ (E) $\dfrac{1}{9}$

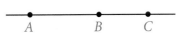

Note: Figure not drawn to scale.

2. The total surface area of a rectangular box is the sum of the areas of the 6 sides. If the length is 3 inches, the width is 4 inches, and the height is 7 inches, what is the box's total surface area, in square inches?

(A) 56   (B) 62   (C) 84   (D) 112   (E) 122

SAT Strategy: Writing Gridded Responses

Your score for the 10 gridded-response questions on the SAT test is based on the number of questions you answer correctly. Your answer must be gridded properly to count as a correct response. There is no penalty for incorrect answers so answer every question, even if you need to make a guess.

EXAMPLE

The figure shown below is a square with sides x and $5x - 6$. What is the area of the square?

Sides x and $5x - 6$ are equal. Solve for x.

$$x = 5x - 6$$
$$x - 5x = 5x - 5x - 6$$
$$-4x = -6$$
$$\frac{-4x}{-4} = \frac{-6}{-4}$$
$$x = \frac{3}{2}$$

$5x - 6$

x

Find the area of the square.

$$\text{Area} = (x)^2$$
$$\text{Area} = \left(\frac{3}{2}\right)^2$$
$$\text{Area} = 2.25$$

To enter the 2.25 into the grid, start at the left placing each digit or symbol into one column only. Then, fill the corresponding circle with the digit or symbol at the top of the column.

Fractions can be placed in the grid. However, mixed numbers should NOT be gridded. For example, 2.25 can be written as $2\frac{1}{4}$. If entered into the grid, $2\frac{1}{4}$ would be interpreted as $\frac{21}{4}$, making the answer incorrect. Mixed numbers must be changed into an improper fraction. $2\frac{1}{4} = \frac{9}{4}$.

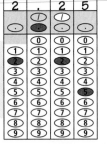

incorrect   **correct**

A decimal answer may have more digits than the grid allows. When recording these decimals, the grid must be filled to the thousandths place. For example, a repeating decimal such as .363636....., can be placed in the grid as either $\frac{11}{4}$, .363, or .364.

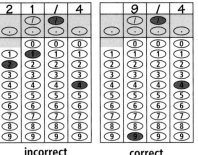

EXERCISES

Rewrite each answer to properly fit the grid.

1. $8\frac{1}{2}$

2. 1.6667

3. 0.745

4. $0.\overline{3}$

ACT Strategy: Answering the Question Asked

Many test questions require more than one step. As you solve these multiple-step problems, make sure you complete all of the steps necessary to solve the final question. As you read a problem, it may be helpful to underline the question being asked. After you solve, check over your work to ensure that you answered the final question.

EXAMPLE

The ratio of length to width of a quilt is 8:5. If the quilt has a length of 84 inches, what is its perimeter?

A. 52.5
B. 136.5
C. 273
D. 436.8
E. 4,410

Write and solve a proportion to find the width of the quilt.	$\dfrac{8}{5} = \dfrac{84}{w}$

$8w = 5(84)$ **Cross-Product Property**

$\dfrac{8w}{8} = \dfrac{420}{8}$ **Divide each side by 8.**

$w = 52.8$

The quilt's width is 52.5 inches. You can now find the quilt's perimeter.	$P = 2w + 2l$

$P = 2(52.5) + 2(84)$

$P = 273$ in.

● The correct answer is (C).

EXERCISES

1. The angles of a triangle are in the ratio 1:2:3. What is the measure of the largest angle?

A. 30° B. 60° C. 90° D. 120° E. 180°

2. In the drawing of the rectangle below, the diameter of the circle is 4 cm. What is the shaded area, in square centimeters?

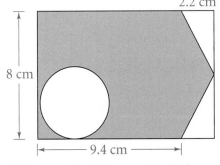

F. 53.8 G. 71.4 H. 73.6 J. 84 K. 92.8

ACT Strategy: Choosing "Cannot Be Determined"

Some test questions have an answer choice of "cannot be determined". When a multiple-choice question has this as an answer choice, try to solve the problem much like any other question. Determine what question needs to be answered and consider the facts provided in the problem. If you were not provided with all of the necessary information to solve the problem, then the correct answer choice is "cannot be determined".

1 EXAMPLE

In the drawing at the right, find the measure of $\angle EHF$.

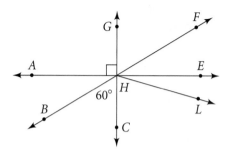

F. 15°
G. 30°
H. 60°
J. 90°
K. Cannot be determined

Consider the given information.	$\rightarrow$	$\angle BHC = 60°$ $\angle AHG = 90°$ Lines BHF and CHG interest at point H. $\angle BHC$ and $\angle BHA$ are complementary angles. $\angle FHG$ and $\angle FHE$ are complementary angles.
Use deductive reasoning.	$\rightarrow$	If lines BHF and CHG intersect, then $\angle BHC$ and $\angle GHF$ are congruent and $\angle GHF = 60°$. If $\angle GHF = 60°$, then $\angle EHF = 30°$.

● The correct answer is G.

2 EXAMPLE

In the figure at the right, $\overline{AB} \cong \overline{AC}$ and $\overline{BC}$ is 15 inches long. What is the perimeter, in inches, of $\triangle ABC$?

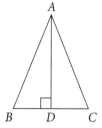

A. 17.5   B. 22.75   C. 30   D. 45   E. Cannot be determined

Consider the given information.	$\rightarrow$	$\overline{AB} \cong \overline{AC}$ and $\overline{BC}$ is 15 inches; $\triangle ABC$ is an isosceles triangle.
Determine whether or not the question can be answered with the given information.	$\rightarrow$	The length of $\overline{AB}$ and $\overline{AC}$ is needed to find the perimeter. It is not possible to calculate that information.

● The correct answer is E.

EXERCISES

1. Equilateral triangle ABC is shown at the right. If $\overline{BD}$ bisects $\angle ABC$, what is the measure of $\angle ABD$?

A. 15   B. 30   C. 45   D. 60   E. Cannot be determined

2. The trapezoid has a height of 115 cm and an area of 17,940 square centimeters. What is the length of the longer base?

115 cm

F. 183 cm   G. 172 cm   H. 166 cm   J. 156 cm   K. Cannot be determined

Tables

Table 1 Math Symbols

Symbol	Meaning	Page
$\ldots$	and so on	p. 4
$=$	is equal to, equality	p. 4
$\cdot, \times$	times (multiplication)	p. 4
n^2	square of n	p. 4
$+$	plus (addition)	p. 5
$-a$	opposite of a	p. 6
$^\circ$	degree(s)	p. 7
$-$	minus (subtraction)	p. 16
$\overleftrightarrow{AB}$	line through points A and B	p. 17
$\overline{AB}$	segment with endpoints A and B	p. 23
$\overrightarrow{AB}$	ray with endpoint A and through point B	p. 23
$\parallel$	is parallel to	p. 24
$(\ )$	parentheses for grouping	p. 30
AB	length of $\overline{AB}$	p. 31
$\lvert a \rvert$	absolute value of a	p. 31
$\cong$	is congruent to	p. 31
$\angle A$	angle with vertex A	p. 36
$\angle ABC$	angle with sides $\overrightarrow{BA}$ and $\overrightarrow{BC}$	p. 36
$m\angle A$	measure of angle A	p. 36
$\llcorner$	right angle symbol	p. 37
$\perp$	is perpendicular to	p. 45
d	distance	p. 53
(a, b)	ordered pair with x-coordinate a and y-coordinate b	p. 53
$\sqrt{x}$	nonnegative square root of x	p. 53
A	area	p. 62
s	length of a side	p. 62
b	base length	p. 62
h	height	p. 62
$\approx$	is approximately equal to	p. 62
d	diameter	p. 62
r	radius	p. 62
P	perimeter	p. 62
π	pi, ratio of the circumference of a circle to its diameter	p. 62
C	circumference	p. 62
$p \rightarrow q$	If p, then q	p. 82
$>$	is greater than	p. 85
$<$	is less than	p. 85
$p \leftrightarrow q$	p if and only if q	p. 88
$\neq$	is not equal to	p. 103
$\measuredangle$	angles	p. 135
$\triangle ABC$	triangle with vertices A, B, and C	p. 147
n-gon	polygon with n sides	p. 158
m	slope of a linear function	p. 166
b	y-intercept of a linear function	p. 166
$[\ ]$	brackets for grouping	p. 168
$\not\cong$	is not congruent to	p. 218
$\triangle$	triangles	p. 237
$\sim$	not	p. 281
$\geq$	is greater than or equal to	p. 288
$\leq$	is less than or equal to	p. 288
$\not>$	is not greater than	p. 291
$\not<$	is not less than	p. 291
$\square ABCD$	parallelogram with vertices A, B, C, and D	p. 312
$\square$	parallelograms	p. 312
$a{:}b, \frac{a}{b}$	ratio of a to b	p. 366
$\pm$	plus or minus	p. 372
$\sim$	is similar to	p. 373
$\overset{?}{=}$	Is this statement true?	p. 419
$\tan A$	tangent of $\angle A$	p. 432
$\sin A$	sine of $\angle A$	p. 439
$\cos A$	cosine of $\angle A$	p. 439
$\overrightarrow{AB}$	vector with initial point A and terminal point B	p. 452
$\langle x, y \rangle$	ordered pair notation for a vector	p. 452
$\vec{v}$	vector $\mathbf{v}$	p. 454
A'	image of A, A prime	p. 471
$\rightarrow$	maps to	p. 471
$\begin{bmatrix} 1 & 2 \\ 3 & 4 \end{bmatrix}$	matrix	p. 504
b_1, b_2	bases of a trapezoid	p. 540
d_1, d_2	lengths of diagonals	p. 541
a	apothem	p. 547
$\odot A$	circle with center A	p. 566
$\%$	percent	p. 566
$\overparen{AB}$	arc with endpoints A and B	p. 567
$\overparen{ABC}$	arc with endpoints A and C and containing B	p. 567
$m\overparen{AB}$	measure of $\overparen{AB}$	p. 567
$P(\text{event})$	probability of the event	p. 582
h	length of an altitude	p. 608
B	area of a base	p. 609
L.A.	lateral area	p. 609
S.A.	surface area	p. 609
ℓ	slant height	p. 617
V	volume	p. 624

Table 2 Formulas

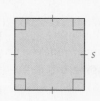

$P = 4s$
$A = s^2$

Square

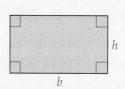

$P = 2b + 2h$
$A = bh$

Rectangle

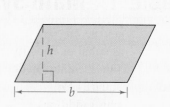

$A = bh$

Parallelogram

$A = \frac{1}{2}bh$

Triangle

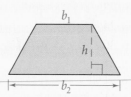

$A = \frac{1}{2}h(b_1 + b_2)$

Trapezoid

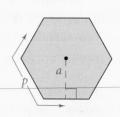

$A = \frac{1}{2}ap$

Regular Polygon

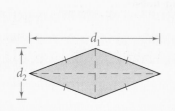

$A = \frac{1}{2}d_1d_2$

Rhombus

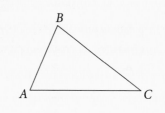

$m\angle A + m\angle B + m\angle C = 180$

Triangle Angle Sum

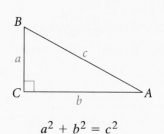

$a^2 + b^2 = c^2$

Pythagorean Theorem

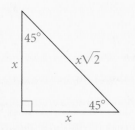

Ratio of sides = $1:1:\sqrt{2}$

45°-45°-90° Triangle

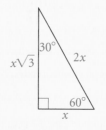

Ratio of sides = $1:\sqrt{3}:2$

30°-60°-90° Triangle

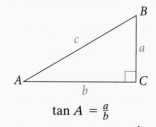

$\tan A = \frac{a}{b}$

$\sin A = \frac{a}{c}$ $\cos A = \frac{b}{c}$

Trigonometric Ratios

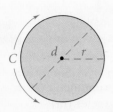

$C = \pi d$ or $C = 2\pi r$
$A = \pi r^2$

Circle

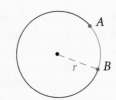

Length of $\overset{\frown}{AB} = \dfrac{m\overset{\frown}{AB}}{360} \cdot 2\pi r$

Arc

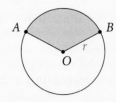

Area of sector $AOB = \dfrac{m\overset{\frown}{AB}}{360} \cdot \pi r^2$

Sector of a Circle

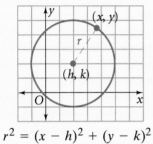

$r^2 = (x - h)^2 + (y - k)^2$

Equation of Circle

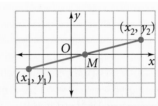

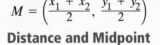

$d = \sqrt{(x_2 - x_1)^2 + (y_2 - y_1)^2}$

$M = \left(\dfrac{x_1 + x_2}{2}, \dfrac{y_1 + y_2}{2} \right)$

Distance and Midpoint

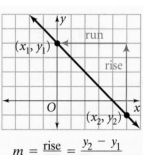

$m = \dfrac{\text{rise}}{\text{run}} = \dfrac{y_2 - y_1}{x_2 - x_1}$

Slope

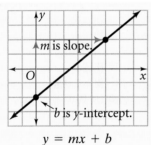

$y = mx + b$

Slope-intercept Form of a Linear Equation

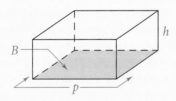

L.A. $= ph$
S.A. $=$ L.A. $+ 2B$
$V = Bh$

Right Prism

L.A. $= 2\pi rh$ or L.A. $= \pi dh$
S.A. $=$ L.A. $+ 2B$
$V = Bh$ or $V = \pi r^2 h$

Right Cylinder

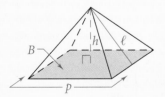

L.A. $= \frac{1}{2} p\ell$
S.A. $=$ L.A. $+ B$
$V = \frac{1}{3} Bh$

Regular Pyramid

L.A. $= \pi r \ell$
S.A. $=$ L.A. $+ B$
$V = \frac{1}{3} Bh$ or $V = \frac{1}{3} \pi r^2 h$

Right Cone

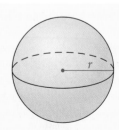

S.A. $= 4\pi r^2$
$V = \frac{4}{3} \pi r^3$

Sphere

Table 3 Measures

United States Customary	Metric

Length

12 inches (in.) = 1 foot (ft)	10 millimeters (mm) = 1 centimeter (cm)
36 in. = 1 yard (yd)	100 cm = 1 meter (m)
3 ft = 1 yard	1000 mm = 1 meter
5280 ft = 1 mile (mi)	1000 m = 1 kilometer (km)
1760 yd = 1 mile	

Area

144 square inches $(in.^2)$ = 1 square foot (ft^2)	100 square millimeters (mm^2) = 1 square centimeter (cm^2)
9 ft^2 = 1 square yard (yd^2)	10,000 cm^2 = 1 square meter (m^2)
43,560 ft^2 = 1 acre	10,000 m^2 = 1 hectare (ha)
4840 yd^2 = 1 acre	

Volume

1728 cubic inches $(in.^3)$ = 1 cubic foot (ft^3)	1000 cubic millimeters (mm^3) = 1 cubic centimeter (cm^3)
27 ft^3 = 1 cubic yard (yd^3)	1,000,000 cm^3 = 1 cubic meter (m^3)

Liquid Capacity

8 fluid ounces (fl oz) = 1 cup (c)	1000 milliliters (mL) = 1 liter (L)
2 c = 1 pint (pt)	1000 L = 1 kiloliter (kL)
2 pt = 1 quart (qt)	
4 qt = 1 gallon (gal)	

Weight or Mass

16 ounces (oz) = 1 pound (lb)	1000 milligrams (mg) = 1 gram (g)
2000 pounds = 1 ton (t)	1000 g = 1 kilogram (kg)
	1000 kg = 1 metric ton

Temperature

32°F = freezing point of water	0°C = freezing point of water
98.6°F = normal body temperature	37°C = normal body temperature
212°F = boiling point of water	100°C = boiling point of water

Time

60 seconds (s) = 1 minute (min)	365 days = 1 year (yr)
60 minutes = 1 hour (h)	52 weeks (approx.) = 1 year
24 hours = 1 day (d)	12 months = 1 year
7 days = 1 week (wk)	10 years = 1 decade
4 weeks (approx.) = 1 month (mo)	100 years = 1 century

Table 4 Properties of Real Numbers

Unless otherwise stated, a, b, c, and d are real numbers.

Identity Properties

Addition $a + 0 = a$ and $0 + a = a$

Multiplication $a \cdot 1 = a$ and $1 \cdot a = a$

Commutative Properties

Addition $a + b = b + a$

Multiplication $a \cdot b = b \cdot a$

Associative Properties

Addition $(a + b) + c = a + (b + c)$

Multiplication $(a \cdot b) \cdot c = a \cdot (b \cdot c)$

Inverse Properties

Addition

The sum of a number and its *opposite*, or *additive inverse*, is zero.

$$a + (-a) = 0 \text{ and } -a + a = 0$$

Multiplication

The reciprocal, or multiplicative inverse, of a rational number $\frac{a}{b}$ is $\frac{b}{a}$ $(a, b \neq 0)$.

$$a \cdot \frac{1}{a} = 1 \text{ and } \frac{1}{a} \cdot a = 1 \ (a \neq 0)$$

Distributive Properties

$a(b + c) = ab + ac$ $(b + c)a = ba + ca$

$a(b - c) = ab - ac$ $(b - c)a = ba - ca$

Properties of Equality

Addition   If $a = b$, then $a + c = b + c$.

Subtraction   If $a = b$, then $a - c = b - c$.

Multiplication   If $a = b$, then $a \cdot c = b \cdot c$.

Division   If $a = b$ and $c \neq 0$, then $\frac{a}{c} = \frac{b}{c}$.

Substitution   If $a = b$, then b can replace a in any expression.

Reflexive $a = a$

Symmetric   If $a = b$, then $b = a$.

Transitive   If $a = b$ and $b = c$, then $a = c$.

Properties of Proportions

$\frac{a}{b} = \frac{c}{d}$ $(a, b, c, d \neq 0)$ is equivalent to

(1) $ad = bc$ (2) $\frac{b}{a} = \frac{d}{c}$

(3) $\frac{a}{c} = \frac{b}{d}$ (4) $\frac{a + b}{b} = \frac{c + d}{d}$

Zero-Product Property

If $ab = 0$, then $a = 0$ or $b = 0$.

Properties of Inequality

Addition   If $a > b$ and $c \geq d$, then $a + c > b + d$.

Multiplication   If $a > b$ and $c > 0$, then $ac > bc$.
If $a > b$ and $c < 0$, then $ac < bc$.

Transitive   If $a > b$ and $b > c$, then $a > c$.

Comparison   If $a = b + c$ and $c > 0$, then $a > b$.

Properties of Exponents

For any nonzero numbers a and b, any positive number c, and any integers m and n,

Zero Exponent $a^0 = 1$

Negative Exponent $a^{-n} = \frac{1}{a^n}$

Product of Powers $a^m \cdot a^n = a^{m+n}$

Quotient of Powers $\frac{a^m}{a^n} = a^{m-n}$

Power to a Power $(c^m)^n = c^{mn}$

Product to a Power $(ab)^n = a^n b^n$

Quotient to a Power $\left(\frac{a}{b}\right)^n = \frac{a^n}{b^n}$

Properties of Square Roots

For any nonnegative numbers a and b, and any positive number c,

Product of Square Roots $\sqrt{a} \cdot \sqrt{b} = \sqrt{ab}$

Quotient of Square Roots $\frac{\sqrt{a}}{\sqrt{c}} = \sqrt{\frac{a}{c}}$

Table 5 Squares and Square Roots

Number n	Square n^2	Positive Square Root $\sqrt{n}$	Number n	Square n^2	Positive Square Root $\sqrt{n}$	Number n	Square n^2	Positive Square Root $\sqrt{n}$
1	1	1.000	51	2601	7.141	101	10,201	10.050
2	4	1.414	52	2704	7.211	102	10,404	10.100
3	9	1.732	53	2809	7.280	103	10,609	10.149
4	16	2.000	54	2916	7.348	104	10,816	10.198
5	25	2.236	55	3025	7.416	105	11,025	10.247
6	36	2.449	56	3136	7.483	106	11,236	10.296
7	49	2.646	57	3249	7.550	107	11,449	10.344
8	64	2.828	58	3364	7.616	108	11,664	10.392
9	81	3.000	59	3481	7.681	109	11,881	10.440
10	100	3.162	60	3600	7.746	110	12,100	10.488
11	121	3.317	61	3721	7.810	111	12,321	10.536
12	144	3.464	62	3844	7.874	112	12,544	10.583
13	169	3.606	63	3969	7.937	113	12,769	10.630
14	196	3.742	64	4096	8.000	114	12,996	10.677
15	225	3.873	65	4225	8.062	115	13,225	10.724
16	256	4.000	66	4356	8.124	116	13,456	10.770
17	289	4.123	67	4489	8.185	117	13,689	10.817
18	324	4.243	68	4624	8.246	118	13,924	10.863
19	361	4.359	69	4761	8.307	119	14,161	10.909
20	400	4.472	70	4900	8.367	120	14,400	10.954
21	441	4.583	71	5041	8.426	121	14,641	11.000
22	484	4.690	72	5184	8.485	122	14,884	11.045
23	529	4.796	73	5329	8.544	123	15,129	11.091
24	576	4.899	74	5476	8.602	124	15,376	11.136
25	625	5.000	75	5625	8.660	125	15,625	11.180
26	676	5.099	76	5776	8.718	126	15,876	11.225
27	729	5.196	77	5929	8.775	127	16,129	11.269
28	784	5.292	78	6084	8.832	128	16,384	11.314
29	841	5.385	79	6241	8.888	129	16,641	11.358
30	900	5.477	80	6400	8.944	130	16,900	11.402
31	961	5.568	81	6561	9.000	131	17,161	11.446
32	1024	5.657	82	6724	9.055	132	17,424	11.489
33	1089	5.745	83	6889	9.110	133	17,689	11.533
34	1156	5.831	84	7056	9.165	134	17,956	11.576
35	1225	5.916	85	7225	9.220	135	18,225	11.619
36	1296	6.000	86	7396	9.274	136	18,496	11.662
37	1369	6.083	87	7569	9.327	137	18,769	11.705
38	1444	6.164	88	7744	9.381	138	19,044	11.747
39	1521	6.245	89	7921	9.434	139	19,321	11.790
40	1600	6.325	90	8100	9.487	140	19,600	11.832
41	1681	6.403	91	8281	9.539	141	19,881	11.874
42	1764	6.481	92	8464	9.592	142	20,164	11.916
43	1849	6.557	93	8649	9.644	143	20,449	11.958
44	1936	6.633	94	8836	9.695	144	20,736	12.000
45	2025	6.708	95	9025	9.747	145	21,025	12.042
46	2116	6.782	96	9216	9.798	146	21,316	12.083
47	2209	6.856	97	9409	9.849	147	21,609	12.124
48	2304	6.928	98	9604	9.899	148	21,904	12.166
49	2401	7.000	99	9801	9.950	149	22,201	12.207
50	2500	7.071	100	10,000	10.000	150	22,500	12.247

Table 6 Trigonometric Ratios

Angle	Sine	Cosine	Tangent		Angle	Sine	Cosine	Tangent
1°	0.0175	0.9998	0.0175		46°	0.7193	0.6947	1.0355
2°	0.0349	0.9994	0.0349		47°	0.7314	0.6820	1.0724
3°	0.0523	0.9986	0.0524		48°	0.7431	0.6691	1.1106
4°	0.0698	0.9976	0.0699		49°	0.7547	0.6561	1.1504
5°	0.0872	0.9962	0.0875		50°	0.7660	0.6428	1.1918
6°	0.1045	0.9945	0.1051		51°	0.7771	0.6293	1.2349
7°	0.1219	0.9925	0.1228		52°	0.7880	0.6157	1.2799
8°	0.1392	0.9903	0.1405		53°	0.7986	0.6018	1.3270
9°	0.1564	0.9877	0.1584		54°	0.8090	0.5878	1.3764
10°	0.1736	0.9848	0.1763		55°	0.8192	0.5736	1.4281
11°	0.1908	0.9816	0.1944		56°	0.8290	0.5592	1.4826
12°	0.2079	0.9781	0.2126		57°	0.8387	0.5446	1.5399
13°	0.2250	0.9744	0.2309		58°	0.8480	0.5299	1.6003
14°	0.2419	0.9703	0.2493		59°	0.8572	0.5150	1.6643
15°	0.2588	0.9659	0.2679		60°	0.8660	0.5000	1.7321
16°	0.2756	0.9613	0.2867		61°	0.8746	0.4848	1.8040
17°	0.2924	0.9563	0.3057		62°	0.8829	0.4695	1.8807
18°	0.3090	0.9511	0.3249		63°	0.8910	0.4540	1.9626
19°	0.3256	0.9455	0.3443		64°	0.8988	0.4384	2.0503
20°	0.3420	0.9397	0.3640		65°	0.9063	0.4226	2.1445
21°	0.3584	0.9336	0.3839		66°	0.9135	0.4067	2.2460
22°	0.3746	0.9272	0.4040		67°	0.9205	0.3907	2.3559
23°	0.3907	0.9205	0.4245		68°	0.9272	0.3746	2.4751
24°	0.4067	0.9135	0.4452		69°	0.9336	0.3584	2.6051
25°	0.4226	0.9063	0.4663		70°	0.9397	0.3420	2.7475
26°	0.4384	0.8988	0.4877		71°	0.9455	0.3256	2.9042
27°	0.4540	0.8910	0.5095		72°	0.9511	0.3090	3.0777
28°	0.4695	0.8829	0.5317		73°	0.9563	0.2924	3.2709
29°	0.4848	0.8746	0.5543		74°	0.9613	0.2756	3.4874
30°	0.5000	0.8660	0.5774		75°	0.9659	0.2588	3.7321
31°	0.5150	0.8572	0.6009		76°	0.9703	0.2419	4.0108
32°	0.5299	0.8480	0.6249		77°	0.9744	0.2250	4.3315
33°	0.5446	0.8387	0.6494		78°	0.9781	0.2079	4.7046
34°	0.5592	0.8290	0.6745		79°	0.9816	0.1908	5.1446
35°	0.5736	0.8192	0.7002		80°	0.9848	0.1736	5.6713
36°	0.5878	0.8090	0.7265		81°	0.9877	0.1564	6.3138
37°	0.6018	0.7986	0.7536		82°	0.9903	0.1392	7.1154
38°	0.6157	0.7880	0.7813		83°	0.9925	0.1219	8.1443
39°	0.6293	0.7771	0.8098		84°	0.9945	0.1045	9.5144
40°	0.6428	0.7660	0.8391		85°	0.9962	0.0872	11.4301
41°	0.6561	0.7547	0.8693		86°	0.9976	0.0698	14.3007
42°	0.6691	0.7431	0.9004		87°	0.9986	0.0523	19.0811
43°	0.6820	0.7314	0.9325		88°	0.9994	0.0349	28.6363
44°	0.6947	0.7193	0.9657		89°	0.9998	0.0175	57.2900
45°	0.7071	0.7071	1.0000		90°	1.0000	0.0000	

Tables

Postulates, Theorems, and Constructions

Chapter 1: Tools of Geometry

Postulate 1-1
Through any two points there is exactly one line. (p. 18)

Postulate 1-2
If two lines intersect, then they intersect in exactly one point. (p. 18)

Postulate 1-3
If two planes intersect, then they intersect in exactly one line. (p. 18)

Postulate 1-4
Through any three noncollinear points there is exactly one plane. (p. 19)

Postulate 1-5
Ruler Postulate
The points of a line can be put into one-to-one correspondence with the real numbers so that the distance between any two points is the absolute value of the difference of the corresponding numbers. (p. 31)

Postulate 1-6
Segment Addition Postulate
If three points A, B, and C are collinear and B is between A and C, then $AB + BC = AC$. (p. 32)

Postulate 1-7
Protractor Postulate
Let $\overrightarrow{OA}$ and $\overrightarrow{OB}$ be opposite rays in a plane. $\overrightarrow{OA}$, $\overrightarrow{OB}$, and all the rays with endpoint O that can be drawn on one side of $\overleftrightarrow{AB}$ can be paired with the real numbers from 0 to 180 so that
a. $\overrightarrow{OA}$ is paired with 0 and $\overrightarrow{OB}$ is paired with 180.
b. If $\overrightarrow{OC}$ is paired with x and $\overrightarrow{OD}$ is paired with y, then $m\angle COD = |x - y|$. (p. 37)

Postulate 1-8
Angle Addition Postulate
If point B is in the interior of $\angle AOC$, then $m\angle AOB + m\angle BOC = m\angle AOC$.
If $\angle AOC$ is a straight angle, then $m\angle AOB + m\angle BOC = 180$. (p. 38)

The Distance Formula
The distance d between two points $A(x_1, y_1)$ and $B(x_2, y_2)$ is $d = \sqrt{(x_2 - x_1)^2 + (y_2 - y_1)^2}$. (p. 53)
• Proof on p. 421, Exercise 34

The Midpoint Formula
The coordinates of the midpoint M of $\overline{AB}$ with endpoints $A(x_1, y_1)$ and $B(x_2, y_2)$ are the following.
$M\left(\dfrac{x_1 + x_2}{2}, \dfrac{y_1 + y_2}{2}\right)$ (p. 55)

The Distance Formula (Three Dimensions)
In a three-dimensional coordinate system, the distance between two points (x_1, y_1, z_1) and (x_2, y_2, z_2) can be found using this extension of the Distance Formula.
$d = \sqrt{(x_2 - x_1)^2 + (y_2 - y_1)^2 + (z_2 - z_1)^2}$ (p. 58)

Postulate 1-9
If two figures are congruent, then their areas are equal. (p. 64)

Postulate 1-10
The area of a region is the sum of the areas of its nonoverlapping parts. (p. 64)

Chapter 2: Reasoning and Proof

Law of Detachment
If a conditional is true and its hypothesis is true, then its conclusion is true. In symbolic form: If $p \rightarrow q$ is a true statement and p is true, then q is true. (p. 95)

Law of Syllogism
If $p \rightarrow q$ and $q \rightarrow r$ are true statements, then $p \rightarrow r$ is a true statement. (p. 95)

Properties of Congruence
Reflexive Property
$\overline{AB} \cong \overline{AB}$ and $\angle A \cong \angle A$
Symmetric Property
If $\overline{AB} \cong \overline{CD}$, then $\overline{CD} \cong \overline{AB}$.
If $\angle A \cong \angle B$, then $\angle B \cong \angle A$.
Transitive Property
If $\overline{AB} \cong \overline{CD}$ and $\overline{CD} \cong \overline{EF}$, then $\overline{AB} \cong \overline{EF}$.
If $\angle A \cong \angle B$ and $\angle B \cong \angle C$, then $\angle A \cong \angle C$. (p. 105)

Theorem 2-1
Vertical Angles Theorem
Vertical angles are congruent. (p. 110)
- Proof on p. 111

Theorem 2-2
Congruent Supplements Theorem
If two angles are supplements of the same angle (or of congruent angles), then the two angles are congruent. (p. 111)
- Proofs on p. 112, Example 2; p. 114, Exercise 27

Theorem 2-3
Congruent Complements Theorem
If two angles are complements of the same angle (or of congruent angles), then the two angles are congruent. (p. 112)
- Proofs on p. 113, Exercise 7; p. 114, Exercise 28

Theorem 2-4
All right angles are congruent. (p. 112)
- Proof on p. 113, Exercise 14

Theorem 2-5
If two angles are congruent and supplementary, then each is a right angle. (p. 112)
- Proof on p. 114, Exercise 21

Chapter 3: Parallel and Perpendicular Lines

Postulate 3-1
Corresponding Angles Postulate
If a transversal intersects two parallel lines, then corresponding angles are congruent. (p. 128)

Theorem 3-1
Alternate Interior Angles Theorem
If a transversal intersects two parallel lines, then alternate interior angles are congruent. (p. 128)
- Proof on p. 129

Theorem 3-2
Same-Side Interior Angles Theorem
If a transversal intersects two parallel lines, then same-side interior angles are supplementary. (p. 128)
- Proof on p. 132, Exercise 29

Theorem 3-3
Alternate Exterior Angles Theorem
If a transversal intersects two parallel lines, then alternate exterior angles are congruent. (p. 130)
- Proof on p. 129, Example 3

Theorem 3-4
Same-Side Exterior Angles Theorem
If a transversal intersects two parallel lines, then same-side exterior angles are supplementary. (p. 130)
- Proof on p. 129, Quick Check 3

Postulate 3-2
Converse of the Corresponding Angles Postulate
If two lines and a transversal form corresponding angles that are congruent, then the two lines are parallel. (p. 134)

Theorem 3-5
Converse of the Alternate Interior Angles Theorem
If two lines and a transversal form alternate interior angles that are congruent, then the two lines are parallel. (p. 135)
- Proof on p. 135

Theorem 3-6
Converse of the Same-Side Interior Angles Theorem
If two lines and a transversal form same-side interior angles that are supplementary, then the two lines are parallel. (p. 135)
- Proofs on p. 138, Exercise 22 and p. 139, Exercise 40

Theorem 3-7
Converse of the Alternate Exterior Angles Theorem
If two lines and a transversal form alternate exterior angles that are congruent, then the lines are parallel. (p. 136)
- Proof on p. 136

Theorem 3-8
Converse of the Same-Side Exterior Angles Theorem
If two lines and a transversal form same-side exterior angles that are supplementary, then the lines are parallel. (p. 136)
- Proof on p. 138, Exercise 27

Theorem 3-9
If two lines are parallel to the same line, then they are parallel to each other. (p. 141)
- Proofs on p. 143, Exercise 3; p. 179, Exercise 37

Theorem 3-10
In a plane, if two lines are perpendicular to the same line, then they are parallel to each other. (p. 141)
- Proofs on p. 141; p. 143, Exercise 12; p. 179, Exercise 38

Theorem 3-11
In a plane, if a line is perpendicular to one of two parallel lines, then it is also perpendicular to the other. (p. 142)
- Proof on p. 143, Exercise 11

Theorem 3-12
Triangle Angle-Sum Theorem
The sum of the measures of the angles of a triangle is 180. (p. 147)
- Proof on p. 147

Theorem 3-13
Triangle Exterior Angle Theorem
The measure of each exterior angle of a triangle equals the sum of the measures of its two remote interior angles. (p. 149)
- Proof on p. 152, Exercise 35
  Corollary
  The measure of an exterior angle of a triangle is greater than the measure of each of its remote interior angles. (p. 290)
  - Proof on p. 290

Parallel Postulate
Through a point not on a line, there is one and only one line parallel to the given line. (p. 154)

Spherical Geometry Parallel Postulate
Through a point not on a line, there is no line parallel to the given line. (p. 154)

Theorem 3-14
Polygon Angle-Sum Theorem
The sum of the measures of the angles of an n-gon is $(n - 2)180$. (p. 159)
- Proof on p. 163, Exercise 54

Theorem 3-15
Polygon Exterior Angle-Sum Theorem
The sum of the measures of the exterior angles of a polygon, one at each vertex, is 360. (p. 160)
- Proofs on p. 156 (using a computer) and p. 162, Exercise 46

Slopes of Parallel Lines
If two nonvertical lines are parallel, their slopes are equal. If the slopes of two distinct nonvertical lines are equal, the lines are parallel. Any two vertical lines are parallel. (p. 174)
- Proofs on pp. 387–388, Exercises 42, 43

Slopes of Perpendicular Lines
If two nonvertical lines are perpendicular, the product of their slopes is -1. If the slopes of two lines have a product of -1, the lines are perpendicular. Any horizontal line and vertical line are perpendicular. (p. 175)
- Proofs on p. 346, Exercise 29 and p. 353, Exercise 41

Chapter 4: Congruent Triangles

Theorem 4-1
If the two angles of one triangle are congruent to two angles of another triangle, then the third angles are congruent. (p. 199)
- Proof on p. 202, Exercise 45

Postulate 4-1
Side-Side-Side (SSS) Postulate
If the three sides of one triangle are congruent to the three sides of another triangle, then the two triangles are congruent. (p. 205)

Postulate 4-2
Side-Angle-Side (SAS) Postulate
If two sides and the included angle of one triangle are congruent to two sides and the included angle of another triangle, then the two triangles are congruent. (p. 206)

Postulate 4-3
Angle-Side-Angle (ASA) Postulate
If two angles and the included side of one triangle are congruent to two angles and the included side of another triangle, then the two triangles are congruent. (p. 213)

Theorem 4-2
Angle-Angle-Side (AAS) Theorem
If two angles and a nonincluded side of one triangle are congruent to two angles and the corresponding nonincluded side of another triangle, then the triangles are congruent. (p. 214)
- Proof on p. 214

Theorem 4-3
Isosceles Triangle Theorem
If two sides of a triangle are congruent, then the angles opposite those sides are congruent. (p. 228)
- Proofs on p. 229; p. 231, Exercise 15
  Corollary
  If a triangle is equilateral, then the triangle is equiangular. (p. 230)
  - Proof on p. 231, Exercise 18

Theorem 4-4
Converse of the Isosceles Triangle Theorem
If two angles of a triangle are congruent, then the sides opposite the angles are congruent. (p. 228)
- Proof on p. 231, Exercise 16
  Corollary
  If a triangle is equiangular, then the triangle is equilateral. (p. 230)
  - Proof on p. 231, Exercise 18

Theorem 4-5

The bisector of the vertex angle of an isosceles triangle is the perpendicular bisector of the base. (p. 228)
- Proof on p. 232, Exercise 29

Theorem 4-6

Hypotenuse-Leg (HL) Theorem

If the hypotenuse and a leg of one right triangle are congruent to the hypotenuse and a leg of another right triangle, then the triangles are congruent. (p. 235)
- Proof on p. 235

Chapter 5: Relationships Within Triangles

Theorem 5-1

Triangle Midsegment Theorem

If a segment joins the midpoints of two sides of a triangle, then the segment is parallel to the third side, and is half its length. (p. 260)
- Proof on p. 260

Theorem 5-2

Perpendicular Bisector Theorem

If a point is on the perpendicular bisector of a segment, then it is equidistant from the endpoints of the segment. (p. 265)
- Proof on p. 269, Exercise 40

Theorem 5-3

Converse of the Perpendicular Bisector Theorem

If a point is equidistant from the endpoints of a segment, then it is on the perpendicular bisector of the segment. (p. 265)
- Proof on p. 269, Exercise 41

Theorem 5-4

Angle Bisector Theorem

If a point is on the bisector of an angle, then the point is equidistant from the sides of the angle. (p. 266)
- Proof on p. 269, Exercise 43

Theorem 5-5

Converse of the Angle Bisector Theorem

If a point in the interior of an angle is equidistant from the sides of the angle, then the point is on the angle bisector. (p. 266)
- Proof on p. 269, Exercise 44

Theorem 5-6

The perpendicular bisectors of the sides of a triangle are concurrent at a point equidistant from the vertices. (p. 273)
- Proof on p. 277, Exercise 29

Theorem 5-7

The bisectors of the angles of a triangle are concurrent at a point equidistant from the sides. (p. 273)
- Proof on p. 277, Exercise 30

Theorem 5-8

The medians of a triangle are concurrent at a point that is two thirds the distance from each vertex to the midpoint of the opposite side. (p. 274)
- Proof on p. 352, Exercise 35

Theorem 5-9

The lines that contain the altitudes of a triangle are concurrent. (p. 275)
- Proof on p. 352, Exercise 36

Comparison Property of Inequality

If $a = b + c$ and $c > 0$, then $a > b$. (p. 289)
- Proof on p. 289

Theorem 5-10

If two sides of a triangle are not congruent, then the larger angle lies opposite the longer side. (p. 290)
- Proof on p. 294, Exercise 33

Theorem 5-11

If two angles of a triangle are not congruent, then the longer side lies opposite the larger angle. (p. 291)
- Proof on p. 291

Theorem 5-12

Triangle Inequality Theorem

The sum of the lengths of any two sides of a triangle is greater than the length of the third side. (p. 292)
- Proof on p. 294, Exercise 41

Chapter 6: Quadrilaterals

Theorem 6-1

Opposite sides of a parallelogram are congruent. (p. 312)
- Proofs on p. 312; p. 317, Exercise 35

Theorem 6-2

Opposite angles of a parallelogram are congruent. (p. 313)
- Proof on p. 317, Exercise 36

Theorem 6-3

The diagonals of a parallelogram bisect each other. (p. 314)
- Proofs on p. 314; p. 350, Exercise 2

Theorem 6-4

If three (or more) parallel lines cut off congruent segments on one transversal, then they cut off congruent segments on every transversal. (p. 315)
- Proof on p. 318, Exercise 52

Theorem 6-5

If both pairs of opposite sides of a quadrilateral are congruent, then the quadrilateral is a parallelogram. (p. 321)
- Proof on p. 321

Theorem 6-6

If both pairs of opposite angles of a quadrilateral are congruent, then the quadrilateral is a parallelogram. (p. 321)
- Proof on p. 325, Exercise 12

Theorem 6-7

If the diagonals of a quadrilateral bisect each other, then the quadrilateral is a parallelogram. (p. 322)
- Proof on p. 322

Theorem 6-8

If one pair of opposite sides of a quadrilateral are both congruent and parallel, then the quadrilateral is a parallelogram. (p. 322)
- Proof on p. 325, Exercise 18

Theorem 6-9

Each diagonal of a rhombus bisects two angles of the rhombus. (p. 329)
- Proof on p. 329

Theorem 6-10

The diagonals of a rhombus are perpendicular. (p. 330)
- Proof on p. 333, Exercise 22

Theorem 6-11

The diagonals of a rectangle are congruent. (p. 330)
- Proof on p. 330

Theorem 6-12

If one diagonal of a parallelogram bisects two angles of the parallelogram, then the parallelogram is a rhombus. (p. 331)
- Proof on p. 334, Exercise 39

Theorem 6-13

If the diagonals of a parallelogram are perpendicular, then the parallelogram is a rhombus. (p. 331)
- Proof on p. 333, Exercise 23

Theorem 6-14

If the diagonals of a parallelogram are congruent, then the parallelogram is a rectangle. (p. 331)
- Proof on p. 334, Exercise 40

Theorem 6-15

The base angles of an isosceles trapezoid are congruent. (p. 336)
- Proof on p. 340, Exercise 38

Theorem 6-16

The diagonals of an isosceles trapezoid are congruent. (p. 337)
- Proofs on p. 337; p. 350, Exercise 3

Theorem 6-17

The diagonals of a kite are perpendicular. (p. 338)
- Proof on p. 338

Theorem 6-18

Trapezoid Midsegment Theorem

(1) The midsegment of a trapezoid is parallel to the bases.
(2) The length of a midsegment of a trapezoid is half the sum of the lengths of the bases. (p. 348)
- Proof on p. 349, Quick Check 1

Chapter 7: Similarity

Postulate 7-1

Angle-Angle Similarity (AA ~) Postulate

If two angles of one triangle are congruent to two angles of another triangle, then the triangles are similar. (p. 382)

Theorem 7-1

Side-Angle-Side Similarity (SAS ~) Theorem

If an angle of one triangle is congruent to an angle of a second triangle, and the sides including the two angles are proportional, then the triangles are similar. (p. 383)
- Proof on p. 383

Theorem 7-2

Side-Side-Side Similarity (SSS ~) Theorem

If the corresponding sides of two triangles are proportional, then the triangles are similar. (p. 383)
- Proof on p. 383

Theorem 7-3

The altitude to the hypotenuse of a right triangle divides the triangle into two triangles that are similar to the original triangle and to each other. (p. 392)

- Proof on p. 392

  Corollary 1

  The length of the altitude to the hypotenuse of a right triangle is the geometric mean of the lengths of the segments of the hypotenuse. (p. 392)

  - Proof on p. 392

  Corollary 2

  The altitude to the hypotenuse of a right triangle separates the hypotenuse so that the length of each leg of the triangle is the geometric mean of the length of the adjacent hypotenuse segment and the length of the hypotenuse. (p. 393)

  - Proof on p. 393

Theorem 7-4

Side-Splitter Theorem

If a line is parallel to one side of a triangle and intersects the other two sides, then it divides those sides proportionally. (p. 398)

- Proof on p. 398

  Converse

  If a line divides two sides of a triangle proportionally, then it is parallel to the third side.

  - Proof on p. 402, Exercise 34

  Corollary

  If three parallel lines intersect two transversals, then the segments intercepted on the transversals are proportional. (p. 399)

  - Proof on p. 402, Exercise 35

Theorem 7-5

Triangle-Angle-Bisector Theorem

If a ray bisects an angle of a triangle, then it divides the opposite side into two segments that are proportional to the other two sides of the triangle. (p. 400)

- Proof on p. 400

Chapter 8: Right Triangles and Trigonometry

Theorem 8-1

Pythagorean Theorem

In a right triangle, the sum of the squares of the lengths of the legs is equal to the square of the length of the hypotenuse.

$a^2 + b^2 = c^2$ (p. 417)

- Proofs on p. 395, Exercise 53; p. 416; p. 422, Exercise 48; p. 545; p. 692, Exercise 36

Theorem 8-2

Converse of the Pythagorean Theorem

If the square of the length of one side of a triangle is equal to the sum of the squares of the lengths of the other two sides, then the triangle is a right triangle. (p. 419)

- Proof on p. 423, Exercise 58

Theorem 8-3

If the square of the length of the longest side of a triangle is greater than the sum of the squares of the lengths of the other two sides, the triangle is obtuse. (p. 419)

Theorem 8-4

If the square of the length of the longest side of a triangle is less than the sum of the squares of the lengths of the other two sides, the triangle is acute. (p. 419)

Theorem 8-5

45°-45°-90° Triangle Theorem

In a 45°-45°-90° triangle, both legs are congruent and the length of the hypotenuse is $\sqrt{2}$ times the length of a leg.

$\text{hypotenuse} = \sqrt{2} \cdot \text{leg}$ (p. 425)

- Proof on p. 425

Theorem 8-6

30°-60°-90° Triangle Theorem

In a 30°-60°-90° triangle, the length of the hypotenuse is twice the length of the shorter leg. The length of the longer leg is $\sqrt{3}$ times the length of the shorter leg.

$\text{hypotenuse} = 2 \cdot \text{shorter leg}$
$\text{longer leg} = \sqrt{3} \cdot \text{shorter leg}$ (p. 426)

- Proof on p. 427

Chapter 9: Transformations

Theorem 9-1

A translation or rotation is a composition of two reflections. (p. 506)

Theorem 9-2

A composition of reflections across two parallel lines is a translation. (p. 507)

Theorem 9-3

A composition of reflections across two intersecting lines is a rotation. (p. 507)

Theorem 9-4

Fundamental Theorem of Isometries

In a plane, one of two congruent figures can be mapped onto the other by a composition of at most three reflections. (p. 508)

Theorem 9-5
Isometry Classification Theorem
There are only four isometries. They are reflection, translation, rotation, and glide reflection. (p. 509)

Theorem 9-6
Every triangle tessellates. (p. 516)

Theorem 9-7
Every quadrilateral tessellates. (p. 516)

Chapter 10: Area

Theorem 10-1
Area of a Rectangle
The area of a rectangle is the product of its base and height.
$A = bh$ (p. 534)

Theorem 10-2
Area of a Parallelogram
The area of a parallelogram is the product of a base and the corresponding height.
$A = bh$ (p. 534)

Theorem 10-3
Area of a Triangle
The area of a triangle is half the product of a base and the corresponding height.
$A = \frac{1}{2}bh$ (p. 535)

Theorem 10-4
Area of a Trapezoid
The area of a trapezoid is half the product of the height and the sum of the bases.
$A = \frac{1}{2}h(b_1 + b_2)$ (p. 540)

Theorem 10-5
Area of a Rhombus or a Kite
The area of a rhombus or a kite is half the product of the lengths of its diagonals.
$A = \frac{1}{2}d_1d_2$ (p. 541)

Theorem 10-6
Area of a Regular Polygon
The area of a regular polygon is half the product of the apothem and the perimeter.
$A = \frac{1}{2}ap$ (p. 547)

Theorem 10-7
Perimeters and Areas of Similar Figures
If the similarity ratio of two similar figures is $\frac{a}{b}$, then
(1) the ratio of their perimeters is $\frac{a}{b}$ and
(2) the ratio of their areas is $\frac{a^2}{b^2}$. (p. 554)

Theorem 10-8
Area of a Triangle Given SAS
The area of a triangle is one half the product of the lengths of two sides and the sine of the included angle.
Area of $\triangle ABC = \frac{1}{2}bc(\sin A)$ (p. 561)
- Proof on p. 560

Postulate 10-1
Arc Addition Postulate
The measure of the arc formed by two adjacent arcs is the sum of the measures of the two arcs. (p. 567)

Theorem 10-9
Circumference of a Circle
The circumference of a circle is π times the diameter.
$C = \pi d$ or $C = 2\pi r$ (p. 568)

Theorem 10-10
Arc Length
The length of an arc of a circle is the product of the ratio $\frac{\text{measure of the arc}}{360}$ and the circumference of the circle.
length of $\overset{\frown}{AB} = \frac{m\overset{\frown}{AB}}{360} \cdot 2\pi r$ (p. 569)

Theorem 10-11
Area of a Circle
The area of a circle is the product of π and the square of the radius. $A = \pi r^2$ (p. 576)

Theorem 10-12
Area of a Sector of a Circle
The area of a sector of a circle is the product of the ratio $\frac{\text{measure of the arc}}{360}$ and the area of the circle.
Area of sector $AOB = \frac{m\overset{\frown}{AB}}{360} \cdot \pi r^2$ (p. 576)

Chapter 11: Surface Area and Volume

Theorem 11-1
Lateral and Surface Areas of a Prism
The lateral area of a right prism is the product of the perimeter of the base and the height.
L.A. $= ph$

The surface area of a right prism is the sum of the lateral area and the areas of the two bases.
S.A. $=$ L.A. $+ 2B$ (p. 610)

Theorem 11-2
Lateral and Surface Areas of a Cylinder
The lateral area of a right cylinder is the product of the circumference of the base and the height of the cylinder.
L.A. $= 2\pi rh$, or L.A. $= \pi dh$

The surface area of a right cylinder is the sum of the lateral area and the areas of the two bases.
S.A. $=$ L.A. $+ 2B$, or S.A. $= 2\pi rh + 2\pi r^2$ (p. 610)

Theorem 11-3

Lateral and Surface Areas of a Regular Pyramid

The lateral area of a regular pyramid is half the product of the perimeter of the base and the slant height.

L.A. $= \frac{1}{2}p\ell$

The surface area of a regular pyramid is the sum of the lateral area and the area of the base.

S.A. $=$ L.A. $+ B$ (p. 618)

Theorem 11-4

Lateral and Surface Areas of a Cone

The lateral area of a right cone is half the product of the circumference of the base and the slant height.

L.A. $= \frac{1}{2} \cdot 2\pi r\ell$, or L.A. $= \pi r\ell$

The surface area of a right cone is the sum of the lateral area and the area of the base.

S.A. $=$ L.A. $+ B$ (p. 619)

Theorem 11-5

Cavalieri's Principle

If two space figures have the same height and the same cross-sectional area at every level, then they have the same volume. (p. 625)

Theorem 11-6

Volume of a Prism

The volume of a prism is the product of the area of a base and the height of the prism.

$V = Bh$ (p. 625)

Theorem 11-7

Volume of a Cylinder

The volume of a cylinder is the product of the area of the base and the height of the cylinder.

$V = Bh$, or $V = \pi r^2 h$ (p. 626)

Theorem 11-8

Volume of a Pyramid

The volume of a pyramid is one third the product of the area of the base and the height of the pyramid.

$V = \frac{1}{3}Bh$ (p. 632)

Theorem 11-9

Volume of a Cone

The volume of a cone is one third the product of the area of the base and the height of the cone.

$V = \frac{1}{3}Bh$, or $V = \frac{1}{3}\pi r^2 h$ (p. 633)

Theorem 11-10

Surface Area of a Sphere

The surface area of a sphere is four times the product of π and the square of the radius of the sphere.

S.A. $= 4\pi r^2$ (p. 638)

Theorem 11-11

Volume of a Sphere

The volume of a sphere is four thirds the product of π and the cube of the radius of the sphere.

$V = \frac{4}{3}\pi r^3$ (p. 640)

Theorem 11-12

Areas and Volumes of Similar Solids

If the similarity ratio of two similar solids is $a : b$, then
(1) the ratio of their corresponding areas is $a^2 : b^2$, and
(2) the ratio of their volumes is $a^3 : b^3$. (p. 647)

Chapter 12: Circles

Theorem 12-1

If a line is tangent to a circle, then the line is perpendicular to the radius drawn to the point of tangency. (p. 662)
• Proof on p. 667, Exercise 36

Theorem 12-2

If a line in the plane of a circle is perpendicular to a radius at its endpoint on the circle, then the line is tangent to the circle. (p. 663)
• Proof on p. 667, Exercise 37

Theorem 12-3

The two segments tangent to a circle from a point outside the circle are congruent. (p. 664)
• Proof on p. 667, Exercise 38

Theorem 12-4

Within a circle or in congruent circles
(1) Congruent central angles have congruent chords.
(2) Congruent chords have congruent arcs.
(3) Congruent arcs have congruent central angles. (p. 670)
• Proofs on p. 674, Exercises 23, 24; p. 675, Exercise 35

Theorem 12-5

Within a circle or in congruent circles
(1) Chords equidistant from the center are congruent.
(2) Congruent chords are equidistant from the center. (p. 671)
• Proofs on p. 671; p. 675, Exercise 37

Theorem 12-6

In a circle, a diameter that is perpendicular to a chord bisects the chord and its arcs. (p. 672)
• Proof on p. 674, Exercise 25

Theorem 12-7

In a circle, a diameter that bisects a chord (that is not a diameter) is perpendicular to the chord. (p. 672)
- Proof on p. 672

Theorem 12-8

In a circle, the perpendicular bisector of a chord contains the center of the circle. (p. 672)
- Proof on p. 678, Exercise 36

Theorem 12-9

Inscribed Angle Theorem

The measure of an inscribed angle is half the measure of its intercepted arc. (p. 679)
- Proofs on p. 679; p. 683, Exercises 40, 41
  Corollary 1
  Two inscribed angles that intercept the same arc are congruent. (p. 680)
  - Proof on p. 684, Exercise 42
  Corollary 2
  An angle inscribed in a semicircle is a right angle. (p. 680)
  - Proof on p. 684, Exercise 43
  Corollary 3
  The opposite angles of a quadrilateral inscribed in a circle are supplementary. (p. 680)
  - Proof on p. 684, Exercise 44

Theorem 12-10

The measure of an angle formed by a tangent and a chord is half the measure of the intercepted arc. (p. 680)
- Proof on p. 684, Exercise 45

Theorem 12-11

The measure of an angle formed by two lines that
(1) intersect inside a circle is half the sum of the measures of the intercepted arcs.
(2) intersect outside a circle is half the difference of the measures of the intercepted arcs. (p. 687)
- Proofs on p. 688; p. 692, Exercises 29, 30

Theorem 12-12

For a given point and circle, the product of the lengths of the two segments from the point to the circle is constant along any line through the point and circle. (p. 689)
- Proofs on p. 689; p. 692, Exercises 31–33

Theorem 12-13

An equation of a circle with center (h, k) and radius r is $(x - h)^2 + (y - k)^2 = r^2$. (p. 695)

Constructions

Construction 1
Congruent Segments
Construct a segment congruent to a given segment. (p. 44)

Construction 2
Congruent Angles
Construct an angle congruent to a given angle. (p. 45)

Construction 3
Perpendicular Bisector
Construct the perpendicular bisector of a segment. (p. 46)

Construction 4
Angle Bisector
Construct the bisector of an angle. (p. 47)

Construction 5
Parallel Through a Point Not on a Line
Construct a line parallel to a given line and through a given point that is not on the line. (p. 181)

Construction 6
Perpendicular Through a Point on a Line
Construct the perpendicular to a given line at a given point on the line. (p. 182)

Construction 7
Perpendicular Through a Point Not on a Line
Construct the perpendicular to a given line through a given point not on the line. (p. 183)

A

EXAMPLES

Acute angle (p. 37) An acute angle is an angle whose measure is between 0 and 90.

Ángulo agudo (p. 37) Un ángulo agudo es un ángulo que mide entre 0 y 90 grados.

Acute triangle (p. 148) An acute triangle has three acute angles.

Triángulo acutángulo (p. 148) Un triángulo acutángulo tiene los tres ángulos agudos.

Adjacent angles (p. 38) Adjacent angles are two coplanar angles that have a common side and a common vertex but no common interior points.

Ángulos adyacentes (p. 38) Los ángulos adyacentes son dos ángulos coplanarios que tienen un lado común y el mismo vértice, pero no tienen puntos interiores comunes.

∠1 and ∠2 are adjacent. ∠3 and ∠4 are *not* adjacent.

Adjacent arcs (p. 567) Adjacent arcs are on the same circle and have exactly one point in common.

Arcos adyacentes (p. 567) Los arcos adyacentes están en el mismo círculo y tienen exactamente un punto en común.

$\overset{\frown}{AB}$ and $\overset{\frown}{BC}$ are adjacent arcs.

Alternate interior (exterior) angles (pp. 127, 129) Alternate interior (exterior) angles are nonadjacent interior (exterior) angles that lie on opposite sides of the transversal.

Ángulos alternos internos (externos) (pp. 127, 129) Los ángulos alternos internos (externos) son ángulos internos (externos) no adyacentes situados en lados opuestos de la transversal.

∠1 and ∠2 are alternate interior angles, as are ∠3 and ∠4. ∠5 and ∠6 are alternate exterior angles.

Altitude *See* **cone; cylinder; parallelogram; prism; pyramid; trapezoid; triangle.**

Altura *Ver* **cone; cylinder; parallelogram; prism; pyramid; trapezoid; triangle.**

Altitude of a triangle (p. 275) An altitude of a triangle is a perpendicular segment from a vertex to the line containing the side opposite that vertex.

Altura de un triángulo (p. 275) Una altura de un triángulo es el segmento perpendicular que va desde un vértice hasta la recta que contiene el lado opuesto a ese vértice.

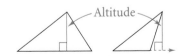

Altitude

English/Spanish Glossary

Angle (p. 36) An angle is formed by two rays with the same endpoint. The rays are the *sides* of the angle and the common endpoint is the *vertex* of the angle.

Ángulo (p. 36) Un ángulo está formado por dos rayos que convergen en un mismo extremo. Los rayos son los *lados* del ángulo y los extremos en común son el vértice.

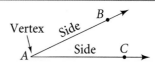

This angle could be named $\angle A$, $\angle BAC$, or $\angle CAB$.

Angle bisector (p. 46) An angle bisector is a ray that divides an angle into two congruent angles.

Bisectriz de un ángulo (p. 46) La bisectriz de un ángulo es un rayo que divide al ángulo en dos ángulos congruentes.

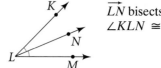

$\overrightarrow{LN}$ bisects $\angle KLM$.
$\angle KLN \cong \angle NLM$.

Angle of elevation or depression (p. 445) An angle of elevation (depression) is the angle formed by a horizontal line and the line of sight to an object above (below) the horizontal line.

Ángulo de elevación o depresión (p. 445) Un ángulo de elevación (depresión) es el ángulo formado por una línea horizontal y la recta que va de esa línea a un objeto situado arriba (debajo) de ella.

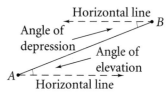

Apothem (p. 546) The apothem of a regular polygon is the distance from the center to a side.

Apotema (p. 546) La apotema de un polígono regular es la distancia desde el centro hasta un lado.

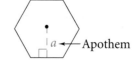

Arc *See* **major arc; minor arc.** *See also* **arc length; measure of an arc.**

Arco *Ver* **major arc; minor arc.** *Ver también* **arc length; measure of an arc.**

Arc length (p. 569) The length of an arc of a circle is the product of the ratio $\frac{\text{measure of the arc}}{360}$ and the circumference of the circle.

Longitud de un arco (p. 569) La longitud del arco de un círculo es el producto del cociente $\frac{\text{medida del arco}}{360}$ por la circunferencia del círculo.

Length of $\overset{\frown}{DE} = \frac{60}{360} \cdot 2\pi(5) = \frac{5\pi}{3}$

Area (pp. 534, 535, 540, 541, 547, 576) The area of a plane figure is the number of square units enclosed by the figure. A list of area formulas is on pp. 764–765.

Área (pp. 534, 535, 540, 541, 547, 576) El área de una figura plana es la cantidad de unidades cuadradas que contiene la figura. Una lista de fórmulas para calcular áreas está en las págs. 764–765.

The area of the rectangle is 12 square units, or 12 units^2.

Axes (p. 53) *See* **coordinate plane.**

Ejes (p. 53) *Ver* **coordinate plane**.

Axiom (p. 18) *See* **postulate.**

Axioma (p. 18) *Ver* **postulate.**

Base(s) *See* **cone; cylinder; isosceles triangle; parallelogram; prism; pyramid; trapezoid; triangle.**

Base(s) *Ver* **cone; cylinder; isosceles triangle; parallelogram; prism; pyramid; trapezoid; triangle.**

Base angles *See* **trapezoid; isosceles triangle.**

Ángulos de base *Ver* **trapezoid; isosceles triangle.**

Biconditional (p. 87) A biconditional statement is the combination of a conditional statement and its converse. A biconditional contains the words "if and only if."

Bicondicional (p. 87) Un enunciado biconditional es la combinación de un enunciado condicional y su recíproco. El enunciado biconditional incluye las palabras "si y solo si".

This biconditional statement is true: Two angles are congruent *if and only if* they have the same measure.

Bisector *See* **segment bisector; angle bisector.**

Bisectriz *Ver* **segment bisector; angle bisector.**

Center *See* **circle; dilation; regular polygon; sphere.**

Centro *Ver* **circle; dilation; regular polygon; sphere.**

Central angle of a circle (p. 566) A central angle of a circle is an angle whose vertex is the center of the circle.

Ángulo central de un círculo (p. 566) Un ángulo central de un círculo es un ángulo cuyo vértice es el centro del círculo.

$\angle ROK$ is a central angle of $\odot O$.

Central angle of a regular polygon (p. 559) A central angle of a regular polygon is an angle formed by two consecutive radii.

Ángulo central de un polígono regular (p. 559) Un ángulo central de un polígono regular es el ángulo formado por dos radios consecutivos.

∠*EFG* is a central angle of the regular pentagon.

Centroid (p. 274) The centroid of a triangle is the point of intersection of the medians of that triangle.

EXAMPLE *P* is the centroid of △*ABC*.

Centroide (p. 274) El centroide de un triángulo es el punto de intersección de las medianas del triángulo.

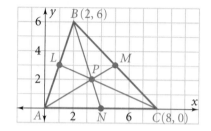

Chord (p. 680) A chord of a circle is a segment whose endpoints are on the circle.

Cuerda (p. 680) Una cuerda de un círculo es un segmento cuyos extremos son dos puntos del círculo.

$\overline{HD}$ and $\overline{HR}$ are chords of ⊙*C*.

Circle (pp. 566, 695) A circle is the set of all points in a plane that are a given distance, the *radius*, from a given point, the *center*. The standard form for an equation of a circle with center (h, k) and radius r is $(x - h)^2 + (y - k)^2 = r^2$.

Círculo (pp. 566, 695) Un círculo es el conjunto de todos los puntos de un plano situados a una distancia dada, el *radio*, de un punto dado, el *centro*. La fórmula normal de la ecuación de un círculo con centro (h, k) y radio r es $(x - h)^2 + (y - k)^2 = r^2$.

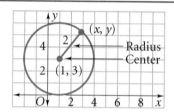

The equation of the circle whose center is $(1, 3)$ and whose radius is 2 is $(x - 1)^2 + (y - 3)^2 = 4$.

Circumcenter (p. 273) A circumcenter is the point of concurrency of the perpendicular bisectors of the sides of a triangle.

Circuncentro (p. 273) El circuncentro es el punto de intersección de las mediatrices perpendiculares a los lados de un triángulo.

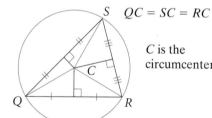

$QC = SC = RC$

C is the circumcenter.

Circumference (p. 568) The circumference of a circle is the distance around the circle. Given the radius r of a circle, you can find its circumference C by using the formula $C = 2\pi r$.

Circunferencia (p. 568) La circunferencia de un círculo es la distancia alrededor del círculo. Dado el radio r de un círculo, se puede hallar la circunferencia C usando la fórmula $C = 2\pi r$.

$$C = 2\pi r$$
$$= 2\pi(4)$$
$$= 8\pi$$

Circumference is the distance around the circle.

Circumference of a sphere (p. 638)   *See* **sphere.**

Circunferencia de una esfera (p. 638)   *Ver* **sphere.**

Circumscribed about (pp. 273, 664)   A circle is circumscribed about a polygon if the vertices of the polygon are on the circle. A polygon is circumscribed about a circle if all the sides of the polygon are tangent to the circle.

Circunscrito en (pp. 273, 664)   Un círculo está circunscrito en un polígono si los vértices del polígono están en el círculo. Un polígono está circunscrito en un círculo si todos los lados del polígono son tangentes al círculo.

⊙*G* is circumscribed about *ABCD*.

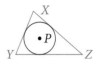

△*XYZ* is circumscribed about ⊙*P*.

Collinear points (p. 17)   Collinear points lie on the same line.

Puntos colineales (p. 17)   Los puntos colineales son los que están sobre la misma recta.

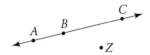

Points *A*, *B*, and *C* are collinear, but points *A*, *B*, and *Z* are noncollinear.

Compass (p. 44)   A compass is a geometric tool used to draw circles and parts of circles, called arcs.

Compás (p. 44)   El compás es un instrumento usado para dibujar círculos y partes de círculos, llamados arcos.

Complementary angles (p. 38)   Two angles are complementary angles if the sum of their measures is 90.

Ángulos complementarios (p. 38)   Dos ángulos son complementarios si la suma de sus medidas es igual a 90 grados.

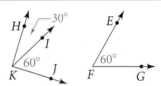

∠*HKI* and ∠*IKJ* are complementary angles, as are ∠*HKI* and ∠*EFG*.

Composite space figures (p. 627)   A composite space figure is the combination of two or more figures into one object.

Figuras geométricas compuestas (p. 627)   Una figura geométrica compuesta es la combinación de dos o más figuras en un mismo objeto.

Composition of transformations (p. 472)   A composition of two transformations is a transformation in which a second transformation is performed on the image of a first transformation.

Composición de transformaciones (p. 472)   Una composición de dos transformaciones es una transformación en la cual una segunda transformación se realiza a partir de la imagen de la primera.

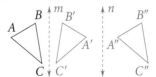

If you reflect △*ABC* across line *m* to get △*A'B'C'* and then reflect △*A'B'C'* across line *n* to get △*A"B"C"*, you perform a composition of transformations.

Concave polygon (p. 158)   *See* **polygon.**

Polígono cóncavo (p. 158)   *Ver* **polygon.**

Concentric circles (p. 568)   Concentric circles lie in the same plane and have the same center.

Círculos concéntricos (p. 568)   Los círculos concéntricos están en el mismo plano y tienen el mismo centro.

The two circles both have center *D* and are therefore concentric.

Conclusion (p. 80)   The conclusion is the part of an *if-then* statement (conditional) that follows *then*.

Conclusión (p. 80)   La conclusión es lo que sigue a la palabra *entonces* en un enunciado (condicional), *si..., entonces....*

In the statement, "If it rains, then I will go outside," the *conclusion* is "I will go outside."

Concurrent lines (p. 273)   Concurrent lines are three or more lines that meet in one point. The point at which they meet is the *point of concurrency*.

Rectas concurrentes (p. 273)   Las rectas concurrentes son tres o más rectas que se unen en un punto. El punto en que se unen es el *punto de concurrencia*.

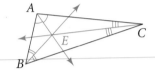

Point *E* is the point of concurrency of the bisectors of the angles of △*ABC*. The bisectors are concurrent.

Conditional (p. 80)   A conditional is an *if-then* statement.

Condicional (p. 80)   Un enunciado condicional es del tipo *si..., entonces...*

If you act politely, *then* you will earn respect.

Cone (p. 619)   A cone is a three-dimensional figure that has a circular *base*, a *vertex* not in the plane of the circle, and a curved lateral surface, as shown in the diagram. The *altitude* of a cone is the perpendicular segment from the vertex to the plane of the base. The *height* is the length of the altitude. In a *right cone*, the altitude contains the center of the base. The *slant height* of a right cone is the distance from the vertex to the edge of the base.

Cono (p. 619)   Un cono es una figura tridimensional que tiene una *base* circular, un *vértice* que no está en el plano del círculo y una superficie lateral curvada (indicada en el diagrama). La *altura* de un cono es el segmento perpendicular desde el vértice hasta el plano de la base. La *altura*, por extensión, es la longitud de la altura. Un *cono recto* es un cono cuya altura contiene el centro de la base. La *longitud de la generatriz* de un cono recto es la distancia desde el vértice hasta el borde de la base.

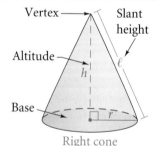

Right cone

Congruence transformation (p. 470)   *See* **isometry.**

Transformación de congruencia (p. 470)   *Ver* **isometry.**

Congruent angles (p. 37)   Congruent angles are angles that have the same measure.

Ángulos congruentes (p. 37)   Los ángulos congruentes son ángulos que tienen la misma medida.

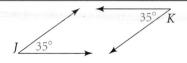

$m\angle J = m\angle K$, so $\angle J \cong \angle K$.

Congruent arcs (p. 569)   Congruent arcs are arcs that have the same measure and are in the same circle or congruent circles.

Arcos congruentes (p. 569)   Arcos congruentes son arcos que tienen la misma medida y están en el mismo círculo o en círculos congruentes.

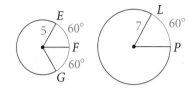

$\overarc{EF} \cong \overarc{FG}$ $\overarc{EF} \not\cong \overarc{LP}$

Congruent circles (p. 566)   Congruent circles are circles whose radii are congruent.

Círculos congruentes (p. 566)   Los círculos congruentes son círculos cuyos radios son congruentes.

$\odot A$ and $\odot B$ have the same radius, so $\odot A \cong \odot B$.

Congruent polygons (p. 198)   Congruent polygons are polygons that have corresponding sides congruent and corresponding angles congruent.

Polígonos congruentes (p. 198)   Los polígonos congruentes son polígonos cuyos lados correspondientes son congruentes y cuyos ángulos correspondientes son congruentes.

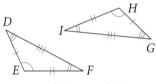

$\triangle DEF \cong \triangle GHI$

Congruent segments (p. 31)   Congruent segments are segments that have the same length.

Segmentos congruentes (p. 31)   Los segmentos congruentes son segmentos que tienen la misma longitud.

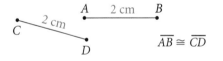

$\overline{AB} \cong \overline{CD}$

Conjecture (p. 5)   A conjecture is a conclusion reached by using inductive reasoning.

Conjetura (p. 5)   Una conjetura es una conclusión obtenida usando el razonamiento inductivo.

As you walk down the street, you see many people holding unopened umbrellas. You conjecture that the forecast must call for rain.

Consecutive angles (p. 312)   Consecutive angles of a polygon share a common side.

Ángulos consecutivos (p. 312)   Los ángulos consecutivos de un polígono tienen un lado común.

In $\square JKLM$, $\angle J$ and $\angle M$ are consecutive angles, as are $\angle J$ and $\angle K$. $\angle J$ and $\angle L$ are *not* consecutive.

Construction (p. 44) A construction is a geometric figure made with only a straightedge and compass.

Construcción (p. 44) Una construcción es una figura geométrica trazada solamente con una regla sin graduación y un compás.

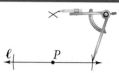

The diagram shows the construction (in progress) of a line perpendicular to a line ℓ through a point P on ℓ.

Contrapositive (p. 280) The contrapositive of the conditional "if p, then q" is the conditional "if not q, then not p." A conditional and its contrapositive always have the same truth value.

Contrapositivo (p. 280) El contrapositivo del condicional "si p, entonces q" es el condicional "si no q, entonces no p." Un condicional y su contrapositivo siempre tienen el mismo valor verdadero.

Conditional: If a figure is a triangle, then it is a polygon.
Contrapositive: If a figure is not a polygon, then it is not a triangle.

Converse (p. 81) The converse of the conditional "if p, then q" is the conditional "if q, then p."

Recíproco (p. 81) El recíproco del condicional "si p, entonces q" es el condicional "si q, entonces p".

Conditional: If you live in Cheyenne, then you live in Wyoming.
Converse: If you live in Wyoming, then you live in Cheyenne.

Convex polygon (p. 158) *See* **polygon.**

Polígono convexo (p. 158) *Ver* **polygon.**

Coordinate(s) of a point (pp. 31, 53) The coordinate of a point is its distance and direction from the origin of a number line. The coordinates of a point on a coordinate plane are in the form (x, y), where x is the x-coordinate and y is the y-coordinate.

Coordenada(s) de un punto (pp. 31, 53) La coordenada de un punto es su distancia y dirección desde el origen en una recta numérica. Las coordenadas de un punto en un plano de coordenadas se expresan como (x, y), donde x es la coordenada x, e y es la coordenada y.

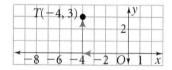

The coordinate of P is -3.

The coordinates of T are $(-4, 3)$.

Coordinate plane (p. 53) The coordinate plane is formed by two number lines, called the *axes*, intersecting at right angles. The x-axis is the horizontal axis, and the y-axis is the vertical axis. The two axes meet at the *origin*, $O(0, 0)$. The axes divide the plane into four *quadrants*.

Plano de coordenadas (p. 53) El plano de coordenadas se forma con dos rectas numéricas, llamadas *ejes*, que se cortan en ángulos rectos. El eje x es el eje horizontal y el eje y es el eje vertical. Los dos ejes se unen en el *origen*, $O(0, 0)$. Los ejes dividen el plano de coordenadas en cuatro *cuadrantes*.

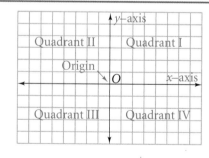

Coordinate proof (p. 260) *See* **proof.**

Prueba de coordenadas (p. 260) *Ver* **proof.**

Coplanar figures (p. 17) Coplanar figures are figures in the same plane.

Figuras coplanarias (p. 17) Las figuras coplanarias son las figuras que están localizadas en el mismo plano.

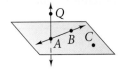

Point C and $\overleftrightarrow{AB}$ are coplanar but points $A, B, C,$ and Q are noncoplanar.

Corollary (p. 229) A corollary is a statement that follows directly from a theorem.

Corolario (p. 229) Un corolario es un enunciado que procede directamente de un teorema.

Theorem: If two sides of a triangle are congruent, then the angles opposite those sides are congruent.
Corollary: If a triangle is equilateral, then it is equiangular.

Corresponding angles (p. 127) Corresponding angles lie on the same side of the transversal t and in corresponding positions relative to ℓ and m.

Ángulos correspondientes (p. 127) Los ángulos correspondientes están en el mismo lado de la transversal t y en las correspondientes posiciones relativas a ℓ y m.

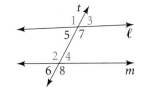

$\angle 1$ and $\angle 2$ are corresponding angles, as are $\angle 3$ and $\angle 4$, $\angle 5$ and $\angle 6$, and $\angle 7$ and $\angle 8$.

Cosine ratio (p. 439) *See* **trigonometric ratios.**

Razón coseno (p. 439) *Ver* **trigonometric ratios.**

Counterexample (pp. 5, 81) A counterexample to a statement is a particular example or instance of the statement that makes the statement false.

Contraejemplo (pp. 5, 81) Un contraejemplo de un enunciado es un ejemplo particular o caso que demuestra que el enunciado no es verdadero.

Statement: If the name of a state begins with W, then that state does not border an ocean.
Counterexample: Washington

CPCTC (p. 221) CPCTC is an abbreviation for "corresponding parts of congruent triangles are congruent."

EXAMPLE By the SAS Congruence Postulate, $\triangle KLM \cong \triangle QPR$. By CPCTC, you also know that $\angle L \cong \angle P, \angle M \cong \angle R,$ and $\overline{LM} \cong \overline{PR}$.

PCTCC (p. 221) PCTCC es la abreviatura para "partes correspondientes de triángulos congruentes son congruentes".

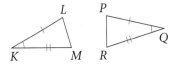

Cross-Product Property (p. 367) The product of the extremes of a proportion is equal to the product of the means.

If $\frac{x}{3} = \frac{12}{21}$, then $21x = 3 \cdot 12$.

Propiedad del producto cruzado (p. 367) El producto de los extremos de una proporción es igual al producto de los medios.

Cross section (p. 600) A cross section is the intersection of a solid and a plane.

The cross section is a circle.

Sección de corte (p. 600) Una sección de corte es la intersección de un plano y un cuerpo geométrico.

Cube (p. 598) A cube is a polyhedron with six faces, each of which is a square.

Cubo (p. 598) Un cubo es un poliedro de seis caras, cada una de las caras es un cuadrado.

Cylinder (p. 610) A cylinder is a three-dimensional figure with two congruent circular *bases* that lie in parallel planes. An *altitude* of a cylinder is a perpendicular segment that joins the planes of the bases. Its length is the *height* of the cylinder. In a *right cylinder,* the segment joining the centers of the bases is an altitude. In an *oblique cylinder,* the segment joining the centers of the bases is not perpendicular to the planes containing the bases.

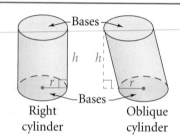

Right cylinder

Oblique cylinder

Cilindro (p. 610) Un cilindro es una figura tridimensional con dos *bases* congruentes circulares en planos paralelos. Una *altura* de un cilindro es un segmento perpendicular que une los planos de las bases. Su longitud es, por extensión, la *altura* del cilindro. En un *cilindro recto,* el segmento que une los centros de las bases es una altura. En un *cilindro oblicuo,* el segmento que une los centros de las bases no es perpendicular a los planos que contienen las bases.

Decagon (p. 158) A decagon is a polygon with ten sides.

Decágono (p. 158) Un decágono es un polígono de diez lados.

Deductive reasoning (p. 94) Deductive reasoning is a process of reasoning logically from given facts to a conclusion.

Based on the fact that the sum of any two even numbers is even, you can deduce that the product of any whole number and any even number is even.

Razonamiento deductivo (p. 94) El razonamiento deductivo es un proceso de razonamiento lógico que parte de hechos dados hasta llegar a una conclusión.

Diagonal (p. 158) *See* **polygon.**

Diagonal (p. 158) *Ver* **polygon.**

Diameter of a circle (p. 566) A diameter of a circle is a segment that contains the center of the circle and whose endpoints are on the circle. The term *diameter* can also mean the length of this segment.

$\overline{DM}$ is a diameter of $\odot C$.

Diámetro de un círculo (p. 566) Un diámetro de un círculo es un segmento que contiene el centro del círculo y cuyos extremos están en el círculo. El término *diámetro* también puede referirse a la longitud de este segmento.

Diameter of a sphere (p. 638) The diameter of a sphere is a segment passing through the center, with endpoints on the sphere.

Diámetro de una esfera (p. 638) El diámetro de una esfera es un segmento que contiene el centro de la esfera y cuyos extremos están en la esfera.

Dilation (p. 498) A dilation, or *similarity transformation,* is a transformation that has *center C* and *scale factor n,* where $n > 0$, and maps a point R to R' in such a way that R' is on $\overrightarrow{CR}$ and $CR' = n \cdot CR$. The center of a dilation is its own image. If $n > 1$, the dilation is an *enlargement,* and if $0 < n < 1$, the dilation is a *reduction.*

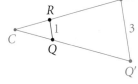

$\overline{R'Q'}$ is the image of $\overline{RQ}$ under a dilation with center C and scale factor 3.

Dilatación (p. 498) Una dilatación, o *transformación de semejanza,* tiene *centro C* y *factor de escala n* para $n > 0$, y asocia un punto R a R' de tal modo que R' está en $\overrightarrow{CR}$ y $CR' = n \cdot CR$. El centro de una dilatación es su propia imagen. Si $n > 1$, la dilatación es un *aumento,* y si $0 < n < 1$, la dilatación es una *reducción.*

Direction of a vector (p. 452) *See* **vector.**

Dirección de un vector (p. 452) *Ver* **vector.**

Distance from a point to a line (p. 266) The distance from a point to a line is the length of the perpendicular segment from the point to the line.

The distance from point P to a line ℓ is $PT.$

Distancia desde un punto hasta una recta (p. 266) La distancia desde un punto hasta una recta es la longitud del segmento perpendicular que va desde el punto hasta la recta.

Dodecagon (p. 158) A dodecagon is a polygon with twelve sides.

Dodecágono (p. 158) Un dodecágono es un polígono de doce lados.

Edge (p. 598) *See* **polyhedron.**

Arista (p. 598) *Ver* **polyhedron.**

Endpoint (p. 23) *See* **ray; segment.**

Extremo (p. 23) *Ver* **ray; segment.**

Enlargement (p. 498) *See* **dilation.**

Aumento (p. 498) *Ver* **dilation.**

Equiangular triangle or polygon (pp. 148, 160) An equiangular triangle (polygon) is a triangle (polygon) whose angles are all congruent.

Triángulo o polígono equiángulo (pp. 148, 160) Un triángulo (polígono) equiángulo es un triángulo (polígono) cuyos ángulos son todos congruentes.

Each angle of the pentagon is a 108° angle.

Equilateral triangle or polygon (pp. 148, 160) An equilateral triangle (polygon) is a triangle (polygon) whose sides are all congruent.

Triángulo o polígono equilátero (pp. 148, 160) Un triángulo (polígono) equilátero es un triángulo (polígono) cuyos lados son todos congruentes.

Each side of the quadrilateral is 1.2 cm long.

Equivalent statements (p. 281) Equivalent statements are statements with the same truth value.

Enunciados equivalentes (p. 281) Los enunciados equivalentes son enunciados con el mismo valor verdadero.

The following statements are equivalent:
If a figure is a square, then it is a rectangle.
If a figure is not a rectangle, then it is not a square.

Euclidean geometry (p. 154) Euclidean geometry is a geometry of the plane in which Euclid's Parallel Postulate is accepted as true.

Geometría euclidiana (p. 154) La geometría euclidiana es una geometría del plano en donde el postulado paralelo de Euclides es verdadero.

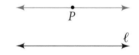

In Euclidean geometry, there is exactly one line parallel to line ℓ through point P.

Extended proportion (p. 367) *See* **proportion.**

Proporción extendida (p. 367) *Ver* **proportion.**

Exterior angle of a polygon (p. 149) An exterior angle of a polygon is an angle formed by a side and an extension of an adjacent side.

Ángulo exterior de un polígono (p. 149) El ángulo exterior de un polígono es un ángulo formado por un lado y una extensión de un lado adyacente.

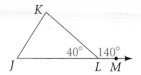

$\angle KLM$ is an exterior angle of $\triangle JKL$.

Face (p. 598) *See* **polyhedron.**

Cara (p. 598) *Ver* **polyhedron.**

Flip (p. 478) *See* **reflection.**

Flow proof (p. 135) *See* **proof.**

Prueba de flujo (p. 135) *Ver* **proof.**

Foundation drawing (p. 11) A foundation drawing shows the base of a structure and the height of each part.

Dibujo de fundación (p. 11) Un dibujo de fundación muestra la base de una estructura y la altura de cada parte.

The first drawing is a foundation drawing, and the second is an isometric drawing based on the foundation drawing.

Geometric mean (p. 392) The geometric mean is the number x such that $\frac{a}{x} = \frac{x}{b}$, where a, b and x are positive numbers.

Media geométrica (p. 392) La media geométrica es el número x tanto que $\frac{a}{x} = \frac{x}{b}$, donde a, b y x son números positivos.

The geometric mean of 6 and 24 is 12.
$$\frac{6}{x} = \frac{x}{24} \rightarrow x^2 = 144 \rightarrow x = 12$$

Geometric probability (p. 582) Geometric probability is a probability that uses a geometric model in which points represent outcomes.

Probabilidad geométrica (p. 582) La probabilidad geométrica es una probabilidad que utiliza un modelo geométrico donde se usan puntos para representar resultados.

English/Spanish Glossary

Glide reflection (p. 508)   A glide reflection is the composition of a translation followed by a reflection across a line parallel to the translation vector.

Reflexión deslizada (p. 508)   Una reflexión deslizada es la composición de una traslación seguida de una reflexión en una recta paralela al vector de traslación.

The blue G in the diagram is a glide reflection image of the black G.

Glide reflectional symmetry (p. 516)   Glide reflectional symmetry is the type of symmetry for which there is a glide reflection that maps a figure onto itself.

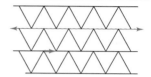

Simetría por reflexión deslizada (p. 516)   La simetría por reflexión deslizada es un tipo de simetría en la que una reflexión deslizada vuelve a trazar una figura sobre sí misma.

The tessellation shown can be mapped onto itself by a glide reflection for the given glide vector and reflection line.

Golden rectangle, Golden ratio (p. 375)   A *golden rectangle* is a rectangle that can be divided into a square and a rectangle that is similar to the original rectangle. The *golden ratio* is the ratio of the length of a golden rectangle to its width. The value of the golden ratio is $\frac{1 + \sqrt{5}}{2}$, or about 1.62.

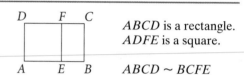

ABCD is a rectangle.
ADFE is a square.

ABCD ~ *BCFE*

Rectángulo áureo, razón áurea (p. 375)   Un *rectángulo áureo* es un rectángulo que se puede dividir en un cuadrado y un rectángulo semejante al rectángulo original. La *razón áurea* es la razón de la longitud de un rectángulo áureo en relación a su ancho. El valor de la razón áurea es $\frac{1 + \sqrt{5}}{2}$ o aproximadamente 1.62.

Great circle (p. 638)   A great circle is the intersection of a sphere and the plane containing the center of the sphere. A great circle divides a sphere into two *hemispheres*.

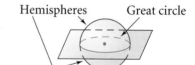

Hemispheres   Great circle

Círculo máximo (p. 638)   Un círculo máximo es la intersección de una esfera y un plano que contiene el centro de la esfera. Un círculo máximo divide una esfera en dos *hemisferios*.

Height   *See* **cone; cylinder; parallelogram; prism; pyramid; trapezoid; triangle.**

Altura   *Ver* **cone; cylinder; parallelogram; prism; pyramid; trapezoid; triangle.**

Hemisphere (p. 638) *See* **great circle.**

Hemisferio (p. 638) *Ver* **great circle.**

Heron's Formula (p. 538) Heron's Formula is a formula for finding the area of a triangle given the lengths of its sides.

Fórmula de Herón (p. 538) La fórmula de Herón se usa para hallar el área de un triángulo, dadas las longitudes de sus lados.

$A = \sqrt{s(s - a)(s - b)(s - c)}$, where s is half the perimeter (semi-perimeter) of the triangle and a, b, and c are the lengths of its sides.

Hexagon (p. 158) A hexagon is a polygon with six sides.

Hexágono (p. 158) Un hexágono es un polígono de seis lados.

Hypotenuse (p. 235) *See* **right triangle.**

Hipotenusa (p. 235) *Ver* **right triangle.**

Hypothesis (p. 80) The hypothesis is the part that follows *if* in an *if-then* statement (conditional).

Hipótesis (p. 80) La hipótesis es lo que sigue a la palabra *si* en un enunciado (condicional), *si..., entonces....*

In the statement "If she leaves, then I will go with her," the hypothesis is "she leaves."

Identity (p. 440) An identity is an equation that is true for all allowed values of the variable.

Identidad (p. 440) Una identidad es una ecuación que es verdadera para todos los valores posibles de las variables.

$\sin x° = \cos(90 - x)°$

Incenter of a triangle (p. 273) The incenter of a triangle is the point of concurrency of the angle bisectors of the triangle.

Incentro de un triángulo (p. 273) El incentro de un triángulo es el punto donde concurren las tres bisectrices de los ángulos del triángulo.

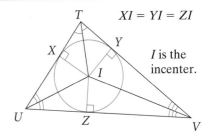

$XI = YI = ZI$

I is the incenter.

Image (p. 470) *See* transformation.

Imagen (p. 470) *Ver* transformation.

Indirect measurement (p. 384)   Indirect measurement is a way of measuring things that are difficult to measure directly.

EXAMPLE   By measuring the distances shown in the diagram and using proportions of similar figures, you can find the height of the taller tower. $\frac{196}{540} = \frac{x}{1300} \rightarrow x \approx 472$ ft

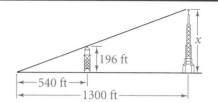

Medición indirecta (p. 384)   La medición indirecta es un modo de medir cosas difíciles de medir directamente.

Indirect proof (p. 281)   *See* **indirect reasoning; proof.**

Prueba indirecta (p. 281)   *Ver* **indirect reasoning; proof.**

Indirect reasoning (p. 281)   Indirect reasoning is a type of reasoning in which all possibilities are considered and then the unwanted ones are proved false. The remaining possibilities must be true.

Razonamiento indirecto (p. 281)   El razonamiento indirecto es un tipo de razonamiento en el que todas las posibilidades se consideran, y luego las no deseadas resultan falsas. Las posibilidades que quedan deben ser verdaderas.

Eduardo spent more than $60 on two books at a store. Prove that at least one book costs more than $30.
Proof: Suppose neither costs more than $30. Then he spent no more than $60 at the store. Since this contradicts the given information, at least one book costs more than $30.

Inductive reasoning (p. 4)   Inductive reasoning is a type of reasoning that reaches conclusions based on a pattern of specific examples or past events.

Razonamiento inductivo (p. 4)   El razonamiento inductivo es un tipo de razonamiento en el cual se llega a conclusiones con base en un patrón de ejemplos específicos o sucesos pasados.

You see four people walk into a building. Each person emerges with a small bag containing hot food. You use inductive reasoning to conclude that this building contains a restaurant.

Initial point of a vector (p. 452)   *See* **vector.**

Punto inicial de un vector (p. 452)   *Ver* **vector.**

Inscribed in (pp. 273, 664)   A circle is inscribed in a polygon if the sides of the polygon are tangent to the circle. A polygon is inscribed in a circle if the vertices of the polygon are on the circle.

Inscrito en (pp. 273, 664)   Un círculo está inscrito en un polígono si los lados del polígono son tangentes al círculo. Un polígono está inscrito en un círculo si los vértices del polígono están en el círculo.

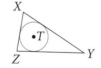

$\odot T$ is inscribed in $\triangle XYZ$.

$ABCD$ is inscribed in $\odot J$.

Inscribed angle (p. 678)   An angle is inscribed in a circle if the vertex of the angle is on the circle and the sides of the angle are chords of the circle.

Ángulo inscrito (p. 678)   Un ángulo está inscrito en un círculo si el vértice del ángulo está en el círculo y los lados del ángulo son cuerdas del círculo.

$\angle C$ is inscribed in $\odot M$.

Intercepted arc (p. 678)   An intercepted arc is an arc of a circle having endpoints on the sides of an inscribed angle, and its other points in the interior of the angle.

Arco interceptor (p. 678)   Un arco interceptor es un arco de un círculo cuyos extremos están en los lados de un ángulo inscrito y los puntos restantes están en el interior del ángulo.

$\overarc{UV}$ is the intercepted arc of inscribed angle $\angle T$.

Inverse (p. 280)   The inverse of the conditional "if p, then q" is the conditional "if not p, then not q."

Inverso (p. 280)   El inverso del condicional "si p, entonces q," es el condicional "si no p, entonces no q".

Conditional: If a figure is a square, then it is a parallelogram.
Inverse: If a figure is not a square, then it is not a parallelogram.

Isometric drawing (p. 10)   An isometric drawing of a three-dimensional object shows a corner view of a figure. It is not drawn in perspective and distances are not distorted.

Dibujo isométrico (p. 10)   Un dibujo isométrico de un objeto tridimensional muestra una vista desde una esquina de la figura. No se muestra en perspectiva y las distancias no aparecen distorcionadas.

Isometry (p. 470)   An isometry, also known as a *congruence transformation,* is a transformation in which an original figure and its image are congruent.

Isometría (p. 470)   Una isometría, conocida también como una *transformación de congruencia,* es una transformación en donde una figura original y su imagen son congruentes.

The four isometries are reflections, rotations, translations, and glide reflections.

Isosceles trapezoid (p. 306)   An isosceles trapezoid is a trapezoid whose nonparallel opposite sides are congruent.

Trapecio isósceles (p. 306)   Un trapecio isósceles es un trapecio cuyos lados opuestos no paralelos son congruentes.

Isosceles triangle (pp. 148, 228)   An isosceles triangle is a triangle that has at least two congruent sides. If there are two congruent sides, they are called *legs.* The *vertex angle* is between them. The third side is called the *base* and the other two angles are called the *base angles.*

Triángulo isósceles (pp. 148, 228)   Un triángulo isósceles es un triángulo que tiene por lo menos dos lados congruentes. Si tiene dos lados congruentes, éstos se llaman *catetos.* Entre ellos se encuentra el *ángulo del vértice.* El tercer lado se llama *base* y los otros dos ángulos se llaman *ángulos de base.*

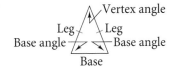

English/Spanish Glossary

Kite (p. 306) A kite is a quadrilateral with two pairs of congruent adjacent sides and no opposite sides congruent.

Cometa (p. 306) Una cometa es un cuadrilátero con dos pares de lados adyacentes congruentes, pero sin lados opuestos congruentes.

Lateral area (pp. 608, 610, 617, 619) The lateral area of a prism or pyramid is the sum of the areas of the lateral faces. The lateral area of a cylinder or cone is the area of the curved surface. A list of lateral area formulas is on p. 765.

Área lateral (pp. 608, 610, 617, 619) El área lateral de un prisma o pirámide es la suma de las áreas de sus caras laterales. El área lateral de un cilindro o de un cono es el área de la superficie curvada. Una lista de las fórmulas de áreas laterales está en la p. 765.

$$\text{L.A. of pyramid} = \tfrac{1}{2}\,p\ell$$
$$= \tfrac{1}{2}(20)(6)$$
$$= 60 \text{ cm}^2$$

Lateral face *See* **prism; pyramid.**

Cara lateral *Ver* **prism; pyramid.**

Leg *See* **isosceles triangle; right triangle; trapezoid.**

Cateto *Ver* **isosceles triangle; right triangle; trapezoid.**

Line (pp. 17, 154) In Euclidean geometry, a line is undefined. You can think of a line as a series of points that extend in two directions without end. In spherical geometry, you can think of a line as a great circle of a sphere.

Recta (pp. 17, 154) En la geometría euclidiana, una recta es indefinida. Puedes imaginarte a una recta como una serie de puntos que se extienden en dos direcciones sin fin. En la geometría esférica, puedes imaginarte a una recta como un círculo máximo de una esfera.

Line symmetry (p. 492) *See* **reflectional symmetry.**

Simetría axial (p. 492) *Ver* **reflectional symmetry.**

Locus (p. 701) A locus is a set of points, all of which meet a stated condition.

Lugar geométrico (p. 701) Un lugar geométrico es un conjunto de puntos que cumplen una condición dada.

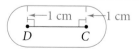

The points in blue are the locus of points in a plane 1 cm from $\overline{DC}$.

Magnitude of a vector (p. 452)   *See* **vector.**

Magnitud de un vector (p. 452)   *Ver* **vector.**

Major arc (p. 567)   A major arc of a circle is an arc that is larger than a semicircle.

Arco mayor (p. 567)   Un arco mayor de un círculo es cualquier arco más grande que un semicírculo.

$\overset{\frown}{DEF}$ is a major arc of $\odot C$.

Map (p. 471)   *See* **transformation.**

Trazar (p. 471)   *Ver* **transformation.**

Matrix (p. 504)   A matrix is a rectangular array of numbers. Each number in a matrix is called an *entry.*

Matriz (p. 504)   Una matriz es un conjunto de números dispuestos en forma de rectángulo. Cada número de una matriz se llama *elemento* de la matriz.

The matrix $\begin{bmatrix} 1 & -2 \\ 0 & 13 \end{bmatrix}$ has dimensions 2×2. The number 1 is the entry in the first row and first column.

Measure of an angle (p. 36)   The measure of an angle is a number of degrees greater than 0 and less than or equal to 180. An angle can be measured with a protractor.

Medida de un ángulo (p. 36)   La medida de un ángulo es un número de grados mayor de 0 y menor o igual a 180. Un ángulo se puede medir con un transportador.

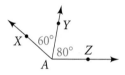

$m\angle ZAY = 80, m\angle YAX = 60,$ and $m\angle ZAX = 140.$

Measure of an arc (p. 567)   The measure of a minor arc is the measure of its central angle. The measure of a major arc is 360 minus the measure of its related minor arc.

Medida de un arco (p. 567)   La medida de un arco menor es la medida de su ángulo central. La medida de un arco mayor es 360 menos la medida en grados de su arco menor correspondiente.

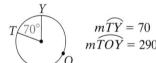

$m\overset{\frown}{TY} = 70$
$m\overset{\frown}{TOY} = 290$

Median of a triangle (p. 274)   A median of a triangle is a segment that has as its endpoints a vertex of the triangle and the midpoint of the opposite side.

Mediana de un triángulo (p. 274)   Una mediana de un triángulo es un segmento que tiene en sus extremos el vértice del triángulo y el punto medio del lado opuesto.

Midpoint of a segment (p. 32)   A midpoint of a segment is the point that divides the segment into two congruent segments.

Punto medio de un segmento (p. 32)   El punto medio de un segmento es el punto que divide el segmento en dos segmentos congruentes.

Midpoint of $\overline{AB}$

$A \qquad M \qquad B$

English/Spanish Glossary

Midsegment of a trapezoid (p. 348) The midsegment of a trapezoid is the segment that joins the midpoints of the nonparallel opposite sides of a trapezoid.

Segmento medio de un trapecio (p. 348) El segmento medio de trapecio es el segmento que une los puntos medios de los lados no paralelos de un trapecio.

Midsegment of a triangle (p. 259) A midsegment of a triangle is the segment that joins the midpoints of two sides of the triangle.

Segmento medio de un triángulo (p. 259) Un segmento medio de un triángulo es el segmento que une los puntos medios de dos lados del triángulo.

Minor arc (p. 567) A minor arc is an arc that is smaller than a semicircle.

$\overset{\frown}{KC}$ is a minor arc of $\odot S$.

Arco menor (p. 567) Un arco menor de un círculo es un arco más corto que un semicírculo.

Negation (p. 280) The negation of a statement has the opposite meaning of the original statement.

Statement: The angle is obtuse.
Negation: The angle is not obtuse.

Negación (p. 280) La negación de un enunciado tiene el sentido opuesto del enunciado original.

Net (p. 12) A net is a two-dimensional pattern that you can fold to form a three-dimensional figure.

Plantilla (p. 12) Una plantilla es una figura bidimensional que se puede doblar para formar una figura tridimensional.

The net shown can be folded into a prism with pentagonal bases.

***n*-gon (p. 158)** An *n*-gon is a polygon with *n* sides.

***n*-ágono (p. 158)** Un *n*-ágono es un polígono de *n* lados.

Nonagon (p. 158) A nonagon is a polygon with nine sides.

Nonágono (p. 158) Un nonágono es un polígono de nueve lados.

Oblique cylinder or prism *See* **cylinder; prism.**

Cilindro oblicuo o prisma *Ver* **cylinder; prism.**

Obtuse angle (p. 37)   An obtuse angle is an angle whose measure is between 90 and 180.

Ángulo obtuso (p. 37)   Un ángulo obtuso es un ángulo que mide entre 90 y 180 grados.

Obtuse triangle (p. 148)   An obtuse triangle has one obtuse angle.

Triángulo obtusángulo (p. 148)   Un triángulo obtusángulo tiene un ángulo obtuso.

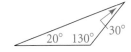

Octagon (p. 158)   An octagon is a polygon with eight sides.

Octágono (p. 158)   Un octágono es un polígono de ocho lados.

Opposite rays (p. 23)   Opposite rays are collinear rays with the same endpoint. They form a line.

Rayos opuestos (p. 23)   Los rayos opuestos son rayos colineales con el mismo extremo. Forman una recta.

$\overrightarrow{UT}$ and $\overrightarrow{UN}$ are opposite rays.

Orientation (p. 478)   Two congruent figures have *opposite* orientation if a reflection is needed to map one onto the other. If a reflection is not needed to map one figure onto the other, the figures have the same orientation.

Orientación (p. 478)   Dos figuras congruentes tienen orientación *opuesta* si una reflexión es necesaria para trazar una sobre la otra. Si una reflexión no es necesaria para trazar una figura sobre la otra, las figuras tiene la misma orientación.

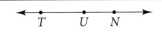

The two R's have opposite orientation.

Origin (p. 53)   *See* **coordinate plane.**

Origen (p. 53)   *Ver* **coordinate plane.**

Orthocenter (p. 275)   The orthocenter of a triangle is the point of intersection of the lines containing the altitudes of the triangle.

Ortocentro (p. 275)   El ortocentro de un triángulo es el punto donde concurren las tres alturas del triángulo.

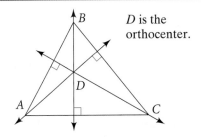

D is the orthocenter.

Orthographic drawing (p. 11) An orthographic drawing is the top view, front view, and right-side view of a three-dimensional figure.

EXAMPLE The diagram shows an isometric drawing (upper right) and the three views that make up an orthographic drawing.

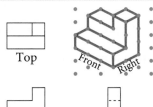

Top

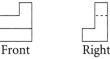

Front   Right

Dibujo ortográfico (p. 11) Un dibujo ortográfico es la vista desde arriba, la vista de frente y la vista del lado derecho de una figura tridimensional.

Paragraph proof (p. 111) *See* **proof.**

Prueba de párrafo (p. 111) *Ver* **proof.**

Parallel lines (p. 24) Two lines are parallel if they lie in the same plane and do not intersect. The symbol ‖ means "is parallel to."

Rectas paralelas (p. 24) Dos rectas son paralelas si están en el mismo plano y no se cortan. El símbolo ‖ significa "es paralelo a".

These symbols indicate parallel lines.

Parallel planes (p. 24) Parallel planes are planes that do not intersect.

Planos paralelos (p. 24) Planos paralelos son planos que no se cortan.

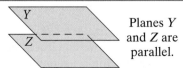

Planes *Y* and *Z* are parallel.

Parallelogram (p. 306) A parallelogram is a quadrilateral with two pairs of parallel sides. You can choose any side to be the *base*. An *altitude* is any segment perpendicular to the line containing the base drawn from the side opposite the base. The *height* is the length of an altitude.

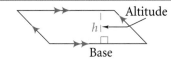

Paralelogramo (p. 306) Un paralelogramo es un cuadrilátero con dos pares de lados paralelos. Se puede escoger cualquier lado como la *base*. Una *altura* es un segmento perpendicular a la recta que contiene la base, trazada desde el lado opuesto a la base. La *altura*, por extensión, es la longitud de una altura.

Pentagon (p. 158) A pentagon is a polygon with five sides.

Pentágono (p. 158) Un pentágono es un polígono de cinco lados.

Perimeter of a polygon (p. 61) The perimeter of a polygon is the sum of the lengths of its sides.

Perímetro de un polígono (p. 61) El perímetro de un polígono es la suma de las longitudes de sus lados.

4 in.

4 in.   3 in.

5 in.

$P = 4 + 4 + 5 + 3$
$= 16$ in.

Perpendicular bisector (p. 45) The perpendicular bisector of a segment is a line, segment, or ray that is perpendicular to the segment at its midpoint.

Mediatriz (p. 45) La mediatriz de un segmento es una recta, segmento o rayo que es perpendicular al segmento en su punto medio.

$\overrightarrow{YZ}$ is the perpendicular bisector of $\overline{AB}$. It is perpendicular to $\overline{AB}$ and intersects $\overline{AB}$ at midpoint M.

Perpendicular lines (p. 45) Perpendicular lines are lines that intersect and form right angles. The symbol ⊥ means "is perpendicular to."

Rectas perpendiculares (p. 45) Las rectas perpendiculares son rectas que se cortan y forman ángulos rectos. El símbolo ⊥ significa "es perpendicular a".

$m \perp n$

Perspective drawing (p. 606) Perspective drawing is a way of drawing objects on a flat surface so that they look the same way as they appear to the eye. In *one-point perspective,* there is one *vanishing point.* In *two-point perspective,* there are two vanishing points.

Dibujar en perspectiva (p. 606) Dibujar en perspectiva es una manera de dibujar objetos en una superficie plana de modo que se vean como los percibe el ojo humano. En la *perspectiva de un punto* hay un *punto de fuga.* En la *perspectiva de dos puntos* hay dos puntos de fuga.

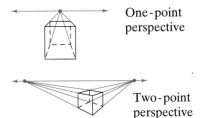

One-point perspective

Two-point perspective

Pi (p. 568) Pi (π) is the ratio of the circumference of any circle to its diameter. The number π is irrational and is approximately 3.14159.

Pi (p. 568) Pi (π) es la razón de la circunferencia de cualquier círculo a su diámetro. El número π es irracional y se aproxima a $\pi \approx 3.14159$.

$\pi = \dfrac{C}{d}$

Plane (p. 17) In Euclidean geometry, a plane is undefined. You can think of a plane as a flat surface that has no thickness. A plane contains many lines and extends without end in the directions of its lines.

Plano (p. 17) En la geometría euclidiana, un plano es indefinido. Puedes imaginarte a un plano como una superficie plana que no tiene grosor. Un plano tiene muchas rectas y se extiende sin fin en la misma dirección que todas las rectas.

Plane ABC or plane Z

Point (p. 17) In Euclidean geometry, a point is undefined. You can think of a point as a location. A point has no size.

Punto (p. 17) En la geometría euclidiana, un punto es indefinido. Puedes imaginarte a un punto como un lugar. Un punto no tiene dimensión.

• P

Point of concurrency (p. 273) *See* **concurrent lines.**

Punto de concurrencia (p. 273) *Ver* **concurrent lines.**

English/Spanish Glossary

Point symmetry (p. 493)   Point symmetry is the type of symmetry for which there is a rotation of 180° that maps a figure onto itself.

Simetría central (p. 493)   La simetría central es un tipo de simetría en la que una figura se ha rotado 180° sobre sí misma.

180°

Point-slope form (p. 168)   The point-slope form for a nonvertical line with slope m and through point (x_1, y_1) is $y - y_1 = m(x - x_1)$.

Forma punto-pendiente (p. 168)   La forma punto-pendiente para una recta no vertical con pendiente m y que pasa por el punto (x_1, y_1) es $y - y_1 = m(x - x_1)$.

$$y + 1 = 3(x - 4)$$

In this equation, the slope is 3 and (x_1, y_1) is $(4, -1)$.

Point of tangency (p. 598)   *See* **tangent to a circle.**

Punto de tangencia (p. 598)   *Ver* **tangent to a circle.**

Polygon (p. 157)   A polygon is a closed plane figure with at least three *sides* that are segments. The sides intersect only at their endpoints and no two adjacent sides are collinear. The *vertices* of the polygon are the endpoints of the sides. A *diagonal* is a segment that connects two nonconsecutive vertices. A polygon is *convex* if no diagonal contains points outside the polygon. A polygon is *concave* if a diagonal contains points outside the polygon.

Polígono (p. 157)   Un polígono es una figura plana cerrada de, por lo menos, tres *lados,* los cuales son segmentos. Los lados se cortan solo en los extremos. No hay dos lados adyacentes que sean colineales. Los *vértices* del polígono son los extremos de los lados. Una *diagonal* es un segmento que une dos vértices no consecutivos. Un polígono es *convexo* si ninguna diagonal contiene puntos fuera del polígono. Un polígono es *cóncavo* si una diagonal contiene puntos fuera del polígono.

Vertices — Diagonal — Sides

Convex   Concave

Polyhedron (p. 598)   A polyhedron is a three-dimensional figure whose surfaces, or *faces,* are polygons. The vertices of the polygons are the *vertices* of the polyhedron. The intersections of the faces are the *edges* of the polyhedron.

Poliedro (p. 598)   Un poliedro es una figura tridimensional cuyas superficies, o *caras,* son polígonos. Los vértices de los polígonos son los *vértices* del poliedro. Las intersecciones de las caras son las *aristas* del poliedro.

Vertices — Faces — Edges

Postulate (p. 18)   A postulate, or *axiom,* is an accepted statement of fact.

Postulado (p. 18)   Un postulado, o *axioma,* es un enunciado que se acepta como un hecho.

Postulate: Through any two points there is exactly one line.

Preimage (p. 470)   *See* **transformation.**

Preimagen (p. 470)   *Ver* **transformation.**

Prime notation (p. 471) *See* **transformation.**

Notación prima (p. 471) *Ver* **transformation.**

Prism (p. 608) A prism is a polyhedron with two congruent and parallel faces, which are called the *bases*. The other faces, which are parallelograms, are called the *lateral faces*. An *altitude* of a prism is a perpendicular segment that joins the planes of the bases. Its length is the *height* of the prism. A *right prism* is one whose lateral faces are rectangular regions and a lateral edge is an altitude. In an *oblique prism,* some or all of the lateral faces are nonrectangular.

Prisma (p. 608) Un prisma es un poliedro con dos caras congruentes paralelas llamadas *bases.* Las otras caras son paralelogramos llamados *caras laterales.* La *altura* de un prisma es un segmento perpendicular que une los planos de las bases. Su longitud es también la *altura* del prisma. En un *prisma rectangular,* las caras laterales son rectangulares y una de las aristas laterales es la altura. En un *prisma oblicuo,* algunas o todas las caras laterales no son rectangulares.

Proof (pp. 111, 129, 135, 260, 281) A proof is a convincing argument that uses deductive reasoning. A proof can be written in many forms. In a *two-column proof,* the statements and reasons are aligned in columns. In a *paragraph proof,* the statements and reasons are connected in sentences. In a *flow proof,* arrows show the logical connections between the statements. In a *coordinate proof,* a figure is drawn on a coordinate plane and the formulas for slope, midpoint, and distance are used to prove properties of the figure. An *indirect proof* involves the use of indirect reasoning.

Prueba (pp. 111, 129, 135, 260, 281) Una prueba es un argumento convincente en el cual se usa el razonamiento deductivo. Una prueba se puede escribir de varias maneras. En una *prueba de dos columnas,* los enunciados y las razones se alinean en columnas. En una *prueba de párrafo,* los enunciados y razones están unidos en oraciones. En una *prueba de flujo,* hay flechas que indican las conexiones lógicas entre enunciados. En una *prueba de coordenadas,* se dibuja una figura en un plano de coordenadas y se usan las fórmulas de la pendiente, punto medio y distancia para probar las propiedades de la figura. Una *prueba indirecta* incluye el uso de razonamiento indirecto.

Given: $\triangle EFG$, with right angle $\angle F$

Prove: $\angle E$ and $\angle G$ are complementary.

Paragraph Proof: Because $\angle F$ is a right angle, $m\angle F = 90$. By the Triangle Angle-Sum Theorem, $m\angle E + m\angle F + m\angle G = 180$. By substitution, $m\angle E + 90 + m\angle G = 180$. Subtracting 90 from each side yields $m\angle E + m\angle G = 90$. $\angle E$ and $\angle G$ are complementary by definition.

Proportion (p. 367) A proportion is a statement that two ratios are equal. An *extended proportion* is a statement that three or more ratios are equal.

Proporción (p. 367) Una proporción es un enunciado en el cual dos razones son iguales. Una *proporción extendida* es un enunciado que dice que tres razones o más son iguales.

$\frac{x}{5} = \frac{3}{4}$ is a proportion.

$\frac{9}{27} = \frac{3}{9} = \frac{1}{3}$ is an extended proportion.

Pyramid (p. 617) A pyramid is a polyhedron in which one face, the *base,* is a polygon and the other faces, the *lateral faces,* are triangles with a common vertex, called the *vertex* of the pyramid. An *altitude* of a pyramid is the perpendicular segment from the *vertex* to the plane of the base. Its length is the *height* of the pyramid. A *regular pyramid* is a pyramid whose base is a regular polygon and whose lateral faces are congruent isosceles triangles. The *slant height* of a regular pyramid is the length of an altitude of a lateral face.

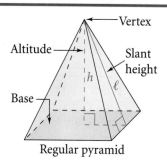

Regular pyramid

Pirámide (p. 617) Una pirámide es un poliedro en donde una cara, la *base,* es un polígono y las otras caras, las *caras laterales,* son triángulos con un vértice común, llamado el *vértice* de la pirámide. Una *altura* de una pirámide es el segmento perpendicular que va del *vértice* hasta el plano de la base. Su longitud es, por extensión, la *altura* de la pirámide. Una *pirámide regular* es una pirámide cuya base es un polígono regular y cuyas caras laterales son triángulos isósceles congruentes. La *apotema* de una pirámide regular es la longitud de la altura de la cara lateral.

Pythagorean triple (p. 417) A Pythagorean triple is a set of three nonzero whole numbers a, b, and c, that satisfy the equation $a^2 + b^2 = c^2$.

The numbers 5, 12, and 13 form a Pythagorean triple because $5^2 + 12^2 = 13^2 = 169$.

Tripleta de Pitágoras (p. 417) Una tripleta de Pitágoras es un conjunto de tres números enteros positivos a, b, and c que satisfacen la ecuación $a^2 + b^2 = c^2$.

Quadrant (p. 53) *See* **coordinate plane.**

Cuadrante (p. 53) *Ver* **coordinate plane.**

Quadrilateral (p. 158) A quadrilateral is a polygon with four sides.

Cuadrilátero (p. 158) Un cuadrilátero es un polígono de cuatro lados.

Radius of a circle (p. 566) A radius of a circle is any segment with one endpoint on the circle and the other endpoint at the center of the circle. *Radius* can also mean the length of this segment.

$\overline{DE}$ is a radius of $\odot D$.

Radio de un círculo (p. 566) Un radio de un círculo es cualquier segmento con un extremo en el círculo y el otro extremo en el centro del círculo. *Radio* también se refiere a la longitud de este segmento.

Radius of a regular polygon (p. 546) The radius of a regular polygon is the distance from the center to a vertex.

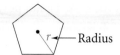

Radio de un polígono regular (p. 546) El radio de un polígono regular es la distancia desde el centro hasta un vértice.

Radius of a sphere (p. 638) The radius of a sphere is a segment that has one endpoint at the center and the other endpoint on the sphere.

Radio de una esfera (p. 638) El radio de una esfera es un segmento con un extremo en el centro y otro en la esfera.

Ratio (p. 366) A ratio is the comparison of two quantities by division.

5 to 7
5 : 7
$\frac{5}{7}$

Razón (p. 366) Una razón es la comparación de dos cantidades por medio de una división.

Ray (p. 23) A ray is the part of a line consisting of one *endpoint* and all the points of the line on one side of the endpoint.

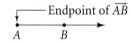

Rayo (p. 23) Un rayo es una parte de una recta que contiene un *extremo* y todos los puntos de la recta a un lado del extremo.

Rectangle (p. 306) A rectangle is a parallelogram with four right angles.

Rectángulo (p. 306) Un rectángulo es un paralelogramo con cuatro ángulos rectos.

Reduction (p. 498) *See* **dilation.**

Reducción (p. 498) *Ver* **dilation.**

Reflection (p. 478) A reflection (*flip*) across line *r* is a transformation such that if a point *A* is on line *r*, then the image of *A* is itself, and if a point *B* is not on line *r*, then its image *B'* is the point such that *r* is the perpendicular bisector of $\overline{BB'}$.

Reflexión (p. 478) Una reflexión en la recta *r* es una transformación tal que si un punto *A* está en la recta *r*, entonces la imagen de *A* es ella misma, y si un punto *B* no está en la recta *r*, entonces su imagen *B'* es el punto tal que *r* es la mediatriz de $\overline{BB'}$.

Reflectional symmetry (p. 492) Reflectional symmetry, or *line symmetry,* is the type of symmetry for which there is a reflection that maps a figure onto itself. The reflection line is the line of symmetry.

A reflection across the given line maps the figure onto itself.

Simetría por reflexión (p. 492) La simetría por reflexión, o *simetría axial,* es un tipo de simetría en la que la reflexión vuelve a trazar la figura sobre sí misma. La recta de reflexión es el eje de simetría.

English/Spanish Glossary

Regular polygon (p. 160)   A regular polygon is a polygon that is both equilateral and equiangular. Its *center* is the center of the circumscribed circle.

Polígono regular (p. 160)   Un polígono regular es un polígono que es equilátero y equiángulo. Su *centro* es el centro del círculo circunscrito.

ABCDEF is a regular hexagon. Point *X* is its center.

Regular pyramid (p. 617)   *See* **pyramid.**

Pirámide regular (p. 617)   *Ver* **pyramid.**

Remote interior angles (p. 149)   Remote interior angles are the two nonadjacent interior angles corresponding to each exterior angle of a triangle.

Ángulos interiores remotos (p. 149)   Los ángulos interiores remotos son los dos ángulos interiores no adyacentes que corresponden a cada ángulo exterior de un triángulo.

∠1 and ∠2 are remote interior angles of ∠3.

Resultant vector (p. 454)   The sum of two vectors is a resultant.

Vector resultante (p. 454)   La suma de dos vectores es el vector resultante.

$\vec{w}$ is the resultant of $\vec{u} + \vec{v}$.

Rhombus (p. 306)   A rhombus is a parallelogram with four congruent sides.

Rombo (p. 306)   Un rombo es un paralelogramo de cuatro lados congruentes.

Right angle (p. 37)   A right angle is an angle whose measure is 90.

Ángulo recto (p. 37)   Un ángulo recto es un ángulo que mide 90.

This symbol indicates a right angle.

Right cone (p. 619)   *See* **cone.**

Cono recto (p. 619)   *Ver* **cone.**

Right cylinder (p. 610)   *See* **cylinder.**

Cilindro recto (p. 610)   *Ver* **cylinder.**

Right prism (p. 608)   *See* **prism.**

Prisma rectangular (p. 608)   *Ver* **prism.**

Right triangle (pp. 148, 235) A right triangle contains one right angle. The side opposite the right angle is the *hypotenuse* and the other two sides are the *legs*.

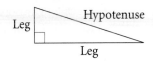

Triángulo rectángulo (pp. 148, 235) Un triángulo rectángulo contiene un ángulo recto. El lado opuesto del ángulo recto es la *hipotenusa* y los otros dos lados son los *catetos*.

Rotation (p. 484) A rotation (*turn*) of $x°$ about a point R is a transformation such that for any point V, its image is the point V', where $RV = RV'$ and $m\angle VRV' = x$. The image of R is itself.

Rotación (p. 484) Una rotación de $x°$ alrededor de un punto R es una transformación de modo que para cualquier punto V, su imagen es el punto V', donde $RV = RV'$ y $m\angle VRV' = x$. La imagen de R es R misma.

Rotational symmetry (p. 493) Rotational symmetry is the type of symmetry for which there is a rotation of 180° or less that maps a figure onto itself.

The figure has 120° rotational symmetry.

Simetría rotacional (p. 493) La simetría rotacional es un tipo de simetría en la que una rotación de 180° o menos vuelve a trazar una figura sobre sí misma.

Same-side interior (exterior) angles (pp. 127, 129) Same side interior (exterior) angles lie on the same side of the transversal t and between (outside of) ℓ and m.

Ángulos internos (externos) del mismo lado (pp. 127, 129) Los ángulos internos (externos) del mismo lado están situados en el mismo lado de la transversal t y dentro de (fuera de) ℓ y m.

$\angle 1$ and $\angle 2$ are same-side interior angles, as are $\angle 3$ and $\angle 4$. $\angle 5$ and $\angle 6$ are same-side exterior angles.

Scalar multiplication (p. 504) Scalar multiplication is the multiplication of each entry in a matrix by the same number, the *scalar*.

$$2 \cdot \begin{bmatrix} 1 & 0 \\ -2 & 3 \end{bmatrix} = \begin{bmatrix} 2(1) & 2(0) \\ 2(-2) & 2(3) \end{bmatrix}$$
$$= \begin{bmatrix} 2 & 0 \\ -4 & 6 \end{bmatrix}$$

Multiplicación escalar (p. 504) La multiplicación escalar es la multiplicación de cada elemento de una matriz por el mismo número, el *escalar*.

Scale (p. 368) A scale is the ratio of any length in a scale drawing to the corresponding actual length. The lengths may be in different units.

1 cm to 1 ft
1 cm = 1 ft
1 cm : 1 ft

Escala (p. 368) Una escala es la razón de cualquier longitud en un dibujo a escala en relación a la longitud verdadera correspondiente. Las longitudes pueden expresarse en distintas unidades.

English/Spanish Glossary

Scale drawing (p. 368) A scale drawing is a drawing in which all lengths are proportional to corresponding actual lengths.

Dibujo a escala (p. 368) Un dibujo a escala es un dibujo en el que todas las longitudes son proporcionales a las longitudes verdaderas correspondientes.

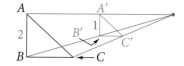

Scale: 1 in. = 30 ft

Scale factor (p. 498) The scale factor of a dilation is the number that describes the size change from an original figure to its image. *See also* **dilation.**

Factor de escala (p. 498) El factor de escala de una dilatación es el número que describe el cambio de tamaño de una figura original a su imagen. *Ver también* **dilation.**

The scale factor of the dilation that maps $\triangle ABC$ to $\triangle A'B'C'$ is $\frac{1}{2}$.

Scalene triangle (p. 148) A scalene triangle has no sides congruent.

Triángulo escaleno (p. 148) Un triángulo escaleno no tiene lados congruentes.

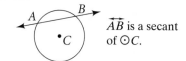

Secant (p. 687) A secant is a line, ray, or segment that intersects a circle at two points.

Secante (p. 687) Una secante es una recta, rayo o segmento que corta un círculo en dos puntos.

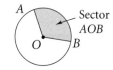

$\overleftrightarrow{AB}$ is a secant of $\odot C$.

Sector of a circle (p. 576) A sector of a circle is the region bounded by two radii and their intercepted arc.

Sector de un círculo (p. 576) Un sector de un círculo es la región limitada por dos radios y el arco abarcado por ellos.

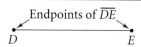

Sector AOB

Segment (p. 23) A segment is the part of a line consisting of two points, called *endpoints,* and all points between them.

Segmento (p. 23) Un segmento es una parte de una recta que consiste en dos puntos, llamados *extremos,* y todos los puntos entre los extremos.

Endpoints of $\overline{DE}$

Segment bisector (p. 32) A segment bisector is a line, segment, ray, or plane that intersects a segment at its midpoint.

Bisectriz de un segmento (p. 32) La bisectriz de un segmento es una recta, segmento, rayo o plano que corta un segmento en su punto medio.

ℓ bisects $\overline{KJ}$.

Segment of a circle (p. 577) A segment of a circle is the part of a circle bounded by an arc and the segment joining its endpoints.

Segmento de un círculo (p. 577) Un segmento de un círculo es la parte de un círculo bordeada por un arco y el segmento que une sus extremos.

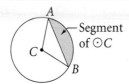

Semicircle (p. 567) A semicircle is half a circle.

Semicírculo (p. 567) Un semicírculo es la mitad de un círculo.

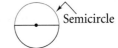

Side *See* **angle; polygon.**

Lado *Ver* **angle; polygon.**

Similar polygons (p. 373) Similar polygons are polygons having corresponding angles congruent and corresponding sides proportional. You denote similarity by ∼.

Polígonos semejantes (p. 373) Los polígonos semejantes son polígonos cuyos ángulos correspondientes son congruentes y los lados correspondientes son proporcionales. El símbolo ∼ significa "es semejante a".

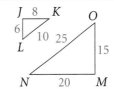

Similar solids (p. 646) Similar solids have the same shape and have all their corresponding dimensions proportional.

Cuerpos geométricos semejantes (p. 646) Los cuerpos geométricos semejantes tienen la misma forma y todas sus dimensiones correspondientes son proporcionales.

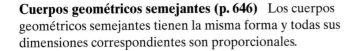

Similarity ratio (pp. 373, 646) The similarity ratio is the ratio of the lengths of corresponding sides of similar figures.

Razón de semejanza (pp. 373, 646) La razón de semejanza es la razón de la longitud de los lados correspondientes de figuras semejantes.

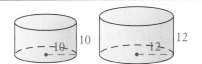

Similarity transformation (p. 498) *See* **dilation.**

Transformación de semejanza (p. 498) *Ver* **dilation.**

Sine ratio (p. 439) *See* **trigonometric ratios.**

Razón seno (p. 439) *Ver* **trigonometric ratios.**

Skew lines (p. 24) Skew lines are lines that do not lie in the same plane.

Rectas cruzadas (p. 24) Las rectas cruzadas son rectas que no están en el mismo plano.

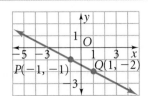

$\overleftrightarrow{AB}$ and $\overleftrightarrow{EF}$ are skew.

Slant height *See* **cone; pyramid.**

Generatriz (cono) o apotema (pirámide) *Ver* **cone; pyramid.**

Slide (p. 471) *See* **translation.**

Traslación (p. 471) *Ver* **translation.**

Slope-intercept form (p. 166) The slope-intercept form of a linear equation is $y = mx + b$, where m is the slope of the line and b is the y-intercept.

$y = \frac{1}{2}x - 3$

In this equation, the slope is $\frac{1}{2}$ and the y-intercept is -3.

Forma pendiente-intercepto (p. 166) La forma pendiente-intercepto es la ecuación lineal $y = mx + b$, en la que m es la pendiente de la recta y b es el punto de intersección de esa recta con el eje y.

Slope of a line (p. 165) The slope of a line is the ratio of its vertical change in the coordinate plane to the corresponding horizontal change. If (x_1, y_1) and (x_2, y_2) are points on a nonvertical line, then the slope is $\frac{y_2 - y_1}{x_2 - x_1}$. The slope of a horizontal line is 0 and the slope of a vertical line is undefined.

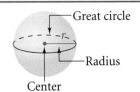

The line containing $P(-1, -1)$ and $Q(1, -2)$ has slope

$\frac{-2 - (-1)}{1 - (-1)} = \frac{-1}{2} = -\frac{1}{2}.$

Pendiente de una recta (p. 165) La pendiente de una recta es la razón del cambio vertical en el plano de coordenadas en relación al cambio horizontal correspondiente. Si (x_1, y_1) y (x_2, y_2) son puntos en una recta no vertical, entonces la pendiente es $\frac{y_2 - y_1}{x_2 - x_1}$. La pendiente de una recta horizontal es 0, y la pendiente de una recta vertical es indefinida.

Space (p. 17) Space is the set of all points.

Espacio (p. 17) El espacio es el conjunto de todos los puntos.

Sphere (p. 638) A sphere is the set of all points in space that are a given distance r, the *radius,* from a given point C, the *center.* A *great circle* is the intersection of a sphere with a plane containing the center of the sphere. The *circumference* of a sphere is the circumference of any great circle of the sphere.

Esfera (p. 638) Una esfera es el conjunto de los puntos del espacio que están a una distancia dada r, el *radio,* de un punto dado C, el *centro.* Un *círculo máximo* es la intersección de una esfera y un plano que contiene el centro de la esfera. La *circunferencia* de una esfera es la circunferencia de cualquier círculo máximo de la esfera.

Spherical geometry (p. 154)   In spherical geometry, a plane is considered to be the surface of a sphere and a line is considered to be a great circle of the sphere. In spherical geometry, through a point not on a given line there is no line parallel to the given line.

In spherical geometry, lines are represented by great circles of a sphere.

Geometría esférica (p. 154)   En la geometría esférica, un plano es la superficie de una esfera y una recta es un círculo máximo de la esfera. En la geometría esférica, a través de un punto que no está en una recta dada, no hay recta paralela a la recta dada.

Square (p. 306)   A square is a parallelogram with four congruent sides and four right angles.

Cuadrado (p. 306)   Un cuadrado es un paralelogramo con cuatro lados congruentes y cuatro ángulos rectos.

Standard form of an equation of a circle (p. 695)   The standard form of an equation of a circle is $(x - h)^2 + (y - k)^2 = r^2$, where (h, k) is the center of the circle.

In $(x + 5)^2 + (y + 2)^2 = 48$, $(-5, -2)$ is the center of the circle.

Forma normal de la ecuación de un círculo (p. 695)   La forma normal de la ecuación de un círculo es $(x - h)^2 + (y - k)^2 = r^2$, donde (h, k) son las coordenadas del centro del círculo.

Standard form of a linear equation (p. 167)   The standard form of a linear equation is $Ax + By = C$, where A, B, and C are integers and A and B are not both zero.

$6x - y = 3$

Forma normal de una ecuación lineal (p. 167)   La forma normal de una ecuación lineal es $Ax + By = C$, donde A, B y C son números reales, y donde A y B no son iguales a cero.

Straight angle (p. 37)   A straight angle is an angle whose measure is 180.

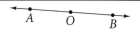

$m\angle AOB = 180$

Ángulo llano (p. 37)   Un ángulo llano es un ángulo que mide 180.

Straightedge (p. 44)   A straightedge is a ruler with no markings on it.

Regla sin graduación (p. 44)   Una regla sin graduación no tiene marcas.

Supplementary angles (p. 38)   Two angles are supplementary if the sum of their measures is 180.

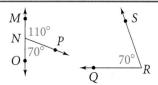

Ángulos suplementarios (p. 38)   Dos ángulos son suplementarios cuando sus medidas suman 180.

$\angle MNP$ and $\angle ONP$ are supplementary, as are $\angle MNP$ and $\angle QRS$.

English/Spanish Glossary

Surface area (pp. 608, 610, 618, 619, 638) The surface area of a prism, cylinder, pyramid, or cone is the sum of the lateral area and the areas of the bases. The surface area of a sphere is four times the area of a great circle. A list of surface area formulas is on p. 765.

Área (pp. 608, 610, 618, 619, 638) El área de un prisma, pirámide, cilindro o cono es la suma del área lateral y las áreas de las bases. El área de una esfera es igual a cuatro veces el área de un círculo máximo. Una lista de fórmulas de áreas está en la p. 765.

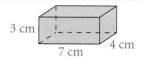

S.A. of prism = L.A. + 2B
= 66 + 2(28)
= 122 cm^2

Symmetry (pp. 492, 493, 516) A figure has symmetry if there is an isometry that maps the figure onto itself. *See* **glide reflectional symmetry; point symmetry; reflectional symmetry; rotational symmetry; translational symmetry.**

Simetría (pp. 492, 493, 516) Una figura tiene simetría si hay una isometría que traza la figura sobre sí misma. *Ver* **glide reflectional symmetry; point symmetry; reflectional symmetry; rotational symmetry; translational symmetry.**

A regular pentagon has reflectional symmetry and 72° rotational symmetry.

Tangent ratio (p. 432) *See* **trigonometric ratios.**

Razón tangente (p. 432) *Ver* **trigonometric ratios.**

Tangent to a circle (p. 598) A tangent to a circle is a line, segment, or ray in the plane of the circle that intersects the circle in exactly one point. That point is the *point of tangency.*

Tangente de un círculo (p. 598) Una tangente de un círculo es una recta, segmento o rayo en el plano del círculo que corta el círculo en exactamente un punto. Ese punto es el *punto de tangencia.*

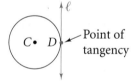

Line ℓ is tangent to ⊙C. Point D is the point of tangency.

Terminal point of a vector (p. 452) *See* **vector.**

Punto terminal de un vector (p. 452) *Ver* **vector.**

Tessellation (p. 515) A tessellation, or *tiling,* is a repeating pattern of figures that completely covers a plane without gaps or overlap. A *pure tessellation* is a tessellation that consists of congruent copies of one figure.

Teselado (p. 515) Un teselado o *reticulado* es un patrón repetitivo de figuras que cubre completamente una superficie plana sin dejar espacios vacíos ni traslaparse. Un *teselado puro* consiste en copias congruentes de una figura.

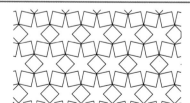

Theorem (p. 110)   A theorem is a conjecture that is proven.

Teorema (p. 110)   Un teorema es una conjetura que se demuestra.

The theorem "Vertical angles are congruent" can be proven by using postulates, definitions, properties, and previously stated theorems.

Tiling (p. 515)   *See* **tessellation.**

Reticulado (p. 515)   *Ver* **tessellation.**

Transformation (p. 470)   A transformation is a change in the position, size, or shape of a geometric figure. The given figure is called the *preimage* and the resulting figure is called the *image*. A transformation *maps* a figure onto its image. *Prime notation* is sometimes used to identify image points. In the diagram, X' (read "X prime") is the image of X.

Transformación (p. 470)   Una transformación es un cambio en la posición, tamaño o forma de una figura. La figura dada se llama la *preimagen* y la figura resultante se llama la *imagen*. Una transformación *traza* la figura sobre su propia imagen. La *notación prima* a veces se utiliza para identificar los puntos de la imagen. En el diagrama de la derecha, X' (leído X prima) es la imagen de X.

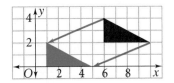

$\triangle XYZ \rightarrow \triangle X'Y'Z'$

Translation (p. 471)   A translation (*slide*) is a transformation that moves points the same distance and in the same direction. A translation in the coordinate plane is described by a vector.

Traslación (p. 471)   Una traslación es la transformación que mueve puntos a la misma distancia y en la misma dirección. Un vector puede describir la traslación en un plano de coordenadas.

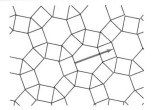

The blue triangle is the image of the black triangle under the translation $\langle -5, -2 \rangle$.

Translational symmetry (p. 516)   Translational symmetry is the type of symmetry for which there is a translation that maps a figure onto itself.

Simetría traslacional (p. 516)   La simetría traslacional es un tipo de simetría en la que la traslación vuelve a trazar la figura sobre sí misma.

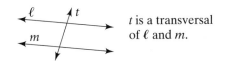

The tessellation shown can be mapped onto itself by the given translation.

Transversal (p. 127)   A transversal is a line that intersects two coplanar lines in two points.

Transversal (p. 127)   Una transversal es una recta que corta dos rectas coplanarias en dos puntos.

t is a transversal of ℓ and m.

Trapezoid (pp. 306, 336)   A trapezoid is a quadrilateral with exactly one pair of parallel sides, the *bases*. The nonparallel sides are called the *legs* of the trapezoid. Each pair of angles adjacent to a base are *base angles* of the trapezoid. An *altitude* of a trapezoid is a perpendicular segment from one base to the line containing the other base. Its length is called the *height* of the trapezoid.

Trapecio (pp. 306, 336)   Un trapecio es un cuadrilátero con exactamente un par de lados paralelos, las *bases*. Los lados no paralelos se llaman los *catetos* del trapecio. Cada par de ángulos adyacentes a la base son los *ángulos de base* del trapecio. Una *altura* del trapecio es un segmento perpendicular que va de una base a la recta que contiene la otra base. Su longitud se llama, por extensión, la *altura* del trapecio.

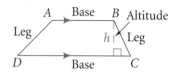

In trapezoid $ABCD$, $\angle ADC$ and $\angle BCD$ are one pair of base angles, and $\angle DAB$ and $\angle ABC$ are the other.

Triangle (pp. 158, 275, 536)   A triangle is a polygon with three sides. You can choose any side to be a *base*. The *height* is the length of the altitude drawn to the line containing that base.

Triángulo (pp. 158, 275, 536)   Un triángulo es un polígono con tres lados. Se puede escoger cualquier lado como *base*. La *altura,* entonces, es la longitud de la altura trazada hasta la recta que contiene la base.

Trigonometric ratios (pp. 432, 439)   In right triangle $\triangle ABC$ with acute angle $\angle A$

sine $\angle A = \sin A = \dfrac{\text{leg opposite } \angle A}{\text{hypotenuse}}$

cosine $\angle A = \cos A = \dfrac{\text{leg adjacent } \angle A}{\text{hypotenuse}}$

tangent $\angle A = \tan A = \dfrac{\text{leg opposite } \angle A}{\text{leg adjacent } \angle A}$

Razones trigonométricas (pp. 432, 439)   En un triángulo rectángulo $\triangle ABC$ con ángulo agudo $\angle A$

seno $\angle A = \text{sen } A = \dfrac{\text{cateto opuesto a } \angle A}{\text{hipotenusa}}$

coseno $\angle A = \cos A = \dfrac{\text{cateto adyecente a } \angle A}{\text{hipotenusa}}$

tangente $\angle A = \tan A = \dfrac{\text{cateto opuesto a } \angle A}{\text{cateto adyecente a } \angle A}$

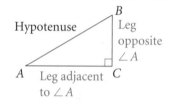

Truth value (p. 81)   The truth value of a statement is "true" or "false" according to whether the statement is true or false, respectively.

Valor verdadero (p. 81)   El valor verdadero de un enunciado es "verdadero" o "falso" según el enunciado sea verdadero o falso, respectivamente.

The truth value of the statement "If a figure is a triangle, then it has four sides" is *false*.

Turn (p. 484)   *See* **rotation.**

Rotación (p. 484)   *Ver* **rotation.**

Two-column proof (p. 129) *See* **proof.**

Prueba de dos columnas (p. 129) *Ver* **proof.**

Vector (p. 452) A vector is any quantity that has *magnitude* (size) and *direction.* You can represent a vector as an arrow that starts at one point, the *initial point,* and points to a second point, the *terminal point.* A vector can be described by *ordered pair notation* ⟨*x, y*⟩, where *x* represents horizontal change from the initial point to the terminal point and *y* represents vertical change from the initial point to the terminal point.

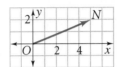

Vector *ON* has initial point *O* and terminal point *N.* The ordered pair notation for the vector is ⟨5, 2⟩.

Vector (p. 452) Un vector es cualquier cantidad que tiene *magnitud* (tamaño) y *dirección.* Se puede representar un vector como una flecha que empieza en un punto, el *punto inicial,* y se dirige a un segundo punto, el *punto terminal.* Un vector se puede describir mediante la *notación de pares ordenados* ⟨*x, y*⟩, donde *x* representa el cambio horizontal desde el punto inicial hasta el punto final, e *y* representa el cambio vertical desde el punto inicial hasta el punto final.

Vertex *See* **angle; cone; polygon; polyhedron; pyramid.** The plural form of *vertex* is *vertices.*

Vértice *Ver* **angle; cone; polygon; polyhedron; pyramid.**

Vertex angle (p. 228) *See* **isosceles triangle.**

Ángulo del vértice (p. 228) *Ver* **isosceles triangle.**

Vertical angles (p. 38) Vertical angles are two angles whose sides form two pairs of opposite rays.

∠1 and ∠2 are vertical angles, as are ∠3 and ∠4.

Ángulos opuestos por el vértice (p. 38) Dos ángulos son ángulos opuestos por el vértice si sus lados son rayos opuestos.

Volume (p. 624) Volume is a measure of the space a figure occupies. A list of volume formulas is on p. 765.

The volume of this prism is 24 cubic units, or 24 units^3.

Volumen (p. 624) El volumen es una medida del espacio que ocupa una figura. Una lista de las fórmulas de volumen está en la p. 765.

English/Spanish Glossary

Chapter 1

Check Your Readiness   p. 2

1. 9 **2.** 16 **3.** 121 **4.** 37 **5.** 78.5 **6.** 13 **7.** 1 **8.** $-\frac{3}{5}$
9. 5 **10.** 8 **11.** 4 **12.** 3 **13.** 3 **14.** 6 **15.** 1

Lesson 1-1   pp. 4–6

Check Skills You'll Need 1. 2, 4, 6, 8, 10, . . . **2.** 1, 3, 5,
7, 9, . . . **3.** $1^2 = 1, 2^2 = 4, 3^2 = 9, 4^2 = 16,$
$5^2 = 25, 6^2 = 36, 7^2 = 49, 8^2 = 64, 9^2 = 81,$
$10^2 = 100$ **4.** It is odd.

Quick Check 1a. 29, 37 **b.** Thursday, Friday

c. 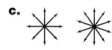   **2.** The sum of the first 35 odd
numbers is 35^2, or 1225.

3.   **4a.** 39 skateboards **b.** Not
confident; December is too
far away.

Lesson 1-2   pp. 10–12

Check Skills You'll Need
1. E B   **2.** H ↗

Quick Check
1. 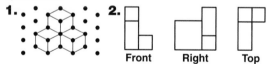 **2.** Front Right Top

3a. 9 cubes **b.** Answers may vary. Sample: The
foundation drawing; you can just add the five
numbers. **4.** E, C [cube image]

Lesson 1-3   pp. 16–19

Check Skills You'll Need 1. (1, 6) **2.** (3, 2) **3.** (5, 10)

Quick Check 1a. no **b.** Answers may vary.
Sample: $\overleftrightarrow{EF}, \overleftrightarrow{FC}, \overleftrightarrow{CE}$ **c.** Arrowheads are used to
show that the line extends in opposite directions
without end. **2.** Answers may vary. Sample: *HEF,*
HEFG, FGH **3.** *ABF* and *CBF* **4a.** *D* **b.** *B*

Lesson 1-4   pp. 23–25, 28

Check Skills You'll Need 1. no **2.** yes **3.** no
4–9. Answers may vary. Samples are given.
4. *NMR* **5.** *PQL* **6.** *NKL* **7.** *PQR* **8.** *PKN* **9.** *LQR*

Quick Check 1. No, they do not have the same
endpoint. **2a.** $\overline{HI}, \overline{DN}$ **b.** $\overline{AB}, \overline{CD}, \overline{CH}$
c. $\overline{DN}, \overline{HI}; \overline{DN}, \overline{HC}$ **3a.** *PSWT* ∥ *RQVU*, *PRUT* ∥
SQVW, PSQR ∥ *TWVU* **b.** Answers may vary.
Sample: $\overleftrightarrow{PS}$

Checkpoint Quiz 1 1. 29, 31.5 **2.** 3.45678, 3.456789
3. For 1: Add 2.5. For 2: Extend the decimal to
one more place with a digit that is 1 more than
the one to its left. **4.** yes, plane *AEF* **5.** yes, plane
DCEF **6.** No; *H, G,* and *F* are in the front plane, *B*
is not. **7.** No; *A, E,* and *B* are in the top plane, *C* is
not. **8.** $\overline{CD}, \overline{AB}, \overline{EF}$ **9.** Answers may vary.
Sample: $\overleftrightarrow{AE}$ and $\overleftrightarrow{BC}$ **10.** [figure]

Lesson 1-5   pp. 31–33

Check Skills You'll Need 1. 6 **2.** 3.5 **3.** 3 **4.** 6 **5.** 2 **6.** 9
7. 4 **8.** 9 **9.** $\frac{1}{3}$

Quick Check 1a. *CD* = *DE*; yes **b.** yes;
$|-5 - (-8)| = |3| = 3$ **2.** 15; *EF* = 40, *FG* = 60
3. 13.5

Lesson 1-6   pp. 36–39

Check Skills You'll Need 1. 80 **2.** 130 **3.** 45 **4.** 135
5. 110 **6.** 45

Quick Check 1a. ∠2, ∠*DEC* **b.** No; 3 ⦞ have *E* for a
vertex, so you need more info. in the name to
distinguish them from one another. **2a.** 30; acute
b. 90; right **c.** 140; obtuse **3.** 35 **4a.** Answers may
vary. Sample: ∠*AFB* and ∠*BFC*; ∠*BFD* and ∠*DFE*
b. 153 **5a.** Yes; the congruent segments are
marked. **b.** No; there are no markings. **c.** No; there
are no markings. **d.** No; there are no markings.
e. Yes; the congruent segments are marked.

Lesson 1-7   pp. 44–47

Check Skills You'll Need 1. [figure with C, D]

2. G H   **3.** A B
4. m   **5.** [figure A, B, C]   **6.** [figure X, Y, S, T]
7. 10 **8.** [60° angle] **9.** [120° angle]

Quick Check **1.** X •———————• Y

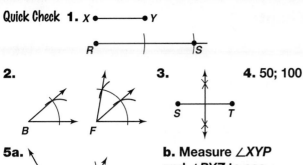

2. 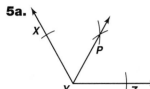 **3.** **4.** 50; 100

5a. **b.** Measure ∠*XYP* and ∠*PYZ* to see that they are ≅.

Chapter 2

Check Your Readiness p. 78

1. 50 **2.** −3 **3.** $25\frac{1}{2}$ **4.** 10.5 **5.** 15 **6.** 11 **7.** 20
8. −5 **9.** −4 **10.** 5 **11.** 6 **12.** 7 **13.** 18 **14.** ∠*ACD*, ∠*DCA* **15.** C **16.** 3 **17.** ∠*ADB* or ∠*BDA* **18.** $\overline{CD}$
19. 45 **20.** 48, 42

Lesson 1-8 pp. 53–55

Check Skills You'll Need **1.** 5.0 **2.** 4.1 **3.** 11.1 **4.** 100
5. 100 **6.** 58 **7.** 196 **8.** 10 **9.** −1

Quick Check **1a.** 8.6 **b.** Yes; the differences are opposites, and the square of a number and the square of its opposite are the same. **2a.** about 8.9 mi **b.** about 3.2 mi **3.** (4, 4) **4.** (6, −9)

Checkpoint Quiz 2 **1.** 17 **2.** 110 **3.** 140 **4a.** 90 **b.** 60
5. ∠*APT* ≅ ∠*RPT* **6.** 45

7. 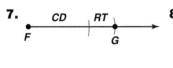 **8.**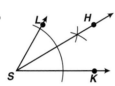

9. 12.2 units **10.** (1, 1.5)

Lesson 1-9 pp. 61–64

Check Skills You'll Need **1.** 4 **2.** 15 **3.** 8 **4.** 6.7 **5.** 3.2
6. 7.8 **7.** 4.5 **8.** 13.0 **9.** 7.8

Quick Check **1a.** 26 in. **b.** 30 in. **2a.** 36π m
b. 56.5 m **3.** 20 units

4. $9\frac{1}{3}$ yd²; $9\frac{1}{3}$ is one-ninth of 84.
5a. $\frac{25}{4}\pi$ ft² **b.** 19.6 ft²
6.

24 cm²

Lesson 2-1 pp. 80–82

Check Skills You'll Need **1.** −1 **2.** −2, 2 **3.** 2 **4.** −4
5. 0 **6.** −1

Quick Check **1.** Hypothesis: $T − 38 = 3$, Conclusion: $T = 41$ **2a.** If an integer ends with 0, then it is divisible by 5. **b.** If a figure is a square, then it has 4 congruent sides. **3.** The conditional is false; New Mexico is a counterexample.

4.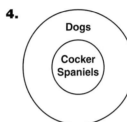

5. If two lines are skew, then they are not parallel and do not intersect.
6a. If two lines are parallel, then they do not intersect. The conditional is false and the converse is true. **b.** If |*x*| = 2, then *x* = 2. The conditional is true and the converse is false. **7.** Answers may vary. Sample: The statement "I breathe when I sleep" can be rewritten as "If I sleep, then I breathe." The statement "I sleep when I breathe" can be rewritten as "If I breathe, then I sleep." The two statements are converses, and do not have the same meaning.

Lesson 2-2 pp. 87–89

Check Skills You'll Need **1.** Hypothesis: $x > 10$, Conclusion: $x > 5$ **2.** Hypothesis: You live in Milwaukee. Conclusion: You live in Wisconsin. **3.** If a figure is a square, then it has four sides. **4.** If something is a butterfly, then it has wings. **5.** If we go on a picnic, then the sun shines. **6.** If two lines do not intersect, then they are skew. **7.** If $x^3 = −27$, then $x = −3$.

Quick Check **1.** If three points lie on the same line, then they are collinear. The converse is also true. Three points are collinear if and only if they lie on the same line. **2.** If a number is prime, then it has only two distinct factors, 1 and itself. If a number has only two distinct factors, 1 and itself, then it is prime. **3.** Conditional: If an angle is a right angle, then its measure is 90. Converse: If an angle has measure 90, then it is a right angle. The two statements are true. An angle is a right angle if and only if its measure is 90. **4.** It is not a good definition because a rectangle has four right angles and is not necessarily a square.

Lesson 2-3   pp. 94–96, 100

Check Skills You'll Need 1. If your grades suffer, then you don't sleep enough. **2. If** you must start early, then you want to arrive on time. **3. If** a year is a leap year, then it has 366 days. **4. If** students do not complete their homework, then they will have lower grades. **5. If** two lines are perpendicular, then they meet to form right angles. **6. If** a person is 16 years old, then that person is a teenager.

Quick Check 1. No, there could be other things wrong with the car, such as a faulty starter.
2. Answers may vary. Sample: Vladimir Nuñez should not pitch a complete game on Tuesday.
3. Not possible; you do not know that the hypothesis is true. **4a. If** a number ends in 0, then it is divisible by 5. **b.** Not possible; the conclusion of one statement is not the hypothesis of the other statement. **5.** The Volga River is less than 2300 miles long. The Volga River is not one of the world's ten longest rivers.

Checkpoint Quiz 1 1. Hypothesis: $x > 5$, Conclusion: $x^2 > 25$ **2. If** something is a rose, then it is a beautiful flower. **3. If** an integer is divisible by 2, then the integer ends with 0. **4.** Answers may vary. Sample: 42 is divisible by 2, but it does not end with 0. **5. If** an angle is an acute angle, then its measure is between 0 and 90. If an angle's measure is between 0 and 90, then it is an acute angle. **6.** Points are collinear if and only if they lie on the same line. **7.** Answers may vary. Sample: A graphing calculator has a keyboard and a memory. **8.** Theresa has passing grades. **9. If** a student studies geometry, then the student's mind is expanded. **10.** not possible

Lesson 2-4   pp. 103–105

Check Skills You'll Need 1. $\angle AOB$, $\angle BOA$ **2.** O **3.** $\overrightarrow{OB}$
4. 45 **5.** $\overrightarrow{OA}$ and $\overrightarrow{OC}$

Quick Check 1. Subst. Prop.; Subtr. Prop. of =; Div. Prop. of = **2.** $AB = 12$; $BC = 9$; $AB + BC = 12 + 9 = 21$ **3a.** Reflexive Prop. of ≅
b. Transitive or Subst. Prop. of ≅

Lesson 2-5   pp. 110–112

Check Skills You'll Need 1. 50 **2.** 90 **3.** 35 **4.** right ∠
5. vertex

Quick Check 1a. 140; 140 **b.** 40; 40 **c.** 140 + 40 = 180 **2.** Subtraction Property of Equality

Chapter 3

Check Your Readiness   p. 124

1. 72 **2.** 1260 **3.** 2700 **4.** 4 **5.** 15 **6.** −24
7. $2x + x + x = 180$; 45, 45, 90
8. $\frac{1}{2}x + x + x = 180$; 36, 72, 72
9.

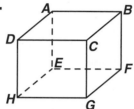

$\overleftrightarrow{AB}$ and $\overleftrightarrow{GH}$; $\overleftrightarrow{AE}$ and $\overleftrightarrow{EF}$

10–12. Check students' work.

Lesson 3-1   pp. 127–130

Check Skills You'll Need 1. 30 **2.** 30 **3.** 60 **4.** 9
5. $m\angle 1 + 2(90 - m\angle 1) = 146$; $m\angle 1 = 34$
6. 72 and 108

Quick Check 1. $\angle 5$ and $\angle 4$; $\angle 6$ and $\angle 2$; $\angle 3$ and $\angle 8$
2. same-side int. ∠ **3.** 1. $a \parallel b$ (Given) 2. $m\angle 3 + m\angle 2 = 180$ (∠ Add. Post.) 3. $m\angle 1 = m\angle 2$ (Corr. ∠ Post.) 4. $m\angle 1 + m\angle 3 = 180$ (Subst.) 5. $\angle 1$ and $\angle 3$ are suppl. (Def. of Suppl. ∠) **4a.** 130; corr. ∠ are ≅. **b.** 130; vert. ∠ are ≅. **c.** 50; alt. int. ∠ are ≅.
d. 50; alt. int. ∠ are ≅. **e.** 130; same-side int. ∠ are suppl. **f.** 50; corr. ∠ are ≅ or vert. ∠ are ≅.
5. $x = 45$, $y = 115$; 90, 90, 115, 65

Lesson 3-2   pp. 134–137

Check Skills You'll Need 1. 11 **2.** 4 **3.** 26 **4.** 6 **5. If** a △ has a 90° ∠, then it is a right △; true. **6. If** two ∠ are ≅, then they are vert. ∠; false. **7. If** two ∠ are suppl., then they are same-side int. ∠; false.

Quick Check 1. $\overleftrightarrow{EC} \parallel \overleftrightarrow{DK}$; Conv. of Corr. ∠ Post.
2. 18; 7 · 18 − 8 = 118, and 62 + 118 = 180.
3. $m\angle 3 = 120$; the lines will be $\parallel$ by the Conv. of Same-Side Ext. ∠ Thm.

Lesson 3-3   pp. 141–142

Check Skills You'll Need 1. sometimes **2.** always
3. sometimes **4.** never

Quick Check 1. Yes; 30° + 60° = 90° **2.** Yes; $b \parallel d$ by Thm. 3-10.

Lesson 3-4   pp. 147–150

Check Skills You'll Need 1. right **2.** acute **3.** acute
4. 60 **5.** 20 **6.** 32 **7.** 58

Quick Check 1. 55

2a. **b.** **c.** Not possible; an equilateral △ has all acute ∠s.

3. 135, 135, 90 **4a.** 130 **b.** Answers may vary. Sample: Find the measure of the third ∠ of the triangle. Subtract this from 180.

Checkpoint Quiz 1 **1.** Corr. ∠s Post. **2.** Conv. of Corr. ∠s Post. **3.** Same-Side Int. ∠s Thm. **4.** Conv. of the Alt. Int. ∠s Thm. **5.** Vert. ∠s Thm. **6.** Alt. Int. ∠s Thm. **7.** Conv. of Alt. Ext. ∠s Thm. **8.** Corr. ∠s Post. **9.** If a line is ⊥ to 1 of 2 parallel lines, it is ⊥ to both. **10.** 38, 55, 87; acute, 55, 26, 99; obtuse

Lesson 3-5   pp. 157–160

Check Skills You'll Need **1.** $m\angle DAB = 77$; $m\angle B = 65$; $m\angle BCD = 131$; $m\angle D = 87$ **2.** $m\angle D = m\angle B = 60$; $m\angle DAB = m\angle DCB = 120$ **3.** $m\angle A = 70$; $m\angle ABC = 85$; $m\angle C = 125$; $m\angle ADC = 80$

Quick Check **1.** ABE; sides: $\overline{AB}, \overline{BE}, \overline{EA}$; ∠s: $\angle A$, $\angle ABE$, $\angle BEA$; $BCDE$; sides: $\overline{BC}, \overline{CD}, \overline{DE}, \overline{EB}$; ∠s: $\angle EBC$, $\angle C$, $\angle D$, $\angle DEB$; $ABCDE$; sides: $\overline{AB}$, $\overline{BC}, \overline{CD}, \overline{DE}, \overline{EA}$; ∠s: $\angle A$, $\angle ABC$, $\angle C$, $\angle D$, $\angle AED$ **2a.** hexagon; convex **b.** octagon; concave **c.** 24-gon; concave **3a.** 1980 **b.** You can solve the equation $(n - 2)180 = 720$. **4.** 108 **5a.** 60 **b.** 30; no, it is not formed by extending one side of the polygon.

Lesson 3-6   pp. 166–168

Check Skills You'll Need **1.** $-\frac{2}{3}$ **2.** $\frac{5}{3}$ **3.** 0 **4.** undefined or no slope **5.** $\frac{2}{3}$ **6.** $-\frac{3}{2}$ **7.** $\frac{4}{5}$ **8.** -1

Quick Check

1. 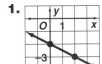 **2.**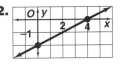

3. **4.** $y + 4 = -1(x - 2)$
5. $y - 0 = -\frac{3}{2}(x - 5)$ or $y + 3 = -\frac{3}{2}(x - 7)$
6. $y = -1; x = 5$

Lesson 3-7   pp. 174–177, 180

Check Skills You'll Need **1.** $\frac{1}{2}$ **2.** $\frac{5}{2}$ **3.** -5 **4.** 2 **5.** -1 **6.** $\frac{2}{3}$ **7.** 1 **8.** 0 **9.** $\frac{2}{3}$

Quick Check 1. No; the slope of $\ell_3 = -\frac{1}{7}$, and the slope of $\ell_4 = -\frac{1}{6}$. **2a.** Yes; both slopes $= -\frac{1}{2}$, and the y-intercepts are different. **b.** No; the lines are the same. **3.** $y = -x + 3$ **4.** No; the slope of $\ell_3 = \frac{4}{9}$, and the slope of $\ell_4 = -\frac{7}{3}$, and $\frac{4}{9} \cdot -\frac{7}{3} \neq -1$. **5.** $y + 4 = -5(x - 15)$ **6.** $y - 8 = \frac{3}{2}(x - 2)$ or $y = \frac{3}{2}x + 5$

Checkpoint Quiz 2 **1.** octagon; $n = 125$ **2.** hexagon; $x = 122$; $w = 90$ **3.** quad.; $a = 105$; $m = 116$

4. 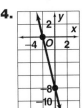 **5.**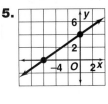

6.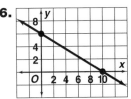

7. slope of $\overleftrightarrow{RS} = -\frac{2}{5}$; slope of $\overleftrightarrow{TV} = \frac{5}{3}$; neither

8. slope of $\overleftrightarrow{RS} = 1$; slope of $\overleftrightarrow{TV} = -1$; ⊥

9. slope of $\overleftrightarrow{RS} = -\frac{5}{4}$; slope of $\overleftrightarrow{TV} = \frac{4}{5}$; ⊥

10. slope of $\overleftrightarrow{RS} = \frac{2}{9}$; slope of $\overleftrightarrow{TV} = \frac{2}{9}$; ∥

Lesson 3-8   pp. 181–183

Check Skills You'll Need **1.** 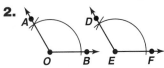 $\overline{AB} \cong \overline{CD}$

2.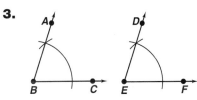
$\angle AOB \cong \angle DEF$

3.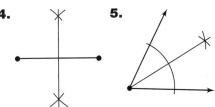
$\angle ABC \cong \angle DEF$

4. **5.**

6.

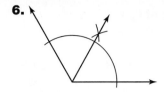

Quick Check 1. If corr. ∠ are ≅, the lines are ∥ by the Conv. of Corr. ∠ Post.

2. **3.** **4.**

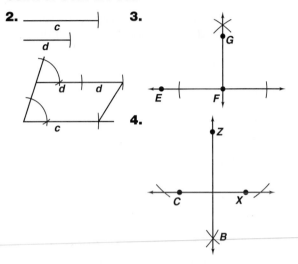

Chapter 4

Check Your Readiness p. 196

1. $AB = 4$, $BC = 3$, $AC = 5$ **2.** $AB = 8$, $BC = \sqrt{265}$, $AC = \sqrt{137}$ **3.** $AB = \sqrt{58}$, $BC = 4\sqrt{2}$, $AC = \sqrt{58}$ **4.** ∠A ≅ ∠C **5.** $m\angle A = 90 = m\angle B$ or ∠A and ∠B are rt. ∠. **6.** ∠B is a rt. ∠. **7.** ∠AFB ≅ ∠DFC **8.** ∠A ≅ ∠D and ∠B ≅ ∠C **9.** ∠ACD ≅ ∠CAB and ∠DAC ≅ ∠BCA **10.** $x = 12$

Lesson 4-1 pp. 198–200

Check Skills You'll Need 1. 19 **2.** 13 **3.** 108 **4.** 10 **5.** 50

Quick Check 1. ∠WSY ≅ ∠MVK; ∠SWY ≅ ∠VMK; ∠WYS ≅ ∠MKV; $\overline{WY}$ ≅ $\overline{MK}$; $\overline{WS}$ ≅ $\overline{MV}$; $\overline{YS}$ ≅ $\overline{KV}$ **2.** $m\angle K = 35$; corr. ∠ are ≅. **3.** No; corr. sides are not necessarily ≅. **4a.** ∠A ≅ ∠D; ∠E ≅ ∠C (Given) **b.** ∠ABE ≅ ∠DBC (Vert. ∠ are ≅.) **c.** $\overline{AE}$ ≅ $\overline{CD}$; $\overline{AB}$ ≅ $\overline{BD}$; $\overline{EB}$ ≅ $\overline{BC}$ (Given) **d.** △ABE ≅ △DBC (Def. of ≅ △)

Lesson 4-2 pp. 205–207

Check Skills You'll Need 1. $\overline{AB}$ ≅ $\overline{DE}$; ∠C ≅ ∠F **2.** ∠Q ≅ ∠S; ∠QPR ≅ ∠SRP; $\overline{PR}$ ≅ $\overline{PR}$ **3.** ∠M ≅ ∠S; ∠MON ≅ ∠SVT; $\overline{TO}$ ≅ $\overline{NV}$; $\overline{MO}$ ∥ $\overline{VS}$

Quick Check 1. You are given that $\overline{HF}$ ≅ $\overline{HJ}$ and $\overline{FG}$ ≅ $\overline{JK}$. H is the midpoint of $\overline{GK}$, so $\overline{GH}$ ≅ $\overline{KH}$ by def. of midpoint. △FGH ≅ △JKH by SSS.

2. ∠DCA ≅ ∠BAC **3.** No; you don't know that ∠E ≅ ∠DBC or that $\overline{AB}$ ≅ $\overline{DC}$.

Lesson 4-3 pp. 213–215, 219

Check Skills You'll Need 1. $\overline{JH}$ **2.** $\overline{HK}$ **3.** ∠L **4.** ∠N **5.** Reflexive Prop. of ≅ **6.** If 2 ∠ of a △ are ≅ to 2 ∠ of another △, the third ∠ are ≅.

Quick Check 1. No; the ≅ side is not the included side. **2.** It is given that $\overline{NM}$ ≅ $\overline{NP}$ and ∠M ≅ ∠P. ∠MNL ≅ ∠PNO because vert. ∠ are ≅. △NML ≅ △NPO by ASA.

3.

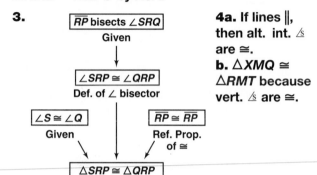

4a. If lines ∥, then alt. int. ∠ are ≅. **b.** △XMQ ≅ △RMT because vert. ∠ are ≅.

Checkpoint Quiz 1 1. $\overline{RS}$ ≅ $\overline{JK}$; $\overline{ST}$ ≅ $\overline{KL}$; $\overline{RT}$ ≅ $\overline{JL}$; ∠R ≅ ∠J; ∠S ≅ ∠K; ∠T ≅ ∠L **2.** ASA **3.** SSS **4.** SAS **5.** not possible **6.** AAS **7.** not possible **8.** If ∥ lines, then alt. int. ∠ are ≅. **9.** Vert. ∠ are ≅. **10.** ASA or AAS

Lesson 4-4 pp. 221–222

Check Skills You'll Need 1. ∠J ≅ ∠H; ∠R ≅ ∠V; ∠C ≅ ∠G **2.** $\overline{JR}$ ≅ $\overline{HV}$; $\overline{RC}$ ≅ $\overline{VG}$; $\overline{JC}$ ≅ $\overline{HG}$ **3.** ∠T ≅ ∠L; ∠I ≅ ∠O; ∠C ≅ ∠K **4.** $\overline{TI}$ ≅ $\overline{LO}$; $\overline{IC}$ ≅ $\overline{OK}$; $\overline{TC}$ ≅ $\overline{LK}$

Quick Check 1. It is given that ∠Q ≅ ∠R and ∠QPS ≅ ∠RSP. $\overline{PS}$ ≅ $\overline{SP}$ by the Reflexive Prop. of ≅. △QPS ≅ △RSP by ASA. $\overline{SQ}$ ≅ $\overline{PR}$ by CPCTC. **2.** 50 ft

Lesson 4-5 pp. 228–230

Check Skills You'll Need 1. ∠C **2.** ∠A **3.** $\overline{BC}$ **4.** $\overline{BA}$ **5.** 105

Quick Check 1. No, point U could be anywhere between R and T. **2.** 6 **3.** 150

Lesson 4-6 pp. 235–237, 240

Check Skills You'll Need 1. yes **2.** yes **3.** no **4.** yes **5.** yes **6.** no **7.** yes; SAS **8.** yes; SAS

Quick Check 1. △LMN ≅ △OQP **2.** In the two rt. △, the hypotenuses are given to be ≅. Legs $\overline{CB}$ and

$\overline{EB}$ are $\cong$ by def. of $\perp$ bis. $\triangle CBD \cong \triangle EBA$ by HL.
3. In the two rt. $\triangle$, hypotenuses $\overline{JZ} \cong \overline{JZ}$ are $\cong$ by the Reflex. Prop. of $\cong$. Two legs are given to be $\cong$. $\triangle JWZ \cong \triangle ZKJ$ by HL.

Checkpoint Quiz 2 1. $\overline{PR} \cong \overline{SQ}$; $\angle P \cong \angle S$; $\angle PRQ \cong \angle SQR$ **2a.** Isosc. $\triangle$ **b.** $\cong$ **c.** Converse of the Isosc. $\triangle$ Thm. **3.** $\triangle AED$; $\angle EAB \cong \angle EDC$ (Given), $\triangle EBC$; $\angle EBC \cong \angle ECB$ (Suppl. of $\cong \triangle$ are $\cong$.) **4.** HL **5.** $\triangle GTW \cong \triangle SWT$ by SAS since $\overline{WT} \cong \overline{WT}$, $\angle WTG \cong \angle TWS$, and $\overline{GT} \cong \overline{SW}$. So $\overline{GW} \cong \overline{ST}$ by CPCTC.

Lesson 4-7   pp. 241–243

Check Skills You'll Need 1. 15; 31 **2a.** yes; SAS **b.** yes; AAS **c.** yes; Trans. Prop. of $\cong$

Quick Check 1a. $\overline{CD}$ **b.** Answers may vary. Sample: $\triangle ABD$ and $\triangle CBD$; $\overline{BD}$ **2. 1.** $\triangle ACD \cong \triangle BDC$ (Given) **2.** $\angle ADC \cong \angle BCD$ (CPCTC) **3.** $\overline{CE} \cong \overline{DE}$ (If base $\triangle$ are $\cong$, the opp. sides are $\cong$.) **3. 1.** $\overline{PS} \cong \overline{RS}$; $\angle PSQ \cong \angle RSQ$ (Given) **2.** $\overline{QS} \cong \overline{QS}$ (Reflexive Prop. of $\cong$) **3.** $\triangle PSQ \cong \triangle RSQ$ (SAS) **4.** $\overline{PQ} \cong \overline{RQ}$ (CPCTC) **5.** $\angle PQT \cong \angle RQT$ (CPCTC) **6.** $\overline{QT} \cong \overline{QT}$ (Reflexive Prop. of $\cong$) **7.** $\triangle PQT \cong \triangle RQT$ (SAS) **4. 1.** $\angle CAD \cong \angle EAD$; $\angle C \cong \angle E$ (Given) **2.** $\overline{AD} \cong \overline{AD}$ (Reflexive Prop. of $\cong$) **3.** $\triangle ACD \cong \triangle AED$ (AAS) **4.** $\overline{CD} \cong \overline{ED}$ (CPCTC) **5.** $\angle BDC \cong \angle FDE$ (Vert. $\triangle$ are $\cong$.) **6.** $\triangle BDC \cong \triangle FDE$ (ASA) **7.** $\overline{BD} \cong \overline{FD}$ (CPCTC)

Chapter 5

Check Your Readiness   p. 256

1. $x \leq 4$ **2.** $x > \frac{15}{2}$ **3.** $x \leq 1$

4.   **5.**

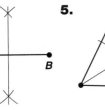

6. 5 **7.** 13 **8.** $4\sqrt{5}$ **9.** (5, 7) **10.** $\left(-3, -\frac{7}{2}\right)$ **11.** $\left(-\frac{9}{2}, \frac{5}{2}\right)$ **12.** -9 **13.** $-\frac{8}{3}$ **14.** 0

Lesson 5-1   pp. 259–261

Check Skills You'll Need 1. (1, 2) **2.** $\left(\frac{3}{2}, \frac{11}{2}\right)$ **3.** $\left(-\frac{1}{2}, 8\right)$ **4.** (1, 1) **5.** $-\frac{2}{5}$ **6.** $\frac{1}{3}$ **7.** $\frac{4}{7}$ **8.** $\frac{3}{2}$

Quick Check 1. $EB = 9$; $BC = 10$; $AC = 20$ **2.** 65; the lines are $\parallel$ so $\angle VUZ$ and $\angle YXZ$ are corr. and $\cong$. **3a.** 1320 ft **b.** $\frac{1}{4}$ mi

---

Lesson 5-2   pp. 265–267

Check Skills You'll Need

1.

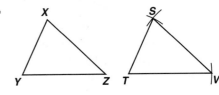

2.

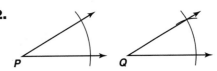

3.   **4.**

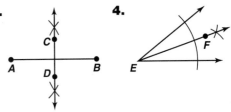

5. 6 **6.** 68

Quick Check 1. $CA = 5$; $DB = 6$; $\overleftrightarrow{CD}$ is the $\perp$ bis. of $\overline{AB}$; therefore $CA = CB$ and $DA = DB$. **2a.** 10; 10 **b.** $\overrightarrow{EK}$ is on the $\angle$ bis. of $\angle DEH$. **c.** 20 **d.** 80

Lesson 5-3   pp. 272–275, 279

Check Skills You'll Need

1–2.   **3.**

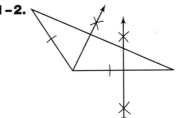 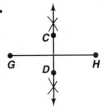

4.

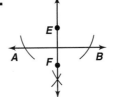

Quick Check 1a. $(-4, 3)$ **b.** Thm. 5-6: All of the $\perp$ bis. of the sides of a $\triangle$ are concurrent. **2a.** Draw segments connecting the towns. Build the library at the inters. pt. of the $\perp$ bisectors of the segments. **b.** The $\perp$ bisectors of the sides of a $\triangle$ are concurrent at a point equidistant from the vertices. **3.** 12 **4.** Median; $\overline{UW}$ is a segment drawn from vertex U to the midpt. of the opp. side.

Checkpoint Quiz 1 **1.** 6 **2.** 3 **3a.** 104 **b.** 228 **4.** right ∠; suppl. to ∠ADB **5.** △ABD ≅ △CBD; HL **6.** $\overline{AD}$ ≅ $\overline{DC}$; CPCTC **7.** $\overrightarrow{XY}$ bisects ∠ZXW; Y is equidist. from $\overrightarrow{XZ}$ and $\overrightarrow{XW}$. **8.** 21; △XYZ ≅ △XYW by HL, so XZ = 21 by CPCTC. **9.** Answers may vary. Sample: Bisect a side of a △. Connect the opp. vertex with the midpt. **10.** Use the procedure for constructing a ⊥ to a line from a point not on the line.

Lesson 5-4 pp. 280–283

Check Skills You'll Need **1.** If we go skiing, then it snows tomorrow. **2.** If 2 lines do not intersect, then they are parallel. **3.** If $x^2 = 1$, then $x = -1$. **4.** If a point is on the bisector of an angle, then it is equidistant from the sides of the angle. If a point is equidistant from the sides of an angle, then it is on the bisector of the angle. **5.** If a point is on the ⊥ bis. of a segment, then it is equidistant from the endpoints of the segment. If a point is equidistant from the endpoints of a segment, then it is on the ⊥ bis. of the segment. **6.** If you will pass a geometry course, then you are successful with your homework. If you are successful with your homework, then you will pass a geometry course.

Quick Check **1a.** The measure of ∠XYZ is not more than 70. **b.** Today is Tuesday. **2a.** If you stand for something, you won't fall for anything. **b.** If you won't fall for anything, then you stand for something. **3a.** Assume that the shoes cost more than $20. **b.** Assume that $m∠A ≤ m∠B$. **4.** I and II **5.** ∠X could be a right ∠.

Lesson 5-5 pp. 289–292

Check Skills You'll Need

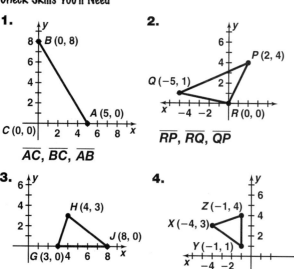

$\overline{AC}$, $\overline{BC}$, $\overline{AB}$

$\overline{RP}$, $\overline{RQ}$, $\overline{QP}$

$\overline{GH}$, $\overline{HJ}$, $\overline{JG}$

$\overline{ZY}$, $\overline{XZ}$, $\overline{XY}$

5. Assume that $m∠A ≤ m∠B$. **6.** AB < AC

Quick Check **1.** $m∠OTY > m∠2$ by the Comparison Prop. of Ineq. Since it was proven that $m∠2 > m∠3$, then by the Trans. Prop. $m∠OTY > m∠3$. **2.** ∠A, ∠C, ∠B **3.** $\overline{YZ}$, $\overline{XY}$, $\overline{XZ}$; $m∠Y = 80$. **4a.** No; $2 + 7 \not> 9$. **b.** Yes; $4 + 6 > 9$; $6 + 9 > 4$; $4 + 9 > 6$. **5.** $9 < x < 15$

Chapter 6

Check Your Readiness p. 304

1. 30 **2.** 42 **3.** 22 **4.** yes **5.** no **6.** yes **7.** parallel **8.** perpendicular **9.** neither **10.** ASA **11.** SAS **12.** AAS

Lesson 6-1 pp. 306–308

Check Skills You'll Need **1.** 10.8 **2.** 7.8 **3.** 12.7 **4.** $\frac{3}{4}$ **5.** $-\frac{8}{3}$ **6.** 1

Quick Check **1a.** quad., ▱, rhombus **b.** Rhombus; it is a ▱ and quad. with 4 sides that are ≅. **2.** square **3.** $a = 2, b = 4$; $LN = ST = NT = SL = 14$

Lesson 6-2 pp. 312–315

Check Skills You'll Need **1.** ASA **2a.** ∠HGE **b.** ∠GHE **c.** ∠HEG **d.** $\overline{GH}$ **e.** $\overline{HE}$ **f.** $\overline{EG}$ **3.** They are ∥.

Quick Check **1.** Yes; by the Converse of the Same-Side Int. ∠ Thm., both pairs of opp. sides are ∥. **2.** 11; $m∠E = 70, m∠G = 70, m∠F = 110, m∠H = 110$ **3.** $a = 16, b = 14$ **4.** 7.5

Lesson 6-3 pp. 321–324, 327

Check Skills You'll Need **1.** $(\frac{5}{2}, \frac{3}{2})$, $(\frac{5}{2}, \frac{3}{2})$; they bisect each other. **2.** Slope of $\overline{BC} = \frac{1}{3}$, slope of $\overline{AD} = \frac{1}{3}$. The slopes are =. **3.** Yes; they are vertical lines. **4.** parallelogram

Quick Check **1.** 70, 2 **2a.** Yes; $\overline{PQ}$ and $\overline{SR}$ are congruent and parallel, so PQRS is a parallelogram. **b.** No; it is possible that the diagonals do not bisect each other, so DEFG would not be a prallelogram. **3.** Once in place, both rulers show the direction and remain ∥. Keep the second ruler in place and move the first ruler to get the compass reading.

Checkpoint Quiz 1 **1.** $m∠1 = 59, m∠2 = 121, m∠3 = 59$ **2.** $m∠1 = 43, m∠2 = 62, m∠3 = 62$ **3.** $m∠1 = 106, m∠2 = 74, m∠3 = 26$ **4.** trapezoid, isos. trapezoid **5.** rectangle, ▱ **6.** rectangle, ▱ **7.** $x = 45, y = 60$ **8.** $x = 1, y = 2$ **9.** 20.6 **10.** kite

Lesson 6-4   pp. 329–332

Check Skills You'll Need 1. 4.5 **2.** 7 **3.** 109 **4.** 4.75
5. 3.5 **6.** 9.5 **7.** 71 **8.** 71

9.

The rhombus is not a square because it has no right $\angle$. The rectangle is not a square because all 4 sides aren't $\cong$.

Quick Check 1. $\angle 1 = 90$, $\angle 2 = 50$, $\angle 3 = 50$, $\angle 4 = 40$
2. $8\frac{1}{2}$ **3.** Sample: No; there is not enough information to conclude that the parallelogram is a rhombus. It cannot be a rectangle because it has no right $\angle$. **4.** Yes; if the ropes are $\perp$ to each other, then the endpoints of the ropes determine a square.

Lesson 6-5   pp. 336–338

Check Skills You'll Need 1. $a = 5.6$, $b = 6.8$; 4.5, 4.2, 4.5, 4.2 **2.** 3; 4.8, 16.4, 18, 18 **3.** $m = 5$, $n = 15$; 15, 15, 21, 21

Quick Check 1. 110, 110, 70 **2.** 85, 95 **3.** 90, 46, 44

Lesson 6-6   pp. 343–344, 347

Check Skills You'll Need

1.

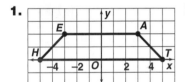

isosc. trapezoid

2.

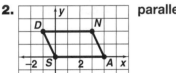

parallelogram

3.

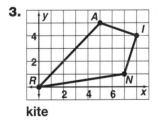

kite

4.

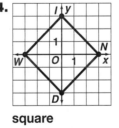

square

Quick Check 1. Both $= \frac{b}{a - e}$, so $\overline{WV} \parallel \overline{TU}$.
2. Q $(s + b, c)$.

Checkpoint Quiz 2 1. $x = 51$, $y = 51$ **2.** $x = 58$, $y = 32$ **3.** $x = 2$, $y = 4$ **4.** 3 **5.** $x = \frac{5}{3}$, $y = \frac{9}{2}$, $b = 90$
6–8. Counterexamples may vary.

6. false;

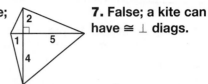

7. False; a kite can have $\cong$ $\perp$ diags.

8. false

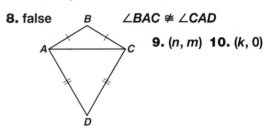

$\angle BAC \neq \angle CAD$

9. (n, m) **10.** $(k, 0)$

Lesson 6-7   pp. 348–349

Check Skills You'll Need 1. The quad. is a rectangle.
2. (a, c) **3.** $(-a, 0)$

Quick Check 1a. $M(b, c)$, $N(a + d, c)$; by starting with multiples of 2 you eliminate fractions when using the midpoint formula. **b.** 0, 0, 0; they are $=$. **c.** $MN = d + a - b$, $TP = 2a$, $RA = 2d - 2b$; so $RA + TP = 2d + 2a - 2b$ which is twice MN. So the midsegment is half the sum of the lengths of the bases. **d.** The base along the x-axis allows us to calculate length by subtracting x–values.
2. Using multiples of 2 in the coordinates for M, N, P, and O eliminates the use of fractions when finding midpoints since finding midpoints requires division by 2.

Chapter 7

Check Your Readiness   p. 364

1. $\frac{2}{3}$ **2.** 5 **3.** $\frac{4}{3a^3}$ **4.** $\frac{3x - 12}{x^2 - x}$ **5.** 108 **6.** 135 **7.** 144
8. $166\frac{2}{3}$ **9.** $\overline{DL}$ **10.** $\angle A$ **11.** $\angle DLH$ **12.** $\triangle APC$
13. SAS **14.** AAS **15.** SSS **16.** 6; 6 **17.** 4.7; 9.4

Lesson 7-1   pp. 366–368

Check Skills You'll Need 1–5. Answers may vary.
Samples are given. **1.** $\frac{1}{2}$ **2.** $\frac{2}{3}$ **3.** $\frac{3}{4}$ **4.** 1 **5.** $\frac{2}{3}$ **6.** 4
7. $\frac{1}{4}$ **8.** $\frac{4}{3}$ **9.** Each side of the smaller $\triangle$ is $\frac{1}{2}$ the length of a side of the larger $\triangle$.

Quick Check 1. 1 : 3 **2.** Answers may vary.
Sample: $\frac{4}{m} = \frac{11}{n}$, $\frac{m + 4}{4} = \frac{n + 11}{11}$ **3a.** 0.75 **b.** 3
4. 10 ft

Lesson 7-2
pp. 373–375, 379

Check Skills You'll Need **1.** $\overline{AB} \cong \overline{HI}$; $\overline{BC} \cong \overline{IJ}$; $\overline{AC} \cong \overline{HJ}$ **2.** 6 **3.** 6 **4.** 3 **5.** 5

Quick Check **1.** $m\angle F = 127$; BC **2.** Yes; corr. $\angle s$ are $\cong$ and corr. sides are prop. **3.** 3.8 **4.** 7.2 in. by 12 in. **5.** about 32.4 cm

Checkpoint Quiz 1 **1.** 1 : 8 **2.** $\frac{b}{10}$ **3.** no; $\frac{4}{6} \neq \frac{8}{10}$ **4.** 2 **5.** $\frac{34}{7}$ **6.** 12.5 ft **7.** $\angle BDF$ **8.** BF **9.** 3 ft by 2 ft **10.** $3\frac{1}{3}$

Lesson 7-3
pp. 382–385

Check Skills You'll Need **1.** SSS **2.** SAS **3.** ASA

Quick Check **1.** No; none of the side lengths are known. **2.** $\frac{AC}{EG} = \frac{CB}{GF} = \frac{AB}{EF} = \frac{3}{4}$, so the $\triangle$ are $\sim$ by SSS $\sim$ Thm.; $\triangle ABC \sim \triangle EFG$. **3.** 9 **4.** 13.5 ft

Lesson 7-4
pp. 391–393

Check Skills You'll Need **1.** 6 **2.** $\frac{14}{3}$ **3.** $\frac{24}{5}$ **4.** 39 **5.** 2 **6.** $\frac{8}{3}$ **7.** $\frac{40}{9}$ **8.** 18
9. Sample: $\triangle ADC$ and $\triangle BCD$

Quick Check **1.** $10\sqrt{3}$ **2.** $x = 8$; $y = 4\sqrt{3}$ **3.** 240 m

Lesson 7-5
pp. 398–400, 404

Check Skills You'll Need **1.** 28 cm **2.** $3\frac{3}{7}$ mm **3.** 9.8 in. **4.** 11.25 ft

Quick Check **1.** 1.5 **2.** $x = \frac{225}{13}$; $y = 28.6$ **3.** 5.76

Checkpoint Quiz 2 **1.** $\triangle ABC \sim \triangle XYZ$; AA $\sim$ Post. **2.** $\triangle WST \sim \triangle HJG$; SAS $\sim$ Thm. **3.** $x = \frac{3\sqrt{13}}{2}$; $w = 4.5$ **4.** 12 **5.** 15 **6.** 7.5 **7.** $4\sqrt{5}$ **8.** 3.6 **9.** 17.5 **10.** 77

Chapter 8

Check Your Readiness
p. 414

1. 4.648 **2.** 40.970 **3.** 6149.090 **4.** -5
5. AA$\sim$ Post. **6.** SSS$\sim$ Thm. **7.** SAS$\sim$ Thm.
8. 12 **9.** 8 **10.** $2\sqrt{13}$ **11.** 9

Lesson 8-1
pp. 417–419

Check Skills You'll Need **1.** $3^2 + 4^2 = 5^2$ **2.** $5^2 + 12^2 = 13^2$ **3.** $6^2 + 8^2 = 10^2$ **4.** $4^2 + 4^2 = (4\sqrt{2})^2$

Quick Check **1.** $5\sqrt{21}$; no **2.** $6\sqrt{3}$ **3.** You want to know the nearest whole number value, which may not be apparent in a radical expression. **4.** no **5.** acute

Lesson 8-2
pp. 425–427, 430

Check Skills You'll Need **1.** 45, 45, 90 **2.** 30, 60, 90 **3.** 45, 45, 90

Quick Check **1.** $5\sqrt{6}$ **2.** $5\sqrt{2}$ **3.** 141 ft **4.** $x = 4$, $y = 4\sqrt{3}$ **5.** about 0.9 m

Checkpoint Quiz 1 **1.** 12 **2.** $x = 10$; $y = 10\sqrt{2}$ **3.** $x = 12\sqrt{3}$, $y = 24$ **4.** 9 **5.** $x = 4\sqrt{3}$; $y = 6$ **6.** $x = 5\sqrt{3}$; $y = 10\sqrt{3}$ **7.** acute **8.** right **9.** obtuse **10.** 28.3 cm

Lesson 8-3
pp. 432–434

Check Skills You'll Need **1.** 0.71, 0.71, 1.00 **2.** 0.71; 0.71; 1.00 **3.** 0.87; 0.5; 1.73 **4.** $\frac{12}{7}$ **5.** $\frac{54}{11}$ **6.** $\frac{15}{2}$ **7.** $\frac{60}{7}$

Quick Check **1a.** $\frac{3}{7}$; $\frac{7}{3}$ **b.** They are reciprocals. **2a.** 13.8 **b.** 1.9 **c.** 3.8 **3.** 68

Lesson 8-4
pp. 439–440

Check Skills You'll Need **1.** 9; 12 **2.** 7; $2\sqrt{78}$ **3.** 10; $3\sqrt{29}$

Quick Check **1a.** $\sin X = \frac{64}{80}$; $\cos X = \frac{48}{80}$; $\sin Y = \frac{48}{80}$; $\cos Y = \frac{64}{80}$ **b.** $\sin X = \cos Y$ when $\angle X$ and $\angle Y$ are complementary. **2a.** about 0.72 AU **b.** 66,960,000 mi; 35,340,000 mi **3a.** 41 **b.** 68

Lesson 8-5
pp. 445–446, 450

Check Skills You'll Need **1.** $\angle 7$ **2.** $\angle 11$ **3.** $\angle 6$ **4.** 90 **5.** 180 **6.** $\angle 8$

Quick Check **1a.** $\angle$ of elevation from hiker to peak **b.** $\angle$ of depression from hiker to hut **2.** about 1179 ft **3.** about 6.2 km

Checkpoint Quiz 1 **1.** $\tan A = \frac{5}{4}$; $\sin A = \frac{25}{32}$; $\cos A = \frac{5}{8}$; $\tan B = \frac{4}{5}$; $\sin B = \frac{5}{8}$; $\cos B = \frac{25}{32}$ **2.** $\tan A = \frac{5}{12}$; $\sin A = \frac{5}{13}$; $\cos A = \frac{12}{13}$; $\tan B = \frac{12}{5}$; $\sin B = \frac{12}{13}$; $\cos B = \frac{5}{13}$ **3.** $\tan A = \frac{57}{40}$; $\sin A = \frac{57}{70}$; $\cos A = \frac{4}{7}$; $\tan B = \frac{40}{57}$; $\sin B = \frac{4}{7}$; $\cos B = \frac{57}{70}$ **4.** 15.0 **5.** 61 **6.** 20.8 **7.** about 13.1 ft **8.** about 393 m

9. Answers may vary. Sample: Identify the unknown you want to find in a right triangle. Then find two pieces of known information that will let you write a trigonometric-ratio equation you can solve for the unknown. **10.** about 22.7 m

Check Skills You'll Need 1. $4\sqrt{41}$ **2.** $\sqrt{13}$ **3.** $10\sqrt{65}$

Quick Check 1. $\langle -21.6, 46.2 \rangle$

2a.

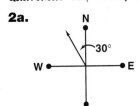

b. 60° north of west
3. about 257 mi at 17° north of east **4.** $\langle -2, 1 \rangle$
5. about 16° north of west

Chapter 9

1. 108 **2.** 135 **3.** 144 **4.** 160 **5.** △RTS **6.** △LJK
7. △ADC **8.** △LHC **9.** always **10.** never
11. sometimes **12.** always **13.** 50 ft **14.** 0.25 in.
15. $\langle 1, 5 \rangle$ **16.** $\langle -6, 2 \rangle$ **17.** $\langle -6, 3 \rangle$

Check Skills You'll Need 1. $\overline{EF}$ **2.** $\overline{AC}$ **3.** $\overline{BC}$ **4.** ∠G
5. ∠A **6.** ∠F **7.** $\overline{GL}$; ∠GRL

Quick Check 1a. Yes; the figures are ≅ by a flip.
b. Yes; the figures are ≅ by a flip and a slide.
2a. ∠U; P **b.** $\overline{NI}$ and $\overline{SU}$; $\overline{ID}$ and $\overline{UP}$; $\overline{ND}$ and $\overline{SP}$
3. $X'(2, 2)$, $Y'(-1, 4)$, $Z'(-5, 4)$
4. $(x, y) \rightarrow (x + 7, y - 1)$ **5.** 1 block east and 3 blocks north of her hotel

Check Skills You'll Need 1. $y = -2$ **2.** $x = -1$
3. $y = -x + 1$

Quick Check 1. $(0, -1)$

2. 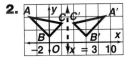 **3.** Yes, the intersection of $\overline{DW'}$ and ℓ is the same point P.

Check Skills You'll Need 1-7. Check students' work.

1. 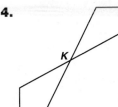   **2.** E **3.** 135°

4.

Checkpoint Quiz 1 1. No; the figures are not ≅.
2. Yes; the figures are ≅ and the transf. is a translation. **3.** Yes; the figures are ≅ and the transf. is a translation, rotation or reflection.
4. a translation of 3 units left and 5 units up.
5. $(x, y) \rightarrow (x - 5, y + 10)$ **6.** a translation 4 units right

7. 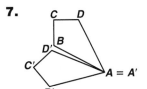   **8.**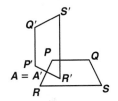

9. 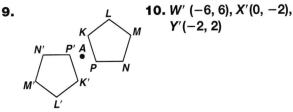   **10.** $W'(-6, 6)$, $X'(0, -2)$, $Y'(-2, 2)$

Check Skills You'll Need 1. G **2.** D **3.** $\overline{AH}$ **4.** $\overline{CB}$

Quick Check

1.   **2a.** yes; 180° **b.** yes
3. rotational and reflectional symmetry

Check Skills You'll Need 1. 3 in. by 4 in. **2.** 2 in. by $2\frac{1}{2}$ in.
3. $1\frac{1}{2}$ in. by $2\frac{1}{4}$ in. **4.** $1\frac{1}{4}$ in. by $1\frac{3}{4}$ in.

Quick Check 1. The dilation is a reduction with center $(0, 0)$ and scale factor $\frac{1}{2}$. **2.** 8 cm
3. $P'(1, 0)$, $Z'(-\frac{1}{2}, \frac{1}{4})$, $G'(\frac{1}{2}, -1)$

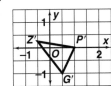

Check Skills You'll Need

1. **2.**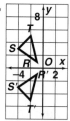

3. 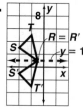 **4.**

5. **6.**

7.
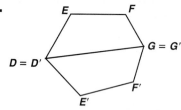

Quick Check 1. Neither; the figures do not have the same orientation.

2.

R is translated the distance and direction shown by the arrow. The length of the arrow is twice the distance between ℓ and *m*.

3. Answers may vary. Sample:

4a.
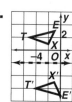

b. Yes; if you reflected it then moved it right, the result would be the same.

5. rotation

Check Skills You'll Need 1. pentagon **2.** octagon **3.** dodecagon

Quick Check 1a. rotation; one fish **b.** translation, glide reflection; horse and rider **2.** The interior ⧔ of an equilateral △ measure 60. 60 divides 360, so it will tessellate. **3.** line symmetry, rotational symmetry, glide reflectional symmetry, translational symmetry

Checkpoint Quiz 2 1. Rotation; the image appears rotated ≈ 90°. **2.** reflection; reverse orientation **3.** enlargement; center (0, 0); scale factor 2 **4.** reduction; center (0, 0); scale factor 5 **5.** line, point **6.** point **7.** line, rotational: 120

8. point **9.** rotational, point, reflectional, glide reflectional, and translational **10.** rotational, point, reflectional, glide reflectional, and translational

Chapter 10

Check Your Readiness

p. 530

1. 9 **2.** 64 **3.** 144 **4.** 225 **5.** 4 **6.** 8 **7.** 10 **8.** 13 **9.** ±6 **10.** ±10.2 **11.** ±6.9 **12.** ±8.1 **13.** $2\sqrt{2}$ **14.** $3\sqrt{3}$ **15.** $4\sqrt{3}$ **16.** $36\sqrt{2}$ **17.** $\frac{3}{5}$ **18.** $\frac{2}{5}$ **19.** 0 **20.** 1 **21.** rhombus **22.** parallelogram **23.** rhombus

Lesson 10-1

pp. 534–536

Check Skills You'll Need 1. 25 cm² **2.** 28 in.² **3.** 11.5 m² **4.** $\frac{3}{2}$ ft² **5.** 6 units² **6.** 2 units² **7.** 8 units²

Quick Check 1. 108 m² **2.** 7.5 cm **3.** 30 cm² **4.** The force is doubled.

Lesson 10-2

pp. 540–542

Check Skills You'll Need 1. $A = bh$ or $A = \ell w$ **2.** $A = \frac{1}{2}bh$ **3.** 9 units² **4.** 7 units² **5.** 13.5 units²

Quick Check 1. 94.5 cm² **2.** 12 m² **3.** 54 in.² **4.** $9^2 + 12^2 = 15^2$

Lesson 10-3 pp. 546–547, 551

Check Skills You'll Need **1.** $25\sqrt{3}$ cm^2 **2.** 50 ft^2
3. $\frac{100\sqrt{3}}{3}$ m^2 **4.** 24 in. **5.** $16\sqrt{3}$ cm

Quick Check **1.** $m\angle 1 = 45$; $m\angle 2 = 22.5$;
$m\angle 3 = 67.5$ **2.** 232 cm^2 **3.** $384\sqrt{3}$ ft^2

Checkpoint Quiz 1 **1.** 84 in.^{2} **2.** 112 cm^2 **3.** 48 m^2
4. 135 in.^{2} **5.** 58.5 m^2 **6.** $72\sqrt{3}$ ft^2 **7.** $27\sqrt{3}$ ft^2
8. 32 yd^2 **9.** $16\sqrt{3}$ in.^{2} **10.** $\frac{81\sqrt{3}}{2}$ m^2; 70.1 m^2

Lesson 10-4 pp. 553–555

Check Skills You'll Need **1.** 28 in.; 49 in.^{2} **2.** 24 m;
32 m^2 **3.** 24 cm; 24 cm^2 **4.** 8 cm; 3 cm^2 **5.** 16 cm;
12 cm^2 **6.** 24 cm; 27 cm^2

Quick Check **1a.** 5 : 7 **b.** 25 : 49 **2.** 54 in.^{2}
3. $6.94 **4.** $5\sqrt{5}$: 3

Lesson 10-5 pp. 559–561

Check Skills You'll Need **1.** 36 m^2 **2.** 4536 in.^{2} **3.** 168 ft^2

Quick Check **1.** 482.8 in.^{2} **2.** It is 4 times as large.
3. 5081 ft^2

Lesson 10-6 pp. 566–569

Check Skills You'll Need **1.** 14 cm **2.** 3.2 m **3.** 5 ft
4. 2.5 in. **5.** 32 **6.** 137 **7.** 180 **8.** 76

Quick Check **1a.** number of hours spent doing an
activity **b.** Each section represents the average
of the 3600 participants' answers.
2. $\overarc{CEA}$, $\overarc{DAE}$, $\overarc{ACD}$, $\overarc{EDC}$ **3.** 58; 180; 122; 270
4. 17 revolutions **5.** 1.3π m

Lesson 10-7 pp. 575–577, 580

Check Skills You'll Need **1.** 4.5 cm **2.** 16 ft **3.** 12π or
about 37.7 in. **4.** 6π or about 18.8 m

Quick Check **1.** about 41 in.^{2} **2.** 181.5 cm^2
3. 13.0 cm^2

Checkpoint Quiz 2 **1.** 60 in. **2.** 45 in.^{2} **3.** 6 : 5
4. about 4.8 ft^2 **5.** 44.8 cm^2 **6.** 100π in.^{2}
7. 27π m^2 **8.** $(16\pi - 32)$ cm^2 **9.** 22.5π ft^2
10. $\frac{9\pi}{2}$ mm

Lesson 10-8 pp. 582–584

Check Skills You'll Need **1.** $\frac{1}{3}$ **2.** $\frac{1}{2}$ **3.** 1 **4.** $\frac{1}{4}$ **5.** $\frac{1}{6}$ **6.** $\frac{1}{2}$
7. $\frac{1}{3}$ **8.** $\frac{1}{2}$

Quick Check **1.** $\frac{2}{5}$ **2.** $\frac{2}{5}$ **3a.** It becomes about 8.7%, or
about 4 times greater. **b.** It becomes 19.6%, or
about 9 times greater. **4.** Yes; theoretically you
should win 1.4 times out of 100.

Chapter 11

Check Your Readiness p. 596

1. 17 **2.** $8\sqrt{2}$ **3.** 6 **4.** $4\sqrt{5}$ **5.** $6\sqrt{2}$ **6.** $4\sqrt{3}$
7. 44 units^2 **8.** $14\sqrt{3}$ units^2 **9.** 234 units^2
10. $54\sqrt{3}$ units^2 **11.** 24 **12.** $2\sqrt{2}$: 5

Lesson 11-1 pp. 598–600

Check Skills You'll Need

1. 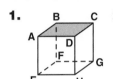 **2.** 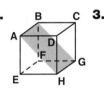 **3.**

4. 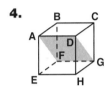 **5.** **6.**

7.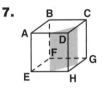

Quick Check
1. R, S, T, U, V; $\overline{RS}$, $\overline{RU}$, $\overline{RT}$, $\overline{VS}$, $\overline{VU}$, $\overline{VT}$, $\overline{SU}$, $\overline{UT}$,
$\overline{TS}$; $\triangle RSU$, $\triangle RUT$, $\triangle RTS$, $\triangle VSU$, $\triangle VUT$ $\triangle VTS$
2. 12 edges
3a. $6 + 8 = 12 + 2$
b.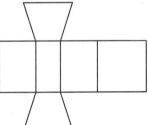
c. Sample:
$6 + 14 = 19 + 1$

4a. 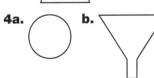 **b.**

5. square

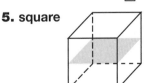

Lesson 11-2 pp. 608–611

Check Skills You'll Need **1.** 96 cm^2 **2.** 40π cm^2
3. $36\sqrt{3}$ m^2

Quick Check

1. 216 cm²

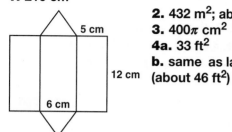

2. 432 m²; about 619 m²
3. 400π cm²
4a. 33 ft²
b. same as large drum (about 46 ft²)

Lesson 11-3   pp. 617–620, 623

Check Skills You'll Need 1. $\sqrt{233}$ in. **2.** $\sqrt{130}$ m
3. $\sqrt{313}$ cm

Quick Check 1. 55 m² **2.** 1,496,511 ft²
3. 704π m² **4.** 1178 in.²

Checkpoint Quiz 1

1.

2.

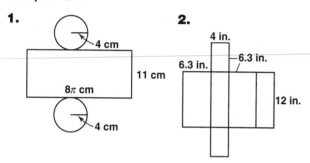

3.

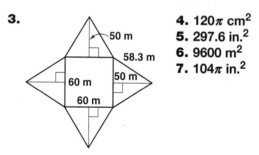

4. 120π cm²
5. 297.6 in.²
6. 9600 m²
7. 104π in.²

8.

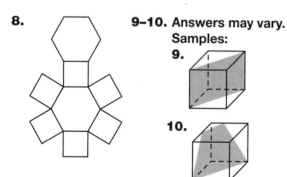

9–10. Answers may vary. Samples:
9.

10.

Lesson 11-4   pp. 624–627

Check Skills You'll Need 1. 49 cm² **2.** 176.7 in.²
3. 314.2 mm² **4.** 3 ft² **5.** 154 in.² **6.** 27.5 cm²
7. 27.7 in.²

Quick Check 1. Answers may vary. Sample:
Multiplication is commutative. **2.** 150 m³
3a. 256π m³ **b.** 804.2 m³ **4.** 12 in.³

Lesson 11-5   pp. 631–634

Check Skills You'll Need 1. 12 cm **2.** 8 in. **3.** 2.5 m
Quick Check 1. 384 in.³ **2.** 960 m³
3a. 144π m³; 452 m³ **b.** 6174π mm³;
19,396 mm³ **4.** 77 ft³

Lesson 11-6   pp. 638–640, 644

Check Skills You'll Need 1. 113.1 in.²; 37.7 in.
2. 78.5 cm²; 31.4 cm **3.** 19.6 ft²; 15.7 ft
4. 4.5 m²; 7.5 m **5.** 706.9 yd²; 94.2 yd
6. 452.4 mm²; 75.4 mm

Quick Check 1. 196π in.²; 616 in.² **2.** 100 in.²
3. 113,097 in.³ **4.** 1258.9 ft²

Checkpoint Quiz 2 1. 60.2 ft²; 22.5 ft³ **2.** 332.9 in.²;
377.0 in.³ **3.** 113.1 m²; 113.1 m³ **4.** 439.8 cm²;
706.9 cm³ **5.** 207 yd²; 144.8 yd³ **6.** 44.8 m²; 16 m³
7. 181.7 m²; 2217.0 m³ **8.** 32 ft²; 12 ft³ **9.** 75.4 cm²;
37.7 cm³ **10.** The balls; the volume of the space
is 2πr³ and the volume of the balls is 4πr³.

Lesson 11-7   pp. 646–648

Check Skills You'll Need 1. Yes; all corr. ∠s are ≅ and
corr. sides are prop.; 3 : 1. **2.** Yes; all corr. ∠s are ≅
and corr. sides are prop.; 3 : $\sqrt{2}$. **3.** 27 in.³
4. 135 m³ **5.** 402.1 cm³

Quick Check 1. yes; 6 : 5 **2.** 2 : 3 **3.** 160 m²
4. 0.01875 lb

Chapter 12

Check Your Readiness   p. 660

1. 82 **2.** $6\frac{2}{3}$ **3.** 15 **4.** 25 **5.** 10 **6.** 5 **7.** 6 **8.** 18
9. 24 **10.** $\sqrt{2}$ **11.** 12 **12.** $4\sqrt{2}$ **13.** 13 **14.** $\sqrt{10}$
15. 6

Lesson 12-1 — pp. 662–665

Check Skills You'll Need **1.** $p^2 + 6p + 9$
2. $w^2 + 20w + 100$ **3.** $m^2 - 4m + 4$ **4.** $8\sqrt{5}$
5. $2\sqrt{30}$ **6.** 12

Quick Check **1.** 52 **2.** about 35.5 in.
3. No; $4^2 + 7^2 \neq 8^2$ **4.** If $\overleftrightarrow{BC}$ and $\overleftrightarrow{GF}$ never
intersect, then *BCFG* is a rectangle. **5.** 12 cm

Lesson 12-2 — pp. 670–673

Check Skills You'll Need **1.** $\frac{11\sqrt{2}}{2}$ **2.** 5 **3.** 28

Quick Check **1.** $\angle O \cong \angle P$; $\overline{BC} \cong \overline{DF}$ **2.** 16
3a. about 11 **b.** 2.8

Lesson 12-3 — pp. 678–681, 685

Check Skills You'll Need **1–4.** Answers may vary.
Samples are given. **1.** $\overparen{STQ}$ **2.** $\overparen{SR}$ **3.** $\overparen{RTQ}$
4. $\angle TPQ$ **5.** 86 **6.** 180 **7.** 121 **8.** 94

Quick Check **1.** 105 **2.** $m\angle 1 = 90, m\angle 2 = 110,$
$m\angle 3 = 90, m\angle 4 = 70$ **3.** $m\angle QJK = m\angle LJK +$
$m\angle QJL = 35 + 90 = 125$; $m\angle QJK + \frac{1}{2}m\overparen{QLJ} =$
$\frac{1}{2}(70 + 180) = 125$

Checkpoint Quiz 1 **1.** 76 cm **2.** 48 in. **3.** 51 m **4.** 24
5. 5 **6.** 8 **7.** $w = 104; x = 22; y = 108$ **8.** $a = 30;$
$b = 42; c = 80; d = 116$ **9.** $w = 105; x = 75; y =$
210 **10.** $a = 140; b = 70; c = 47.5$

Lesson 12-4 — pp. 687–690

Check Skills You'll Need **1.** 57 **2.** 180 **3.** 303 **4.** 28.5
5. 28.5 **6.** 4 **7.** 2 **8.** about 4.5 **9.** 123

Quick Check **1a.** 250 **b.** 40 **2.** away; 20°
3a. 13.8 **b.** 3.2 **4a.** $\left(8 - 4\sqrt{3}\right)$ in. **b.** $\frac{16}{8 + 4\sqrt{3}}$ in.

Lesson 12-5 — pp. 695–697, 700

Check Skills You'll Need **1.** 5.8 **2.** 12.8 **3.** 5.8

Quick Check **1a.** $(x - 3)^2 + (y - 5)^2 = 36$
b. $(x + 2)^2 + (y + 1)^2 = 2$ **2.** $(x - 2)^2 +$
$(y - 3)^2 = 13$ **3.** center: (2, 3); radius: 10

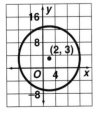

4. $(x - 0)^2 + (y - 0)^2 = x^2 + y^2 = 144$

Checkpoint Quiz 2 **1.** 58 **2.** 226 **3.** 30 **4.** about 3.0
5. about 15.7 **6.** 40 **7.** A chord is a segment
whose endpoints are on the circle. A secant is a
line, ray, or segment that intersects a circle at
two points.

$\overline{AB}$ is a chord
$\overleftrightarrow{CD}$ is a secant

8. $(x - 1.5)^2 + (y - 0.5)^2 = 2.5$
9. $(x - 3.5)^2 + (y - 1)^2 = 46.25$
10. $(x + 1.5)^2 + (y + 4)^2 = 22.25$

Lesson 12-6 — pp. 701–702

Check Skills You'll Need

1.
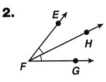

2.
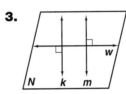

3.
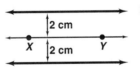

Quick Check **1.** Two lines ∥ to $\overleftrightarrow{XY}$, each 2 cm
from $\overleftrightarrow{XY}$.

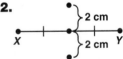 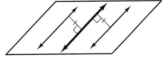

3a. Consider two ∥ lines in a plane. The locus is
the line ∥ to and equidist. from the ∥ lines.

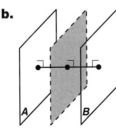

b. Consider a segment ⊥ to
both planes, with one
endpoint on plane *A* and the
other on plane *B*. The locus
is the plane ⊥ to this
segment at the segment's
midpoint.

Selected Answers

Chapter 1

Lesson 1-1   pp. 6–9

EXERCISES 1. 80, 160 **3.** −3, 4 **17.**

19. The sum of the first 6 pos. even numbers is 6·7, or 42. **21.** The sum of the first 100 pos. even numbers is 100·101, or 10,100. **23.** 555,555,555 **25–27.** Answers may vary. Samples are given. **25.** 8 + (−5) = 3 and 3 ≯ 8 **27.** −6 − (−4) ≮ −6 and −6 −(−4) ≮ −4 **29.** 75°F **31.** 31, 43

33. 0.0001, 0.00001 **43.**   **45.**

Lesson 1-2   pp. 13–15

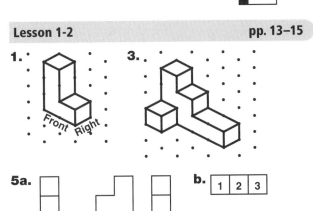

5a.   **b.**

Front   Right   Top

7. 6 **9.** 8 **11.** C **13.** B **15.** Answers may vary.
Sample:

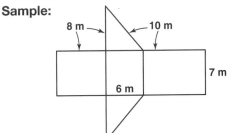

19.a-b.

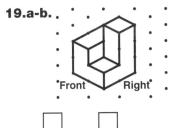

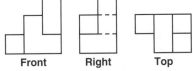

Front   Right   Top

21. orthographic top view **23.** blue **25.** orange
31. Answers may vary.
Sample:

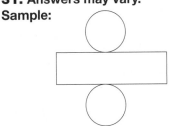

Lesson 1-3   pp. 19–22

EXERCISES 1. no **3.** yes; line n **11.** *ABCD*
13. *ABHF* **17.** $\overleftrightarrow{RS}$ **19.** $\overleftrightarrow{UV}$ **21.** planes *QUX* and
QUV **23.** planes *UXT* and *WXT*

25.   **27.**

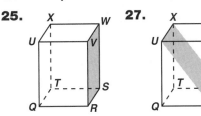

31. *X* **33.** *Q* **35.** no **37.** no **39.** coplanar
41. coplanar **47.** not possible **49.** not possible
51.   no **55.** always **57.** always

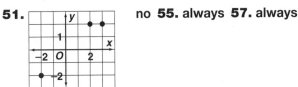

63. Answers may vary.
Sample:

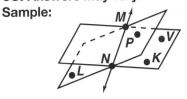

69.   yes **71.**   no

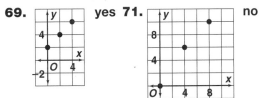

Lesson 1-4   pp. 25–28

EXERCISES 1. $\overline{RS}, \overline{RT}, \overline{RW}, \overline{ST}, \overline{SW}, \overline{TW}$
3.a. $\overrightarrow{TS}$ or $\overrightarrow{TR}, \overrightarrow{TW}$ **b.** $\overrightarrow{SR}, \overrightarrow{ST}$ **5.** $\overline{BC}$ **7.** $\overline{DE}, \overline{EF}, \overline{BE}$
9. $\overline{BC}, \overline{EF}$ **11.** *DEF*, $\overleftrightarrow{BC}$ **13.** Answers may vary.
Sample: $\overleftrightarrow{CD}, \overleftrightarrow{AB}$ **15.** $\overleftrightarrow{AF}$ **17.** False; they are skew.

Selected Answers   **855**

19. False; they intersect above $\overline{CG}$. **25.** always
27. always **35.** Answers may vary. Sample: (0, 0)
39.a. The lines of intersection are parallel **b.**
Examples may vary. Sample: The floor and ceiling
are parallel. A wall intersects both. The lines of
intersection are parallel.

Algebra 1 Review	p. 30

1. 6 **3.** $-\frac{7}{3}$ or $-2\frac{1}{3}$ **5.** 4 **7.** −6

Lesson 1-5	pp. 33–35

EXERCISES 1. 9; 9; yes **3.** 11; 13; no **5.** $XY =$
$ZW = 4$; yes **7.** $YZ = 4$, $XW = 12$; no **9.** 25
11.a. 7 **b.** $RS = 60$, $ST = 36$, $RT = 96$ **13.** 33
15. 130 **17.** 6 **19.** 1 **21.** −3.5, 3.5 **23.a.** 114 miles
b. Conway **29.** true; $AB = 2$, $CD = 2$ **31.** false;
$AC = 9$, $BD = 9$, $AD = 11$, and $9 + 9 \neq 11$
35. $ED = 10$, $DB = 10$, $EB = 20$

Lesson 1-6	pp. 40–42

1. $\angle XYZ$, $\angle ZYX$, $\angle Y$ **3.** $\angle ABC$, $\angle CBA$

5. 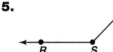 **7.**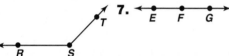

9. 60; acute **11.** 90; right **15.** $\angle AOB$ or $\angle DOC$
17. $\angle EOC$ **21.** 30 **23.** 30 **25.** No; there are no
markings. **27.** No; there are no markings.
33. 115 **35.** 180 **37.** 30 **41.** 45, 75 and 165, or 135,
105, and 15 **43.** 8; $m\angle AOB = 30$, $m\angle BOC = 50$,
$m\angle COD = 30$ **45.** 7; $m\angle AOB = 31$, $m\angle BOC = 49$,
$m\angle AOD = 111$ **47.a.** 19.5; **b.** 43; 137 **c.** Answers
may vary. Sample: The sum of the $\angle$ measures
should be 180.

Lesson 1-7	pp. 47–50

EXERCISES 1.
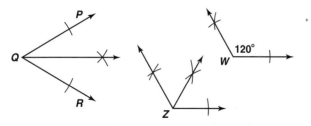

3.

9.a. 11; 30 **b.** 30 **c.** 60 **11.** 15; 48

13. **15.**

17. Find a segment on $\overleftrightarrow{SQ}$ so that you can
construct $\overleftrightarrow{SP}$ as its $\perp$ bisector. Then bisect $\angle PSQ$.

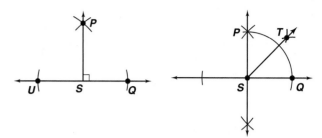

19.a–b.

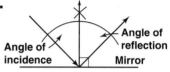

23. **29.**

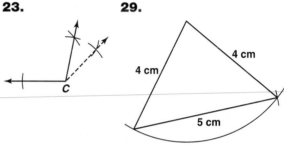

31. impossible; the short segments are not long
enough to form a $\triangle$.

Lesson 1-8	pp. 56–59

EXERCISES 1. 6 **3.** 8 **11.** about 4.5 mi **13.** 6.4
15. 15.8 **19.** (3, 1) **21.** (6, 1) **25.** (5, −1)
27. (12, −24) **31.** (4, −11) **33.** 5.8; (1.5, 0.5) **41.** IV
45. 10.8 units; (3, −4) **49.** 934 mi **51.** 2693 mi
53–55. Answers may vary. Samples are given.
53. (3, 6), (0, 4.5) **55.** (1, 0), (−1, 4)

Lesson 1-9	pp 65–68

EXERCISES 1. 22 in. **3.** 56 in. **5.** 120 m **9.** 10π ft
11. $\frac{1}{2}\pi$ m **13.** 22.9 m **15.** 351.9 cm
17. 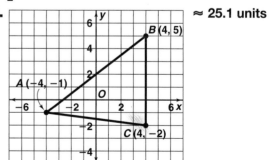 $\approx$ 25.1 units

21. 4320 in.^{2}, or $3\frac{1}{3}$ yd^2 **23.** 8000 cm^2, or 0.8 m^2
27. 400π m^2 **29.** $\frac{9}{64}\pi$ in.^{2} **33.** 153.9 ft^2

35. 452.4 cm² **37.** 310 m² **39.a.** 144 in.² **b.** 1 ft²
c. There are 144 square inches in one square
foot. A square whose sides are 12 in. long and a
square whose sides are 1 ft long are the same
size. **43–45.** Answers may vary. Check students'
work. Samples are given. **43.** 39 in.; 93.5 in.²
45. 8 ft; 3.75 ft²

53.

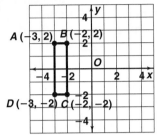

perimeter = 10 units, area = 4 units²
55. 38 units **57.** 1,620,000 m² **59.** Area; the wall
is a surface. **61.** Perimeter; the fence must fit the
perimeter of the garden. **63.** 6.25 π units²

Chapter Review   pp. 71–73

1. coplanar **2.** segment **3.** congruent **4.** midpoint
5. angle bisector **6.** conjecture **7.** postulate or
axiom **8.** Parallel lines **9.** obtuse angle
10. subtract 5; 20, 15 **11.** Answers may vary.
Sample: mult. by −1; 5, −5 **12.** subtr. 7; −1, −8
13. mult. by 4; 1536, 6144 **14.** mult. by 2; 64, 128
15. alternate adding 1 and 3; 10, 13

16.   **17.** 4, 6, 11 **18.**

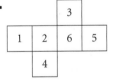

19–22. Answers may vary. Samples are given.
19. $\overleftrightarrow{AQ}$ and $\overleftrightarrow{QR}$ **20.** A, Q, R, **21.** AQTD and BRSC
22. $\overleftrightarrow{AD}$, $\overleftrightarrow{TD}$, $\overleftrightarrow{CD}$ **23.** always **24.** sometimes
25. never **26.** always **27.** −7, 3 **28.** 0.5
29. 15 **30.** 31 **31.** Answers may vary. Samples
are given. **a.** ∠ADB, ∠BDC **b.** ∠ADE, ∠EDF
c. ∠ADC, ∠EDF

32.   **33.a–b.**

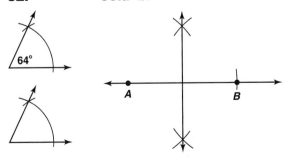

34. 1.4 units **35.** 7.6 units **36.** 14.4 units **37.** (0, 0)
38. 7.2 units **39.** P = 32 cm, A = 64 cm² **40.** P =
38 ft, A = 78 ft² **41.** P = 32 in., A = 40 in.²
42. C = 18.85 in., A = 28.27 in.² **43.** C = 47.12 m,
A = 176.71 m² **44.** C = 163.36 m, A = 2123.72 m²

Chapter 2

Lesson 2-1   pp. 83–86

EXERCISES 1. Hypothesis: You send in the
proof-of-purchase. Conclusion: They send you a
get-well card. **3.** Hypothesis: x + 20 = 32,
Conclusion: x = 12 **9.** If an object is glass, then it
is fragile. **11.** If a whole number has 2 as a factor,
then it is even. **15.** Sunday **17.** Mexico

19.    **23.** If you grow, then you
eat your vegetables.

25. If two segments have the same length, then
they are congruent. **27.** Converse: If you have a
passport, then you travel from the United States to
Kenya. The original conditional is true and the
converse is false. **29.** Converse: If the chemical
formula for a substance is H_2O, then it is water.
Both statements are true. **33.** If a person is an
Olympian, then that person is an athlete.
35. If something is a whole number, then it is an
integer. **37.** A **39.** If a work is great, then it is
made out of a combination of obedience and
liberty. **41.** If x = 18, then x − 3 = 15; true.
43. If |x| = 6, then x = −6; 6. **49.** If a figure has
four congruent angles, then it is a square; false; a
rectangle that is not a square. **51.** If a figure has
four congruent angles and four congruent sides,
then it is a square; true. **55–57.** Answers may
vary. Samples are given. **55.** If two planes
intersect, then they meet in exactly one line.
57. If two points are given, then there is exactly
one line through them.

Lesson 2-2   pp. 90–93

EXERCISES 1. If two segments are congruent,
then they have the same length. It is true. Two
segments have the same length if and only if they
are congruent. **3.** If a number is even, then it is
divisible by 20. It is false since 4 is even but not
divisible by 20. **7.** If a line bisects a segment,
then the line intersects the segment only at its
midpoint. If a line intersects a segment only at its

midpoint, then it bisects the segment. **9.** If you live in Washington, D.C., then you live in the capital of the United States. If you live in the capital of the United States, then you live in Washington, D.C. **13.** A line, segment, or ray is a perpendicular bisector of a segment if and only if it is perpendicular to the segment at its midpoint. **15.** not reversible **19–21.** Answers may vary. Samples are given. **19.** No; it is not reversible; a cat is a counterexample. **21.** No; it is not reversible; skew lines are not parallel. **29.** Yes; ∠1 and ∠2 share a side and a vertex, and are suppl. **31.** No; ∠1 and ∠2 do not share a side, and are not suppl. **33.** good definition **35.** L is a counter example. **39.** The sum of the digits of an integer is divisible by 9 if and only if the integer is divisible by 9. **41.** If ∠A and ∠B are right angles, then ∠A and ∠B are supplementary angles. **43.** ∠A and ∠B are right angles if and only if ∠A and ∠B are supplementary angles.

Lesson 2-3 pp. 96–99

EXERCISES 1. Felicia will pass the music theory course. **3.** Line ℓ and line m do not intersect. **5.** Figure ABCD has two pairs of parallel sides. **7.** Points X, Y, and Z are collinear. **11.** If two planes are not parallel, then they intersect in a line. **13.** If you are studying botany, then you are studying a science. **15.** Answers may vary. Sample: If you live in Texas, then you live in the 28th state to enter the Union. Levon lives in the 28th state to enter the Union. **17.** Must be true; by (E) and (A), it is breakfast time. Then by (C), Curtis drinks water. **19.** Is not true; by (E) and (A), it is breakfast time. By (C), Curtis drinks water and nothing else. **23.** If you are in Key West, Florida, then the temperature is always above 32°F; not possible. **25.** If a figure is a square, then it is a rectangle; ABCD is a rectangle. **27.** No; red cars can never park. **29.** yes

Lesson 2-4 pp. 105–108

EXERCISES 1.a. ∠ Add. Post. **b.** Subst. Prop. **c.** Simplify. **d.** Subtr. Prop. of = **e.** Div. Prop. of = **3.a.** Mult. Prop. of = **b.** Distr. Prop. **c.** Add. Prop. of = **5.** Reflexive Prop. of ≅ **7.** Div. Prop. of = **17.** 5x **19.** ∠K **27.a.** Given **b.** Def. of midpoint **c.** Subst. Prop. of = **d.** Subtr. Prop. of = **e.** Div. Prop. of = **29.a.** m∠GFE + m∠EFI = m∠GFI (∠ Add. Post.) 9x − 2 + 4x = 128 (Subst. Prop.) 13x − 2 = 128 (Simplify.) 13x = 130 (Add. Prop. of =) x = 10 (Div. Prop. of =) **b.** 40

Lesson 2-5 pp. 112–115

1. 20 **3.** 30 **5.** 75, 105 **7.a.** 90 **b.** 90 **c.** Subst. **d.** m∠3 **9.** Answers may vary. Sample: scissors. **11.** The two acute ⦞ have measure 72. The two obtuse ⦞ have measure 108. **13.** x = 14, y = 15; 50, 50, 130 **15.** C **17.** ∠EIG ≅ ∠FIH since all rt. ⦞ are ≅; ∠EIF ≅ ∠HIG since they are compl. of the same ∠. **19.** Answers may vary. Sample: (−5, −1) **21.a.** V **b.** 180 **c.** Division **d.** right **23.** m∠A = 60; m∠B = 30 **25.** m∠A = 120; m∠B = 60

Chapter Review pp. 117–119

1. Reflexive **2.** hypothesis **3.** Transitive **4.** biconditional **5.** converse **6.** Symmetric **7.** conclusion **8.** truth value **9.** deductive reasoning **10.** theorem **11.a.** If you are younger than 20, then you are a teenager. **b.** conditional: true, converse: false **12.a.** If an angle has measure greater than 90 and less than 180, then it is obtuse. **b.** conditional: true, converse: true **c.** An angle is obtuse if and only if it has measure greater than 90 and less than 180. **13.a.** If a figure has four sides, then it is a square. **b.** conditional: true, converse: false **14.** If something is a flower, then it is beautiful. **15.** Rico's definition is not reversible. A magazine is a counterexample. You read a magazine, but it is not a book. **16.** A phrase is an oxymoron if and only if it contains contradictory terms. **17.** If two angles are complementary, then the sum of their measures is 90. If the sum of the measures of two angles is 90, then the angles are complementary. **18.** Lucy will become a better player. **19.** Lines ℓ and m intersect to form right angles. **20.** The sum of the measures of ∠1 and ∠2 is 180. **21.** If Kate studies, then she will graduate. **22.** If a, then c. **23.** If the weather is wet, then Nathan can stop at the ice cream shop. **24.a.** Segment Add. Post. **b.** Subst. Prop. **c.** Simplify. **d.** Subtr. Prop. of Equality **e.** Div. Prop. of Equality **25.** 8 **26.** BY **27.** m∠Y **28.** RS = XY **29.** y **30.** 10 **31.** p − 2q **32.** $\overline{NM}$ **33.** 18 **34.a.** 74 **b.** 74 **c.** 106 **35.** ∠2, ∠3, transitive

Chapter 3

Lesson 3-1 pp. 131–133

EXERCISES 1. $\overleftrightarrow{PQ}$ and $\overleftrightarrow{SR}$ with transversal $\overleftrightarrow{SQ}$; alt. int. ⦞ **3.** $\overleftrightarrow{PS}$ and $\overleftrightarrow{QR}$ with transversal $\overleftrightarrow{PQ}$; same-side int. ⦞ **5.** ∠1 and ∠2: corr. ⦞, ∠3 and ∠4: alt. int. ⦞, ∠5 and ∠6: corr. ⦞ **7.** ∠1 and ∠2: corr. ⦞, ∠3 and ∠4: same-side int. ⦞, ∠5 and ∠6: alt. int. ⦞ **9.** 2. Same-Side Int. ⦞ Thm., 4. Same-Side Int. ⦞ Thm., 5. Congruent Supplements Thm.

11. $m\angle 1 = 75$ because corr. ∠s of ∥ lines are ≅; $m\angle 2 = 105$ because same-side int. ∠s of ∥ lines are suppl. **13.** $m\angle 1 = 100$ because same-side int. ∠s of ∥ lines are suppl.; $m\angle 2 = 70$ because alt. int. ∠s of ∥ lines have = measure. **15.** 25; 65, 65 **17.** $m\angle 1 = m\angle 3 = m\angle 6 = m\angle 8 = m\angle 9 = m\angle 11 = m\angle 13 = m\angle 15 = 52$; $m\angle 2 = m\angle 4 = m\angle 5 = m\angle 7 = m\angle 10 = m\angle 12 = m\angle 14 = 128$ **19.** two **21.** two **23.** 32 **25.** $x = 135$, $y = 45$ **29.** Since $a \parallel b$, $\angle 1 \cong \angle 3$ because they are corr. ∠s. Also, $\angle 3$ and $\angle 2$ are supplementary by the ∠Add. Post. So by Subst., $\angle 1$ and $\angle 2$ are supplementary.

Lesson 3-2   pp. 137–140

EXERCISES **1.** $\overleftrightarrow{BE} \parallel \overleftrightarrow{CG}$; Conv. of Corr. ∠s Post. **3.** $\overline{JO} \parallel \overline{LM}$; if two lines and a transversal form same-side int. ∠s that are suppl., then the lines are ∥. **5.** 30 **7.** 59 **11.** $a \parallel b$; if two lines and a transversal form same-side int. ∠s that are suppl., then the lines are ∥. **23.** The corr. ∠s he draws are ≅. **25.** 20 **29.** 5; $m\angle 1 = m\angle 2 = 50$ **31.** 1.25; $m\angle 1 = m\angle 2 = 10$ **33.** Answers may vary. Sample: $\angle 3 \cong \angle 9$; $j \parallel k$ by Conv. of the Alt. Int. ∠s Thm.

Lesson 3-3   pp. 143–144

1.a. $\angle 1 \cong \angle 2 \cong \angle 3$ **b.** Slat D is perp. to slats B and C. Explanations may vary. Sample: Slat D is perp. to slat A. Slats A, B, and C are parallel, so by Theorem 3-11, slat D is also perp. to B and C. **3.a.** corr. ∠s **b, c.** $\angle 1$, $\angle 3$ (any order) **d.** Converse of Corr. ∠s **5.** All of the rungs are perpendicular to one side. The side is perp. to the top rung, and because all of the rungs are parallel to each other, the side is perp. to all of the rungs. **7.** The rungs are parallel to each other because they are all perpendicular to one side. The sides are parallel because they are both perpendicular to one rung. **9.** All of the rungs are parallel. All of the rungs are parallel to one rung, so they are all parallel to each other.

Lesson 3-4   pp. 150–153

EXERCISES **1.** 30 **3.** 90 **5.** $x = 80$; $y = 80$ **7.** right, scalene **9.** obtuse, isosceles **11.** Not possible; a right △ will always have one longest side opp. the right ∠.

13.
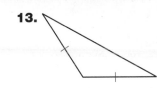

17.a. 2 **b.** 6 **19.** 115.5 **21.** $x = 147$, $y = 33$ **23.** $x = 7$; 55, 35, 90; right **25.** $x = 38$, $y = 36$, $z = 90$; △ABD: 36, 90, 54; right; △BCD: 90, 52, 38; right; △ABC: 74; 52, 54; acute.

Lesson 3-5   pp. 161–164

EXERCISES **1.** yes **3.** No; it is not a plane figure. **5.** $MWBFX$; sides: $\overline{MW}, \overline{WB}, \overline{BF}, \overline{FX}, \overline{XM}$; ∠s: $\angle M$, $\angle W$, $\angle B$, $\angle F$, $\angle X$ **7.** $HEPTAGN$; sides: $\overline{HE}, \overline{EP}, \overline{PT}, \overline{TA}, \overline{AG}, \overline{GN}, \overline{NH}$; ∠s: $\angle H$, $\angle E$, $\angle P$, $\angle T$, $\angle A$, $\angle G$, $\angle N$ **9.** decagon; concave **11.** 1080 **13.** 1440 **17.** 103 **19.** 37 **23.** 150; 30 **25.** 176.4; 3.6

29.   **31.**
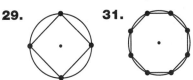

33. 8 **35.** 18 **37.** octagon; $m\angle 1 = 135$; $m\angle 2 = 45$ **41.** 144; 10 **43.** 150; 12 **45.** $\frac{4}{5}$ **47.** $y = 103$; $z = 70$; quad. **49.** $x = 36$, $2x = 72$, $3x = 108$, $4x = 144$; quad.

Algebra 1 Review   p. 165

1. $\frac{8}{3}$; it is steeper than line ℓ, but has the same tilt.

3. $-\frac{3}{2}$; it is steeper than line r, but has the same tilt.

5. -1; it is the same steepness and tilt as line r.

7. -5; it is steeper than line r, but has the same tilt.

11.

The line is steeper than line r, but has the same tilt.

13. The line is horizontal.

Lesson 3-6   pp. 169–171

EXERCISES

1.   **3.**

5.
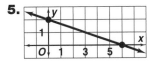

11.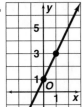

$y = 2x + 1$

13. $y = -2x + 4$
17. $y - 3 = 2(x - 2)$
19. $y - 5 = -1(x + 3)$
23–25. Equations may vary from the pt. chosen. Samples are given.

23. $y - 5 = \frac{3}{5}(x - 0)$
25. $y - 6 = 1(x - 2)$ **29.a.** $y = 7$
b. $x = 4$ **31.a.** $y = -1$ **b.** $x = 0$

33. **35.**

39. No; a line with no slope is a vertical line. 0 slope is a horizontal line. **41.a.** Undefined; it is a vertical line. **b.** $x = 0$ **43.** Answers may vary. Sample: The eq. is in slope-int. form; use slope-int. form because the eq. is already in that form.

49. **51.**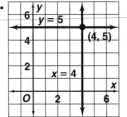

Lesson 3-7 pp. 177–180

EXERCISES 1. Yes; both slopes $= -\frac{1}{2}$. **3.** No; the slope of $\ell_1 = \frac{3}{2}$, and the slope of $\ell_2 = 2$. **7.** Yes; the lines both have a slope of $\frac{3}{4}$ but different y-intercepts. **9.** No; one slope $= 7$ and the other slope $= -7$. **13.** $y - 0 = \frac{1}{3}(x - 6)$ or $y = \frac{1}{3}(x - 6)$ **15.** $y + 2 = -\frac{3}{2}(x - 6)$ **17.** Yes; the slope of $\ell_1 = -\frac{3}{2}$, and the slope of $\ell_2 = \frac{2}{3}$; $-\frac{3}{2} \cdot \frac{2}{3} = -1$. **19.** Yes; the slope of $\ell_1 = -1$, and the slope of $\ell_2 = 1$; $-1 \cdot 1 = -1$. **21–23.** Answers may vary. Samples are given. **21.** $y = -2(x - 4)$ **23.** $y = \frac{4}{5}x$ **25.** Yes; $1 \cdot (-1) = -1$ **27.** No; $\frac{2}{7} \cdot (-\frac{7}{4}) \neq 1$ **29.** slope of $\overline{AB}$ = slope of $\overline{CD} = \frac{2}{3}$; $\overline{AB} \parallel \overline{CD}$ slope of $\overline{BC}$ = slope of $\overline{AD} = -3$; $\overline{BC} \parallel \overline{AD}$ **31.** slope of $\overline{AB} = \frac{1}{2}$; slope of $\overline{CD} = \frac{1}{4}$; $\overline{AB} \nparallel \overline{CD}$; slope of $\overline{BC} = -1$; slope of $\overline{AD} = -\frac{1}{2}$; $\overline{BC} \nparallel \overline{AD}$ **37.** The lines will have the same slope. **41.** $\perp$ **43.** $\perp$

Lesson 3-8 pp. 184–186

1.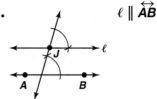

$\ell \parallel \overleftrightarrow{AB}$

For Exercise 5, constructions may vary. Sample using the following segments is shown:

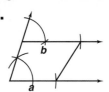

5.

9. 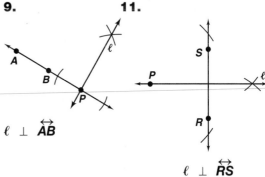 **11.**

$\ell \perp \overleftrightarrow{AB}$

$\ell \perp \overleftrightarrow{RS}$

For Exercises 17–19, constructions may vary. Samples are given.

17. 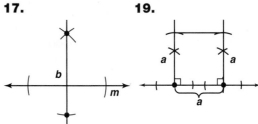 **19.**

23. 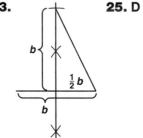 **25.** D

Chapter Review pp. 189–191

1. acute **2.** obtuse **3.** corr. ⚓ **4.** exterior
5. convex **6.** equiangular **7.** regular
8. point-slope **9.** slope-int. **10.** alt. int. ⚓
11. $m\angle 2 = 121$, $m\angle 3 = 59$, $m\angle 4 = 59$ **12.** $m\angle 1 = 120$; corr. ⚓ are ≅. $m\angle 2 = 120$; vert. ⚓ are ≅ or alt. ext. ⚓ are ≅.

13. $m\angle 1 = 75$; same side int. ⓢ are suppl. $m\angle 2 = 105$; alt. int. ⓢ are ≅ or two ⓢ that form a straight ∠ are suppl. **14.** $m\angle 1 = 55$; same side int. ⓢ are suppl. $m\angle 2 = 90$; alt. int. ⓢ are ≅. **15.** 20 **16.** 20 **17.** 24

18. **19.**

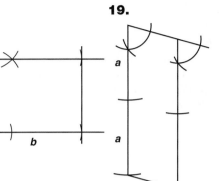

20. If 2 lines are ∥ to the same line, then the lines are ∥ (Thm. 3–10). **21.** If a line is ⊥ to one of two ∥ lines, then it is also ⊥ to the other (Thm. 3–11). **22.** 61; scalene, acute **23.** $x = 60$; $y = 60$; equilateral, acute **24.** $x = 45$; $y = 45$; isosc., right **25.** 55; acute **26.** 30; right **27.** 3; acute **28.** 8; obtuse **29.** One ∠ is 90; the remaining 2 ⓢ are compl. **30.** 120; 60 **31.** 135; 45 **32.** 144; 36 **33.** 165; 15 **34.** 360

35. $m = 2$; **36.** $m = -2$;
y-int. $= -1$ point $= (-5, 3)$

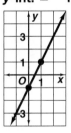 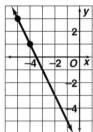

37. **38.**

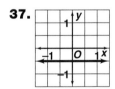 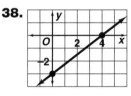

39. $x = 6$ **40.** neither **41.** ∥ **42.** ⊥ **43.** ∥ **44.** 0; the difference of y-coordinates is always zero.

Chapter 4

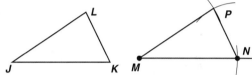

Lesson 4-1   pp. 200–203

EXERCISES **1.** $\angle CAB \cong \angle DAB$; $\angle C \cong \angle D$; $\angle ABC \cong \angle ABD$; $\overline{AC} \cong \overline{AD}$; $\overline{AB} \cong \overline{AB}$; $\overline{CB} \cong \overline{DB}$. **3.** $\overline{BK}$ **5.** $\overline{ML}$ **9.** $\triangle KJB$ **11.** $\triangle JBK$ **15.** $\angle P \cong \angle S$; $\angle O \cong \angle I$; $\angle L \cong \angle D$; $\angle Y \cong \angle E$ **17.** 54 in. **19.** 77

25. No; the corr. sides are not ≅. **27.** Yes; all corr. sides and ⓢ are ≅. **29.** B **31.** 5 **33.** $m\angle B = m\angle E = 21$ **35.** $AC = DF = 19$ **39.** $\triangle BCE \cong \triangle ADE$ **41.** $\triangle JLM \cong \triangle NRZ$; $\triangle JLM \cong \triangle ZRN$

Lesson 4-2   pp. 208–211

EXERCISES **1.a.** Given **b.** Reflexive **c.** $\triangle JKM$ **d.** $\triangle LMK$ **3.** It is given that $\overline{WZ} \cong \overline{ZS} \cong \overline{SD} \cong \overline{DW}$. $\overline{ZD} \cong \overline{ZD}$ by the Reflexive Prop. of Congruence. Therefore, $\triangle WZD \cong \triangle SDZ$ by SSS. **5.** Yes; the legs have equal lengths and are joined at midpts. So $\overline{AE} \cong \overline{CE}$ and $\overline{BE} \cong \overline{DE}$; $\angle AEB \cong \angle CED$ by vert. ⓢ are ≅; $\triangle AEB \cong \triangle CED$; by SAS. **11.** $\angle W$ **13.** $\overline{WU}$ **15.** $\overline{XZ}$, $\overline{YZ}$ **17.** Yes; $\triangle PVQ \cong \triangle STR$ by SSS. **21.** $\triangle KLJ \cong \triangle MON$; SSS **23.** $\triangle JEF \cong \triangle SVF$ or $\triangle JEF \cong \triangle SFV$; SSS **29.** No; you would need $\angle H \cong \angle K$ or $\overline{GI} \cong \overline{JL}$.

31.

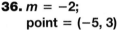

35. $\overline{IP} \cong \overline{PO}$; $\triangle ISP \cong \triangle OSP$ by SSS. **37.** Yes; $\triangle ABC \cong \triangle CDA$ by SAS; $\angle DAC \cong \angle ACB$ because if ∥ lines, then alt. int. ⓢ are ≅.

Lesson 4-3   pp. 215–219

EXERCISES **1.** $\triangle PQR \cong \triangle VXW$ **3.** $\overline{RS}$ **5.** Reflexive, ASA **7.** $\overline{QR} \cong \overline{TS}$: given, $\overline{QR} \parallel \overline{ST}$: given, $\angle TQR \cong \angle QTS$: Alt. Int. ⓢ Thm., $\angle QTR \cong \angle TQS$: Alt. Int. ⓢ Thm., $\triangle QRT \cong \triangle TSQ$: AAS **13.** $\triangle PMO \cong \triangle NMO$; ASA **15.** $\triangle ZVY \cong \triangle WVY$; AAS **21.** $\triangle FGJ \cong \triangle HJG$ by AAS since $\angle FGJ \cong \angle HJG$ because when lines are ∥, then alt. int. ⓢ are ≅ and $\overline{GJ} \cong \overline{GJ}$ by the Reflexive Prop. of ≅.

Lesson 4-4   pp. 222–225

EXERCISES **1.** SAS; $\triangle KLJ \cong \triangle OMN$; $\angle K \cong \angle O$, $\angle J \cong \angle N$, $\overline{JK} \cong \overline{NO}$ **3.** $\triangle MOE \cong \triangle REO$ by SSS because $\overline{OE} \cong \overline{OE}$ by Reflexive Prop. of ≅; $\angle M \cong \angle R$ by CPCTC. **5.** The ⓢ are ≅ by SAS so the distance across the sinkhole is 26.5 yd by CPCTC. **7.** $\overline{YT} \cong \overline{YP}$, $\angle C \cong \angle R$, $\angle T \cong \angle P$ (Given), $\angle CYT \cong \angle RYP$ (If 2 ⓢ of a △ are ≅ to 2 ⓢ of another, the 3rd ⓢ are ≅.), $\triangle CYT \cong \triangle RYP$ (ASA), $\overline{CT} \cong \overline{RP}$, (CPCTC) **9.** $\overline{KL} \cong \overline{KL}$ by Reflexive Prop. of ≅; $\overline{PL} \cong \overline{LQ}$ by def. of ⊥ bis.; $\angle KLP \cong \angle KLQ$ by def. of ⊥; the ⓢ are ≅ by SAS. **17.** $\triangle ABX \cong \triangle ACX$ by SSS, so $\angle BAX = \angle CAX$ by CPCTC. Thus $\overrightarrow{AX}$ bisects $\angle BAC$ by the def. of ∠bis.

Lesson 4-5 pp. 230–233

EXERCISES 1. $\overline{VX}$; Conv. of the Isosc. △ Thm.
3. $\overline{VY}$; $VT = VX$ (Ex. 1) and $UT = YX$ (Ex. 2), so $VU = VY$ by the Subtr. Prop. of =. **7.** $x = 38$; $y = 4$
9. 24, 48, 72, 96, 120 **11.** $2\frac{1}{2}$ **13.** 35

17a.
30, 30, 120 **b.** 5; 30, 60, 90, 120, 150 **21.** $x = 36$; $y = 36$
23. Two sides of a △ are ≅ if and only if the ∡ opp. those sides are ≅.
25. a. isosc. ∡ **b.** 900 ft;
1100 ft **c.** The tower is the ⊥ bis. of the base of each △. **31.** $m = 60$; $n = 30$

Algebra Review p. 234

1. $(-3, -7)$ **3.** no sol. **5.** inf. many sol. **7.** inf. many sol.

Lesson 4-6 pp. 237–240

EXERCISES 1. $\triangle ABC \cong \triangle DEF$ by HL. Both ∡ are rt. ∡, $\overline{AC} \cong \overline{DF}$, and $\overline{CB} \cong \overline{FE}$. **3.** ∠T and ∠Q are rt. ∡. **5a.** ≅ supp. ∡ are rt. ∡ **b.** Def of rt. △ **c.** Given **d.** Reflexive Prop. of ≅ **e.** HL **9.** HL; each rt. △ has a ≅ hyp. and side. **11.** $x = -1$; $y = 3$

17. 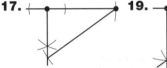 **19.**

21.1. $\overline{LO}$ bisects ∠MLN, $\overline{OM} \perp \overline{LM}$, $\overline{ON} \perp \overline{LN}$ (Given) 2. ∠M and ∠N are rt. ∡ (Def. of ⊥) 3. ∠MLO ≅ ∠NLO (Def. of ∠ bis.) 4. ∠M ≅ ∠N (All rt. ∡ are ≅.) 5. $\overline{LO} \cong \overline{LO}$ (Reflexive Prop. of ≅) 6. $\triangle LMO \cong \triangle LNO$ (AAS)

Lesson 4-7 pp. 243–246

EXERCISES 1. ∠M **3.** $\overline{XY}$

5.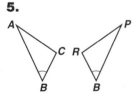
7.a. Given **b.** Reflexive Prop. of ≅ **c.** Given **d.** AAS **e.** CPCTC

11. $\triangle ADC \cong \triangle EDG$; by ASA if ∠A ≅ ∠E. ∠A and ∠E are corr. parts in $\triangle ADB$ and $\triangle EDF$, which are ≅ by SAS. **17.** 1. $\overline{AC} \cong \overline{EC}$; $\overline{CB} \cong \overline{CD}$ (Given), 2. ∠C ≅ ∠C (Reflexive Prop. of ≅) 3. $\triangle ACD \cong \triangle ECB$ (SAS), 4. ∠A ≅ ∠E (CPCTC) **19.** $m\angle 1 = 56$; $m\angle 2 = 56$; $m\angle 3 = 34$; $m\angle 4 = 90$; $m\angle 5 = 22$; $m\angle 6 = 34$; $m\angle 7 = 34$; $m\angle 8 = 68$; $m\angle 9 = 112$

Chapter Review pp. 249–251

1. legs **2.** vertex angle **3.** CPCTC **4.** hypotenuse **5.** base angles **6.** corollary **7.** legs **8.** Congruent polygons **9.** base **10.** $\overline{ML}$ **11.** ∠U **12.** $\overline{ST}$
13. ONMLK **14.** 80 **15.** 3 **16.** 5 **17.** 35 **18.** 100
19. SSS **20.** not possible **21.** SAS **22.** not possible **23.** AAS **24.** ASA **25.** $\triangle AWC \cong \triangle RCW$; AAS **26.** $\triangle JKL \cong \triangle UVT$; SAS **27.** $\triangle RGB \cong \triangle DCS$; ASA **28.** $\triangle VTY \cong \triangle WYX$ by AAS so $\overline{TV} \cong \overline{YW}$ by CPCTC. **29.** $\triangle BCE \cong \triangle DCE$ by ASA so $\overline{BE} \cong \overline{DE}$ by CPCTC. **30.** $\triangle KNM \cong \triangle MLK$ by SAS so $\overline{KN} \cong \overline{ML}$ by CPCTC. **31.** $x = 4$, $y = 65$
32. $x = 55$, $y = 62.5$ **33.** $x = 65$, $y = 90$
34. Since $\overline{PS} \perp \overline{SQ}$ and $\overline{RQ} \perp \overline{QS}$, $\triangle PSQ$ and $\triangle RQS$ are rt. ∡. $\overline{PQ} \cong \overline{RS}$ and $\overline{QS} \cong \overline{SQ}$ so $\triangle PSQ \cong \triangle RQS$ by HL. **35.** Since $\overline{LN} \perp \overline{KM}$, $m\angle LNK = m\angle LNM = 90$. $\overline{KL} \cong \overline{ML}$ and $\overline{LN} \cong \overline{LN}$ so $\triangle KLN \cong \triangle MLN$ by HL. **36.** $\triangle AEC \cong \triangle ABD$ by SAS. **37.** $\triangle FIH \cong \triangle GHI$ by SAS. **38.** $\triangle PTS \cong \triangle RTA$ by ASA. **39.** $\triangle CFE \cong \triangle DEF$ by ASA.

Chapter 5

Lesson 5-1 pp. 262–264

EXERCISES 1. 9 **3.** 14 **7.** 40 **9.** 160 **11.** $\overline{UW} \parallel \overline{TX}$; $\overline{UY} \parallel \overline{VX}$; $\overline{YW} \parallel \overline{TV}$ **13a.** $\overline{ST} \parallel \overline{PR}$; $\overline{SU} \parallel \overline{QR}$; $\overline{UT} \parallel \overline{PQ}$ **b.** $m\angle QPR = 40$ **15.** $\overline{FG}$ **17.** $\overline{EG}$
21.a. 114 ft 9 in. **b.** Answers may vary. Sample: The highlighted segment is a midsegment of the triangular face of the building. **23.** 45 **25.** 55
27. $18\frac{1}{2}$ **29.** C **31.** 50 **33.** $x = 6$; $y = 6\frac{1}{2}$
35. $x = 3$; $DF = 24$

Lesson 5-2 pp. 267–270

EXERCISES 1. $\overline{AC}$ is the ⊥ bis. of $\overline{BD}$. **3.** 18
7. $y = 3$; $ST = 15$; $TU = 15$ **9.** $\overrightarrow{HL}$ is the ∠bis. of ∠KHF because a point on $\overrightarrow{HL}$ is equidistant from $\overrightarrow{HK}$ and $\overrightarrow{HF}$. **11.** 54; 54 **13.** 10 **15.** Isosceles; it has 2 ≅ sides. **19.** 4 **21.** 16 **27.** Answers may vary. Sample: The student needs to know that $\overline{QS}$ bisects $\overline{PR}$. **31.** the pitcher's plate **35.** $C(3, 2)$, $D(3, 0)$; $AC = BC = 3$, $AD = BD = \sqrt{13}$
37. $C(0, 0)$, $D(1, 1)$; $AC = BC = 3$, $AD = BD = \sqrt{5}$
41. $\triangle ABP$ and $\triangle ABQ$ are right ∡ with a common leg and ≅ hypotenuses. Thus, $\triangle BAP \cong \triangle BAQ$ by the HL Thm. $\overline{PB} \cong \overline{BQ}$ using CPCTC, so $\overline{AB}$ bisects $\overline{PQ}$ by the def. of bis. Hence, $\overline{AB}$ is the ⊥ bis. of $\overline{PQ}$. **45.** D **47.** $y = -(x - 2)$

EXERCISES 1. $(-2, -3)$ **3.** $\left(1\frac{1}{2}, 1\right)$ **5.** $\left(-3, 1\frac{1}{2}\right)$
9. Z **11.** $TY = 18$; $TW = 27$ **13.** $VY = 6$; $YX = 3$
15. Neither; it's not a segment drawn from a vertex.
17.

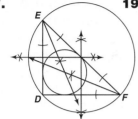

19. $\overline{BE}$ **21.** $\overrightarrow{CA}$ **27.** D
29.a. $\overline{AB}$ **b.** $\overline{BC}$
c. XC **d.** $\perp$ bis.

EXERCISES 1. Two angles are not congruent.
3. The angle is obtuse. **7.a.** If you don't eat all of
your vegetables, then you won't grow. **b.** If you
won't grow, then you don't eat all of your
vegetables. **9.a.** If a figure isn't a rectangle, then
it doesn't have four sides. **b.** If a figure doesn't
have four sides, then it isn't a rectangle.
11. Assume that $\angle J$ is a right angle. **13.** Assume
that none of the angles is obtuse. **17.** I and II
19. II and III **21.a.** right angle **b.** right angles
c. 90 **d.** 180 **e.** 90 **f.** 90 **g.** 0 **h.** more than one
right angle **i.** at most one right angle
23. Assume one base $\angle$ is a right $\angle$. Then the
other base $\angle$ is also a right $\angle$ since the base $\angle$s of
an isosc. $\triangle$ are $\cong$. But a $\triangle$ can have at most one
right $\angle$. So neither base $\angle$ is a right $\angle$. **25.a.** If
four points aren't collinear, then they aren't
coplanar; false **b.** If four points aren't coplanar,
then they aren't collinear; true **35.** If the $\angle$
measures 120, then it is obtuse. If the $\angle$ isn't
obtuse, then it doesn't measure 120. **37.** Angie
assumed that the inverse of the statement was
true, but a conditional and its inverse may not
have the same truth value. **39.** The culprit
entered the room through a hole in the roof; the
other possibilities were eliminated.

1. $x \leq -1$ **3.** $x > -\frac{21}{2}$ **5.** $a \leq 18$ **7.** $z > \frac{1}{2}$

EXERCISES 1. $\angle 3 \cong \angle 2$ because they are
vertical $\angle$s and $m\angle 1 > m\angle 3$ by Corollary to the
Ext. $\angle$ Thm. So, $m\angle 1 > m\angle 2$ by subst. **3.** $m\angle 1 >$
$m\angle 4$ by Corollary to the Ext. $\angle$ Thm. and $\angle 4 \cong$
$\angle 2$ because if $\parallel$ lines, then alt. int. $\angle$s are $\cong$.
5. $\angle D$, $\angle C$, $\angle E$ **7.** $\angle A$, $\angle B$, $\angle C$ **9.** $\angle Z$, $\angle X$, $\angle Y$

11. $\overline{FH}$, $\overline{GF}$, $\overline{GH}$ **13.** $\overline{AC}$, $\overline{AB}$, $\overline{CB}$ **17.** Yes;
$11 + 12 > 15$; $12 + 15 > 11$; $11 + 15 > 12$.
19. Yes; $1 + 15 > 15$; $15 + 15 > 1$.
23. $11 < s < 21$ **25.** $5 < s < 41$ **31.** Answers
may vary. Sample: The shortcut across the grass
is shorter than the sum of the two paths. **35.** $\overline{RS}$
37. $\overline{XY}$

1. median of a $\triangle$ **2.** distance from the point to
the line **3.** $\perp$ Bis. Thm. **4.** altitude
5. contrapositive **6.** indirect proof **7.** $\triangle$ Ineq.
Thm. **8.** incenter **9.** $\angle$ Bis. Thm. **10.** point of
concurrency **11.** 15 **12.** 11 **13.** 40 **14.** 7 **15.** 14
16. 80 **17.** $(-1, 0)$ **18.** $(0, -1)$ **19.** $(2, -3)$
20. $\angle$ bisector; it bisects an $\angle$. **21.** altitude; it is
$\perp$ to a side. **22.** median; it goes through a
midpoint. **23.** Inverse: If it is not snowing, then it
is not cold outside. Contrapositive: If it is not cold
outside, then it is not snowing. **24.** Inverse:
If an angle is not obtuse, then its measure is not
greater than 90 and less than 180. Contrapositive:
If an angle's measure is not greater than 90 and
less than 180, then it is not obtuse. **25.** Inverse:
If a figure is not a square, then its sides are not
congruent. Contrapositive: If a figure's sides are
not congruent, then it is not a square. **26.** Inverse:
If you are not in Australia, then you are not south
of the equator. Contrapositive: If you are not
south of the equator, then you are not in
Australia. **27.** Assume that both numbers are
odd. The product of 2 odd numbers is always
odd, which contradicts that the product is even.
Therefore, at least one number must be even.
28. Assume a right $\angle$ can be formed by non-
perp. lines. Then by the def. of $\perp$, the lines are $\perp$.
Therefore, the assumption is false. **29.** Assume
that an $\triangle$ has 2 obtuse $\angle$s. Then these $\angle$s by def.
are greater than 90, which makes their sum
greater than 180. But the sum of the measures of
the $\angle$s of a $\triangle$ = 180, so the assumption must be
false. **30.** Assume an $\angle$ is obtuse, and therefore
has measure greater than 90. Since the $\triangle$ is
equilateral, it is equiangular, and each $\angle$ measures
60. **31.** $\angle T$, $\angle R$, $\angle S$; $\overline{RS}$, $\overline{TS}$, $\overline{TR}$ **32.** $\angle G$, $\angle O$,
$\angle F$; $\overline{OF}$, $\overline{FG}$, $\overline{OG}$ **33.** No; $5 + 8 \ngtr 15$. **34.** Yes;
each pair > 3rd. **35.** Yes; each pair > 3rd.
36. Yes; each pair > 3rd. **37.** No; $1 + 1 \ngtr 3$.
38. Yes; each pair > 3rd. **39.** $3 < x < 11$
40. $7 < x < 23$ **41.** $6 < x < 10$ **42.** $1 < x < 25$

Chapter 6

EXERCISES 1. $\square$, rectangle, rhombus, square
3. trapezoid **7.** rhombus **9.** rhombus

13. rhombus

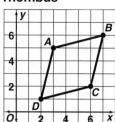

15. trapezoid
19. $x = 11, y = 29$; 13, 13, 23, 23 **21.** $x = 2$, $y = 6$; 2, 7, 7, 2 **25.** 40, 40, 140, 140; 11, 11, 15, 32 **27.** rectangle, square, trapezoid

29–31. Answers may vary. Samples are given.

29.

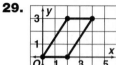

31. Impossible; a trapezoid with one rt. ∠ must have another, since two sides are ∥.

37. False; a trapezoid only has one pair of ∥ sides. **39.** True; all squares are ⌷.
45. Some isosc. trapezoids, some trapezoids
47. rectangle, square **51.** rhombus, ▱
53. rhombus, ▱, kite

| Lesson 6-2 | pp. 315–318 |

EXERCISES 1. 127 **3.** 76 **7.** 3; 10, 20, 20 **9.** 20
11. 17 **13.** 6; $m\angle H = m\angle J = 30, m\angle I = m\angle K =$
150 **15.** $x = 5, y = 7$ **17.** $x = 6, y = 9$ **21.** 3,
23. 6 **31.** $BC = AD = 14.5$ in.; $AB = CD = 9.5$ in.
33. A **35.a.** $\overline{DC}$ **b.** $\overline{AD}$ **c.** ≅ **d.** Reflexive **e.** ASA
f. CPCTC **37.** 38, 32, 110 **39.** 95, 37, 37 **41.** 18,
162 **43.** Answers may vary. Sample: In ⌷ *LENS*
and *NGTH*, $\overline{GT} \parallel \overline{EH}$ and $\overline{EH} \parallel \overline{LS}$ by the def. of a
▱. Therefore $\overline{LS} \parallel \overline{GT}$ because if 2 lines are ∥ to
the same line then they are ∥ to each other.
45. $x = 12, y = 4$ **47.** $x = 9, y = 6$

| Lesson 6-3 | pp. 324–327 |

EXERCISES 1. 5 **3.** $x = 1.6, y = 1$ **7.** Yes; both
pairs of opp. sides are ≅. **9.** Yes; both pairs of
opp. ∠ are ≅. **11.** A quad. is a ▱ if and only if
opp. sides are ≅ (6-1 and 6-5); opp. ∠ are ≅ (6-2
and 6-6); diags. bisect each other (6-3 and 6-7).
15. $x = 3, y = 11$ **17.** $k = 9, m = 23.4$
21. $\triangle TRS \cong \triangle RTW$ (given), so $\overline{ST} \cong \overline{RW}$ and
$\overline{SR} \cong \overline{TW}$. *RSTW* is a ▱ because opp. sides
are ≅ (Thm. 6-5) **23.** (6, 6)

| Lesson 6-4 | pp. 332–335 |

EXERCISES 1. 38, 38, 38, 38 **3.** 118, 31, 31
11. 3; $LN = MP = 7$ **13.** 9; $LN = MP = 67$
17. rhombus; the diags. are ⊥ **19.** The pairs of
opp. sides of the frame remain ≅, so the frame
remains a ▱. **25–27.** Symbols may vary. Sample:
parallelogram: ▱; rhombus: ▱; rectangle: ▯;
square: ⌷ **25.** ▱, ⌷ **27.** ▱, ▱, ▯, ⌷

37.

Diag. are ≅, diag. are ⊥.
39. 1. *ABCD* is a parallelogram
(Given) $\overline{AC}$ bisects ∠*BAD* and
∠*BCD* (Given). **2.** ∠1 ≅ ∠2,
∠3 ≅ ∠4 (Def. of bisect)

3. $\overline{AC} \cong \overline{AC}$ (Refl. Prop. of ≅) **4.** $\triangle ABC \cong \triangle ADC$
(ASA) **5.** $\overline{AB} \cong \overline{AD}$ (CPCTC) **6.** $\overline{AB} \cong \overline{DC}$,
$\overline{AD} \cong \overline{BC}$ (Opp. sides of a ▱ are ≅.) **7.** $\overline{AB} \cong$
$\overline{BC} \cong \overline{CD} \cong \overline{AD}$ (Trans. Prop. of ≅) **8.** *ABCD* is a
rhombus. (Def. of rhomb.) **45.** $x = 5, y = 32, z =$
7.5 **51.** 2, 2 **53.** 1, 1 **55.** Construct a rt. ∠, and
draw. diag. 1 from its vertex. Construct the ⊥
from the opp. end of diag. 1 to a side of the rt. ∠.
Repeat to the other side.

| Lesson 6-5 | pp. 338–341 |

EXERCISES 1. 77, 103, 103 **3.** 49, 131, 131
7.a. isosc. trapezoids **b.** 69, 69, 111, 111 **9.** 90,
45, 45 **11.** 90, 26, 90 **21.** 15 **23.** 3 **27.** $x = 35$,
$y = 30$ **29.** isosc. trapezoid; all the large rt. ∠
appear to be ≅. **39.** Answers may vary. Sample:
Draw $\overline{TA}$ and $\overline{RP}$. **1.** isosc. trapezoid *TRAP* (Given)
2. $\overline{TA} \cong \overline{PR}$ (Diags. of an isosc. trap. are ≅.)
3. $\overline{TR} \cong \overline{PA}$ (Given) **4.** $\overline{RA} \cong \overline{RA}$ (Reflexive Prop.
of ≅) **5.** $\triangle TRA \cong \triangle PAR$ (SSS) **6.** ∠*RTA* ≅ ∠*APR*
(CPCTC)

| Lesson 6-6 | pp. 344–346 |

EXERCISES 1.a. $(2a, 0)$ **b.** $(0, 2b)$ **c.** (a, b)
d. $\sqrt{b^2 + a^2}$ **e.** $\sqrt{b^2 + a^2}$ **f.** $\sqrt{b^2 + a^2}$ **g.** *MA*
= *MB* = *MC* **3.** $W(a, a); Z(a, 0)$ **5.** $W(0, b); Z(a, 0)$
11. *B, D, H, F* **13.** *A, C, G, E* **17.** $W(2a, 2a)$;
$Z(2a, 0)$ **19.** $W(0, b); Z(2a, 0)$ **23.** $(c-a, b)$
25. $(-b, 0)$ **27.a.** **b.**

c. $\sqrt{b^2 + 4c^2}$ **d.** $\sqrt{b^2 + 4c^2}$ **e.** The lengths are =.

| Lesson 6-7 | pp. 349–353 |

EXERCISES 1.a. $W\left(\frac{a}{2}, \frac{b}{2}\right); Z\left(\frac{c + e}{2}, \frac{d}{2}\right)$ **b.** $W(a, b)$;
$Z(c + e, d)$ **c.** $W(2a, 2b); Z(2c + 2e, 2d)$; c; it
uses multiples of 2 to name the coordinates of
W and *Z*. **3.a.** *y*-axis **b.** Distance **5.a.** isos.
b. *x*-axis **c.** *y*-axis **d.** Midpoints **e.** ≅ sides
f. slopes **g.** the Distance Formula **7.a.** $\sqrt{a^2 + b^2}$
b. $2\sqrt{a^2 + b^2}$ **9.a.** (a, b) **b.** (a, b) **c.** the same
point **11.** The vertices of *KLMN* are *L* $(b, a + c)$,
M (b, c), *N* $(-b, c)$, and *K* $(-b, a + c)$. The slopes
of $\overline{KL}$ and $\overline{MN}$ are zero, so these segments are

horizontal. The endpoints of $\overline{KN}$ have $=$
x-coordinates, and so do the endpoints of $\overline{LM}$.
So these segments are vertical. Hence, opp. sides
of *KLMN* are $\parallel$ and consecutive sides are $\perp$.
It follows that *KLMN* is a rectangle.
13–15. Answers may vary. Samples are given.
13. yes; same slope **15.** no; may not have
intersection pt. **25.** 1, 4, 7 **27.** −0.8, 0.4, 1.6, 2.8,
4, 5.2, 6.4, 7.6, 8.8 **31.** $\left(-1, 6\frac{2}{3}\right)$, $\left(1, 8\frac{1}{3}\right)$, (3, 10),
$\left(5, 11\frac{2}{3}\right)$, $\left(7, 13\frac{1}{3}\right)$ **33.** (−2.76, 5.2), (−2.52, 5.4),
(−2.28, 5.6), . . . , (8.52, 14.6), (8.76, 14.8)

1. 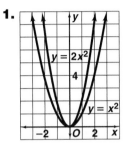 **3.**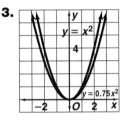

5. 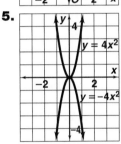 **7.**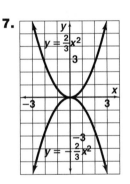

13. 0,4 **15.** −5, 5

1. F **2.** H **3.** G **4.** B **5.** I **6.** J **7.** A **8.** C **9.** E **10.** D

11. 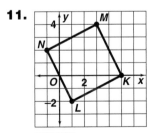 **12.**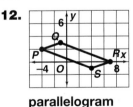

parallelogram

square

13. $x = 8$; 9, 14, 9, 7 **14.** $m = 4, t = 5$; 7, 14, 14,
7 **15.** 101, 79, 101 **16.** 38, 43, 99 **17.** 37, 26, 26
18. yes **19.** yes **20.** no **21.** yes **22.** $x = 29$,
$y = 28$ **23.** $x = 4, y = 5$ **24.** 124, 28, 62 **25.** 60,
90, 30 **26.** 90, 25 **27.** 26 in. **28.** 20 cm **29.** 19 ft
30. (a, b) **31.** $(0, c)$ **32.** $(a - b, c)$ **33.a.** −1
b. 1 **c.** The prod. of the slopes is −1. **34.a.** a
b. $(0, b)$ **c.** $\sqrt{a^2 + b^2}$ **d.** $\sqrt{a^2 + b^2}$ **e.** *BD*

Chapter 7

EXERCISES 1. 1 : 1000 **3.** 3*b* **5.** $\frac{b}{4}$ **13.** $1\frac{2}{3}$
15. 6.875 **21.** 125 mi **23.** about 135 mi
27. 5 : 4 **29.** A **31.** $\frac{9}{4}$ **33.** $\frac{b}{2}$ **35.** 6 **37.** 16.5
45. 9; 18 **47.** 8; 21
49–51. Answers may vary. Samples are given.

49.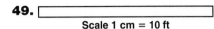
Scale 1 cm = 10 ft

51. 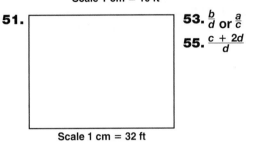 **53.** $\frac{b}{d}$ or $\frac{a}{c}$
  55. $\frac{c + 2d}{d}$

Scale 1 cm = 32 ft

1. −7, 2 **3.** −0.5, −3

EXERCISES 1. $\angle JHY$ **3.** $\angle JXY$ **7.** no; $\frac{20}{30} \neq \frac{36}{52}$
9. yes; $KLMJ \sim PQNO$; $\frac{3}{5}$ **11.** No; corr. $\triangle$ are not
$\cong$. **13.** $x = 4; y = 3$ **15.** $x = 16; y = 4.5; z = 7.5$
17. 6.6 in. by 11 in. **19.** 70 mm **21.** 2 : 3 **23.** 50
29. Yes; corresponding $\triangle$ are $\cong$ and corresponding sides are proportional with a ratio of 1:1 **31.** C
33. 2.6 cm **35.** 3 : 1 **37.** 1 : 2 **39.** 2 : 3 **41.** sides
of 2 cm; $\triangle$ of 60° and 120° **43.** sides of 0.8 cm; $\triangle$
of 60° and 120° **47.** 6.2 in.

1. 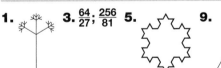 **3.** $\frac{64}{27}$; $\frac{256}{81}$ **5.** **9.**

EXERCISES 1. Yes; $\triangle ABC \sim \triangle FED$; SSS $\sim$ Thm.
3. Ex. 1: $\frac{2}{3}$ (for $\triangle ABC$ to $\triangle FED$); Ex. 2: Not
possible; the $\triangle$ aren't similar. **5.** No; $\frac{6}{3} \neq \frac{10}{4}$
7. Yes; $\triangle APJ \sim \triangle ABC$; SSS $\sim$ Thm. or SAS $\sim$
Thm. **11.** AA $\sim$ Post.; 2.5 **13.** AA $\sim$ Post.; 12
17. AA $\sim$ Post.; 220 yd **19.** AA $\sim$ Post.; 90 ft
21. 151 m **23.a.** No; the corr. $\triangle$ may not be $\cong$.
b. Yes; every isosc. rt. $\triangle$ is a 45°-45°-90° $\triangle$.

Selected Answers

Therefore, by AA ~ Thm. they are all ~. **25.** Yes; $\triangle AWV \sim \triangle AST$; SAS ~ Thm. **27.** No; there is only one pair of $\cong$ &. **31.** 2:1 **33.** 4:3

Algebra 1 Review   **pp. 390**

1. $5\sqrt{2}$ **3.** 8 **5.** $4\sqrt{3}$ **7.** $6\sqrt{2}$

Lesson 7-4   **pp. 394–396**

EXERCISES 1. 6 **3.** $4\sqrt{3}$ **9.** s **11.** c **15.** 9
17. 10 **21.a.** 18 mi **b.** 24 mi **23.a.** 4 cm
23.b.

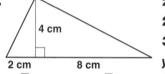

25. (10, 6), (−2, 6)
27. 14 **29.** $\sqrt{14}$
35. $x = 12\sqrt{5}$;
$y = 12$; $z = 6\sqrt{5}$
37. $12\sqrt{2}$ **39.** $\ell_1 = \sqrt{2}, \ell_2 = \sqrt{2}, a = 1, h_2 = 1$
41. $\ell_1 = \ell_2 = 6\sqrt{2}, h = 12, h_2 = 6$ **49.** 3 **51.** 4.5

Lesson 7-5   **pp. 400–404**

EXERCISES 1. 7.5 **3.** 5.2 **5.** c **7.** d **9.** $3\frac{1}{3}$ **11.** 6
13. 35 **17.** KS **19.** JP **25.** 559 ft **27.** 2.4 cm and
2.6 cm; 3.3 cm and 8.7 cm; 3.8 cm and 9.2 cm
29. $x = 18$ m; $y = 12$ m **31.** 20 **33.** $\frac{2}{7}$, 3
35.a. $\frac{AB}{BC}$ **b.** $\frac{WX}{XY}$ **c.** $\frac{AB}{BC} = \frac{WX}{XY}$ **37.** No; $\frac{28}{12} \neq \frac{24}{10}$
39. Measure $\overline{AC}$, $\overline{CE}$ and $\overline{BD}$. Use the Side-Splitter Thm. Write the proport. $\frac{AC}{CE} = \frac{AB}{BD}$ and
solve for AB. **41.** 6 **43.** 19.5

Chapter Review   **pp. 407–409**

1. similar **2.** Cross-Product Property **3.** golden
rectangle **4.** similarity ratio **5.** proportion
6. indirect measurement **7.** golden ratio **8.** 1 : 48
9. 1 : 24 **10.** True; use the Cross-Product Prop.
11. True; the cross product is equivalent to the
original proportion. **12.** False; the cross product
is *not* equivalent to the given equation.
13. True; the cross product is equivalent to the
original proportion. **14.** $\angle M \cong \angle R, \angle N \cong \angle S$,
$\angle P \cong \angle T$; $\frac{MN}{RS} = \frac{MP}{RT} = \frac{NP}{ST}$ **15.** 39 in. or 15 in.
16. 2 : 3 **17.** 2.5 : 1 or 5 : 2 **18.** 9 **19.** $x = 12$;
$y = 15$ **20.** $\triangle XYZ \sim \triangle JKL$; SAS ~ Thm. **21.** No;
corr. sides are not in prop. **22.** AA ~ Post.
23. 13.5 ft **24.** 10 **25.** 30 **26.** $2\sqrt{15}$ **27.** $\frac{2\sqrt{7}}{3}$
28. 1.2 **29.** $\sqrt{12}$ **30.** $x = 15$; $y = 12$; $z = 20$
31. $x = 2\sqrt{21}$; $y = 4\sqrt{3}$; $z = 4\sqrt{7}$ **32.** $x = 2\sqrt{3}$;
$y = 6$; $z = 4\sqrt{3}$ **33.** 7.5 **34.** 5.5 **35.** 37.5
36. 11.25 **37.** 17.5 **38.** 12

Chapter 8

Lesson 8-1   **pp. 420–423**

EXERCISES 1. 10 **3.** 34 **5.** 65 **7.** No;
$4^2 + 5^2 \neq 6^2$. **9.** Yes; $15^2 + 20^2 = 25^2$. **11.** $\sqrt{33}$
13. $2\sqrt{89}$ **17.** 17.0 m **19.** no; $8^2 + 24^2 \neq 25^2$
21. acute **23.** acute **27.** 10 **29.** $2\sqrt{2}$ **31.** B
33. Yes; $7^2 + 24^2 = 25^2$, so $\angle RST$ is a rt. $\angle$.
37. 50 **39.** 35 **41–47.** Answers may vary.
Samples are given. **41.** 4; 5 **43.** 11; 12
49. 2830 km **51.** 12.5 cm

Lesson 8-2   **pp. 428–430**

EXERCISES 1. $x = 8$; $y = 8\sqrt{2}$ **3.** $y = 60\sqrt{2}$
5. $4\sqrt{2}$ **7.** 14.1 cm **9.** $x = 20$; $y = 20\sqrt{3}$
11. $x = 5$; $y = 5\sqrt{3}$ **15.** 50 ft; **17.** $a = 7$; $b = 14$;
$c = 7$; $d = 7\sqrt{3}$ **19.** $a = 10\sqrt{3}$; $b = 5\sqrt{3}$; $c = 15$;
$d = 5$ **25.a.** 28 ft **b.** 0.28 min **27.a.** $\sqrt{3}$ units
b. $2\sqrt{3}$ units **c.** $s\sqrt{3}$ units

Lesson 8-3   **pp. 434–437**

EXERCISES 1. $\frac{1}{2}$; 2 **3.** 1; 1 **5.** 12.3 **7.** 2.5 **11.** 32
13. 48 **17.** 74.1 **19.** 114.5 **21.** 44 and 136 **23.** D
27. $w = 5$; $x \approx 4.7$ **29.** $w \approx 59$; $x \approx 36$
31. about 51° **33.** about 296 ft **35.** 71.6 **37.** 45.0

Lesson 8-4   **pp. 441–443**

EXERCISES 1. $\frac{7}{25}$; $\frac{24}{25}$ **3.** $\frac{1}{2}$; $\frac{\sqrt{3}}{2}$ **5.** 8.3 **7.** 17.0
11. 21 **13.** 46 **17.** about 17 ft 8 in. **19.** $\cos X \cdot \tan$
$X = \frac{\text{adj.}}{\text{hyp.}} \cdot \frac{\text{opp.}}{\text{adj.}} = \frac{\text{opp.}}{\text{hyp.}} = \sin X$ **21.** No; the & are ~
and the sine ratio for 35° is constant. **23.** $w \approx 37$;
$x \approx 7.5$ **25.a.** They are equal; yes; The sine and
cosine of complementary & are equal. **b.** $\angle B$; $\angle A$
c. Sample: cosine of $\angle A$ = sine of the compl. of $\angle A$.

Lesson 8-5   **pp. 447–450**

EXERCISES 1. $\angle$ of elevation from sub to boat
3. $\angle$ of elevation from boat to lighthouse **9.** 34.2 ft
11. about 986 m **13.** 0.6 km **15.** 64° **17.** about
193 m **19.** 72, 72 **21.** 27, 27 **23.** B **25.** 5 **27.** 0.5;
about 84.9

Lesson 8-6   **pp. 455–459**

EXERCISES 1. $\langle 602.2, 668.8 \rangle$ **3.** $\langle 37.5, -65.0 \rangle$
5. 20° west of south

7.

13. about 97 mi at 41° south of west **15.** about 54 mi/h; 22° north of east

17.a. $\langle -9, -9 \rangle$

b.

23. $\langle -1, 3 \rangle$ **25.** $\langle -2, 3 \rangle$
27. about 13.2° north of west
29. Yes; both vectors have the same direction, but could have diff. mag.

31.a.

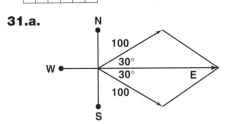

b. about 173 due east **33.** Vectors are ∥ if they have the same or opp. directions. **35.a.** $\langle 0, 0 \rangle$
b. $\vec{a}$ and $\vec{c}$ have = mag. and opp. direction.

37. $\begin{bmatrix} -1 \\ -2 \end{bmatrix}$ **39.** $\begin{bmatrix} -1 \\ 0 \end{bmatrix}$

43. $\langle 0, -4 \rangle$

45. The vectors have the same mag.; the vectors have opp. directions. **47.a.** 15° south of west
b. about 6.7 h

Chapter Review
pp. 461–463

1. vector **2.** angle of elevation **3.** cosine **4.** sine
5. identity **6.** resultant vector **7.** Pythagorean triple **8.** 16 **9.** $2\sqrt{113}$ **10.** 17 **11.** $x = 9\sqrt{3}$; $y = 18$;
12. $12\sqrt{2}$ **13.** $x = \frac{20\sqrt{3}}{3}$; $y = \frac{40\sqrt{3}}{3}$; **14.** 42 **15.** 2
16. 2 **17.** 67 **18.** 23 **19.** 42 **20.** 12 **21.** 8
22. $\frac{\sqrt{3}}{2}, \frac{1}{2}, \sqrt{3}$ **23.** $\frac{\sqrt{19}}{10}, \frac{9}{10}, \frac{\sqrt{19}}{9}$ **24.** $\frac{4}{5}, \frac{3}{5}, \frac{4}{3}$ **25.** 51
26. 16.5 **27.** 33 **28.** 1410 ft **29.** 280 ft **30.** about 38.2 yd **31.** $\langle 125.8, 81.7 \rangle$ **32.** $\langle 37.5, -92.7 \rangle$
33. $\langle -21.8, 33.5 \rangle$ **34.** about 167.7 mi; about 26.6° east of south **35.** about 206.2 km; about 14.0° west of south **36.** about 503.1 mi/h; about 26.6° north of west **37.** $\langle 6, 8 \rangle, \langle 9, 12 \rangle, \langle 30, 40 \rangle$; $\langle x, y \rangle$ and $\langle nx, ny \rangle$ have the same direction for $n > 0$.

38.

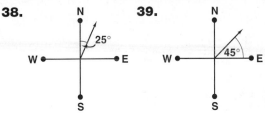

39.

40.

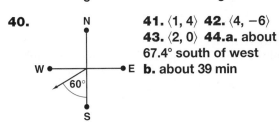

41. $\langle 1, 4 \rangle$ **42.** $\langle 4, -6 \rangle$
43. $\langle 2, 0 \rangle$ **44.a.** about 67.4° south of west
b. about 39 min

Chapter 9

Lesson 9-1
pp. 473–476

EXERCISES 1. Yes; the trans. is a slide. **3.** No; the figures are not ≅. **5.a.** Answers may vary.
Sample: $\angle R \rightarrow \angle R'$ **b.** $\overline{RI}$ and $\overline{R'I'}$; $\overline{IT}$ and $\overline{I'T'}$; $\overline{RT}$
and $\overline{R'T'}$ **7.** $(-6, 5), (2, 4), (-3, 1)$ **9.** $(-7, 5), (-7, 8)$,
$(-4, 8), (-1, 2)$ **11.** $(x - y) \rightarrow (x + 1, y - 3)$
13. $(x, y) \rightarrow (x - 5, y - 2)$
15.a.

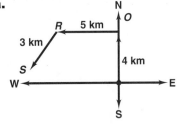

b. about 7.1 km west, 1.9 km north

17. $(x, y) \rightarrow (x + 2, y + 2)$ **19.** D **23.a.** at least 5 ft east, 10 ft north **b.** Sample: $(x, y) \rightarrow (x + 5, y + 10)$
25. U' (1, 16) G' (2, 12) **27.** No; $\triangle HYP \rightarrow \triangle Y'H'P'$ is the translation. **29.** $(x, y) \rightarrow (x + 13, y - 2.5)$

31.

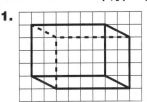

33.

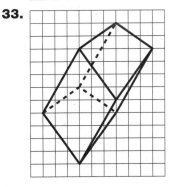

Lesson 9-2
pp. 480–482

1. (−1, 2) **3.** (3, −2)

7. **9.**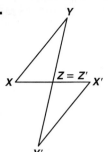

15. Reflect point *D* over the mirrored wall. Connect this point and *C*. The intersection of the segment and the wall is the point to focus the camera.

17.  **19.** A **21.**

23. Answers may vary. Sample: scissors, a baseball glove, a guitar **25.** (*x*, *y*) has image (−*x*, *y*).
29. (4, 0) **31.** (−4, 6)

Lesson 9-3
pp. 485–488

EXERCISES

1. 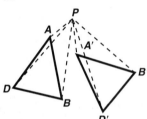 **3.**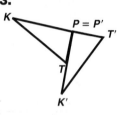

5. 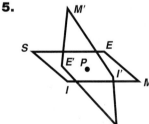 **7.**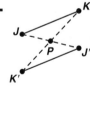

9.  **11.** *M* **13.** $\overline{BC}$ **19.** 108°; 252°
21.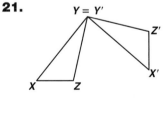

23. 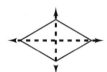 **27.** 180° rotation about its center **29.** Answers may vary. Sample: 110
33. Answers may vary. Sample: a 90° a 270° rotation

Lesson 9-4
pp. 494–496

EXERCISES

1. line; rotational 180° **3.** rotational: 90°

 **13.**

15. 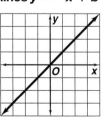 **17.** rotational and reflectional
19. Answer may vary. Samples: CODE, HOOD, DOCK

21a.

Language	Horiz. line	Vert. line	Point
English	B, C, D, E, H, I, K, O, X	A, H, I, M, O, T, U, V, W, X, Y	H, I, N, O S, X, Z
Greek	B, E, H, Θ, I, K, Ξ, O, Σ, Φ, X	A, Δ, H, Θ, I, Λ, M, Ξ, O, Π, T, Υ, Φ, X, Ψ, Ω	Z, H, Θ, I, N, Ξ, O, Φ, X

b. Sample: Greek; Greek alphabet has more letters with at least one kind of symmetry and more letters with multiple symmetries.
23. rotational: 90°; reflectional **25.** reflectional; rotational **27.** point **33.** Yes; the bisector divides the ∠ into 2 ≅ △s with one side of the ∠ being the reflection of the other. **35.** Not necessarily; the bisector divides the segment into 2 ≅ parts but one part cannot be the reflection of the other unless the bisector is the ⊥ bisector.
37. (−3, 4) **39.** (−3, −4)
41. point symmetry about any pt. on the line; reflectional in any member of the family of lines *y* = −*x* + *b*
43. reflectional in *x*-axis

EXERCISES 1. enlargement; center *A*, scale factor $\frac{3}{2}$ **3.** enlargement; center *R*, scale factor $\frac{3}{2}$
11. 512 in. **13.** 1.25 ft
15. *P*′(6, −3), *Q*′(6, 12), *R*′(12, −3)

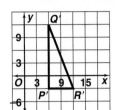

19. *L*′(−15, 0)
21. *T*′(0, 18)

25. $B'\left(\frac{1}{8}, -\frac{1}{15}\right)$ **27.** *Q*′(−9, 12), *W*′(9, 15), *T*′(9, 3), *R*′(−6, −3) **29.** $Q'\left(-\frac{3}{2}, 2\right)$, $W'\left(\frac{3}{2}, \frac{5}{2}\right)$, $T'\left(\frac{3}{2}, \frac{1}{2}\right)$, $R'\left(-1, -\frac{1}{2}\right)$ **35.** The image has side lengths 10 in. and ∠measures 60.

37.

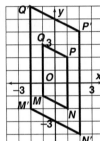

43. *I*′*J*′ = 10; *H*′*J*′ = 12
45. *HI* = 32; *I*′*J*′ = 7.5
47. *x* = 3; *y* = 60
49.

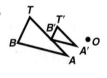

55. 60 cm **57.** False; a dilation doesn't map a segment to a ≅ segment unless the scale factor is 1. **59.** False; a dilation with a scale factor greater than 1 is an enlargement.

EXERCISES 1. rotation **3.** Neither; the figures do not have the same orientation.

5.

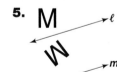

M is translated across line *m* twice the distance between ℓ and *m*.

7.

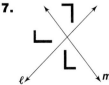

L is rotated clockwise about 180°.

9.

N is rotated clockwise about 160°.

11.

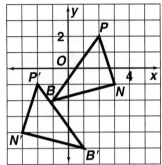

13.

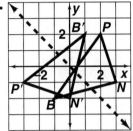

17. opp.; glide reflection **19.** same; rotation
25. rotation; 180° about the point $\left(0, \frac{1}{2}\right)$ **29.** Yes; a rotation of *x*° followed by a rotation of *y*° is equivalent to a rotation of (*x* + *y*)°. **31.** 60°
33. $51\frac{3}{7}°$ **35.** rotation; center *C*, ∠of rotation 180°
37. translation; (*x*, *y*) → (*x* − 9, *y*)

EXERCISES 1–3. Answers may vary. Samples are given. **1.** yes; translation; two ⊥ rectangles **3.** yes; translation; four rectangles in a square shape **5.** yes **7.** no **11.** rotational, reflectional, glide reflectional, and translational **13.** rotational, reflectional, glide reflectional, and translational

15.

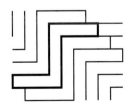

17.

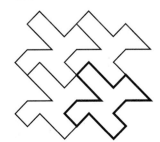

19.

21.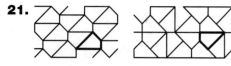

23. no **25.** yes;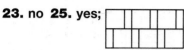

27. reflectional, rotational, glide reflectional, translational

Chapter Review   pp. 523–525

1. F **2.** D **3.** E or B **4.** G **5.** B **6.** C **7.** A **8.** $A'(7, 12)$, $B'(8, 6)$, $C'(3, 5)$ **9.** $R'(-4, 3)$, $S'(-6, 6)$, $T'(-10, 8)$ **10.** $(x, y) \rightarrow (x - 2, y + 1)$
11. $(x, y) \rightarrow (x + 11, y - 4)$

12. 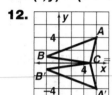 **13.**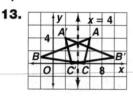

14. 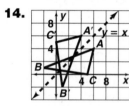 **15.**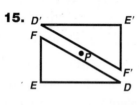

16. 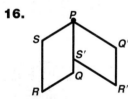 **17.**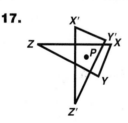

18. $(-2, 5)$ **19.** $(-3, 0)$ **20.** $(-1, -4)$ **21.** $(0, 7)$
22. $(8, -2)$ **23.** reflectional **24.** rotational; 72°
25. $A'(0, 12)$ **26.** $B'(-1, 3)$ **27.** $C'(15, -20)$
28. $M'(-15, 20)$, $A'(-30, -5)$, $T'(0, 0)$, $H'(15, 10)$
29. $F'(-2, 0)$, $U'\left(\frac{5}{2}, 0\right)$, $N'\left(-1, -\frac{5}{2}\right)$ **30.** same;
rotation **31.** opposite; reflection **32.** same;
translation **33.** opposite; glide reflection
34. $T'(-4, -9)$, $A'(0, -5)$, $M'(-1, -10)$

35.a. 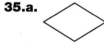 **b.** rotational, point, reflectional, translational, glide reflectional

36.a. and ◯ **b.** rotational, point, reflectional, translational, glide reflectional

Chapter 10

Lesson 10-1   pp. 536–539

EXERCISES 1. 240 cm^2 **3.** 20.3 m^2 **5.** 0.24
7. 14 m^2 **9.** 3 ft^2 **11.** 4 in. **13.** 14 cm **15.** The
area does not change; the height and base AB do
not change. **17.** 15 units^2 **19.** 6 units^2

25.a. 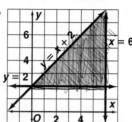 **b.** 18 units^2

27.b. 6 units^2 **29.a.** Blank grid; area is 84 units^2
while figures are 36 units^2. **b.** No; the figures
have the same area. **31.** 28 units^2 **33.** 88 units^2
35. 525 cm^2

Lesson 10-2   pp. 542–545

EXERCISES 1. 472 in.^{2} **3.** 108 ft^2 **5.** 150 cm^2
7. about 43,290 mi^2 **9.** 72 m^2 **11.** 80 in.^{2}
13. 24 ft^2 **15.** 96 in.^{2} **17.** 20 in.^{2} **21.** 19.5 cm^2
23. 49.9 ft^2 **25.** 18 units^2 **27.** 15 units^2
31. $32\sqrt{3}$ m^2 **33.a.** $A = \frac{1}{2}b_1 h$; $A = \frac{1}{2}b_2 h$; **b.** The
area of the trapezoid is the sum of the areas of the
triangles, so $A = \frac{1}{2}b_1 h + \frac{1}{2}b_2 h = \frac{1}{2}h(b_1 + b_2)$.

Lesson 10-3   pp. 548–551

EXERCISES 1. $m\angle 1 = 120$; $m\angle 2 = 60$;
$m\angle 3 = 30$ **3.** $m\angle 7 = 60$; $m\angle 8 = 30$; $m\angle 9 = 60$
5. 2851.8 ft^2 **7.** 2475 in.^{2} **11.** 27.7 in.^{2} **13.** 210 in.^{2}
15. $384\sqrt{3}$ in.^{2} **17.** $75\sqrt{3}$ m^2 **19.a.** 72 **b.** 54
21.a. 40 **b.** 70 **23.** 73 cm^2 **25a.** 9.1 in. **b.** 6 in.
c. 3.7 in. **d.** Answers may vary. Sample: About 4.6
in.; the length of a side of a pentagon should be
between 3.7 in. and 6 in. **27.** The apothem is one
leg of a rt $\triangle$, and the radius is the hypotenuse.
29. $600\sqrt{3}$ m^2 **31.** 128 cm^2 **33.** $900\sqrt{3}$ m^2;
1558.8 m^2

Lesson 10-4   pp. 555–558

EXERCISES 1. 1 : 2; 1 : 4 **3.** 2 : 3; 4:9 **5.** 24 in.^{2}
7. 59 ft^2 **9.** $384 **11.** 1 : 2; 1 : 2 **13.** 7 : 3; 7 : 3
17. 3 : 1; 9 : 1 **19.** 2 : 3; 4 : 9 **23.** While the ratio
of lengths is 2 : 1, the ratio of the areas is 4 : 1.
25. 252 m^2 **27.** $x = 2\sqrt{2}$ cm, $y = 3\sqrt{2}$ cm
29. $x = \frac{8\sqrt{3}}{3}$ cm, $y = 4\sqrt{3}$ cm **35.a.** $\frac{5}{2}$ **b.** $\frac{25}{4}$

37.a. $\frac{2}{1}$ **b.** $\frac{4}{1}$ **39.** Answers may vary.
a. Sample:

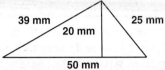

b. 114 mm; 475 mm²; **c.** 456 yd; 7600 yd²

Lesson 10-5   pp. 561–564

EXERCISES 1. 173.8 cm² **3.** 259.8 m² **7.** 47.0 in.²
9. 8 ft² **11.** 27.7 m² **13.** 7554.0 m² **17.** 5523 yd²
21. 20.8 m; 20.8 m² **23.** 17.6 ft; 21.4 ft² **27.** D
29. (area of pent. A) ≈ 1.53 · (area of pent. B)
31. (area of oct. B) ≈ 1.17 · (area of oct. A)
33. 5.0 ft²

Lesson 10-6   pp.569–573

EXERCISES 1. 18 **3.** 40 **9–11.** Answers may
vary. Samples are given. **9.** $\overset{\frown}{ED}$ **11.** $\overset{\frown}{BFE}$
15. 128 **17.** 218 **27.** 20π cm **29.** 8.4π m **33.** 25 in.
35. 8π ft **37.** 33π in. **43.** 180 **45.** 55 **49.a.** 6
b. 30 **c.** 120 **51.** 100 **53.** 40 **55.** 50 in.
57. B. **59.** 105 ft **61.** (2.5, 5) **63.** 5.125π ft
65. 3π m **69.** 12.6 units

Lesson 10-7   pp. 577–580

EXERCISES 1. 9π m² **3.** 0.7225π ft² **5.** about
86,394 ft² **7.** 40.5π yd² **9.** $\frac{169\pi}{6}$ m² **13.** $\frac{25\pi}{4}$ m²
15. 24π in.² **17.** 22.1 cm² **19.** 3.3 m²
21. 120.4 cm² **23.** (54π + 20.25$\sqrt{3}$) cm²
25. (4 − π) ft² **31.** 15.7 in.²

Lesson 10-8   pp. 584–587

EXERCISES 1. $\frac{1}{2}$ **3.** $\frac{3}{5}$ **7.** $\frac{2}{5}$ or 40%

         9. $\frac{4}{15}$ or about 27%

11. $\frac{1}{3}$ or about 33% **15.** $\frac{1}{4}$ or 25% **17.** $\frac{2}{5}$ or 40%
21. 4% **23.** $\frac{\pi}{4}$ **25.** $\frac{\pi}{4}$ **27a.** 14 prizes **b.** $110
29. 36 s **31.a.**
If it starts after 45 min, you cannot erase 15 min
of a 60 min tape. **b.** $\frac{7}{45}$ or about 16% **33.** $\frac{3}{10}$ **35.** 0
41. about 36% **43.** about 46%

Chapter Review   pp. 589–591

1. base **2.** sector **3.** diameter **4.** apothem
5. adjacent arcs **6.** 10 m² **7.** 90 in.² **8.** 33 ft²
9. 96$\sqrt{3}$ mm² **10.** 96 ft² **11.** 117 cm²

12. 20.8 in.²   **13.** 128 mm²

14. 127.3 cm²

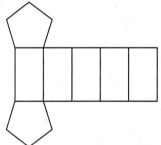

15. 4 : 9 **16.** 9 : 4
17. 1 : 4 **18.** 2$\sqrt{2}$: 5
19. 73.5 ft² **20.** 232.5 cm²
21. 124.7 in.² **22.** 8 m²
23. 100.8 cm² **24.** 88.4 ft²
25. 70.4 m² **26.** 30 **27.** 120
28. 330 **29.** 120 **30.** $\frac{22}{9}\pi$ in.

31. π mm **32.** 18.3 m² **33.** 41.0 cm² **34.** $\frac{1}{2}$ or 50%
35. $\frac{3}{8}$ or 37.5% **36.** $\frac{1}{6}$ or about 16.7%

Chapter 11

Lesson 11-1   pp. 601–604

1. 4, 6, 4; *M, N, O, P*; $\overline{MN}$, $\overline{MP}$, $\overline{MO}$, $\overline{NO}$, $\overline{NP}$, $\overline{OP}$;
△*MNP*, △*MOP*, △*MNO*, △*PNO* **5.** 12 **7.** 8 **9.** 9
11. 7 + 10 = 15 + 2;
13. two concentric
circles
15. rectangle

17. rectangle   **19.** rectangle

21. rectangle   **23.** triangle

25. circle **27.** cone **29.** cylinder attached to a
cone **31.** 18 + 32 = 48 + 2 **33.** 6 + 4 = 9 + 1
37. A

Extension   p. 606

1. two-point **3.** one-point

7.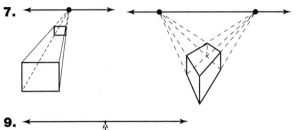

9.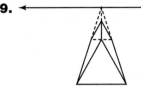

15. Answer may vary. Sample: The horizontal lines appear to be curved; the slanted lines that would meet at the vanishing pt. create a cylinder effect.

Algebra 1 Review   p. 607

1. $r = \frac{C}{2\pi}$ **3.** $r = \sqrt{\frac{A}{\pi}}$ **5.** $y = x \tan A; x = \frac{y}{\tan A}$

7. $C = 2\sqrt{\pi A}$ **9.** $a = \frac{\sqrt{6A}\sqrt{3}}{6}$ or $\frac{\sqrt{6A} \cdot \sqrt[4]{3}}{6}$

Lesson 11-2   pp. 611–614

EXERCISES 1. 1726 cm^2

3. $(80 + 32\sqrt{2})$ in.^{2} or about 125.3 in.^{2}

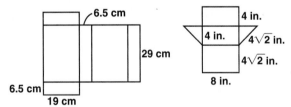

5. 120 ft^2; 220 ft^2 **7.** 880 cm^2; 1121 cm^2
9. 16.5π cm^2 **11.** 36.8 cm^2 **13.** 107 in.^{2}
15. 1407 cm^2 **17.** 150 cm^2 **19.** 4080 mm^2
23. 47.5 in.^{2} **25.a.** 7 units **b.** 196π units^2
27. cylinder of radius 4 and height 2; 48π units^2
29. cylinder of radius 2 and height 4; 24π units^2

Lesson 11-3   pp. 620–623

EXERCISES 1. 408 in.^{2} **3.** 179 in.^{2} **5.** 354 cm^2
7. 834,308 ft^2 **9.** 33π ft^2 **11.** 1044 in.^{2} **13.** 47 cm^2
17. 8 ft **19.** 62 cm^2 **21.** 4 in. **23.** 1580.6 ft^2 **27.** A
29. 471 ft^2 **31.** L.A. = 30 in.^{2}, $h \cong 4.8$ in.; $\ell = 5$ in.
33. $s = 2$ ft, $h = 4.9$ ft, S.A. = 24 ft^2 **35.** 21.2 cm
37. cone with $r = 4$ and $h = 3$; 36π

Lesson 11-4   pp. 627–630

EXERCISES 1. 216 ft^3 **3.** 180 m^3 **5.** about
280.6 cm^3 **7.** 720 mm^3 **9.** 288π in.^{3}, 904.8 in.^{3}
11. 37.5π m^3, 117.8 in.^{3} **13.** 3445 in.^{3} **15.** 501 in.^{3}
17. $\frac{26}{9}$ cm **19.** 6 ft **21.** 28–42 pots **25.** 80 units^3
27. 3 cm **29.** Bulk; cost of bags ≈ $1167, cost of
bulk ≈ $1161. **31.** cylinder with $r = 4$ and $h = 2$;
32π units^3 **33.** cylinder with $r = 5$ and $h = 2$, and
a hole of radius 1; 48π units^3 **35.** 140.6 in.^{2}

Lesson 11-5   pp. 634–637

EXERCISES 1. about 233,333 ft^3 **3.** 1296 in.^{3}
5. about 443.7 cm^3 **7.** 2048 m^3 **9.** about
3714.5 mm^3 **11.** $\frac{16}{3}\pi$ ft^3; 17 ft^3 **13.** 36.75π in.^{3};
115 in.^{3} **15.** ≈ 4.7 cm^3 **17.** 312 cm^3 **19.** They
are equal; both volumes are $\frac{1}{3}\pi r^2 h$. **21.** 6
23. $3\sqrt{2}$ **25.** B **27.a.** 120π ft^3 **b.** 60π ft^3
c. 240π ft^3 **29.** cone with $r = 4$ and $h = 3$; 16π

Lesson 11-6   pp. 640–644

EXERCISES 1. 900π m^2 **3.** 1024π mm^2
5. 4624π mm^2 **7.** $\frac{121}{16}\pi$ in.^{2} **9.** 232 in.^{2} **11.** 154 in.^{2}
13. 288π cm^3; 905 cm^3 **15.** $\frac{2048}{3}\pi$ cm^3; 2145 cm^3
19. 1006 m^2 **21.** S.A. ≈ 108 cm^2, V ≈ 108 cm^3
27. 1.7 lb **29.** $\frac{4}{3}\pi$ m^3 **31.** $\frac{9}{2}\pi$ ft^3 **39.** Answers may
vary. Sample: sphere radius 3 in.; cylinder radius
3 in., height 4 in. **41.** 26π cm^2; $\frac{62}{3}\pi$ cm^3

Lesson 11-7   pp. 648–651

EXERCISES 1. no **3.** yes; 2 : 3 **7.** 5 : 6 **9.** 3 : 4
11. 240 in.^{3} **13.** 24 ft^3 **15.** 112 m^2
17. 6000 toothpicks **19.a.** It is 64 times the
smaller prism. **b.** It is 64 times the smaller prism.
23. about 1000 cm^3 **27.a.** 3 : 1 **b.** 9 : 1
29. 864 in.^{3} **31.** 9 : 25; 27 : 125 **33.** 5 : 8; 25 : 64

Chapter Review   pp. 653–655

1. sphere **2.** altitude **3.** pyramid **4.** cross
section **5.** right

6.   **7.**

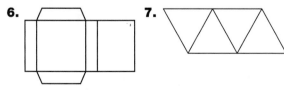

8.   **9.** 8 **10.** 8

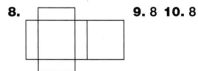

11. **12.** 36 cm², 12 cm³ **13.** 66π m²; 72π m³ **14.** 208 in.²; 192 in.³
15. 186.0 ft²; 187.6 ft³
16. 576 in.²; 512 in.³
17. 50.3 in.²; 23.7 in.³
18. 391.6 in.²; 443.7 in.³ **19.** 282.8 yd²; 410.5 yd³
20. 36.5 in.²; 13.9 in.³ **21.** $B = \frac{S.A. - L.A.}{2}$.
22. $r = -\frac{1}{2}h + \sqrt{\frac{1}{4}h^2 + \frac{S.A.}{2\pi}}$ **23.** 314.2 in.²; 523.6 in.³
24. 153.9 cm²; 179.6 cm³ **25.** 50.3 ft²; 33.5 ft³
26. 8.0 ft²; 2.1 ft³ **27.** 8.6 in.³ **28.** Answers may vary. Sample:

 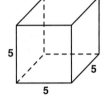

29. 27 : 64 **30.** 64 : 27

Chapter 12

Lesson 12-1   pp. 665–668

EXERCISES 1. 120 **3.** 30 **7.** 20.0 in. **9.** Yes; 2.5² + 6² = 6.5² **11.** 78 cm **13.** 13 **15.** 8 in. **17.** 35.8 km **19.** 113.1 km **25.** 35 **29.a.** ⊥; **b.** *LK*; **c.** SAS; **d.** CPCTC; **e.** tangent; **f.** false

Lesson 12-2   pp. 673–676

EXERCISES 1. $\overset{\frown}{BC} \cong \overset{\frown}{YZ}$; $\overline{BC} \cong \overline{YZ}$ **3.** 14 **5.** 7
9. Answers may vary. Samples are given. **a.** $\overline{CE}$
b. $\overline{DE}$ **c.** ∠CEB **d.** ∠DEA **11.** 6 **13.** 8.9 **17.** 108
19. about 123.9 **21.** 12 cm **31.** 10 cm **33.** C

Lesson 12-3   pp. 681–685

EXERCISES 1. ∠ACB; $\overset{\frown}{AB}$ **3.** ∠MPN; $\overset{\frown}{MN}$ **5.** 58
7. a = 218; b = 109 **11.** x = 36; y = 36 **13.** a = 50;
b = 90; c = 90 **15.** w = 123 **17.** e = 65; f = 130
21.a. 96 **b.** 55 **c.** 77 **d.** 154 **23.** B **27.a.** $77\frac{1}{7}$
b. 36 **29.** about 7.1 cm by 7.1 cm **31.** about 7.1 cm legs, and a 10 cm base

Lesson 12-4   pp. 691–693

EXERCISES 1. 46 **3.** x = 60; y = 70 **7.** 160° **9.** 15
11. 13.2 **15.** about 270.8 ft **17.** 360 − x
19. 180 − y **21.** 16.7 **23.** x ≈ 10.7; y = 10
25. x ≈ 10.9; y ≈ 2.3 **27.** 95, 104, 86, 75
31. Given: a circle with secant segments $\overline{XV}$ and $\overline{ZV}$; Prove: $XV \cdot WV = ZV \cdot YV$.

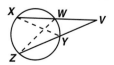

1. Construct $\overline{XY}$ and $\overline{ZW}$. **2.** ∠XVY ≅ ∠ZVW
(Reflexive Prop. of ≅) **3.** ∠VXY ≅ ∠WZV
(2 inscribed ∡ that intercept the same arc are ≅.)
4. △XVY ~ △ZVW (AA~) **5.** $\frac{XV}{ZV} = \frac{YV}{WV}$ (In similar
figures, corr. sides are proport.) **6.** $XV \cdot WV = YV \cdot ZV$ (Prop. of Proport.)

Extension   p. 694

1. *ACD*; *HGA*; *EFA* **3.** The radius of the circle is 1, which is the denominator in each of the ratios.
5. On the unit circle, secant $A = \frac{hyp.}{1}$ = hyp. =
length of $\overline{DA}$, the secant segment.
7. $(\text{tangent } A)^2 = \left(\frac{DC}{CA}\right)^2 = \left(\frac{EB}{BA}\right)^2 =$
$\frac{(EB)^2}{(BA)^2} = \frac{(EA)^2 - (BA)^2}{(BA)^2} = \frac{(EA)^2}{(BA)^2} - 1 = (\text{secant } A)^2 - 1$
9. cotangent $A = \frac{HG}{GA} = \frac{CA}{CD} = \frac{1}{\frac{CD}{CA}} = \frac{1}{\text{tangent } A}$
11. secant $A = \frac{EA}{AB} = \frac{EA}{EF} = \frac{1}{\frac{EF}{EA}} = \frac{1}{\text{cosine } A}$

13. Consider the possible lengths of each segment in the diagrams: cosecant *A* and secant *A* can be any value > 1; cotangent *A* can be any value > 0.

Lesson 12-5   pp. 697–700

EXERCISES 1. $(x - 2)^2 + (y + 8)^2 = 81$
3. $(x - 0.2)^2 + (y - 1.1)^2 = 0.16$ **11.** $(x - 1)^2 + (y - 2)^2 = 17$ **13.** $(x + 10)^2 + (y + 5)^2 = 125$
17. center: (3, −8); radius: 10

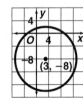

23. $(x - 4)^2 + (y + 4)^2 = 4$
25. position: (5, 7); range: 9 units
27. $x^2 + y^2 = 4$ **29.** $x^2 + (y - 3)^2 = 4$
33. $(x - 4)^2 + (y - 3)^2 = 25$

35. $(x - 3)^2 + (y - 3)^2 = 8$ **39.** $x^2 + y^2 = 1$
43. No; the x and y terms are not squared.
45. circumference: 16π; area: 64π

49. 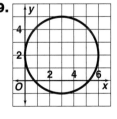 **53.** (3, 2); (2, 3)
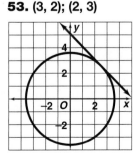

EXERCISES 1. a circle of radius 4 cm with center X

3. two distinct lines ∥ to $\overset{\leftrightarrow}{LM}$ and 3 mm from $\overset{\leftrightarrow}{LM}$

9. the single pt. L **11.** the single pt. N

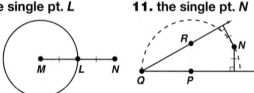

15. an endless cylinder with radius 4 cm and center-line $\overset{\leftrightarrow}{DE}$

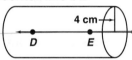

17. an endless cylinder with radius 5 mm and center-ray $\overset{\rightarrow}{PQ}$ and a hemisphere of radius 5 mm centered at P

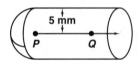

19. the set of all points 2 units from the origin
23. $y = x$ **25.** $y = -x + 3$

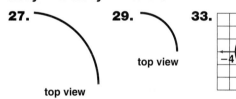

27.   **29.**   **33.**

top view   top view

35.   **41.** a circle

1. tangent to **2.** chord **3.** inscribed in **4.** inscribed angle **5.** locus **6.** 57 in. **7.** 72 mm **8.** 9.8 cm **9.** 4.3 **10.** 19.5 **11.** 6.4 **12.** 4.5 **13.** $a = 40$; $b = 140$; $c = 90$ **14.** $a = 118$; $b = 49$; $= 144$; $d = 98$ **15.** $a = 90$; $b = 90$; $c = 70$; $d = 65$ **16.** $a = 95$; $b = 85$ **17.** 37 **18.** $x = 57$; $y = 44.5$; $z = 129$; $v = 51$ **19.** 15.2 **20.** 4 **21.** 13.1

22. $(x - 2)^2 + (y - 5)^2 = 12.25$ **23.** $(x + 3)^2 + (y - 1)^2 = 5$ **24.** $(x - 9)^2 + (y + 4)^2 = 16$ **25.** $x^2 + (y - 1)^2 = 80$ **26.** $(x + 2)^2 + (y - 3)^2 = 85$ **27.** $(x - 10)^2 + (y - 7)^2 = 468$ **28.** center $= (0, 8)$; $r = 7$ **29.** center $= (5, -9)$; $r = 2\sqrt{10}$ **30.** center $= (-1, 0)$; $r = 3$ **31.** a circle of **32.** the ⊥ bis. of the radius 3 cm   segment between the pts.

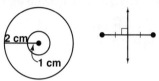

33. a cylinder with hemispherical ends

Extra Practice

Chapter 1

1. 37, 42 **3.** 8, $\frac{8}{5}$ **9.** cube **11.** cylinder **19.** false **21.** true **31.** $\overline{CG}$, $\angle BCA$ **33.** 8 **43.** 7.3; $\left(\frac{3}{2}, 3\right)$ **45.** 10.4; $\left(-\frac{13}{2}, -6\right)$ **49.** 42 in., 98 in.^{2} **51.** 3π m, $\frac{9}{4}\pi$ m^2

Chapter 2

1. hyp: you can predict the future; concl: you can control the future **3.** hyp: lines k and m are skew; concl: lines k and m are not perpendicular **5.** If a number is one, then it is the smallest positive square. If a number is the smallest positive square, then it is one. A number is one if and only if it is the smallest positive square. **11.** yes **13.** yes **17.** Shauna will gain a promotion. **19.** Brendan will be selected for the chorus. **27.** 15

Chapter 3

1. $m\angle 1 = 134$; Same-Side Int. $\angle$ Thm. $m\angle 2 = 46$; Alt. Int. $\angle$ Thm. **3.** $m\angle 1 = 58$; Alt. Int. $\angle$ Thm. $m\angle 2 = 122$; Same-Side Int. $\angle$ Thm. **7.** $r \parallel s$, Conv. of Corr. $\angle$ Post. **9.** none **15.** obtuse; scalene **17.** acute; isosceles **21.** 100 **23.** $x = 110$; $y = 102$; $z = 82$ **27.** $y + 1 = (x + 1)$ or $y - 1 = (x - 1)$ **29.** $y = -\frac{1}{5}(x - 5)$ or $y - 2 = -\frac{1}{5}(x + 5)$ **35.** neither; not same slope and $m_1 \cdot m_2 \neq -1$ **37.** ⊥; $m_1 \cdot m_2 = -1$

Chapter 4

1. $\angle G$ **3.** $\angle T$ **13.** $\angle T \cong \angle S$, $\angle Y \cong \angle W$ and included sides $\overline{TY} \cong \overline{SW}$; ASA **15.** not possible
21. $\overleftrightarrow{OL} \parallel \overleftrightarrow{MN}$, so $\angle OLN \cong \angle MNL$. $\overline{LN} \cong \overline{LN}$ by the Reflexive Prop. of $\cong$. Since $\overline{LO} \cong \overline{MN}$, $\triangle MLN \cong \triangle ONL$ by SAS, and $\angle MLN \cong \angle ONL$ by CPCTC.
23. $\overline{BI} \cong \overline{BI}$ by the Reflexive Prop. of $\cong$. Since $\angle MBI \cong \angle RIB$ and $\angle MIB \cong \angle RBI$, $\triangle MBI \cong \triangle RIB$ by ASA, and $\overline{MB} \cong \overline{RI}$ by CPCTC.
27. $x = 57$; $y = 66$ **31.** $\triangle ARO \cong \triangle RAF$; HL
33. $\triangle AON \cong \triangle MOP$; AAS

Chapter 5

1. $\frac{25}{7}$ **3.** 7 **7.** 7 **9.** $\frac{15}{2}$ **13.** (1, 5) **15.** (0, 0)
25.a. If figures are not similar, then their side lengths are not prop. **b.** If their side lengths are not prop., then figures are not similar.
29. Assume $\triangle ABC$ is not a right $\triangle$.
31. Assume lines ℓ and m are $\parallel$.
33. $\overline{RN}, \overline{RS}, \overline{NS}$ **35.** $\overline{MQ}, \overline{QD}, \overline{MD}$

Chapter 6

1.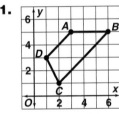
trapezoid

5. $x = 12$, $y = 84$ **7.** $x = 8$, $y = 25$ **11.** yes **13.** no
17. square; $m\angle 1 = 45$, $m\angle 2 = 45$ **19.** $\square$; $m\angle 1 = 45$, $m\angle 2 = 45$, $m\angle 3 = 80$, $m\angle 4 = 55$ **23.** $m\angle 1 = 110$, $m\angle 2 = 25$
25. $m\angle 1 = 110$, $m\angle 2 = 70$ **29.** $D(0, b)$; $S(-a, 0)$
31. $D(0, 2a)$; $S(0, -2a)$

Chapter 7

1. 10 **3.** 2 **7.** $x = \frac{80}{3}$; $y = 6$; $z = \frac{16}{3}$ **9.** $x = 30$; $y = 4$ **13.** Yes; $\triangle XZY \sim \triangle EWN$ by AA$\sim$.
19. $x = \sqrt{5}$; $y = 2$; $z = 2\sqrt{5}$ **21.** $x = 65$; $y = 60$; $z = 156$

Chapter 8

1. 15 **3.** $3\sqrt{5}$ **11.** 29 **13.** 9.4 **23.** 139 ft
27.a. $\langle -49, 142 \rangle$, $\langle 38, 47 \rangle$ **b.** $\langle -11, 189 \rangle$
29.a. $\langle -54, 72 \rangle$, $\langle -95, -33 \rangle$ **b.** $\langle -149, 39 \rangle$

Chapter 9

1. E **3.** C **7.** $A'(-5, 9)$, $B'(-3, 3)$, $C'(-1, 10)$
9. $P'(-15, -11)$, $Q'(-11, -6)$, $R'(-4, 1)$

11.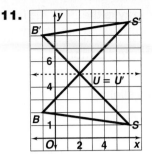

21. 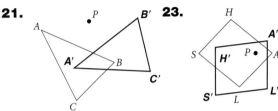 **23.**

27.a. line, rotation, point **b.** yes **29.a.** line
b. yes **33.** dilation **35.** glide reflection

Chapter 10

1. 15 ft; 10.825 ft^2 **3.** 50 ft; 143 ft^2 **7.** 72 cm^2
9. $\frac{25}{4}\sqrt{3}$ mm^2 **15.** 5 : 8; 25 : 64 **17.** 5 : 16; 25 : 256
21. 78.0 in.^{2} **23.** 20 m^2

Chapter 11

1. equilateral triangle; $7 + 10 = 15 + 2$
3. equilateral triangle; $7 + 7 = 12 + 2$
7. 28π cm^2; 36π cm^2 **9.** $108\sqrt{3}$ in.^{2}; $144\sqrt{3}$ in.^{2}
17. 175 mm^3 **19.** 45π in.^{3} **27.** $\frac{500\pi}{3}$ cm^3, 524 cm^3; 100π cm^2, 314 cm^2 **29.** $\frac{256\pi}{3}$ in.^{3}, 268 in.^{3}; 64π in.^{2}, 201 in.^{2} **37.** 4 : 9; 8 : 27 **39.** 3 : 4; 9 : 16

Chapter 12

1. 65 **3.** 6 **7.** 5.2 **9.** 20 **17.** $x = 193$; $y = 60.5$
19. ≈ 10.4 **27.** $x^2 + y^2 = 16$
29. $(x - 9)^2 + (y + 3)^2 = 49$

35. two rays $\parallel$ to and 2 cm from $\overrightarrow{AB}$, and the semicircle of radius 2 with center A, opp. pt. B.
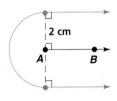

Skills Handbook

p. 740 **1.** 10 handshakes **3.** 16 regions

p. 741 **1.** 6, 8, 10 **3.** any pos. number less than 4 **5.** –2 **7.** 54

p. 742 **1.** 165 toothpicks **3.** 3280 triangles

p. 743 **1.** 28 posts **3.** 84 paths

p. 744 1. J.T. is the parrot; Izzy is the dog; Arf is the goldfish; Blinky is the hamster. **3.** 14 students

p. 745 1. Route 90 east, Route 128 north, Route 4 north **3.** 3 tickets

p. 746 1. Answers may vary slightly. Sample: 35 mm; 46 mm

3.

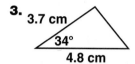

3.7 cm
34°
4.8 cm

p. 747 1. 0.4 **3.** 600

p. 748 1. $23\frac{1}{2}$ ft to $24\frac{1}{2}$ ft **3.** $339\frac{1}{2}$ mL to $340\frac{1}{2}$ mL

p. 749 1. 18% **3.** 8% **9.** ≈7% **11.** ≈2%

p. 750 1. 353.6; 301; no mode **3.** $40,533; $28,150; $18,000

p. 751 1.

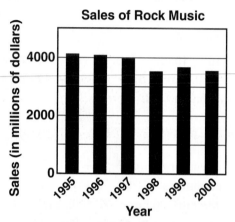

Sales of Rock Music

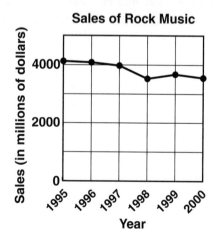

Sales of Rock Music

3. 8 A.M.–2 P.M.

p. 752 1.a. 16.6 **b.** 15.1 **c.** 19.4

3.

Pages in Books

185   205   258 297   356

p. 753 1. 121 **3.** 196 **9.** 10 **11.** 8.6 **17.** ±7 **19.** ±1
p. 754 1. −50 **3.** 15 **19.** $2\ell + 2w$ **21.** $-4x^2 + 8x$
p. 755 1. $3\sqrt{3}$ **3.** $5\sqrt{6}$ **11.** $5\sqrt{10}$ **13.** $2\sqrt{39}$

p. 756 1. $\frac{5}{3}$ **3.** $\frac{2}{3}$ **19.** $\frac{5}{12}$ **21.** $\frac{5}{13}$ **25.** $\frac{2}{\pi}$ **27.** $\frac{1}{9}$

p. 757 1. 8 **3.** 16 **13.** −16 or 16 **15.** −20 or 20
p. 758 1. 5 **3.** 3 **13.** $35 - 2x = 9$; 13
15. $\$9.95 + \$0.035m = \$12.75$; 80 min

p. 759 1. $w = \frac{P - 2\ell}{2}$ **3.** $r = \frac{1}{2}\sqrt{\frac{S}{\pi}} = \frac{\sqrt{\pi S}}{2\pi}$

p. 760 1. (4, −1) **3.** (4, 1)
p. 761 1. 0.5 **3.** 0.27 **9.** 8.4 **11.** 7.2
17–19. Answers may vary. Samples are given.
17. 7 **19.** 45

p. 762 1. $\frac{1}{2}$ **3.** $\frac{2}{7}$ **7.** $\frac{1}{12}$ **9.** $\frac{1}{2}$ **13.** $\frac{3}{8}$ **15.** $\frac{7}{8}$

Index

A

AA~ (Angle-Angle) Similarity Postulate, 382–383, 384–385, 408, 433

AAA (Angle-Angle-Angle), 220

AAS (Angle-Angle-Side) Theorem, 214–215, 250

Absolute value, 757

ACT preparation. *See* Test Prep

Active Math Online, 45, 54, 61, 80, 89, 149, 160, 182, 207, 230, 236, 266, 275, 292, 307, 323, 331, 374, 393, 419, 434, 440, 454, 479, 535, 555, 583, 625, 633, 670, 680

Activity. *See also* Technology

Active Math Online, 45, 54, 61, 80, 89, 149, 160, 182, 207, 230, 236, 266, 275, 292, 307, 323, 331, 374, 393, 419, 434, 440, 454, 479, 535, 555, 583, 625, 633, 670, 680

Exploring Inscribed Angles, 678

Exploring the Area of a Circle, 575

Finding Perimeter and Area, 61

Finding Volume, 624, 631

Hands-On, 16, 61, 259, 272, 391, 483, 517, 575, 624, 631, 678

How Many Lines Can You Draw?, 16

Making Designs with Rotations, 483

Making Tessellations, 517

Midsegments of Triangles, 259

Paper Folding Bisectors, 272

Perimeters and Areas of Similar Rectangles, 553

Similarity in Right Triangles, 391

Sum of Polygon Angle Measures, 159

Tangent Ratios, 432

Triangles with Two Pairs of Congruent Angles, 382

Writing a Definition, 88–89

Activity Lab

Accuracy and Precision in Measuring, 412–413

Angle Dynamics, 146

Applying Parallel Lines, 194–195

Applying Reasoning, 122–123

Applying the Pythagorean Theorem, 466–467

Applying Theorems about Triangles, 302–303

Applying Translations and Rotations, 528–529

Applying Volume, 658–659

Building Congruent Triangles, 204

Colossal Task, 658–659

Comparing Perimeters and Areas, 69

Compass Designs, 43

Data Analysis, 76–77, 254–255, 362–363, 412–413, 594–595, 714–715

Delicious Side of Division, 122–123

Diagonals of Parallelograms, 328

Distance in the Coordinate Plane, 52

Dorling Kindersley, 122–123, 194–195, 302–303, 466–467, 528–529, 658–659

Exploring AAA and SSA, 220

Exploring Area and Circumference, 581

Exploring Chords and Secants, 686

Exploring Constructions, 51

Exploring Proportions in Triangles, 397

Exploring Spherical Geometry, 154–155

Exploring Surface Area, 616

Exploring Trigonometric Ratios, 438

Exterior Angles of Polygons, 156

Frieze Patterns, 514

Geo-Models, 320

graphing calculator, 69, 172–173, 187

Hands-On, 43, 52, 102, 146, 204, 227, 320, 416, 431, 444, 477, 490–491, 514, 532–533, 669

How Far Can You See?, 466–467

How'd They Do That?, 528–529

Interpreting Data, 362–363

Investigating Midsegments, 258

Kaleidoscopes, 513

Linear Regression, 76–77

Market Research, 714–715

Mathematical Systems, 101

Measuring from Afar, 444

Misleading Graphs, 594–595

Paper Folding and Reflections, 477

Paper Folding With Circles, 669

Paper-Folding Conjectures, 227

Paper-Folding Constructions, 102

Parallel Lines and Related Angles, 126

Point of Balance, 302–303

Probability, 254–255

Pythagorean Theorem, 416

Quadrilaterals in Quadrilaterals, 342

The Science of Reflection, 194–195

software, 51, 126, 156, 220, 258, 271, 328, 342, 397, 438, 513, 581

Solving Linear Equations with Graphs and Tables, 172–173

Special Segments in Triangles, 271

Staff and Stadiascope, 431

Technology, 51, 69, 126, 156, 172–173, 187, 220, 258, 271, 328, 342, 397, 438, 513, 581, 616, 645, 686

Tracing Paper Transformations, 490–491

Transforming to Find Area, 532–533

Using Tables and Lists, 187

Acute angle, 37, 229

Acute triangle, 148

Addition

of angles, 38

of arcs, 567–568

of segments, 32

of vectors, 454–455, 463

Addition Property

of Equality, 103, 104, 118

of Inequality, 288

Adjacent angles, 38, 116

Adjacent arcs, 567, 591

Algebra. *See also* Algebra 1 Review; Equations

angle measures, 130

coordinate plane. *See* Coordinate plane

dimensional analysis, 574

examples, 32, 33, 46, 55, 63, 88, 96, 104, 136, 148, 149, 160, 167, 168, 176, 229, 234, 292, 308, 313, 314, 349, 367, 392, 393, 418, 427, 672, 688, 690, 745

exercises, 2, 9, 33, 34, 41, 47, 48, 53, 56, 59, 67, 72, 74, 78, 83, 85, 90, 97, 99, 105, 106, 107, 109, 113, 114, 115, 132, 137, 138, 139, 144, 150, 151, 153, 161, 162, 163, 169, 171, 178, 180, 192, 201, 202, 230, 232, 238, 252, 263, 264, 267, 268, 270, 279, 288, 293, 294, 296, 300, 304, 309, 315, 316, 317, 324, 325, 327, 332, 334, 335, 339, 340, 341, 344, 346, 347, 360, 368, 369, 370, 376, 377, 378, 379, 386, 394, 395, 396, 400, 401, 402, 404, 410, 414, 420, 421, 428, 430, 448, 450, 464, 465, 502, 512, 520, 537, 544, 557, 558, 571, 586, 587, 613, 621, 630, 635, 651, 665, 666, 668, 673, 685, 691, 709, 710, 753, 754, 755, 756, 757, 758, 759, 760

inequalities, 288, 289–292, 299

linear equations, 30, 166–167, 168, 172–173, 174–177, 191, 234, 758, 760, 765

literal equations, 607, 759

matrices, 504, 524

proofs using coordinate geometry, 348–349

proportions, 366–368, 397, 398–400, 407

radicals, 390, 755

ratios, 366–368, 373–375, 756

reasoning in, 103–105

slope, 165, 174–177, 191, 765

systems of linear equations, 234, 760

Algebra 1 Review

Dimensional Analysis, 574

Literal Equations, 607

Quadratics, 354–355

Simplifying Radicals, 390

Slope, 165

Solving Inequalities, 288

Solving Linear Equations, 30

Solving Quadratic Equations, 372

Systems of Linear Equations, 234

Ali, Muhammad, 83

Allen, George, 83

Alternate exterior angle, 130, 136, 137

Alternate Exterior Angles Theorem, 130 converse of, 136, 137

Alternate interior angle, 127, 128, 129, 135, 189

Alternate Interior Angles Theorem, 128, 129, 135

Altitude
of cone, 619
of cylinder, 610
defined, 534
of parallelogram, 534
of prism, 608
of pyramid, 617
of trapezoid, 541
of triangle, 271, 275, 298

Analyzing data. *See* Data analysis

Anemometer, 704

Angelou, Maya, 281

Angle(s)
acute, 37, 229
addition of, 38
adjacent, 38, 116
alternate exterior, 130, 136, 137
alternate interior, 127, 128, 129, 135, 189
base, 228, 336, 358–359
bisector of, 46–47, 73, 266–267, 400
central, 566–567
classifying, 37
complementary, 38, 39, 229
congruent, 37, 45, 72, 110–112
consecutive, 312–313
constructing, 45
corresponding, 127, 128, 130, 134, 189
defined, 36, 72
of depression, 445–446, 462
dynamics of, 146
of elevation, 445–446, 462
exterior, 149–150, 156, 190, 290
formed by a tangent and a chord, 680–681
identifying, 127–128
of incidence, 385
inequalities involving angles of triangles, 289–290, 299
inscribed, 678–681, 708, 710
naming, 36
obtuse, 37
with parallel lines, 126
of parallelogram, 312–313, 358
of reflection, 385
remote interior, 147, 149, 190
of rhombus, 330, 332
right, 37
of rotation, 493
same-side exterior, 130, 136, 189
same-side interior, 127, 128, 135, 189
segment bisecting, 329
sides of, 36
straight, 37
sum in polygon, 159–160, 191
sum in triangle, 147–148
supplementary, 38, 39, 112
theorems about, 110–112. *See also* Theorems
vertex of, 36, 228
vertical, 38, 39, 110, 119

Angle Addition Postulate, 38, 104, 130

Angle-Angle-Angle (AAA), 220

Angle-Angle-Side (AAS) Theorem, 214–215, 250

Angle-Angle Similarity (AA~) Postulate, 382–383, 384–385, 408, 433

Angle bisector, 266–267, 400
constructing, 47
defined, 46, 73
finding angle measures, 46

Angle Bisector Theorem, 266, 267

Angle measures, 36–38, 46, 130, 159, 330, 546, 591, 680, 687–688

Angle pairs, 38, 39

Angle-Side-Angle (ASA) Postulate, 213–215, 250–251

Angle-Sum Theorems
Polygon, 159–160, 191
Triangle, 147–150, 154, 159, 291, 764

Answers
answering the question asked, 522
eliminating, 460
multiple correct, 588

Apothem, 546, 589

Applications. *See* Careers; Geometry at Work; Interdisciplinary Connections; Real-World Connections

Arc(s)
addition of, 567–568
adjacent, 567, 591
congruent, 569
defined, 44, 567
drawing, 43
identifying, 567
intercepted, 678–679, 708
major, 567, 591
measures of, 568, 591
minor, 567, 591
semicircle, 567

Arc length, 569, 591, 765

Arc Length Theorem, 569

Archimedes, 643

Area(s). *See also* Lateral area; Surface area
of circle, 62, 64, 73, 575–577, 765
comparing, 69
defined, 61, 73
finding, 63–64, 73
finding probability using, 583
of irregular shape, 64
of kite, 541, 589, 764
measuring, 62
of parallelogram, 533, 534–535, 589, 764
of pentagon, 559
of quadrilateral, 540–542
of rectangle, 63, 534, 553, 589, 764
of regular polygon, 546–547, 559–560, 581, 589, 764
of rhombus, 541, 542, 589, 764
of sector of circle, 576, 591, 765
of segment of circle, 577, 591
of similar figures, 553–555
of similar solids, 646–648
of square, 460, 607, 764
transforming to find, 532–533
of trapezoid, 540–541, 589, 764
of triangle, 535–536, 549, 560–561, 589, 764
trigonometry and, 559–561

Arrow notation, 471

ASA (Angle-Side-Angle) Postulate, 213–214, 250–251

Assessment. *See also* Instant Check System; Open-Ended; Test-Taking Strategies
Chapter Review, 71–73, 117–119, 189–191, 249–251, 297–299, 357–359, 407–409, 461–463, 523–525, 589–591, 653–655, 707–709
Chapter Test, 74, 120, 192, 252, 300, 360, 410, 464, 526, 592, 656, 710
Check Your Readiness, 2, 78, 124, 196, 256, 304, 364, 414, 468, 530, 596, 660
Checkpoint Quiz, 28, 59, 100, 153, 180, 219, 240, 279, 327, 347, 379, 404, 430, 450, 488, 521, 551, 580, 623, 644, 685, 700
Cumulative Review, 121, 253, 361, 465, 593, 711–713
Mixed Review, 9, 15, 22, 28, 35, 42, 50, 59, 68, 86, 93, 99, 108, 115, 133, 140, 144, 153, 164, 171, 180, 186, 203, 211, 219, 225, 233, 240, 246, 264, 270, 279, 286, 295, 311, 318, 327, 335, 341, 346, 353, 371, 379, 388, 396, 404, 423, 430, 437, 443, 449–450, 459, 476, 482, 488, 496, 503, 512, 520, 539, 545, 551, 558, 564, 573, 580, 587, 604, 614, 623, 630, 637, 644, 651, 668, 676, 685, 693, 700, 705
Quick Check, 4, 5, 6, 10, 11, 12, 17, 19, 23, 24, 25, 32, 33, 36, 37, 38, 39, 44, 45, 46, 47, 54, 55, 62, 63, 64, 80, 81, 82, 87, 88, 89, 94, 95, 96, 104, 105, 111, 112, 127, 128, 129, 130, 135, 136, 137, 142, 148, 149, 150, 157, 158, 159, 160, 166, 167, 168, 174, 175, 176, 177, 181, 182, 183, 198, 199, 200, 206, 207, 213, 214, 215, 221, 222, 229, 230, 236, 237, 241, 242, 243, 260, 261, 266, 267, 273, 274, 275, 280, 281, 282, 283, 290, 291, 292, 307, 308, 313, 314, 315, 323, 324, 330, 331, 332, 337, 338, 343, 344, 349, 366, 367, 368, 373, 374, 375, 383, 384, 385, 392, 393, 399, 400, 418, 419, 425, 426, 427, 433, 434, 439, 440, 445, 446, 453, 454, 455, 470, 471, 472, 479, 484, 485, 492, 493, 499, 506, 507, 508, 509, 515, 516, 517, 535, 536, 540, 541, 542, 546, 547, 554, 555, 559, 560, 561, 567, 568, 569, 576, 577, 582, 583, 584, 598, 599, 600, 609, 611, 618, 619, 620, 625, 626, 627, 632, 633, 634, 639, 640, 646, 647, 648, 663, 664, 665, 670, 671, 673, 679, 680, 681, 688, 689, 690, 695, 696, 697, 701, 702
Reading Comprehension, 75, 193, 301, 411, 527, 657
Test Prep, 9, 15, 22, 27–28, 35, 42, 49–50, 58, 68, 75, 86, 93, 99, 108, 115, 121, 133, 140, 144, 153, 164, 171, 180, 186, 193, 203, 211, 218, 225, 233, 239–240, 246, 253, 264, 270, 278, 285–286, 295, 301, 311, 318, 326, 335, 341, 346, 353, 361, 370, 378, 388, 396, 403, 411, 423, 429–430, 437, 443, 449, 459, 465, 476, 482, 487, 496, 503, 512,

Extended proportion, 367, 407

Extended Response exercises, 9, 15, 50,
93, 99, 121, 126, 140, 188, 218, 246,
253, 270, 278, 326, 353, 361, 388, 403,
459, 465, 512, 520, 550, 563, 587, 593,
623, 684, 700

Extensions. *See also* Dimensionality
Fractals, 380–381
Laws of Sines and Cosines, 565
Perspective Drawing, 605–606
Tangent Lines, Tangent Ratios, 694
Transformations Using Vectors and
Matrices, 504–505
Writing Flow Proofs, 247

Exterior angle
of polygon, 149, 156, 190
of triangle, 149–150, 190, 290

Extra Practice, 716–739

F

Face
lateral, 608, 617
of polyhedron, 598, 599, 653

Factors
conversion, 574, 747
repeated, 96
scale, 498–499

Fibonacci, Leonardo (Leonardo of Pisa),
8

Fibonacci sequence, 8, 378

Flip, 198, 470. *See also* Reflection(s)

Flow proof, 135, 247, 674

Formulas. *See also* Area(s); Circle(s)
for arc length, 765
for area of circle, 62, 64, 73, 575–577,
765
for area of kite, 541, 589, 764
for area of parallelogram, 533, 534,
589, 764
for area of rectangle, 534, 589, 764
for area of regular polygon, 547, 559,
764
for area of rhombus, 541, 589, 764
for area of sector of circle, 576, 591, 765
for area of square, 607, 764
for area of trapezoid, 540, 589, 764
for area of triangle, 535, 549, 560, 561,
589, 764
for circumference, 62, 73, 568, 765
for distance, 53, 60, 73, 260, 421, 695,
696
for equation of circle, 695–696, 765
Euler's, 599, 601, 602, 653
Heron's, 538
for lateral area of cone, 619, 620, 655
for lateral area of cylinder, 610, 654
for lateral area of prism, 609, 610,
654
for lateral area of pyramid, 617, 618,
654
literal equations as, 759
for midpoint, 55, 70, 73, 260, 765
for perimeter, 62, 73, 553, 607, 764
polygon angle-sum, 159

quadratic, 372
for slope, 165, 765
for surface area of cone, 619, 655
for surface area of cylinder, 610, 611,
654
for surface area of prism, 609, 610,
654
for surface area of pyramid, 618, 654
for surface area of sphere, 638–639,
640, 655, 765
table of, 764–765
for volume of cone, 633–634, 765
for volume of cylinder, 607, 626
for volume of prism, 616, 624–626, 765
for volume of pyramid, 632, 654
for volume of sphere, 639–640, 765

45°-45°-90° triangles, 425–426, 764

Foundation drawing, 11

Fractals, 380–381

Frieze Patterns, 514

Frustum, of cone, 636

Functions, 248. *See also* Equations;
Graph(s)

Fundamental Theorem of Isometries, 508

Funnel, 620

G

Garfield, James, 545

Gauss, Karl, 9

Geoboard, 151, 202, 320

Geometric mean, 392, 409

Geometric probability, 582–584, 591

Geometry
coordinate. *See* Coordinate geometry
Euclidean, 154–155
reasoning in, 103–105
spherical, 154–155

Geometry at Work. *See also* Careers
aerospace engineer, 693
cabinetmaker, 50
die casting, 203
industrial designer, 286
package designer, 637
surveyor, 564

Geometry in 3 dimensions, 58, 239, 403,
422, 429, 458, 475, 613, 641, 674, 699

Geo-models, 320

Geopaper, 320

Glide reflection, 508–509, 525

Glide reflectional symmetry, 516–517,
524, 525

Golden ratio, 375, 408

Golden rectangle, 375, 408

Graph(s)
bar, 751
box-and-whisker, 714, 752
circle, 566
of circle given its equation, 696
of dilation images, 499
line, 751
of line in coordinate plane, 166–168
of linear equations, 166–167

solving linear equations with, 172–173
of systems of equations, 18

GRAPH feature, of graphing calculator,
170

Graph paper, 303, 416, 519

Graphing calculator. *See also* Calculator
Activity Lab, 69, 172–173, 187
comparing perimeters and areas, 69
cosine, 440
exercises that use, 67, 69, 163, 170, 179,
436, 442, 563, 586, 636, 642, 682, 699
GRAPH feature of, 170
sine, 440
solving linear equations with graphs
and tables, 172–173
STAT feature of, 163
STAT PLOT features of, 163, 187
surface area, 616
TABLE feature of, 187, 563
tangent, 433, 434
TBLSET feature of, 187, 436, 442, 563
using tables and lists, 187
Y= window of, 170, 187

Great circle, 638

Gridded Response examples, 70, 111, 130,
274, 313, 399, 418, 485, 566

Gridded Response exercises, 68, 70, 115,
121, 164, 203, 253, 264, 318, 361, 423,
437, 465, 503, 558, 593, 693

Guided Problem Solving
Analyzing Errors, 226
Understanding Math Problems, 109,
489, 677
Understanding Proof Problems, 145,
287, 319
Understanding Word Problems, 60,
405, 424, 451, 552, 615

Gyroscopes, 302, 303

H

Hands-On
Activity, 16, 259, 272, 391, 483, 517,
624, 631
Activity Lab, 43, 52, 102, 146, 204, 227,
320, 416, 431, 444, 477, 490–491, 514,
532, 533, 669

Height
of cone, 619
of cylinder, 610
defined, 534, 589
measuring, 444, 446
of parallelogram, 534, 589
of prism, 608
of pyramid, 617
slant, 617, 619
of trapezoid, 540, 589
of triangle, 536, 589

Hemispheres, 638

Heron, 538

Heron's formula, 538

Hexagon, 158, 492, 513, 599

Hexagonal prism, 612

Hinge Theorem, 293

Index

even, 510
　Fundamental Theorem of, 508
　identifying, 470–471
　odd, 510
Isometry Classification Theorem, 509
Isosceles (45°-45°-90°) right triangle,
　425–426, 764
Isosceles trapezoid, 306, 336–337, 357,
　359, 496
Isosceles triangle, 148, 227, 228–229
Isosceles Triangle Theorems
　Converse of, 228, 230, 251
　Corollary to, 230
　proof of, 229
　statement of, 228, 251
　using, 229
Iteration, 380

J

Justifying Steps, 104

K

Kaleidoscopes, 511, 513
Kettering, Charles F., 85
Kites
　angle measures in, 338
　area of, 541, 589, 764
　defined, 306, 357
　diagonals of, 337–338, 541
　properties of, 338
Koch, Helge von, 380
Koch Curve, 380, 381
Koch Snowflake, 380–381
Kraus, Hannelore, 387

L

Landers, Ann, 107
Lateral area
　of cone, 619, 620, 654
　of cylinder, 610, 654
　defined, 608, 654
　of prism, 608, 609–610, 654
　of pyramid, 617, 618, 654
Lateral faces
　of prism, 608
　of pyramid, 617
Latitude, 154
Law
　of Cosines, 565
　of Detachment, 94–95, 118
　of Sines, 565
　of Syllogism, 95–96, 118
Legs
　of isosceles triangle, 228
　length of, 426
　of right triangle, 235
　of trapezoid, 336
Length
　of arc, 569, 591, 765
　of diagonal, 331
　of hypotenuse, 425

of leg, 426
of segment, 31–33, 689–690
Leonardo of Pisa, 8
Lesson Quiz Online, 9, 15, 21, 27, 35, 41,
　49, 59, 67, 85, 93, 99, 107, 115, 133,
　139, 143, 153, 163, 171, 179, 185, 201,
　211, 219, 225, 233, 239, 245, 263, 269,
　279, 285, 295, 311, 317, 327, 335, 341,
　345, 353, 369, 379, 387, 395, 403, 423,
　429, 437, 443, 449, 459, 475, 481, 487,
　495, 503, 511, 519, 539, 543, 549, 557,
　563, 573, 579, 587, 603, 613, 621, 629,
　635, 643, 651, 667, 675, 685, 691, 699,
　705
Light, speed of, 371
Line(s). *See also* Linear equations
　center of a circle and, 672–673
　concurrent, 273, 298
　constructing, 126, 181–183, 246
　in coordinate plane, 166–168
　coplanar, 17, 20, 72
　defined, 17
　distance from a point to, 266, 297
　drawing, 16
　graphing, 166–167
　horizon, 605
　horizontal, 168
　intersecting, 507
　number, 31, 32, 33, 34
　parallel. *See* Parallel lines
　perpendicular. *See* Perpendicular lines
　skew, 24, 72
　slope of, 165
　of symmetry, 492, 524
　tangent, 662–665, 694
　vertical, 168
Line graph, 751
Line symmetry, 492, 524
Linear equations
　graphing, 166–167
　for parallel lines, 175
　for perpendicular lines, 176
　point-slope form of, 168, 175, 191
　slope-intercept form of, 166, 167, 191,
　765
　solving, 30, 172–173, 758
　solving with graphs and tables,
　172–173
　standard form of, 167, 191
　systems of, 234, 760
　writing, 168, 758
Linear function, 248
Linear regression, 76
Lists, 187
Literal equations, 607, 759
Locus, 701, 702, 709, 710
Logical reasoning, 744
Longitude, 154

M

MacArthur, Ellen, 699
Magnification, 374
Magnitude, 452

Major arc, 567, 591
Make a Conjecture, 5, 6, 27, 49, 69, 185,
　204, 220, 227, 258, 259, 269, 272, 320,
　397, 424, 432, 438, 442, 477, 520, 581,
　645, 669, 682, 686
Manipulatives. *See also* Calculator;
　Computers; Graphing calculator
　centimeter ruler, 150, 192
　compass, 43, 44, 45, 46, 47, 73, 182–183,
　303, 387, 477, 539, 549, 570, 575, 669,
　678
　geoboard, 151, 202, 320
　geopaper, 320
　graph paper, 303, 416
　grid paper, 532–533
　index cards, 204
　isometric dot paper, 10, 14, 518
　number cubes, 72
　parallel rule, 324
　protractor, 37, 40, 48, 49, 50, 102, 150,
　162, 192, 204, 425, 444, 570, 678, 746
　ruler, 31, 49, 368, 411, 416, 477, 582,
　746
　scissors, 303
　straightedge, 44, 48, 49, 50, 59, 73,
　182–183, 186, 204, 210, 303, 315, 387,
　477, 539, 669
　tracing paper, 52, 102, 204, 227, 477,
　483, 490–491, 669
　unit cubes, 624
Math in the Media, 8
Mathematical modeling
　exercises that use, 368, 379
　finding probability using, 582–584
　geo-models, 320
　of quadrilaterals, 320
Mathematical Systems, 29, 101
Matrices
　column, 458
　defined, 504
　elements in, 504
　transformations using, 504–505, 524
　using to find translation images, 504,
　524
Mean
　defined, 750
　geometric, 392, 409
Measurement, 748
　of angles, 36–39, 41, 46, 130, 159, 330,
　546, 591, 680, 687–689
　of arcs, 568
　of area, 62–63
　of circumference, 62
　conversion of, 747
　customary units of, 766
　of distance, 195, 440
　effect of errors in, 749
　of height, 444, 446
　in history, 431
　indirect, 261, 262, 384, 386, 387, 408,
　409, 410, 431, 447, 459, 463, 464, 685
　metric units of, 766
　of parallelogram, 315
　of perimeter, 62
　of segments, 31–33
　of time, 8

Measures of central tendency, 750. *See also* Mean; Median; Mode

Median
defined, 750
length of, 274
of triangle, 271, 274, 275, 298

Mental Math, 641

Midpoint
coordinate of, 54–55
defined, 32, 72
finding length using, 33
of segment, 32–33, 54–55, 72

Midpoint Formula, 55, 70, 73, 260, 765

Midsegments
constructing, 258
defined, 259, 297
properties of, 259–261
of trapezoids, 348, 359
of triangles, 259–261, 297

Minor arc, 567, 591

Mirrors, 194–195, 478

Mixed Review, 9, 15, 22, 28, 35, 42, 50, 59, 68, 86, 93, 99, 108, 115, 133, 140, 144, 153, 164, 171, 180, 186, 203, 211, 219, 225, 233, 240, 246, 264, 270, 279, 286, 295, 311, 318, 327, 335, 341, 346, 353, 371, 379, 388, 396, 404, 423, 430, 437, 443, 449–450, 459, 476, 482, 488, 496, 503, 512, 520, 539, 545, 551, 558, 564, 573, 580, 587, 604, 614, 623, 630, 637, 644, 651, 668, 676, 685, 693, 700, 705

Mode, 750

Models. *See* Mathematical modeling

Mosaics, 521

Multiple Choice examples, 12, 64, 82, 149, 175, 213, 229, 267, 291, 323, 331, 368, 384, 406, 426, 446, 479, 499, 541

Multiple Choice exercises, 7, 9, 15, 20, 22, 27, 35, 42, 49–50, 57, 58, 67, 84, 86, 91, 93, 99, 106, 108, 113, 121, 133, 138, 140, 144, 151, 153, 171, 180, 185, 186, 201, 209, 211, 217, 218, 225, 231, 233, 239–240, 245, 246, 253, 263, 269, 270, 277, 278, 285–286, 295, 309, 311, 316, 325, 326, 333, 335, 341, 345, 346, 351, 353, 361, 369, 370, 377, 378, 388, 395, 396, 403, 406, 421, 429, 435, 443, 448, 449, 457, 459, 465, 466, 474, 476, 480, 482, 487, 495, 510, 512, 519, 520, 537, 539, 543, 549, 550, 554, 556, 562, 563, 571, 573, 578, 580, 587, 603, 604, 613, 614, 621, 622–623, 629, 630, 635, 636–637, 641, 643, 668, 676, 684, 705

Multiplication, scalar, 505, 525

Multiplication Property
of Equality, 103, 118
of Inequality, 288

in finding volume, 631
for three-dimensional figures, 12, 609

New Vocabulary. *See* Vocabulary

Newton, Sir Isaac, 194

n-gon, 159

Noncoplanar points, 20

Notation. *See also* Symbols
arrow, 471
prime, 471

Number(s)
real, 767
squaring, 753

Number cubes, 72

Number line, 31, 32, 33, 34

Nuñez, Vladimir, 95

O _____

Oblique cone, 633

Oblique cylinder, 610

Oblique prism, 608, 610, 625

Obtuse angle, 37

Obtuse triangle, 148

Octagons, 158, 550, 560, 581

Octahedron, 603

Odd isometry, 510

O'Gorman, Juan, 521

O'Keeffe, Georgia, 83

One-point perspective, 605, 606

Online Active Math, 45, 54, 61, 80, 89, 149, 160, 182, 207, 230, 236, 266, 275, 292, 307, 323, 331, 374, 393, 419, 434, 440, 454, 479, 535, 555, 583, 625, 633, 670, 680

Online Chapter Test, 74, 120, 192, 252, 300, 360, 410, 464, 526, 592, 656, 710

Online End-of-Course Test, 711

Online Geometry at Work. *See also* Careers
aerospace engineer, 693
cabinetmaker, 50
die casting, 203
industrial designer, 286
package designer, 637
surveyor, 564

Online Homework Video Tutor, 8, 14, 21, 26, 34, 41, 48, 56, 67, 85, 91, 97, 106, 113, 132, 138, 144, 152, 162, 170, 178, 185, 202, 209, 217, 223, 231, 238, 244, 263, 268, 277, 284, 294, 310, 317, 325, 333, 339, 345, 351, 369, 377, 387, 395, 402, 421, 429, 435, 442, 447, 458, 475, 480, 487, 495, 502, 510, 519, 537, 543, 549, 556, 562, 570, 578, 585, 602, 612, 622, 628, 635, 642, 650, 666, 675, 683, 692, 698, 704

Online Lesson Quiz, 9, 15, 21, 27, 35, 41, 49, 59, 67, 85, 93, 99, 107, 115, 133, 139, 143, 153, 163, 171, 179, 185, 201, 211, 219, 225, 233, 239, 245, 263, 269, 279, 285, 295, 311, 317, 327, 335, 341, 345, 353, 369, 379, 387, 395, 403, 423, 429, 437, 443, 449, 459, 475, 481, 487,

495, 503, 511, 519, 539, 543, 549, 557, 563, 573, 579, 587, 603, 613, 623, 629, 635, 643, 651, 667, 675, 685, 691, 699, 705

Online Point in Time, 100, 347, 371, 521, 545

Online Video Tutor Help, 37, 39, 63, 127, 148, 159, 207, 215, 535, 542, 626

Online Vocabulary Quiz, 71, 117, 189, 249, 297, 357, 407, 461, 523, 589, 653, 707

Open-Ended, 8, 13, 20, 22, 26, 48, 57, 68, 74, 85, 90, 132, 139, 152, 163, 165, 170, 178, 192, 202, 210, 218, 239, 244, 252, 264, 284, 285, 300, 318, 325, 334, 339, 351, 360, 378, 387, 394, 401, 410, 429, 458, 475, 481, 494, 495, 501, 510, 528, 537, 543, 549, 557, 579, 586, 592, 602, 606, 612, 621, 628, 642, 655, 656, 674, 682, 698, 703, 710

Opposite rays, 23

Opposite reciprocals, 176

Optical illusion, 21, 606

Orthocenter of triangle, 275, 298

Orthographic drawing, 11

Overlapping triangles, 241–242, 243, 251

P _____

Paper folding, 110, 114, 185, 227, 243, 310, 669
bisectors, 272, 277
conjectures, 227, 669
constructions, 102, 114, 185, 277, 310, 477

Paragraph proof, 111–112, 119, 141, 142, 145, 317, 674

Parallel lines
applying, 194–195
composition of reflections across, 507
conditions resulting in, 134–137
constructing, 126, 181–182, 190
defined, 24, 72
determining, 175
equations for, 175
properties of, 127–130
proving, 134–137
related angles and, 126
relating to perpendicular lines, 141–142, 190
slope of, 174–175, 191
Triangle Angle-Sum Theorem and, 147–150
writing equations of, 175

Parallel planes, 24–25, 72

Parallel rays, 24

Parallel rule, 324

Parallel segments, 24, 261

Parallel vectors, 457

Parallelogram(s)
altitude of, 534
angles of, 312–313, 358
area of, 532–533, 534–535, 589, 764
base of, 534, 589

Acknowledgments

Staff Credits

The people who made up the High School Mathematics team—representing design services, editorial, editorial services, education technology, image services, marketing, market research, production services, publishing processes, and strategic markets—are listed below. Bold type denotes the core team members.

Leora Adler, **Scott Andrews,** Carolyn Artin, Stephanie Bradley, Amy D. Breaux, Peter Brooks, Judith D. Buice, Ronit Carter, Lisa J. Clark, Bob Cornell, Sheila DeFazio, Marian DeLollis, Kathleen Dempsey, Jo DiGiustini, Delphine Dupee, Emily Ellen, Janet Fauser, Debby Faust, Suzanne Feliciello, Frederick Fellows, Jonathan Fisher, Paula Foye, Paul Frisoli, **Patti Fromkin,** Paul Gagnon, Melissa Garcia, Jonathan Gorey, Jennifer Graham, **Ellen Granter,** Barbara Hardt, Daniel R. Hartjes, Richard Heater, Kerri Hoar, Jayne Holman, Karen Holtzman, Angela Husband, Kevin Jackson-Mead, Al Jacobson, Misty-Lynn Jenese, **Elizabeth Lehnertz,** Carolyn Lock, Diahanne Lucas, Catherine Maglio, **Cheryl Mahan,** Barry Maloney, Meredith Mascola, Ann McSweeney, Eve Melnechuk, Terri Mitchell, Sandy Morris, Cindy Noftle, Marsha Novak, Marie Opera, Jill Ort, **Michael Oster,** Steve Ouellette, Dorothy M. Preston, Rashid Ross, Donna Russo, Suzanne Schineller, Siri Schwartzman, Malti Sharma, Dennis Slattery, Emily Soltanoff, Deborah Sommer, Kathryn Smith, Lisa Smith-Ruvalcaba, Kara Stokes, Mark Tricca, Paula Vergith, Nate Walker, Diane Walsh, Merce Wilczek, **Joe Will,** Amy Winchester, Mary Jane Wolfe, Helen Young, Carol Zacny

Additional Credits: J. J. Andrews, Sarah Aubry, Jonathan Kier, Lucinda O'Neill, Sara Shelton, Ted Smykal, Michael Torocsik

Cover Design

Brainworx Studio

Cover Photos

Zebras, Art Wolfe/Stone/Getty Images, Inc.; Glass Dome, John McAnulty/Corbis.

Technical Illustration

Network Graphics

Illustration

Andrea G. Maginnis: 41
John Edwards: 24, 132, 133t, 200, 207, 290
JB Woolsey: 21, 88, 162, 201, 230, 332, 542, 385, 386, 445
Dennis Harms: 8, 57, 222, 223, 239, 274, 556, 566
Function through Form: 52, 266, 611, 641, 472
Leo Abbett: 92, 606 all
Jim Delapine: 579, 638, 639
Seymour Levy: 643
Kenneth Batelman: 395, 433, 650, 701, 704
Christine Graham: 487, 682
Peter Bollinger: 469, 528, 529
Brucie Rosch: 69, 81, 269, 472, 479
Gary Torrisi: 447 all, 448, 449
Lois Leonard Stock: 517

Ortelius Design, Inc.: 540
Gary Phillips: 37, 41

Photography

Page viii, Johnny Johnson/DRK Photo; **x,** Peter Menzel/Stock Boston; **xi,** Russ Lappa; **xii,** Timothy Hursley/SuperStock, Inc.; **xiii,** Kunio Owaki/Corbis; **xiv,** David Young-Wolff/PhotoEdit; **xix,** Reza Estakhrian/Getty Images, Inc.; **xv,** Scott T. Smith/Corbis; **xvi,** Roy Ooms/Masterfile; **xvii,** David Young-Wolff/PhotoEdit; **xviii,** Felicia Martinez/PhotoEdit

Chapter 1: Pages 2, 3, Courtesy of SeaWorld; **4,** Julie Houck/Stock Boston; **6,** Brett Froomer/Getty Images, Inc.; **7,** Dennis O'Clair/Getty Images, Inc.; **9,** PhotoDisc, Inc./Getty Images, Inc.; **11,** Barry Durand/Odyssey/Chicago; **14,** Reprinted by Permission of Tribune Media Services.; **19,** Russ Lappa; **20,** Johnny Johnson/DRK Photo; **23,** Daryl Benson/Masterfile; **24,** Stacy Pick/Stock Boston; **26,** Tim O'Hara/Index Stock Imagery, Inc.; **33,** AP Photo/Cliff Schiappa; **36,** Bill Nation/Corbis, **36** inset, Chad Slattery/Getty Images, Inc.; **37,** Corbis; **40,** William Sallaz/Duomo; **41,** Cameramann/The Image Works; **45,** Jon Feingersh/Corbis; **46,** Bob Daemmrich/The Image Works; **48 t,** ©1997 Richard Megna/Fundamental Photographs, NYC; **48 b,** Peanuts reprinted by permission of United Feature Syndicate, Inc.; **50,** Dan McCoy/Corbis; **55,** Michael Newman/PhotoEdit; **57,** Laura Dwight/PhotoEdit; **58,** Richard Eller/Aerial Images Photography; **63,** Gail Mooney/Corbis; **66,** Hans Georg Roth/Corbis; **67,** Michael Rosenfeld/Getty Images, Inc.; **76,** Motoring Picture Library/Alamy

Chapter 2: Pages 78, 79, Stewart Cohen/Getty Images, Inc.; **82,** Jon Chomitz; **83 t,** Frank and Ernest reprinted with permission of Newspaper Enterprise Association, Inc.; **83 bl,** AP Photo/Richard Drew; **83 br,** Tony Vaccaro/Archive Photos/Getty Images, Inc.; **85 t,** Hulton-Deutsch Collection/Corbis; **85 b,** Mug Shots/Corbis; **89,** Richard Haynes; **91 l all,** Russ Lappa; **91 r,** National Association for the Deaf; **92 l,** Andy Sacks/Getty Images, Inc.; **92 r,** C. Squared Studios/PhotoDisc, Inc./PictureQuest; **94,** Tony Freeman/PhotoEdit; **96,** InterNetwork Media/PhotoDisc, Inc./PictureQuest; **97,** Damian Strohmey/Sports Illustrated; **98 t,** Fotopic/Omni-Photo Communications, Inc.; **98 m,** Close to Home by John McPherson. Reprinted with permission of Universal Press Syndicate. All rights reserved.; **98 b,** Gregory Scott/Index Stock Imagery, Inc.; **100 l,** Courtesy of WGBH; **100 r,** King Collection/Retna, Ltd.; **106 l,** PhotoDisc, Inc./Getty Images, Inc.; **106 r,** Tony Freeman/PhotoEdit; **107 t,** The Granger Collection, NY; **107 m,** John Lopinot/Black Star Publishing/PictureQuest; **107 b,** Neal Preston-Andy Kent/Still Bill Productions; **113,** Corbis Digital Images/PictureQuest; **114,** Paul A. Souder/Corbis; **122 t,** David Murray and Jules Selmes/Dorling Kindersley; **122 m,** Dorling Kindersley; **122 b both,** Dorling Kindersley; **122-123,** Dorling Kindersley; **123 tl,** Russ Lappa; **123 tr,** Dorling Kindersley; **123 m,** Dorling Kindersley; **123 b all,** Christ Graham and Nick Nichols/Dorling Kindersley

Chapter 3: Pages 124, 125, Imtek Imagineering/Masterfile; **128** composite, H.P. Merten/Corbis, Joe Towers/Corbis; **131,** Peter Menzel/Stock Boston; **134,** Gary Kufner/Corbis; **138 both,** Jon Chomitz; **139,** Richard Pasley/Stock Boston; **142,** Jon Chomitz; **143,** John M. Roberts/Corbis; **149,** Stewart Cohen/Getty Images, Inc.; **151,** John Coletti/The Picture Cube/Index Stock Imagery, Inc.; **152 both,** Jerry Jacka Photography; **157,** Jim Cummins/

Getty Images, Inc.; **158,** Arthur Thevenart/Corbis; **160,** Ken O'Donoghue; **161 l,** Russ Lappa; **161 m,** Nawrocki Stock Photo/Picture Perfect; **161 r,** Russ Lappa; **162,** Courtesy of the Theatre in the Round Players, Inc., Minneapolis, MN; **163,** Michael Newman/PhotoEdit; **169,** NASA; **170,** Michael Newman/PhotoEdit; **174,** Bob Daemmrich/Stock Boston; **177,** PhotoDisc, Inc./Getty Images, Inc.; **179,** Getty Images, Inc.; **181,** Michael Newman/PhotoEdit; **183,** Sol Lewitt, Installation view of photo of students with Sol Lewitt, from Sol Lewitt: Twenty-Five Years of Wall Drawings, 1968-1993, Addison Gallery of American Art, Phillips Academy, Andover, Massachusetts; **185 all,** Jon Chomitz; **194 tl,** NASA; **194 r,** Dorling Kindersley; **194 bl,** Science Museum Photo Library/Dorling Kindersley; **195 all,** Dorling Kindersley

Chapter 4: Pages 196, 197, Chateau de Versailles, France/Peter Willi/SuperStock, Inc.; **199,** NASA; **200,** John Elk III/Stock Boston; **202,** Richard Haynes; **203,** Hans Gregory Roth/Corbis; **206,** Tom Alexander/Getty Images, Inc.; **208 l,** Tony Freeman Photographs; **208 r,** Russ Lappa; **209 both,** Prentice Hall; **210,** Chuck Pefley/Getty Images, Inc.; **214,** Mike Greenlar Photography; **217,** Richard Haynes; **218,** Photo Researchers, Inc.; **221,** Russ Lappa; **223,** Leif Skoogfors/Woodfin Camp & Associates; **229,** Pat O'Hara/Corbis; **231,** Fernando Serna/Department of the Air Force; **232,** Richard Hutchins/PhotoEdit; **233,** Corbis; **236,** Tony Freeman Photographs; **238,** Jeff Greenberg/Omni-Photo Communications, Inc.; **241,** Jim Rudnick/Corbis; **242,** Miriam Nathan-Roberts; **243 t,** Russ Lappa; **243 b,** Walter Hodges/Getty Images, Inc.; **245,** Corbis

Chapter 5: Pages 256, 257, Jeff Greenberg/PhotoEdit; **261,** Clyde Lockwood/Animals Animals/Earth Scenes; **263,** Timothy Hursley/SuperStock, Inc.; **268,** Jim Cummins/Corbis; **281,** AP Photo/Dean Times Hearld; **282,** Michael Newman/PhotoEdit; **284,** AP/Wide World Photos; **285,** Corbis; **286,** Tom Pantages; **289,** Russ Lappa; **290,** David Young-Wolff/PhotoEdit; **293,** Dave Schiefelbein/Getty Images, Inc.; **294,** Fred Wood/Summer Productions; **302 t and b,** Dorling Kindersley; **302-303,** Index Stock Imagery, Inc./PictureQuest; **303 t and b,** Dorling Kindersley

Chapter 6: Pages 304, 305, Complete Sportswear; **306,** The Image Works; **309,** Charles Demuth, My Egypt, 1927 Oil on composition board 35 3/4 x 30 in. Whitney Museum of American Art, New York; Purchase, with funds from Gertrude Vanderbilt Whitney 31.172; **310,** Russ Lappa; **313,** Richard Haynes; **314,** Tony Freeman/PhotoEdit; **316,** Richard Haynes; **320,** Ken O'Donoghue; **321,** Tony Freeman Photographs; **322 t,** Photo of Russell W.M. and Ruth Kraus residence, S.340 in The Architecture of Frank Lloyd Wright, A Complete Catalog, 3rd Edition, ©2002/William Allin Storrer, PhD; **322 b,** Plan of Russell W.M. and Ruth Kraus residence, S.340 in The Frank Lloyd Wright Companion, ©1973/William Allin Storrer, PhD; **324 l,** Dave Bartruff/Stock Boston; **324 r,** Mark Thayer; **325,** Russ Lappa; **330,** Walter Bibikow/Index Stock Imagery, Inc.; **332,** Ron Sherman/Stone/Getty Images, Inc.; **333,** Tony Freeman Photographs; **333 inset,** Russ Lappa; **336,** archivebroehandesign.com, New York; **337,** The Purcell Team/Corbis; **339,** Mark E. Gibson/Visuals Unlimited; **340,** Wallace Garrison/Index Stock Imagery, Inc.; **343,** Richard Haynes; **345,** ©National Park Service, photo by Bill Hudson, Biscayne National Park; **347 both,** Geoffrey Clifford/Woodfin Camp & Associates; **349,** SuperStock, Inc.; **352,** Kunio Owaki/Corbis; **362,** Big Cheese Photo LLC/Alamy; **363,** Tim Wright/Corbis

Chapter 7: Pages 364, 365, David Maenza/SuperStock, Inc.; **366 l,** David Young-Wolff/PhotoEdit; **366 r,** David Young-

Wolff/PhotoEdit; **368,** Prentice Hall; **369,** ©AAA reprinted with permission.; **370,** Michael Dwyer/Stock Boston; **371,** NASA; **374,** Miro Vintoniv/Stock Boston; **375,** Telegraph Colour Library/Getty Images, Inc.; **377 t,** Peanuts reprinted by permission of United Feature Syndicate, Inc.; **377 m,** Ralf-Finn Hestoft/Index Stock Imagery, Inc.; **377 b,** Russ Lappa; **382,** Lee Snider/Photo Images; **385,** Francis Lepine/Animals Animals/Earth Scenes; **387,** AP/Wide World Photos; **393,** Paul Chesley/Getty Images, Inc.; **395,** Kevin Miller/Getty Images, Inc.; **399,** Corbis; **401,** Vincent Hobbs/SuperStock, Inc.; **402,** Mark Richards/PhotoEdit; **403,** David Woodfall/Getty Images, Inc.; **412,** image100/Getty Images, Inc.; **413,** Hemera/AGE Fotostock

Chapter 8: Pages 414, 415, Peter Hendrie/Getty Images, Inc.; **419,** Tom Tracy/Getty Images, Inc.; **422,** The Observatories of the Carnegie Institution of Washington; **426,** Rhoda Sidney/PhotoEdit; **427,** Stephen Saks/ Index Stock Imagery, Inc./PictureQuest; **428,** Russ Lappa; **429,** Macduff Everton/Corbis; **433,** Scott T. Smith/Corbis; **436,** Courtesy of Katoomba Scenic Railway; **439,** Richard Hamilton/Corbis; **441,** David Young Wolff/PhotoEdit; **443 both,** Tom Pantages; **446,** Ron Dahlquist/Getty Images, Inc.; **448,** Dwayne Newton/PhotoEdit; **450,** Giulio Andreini/Getty Images, Inc.; **452,** Lowell Georgia/Corbis; **453,** Getty Images, Inc.; **454,** Bob Daemmrich/Stock Boston; **455,** William J. Weber/Visuals Unlimited; **457,** NOAA; **458,** The Far Side ® by Gary Larson © 1993 Farworks, Inc. All rights reserved. Used with permission.; **466 both,** Dorling Kindersley; **467 t,** Corbis; **467 m,** Ryan McVay/PhotoDisc, Inc./PictureQuest; **467 bl,** Dorling Kindersley; **467 br,** Corbis Images/PictureQuest

Chapter 9: Pages 468, 469 cave, Tom Till/Getty Images, Inc.; **470 tl,** Tom Rosenthal/SuperStock, Inc.; **470 bl,** Tom Rosenthal/SuperStock, Inc.; **470 r all,** Russ Lappa; **471,** Adam Smith Productions/Corbis; **475,** Roy Ooms/Masterfile; **478,** Gabe Palmer/Corbis; **481,** Seth Joel/Corbis; **485,** Jerry Jacka Photography; **486 both,** Jerry Jacka Photography; **492,** Brian Parker/Tom Stack & Associates, Inc.; **493 both,** Prentice Hall; **494 l,** Andrew J. Martinez © 1993/Photo Researchers, Inc.; **494 r,** Guido Alberto Rossi/Image Bank/Getty Images, Inc.; **495 l,** Tom Salyer/ Stock Connection/ PictureQuest; **495 r,** Getty Images, Inc.; **498,** Frank T. Awbrey/Visuals Unlimited; **502,** David Young/Wolff/PhotoEdit; **507,** Michael Newman/PhotoEdit; **508,** PhotoDisc, Inc./Getty Images, Inc.; **511 tl,** Viviane Moos/Corbis; **511 tr,** Alfred Pasieka/Science Photo Library/Photo Researchers, Inc.; **511 bl,** Paul Jablonka/International Stock Photo/ImageState; **511 br,** Adam Peirport/Corbis; **515 t,** ©1996 M.C. Escher Heirs/Cordon Art, Baarn, Holland. All rights reserved.; **515 bl,** ©2001 Cordon Art , Baarn, Holland. All rights reserved.; **515 br,** Symmetry Drawing E67 by M.C. Escher. ©2002 Cordon Art, Baarn, Holland. All rights reserved.; **516,** Syracuse Newspapers/The Image Works; **518 tl,** Ira Kirschenbaum/Stock Boston; **518 tr,** Suzanne Murphy-Larronde; **518 bl,** Margaret Courtney-Clarke/Corbis; **518 br,** M. Angelo/Corbis; **519,** From The Grammar of Ornament by Owen Jones, 1856 Edition/1998 Octavo Corporation; **521,** Robert Frerck/Odyssey/Chicago; **524 l,** Patti Murray/Animals Animals/Earth Scenes; **524 r,** Russ Lappa; **526 l,** Don & Pat Valenti/DRK Photo; **526 r,** Jeff Foott/DRK Photo; **529,** Tom Till/Getty Images, Inc.

Chapter 10: Pages 530, 531 AP Photo/NASA; **536,** Mick Roessler/SuperStock, Inc.; **538,** Larry Lefever/Grant Heilman Photography, Inc.; **543,** Simon Battensby/Getty Images, Inc.; **544,** Tony Donaldson/SportsChrome-USA; **545,** The Granger Collection, NY; **547,** Photo courtesy of NidaCore; **548,** Richard Cummins/Corbis; **549,** Howard G. Ross; **550,** F. Dewey Webster/Sovfoto/Eastfoto/ PictureQuest; **553,** Russ Lappa; **555,**

David Young-Wolff/PhotoEdit; **557,** Bob Daemmrich/Stock Boston/PictureQuest; **560,** George Gerster/Photo Researchers, Inc.; **562,** Corbis; **563,** PhotoDisc, Inc./Getty Images, Inc.; **564,** Ralph Cowan/Getty Images, Inc.; **567,** PhotoDisc, Inc./Getty Images, Inc.; **571,** Bill Horsman/Stock Boston; **572,** Calvin and Hobbs ©Watterson. Reprinted with permission of Universal Press Syndicate. All rights reserved.; **573,** Bob Daemmrich/Stock Boston; **576,** Russ Lappa; **577,** Russ Lappa; **579,** Bill Bachmann/Photo Network/PictureQuest; **583,** David Young-Wolff/PhotoEdit; **585,** Frank Fournier/Corbis; **586,** AP/Wide World Photos

Chapter 11: 596, 597, Dorling Kindersley; **599,** Michael Kevin Daly/Corbis; **599 inset,** Prentice Hall; **602 t all,** Russ Lappa; **602 b,** PhotoDisc, Inc./Getty Images, Inc.; **603,** ©Breck P. Kent; **608,** Uniphoto, Inc./Pictor International/ImageState; **610,** Russ Lappa; **611,** Tony Freeman/PhotoEdit; **612,** Russ Lappa; **618,** Will & Deni McIntyre/Getty Images, Inc.; **620,** Roger Allyn Lee/SuperStock, Inc.; **621,** Michael Busselle/Corbis; **624 both,** Russ Lappa; **628,** James King-Holmes/Science Photo Library/Photo Researchers, Inc.; **631,** Ken O'Donoghue; **632,** Pictor International/ImageState; **634,** Carolyn Ross/Index Stock Imagery, Inc.; **635,** John Elk/Stock, Boston/PictureQuest; **637,** Alan Klehr/Getty Images, Inc.; **638,** Rafael Macia/Photo Researchers, Inc.; **641 l,** Richard Hutchings/Photo Researchers, Inc.; **641 m,** Mark C. Burnett/Photo Researchers, Inc.; **641 r,** Tony Freeman/PhotoEdit; **642,** David Young-Wolff/PhotoEdit; **642,** Gene Moore/Phototake/PictureQuest; **644,** Ken O'Donoghue; **646,** Felicia Martinez/PhotoEdit; **647,** Russ Lappa; **649,** U.S. Department of Commerce; **659 t,** Dorling Kindersley; **659 b,** Mick Roessler/Index Stock Imagery, Inc./PictureQuest

Chapter 12: Page 660, 661, Ray Ooms/Masterfile; **663,** Patrick Ward/Corbis; **666 t,** Carl F. Romney; **666 b,** Russ Lappa; **667 both,** Russ Lappa; **671,** Steve Vidler/SuperStock, Inc.; **672,** Tony Freeman Photographs; **674,** Robert Frerck/Odyssey/Chicago; **675,** Grant Heilman Photography, Inc.; **683,** AP/Wide World Photos; **688,** Shaun Egan/Getty Images, Inc.; **690,** Carol Simowitz; **691,** Getty Images, Inc.; **693,** Rosenfeld Images Ltd/Science Photo Library/Photo Researchers, Inc.; **697,** Tony Freeman Photographs; **699,** AFP/Corbis; **702,** Reza Estakhrian/Stone/Getty Images, Inc.; **703,** Steven E. Sutton/Duomo; **704,** Christian Grzimek / Okapia/Photo Researchers, Inc.; **714,** Russ Lappa